Lecture Notes in Networks and Systems

Volume 68

Series Editor

Janusz Kacprzyk, Systems Research Institute, Polish Academy of Sciences, Warsaw, Poland

The series "Lecture Notes in Networks and Systems" publishes the latest developments in Networks and Systems—quickly, informally and with high quality. Original research reported in proceedings and post-proceedings represents the core of LNNS.

Volumes published in LNNS embrace all aspects and subfields of, as well as new challenges in, Networks and Systems.

The series contains proceedings and edited volumes in systems and networks, spanning the areas of Cyber-Physical Systems, Autonomous Systems, Sensor Networks, Control Systems, Energy Systems, Automotive Systems, Biological Systems, Vehicular Networking and Connected Vehicles, Aerospace Systems, Automation, Manufacturing, Smart Grids, Nonlinear Systems, Power Systems, Robotics, Social Systems, Economic Systems and other. Of particular value to both the contributors and the readership are the short publication timeframe and the world-wide distribution and exposure which enable both a wide and rapid dissemination of research output.

The series covers the theory, applications, and perspectives on the state of the art and future developments relevant to systems and networks, decision making, control, complex processes and related areas, as embedded in the fields of interdisciplinary and applied sciences, engineering, computer science, physics, economics, social, and life sciences, as well as the paradigms and methodologies behind them.

**** Indexing: The books of this series are submitted to ISI Proceedings, SCOPUS, Google Scholar and Springerlink ****

More information about this series at http://www.springer.com/series/15179

Igor Kabashkin · Irina Yatskiv (Jackiva) ·
Olegas Prentkovskis
Editors

Reliability and Statistics in Transportation and Communication

Selected Papers from the 18th International Conference on Reliability and Statistics in Transportation and Communication, RelStat'18, 17–20 October 2018, Riga, Latvia

Springer

Editors
Igor Kabashkin
Transport and Telecommunication Institute
Riga, Latvia

Irina Yatskiv (Jackiva)
Transport and Telecommunication Institute
Riga, Latvia

Olegas Prentkovskis
Vilnius Gediminas Technical University
Vilnius, Lithuania

ISSN 2367-3370 ISSN 2367-3389 (electronic)
Lecture Notes in Networks and Systems
ISBN 978-3-030-12449-6 ISBN 978-3-030-12450-2 (eBook)
https://doi.org/10.1007/978-3-030-12450-2

Library of Congress Control Number: 2018968530

This Springer imprint is published by the registered company Springer Nature Switzerland AG
The registered company address is: Gewerbestrasse 11, 6330 Cham, Switzerland

Preface

In this volume of "Lecture Notes in Networks and Systems", we are pleased to present the proceedings of the *18th International Multidisciplinary Conference on Reliability and Statistics in Transportation and Communication* (*RelStat 2018*), which took place in Riga in Latvia, from October 17 to October 20, 2018. This event belongs to a conference series started in 2001 and organized annually by the Transport and Telecommunication Institute (TTI) in Riga, Latvia. The mission of RelStat is to promote a more comprehensive approach supporting new ideas, theories, technologies, systems, tools, applications, as well as work in progress and activities on all theoretical and practical issues arising in transport, information, and communication technologies. Results of previous editions of RelStat were published by TTI Publishing House (RelStat 2001–2015) in the journal "Transport and Telecommunication" (ISSN 1407-6160), by Elsevier in the "Procedia Engineering" (RelStat 2016) and by Springer in "Lecture Notes in Networks and Systems" volume no. 36 (RelStat 2017).

Design, implementation, operation, and maintenance of contemporary complex systems have brought many new challenges to "classic" reliability theory. We define complex systems as integrated unities of assets: technical, information, organization, economical, software, and human (users, administrators, and management) ones. Their complexity comes not only from their technical and organizational internal structure, which is built upon diverse hardware and software resources, but also from the complexity of information processes (data processing, monitoring, management, etc.) that must be executed in their specific environment.

A system approach to the evaluation of the efficiency of complex systems at all phases of their life cycle is the contemporary answer to new challenges in the use of such systems. The dependability approach in theory and engineering of complex systems (not only computer systems and networks) is based on a multidisciplinary approach to system theory, technology, and maintenance of the systems working in real, very often unfriendly, environment. Usability and dependability concentrates on efficient realization of tasks, services, and jobs by a system considered as a unity of all technical, information, and human assets, in contrast to "classical" reliability, which is more restrained to analysis of technical resources. This difference has caused a natural evolution in the topical range of subsequent RelStat conferences, with an increased focus on dependability approaches over the classical reliability approach. Efficiency of different modes of transport; transport for smart city; reliability, safety, and risk management for transport applications; statistics, modeling, and multicriteria decision making in transportation and logistics; smart solutions, telematics, intelligent transport systems, innovative economics and education and training in engineering are the main topics of RelStat.

Second year the RelStat conference was supported by the HORIZON2020 funded project ALLIANCE (Enhancing Excellence and Innovation Capacity in Sustainable Transport Interchanges), led by the TTI, in collaboration with University of Thessaly (Greece) and the Fraunhofer Institute for Factory Operation and Automation (Germany). The special session entitled "Sustainable Transport Interchanges" was organized to allow collaborative research teams from Latvia, Greece, and Germany to present and discuss their findings in the areas of governance and policy development, smart solutions, and decision making.

The program committee of the 18th International RelStat Conference, the organizers, and the editors of these proceedings would like to acknowledge participation of all reviewers who helped to refine contents of this volume and evaluated conference submissions. Our thanks go to all members of program committee:

- Prof. **Igor Kabashkin**, Transport and Telecommunication Institute, Latvia—Chairman
- Prof. **Irina Yatskiv (Jackiva)**, Transport and Telecommunication Institute, Latvia—Co-Chairman
- Prof. **Irina Kuzmina-Merlino**, Transport and Telecommunication Institute, Latvia—Co-Chairman
- Prof. **Lutfihak Alpkan**, Gebze Institute of Technology, Turkey
- Prof. **Liudmyla Batenko**, Kyiv National Economic University named after Vadym Hetman, Ukraine
- Prof. **Maurizio Bielli**, Institute of System Analysis and Informatics, Italy
- Dr. **Brent D. Bowen**, Purdue University, USA
- Prof. **Inta Bruna**, University of Latvia, Latvia
- Dr. **Ilia B. Frenkel**, Industrial Engineering and Management Department, Sami Shamoon College of Engineering, Israel
- Prof. **Alexander Grakovski**, Transport and Telecommunication Institute, Latvia
- Prof. **Stefan Hittmar**, University of Zilina, Slovakia
- As. Prof. **Ishgaly Ishmuhametov**, Transport and Telecommunication Institute, Latvia
- Prof. Dr. **Nicos Komninos**, Aristotle University of Thessaloniki, Greece
- Prof. **Vulfs Kozlinskis**, Riga International School of Economics and Business Administration, Latvia
- Dr. **Gatis Krumins**, Vidzemes Augstskola, University of Applied Sciences, Latvia

- Prof. **Zohar Laslo**, Sami Shamoon College of Engineering, Israel
- Prof. **Agita Livina**, Vidzemes Augstskola, University of Applied Sciences, Latvia
- As. Prof. **Jacek Mazurkiewicz**, Wroclaw University of Technology, Poland
- Prof. **Massimo Merlino**, University of Bergamo, Italy
- Prof. **Boriss Misnevs**, Transport and Telecommunication Institute, Latvia
- Prof. Dr. **Andres Monzon de Caceres**, Universidad Politécnica de Madrid, Spain
- As. Prof. **Eftihia Nathanail**, University of Thessaly, Greece
- Prof. **Andrzej Niewczas**, Lublin University of Technology, Poland
- Prof. **Lauri Ojala**, Turku School of Economics, Finland
- Prof. **Ramunas Palšaitis**, Vilnius Gediminas Technical University, Lithuania
- Asst. Prof. **Dmitry Pavlyuk**, Transport and Telecommunication Institute, Latvia
- Prof. **Gunnar Prause**, Tallinn Technical University, Estonia
- Prof. **Olegas Prentkovskis**, Vilnius Gediminas Technical University, Lithuania
- Prof. **Svetlana Saksonova**, University of Latvia, Latvia
- Prof. **Natalia Salienko**, Bauman Moscow State Technical University, Russia
- As. Prof. **Mihails Savrasovs**, Transport and Telecommunication Institute, Latvia
- As. Prof. **Julia Stukalina**, Transport and Telecommunication Institute, Latvia
- Prof. **Juri Toluyew**, Transport and Telecommunication Institute, Latvia
- Prof. **Tatjana Volkova**, BA School of Business and Finance, Latvia
- Prof. **Edmundas Zavadskas**, Vilnius Gediminas Technical University, Lithuania

Thanking all the authors who have chosen RelStat as the publication platform for their research, we would like to express our hope that their papers will help in further developments in design and analysis of complex systems, offering a valuable and timely resource for scientists, researchers, practitioners, and students who work in these areas.

Igor Kabashkin
Irina Yatskiv (Jackiva)
Olegas Prentkovskis

Contents

Innovative Economics

Education and Training in Engineering

Transport for Smart City

Socio-Economic Aspects of Free Public Transport

Tarmo Tuisk[1(✉)] and Gunnar Prause[1,2]

[1] School of Business and Governance, Tallinn University of Technology, Akadeemia Tee 3, 12618 Tallinn, Estonia
{Tarmo.Tuisk,Gunnar.Prause}@taltech.ee

[2] Wismar Business School, Wismar University, Philipp-Müller-Str. 14, 23966 Wismar, Germany

Abstract. Since the 1st of January 2013 the public transport is free for residents of the City of Tallinn. People registered outside Tallinn who want to use the public transport have to pay for their trips. Parallel, Tallinn City Governance raised the parking tariffs in the city centre which caused an increase of using of the public transport by 10% and a decrease of car traffic in the city centre by 6%. Currently, the greening of urban transport enjoys a high rank on the political agenda in a lot of European and even some cities are thinking to follow the example of Tallinn. Literature review reveals that scientific papers are focussing mainly on ecological aspects whereas socio-economic studies are neglected. The paper will present insights in social, economic and political aspects which are related to the experience of free public transport in Tallinn. Furthermore, the paper will highlight the role of digitalisation in public transport and discuss the influence of the Estonian e-governance system for the success of the Tallinn case.

Keywords: Digitalisation · Free public transport · Social-economic analysis · Sustainability

1 Introduction

Greening of urban transport is high on political agenda after increasing pollution problems in cities all around the world. Different strategies for sustainable city transport are discussed comprising e-mobility, reduction of vehicles in city centres or free public transport. All these approaches have their pros and cons, but with the integration of digital technologies into transportation systems and infrastructures free public transport has gained importance.

City planners and transportation administrators have been looking for years on how to minimize the share of driving in cities in order to reduce the impacts of public space consuming, polluting and fuel-demanding cars. Knoflacher et al. [13] argue that there is a link between urban transport planning and the ranking of a city in "quality of living" by investigating Vienna as a case. Vienna kept the first rank in the Mercer Quality of Living Survey for the last 8 consecutive years. The survey compares the political,

I. Kabashkin et al. (Eds.): RelStat 2018, LNNS 68, pp. 3–13, 2019.
https://doi.org/10.1007/978-3-030-12450-2_1

social and economic climate, medical care, education, and infrastructural conditions such as public transport, power and water supply of 221 cities worldwide. Buehler et al. [5] highlighted that the roots of Vienna's position is much older than the history of Mercer studies. Vienna's government decided already in the 1970's to take away space from cars which caused a public outcry. An the complaints still goes on despite the fact that this reorganisation of public space and restriction of private motorised cars were the foundation of the attractive, healthy, fair and smart city that Vienna is today. Consequently, Knoflacher et al. [13] conclude that the experience from the past shows that unpopular decisions succeed on the long run. Parallel to the restriction of urban car traffic Vienna developed a high performing public transport system and a competitive tariff system for using the public transport. Currently, Vienna citizens can buy an annual public transport ticket for 365 Euros and with the lowering of the prices for the public transport ticket the number of owners of the annual public transport ticket increased from 373,000 in 2012 to 780,000 owners today with the consequence that more citizens of Vienna possess an annual public transport ticket than those who own a private car [2]. It has to be mentioned that the City of Vienna support annually the public urban transport with 400 million Euros. Despite the cases of Vienna and other cities a view to the statistics of the European Union reveals that the number of cars is still increasing.

So the question rises to which extent policy changes can improve the traffic situation in cities and what are the appropriate measures to force a modal shift in urban transportation. Pricing of these public transportation rides is often considered as a policy instrument that can be devised to influence or make the modal shift a reality. A Special Eurobarometer was carried out in May–June 2013 by European Commission [12]. The results showed that the Europeans are convinced about the two measures that can facilitate urban transport. These are lower fares (supported by 59%) and "better" public transport (56%). These measures were pointed out by all travel mode users. These two options were high in particular for those who considered road congestion to be an important problem. For 50% of the member states participating in the study namely lowering the fares was the most often supported instrument. At the same time respondents from Hungary (28%), Czech Republic (27%), Estonia (23%), Latvia (22%) and Romania (22%) reported higher levels of daily public transport use than the EU average (16%). In Austria the same indicator is 19%, in Germany 13% and in Netherlands just 8%.

The paper addresses the research question to determinate the socio-economic impact of free urban transportation and to distinguish it from low tariff urban public transport models. The research is based on expert interviews, secondary data research, statistical analysis and the case study of the Tallinn free public transport. The paper is organised as follows: after the introduction, the theoretical background of urban public transport are given. Then, the system of methods used for this research will be described followed the discussion of the findings. Finally, the results are summed up in the conclusions.

2 Theoretical Background

Despite the fact that the most of public transport systems need subsidies for operation, a typical financing process includes calculating a cost per rider and passenger fare. The price that all passengers have to pay to cover the cost of his/her personal use of this transport service (so-called passenger fare) can be obtained as one of three following ways. First, the passenger pays the total fare that is very rare in Western countries, to cover the operating costs of public transport. In the sequel we will call such a system a fare free public transport model and abbreviate it by FFPT. Second, the provider i.e. the public transport company pays the full passenger operating cost. However, both these kind of free public transport systems are not common in practice [7]. But the third way, where most passenger related costs are subsidized, and only a share of the cost is paid by the passenger (called 'partial cost recovery') has been largely used in practice (i.e. mixture of the first two options). This partly cost recovery model is realised in the City of Vienna with a very low share of only 1 € per day when buying an annual public transport ticket [2, 13]. A slight modification of the mentioned passenger fare models appears if a third party pays the total cost of the ride, based on an agreement between the external party and the transport system or the passenger. This has been applied usually in the cases when employers pay for their employees, also when social security institutions cover tickets for their clients, and when schools or universities cover for their students [4].

Historically FFPT is not a totally new idea. For decades there has existed public debate that for various reasons public goods and services should be free, granting universal access for everybody, also handling this way road congestion, and reducing negative environmental impact caused by various modes of urban transport. The idea of fare-free public services and goods addresses to the principle that access to the services like schools, libraries, museums, roads, green areas and Wi-Fi are free to use for everyone. By applying the same idea to public transport underlines that mobility as a "service", should be also fare-free as the rapid growth of cities forces people to settle more and more away from the city centre, further from the locations of schools and work places. Thereby, when the transport costs are high, this can be an obstacle for employees in approaching and participating in the labour market. In addition, an important argument that supports introducing of FFPT is improving social inclusion within the society. In respect of considering the modal shift, an analysis based on earlier research has demonstrated that urban dwellers are more likely to replace using of their private car to public transport when the price of car usage increases and not likely due to the decrease of the cost of using the public transport [6].

A sustainability check of the different urban transport modes leads to two main characteristics. First, the required urban space for the transport has been considered by several institutions. The required space is important to investigate when it comes to the aim to reduce traffic jams and to minimize the needed space for urban transport. The second important characteristic of the different urban transport modes concerns the energy consumption and related to that the linked emissions which are heavily influencing the air and life quality in the cities. By taking into account the results about space and sustainability in urban transport one can assume that a tramway requires

85 m^2 and is able to transport 145 persons which equals 124 cars that would need 950 m^2 during transport and in addition to that they require parking lots [16]. A closer look to the needed transport space for different transport modes shall be considered for the case of 50 persons. If 50 persons want to move from one destination to another they need by walking 50 m^2, by using a bike 580 m^2, by using a bus 70 m^2 and by using an average filled cars with about 1.3 persons per car 2375 m^2. Even by assuming a fully charged car the transport of 50 persons requires 610 m^2 of street space plus parking lots. In addition to that, the calculation of required public space is expressed by m^2h taking under account the number of hours the transport device participates in the public. This calculation yields very poor values for cars whereas public transport as well as car sharing delivers better results.

An analysis of emissions reveals that ca. 20% of EU emissions are related to transport and 19% are related to CO_2 emissions from transport. There are different studies concerning the emissions of transport so that in this paper we rely on the results of the German Federal Ministry of Environment which states that a private car emits in the average 206 g CO_2 per km [3]. Additional emissions are related to NOx, micro particles and noise. The pollution in big cities forced already the municipalities to reduce or even to stop the traffic in urban areas. This issue gained rising interest after the Diesel scandal of European diesel car builders. The mentioned study further reveals that a public bus in urban traffic with an average utilization consumes per 100 person km only half of fuel compared to a private car, i.e. public urban busses cause only about half of the CO_2 emissions for the same distance like a private car. Busses and trains are even more efficient in long distance traffic when it comes to the CO_2 emissions per person and trip. Table 1 gives an overview about the emissions of different modes in public transportation [3].

Table 1. Emissions of different modes of transportation [3].

		Private car	Long distance			Urban/short distance		
			Bus	Train	Plane	Bus	Train	Metro/tram
Greenhouse gases	g/Pkm	140	32	38	214	75	63	65
CO	g/Pkm	0.61	0.04	0.02	0.14	0.05	0.04	0.04
Volatile organic compounds (VOC)	g/Pkm	0.14	0.02	0.00	0.04	0.03	0.01	0.00
NOx	g/Pkm	0.35	0.18	0.05	0.57	0.32	0.18	0.06
Particulate matter	g/Pkm	0.004	0.003	0.00	0.005	0.002	0.002	0.00
Utilisation rate	g/Pkm	1.5 pers./car	60%	53%	80%	21%	27%	19%

3 Methodology

The paper addresses the research question to determinate the socio-economic impact of free urban transportation and to distinguish it from low tariff urban public transport models. For this purpose, the research design constitutes as follows. The empirical evidence in this paper is based on a mixture of the qualitative and quantitative research. The complexity of the research question requires on the one hand expert interviews and a qualitative approach to acquaint an in-depth view into the structures and the social-economic consequences of free public transport. The quantitative analysis is based on socio-economic data from municipal statistics for 2012, i.e. the year before FFPT was implemented in Tallinn, and the time after FFTP. The statistics data were complemented by information about cultural differences, distances, political voting and transport mode choice for different districts of Tallinn. The statistical analysis is using different forms of ANOVA and regression analysis. The paper targets on the investigation of the following research questions

- RQ1: What is the socio-economic impact of free public transport in Tallinn?
- RQ2: Is FFPT in Tallinn influencing the voting behaviour?
- RQ3: Is fare-free public transport model more feasible than low tariff model?

The empiric measures are combined with secondary data analysis, expert interviews and case studies.

4 Case Study: Free Public Transport in Tallinn

By January 2019 Estonia's capital Tallinn has reached the population of 453,000. This indicates that Tallinn is world's largest city that has implemented fare-free public transportation for its citizens. Still we have to note that the ridership of public transportation decreased by one third during first 20 years of Estonian re-independence from 1991 to 2012 [9].

The societal transformation that took place after collapse of Soviet Union's and its command economy being replaced by modern market economy caused the significant change that enabled people with new means possibilities that facilitated them to acquire more and more private cars and caused them to decrease consuming broadly exploited public transportation services which dominated under Soviet rule [10].

The municipality of Tallinn carried out a plebiscite for City residents during 19–25 March, 2012. The results showed that 75.5% of the voters supported fare-free public transport for the residents while 24.5% of the participants were against it. By informing this way—as the municipality leaders later admitted—the community was involved into decision-making and political decision was locked. The fact that the voters' turnout was just 20% was considered as unimportant as the decision that was expressed by the voters was not legally binding for the municipality. The City Council decided based on this vote that Fare Free Public Transport (FFPT) will be implemented since 1 January 2013. Already during the first year of FFPT it was extended to national train traffic within the borders of the City of Tallinn. The bus lines were prolonged 19–23 km after this change, but more importantly, new lines were implemented which trespassed the City centre where are/were the traffic congestions. Since 2013 Tallinn became the first capital in the EU providing free public transport to its citizens. To be entitled for FFPT the citizen must buy and own a contact free 'green card' (2 Euros). People registered outside Tallinn who wants to use the same card need to load money to that. Monthly ticket for them costs 23 Euros.

Since the collapse of the Soviet Union the number of the cars per 1,000 people increased twice. This was not only based on the availability of the means and a result of the change to market economy, but also due to the need to commute to these suburbs which were developed since 1991 and not conveniently accessible due to less developed public transport for these new areas [14]. Also, when comparing years 2003 and 2015 the mode shift that increased more even from the beginning of 2000s can be detected. While in 2003 24% of Tallinners reported driving car while bus usage was 41%, then for 2015 the share of driving was already 41% and public transport had dropped to 23% [15]. When comparing increase in private cars ownership between 2012 and 2013 the statistics show in Tallinn it dropped to 0.6%. In contrast, the statistics showed a continuing growth of annually 10% of private car ownerships in neighbouring municipalities like it was before also in Tallinn. Despite this quick change after implementation of FFPT the longer trend shows that the number of cars has not dropped in general. According to data from Estonian Road Administration (Eesti Maanteeamet) in 2013 Harju County (the area that surrounds the capital city Tallinn) had 254,222 private cars and of these 167,553 were registered in Tallinn. In 2018 Harju county had already 310,686 cars and of those 197,922 belonged Tallinn. So, the growth in ownership in Tallinn was 15% during the first 5.5 years of FFPT while for Harju County it makes 23%. This difference in car ownership is clearly visible and can be at least partly attributed to FFPT in Tallinn.

Table 2. Mode share by district before (2012) and after FFPT (2013).

District of Tallinn	Car		Bike		By foot		PT*		Other	
	2012	2013	2012	2013	2012	2013	2012	2013	2012	2013
Haabersti	39	38	0	1	5	3	56	58	0	0
Kesklinn	27	24	1	0	19	22	52	53	1	1
Kristiine	34	30	1	0	5	7	60	62	1	0
Lasnamäe	29	31	1	0	14	6	55	61	1	2
Mustamäe	26	31	0	1	10	5	63	62	1	1
Nõmme	48	28	1	0	8	3	43	69	0	0
Pirita	56	50	0	0	3	2	41	48	0	0
Põhja-Tallinn	27	23	1	2	16	7	56	68	0	0

*Public transportation.

Note: This comparative secondary data describing modal shifts across districts of Tallinn was first presented in Fig. 3 in Cats, Susilo and Reimal in 2017 [6]. The authors of the current chapter used this data as input to their ANOVA and regression analysis.

For deeper analysis of the socio-economic impact, official statistical data were used to execute multivariate methods. Therefore, for eight districts of Tallinn, i.e. for Põhja-Tallinn, Haabersti, Mustamäe, Nõmme, Kristiine, Kesklinn, Lasnamäe and Pirita, the use of private cars and public transport before and after the implementation of free public transport were taken (data source from Table 2).

In addition, the share of foreigners which mainly means Russian-speaking citizens, the share of vote for the Centre Party in the district as well as the average distance of the distance to the city centre was added. The political dimension, i.e. share of the Centre Party before and the after the introduction of free public transport is of special importance for Tallinn because the Centre Party holds the majority in Tallinn city government and before the implementation of the free public transport the opposition parties in Tallinn accused the Centre Party for having decided for FFPT only due to the fact that mainly the voters of their party will benefit. The statistical analysis used mainly ANOVA and regression analysis where the impact of the underlying free variables is measured by the β-coefficient.

5 Findings and Discussion

As Mr. Allan Alaküla (Representative of Tallinn Municipality to EU) has explained several times in public—the impact can be understood in terms of indirect indicators. According to him there is no proper technology to estimate and measure the impact. He has claimed that people's mobility increased, not only for unemployed, but also for those who looked better jobs use public transport more often. People started to go out more in the evenings and weekends, for consumption of local goods and services. They save from paying taxi services, emission of CO_2 is decreasing. Families go out more on weekends, leaving their cars at home [1].

Municipality's calculations show that during 2013–2017 the number of Tallinners increased by 35,000 residents (from 410,000 to 445,000). Every 1000 of new residents brings about 1 million Euros to the city budget (as part of their income tax goes directly to the municipality where they are registered). When comparing the funds made available for public transport in 2012 and 2017, we can see that there was an increase from 53 million Euros to 63 million Euros while at the same time the total share spent for public transport makes now 2.5% of yearly municipal budget. In 2012 tickets revenue was 17 million Euros (12 million Euros from residents and 5 million Euros from non-residents of Tallinn). By making the city transport free, the government lost 12 million Euros per year, but by attracting new taxpayers to register in Tallinn the win was about 20 million Euros in a year. In total change added to the budget 8 million Euros. How the increase in these tax-payers number was gained? Actually, most of these people who registered themselves where already living in Tallinn. When they registered themselves in Tallinn they transformed themselves also to taxpayers. The timing of the implementation of the FFPT was also important to the municipality that was led by Centre Party over 20 years. Interestingly the promotion of FFPT in Tallinn started just 18 months before the municipal elections took place. The elections were in October 2013 and the Centre Party won 58.2% of votes. After seeing this result the shift to FFPT was explained as an instrument to ensure Centre Party's re-election in

Tallinn. Also, it became evident that FFPT was implemented by the municipality without consultations with urban planning experts and transport administrators. Although a special urban transport visionary plan for 10 upcoming years was under development and in a stage to be nearly ready and completed, it was discarded in the decision-making. This plan itself did not propose any options for FFPT anyway. According to critics the municipality artificiallycreated a demand for FFPT that was not never seen as any measure necessary for transport problems solving [10].

Research has even shown that the municipality leaders had considered tax revenue to be as unimportant aspect and the highest importance of this shift to FFPT was attributed just for ensuring re-election of the Centre Party. In short, the implementation of FFPT can be called as "administratively created tax incentive scheme" [9].

Neighbouring counties and municipalities of Tallinn are not impressed about Tallinn's FFPT. The tax revenue (to the municipalities) of these 35,000 people who had registered to Tallinn has been taken from their budget. To support the citizens of these neighbouring counties and their mobility in Tallinn, special agreements for transportation discounts have been made between these areas and the municipality of Tallinn.

The goals of the implementation of FFPT in Tallinn were not easily found in official public documentation dated by municipality as of 2012 or earlier i.e. before FFPT started, but these can be found in different presentations given after the change was implemented [10]. Nevertheless, according to studies one important result of the implementation of FFPT in Tallinn was the increase of use of the public transport by +10% which was accompanied by a slight decrease of car traffic in the city centre by −6% and a slight increase of car traffic around the City centre by +4%. One important parallel to the start of the FFPT in Tallinn was the increase of parking tariffs in the city centre which are currently 6 Euros per hour in the old town and 4.80 Euros in the city centre so that decrease of car traffic in the city centre can be considered as a combination of FFPT together with the increase of parking tariffs.

But more public transportation means also less wasted space on streets and less emission of GHG. 167,553 cars in Tallinn emitted 140 g CO_2/Pkm in 2012 while after the modal shift (2012/2013) from where car usage dropped 3%, the emission of the CO_2 dropped 703 kg/Pkm in total. This means that 7540 persons who had previously used cars started to use public transportation. The same number of persons saved $151 \times 610 = 92{,}110\ m^2$ of space on the streets of Tallinn if their cars were full before their modal shift, but if they used 1.3 people in cars in an average car, then there was saved $151 \times 2375 = 358{,}625\ m^2$ of street space. The savings are remarkable despite a small drop of car passengers. The issue is not only about saving public street space during congestions, but in combination with emission gases and psychological factors which concern all city inhabitants (e.g. causing stress) who participate in traffic not depending what type transport they use.

The analysis of the statistical data revealed a couple of important results. Firstly, the 2-factorial ANOVA showed that in districts with higher share of foreigners the Centre Party significantly higher votes than in districts with low percentage. But it also turned out that there were no significant changes in the share of votes that the Centre Party got before and after the start of FFPT. Consequently, the decision for FFPT did not significantly influence the share of votes on the political level, neither in districts with high share of foreigners nor in those with low share.

For analysing the changes in used transport mode before and after the start of FFPT, the share of use of public transport was considered. Therefore, we analysed the usage rate before and after the FFTP by ANOVA. Here we found a light significant increase for the usage rate from 53% before the FFPT to 60% after the implementation. It has to be added that the slight significant means a p-value of 7.5% which is slightly higher than the usual 5%. Nevertheless, it became visible that the introduction of FFPT impacted positively the use of public transport in Tallinn. The same ANOVA approach for the change in car use before and after the FFPT did not show a significant change, i.e. the car usage rate went slightly down, but there is no significant statistical evidence for the decrease after FFPT.

Finally, by using regression analysis the impact of the four independent variables, share of Centre Party voters and foreigners as well as the distance from the city centre and the start of the FFPT, was analysed. It turned out by considering the β-coefficient that the distance to city centre represented the most important variable with 60% to explain the change in bus usage, followed by the possibility to use the public transport freely with 19% and the share of foreigners in the district with 14%. The share of votes of the Centre Party in the district only explained 7% of the changes in bus usage. Consequently, the political impact of the implementation of the FFPT in Tallinn is negligible and other topics like distance from the city centre, the offer to use public transport freely and the social composition of a district are more important for explaining the results.

Despite the contradictory findings data show that the ridership of public transportation in Tallinn has increased 14% after implementation of FFPT and modal shift of in the city towards using public transportation has increased from 55% to 63%. At the same time this shift is largely based on increase from walking mode to using public transport (as average passenger trip became shorter by 10%). The impact of the FFPT implementation was studied also by using data gathered via detailed automatic vehicle location (AVL) and from aggregate automatic passenger counter (APC). In average, an increase of only 3% in passenger utilisation of transport took place, and from this 1.2% can be directly understood as effect of FFPT when eliminating simultaneous changes which had been applied to the transport system [6].

According to Fig. 1 the increase of using public transportation took place in 2013 after the implementation of FFPT. Still, low fuel price at the world market most probably encourage the use of car transport, and after oil price started to raise the car usage turned slightly down by 2017. This data also supports the finding that people stopped to walk small distances in 2013 when they could use public transport instead. Later the variation in walking to work or school has been 14–16% that shows quite stable trend. The same study also mentioned that in 2017 women were more likely to use public transport when compared to men (54% versus 37%) and males are more likely to use car (47% versus 26%). Going by foot is more preferred by females than males (18% versus 14%). The study shows also that 57% of the PT (public transportation) users are 60 years old or older. These patterns are also supported by Hofstede's national cultural indicators on Estonia which underline the country with dominating masculine values like aspirations towards high competitiveness and success [11]. Thereby it can be said that ecological aspects and sustainability as values are

pl-
ac-
ed
lo-

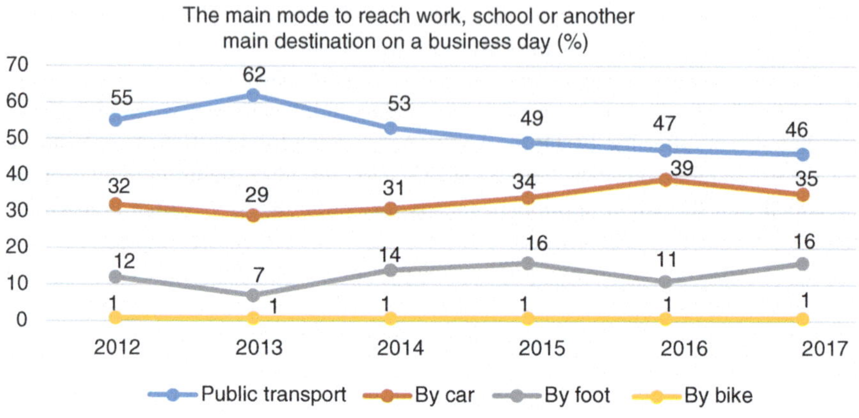

Fig. 1. Source data adapted from *Satisfaction of inhabitants with public services in Tallinn in 2017.* A study carried out by Eesti Uuringukeskus, ordered by Tallinn Municipal Office [8].

wer than material values and desire for being young and successful, accompanied by showing off your property (e.g. cars on the street).

Despite these above presented contradictory data and intertwining of municipal decisions and politics, Tallinn's fare-free public transportation has been vital for five and half years, and there is no evidence of major setbacks. Additional more systematic studies will be necessary to investigate the phenomenon of FFPT in Tallinn.

6 Conclusion

As the results show the implementation of FFPT has been an instrument to ensure re-election of the leading party into city government. Although the criticism been expressed by academic researchers and the press, this has been not sufficient. The impact of FFPT in Tallinn has been also minimal when comparing car ownership in 2013 and 2018 in Tallinn. Increasing number of car owners seems to be inevitable, and most probably raising maintenance, taxes or fuel costs will make these car drivers much more knowledgeable of using FFPT than just changing their habits based on ecologic thinking and fare-free rides. It is possible to conclude that from financial point of view the shift of 2013 to FFPT has been successful for the City of Tallinn, but not from ecological point of view as the total number of car ownership increasing, though not as fast as in the neighbouring Harju county.

As our analysis showed, also political influence to the voters has been negligible. On the other hand, the longer FFPT lasts the more Tallinn citizens will be "spoilt" by using the public transport for free, and once there is an emergency to charge again a fee it will be complicated to predict the political consequences since it will be sensitive for electorate of any party backing this kind of reversal. Furthermore, the findings of this article show that FFPT has been accepted as public benefit for majority of population, but the male-dominated car owners have not demonstrated any conscious approach changing their attitudes towards greener thinking. In the current situation it is hard to argue if a low tariff model like in the City of Vienna would have changed the attitudes of Tallinn citizens since a lowering of the tariffs or the implementation of annual tickets was never offered for them.

References

1. Alaküla, A.: Forum Smart City du Grant Paris, 29 November 2017. https://www.tallinn.ee/eng/freepublictransport/Forum-Smart-City-du-Grand-Paris. Accessed 14 April 2018
2. Berger, M.: 365 Euro für ein Jahresticket—deutsche Städte wollen Wiener Modell übernehmen, Tagesspiegel, 29 June 2018, Berlin (2018)
3. Bundesministerium für Umwelt, Naturschutz, Bau und Reaktorsicherheit: CO_2 Emissionen im Verkehr in Deutschland, Bundesumweltamt Deutschland, Berlin (2016)
4. Brown, J., Hess, D.B., Shoup, D.: Unlimited access. Transportation **28**(3), 233–267 (2001)
5. Buehler, R., Pucher, J., Altshuler, A.: Vienna's path to sustainable transport. Int. J. Sustain. Transp. **11**(4), 257–271 (2017)
6. Cats, O., Susilo, Y.O., Reimal, T.: The prospects of fare-free public transport: evidence from Tallinn. Transportation **44**(5), 1083–1104 (2017)
7. De Witte, A., Macharis, C., Lannoy, P., Polain, C., Steenberghen, T., Van de Walle, S.: The impact of "free" public transport: the case of Brussels. Transp. Res. Part A **40**, 671–689 (2006)
8. Eesti Uuringukeskus.: Satisfaction of inhabitants with public services in Tallinn in 2017. https://aktal.tallinnlv.ee/static/Koosolekud/Dokumendid/pkldok18990.pdf. Accessed 25 June 2018
9. Hess, D.B.: Decrypting fare-free public transport in Tallinn, Estonia. Case Stud Transp Policy, 690–698 (2017)
10. Hess, D.B.: Transport in mikrorayons: accessibility and proximity to centrally planned residential districts during the socialist era, 1957–1989. J. Plan. Hist. **17**(3), 184–204 (2018)
11. Hofstede, G., Hofstede, G.J.: Cultures and Organizations: Software of the Mind. McGraw-Hill, New York (2005)
12. European Commission: Directorate-General for Mobility and Transport (2013). Attitudes of Europeans Towards Urban Mobility. Special Eurobarometer 406. http://ec.europa.eu/public_opinion/archives/ebs/ebs_406_en.pdf. Accessed 14 April 2018
13. Knoflacher, H., Freya, H., Letha, U.: Smartness of urban mobility and "quality of life" in Vienna. In: Proceedings of 7th Transport Research Arena TRA 2018, 16–19 April 2018, Vienna, Austria, Paper #10342. TRA, Vienna (2018)
14. Poltimäe, H., Jüssi, M.: Factors affecting choice of travel mode in Tallinn. In: Swain, R.B. (ed.) Environmental Challenges in the Baltic Region, pp. 135–153. MacMillan, Palgrave (2017)
15. Põldemaa, K.: Tallinna elanike liikumisviisid ja selle muutused. Modal split and it's changes of Tallinn residents. Tallinn University of Technology. Bachelor's thesis, Tallinn (2016). https://digi.lib.ttu.ee/i/file.php?DLID=6518&t=1. Accessed 23 Jan 2019
16. Verkehrsclub Österreich: Urbaner Verkehr der Zukunft. Schriftenreihe Mobilität mit Zukunft 1/2016. Verkehrsclub Österreich, Wien (2016)

Evaluation of the Quality of Public Transport Stops

Ivana Olivková(✉)

Faculty of Mechanical Engineering, VŠB - Technical University of Ostrava, 17. Listopadu 15, 708 33 Ostrava-Poruba, Czech Republic
ivana.olivkova@vsb.cz

Abstract. The role of the public transport is to secure transportation requirements at the required qualitative level. Current level of the public transport capacity is first and foremost the reflection of the transport services quality level. The particular offers, benefits and disadvantages within the network of the transport system have proven complicated to follow for passengers. Passengers' decision depends on the offered transport services' quality. Stops are the gateway to public transport, a place where citizens hesitate if to use public transport services. It depends on the quality of the stops whether the public transport offer is attractive to passengers. According to the quality analysis of the provided services in public transport, which has been carried out in previous years, the quality of public transport stops is one of the most problematic. The article deals with public transport quality evaluation. Methods of the public transport quality evaluation are theoretically described in detail. Methodology of the public transport stops quality evaluation was evolved. The methodology was experimentally verified by conducting a transportation passenger satisfaction survey, which was organized in 2017. The experiment was based on a questionnaire, which was used for a traffic survey of the passengers. This article presents the results of implemented methodology applications, including analysis of the survey using SWOT analysis. Finally, the advantages and opportunities of the practical application of the methodology are assessed.

Keywords: Public transport · Methodology · Evaluation of the quality · Public transport stops

1 Introduction

The issue of measuring the quality of services from a theoretical and practical perspective has already been processed with in many areas [1–3]. In transport this problem was solved later than in other service sectors [4, 5].

Therefore methodology of the public transport quality evaluation has been evolved [6]. In the following years, this methodology was applied to assess the quality of public transport stops. The original findings from the practical application of this methodology are presented in this article. The goal is to present both the theoretical and practical findings. The methodology presented and its experimental verification is the original work of the author.

I. Kabashkin et al. (Eds.): RelStat 2018, LNNS 68, pp. 14–23, 2019.
https://doi.org/10.1007/978-3-030-12450-2_2

2 Public Transport Stops

The stop means any point other than a terminal where according to the specified route a regular service is scheduled to stop for passengers to board or alight. In urban public transport (PT) a stop is usually established for each direction separately. One stop name typically has a pair of stops, sometimes several stops in a point of transfer within or between different transport modes.

There are various PT stops on the PT network. The equipment of the stops varies depending on several factors. Therefore, 3 categories of stops are defined according to their importance. The main parameter determining this importance is the location of the stop, respectively its transport significance.

The stops are divided into three categories: category 1—significant interchange stops, category 2—stops in city centres, major stops in the city and category 3 – urban stops, minor stops in cities.

3 Methodology

The methodology for evaluation of the quality of public transport (PT) can be divided into four steps detailed in the following chapters.

3.1 Defining Quality Criteria

The definition of the criteria is important step of evaluation of the quality. Quality is a multi-criteria phenomenon for which the number of criteria is infinite. The criteria for evaluation PT quality can be separated into groups according to the various viewpoints. The EN 13816 standard [4] lists quality criteria for public transportation that are only general in nature. This is why were created a criteria set that can be used for evaluation of the quality of PT stops.

Following the described previous analysis, eight criteria were defined for evaluation of the quality of PT stops. These eight criteria fulfil and represent the concept of "the quality of PT stops" from the viewpoint of the passengers (see Table 1).

Table 1. Defining quality criteria.

Number of criteria	Criteria
1	Stopover shelter
2	Ticket machine
3	Trash bin
4	Information panel
5	Notice board with timetable
6	Sign stop
7	Bench
8	Stop lighting

All criteria listed in Table 1 are qualitative in nature (values are expressed on an ordinal scale).

3.2 Establishing Quality Criteria Weight

The evaluation method must first establish the weight of the individual criteria that express the numeric meaning of these criteria (the significance of the criteria from the evaluator's standpoint). The following process of calculating criteria weight was applied:

1. The respondents must prioritise the order of criteria based on their own subjective opinion.
2. Based on this criteria ranking, the non-normalised weight of individual criteria is calculated, and is then normalised so that the sum of the weights is equal to one.

The following relationship is applied for establishing the non-normalised weight:

$$k_i = n + 1 - p_i, \tag{1}$$

k_i—non-normalised weight of i-criteria; n—quantity of criteria; p_i—ranking of i-criteria in its preferential order.

Due to the requirements of the comparability of criteria weights established by various methods, it is necessary to normalise these weights (the sum of the normalised weights of the set is equal to one). Criteria weight normalisation is carried out according to the following relationship:

$$v_i = \frac{k_i}{\sum_{i=1}^{n} k_i}, \tag{2}$$

v_i—normalised weight of i-criteria; k_i—non-normalised weight of i-value criteria; n—quantity of criteria.

3.3 Evaluation of the Quality Criteria

In the evaluation of public transport quality criteria, there is the problem, because the criteria are qualitative in nature (values are expressed on an ordinal scale). The means to achieve a statistical assessment typical for metrical scales while using ordinal rankings is through metrisation, i.e. assigning utility value. For every evaluation (descriptor), each quality criteria is given a precisely defined utility value. By assigning a descriptor, passengers determine the level at which the given criterion meets his/ her level of evaluation (Table 2).

Through the use of a descriptor for evaluation of quality criteria, where "very satisfied" represents the best score and "very dissatisfied" is the worst, passengers assign the utility value $u_i = 1$, $u_i = 0.75$, $u_i = 0.5$, $u_i = 0.25$ or $u_i = 0$.

Table 2. Determination of the utility value of the criteria.

Descriptor	Utility value
Very satisfied	1
Satisfied	0.75
Neither satisfied nor dissatisfied	0.5
Dissatisfied	0.25
Very dissatisfied	0

3.4 Comprehensive Evaluation of the Quality Criteria

For a comprehensive evaluation of the quality of PT stops can be applied comprehensive utility value, which is set for various categories of PT stops. The comprehensive utility value is the expression of a comprehensive evaluation of the quality of PT stops of the relevant category.

The utility values for the evaluated categories of PT stops (j = 1, 2, 3) can be calculated for selected criteria (i = 1, 2, …, 8). In this case the following relations can be used.

$$U_j = \sum\nolimits_{j=1}^{m} \sum\nolimits_{i=1}^{n} u_{i,j} \cdot v_{i,j}, \tag{3}$$

U_j—comprehensive utility of j-categories; $u_{i,j}$—partial utility of i-criteria of j-categories; $v_{i,j}$—normalised weight of i-criteria of j-categories; n—quantity of criteria; q—quantity of categories.

4 Passenger Survey Questionnaire

In order to come to all of the necessary conclusions, a questionnaire has been created to surveys urban PT users. To ensure maximum clarity, a transportation psychologist was consulted for the formulation of the questions contained in the questionnaire. The questionnaire is comprised of an introduction and four sections.

The introduction outlines the purpose and aim of the questionnaire for the passengers. Anonymity ensures greater objectivity and respondents' willingness to supply the required information.

The first section of the questionnaire, "Respondent information" contains questions regarding: sex, age, intensity of urban PT use. This information can help divide respondents into categories, because they can affect the preferences or ratings of the individual criteria.

The second section of the questionnaire, "Information on the stops" contains questions regarding names of the departure, transfer, and destination stops of the urban PT the respondent usually uses. Questions regarding the names of stops serve to determine the category of the stops.

The third section of the questionnaire: "Establishing urban PT quality criteria weight." This section includes an easy-to-understand method for establishing the preferential order of criteria. The traveller indicates the order of criteria according to their preferences directly into the respective column of the questionnaire. Due to the number of surveyed criteria (8) and categories of the stops (3), which the respondent usually uses, it was necessary to prepare eight rows and three columns of the questionnaire. Based on their subjective opinion, respondents were asked to establish a preferential order: PT stops quality criteria from the most important (this criterion then occupies the first order position) to the least important (eighth in the preferential order) for category 1—significant interchange stops, category 2—stops in city centres, major stops in the city and category 3—urban stops, minor stops in cities.

The fourth section: "Evaluation of the quality PT stops." By marking the descriptor, the respondent evaluates the quality criteria of PT stops. The descriptors (descriptions of the meaning of the individual levels of the quality) is sufficient for a verbal description of the gradation of criteria intensity. Its metrisation is carried out so that for every evaluation (descriptor), each quality criteria is given a precisely defined utility value.

An odd number eliminates the problem of respondents who cannot decide which ranking to give to quality criteria and can choose the centre. Due to the number of surveyed criteria (8), the number of descriptors (5) and categories of the stops (3), which the respondent usually uses, it was necessary to prepare eight rows and five columns of the questionnaire for every category of the PT stops. Based on their subjective opinion, respondents were asked to evaluate the quality criteria of PT stops for category 1—significant interchange stops, category 2—stops in city centres, major stops in the city and category 3—urban stops, minor stops in cities.

5 Methodology Applications Results

The methodology for evaluation of the quality of PT stops described above was implemented in Ostrava from September 2017 to December 2017. A total of 576 respondents were surveyed. These surveys were carried out in the form of personal interviews. Respondents filled in the questionnaire. The appearance of the questionnaire has been modified so that it can be filled out quickly and easily. It is based on the condition that the interviewers are trained students that supervise the course of the completion of the questionnaire by the respondent. The clarity of the questionnaire was verified through a trial survey.

One specific component of the passenger survey was the fact that respondents were surveyed at their place of employment (or at school), which was chosen due to the time needed to complete the questionnaire. The scope of the questionnaire corresponds to a polling period of approximately 10 minutes to ensure a comprehensive view of PT services. An analysis of the mass transit timetable in Ostrava determined that, during rush hour, the majority of passengers are available for a maximum of five minutes while waiting for transportation on their way to work (school), which is not sufficient time for accurate and complete participation in this survey. Conducting a poll in the transportation vehicle during rush hour is practically impossible. In these cases, passenger motivation for completing a survey decreases, there would be a decreased number of respondents, and this, in turn, would result the reduced quality of the entire survey. On the other hand, truncating the survey would be to the detriment of the assessment and goal of the survey itself. There have been almost none regular PT surveys, and the impact of individual quality components on overall quality of PT stops are unknown. For this reason, the scope of the surveyed PT quality components in the questionnaire must be preserved.

5.1 Evaluation of Respondent Information

Evaluation of respondent information is depicted in Table 3, which presents both absolute and relative frequencies, expressed in percent.

Table 3. Evaluation of respondent information.

Resp. data	Class	Absolute frequency [person]	Relative frequency [%]
Sex	Man	248	43
	Woman	328	57
Age	till 26	92	16
	26–44	216	38
	45–59	195	34
	from 60	73	8
Frequency of use of urban PT	Daily	386	67
	3–4 times a week	114	20
	1–2 times a week	76	13

The city of Ostrava and the Czech Republic overall demonstrates a high frequency of urban PT (approximately 70% of Ostrava residents utilize urban PT daily for regular travel). This is why the survey focused on these regular travelers.

5.2 Evaluation of Information on the Stops

In the section of the questionnaire, "Information on the stops" the respondents answered on the questions regarding names of the departure, transfer, and destination stops of the urban PT the respondents usually uses. Questions regarding the names of stops serve to determine the category of the stops. Evaluation of information on the stops is depicted in Table 4, which presents both absolute and relative frequencies of categories of evaluated stops and percentage overall utilization these PT stops.

Table 4. Evaluation of information on the stops.

Category	Absolute frequency [stop]	Relative frequency [%]	Overall utilization [%]
1	158	21	80
2	250	33	15
3	348	46	5

The proposed categorization of stops on the 3 groups can be described as analogous to the ABC analysis used in inventory management. Similarly, there is an analogy to Paret's addiction, which claims that approximately 20% of the most important storage items account for 80% of the turnover. According to the above model is valid, that at 20% most important PT stops are 80% of passengers of total passengers which there getting in, getting off or changes vehicle. This proves that there is a need to survey of PT stops equipment depending on their importance.

5.3 Evaluation of Criteria in Terms of Subjective Importance

The process described in the previous chapters was used to calculate the weights of individual criteria of the quality of public transport stops. From the collected data, average percentage representations of weight (level of relative importance) can be determined for significant interchange stops—category 1 (see Fig. 1), stops in city centres, major stops in the city—category 2, urban stops, minor stops in cities—category 3.

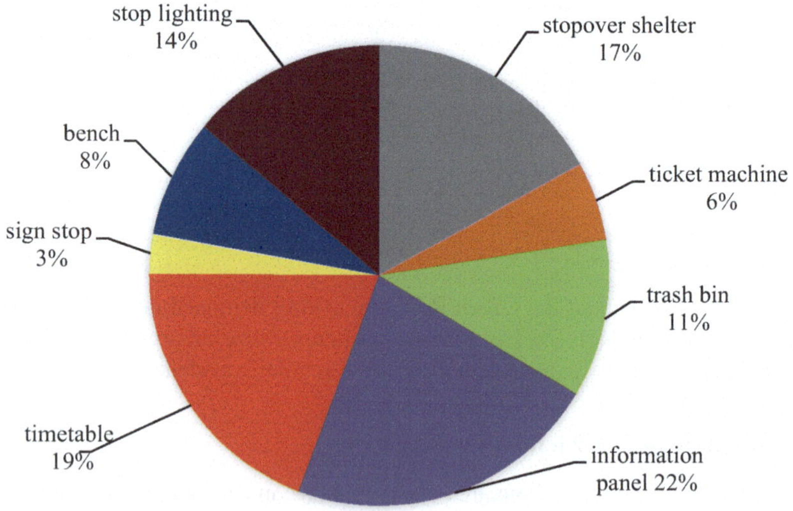

Fig. 1. Average evaluation of the importance of quality criteria of PT stop (category 1).

Figure 1 indicate weight ranking of quality criteria of PT stops (category 1). Other categories of stops (categories 2 and 3) are further evaluated in Table 5. The most important quality criteria of PT stops are information panel, timetable, stopover shelter and stop lighting. From the point of view respondents these criteria are the most important in the case of the PT stops of category 1 as well as category 2.

Table 5. Average values of individual criteria for the categories of PT stops.

Criterion(i)	Weight of i-criteria of j-categories (v_{ij})			Partial utility of i-criteria of j-categories (u_{ij})		
	Category(j)					
	1	2	3	1	2	3
Stopover shelter	0.17	0.14	0.19	0.94	0.68	0.55
Ticket machine	0.06	0.08	0.11	0.62	0.33	0.05
Trash bin	0.11	0.06	0.03	0.92	0.74	0.62
Information panel	0.22	0.22	0.08	0.78	0.54	0.08
Timetable	0.19	0.19	0.22	0.99	0.88	0.71
Sign stop	0.03	0.03	0.06	0.83	0.64	0.43
Bench	0.08	0.11	0.14	0.93	0.67	0.38
Stop lighting	0.14	0.17	0.17	0.96	0.72	0.21
Comprehensive utility of j-categories (U_{ij})	0.89	0.62	0.41			

5.4 Evaluation of the Quality of PT Stops

The procedure for evaluation of the quality of PT stops is described in the previous chapter. Table 5 lists the average values of weight (level of relative importance) and the average values of utility of individual criteria (i = 1, 2, …, 8) for the evaluated categories of PT stops (j = 1, 2, 3), that were calculated overall for all passengers (respondents) who participated in the survey.

It can be stated that all criteria for the evaluated category 1 and seven criteria for category 2, have partial utility values scored on average, above 0.5, i.e. passengers are satisfied with them. Below this threshold are 5 criteria for the category 3. The passengers are dissatisfied with these criteria and that means the potential for improvement for the public transport company. The following section outlines the criteria of quality PT stops, which should improve as soon as possible.

5.5 Comprehensive Evaluation of the Quality Public Transport Stops

The procedure for comprehensive evaluation of the quality of PT stops is described in the previous chapter. The comprehensive utility of each category of PT stops can be calculated based on the weight of the criteria and the partial utility of each criterion. The comprehensive utility of the category of the stops is then determined in accordance with Table 5 as the weighted sum of partial utility.

Comprehensive evaluation of quality was done in the form of a SWOT analysis. Specifically SWOT analysis was used to evaluate the carrier's position by identifying its strengths, weaknesses, opportunities and threats. In the case of SWOT analysis, this is a two-dimensional graph. In the graph are graphically shows the utility values of the given criteria (vertical axis) and their real significance (horizontal axis). In order to be possible to interpret and evaluate the importance of individual criteria for another carrier decision-making, the each SWOT table is divided by a horizontal and vertical line into four quadrants. The horizontal dividing line represents the average value of utility, the vertical dividing line represents of the position of the real significance of all the criteria—the median of the weights of the criteria. This is created a SWOT table with containing 4 areas.

For a comprehensive evaluation of quality was chosen for this article Category 2—stops in the city center, the main station in the city (see Fig. 2).

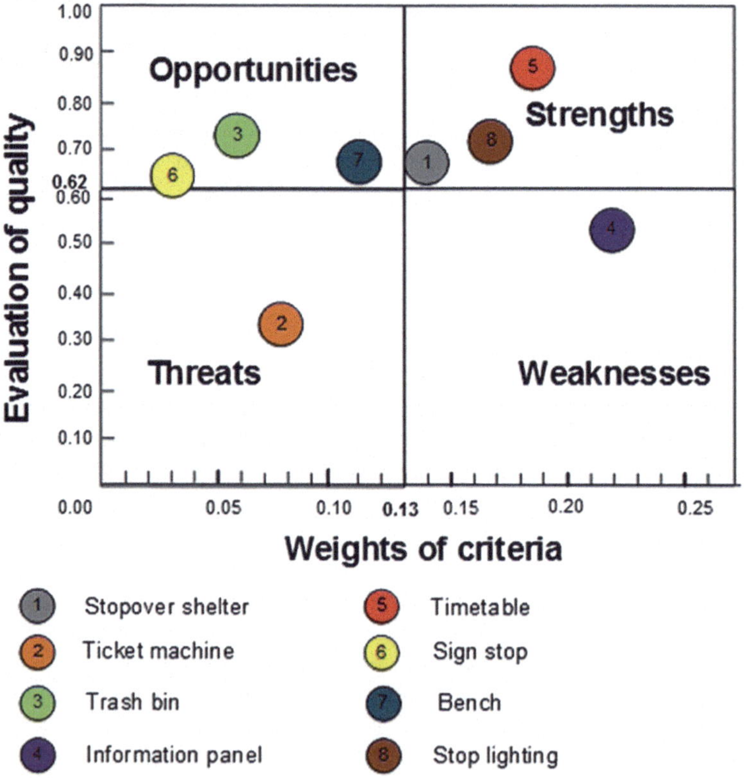

Fig. 2. Evaluating the results for the PT stops for the category 2 (SWOT).

In the lower right corner called "Weaknesses" is the criterion that has a significant effect on evaluation of quality and it is negatively evaluated. This criterion represents a significant threat for the public transport company, is urgently needed to correct. Only 1 criterion of 8 evaluated is located in the "Weaknesses": information panel.

In the upper right corner called "Strengths" are the criteria that have a significant effect on evaluation of quality and are also positively evaluated. These criteria indicate the quality of services and the carrier may be satisfied with their evaluation. Three criteria are located in the "Strengths": notice board with timetable, stop lighting and stopover shelter.

In the upper left corner called "Opportunities" are the criteria that have a relatively small effect on evaluation of quality, but they are positively evaluated. Three criteria are located in the "Opportunities": the layout of the bench, trash bin, and sign stop.

In the lower left corner called "Threats" is the criterion that has a small effect on evaluation of quality and it is negatively evaluated. Only one criterion is located in the "Threats": ticket machine. But attention should be paid to this criterion. In the case of increases in its importance is transferred to the critical area "Weaknesses".

6 Conclusions

This article studies the issues of evaluation of the quality of public transport. It focuses on the methodology of the public transport stops quality evaluation and its experimental validation by the traffic survey of the passengers in Ostrava urban PT system.

The experimental verification indicated the following conclusions:

1. The advantage of the methodology is his theoretical base.
2. Created methodology has benefit for transport science, because presently does not exist any complex model.
3. PT quality can be evaluated by the methodology with using qualitative criteria but does not limit their explanatory value.
4. The results of the traffic survey are very important for the methodology evaluation. The results show suitability of the methodology for practical exploitation, due the following reasons: passengers' expectations identification possibility and present quality level identification possibility. The methodology shows reasons of passengers' dissatisfactions and a benefits and disadvantages of the carrier.

Acknowledgements. This paper was supported by the internal research project SP2018/110 Research and Development in Transport. This study, survey and questionnaire is surely carried out according to ethical policies.

References

1. Hayes, B.E.: Measuring Customer Satisfaction. ASQ Quality Press, Milwaukee (1998)
2. Hill, N., Roche, G., Allen, R.: Customer Satisfaction: The Customer Experience Through the customer's eyes Paperback. Cogent Publishing, London (2003)
3. Nenadál, J., Petříková, R., Hutyra, M., Balcarová, P.: Modely měření a zlepšování spokojenosti zákazníků. Národní informační středisko pro podporu jakosti, Praha (2004)
4. European standard EN 13816: Transportation—Logistics and services—Public passenger transport—Service quality definition, targeting and measurement. European Committee of standardization, Brussels (2002)
5. European Standard EN 151s40: Public Passenger Transport—Basic Requirements and Recommendations for Systems that Measure Delivered Service Quality. European Committee for Standardization, Brussels (2006)
6. Olivkova, I.: Evaluation of quality public transport criteria in terms of passenger satisfaction. Transp. Telecommun. **17**(1), 18–27 (2016). https://doi.org/10.1515/ttj-2016-0003. ISSN 1407-6179

Environmental Safety of City Transport Systems: Problems and Influence of Infrastructure Solutions

Irina Makarova, Ksenia Shubenkova(✉), Vadim Mavrin, and Kirill Magdin

Kazan Federal University, Naberezhnye Chelny, Russian Federation
kamIVM@mail.ru, ksenia.shubenkova@gmail.com,
vadim_mmite@rambler.ru, irina.sippel@yandex.ru

Abstract. Improving the economy and environmental friendliness of transport is one of the main directions of transition to a low-carbon economy. The article indicates that road transport has a significant negative impact on the environment. An assessment of the negative impact and benefits of motor vehicles' transition to alternative fuels is provided. The main directions of transition to the vehicles run on alternative fuels are overviewed in the article. Approaches to reduce environmental damage from motor vehicles are analyzed. The article presents the research results on the example of the bottlenecks of the city road network. The developed simulation model for choosing the optimal configuration of the road network is presented. When constructing the simulation model, the data of full-scale surveys of the structure and intensity of the city traffic flows were used. These surveys were performed together with measurements of the quality of atmospheric air. The ecological effect of the proposed measures is calculated. It is noted that the improvement of the road infrastructure provides the greatest effect of reducing the negative impact on the environment at comparable costs.

Keywords: Transport emissions · Sustainable transport · Simulation

1 Introduction

Modern human civilization has entered the third Millennium and has faced with global challenges. Urbanization is one of the causes of majority problems of our Millennium. According to a new UN DESA report "World Population Prospects: The 2015 Revision" [1], the current world population of 7.3 billion is expected to reach 8.5 billion by 2030, 9.7 billion in 2050 and 11.2 billion in 2100. The world has experienced unprecedented urban growth in recent decades. As the map "Largest Urban Agglomerations" [2] shows, just three cities had populations of 10 million or more in 1975, one of them in a less developed country. There were 16 megacities in 2000. By 2025, 27 megacities will exist, 21 in less developed countries. Though the world's cities occupy just 3 per cent of the Earth's land, they consume 60–80% of the whole world's energy and produce 75% of carbon emissions. Rapid urbanization is exerting pressure on fresh water supplies, sewage, the living environment, and public health. However, high

I. Kabashkin et al. (Eds.): RelStat 2018, LNNS 68, pp. 24–34, 2019.
https://doi.org/10.1007/978-3-030-12450-2_3

density of cities can bring efficiency gains and technological innovation while reducing resource and energy consumption.

Mobility is a key dynamic of urbanization, and associated infrastructure (roads, transport systems, spaces, architectural solutions) invariably shapes the urban environment. Nowadays road transport is one of the main sources of pollutant emissions into the atmosphere (Fig. 1). At the same time, the percentage of cargos and passengers transported by motor transport increases every year [3]. By 2005, approximately 7.5 billion trips were made in cities worldwide every day. In 2050, there may be three to four times as many passenger-kilometers travelled as in the year 2000. Freight movement could also rise more than threefold during the same period.

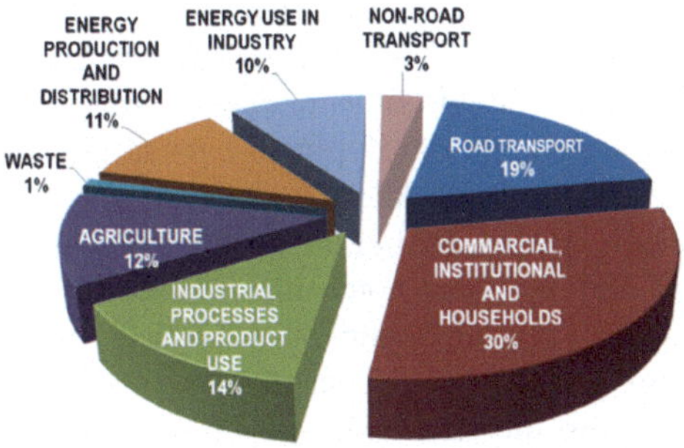

Fig. 1. Emissions in the EU-28: share by sector group.

The need to solve these problems is formulated in "Millennium Development Goals". The key findings and policy messages stemming from the Global Environment Outlook (GEO-6) assessments conducted for the six United Nations Environment Programme regions are summarized and presented in information note by Achim Steiner (Executive Director, United Nations Environment Programme) [4]. Each GEO-6 regional assessment includes review of regional priorities, the state of environment in the region and the main trends that will affect it in the future and an analysis of the actions needed for the region to achieve more sustainable future. Poor air quality, climate change, unhealthy lifestyles and the disconnection between society and natural environments increasingly affect human health and give rise to new risks. Thus, each year, air pollution causes 570,000 deaths in children under 5 [5].

Transition to an inclusive green economy should be based on viable ecosystems, cleaner production, healthy consumer preferences. There is no doubt that achieving a healthy planet and healthy people requires urgent transformation of the current production and consumption systems that most contribute to environmental degradation and inequalities in human health and well-being.

Given the fact that a car with an internal combustion engine on hydrocarbon fuels from non-renewable fossil fuels will remain a main part of the car fleet for a long time, the purpose of the article is to develop methods to improve the environmental efficiency of road transport, believing that the structure of the car fleet will not change significantly.

2 Problem's State

The Global Report on Urban Health [6] presents a comprehensive set of interventions that can reduce the urban burden of non-communicable diseases. These include altering the built environment and promoting alternative transport options to foster greater physical activity and reduce air pollution.

Delegates at the 69th World Health Assembly approved a road map for responding to the adverse health effects of air pollution—the world's largest single environmental risk. Every year 4.3 million deaths occur from exposure to indoor air pollution and 3.7 million deaths are attributable to outdoor air pollution. Most large cities globally, and 98% of low and middle-income cities, have air pollution exceeding WHO's guideline levels. WHO Director General Dr. Margaret Chan described air pollution as "the new epidemic demanding urgent attention". She further stressed the need for air pollution to be combatted jointly from the environment, energy, transport and finance sectors especially as part of the sustainable development agenda [7].

The type-age structure of motor vehicles operating in Russia and the quality of fuels and lubricants used are still lagging behind the level of developed countries. There is a lag in the development and technical condition of the transport infrastructure. This against the background of increased motorization creates environmental problems. In Russia, the contribution of different industries to total emissions differs from the European one. This is due to the territorial disunity of industrial enterprises and the need for transport communications between them. The growth in the share of road transport in transportation increases its contribution to environmental pollution [8].

Naberezhnye Chelny city is center of the one leading agglomerations. Together with Nizhnekamsk and Elabuga cities, it forms the Kamsky innovative territorial-production cluster. The geographical position and the developed transport and production infrastructure of Naberezhnye Chelny make it the center, where material and transport flows formed. Such concentration of the main trunk transport's types makes the Tatarstan Republic and Naberezhnye Chelny a powerful transport hub of both regional and international importance. Therefore, the indices of the contribution of motor vehicles to environmental pollution in the city are higher than in the Tatarstan Republic and the whole country that are comparable with the data typical for megacities (Fig. 2) [9].

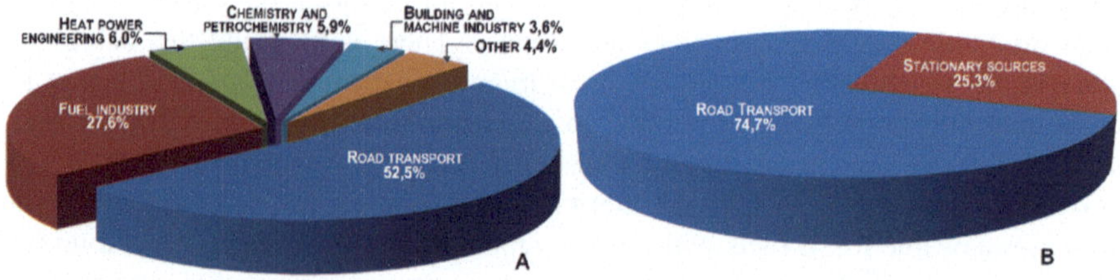

Fig. 2. Emissions in (a) Republic of Tatarstan, (b) Naberezhnye Chelny city.

There is a number of ways to reduce energy consumption and, accordingly, emissions from the transport sector. Basically, researches are performed in three ways:

- new technical and technological decisions aimed at a fuel quality improving, transition to low-carbon kinds of fuel, changing of power generation methods, a principally new types of energy-power plants, and usage of new materials and structural arrangements in order to reduce the weight of vehicle body;
- organizational and administrative decisions aimed at the increase of vehicle maintenance efficiency, including organization of qualitative and opportune service and repair, failure prevention, maintainability and safety increasing;
- application of scientifically based methods to improve the efficiency of traffic management in urban areas at the design and reconstruction stage of transport infrastructure facilities.

Also, to reduce the level of air pollution, it is necessary to regulate transport loads on the city streets by developing public transport. When developing the public transport system, it is necessary to use scientifically based algorithms, methods, approaches and managerial systems, and, if possible, give priority to more environmentally friendly transport, for example, trams, and develop infrastructure and popularize the transition from cycling to bicycle transport. In some developed countries, up to 40% of all trips to work are done using bicycle, even in the winter. In unfavorable climatic conditions and snowfalls in a number of cities, maintenance of bicycle paths is more priority than maintenance of the carriageway [10].

To reduce urban emissions, the managerial decisions are used both to regulate the density of traffic flow and to optimize the vehicle fleet. To solve these issues the measures of state regulation are applied and intelligent transport systems are created.

To regulate type and age structure of the fleet, the upgrading or replacement of vehicles by more ecofriendly ones is performed. Considering the possibilities of using alternative types of fuel in road transport, it is necessary to take into account the long-term demand among consumers. Since the consumer is not always ready to abandon the usual vehicles, perspective engines should be used in those segments of the market where the state can most effectively generate demand using various incentive methods. In this respect, the most promising groups are logistics and transportation companies, which can renew the fleet under appropriate government incentives in optimal for solving of environmental problems way. Furthermore, taking into consideration that buses, light-duty trucks and community services vehicles using mainly the diesel fuel have a big share in the urban traffic flow, the comparison of used vehicles with their alternative counterparts is based on economic, environmental and social criteria.

The authors of the research [11] have analyzed the costs of transition to buses using alternative fuel of various types, and have come to conclusion that most important prerequisite for such a transition is the willingness of cities and bus operators to use such new technologies. The decision-maker must understand the importance of the environmental benefits of such vehicles because from economic point of view these buses are more costly. The estimation of transport effectiveness can be performed using such indicator as energy efficiency using the formula: EE in % = 100/(fuel rate in kg/KWh*h * fuel calorific value (specific heat of combustion) in kWh*h/kg). Wherein, this indicator may be regulated by the government [12].

The author of the article [13] marks two main ways to estimate the transport efficiency: fuel efficiency and fleet efficiency. Herewith, the fuel efficiency is the form of thermal efficiency, which depends on the unique parameters of the engine, aerodynamic resistance, weight and rolling resistance of vehicle, while the fleet efficiency describes the usage of fuel by a group of transport vehicles that can be increased both by improvement of an individual car characteristics and route optimization or behavior modification. According to [13], the environmental impact of large fleets is higher than personal vehicles because of the large annual mileage. The personal car average mileage is 12,000 miles/year, while the average vehicle of the fleet passes 23,000 miles/year. The percentage of new vehicle of the entire fleet is significant, because updating is more common than in individual owners. In addition, to stimulate the owners of vehicle fleets for rational operation is easier than individual owners. Therefore, to analyze the potential benefits from fleet management the subsystem of public passenger vehicles was selected.

3 Methodology

Efficiently functioning transport infrastructure allows reduction of transportation costs, affects the speed of passenger and cargo transportation, reduces capacity limitations and increases access to transport services for the population. For planning activities to reduce emissions from vehicles and road network state analyzing from the point of environmental safety view, to study behavior and evaluate various strategies that ensure the transport system's functioning, simulation modeling is used. Macromodels are used to analyze the environmental situation in city's whole, and micromodels are used to study more problematic zones. Wherein the following factors must be taken into account:

- geometry of the road section;
- traffic density;
- intensity of pedestrian traffic of the road section with the distribution lanes;
- emissions of pollutants from motor vehicles and emissions' quotas;
- modes of traffic lights on the previous and subsequent road sections.

The models development is based on field observation, which are also used in the model's verification and validation. Vehicle emissions exacerbate air pollution, because they are superimposed on the general background, therefore, pollutants industrial plants dispersion's maps must be pre-compiled. This helps to identify the most unfavorable parts of the road network, which require measures to reduce the environmental load.

Total emission of pollutants, M, g/km, is calculated by the formula (1) [14].

$$M = \sum_{l=1}^{n}\left(M^{l}_{P_1} + M^{l}_{P_2}\right) + \sum_{l=1}^{n_1}\left(M^{l}_{L_3} + M^{l}_{L_4}\right) + \sum_{l=1}^{m}\left(M^{l}_{P_3} + M^{l}_{P_4}\right) + \sum_{l=1}^{m_1}\left(M^{l}_{L_1} + M^{l}_{L_2}\right), \tag{1}$$

where $M_{P_1}^l$, $M_{P_2}^l$—emission of pollutants into the atmosphere by cars in the queue before the intersection of the road with greater intensity by l-measurement, $M_{P_3}^l$, $M_{P_4}^l$—emission of pollutants into the atmosphere by cars in the queue before the intersection of the road with lower intensity by l-measurement, g/km; $M_{L_1}^l$, $M_{L_2}^l$—emission of pollutants into the atmosphere by cars moving along the road with greater intensity during the period under consideration, $M_{L_3}^l$, $M_{L_4}^l$—emission of pollutants into the atmosphere by cars moving along the road with lower intensity during the period under consideration, g/km; n, m –number of stops of the traffic flow on the roads that form the intersection during 20 min; n_1, m_1—number of periods of traffic flow movement during 20 min.

Emission of the i-th pollutant by the traffic flow on the motorway with a fixed length M_{L_i}, g/km, is calculated by formula (2):

$$M_{L_i} = \frac{L}{1200}\sum_{k=1}^{K} M_{k,i}^L \cdot G_k \cdot r_{v_{k,i}}, \tag{2}$$

where L—length of the road, from which the length of the queue of cars before the red signal is excluded, km; $M_{k,i}^L$—specific emissions of the i-th pollutant by cars of the k-th group, g/km; K—number of car groups, pcs.; G_k—the actual maximum traffic intensity, i.e. the number of vehicles of each of the k groups passing through a fixed cross-section of the selected road section per time unit in both directions across all lanes; $r_{v_{k,i}}$ –coefficient that takes into account the average velocity of the flow $v_{k,i}$ (km/h).

The emission of the i-th pollutant by cars in the intersection area in the queue during the 20-min period M_{P_i}, g/km, is calculated by the formula (3):

$$M_{P_i} = \frac{P}{60}\sum_{s=1}^{S}\sum_{k=1}^{K}\left(M_{Pi,k}^s \cdot G_k'\right), \tag{3}$$

where P—average stop time during 20 min, seconds; S—number of vehicle stops for a 20-min time period; $M_{P_i,k}^s$—specific emission of the i-th pollutant by cars of the k-th group that are in the queue before the intersection; G_k'—number of cars of the k-th group in the queue at the crossroad [15].

Thus, in order to achieve the stability of the ecological situation in the city, it is necessary that the total volume of emissions from industrial enterprises and vehicles does not exceed the emissions' quota (formula 4):

$$Q_{veh} = \frac{M_{L_i} + M_{P_i} + M_{IE}}{M_{EQ}} \le 1, \tag{4}$$

where M_{IE}—background pollution; M_{EQ}—emission quota for this sector.

4 Results and Discussions

Contaminant dispersion maps received by results of field surveys (Fig. 3a) has allowed highlighting problem areas, one of which is a complex junction formed by the intersection of Mira, Druzhby Narodov and Syuyumbike avenues (Fig. 3b). Analysis of the road traffic accidents statistics has showed that this road section is a place with a high concentration of road accidents. It causes problems for the normal transport system's operation, as well as environmental problems, since the frequent occurrence of the congestion exacerbates the negative impact of transport on the environment.

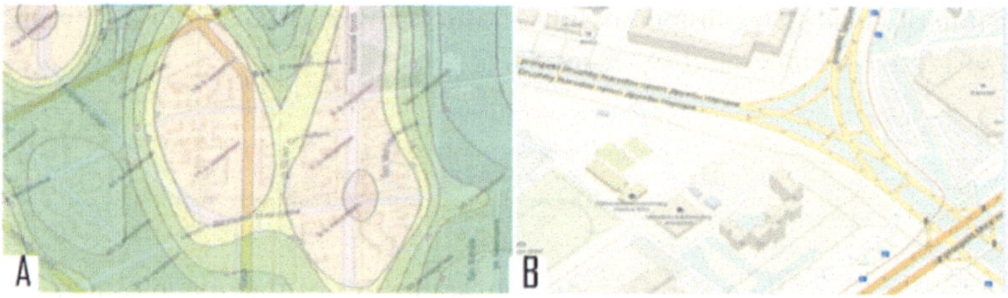

Fig. 3. (a) CO dispersion map; (b) Scheme of the simulated section of road network.

Simulations with different traffic flow's characteristics, which are determined as a result of field surveys, has showed that there was an excess of carbon monoxide and nitrogen dioxide emissions on the considered road section during the congestion situations (Fig. 4). In our case, the constraint of the model corresponded to formula (4). The first stage of optimization experiment allowed us to determine optimal parameters of the traffic flow (the density, the intensity and the speed) that provide not exceeding quota emissions of pollutants [16].

Fig. 4. The simulation model visualization of the road network's section.

Since the explored road section is the confluence of the two main avenues of the city, it forms a significant number of public transport routes, which connect the new and old parts of the city. One of the methods to optimize traffic flow parameters of this road section is the improvement of the route network. The second way is connected to the use of the buses of greater capacity, which will reduce the density of traffic flow and emissions of harmful substances.

In the second stage of the optimization the replacement of the part of public transport buses with the more environmentally friendly ones was simulated and the emissions of this traffic flow were determined. Such replacement leads to considerable decrease in volumes of pollutants' emissions (Table 1).

Table 1. Change of polluting emissions volumes of public transport when using gas motor fuel.

The volume of the maximum allowable emissions	Contaminants						
	CO	NO_x	CH	Soot	SO_2	Formaldehyde	Benzapiren
100% of fleet on diesel fuel	1.04	0.97	0.5	0.58	0.58	0.681	0.579
50% of gas motor fuel fleet	0.87	0.91	0.44	0.47	0.58	0.69	0.587
100% of gas motor fuel fleet	0.69	0.83	0.39	0.34	0.59	0.692	0.589

The state of the transport infrastructure has a great importance for the development of transport and its impact on the environment. Efficiently functioning infrastructure facilities allow reducing transport costs, increasing the speed of passenger and cargo transportation, reducing the limitations of the road network capacity, increasing the availability of transport services to the population and improving the investment climate in the country as a whole. At present, only 37% of federal and 24% of regional roads in Russia meet international standards [17].

Thus, during the peak hours, explored section of the road network does not cope with the transport flow that causes congestions. Drivers spend a lot of time for maneuver on turns because of a huge traffic flow. These dangerous problems arise due to the non-optimal organization of transport flows. The situation was aggravated in connection with input of a new section of a tram route network.

Simulation results have shown that the geometry of the explored road section does not correspond to the traffic flow's parameters, and, therefore, negatively affects traffic characteristics. To improve the situation, we have proposed optimizing configuration of the intersection of Syuyumbike and Druzhby Narodov avenues by implementing a roundabout that will allow reducing the number of conflict points (elimination of turns) on this road section (Fig. 5).

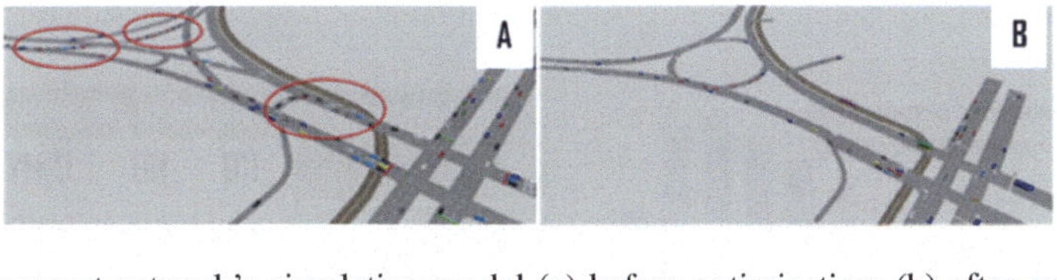

Fig. 5. Transport network's simulation model (a) before optimization; (b) after optimization.

Besides, we have suggested applying the traffic light regulation with alternative number of phases (i.e. the set of the main and intermediate traffic light's cycles) at the adjacent intersection (Druzhby Narodov and Mira avenues). The analysis of the research results has shown that the best option is implementing a roundabout on this road section. Table 2 summarizes changes of the adaptive traffic light's phases, depending on the traffic density. The research results of the explored road section are shown in Table 3. These data demonstrate that traffic parameters of the explored road section could be greatly improved. In addition, organization of the circular motion at this road section will lead to a significant reduction in pollutant emissions of road transport.

Table 2. The change of traffic light's phase depending on the traffic density.

Traffic density of the road section (%)	Total stage duration, sec.	Red signal (basic tact), sec.	Green signal (basic tact), sec.	Red and yellow signal, sec.	Yellow signal, sec.
95	85	41	38	3	3
82	83	37	40	3	3
74	82	35	41	3	3
61	81	32	43	3	3

Table 3. Calculated parameters of the explored road section.

Designation of parameter	Value	
	Prior to any changes	After changes
Average speed along the explored road section, km/h	35	42
The number of stops per time unit, pcs.	6	2
Traffic flow density, % of road area	92	67
Average travel time on the explored road section, min.	4	1.4

To determine the coefficient $r_{v_{k,i}}$ taking into account the average speed of movement, we have used data obtained during simulation (Fig. 6a). Based on the simulation results, the following calculated emission values at the road section were obtained (Fig. 6b).

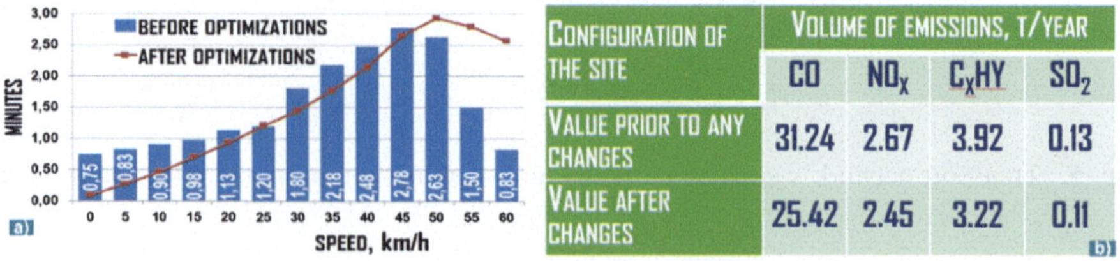

Fig. 6. (a) Average travel time at different speeds (in terms of 20 min interval); (b) Volumes of pollutant emissions.

5 Conclusions

Thus, the change in the configuration of the road section, according to calculations, will cause reduction in CO emissions by 18.6%, NO_X by 8.2%, C_XH_Y by 17.9%, SO_2 by 15.4%. It is necessary to continue research of other problem road sections that have their own specifics and factors that negatively affect the parameters of traffic flow. For example, there are speed bumps before unregulated pedestrian crossings of some city avenues. This leads to inevitable unnecessary cycles of braking of traffic flow even in the absence of pedestrians crossing the road. In the case of implementation of pedestrian traffic lights with a push button instead of speed bumps, traffic flow parameters will significantly improve. And optimization of phases of traffic light regulation with the use of simulation models allows finding a balance between mobility, safety and the environment.

References

1. World Population Prospects (2017). https://esa.un.org/unpd/wpp/publications/files/key_findings_wpp_2015.pdf. Accessed 10 Apr 2018
2. Major Agglomerations of the World. https://www.citypopulation.de/world/Agglomerations.html. Accessed 10 Apr 2018
3. EEA, Air quality in Europe—2016 report. https://www.eea.europa.eu/publications/air-quality-in-europe-2016. Accessed 10 Apr 2018
4. United Nations, Summary of the Sixth Global Environment Outlook GEO—6. http://web.unep.org/geo/assessments/regional. Accessed 10 Apr 2018
5. World Health Organization: The cost of a polluted environment: 1.7 million child deaths a year. http://www.who.int/mediacentre/news/releases/2017/pollution-child-death/en/. Accessed 10 Apr 2018
6. Hosking, J., Mudu, P., Dora, C.: Health co-benefits of climate change mitigation—Transport sector. WHO Document Production Services, Geneva, Switzerland (2011). http://apps.who.int/iris/handle/10665/70913?locale-attribute=ru. Accessed 10 Apr 2018
7. Chan, M.: Address to the Sixty-ninth World Health Assembly. http://www.who.int/dg/speeches/2016/wha-69/en/. Accessed 10 Apr 2018
8. Ministry of Natural Resources and Environment of the Russian Federation: State report "On the state and on the protection of the environment of the Russian Federation in 2015", 639 p. (2016)
9. Ministry of Ecology and Natural Resources of the Republic of Tatarstan: State report "On the state of natural resources and on the protection of the environment of the Republic of Tatarstan in 2015. http://eco.tatarstan.ru/gosdoklad.htm. Accessed 10 Apr 2018
10. Bredal, F.: The case of copenhagen, changing urban traffic and the role of bicycles. In: Proceedings of the conference on Russian and International Experiences, pp. 24–28 (2014)
11. Straten, P., Wiegmans, B.W., Schelling, A.B.: Enablers and barriers to the adoption of alternatively powered buses. Transp. Rev.: Transnatl. Transdiscipl. J. **27**(6), 679–698 (2007)
12. The resolution of Kazakhstan Republic government: About establishment of requirements for energy efficiency of transport. http://online.zakon.kz. Accessed 10 Apr 2018
13. Reich, A.A.: Transportation efficiency. Strateg. Plan. Energy Environ. **32**(2), 32–43 (2012)
14. GOST R 56162-2014. Emissions of Pollutants in the Atmosphere. Method for calculating emissions from road transport during consolidated calculations for urban settlements (2014)

15. Makarova, I., Khabibullin, R., Mavrin, V., et al.: Simulation modeling in improving pedestrians' safety at non-signalized crosswalks. Transp. Probl. **11**(4), 139–150 (2016)
16. Makarova, I., et al.: Transition to green economy in Russia: current and long-term challenges. J. Appl. Eng. Sci. **13**(1), 1–10 (2015). https://doi.org/10.5937/jaes13-7566
17. Popadyuk, T.G., Shevyolkina, K.L.: Possibilities and perspectives of innovative technology of public-private partnership in financing of transport infrastructure. Manage. econ. syst. sci. electron. j. **12** (2017). http://www.uecs.ru/finansi-i-kredit/item/3161-2014-11-17-11-52-25

Tram Service Quality and Its Impact on the Passengers' Modal Choice in Constantine City (Algeria)

Mouloud Khelf[1(✉)], Salim Boukebbab[1], Neila Bhouri[2], and Mohamed Salah Boulahlib[1]

[1] Laboratory of Transportation Engineering and Environment, Department of Transport Engineering, University of Constantine 1, 25017 Constantine, Algeria
mouloud.khelf@umc.edu.dz

[2] Laboratoire Génie des Réseaux de Transport Terrestre et Informatique Avancée (GRETTIA), Université Paris-Est, IFSTTAR/COSYS/GRETTIA, 14-20 Boulevard Newton - Cité Descartes, Champs-sur-Marne, 77447 Marne-la-Vallée Cedex 2, France

Abstract. Nowadays, the tram knows a new golden age since that modern cities prefer a sustainable mobility. It can give a decent image to the urban area and contributes to meet the increasing mobility demand. Without an optimal operation and a better service, the tram can't be an efficient means of transport that may help in decongesting road traffic. Passengers would prefer to use vehicles than trams, therefore road traffic will increase and this, will complicate more the state of traffic in cities. This study uses the opinion of tram users. The main objective is to show the strong characteristics that attract the population to use tram and also its weaknesses. This last, may be improved to reach the maximum satisfaction of users. In addition, this article attempts to show the preferred means of transport for passengers who travel on the same line as the tram. The results show that despite the different advantages of the tram, the population is not all satisfied about its exploitation and about its service quality. Almost 40% of passengers, declare that they prefer to use their cars or taxis than a tram; as a result, the road traffic situation may get complicated in the future. Finally, solutions to ameliorate the performance of the tram and its services' quality are presented, in order to attract more tram users and to reduce traffic congestion in urban areas.

Keywords: Tram · Service quality · Passengers · Modal choice · Traffic congestion

1 Introduction

Traffic congestion is a major issue in the daily life of commuters, especially for those who live in big cities [1]. Although the technical and the technological progress realized by humans in all domains. The road traffic remains a victim of the increasing congestion when the demand exceeds the capacity of the infrastructure [2].

I. Kabashkin et al. (Eds.): RelStat 2018, LNNS 68, pp. 35–44, 2019.
https://doi.org/10.1007/978-3-030-12450-2_4

One of the solutions to this problem is to ameliorate the offered services of public transport [3]. Therefore, the systems of public transport are widely used to help in relieving congestion. Dependability and easy accessibility are the main qualities that make public transport more popular; that, led to the restoration of trams around the world [4–6]. In the last five years, urban railways built in Algeria have been tripled.

A tram network has been set up in five Algerian provinces; in addition to other projects that are on the way of construction [7]. Constantine is one of these provinces, along 8 km, the tramline links the downtown with the suburban zone of Zouaghi Slimane, the tram is designed to a lifetime of 30 years [8]. This mean of transport must be evaluated through surveys [9, 10]. The capacity of the line and the trams operation have a direct impact on the user satisfaction [11].

This paper focuses on passengers' perception about Constantine tram performance. The first part of this work presents the type of the rolling stock. i.e. (The tram's type, length, capacity… etc.). The second part, focuses mainly, on users' opinions towards the comfort and the offered services. The last part, includes questions on the preferred mode of transport for tram users, to show the main factor that affects their modal choice.

Finally, the study aims at knowing what are the most important services that need improvement in order to attract more population. Furthermore, we're going to discover if the clients will use another mean of transport if it is possible. Or, they are globally, satisfied and they prefer always to use the tram.

2 The Characteristics and the Offered Service

2.1 The Architecture of the Tram and Its Characteristics

The tram is a low-floor tram from an Alstom CITADIS family. The tram is composed of seven articulated sections and it includes the following modules:

- 2 motorized cabin (M1 et M2), 1 motorized nacelle (NM) and 1carrier nacelle (NP);
- 3 suspended central section (cars-bodies) (C1, C2 and CC) [12] (Fig. 1).

Fig. 1. The architecture of Constantine's tram.

2.2 The Capacity of the Tram

The length of the vehicle is approximately 44 m, each side of each vehicle is composed of 6 doors with a double access, in addition to two other doors with a simple access. The tram passengers' capacity is as follows [12] (Table 1):

Table 1. The capacity of Constantine trams.

	The average charge (4 p/m^2)	The maximal charge (6 p/m^2)
Users sitting	78	78
Users standing	224	336
The sum of users/tram	302	414

2.3 The Operational Time and the Tram Fleet

The operating hours of the fleet of Constantine tram are as follows [13] (Tables 2, 3, 4 and 5):

Table 2. The operational time of Constantine's tram.

	All year except Ramadan	In Ramadan
The exploitation start	5:00	5:00
The exploitation end	22:00	01:00
The daily exploitation duration	17 h	20 h
Peak hour	7:00 to 9:00	7:00 to 9:00
	14:00 to 18:00	19:00 to 00:00

Table 3. The Nominal service of Constantine's tram.

From sunday to thursday	Time interval between trams (minutes)	Passengers per hour per direction for 4 p/m^2
05:00 to 07:00	5	3624
07:00 to 09:00	3	6040
09:00 to 14:00	5	3624
14:00 to 18:00	3	6040
18:00 to 22:00	5	3624

Table 4. Fridays service of Constantine's tram.

Friday service	Time interval between trams (minutes)	Passengers per hour per direction for 4 p/m^2
05:00 to 07:00	7	2588
07:00 to 09:00	4	4530
09:00 to 14:00	7	2588
14:00 to 18:00	4	4530
18:00 to 22:00	7	2588

Table 5. Saturdays service of Constantine's tram.

Saturday service	Time interval between trams (minutes)	Passengers per hour per direction for 4 p/m^2
05:00 to 07:00	10	1812
07:00 to 09:00	6	3020
09:00 to 14:00	10	1812
14:00 to 18:00	6	3020
18:00 to 22:00	10	1812

2.4 The Travel Time, the Average Speed and the Station Stopping Time

The theoretical travel time for itinerary 1 (from Zouaghi to Benabdelmalek Station) is 1251.5 s, for itinerary 2 (from Benabdelmalek station to Zouaghi Station) the travel time is 1282.5 s. The average speed of Constantine tram for itinerary 1, is approximately 23.0 km/h, for itinerary 2 is almost 22.5 km/h. The average stopping time during peak hours for both directions, should be at least 20 s. The reference stopping time at stations for both directions is 30 s [13].

3 The Survey Methodology

The investigation targets the perception of users about the services' quality and its exploitation. The research methodology used is based on a questionnaire. A sample of 250 of tram users were randomly selected at the level of trams, stations and near the tram lines. The Survey was conducted at several times of the day. The study is based on data collected between the end of January and the beginning of February 2018.

The questionnaire is a set of 25 questions and divided into four parts, where some blanks were left to collect remarks and propositions suggested by respondents.

In the first part of the survey, the socio-economic characteristics of the respondents such as gender, age, occupation and also information about the particularities of using the tram have been recorded. In the second part the collected information is about users' opinions on the offered service. Respondents were asked to evaluate the following indicators (the ticket price, comfort, safety, cleanliness…etc.) on the categorical scale from 1 (Bad), represents the lowest satisfaction with the respective service throw 2 (Average) to 3 (Good) the highest.

The third section is based on passengers' opinions towards the regularity of the tram, such as: punctuality, waiting times, Tram stations numbers and locations and so on. Respondents were asked about their satisfaction of this services (Yes or Not).

Finally, we conclude the work considering the opinions of passengers on what attracts them to use the tram as well as knowing their preferred means of transport. This aims at knowing, if the tram is efficient, or it lacks improvement to achieve the maximal attractiveness.

4 Results

Tram Data of the Number of Passengers Per Year. Figure 2 represents the annual number of passengers at the level of Constantine trams. We observe from the figure, that since 2014, each year the tram users increase, but in the last year (2017) the number of passengers decreased. This drives us to know throughout this study, the main causes of this decrease in the tram usage.

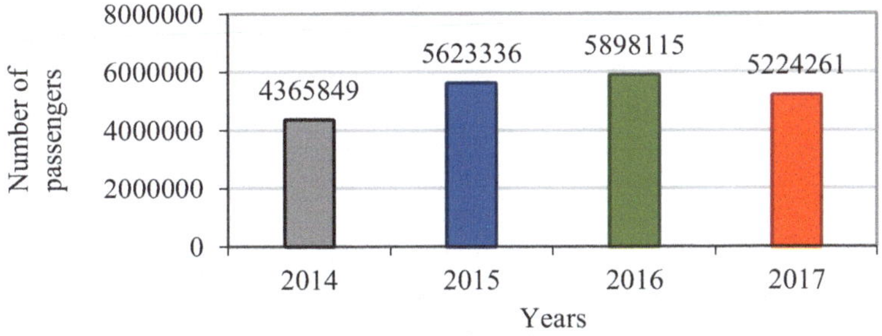

Fig. 2. Number of passengers per years of Constantine's tram. (Source Entreprise Metro Alger)

The Socioeconomic Characteristics of the Sample. Table 6 shows that tram users' rates of both genders are quite close. The dominating category is between 18 and 36 years old with 56.40%, that corresponds to the category of young people, by function, it is the category of students and workers with 82.80%. Since, the tram starts from the downtown of Constantine, coming to Zouaghi Slimane and passing by several university stations. Table 6 also indicates that 78.80% of passengers take the tram between once a week to a daily use, for diverse reasons. The table denotes that there are three main reasons for moving by tram. 38% use it for studies, 22.40% to go to work while 21.20% to shop. Finally, we notice that over 52.80% of users live near the tram lines.

Table 6. The socioeconomic characteristics of the sample

		N	%
Gender	Men	136	54.40
	Women	114	45.60
Age	18–36	141	56.40
	37–55	76	30.40
	56–75	33	13.20
Function	Student	114	45.60
	Employee	93	37.20
	Pensioner	24	9.60
	Unemployed	19	7.60

(*continued*)

Table 6. (*continued*)

		N	%
The tram use Frequency	Daily	63	25.20
	2–3 time a week	86	34.40
	1 time a week	48	19.20
	As necessary	53	21.20
Reason for traveling	Study	95	38.00
	Work	56	22.40
	Shop	53	21.20
	Other	46	18.40
The time between passengers domicile and the closest station	$\leq$ 15 min	132	52.80
	15–30 min	58	23.20
	>30 min	60	24.00

The Service Quality. Table 7 shows that the service quality is acceptable in general. However, it has to be ameliorated, passengers say that agents are good, but they need to be more comprehensive. An electronic information panel is needed at stations to indicate the remaining time of the next tram. The results indicate also, that more security is needed in stations and inside trams, especially in the evening. The comfort and cleanliness ratios are good, but the daily maintenance is still necessary for more tidiness and comfort. Users find the ticket and the subscription price too expensive.

Table 7. The evaluation of the service quality

		N	%
Trams agents' behavior	Good	137	54.8
	Average	83	33.2
	Bad	30	12
Information quality in trams and stations	Good	160	64
	Average	76	30.4
	Bad	14	5.6
Security in stations and inside trams	Good	108	43.2
	Average	68	27.2
	Bad	74	29.6
Air quality inside trams	Good	160	64
	Average	77	30.8
	Bad	13	5.2
Tidiness inside trams	Good	162	64.8
	Average	72	28.8
	Bad	16	6.4

(*continued*)

Table 7. (*continued*)

		N	%
Tidiness at stations	Good	154	61.6
	Average	75	30
	Bad	21	8.4
Seats' comfort	Good	180	72
	Average	64	25.6
	Bad	6	2.4
Ticket's price	Good	33	13.2
	Average	66	26.4
	Bad	151	60.4

The Tram's Operation. Table 8 reveals the opinion of users about the suggested and the actual tram's operation. Travelers think that the stations' number and the location are good. However, they suggest to add a station in front of the para-medical Institute. the table denotes, that only 23.60% find that the tram comes in 3 min at peak hours, against 76.40% who wait between 4 and 12 min. The data show that almost half of the users find the interval between trams disrespected, they're not satisfied of the tram management. Moreover, they see that some trams aren't filled with the required users. Considering these data, we notice that the tram operation doesn't respond to the demand, in addition to the average management of the tram disturbances. Furthermore, users say that the stopping time isn't respected and mostly starts within 10–15 s.

Table 8. The evaluation of the tram's operation

		N	%
Tram stations	Satisfied	178	71.2
Numbers and locations	Not satisfied	72	28.8
Waiting duration at stations in peak hours	3 min	59	23.6
	4–6 min	126	50.4
	7–9 min	34	13.6
	10–12 min	31	12.4
Time respect between trams	Respected	115	46
	Not respected	135	54
Time interval satisfaction between trams	Satisfied	137	54.8
	Not satisfied	113	45.2
Time interval acceptance between trams	3 min	63	25.2
	4–6 min	88	35.2
	7–9 min	24	9.6
	10–12 min	75	30

(*continued*)

Table 8. *(continued)*

		N	%
Trams service amplitude 5–22 h	Good	172	68.8
	Average	66	26.4
	Bad	12	4.8
Trams disturbance situations management	Good	73	29.2
	Average	132	52.8
	Bad	45	18
Speed, travel time and stopping time	Good	147	58.8
	Average	84	33.6
	Bad	19	7.6

Reasons Why People Use the Tram. From the Fig. 3 it can be seen that the first reason for passengers to use the tram, is its comfort and rapidity. Therefore, if the operation of the tram is bad, or if passengers do not feel comfortable and safe, they will certainly change the mode of transport. As a result, they'll use taxis or private vehicles.

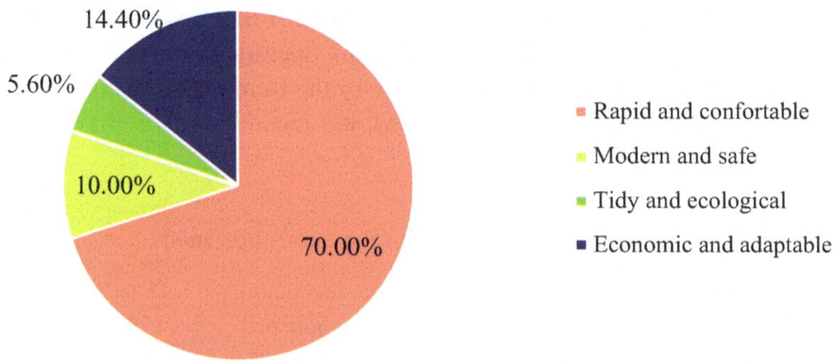

Fig. 3. Reasons why people use the tram.

Tram Users' Satisfaction and Preferred Mode. We observe from Fig. 4, that 39.20% are not satisfied with the operation and the service quality, this rate is quite high. This category of passengers claim that they prefer to use other means of transport, because the exploitation of the tram is not good compared to the previous years. 91.84% of the unsatisfied passengers argue that they prefer to take taxis or private vehicles, while 8.16% are opted for buses.

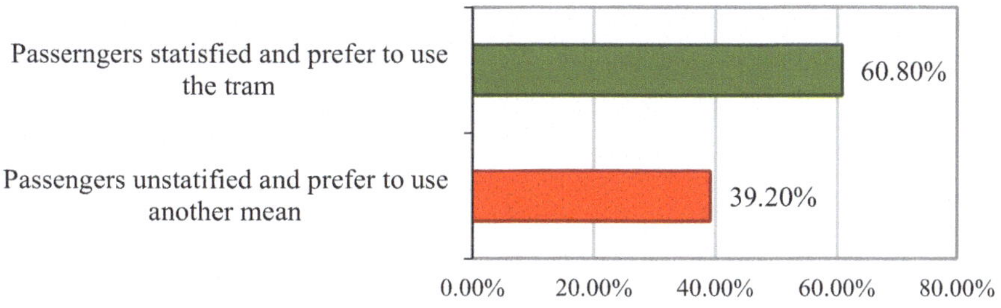

Fig. 4. Tram users' satisfaction on the operation and the service quality, as well the possibility of changing to other means.

5 Conclusion

The aim of the present study is to determine the impact of the proposed service quality on the modal choice of Constantine's tram passengers. In this investigation we try to evaluate the satisfaction of passengers about the service quality, especially the tram operation, and to show the weak characteristics.

It is shown that the offered service is acceptable in general, but it needs some amelioration. 60.40% of passengers argue that the ticket's price is expensive, especially for those who travel on 500 m or 2000 m, since that they pay the same price as those who travel for 8 km. As a result, they choose other means. This survey has shown also that 29.60% of users are unsatisfied about the security at stations and inside trams. In addition, 12.00% of users find that trams agents' behavior is not good and told us that it is an important factor for them.

For the tram's operation, only 23.60% of users say that the tram comes in 3 min at peak hours; while 54.00% find that the time interval between trams is not respected. For that, 45.20% of users aren't satisfied about this problem. For the time acceptance between trams, 25.20% agree that it's necessary for trams to come every 3 min, and 74.80% can wait for 4 to 12 min.

The most obvious finding to emerge from this study, is to know the opinion of passengers about the actual service quality, especially the tram operation, since the particularity of this work, is to find out if the service quality impacts the modal choice of tram users. It has been revealed that 70% of people use the tram for its rapidity and comfort. Because of the disrespect of the proposed operation, 39.20% of passengers are not satisfied and they prefer to take another mean of transport, when it is possible. This may influence negatively the situation of traffic in the urban areas in the future and cause congestion.

That's why it is recommended to optimize the trams' exploitation, as well as to ameliorate both the comfort and the security since that these factors have a direct impact on their modal choice. Furthermore, the ticket's price has to be adapted for the reasons mentioned above. Finally, a daily check out of trams and stations is necessary. The quality of the information is ought to be improved, especially in the disturbances. This may attract more users to take always the tram.

The future research should analyze the actual operation of Constantine trams and try to propose an optimal exploitation, to satisfy the demand of passengers and to encourage the population to always travel by tram.

References

1. Nguyen-Phuoc, D.Q., Currie, G., De Gruyter, Ch., Young, W.: Exploring the impact of public transport strikes on travel behavior and traffic congestion. Int. J. Sustain. Transp. 1–11 (2018)
2. Khelf, M., Boukebbab, S., Boulahlib, M.S.: A mathematical model to regulate roads traffic in order to decongest the urban areas of Constantine city. In: Zamojski, W., Mazurkiewicz, J., Sugier, J., Walkowiak, T., Kacprzyk, J. (eds.) Proceedings of the Eleventh International Conference on Dependability and Complex Systems DepCoS-RELCOMEX, AISC, vol. 470, pp. 279–289. Springer, Heidelberg (2016)
3. Pavkova, K., Currie, G., Delbosc, A., Sarvi, M.: Selecting tram links for priority treatments-The Lorenz Curve approach. J. Transp. Geogr. **55**, 101–109 (2016)
4. Khelf, M., Boukebbab, S.: The effect of noise on the comfort of passengers inside the tramway and its impact on traffic congestion in the urban area. J. VibroEng. **20**(1), 530–540 (2018)
5. Khelf, M., Boukebbab, S.: The effect of vibration inside the Constantine's tramway on the comfort of passengers. In: Proceedings of the 23rd International Conference Vibroengineering, Istanbul, Turkey, pp. 33–38 (2016)
6. Zeng, A.Z., Durach, ChF, Fang, Y.: Collaboration decisions on disruption recovery service in urban public tram systems. Transp. Res. Part E: Logistics Transp. Rev. **48**(3), 578–590 (2012)
7. Segaudi, L., Ecotiere, D., Dutilleux, G., Puymérail, J.-F.: A local directivity for 3 French tramways. In: Online Proceedings of the 10th French Acoustic Congress, Lyon, France (2010)
8. Entreprise Metro Alger (EMA) Alstom: General presentation of the transportation system, Algeria (2013)
9. Ahac, M., Lakušić, S.: Track Gauge Degradation Modelling on Small Urban Rail Networks: Zagreb Tram System Case Study. Urban Transport Systems. In TechOpen (2017)
10. Abramović, B.: Analysis of the mobility of railway passenger transport in small urban areas. In: Proceedings of the 21st International Conference on Urban Transport and the Environment, Valencia, Spain, pp. 665–674 (2015)
11. Molecki, A., Gąska, D.: Calculations of tramway track capacity in the wide area networks. Transport **27**(4), 428–433 (2012)
12. Entreprise Metro Alger (EMA) Alstom: Functional Specification and general technics for rolling stock, Algeria (2010)
13. Entreprise Metro Alger (EMA) Alstom: Principle of tramway exploitation, Algeria (2010)

Problems of the Warsaw Metropolitan Area in Shaping the Principles of Sustainable Transport Development

Cezary Krysiuk[1], Gabriel Nowacki[2(✉)], and Jacek Brdulak[3]

[1] Motor Transport Institute, Jagiellońska 80, 03-301 Warsaw, Poland
cezary.krysiuk@its.waw.pl
[2] Cybernetics Faculty, Military University of Technology, Gen. Witolda Urbanowicza 2, 00-908 Warsaw, Poland
rafal.kopczewski@wat.edu.pl
[3] Warsaw School of Economics, Niepodleglosci 162, 02-554 Warsaw, Poland
jacek.brdulak@sgh.waw.pl

Abstract. The paper refers sustainable transport development in the European Union especially definitions, principles and guidelines of individual member states. Practice shows that implementation of the principles is not easy, as indicated by various available analyzes and studies of transport systems operating in EU member states. In Poland, also activities are undertaken related to the implementation of these principles into the transport systems of the country, large urban centres, metropolitan areas or cities. Due to the continuous development of transport and ways to implement it, there are still many problems that constitute a barrier that is often not easy to overcome (even financial). The paper points out the problems of implementing the principles of sustainable transport development in the Warsaw Metropolitan Area (WMA). To resolve the problem, research methods were used: analogy, analysis, synthesis, induction, deduction.

Keywords: Sustainable development of transport · Transport problems

1 Introduction

The aim of paper is to discuss the problems of the Warsaw Metropolitan Area in shaping the principles of sustainable transport development.

Sustainable urban mobility plan (SUMP) is an integrated transport strategy in the form of a strategic plan that builds on existing planning practices and takes due account of integration, cooperation, and valuation principles in order to meet the mobility needs of people—today and tomorrow—to improve the quality of life in cities and their surroundings. The process followed guidelines set by the European Union as part of their ELTIS program.

The main characteristics of ELTIS program and the reference project for SUMP are as follows [1–3]:

– Active involvement of all stakeholders and the engagement of citizens,

I. Kabashkin et al. (Eds.): RelStat 2018, LNNS 68, pp. 45–56, 2019.
https://doi.org/10.1007/978-3-030-12450-2_5

- Commitment to sustainability, i.e. balancing social equity, environmental quality and economic development,
- Looking "beyond the borders"—an integrated approach between policy sectors—cooperation between authority levels—coordination across neighboring authorities,
- Focus on achieving ambitious, measurable targets,
- Targeting cost internalization i.e. reviewing transport costs and benefits for society,
- Including all steps of the life cycle of policy making and implementation.

When talking about strategies for sustainable urban mobility, most of the model assumptions built upon a properly working relationship between urban local governments, city administration and the private sector [3–5].

The article was based on research carried out at the Motor Transport Institute in Warsaw, entitled: "Formation the transport system of the Warsaw Metropolitan Area" and research carried out for the Ministry of Development under the Operational Program Technical Assistance No. POPT.02.01.00-00-0021/15-00, Support for the strategic coordination institution of the Partnership Agreement in 2015–2016, entitled: "Analysis of the impact of the built road infrastructure on the level of economic activity in the surrounding territorial units" [6].

It required realization of project on search of answer following questions investigative:

1. What is on economic activity in its close enclosing influence of new road infrastructure and if it causes new investments?
2. What is area of interaction of built road infrastructure?
3. If productiveness of economic unit accrues near road infrastructure acting?
4. How to have will influence (income) on local economy growing living costs infrastructure road?

The results of surveys conducted as a part of the research work [7] among both road users and authorities responsible for making decisions regarding the organization and construction of road infrastructure at WMA indicate that many of the most modern transport solutions conducive to the principles of sustainable development have been introduced, but very there is still a lot to be done. Very often it is associated with high costs, and Poland is still building only the basic transport network.

2 Characteristics of the Warsaw Metropolitan Area

The Warsaw Metropolitan Area (WMA) is located in the central eastern part of Poland—Fig. 1. It consists of 72 local government units with a total area of about 6206 km^2, (17.5% of the total area of the Mazowieckie voivodship), of which the area of the Capital City of Warsaw constitutes 8.3% of the total area of the designated area. According to the data of the Central Statistical Office, the number of inhabitants of the WMA is systematically growing, currently the area is inhabited by over 3 million people, which is about 60% of the population of the Masovia Province.

The population density determined by the number of people per 1 km^2 at the WMA in 2016, amounted to 507 people to 151 people in the province. Statistics show that the

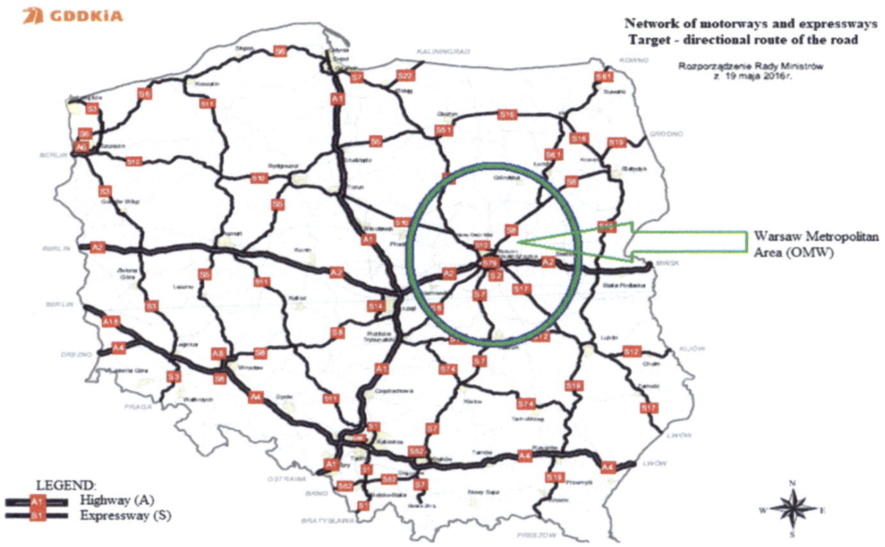

Fig. 1. The road network of Poland, situation as of 05/07/2018, https://www.gddkia.gov.pl.

highest population density was found in the municipalities: Legionowo (3998 people/km^2), urban Piastow (3946 people/km^2) and the Capital City of Warsaw. Warsaw (3391 people/km^2) [7].

Very important communication routes pass through the WMA area not only for the analyzed region, but also for the country or Europe (Fig. 1).

Some of them are part of the priority TEN-T axis. One of the main communication routes passing through the WMA is the A2 motorway, the extension of which is the S2 expressway that forms the Southern Bypass of Warsaw (part of the express bypass). On sections where no motorway or express road was created, there is still DK2 running from Warsaw towards the border with Belarus. In the WMA it runs from Warsaw to the intersection with DK50 in the area of Stojadla. In the southern direction, the traffic is carried on the national road No. 7/S7, with a short northern section from the Nowy Dwor Mazowiecki area having the status of an expressway, whereas the south of Czosnow is a GP class road. In the southern part of the WMA DK7 is also the main road of the accelerated movement outside the Grojec bypass, where the next section of S7 begins. In parallel to DK7, DK79 runs within the limits of the WMA, connecting Warsaw with Gora Kalwaria through Piaseczno [7].

The second important communication route running from north to south of the WMA is DK8 on the greater part of its course in the WMA having the standard of the S8 expressway. It runs from the Wyszkow beltway on the beltway of Marek and Radzymin, and then in Warsaw is an expressway (northern section of the express bypass). DK17 runs from Warsaw to the southeast, bypassing larger towns, however, its intersection with DK50 is one of the most important points on the road network of the area, due to the large congestion. The network of national roads at the WMA is also created by DK61 connecting Warsaw with Legionowo and Serock, DK62 running in the northern part of the WMA from Zakroczym through Serock to Myszków and DK50 passing by Sochaczew through the beltway of Zyrardow, Grojec, Gora Kalwaria and Minsk Mazowiecki towards Łochów. Roads DK 50 and 62 form the so-called a large Warsaw Bypass, or a junction of the Warsaw junction for truck transit [7].

Provincial roads (DW) in the WMA constitute important connections within the area, simultaneously creating bottlenecks, especially at the inlets to Warsaw. DW580 connects Warsaw with Sochaczew and a series of municipalities south of the Kampinos Forest. DW719, until the creation of A2, a key road route in the Pruszkow range, leads traffic from Zyrardow, Grodzisk Mazowiecki, Brwinow, Milanowek and Pruszkow to the centre of Warsaw. Within the city itself, it is one of the most important road routes (Aleje Jerozolimskie). The same is true for DW724 connecting the capital with Konstancin-Jeziorna and Gora Kalwaria, DW801 introducing traffic from Jozefow, Otwock and Karczew, DW634 connecting Warsaw with the towns of the Wolomin-range: Wołomin, Kobyłka, Zielonka and Zabki, as well as Tłuszcz. The DW 630 is of great importance, being the shortest route from Nowy Dwor Mazowiecki to the capital (via Jablonna). The connection between Warsaw and the area of Zalew Zegrzyński and Nowy Dwor Mazowiecki is DW631, which is also one of the routes introducing traffic from the Wolomin band (from the side of Zabki). An alternative to her is in this case DW 629. The role of DW637 in WMA is limited because, apart from connecting its eastern districts—Rembertowand Wesola, from the centre of Warsaw, it introduces traffic to the city only from Sulejowek and smaller towns in the commune. Halinow. The supplementary role is played by the road indirectly directing traffic to Warsaw—DW722 connecting Grojec with Piaseczno. In addition to the above-mentioned routes of provincial roads with a radial connection connecting Warsaw with the towns of the WMA, there are connections linking various satellite towns. A special role is played here by DW721, connecting villages near Warsaw, starting from Pruszkow to villages located on the Vistula River in the Konstancin-Jeziorna commune and towns on the right bank of the Vistula from Jozefow to Wiazowna. In the west of the WMA, an important combination is DW579 running from Mszczonow through Grodzisk Mazowiecki and Błonie to S7 in the Czosnow area, while its northern section, running across the Kampinos National Park, is of limited importance due to the ban on heavy goods vehicles [7].

A certain significance for road traffic in the Warsaw agglomeration is the course of railway lines in the area of the WMA, which reflects both on the issues of traffic flow at intersections of roads and railways, as well as on the shaping of interchange nodes between individual transport and public transport.

There are two airports on the WMA that serve national and international connections that are included, which are a component of the transport system not only of the country, but also of the region.

Such a large area and population entails many problems in the functioning of the transport system for the WMA, including Warsaw, which face serious challenges in shaping a sustainable development policy in this area, so that transport processes do not interfere with the functioning of the area, run smoothly and safely, they did not hamper its development.

3 Problems of Sustainable Development in Polish Cities

The problem of implementation of the principles of sustainable development in the WMA is associated not only with its size in terms of area or population, but also with the lack of an appropriate number of modern solutions ensuring the WMA transport system efficient operation or lack of good organizational solutions supported by ITS solutions, as well as on administrative decisions, or conducting a coherent and comprehensive urban policy related to investments in transport infrastructure (car parks at the boundaries of the area, modern interchanges, bridges, viaducts, by-passes).

In addition, it should be returned to the service character of the city transport system, which means that its development is stimulated by needs [8, 9] including planned social and economic needs of the city (or the functional area, metropolis or region where the city plays a decisive role). It follows that there are close links between the socio-economic development of the city and the development of the transport system. They have a different nature, scope and impact on particular socio-economic areas, which are the reason for the emergence of diversified direct and indirect effects of the development of the city's transport system. Learning about all connections occurring in the city's transport system is one of the conditions for its harmonious development [10, 11]. Consequently, the city's needs should be carefully analyzed, because it is a good idea to design a transport system by selecting the right transport infrastructure investments.

In the case of the WMA, the difficulty in implementing the principles of sustainable development is related to the excessive burden on road networks, which in turn result in congestion. The problem of congestion is a universal problem of cities, which in particular affects large urban agglomerations, metropolises, and a part of the metropolitan area. During holiday periods it is also a nuisance for smaller towns, through which the area leads to attractive tourist areas (sea, mountains, lakes). Problems that should be removed, reduced, and which intensify congestion in cities such as Warsaw and its metropolitan area belong [12–15]:

- increase in the number of population living in cities—not only the increase in the number of city residents is important, but also the resulting phenomenon of the spread of cities,
- lack of unloading/loading infrastructure, which causes that cars delivering goods to shops stop on the street and block traffic,
- development of e-commerce and m-commerce contributing to the increase in the number of deliveries made to the homes of individual clients,
- modernization and development of transport infrastructure causing significant difficulties in traffic and the need for changes in the communication system,
- lack of an adequate number of car parks with convenient interchanges at the WMA, enabling transfers from a passenger car to fast rail, bus, tram, metro),
- implementation of dangerous transport, often through urban areas, (such transport should be based on the safest solutions used in the EU, based on the latest ITS systems [16, 17].

Sustainable transport according to the recommendations of the European Commission experts [18] should meet the ensuring the accessibility of communication objectives in a safe manner, improving the entire transport system of a given city, area or region. However, it should also be taken into account that many countries are just making up for infrastructural backwardness, in terms of the length and quality of roads, which are the basis of the transport processes currently underway. The creation of a comprehensive and modern skeleton of road networks (but also rail, water, rope—along with parking lots, interchange nodes, etc.) will allow to eliminate many of the problems that currently occur and make it difficult to implement the principles of sustainable development.

In the WMA many solutions are already implemented, and many planned for implementation, while planning changes to the transport system, it is worth bearing in mind the best solutions from the sphere of "good practices"—sustainable development should be tested.

Information and educational campaigns alone are not enough to convince a society inhabiting areas such as the WMA to give up individual car transport. Lack of a comprehensive approach to the adaptation of "good practices", land use patterns (e.g. location of shopping and service centres or technological and industrial clusters on the outskirts of cities, but without providing access to public transport there, construction of new housing estates in the suburbs, but without creating local centres service and railway/tram lines providing fast connection with other parts of the city) results in an additional increase in the use of private cars.

The target the WMA transport system should be characterized by availability and meet the basic needs of all users as to their mobility, balance and meet various needs for mobility and transport services for both residents and industrial and service companies [19]. Which means that it should be equipped with the appropriate quality and "number" of transport infrastructure, i.e. should invest in the modernization or construction of this infrastructure.

Taking into account the specificity of Polish cities, it was necessary to face a series of activities that were necessary and are necessary for introducing the principles of sustainable development, such as: stimulating local authorities to take action; creating sustainable transport behaviours; improving the efficiency and performance of urban transport systems; reduction of negative external effects of transport; balancing mobility.

4 The Infrastructure Investments of WMA

Poland is constantly catching up in the construction of transport infrastructure, and in principle in every branch of transport. In the case of roads, there is a goal to build a network of motorways and expressways, the target layout of this network is presented in the Fig. 1. The construction of a comprehensive network of roads (main communication routes) will result in the patency not only of the WMA, but also of our country.

Some regions in Poland are still difficult to access by public transport, especially the regions of south-eastern and north-eastern Poland. The implementation of the target road network system will consequently give a new transport quality and create development opportunities for these regions.

In the case of the WMA, many road and transport investments have been implemented and are currently underway, which will definitely improve the transport system of the area, which will positively affect the introduction of elements of sustainable transport development.

As part of the research conducted at the Motor Transport Institute and carried out for the Ministry of Investments and Development, 52 sections of road infrastructure in Poland were analyzed, some of them were located in the area of the WMA. The economic calculation for these sections was calculated, its task was to determine the payback time of a given investment.

To implement this objective, the method of determining the so-called traffic of excited vehicles on constructed or significantly modernized sections of the main road network (motorways and so-called express roads) as well as important bypasses and roads of the border area. The dynamic approach to traffic induced by the investment enables not only a functional analysis, but first of all a calculation of the economic efficiency of the given infrastructure investment.

As part of the article, two road investments were presented, which are extremely important for both the WMA and the restoration of Warsaw itself. These investments are: Southern Bypass of Warsaw (POW) and Beltway of Marki and Radzymin (OMR). Both investments are part of the creation of the target urban transport system, which should be characterized by availability and meet the basic needs of all users as to their mobility, transport processes, reduce travel time, increase traffic flow, reduce noise and emissions, move truck traffic away from built-up areas.

Warsaw has had serious problems due to traffic caused by an increasing number of cars for a long time. In Warsaw alone there are already over 600 passenger cars per 1000 inhabitants, which is twice as many as in, for example, in Berlin. In Warsaw, the movement takes place mainly from outside the city centre, often from distant places even by several dozen kilometres. In addition, this involves the transport of goods to stores, various types of investments generating traffic, public transport, transit, tourist traffic, etc., generating all kinds of problems related to urban transport, lack of traffic flow, noise or air pollution.

Analyzes of forecasted loads show that traffic associated with Warsaw accounts for 80% of all traffic on the route, of which 50–60% is incoming and outgoing traffic from the city, and 20–30% is traffic inside Warsaw. The remaining traffic is related to the Mazowieckie Voivodship (on average 17%) and long-distance traffic, which is only 3% (http://siskom.waw.pl/s2.htm).

These two investments together with other sections will in the future create a closed ring, which will enable, among others, ensuring the integrity of the WMA and linking the Warsaw's target road system to the external (national) road network, including the A2 motorway at the "Konotopa" junction (in the west) and "Konik" route (in the east). At the "Konotopa" junction, the entry of the A2 motorway to Warsaw branches out on routes towards the POW and expressway no. S7/S8 constituting the extension of the existing Torunska route towards the west, ensuring efficient distribution of traffic from the motorway to the northern and southern districts of Warsaw.

The factors adopted for calculating the economic efficiency calculation for POW and OMR were: increasing capacity, air pollution, Road Traffic Safety, noise and maintenance costs of this section.

Southern Bypass of Warsaw (POW)—road route, along which the route was marked out (POW) uses the former corridor reserved in the development plans for the A2 motorway. This part of the Warsaw bypass will act as a city route with the parameters of the main motorway route and will be characterized by connections with the street layout that are more numerous than the motorway. The corridor of the route between Puławska street and the road to Lublin will run through the districts of Ursynow, Wilanow, Wawer and the Wiazowna commune.

Table 1 presents the results of tests carried out with the modified method of the excited vehicle movement applied in the implemented project [6].

Table 1. Economic calculation POW [6].

No	Topic	Value in PLN/year
1.	Accident including – the benefits of fewer accidents – the benefits of fewer fatalities – the benefits of fewer serious injuries – the benefits of fewer people slightly injured	4 233 497.00 1 846 200.00 615 000.00 1 748 000.00 24 297.00
2.	Smoothness of movement including – cars – lorries	1 330 432 180.92 987 678 011.24 342 754 169.68
3.	Environmental Protection – costs of environmental pollution including • cars • lorries – noise costs including • cars • lorries	94 462 815.05 43 644 336.45 20 091 090.55 23 553 245.90 50 818 478.60 26 520 239.08 24 298 239.08
4.	Costs of using the road surface, maintaining repairs etc. of the motorway section built including – cars – lorries	−10 035 470.71 −4 018 218.11 −6 017 251.60
5.	Total—annual profits	1 420 039 562.25
6.	Total investment cost	2 541 549 492.51
7.	Investment return time (in years)	1.79

The total investment cost will amount to PLN 2 541 549 492.51, the total length of the section—18.5 km and annual profits resulting from the construction of the motorway—approx. PLN 1 427 719 356.24, the investment return time is 1.79 years old.

Recommendations for the analyzed section can be summarized in two points:

- complete the entire investment of the Warsaw Southern Bypass and complete it with the city exit road system. All elements of the investment should have the parameters of the city motorway. The old road system should be preserved in order to service the growing local needs;
- start research work on the creation of the motorway Warsaw bypass (for the time being on the southern and south-eastern sections) in the trail intended for heavy truck transport DK50.

Bypass of Marki and Radzymin—implementation of the expressway, on the route of which the beltway Marki and Radzymin is located is an investment of European significance. It has been included in very important government tasks. The necessity of its construction results from the need to create a transit road system on the territory of the country. The completed road investment is located in Marki and runs towards Radzymin. It is needed due to the fact that the existing road communication system in this area is not able to cope with the rapidly increasing traffic of vehicles.

S8 connects agglomerations: Wroclaw, Łódź, Warsaw and Bialystok. The route on the section Wroclaw–Ostrow Mazowiecka is the Polish part of the E67 route, while the section Warsaw–Ostrow Mazowiecka is the Polish section of the Via Baltica route. The implementation of the S8 express road, including the section under analysis, is an investment of European importance. It is a transit route from Warsaw to Bialystok, towards the eastern border of Poland.

The effects that can be expected and in an immediate time, after putting the investment into use, will be: taking over part of the traffic from existing national and provincial roads; removal of heavy traffic from built-up areas; reduction of travel time; fuel economy; providing driving comfort; reducing the risk of accidents; reduction of exhaust and noise emissions in relation to currently used roads; acceleration of development of adjacent areas.

Table 2 presents the results of the research carried out using the modified method of motion of excited vehicles used in the implemented project [6].

The return on investment is 0.91 years. This means, as with many other analyzed sections of road investments, an "infrastructure disaster" expressing the urgent need to open the communication chain, in this case running through the brands.

Recommendations for the analyzed section can be summarized in three points:

- investments entering communication routes constituting bypasses of large urban agglomerations (in particular, when rail transport is not functioning in a given area) should be implemented in the first place;
- after commissioning the beltway for use, Marki should not narrow the previous communication routes; they should ensure free movement for local traffic;
- future road bottlenecks that contradict the network principle of road infrastructure can not be allowed in the future.

Table 2. Economic calculation OMR [6].

No	Topic	Value in PLN/year
1.	Accident including – the benefits of fewer accidents – the benefits of fewer fatalities – the benefits of fewer serious injuries – the benefits of fewer people slightly injured	4 297 564.00 1 887 000.00 615 000.00 1 771 000.00 24 564.00
2.	Smoothness of movement including – cars – lorries	751 291 173.38 449 635 277.05 301 655 896.33
3.	Environmental protection including – costs of environmental pollution including • cars • lorries – noise costs including • cars • lorries	63 333 337.98 29 875 412.97 9 146 354.46 20 729 058.51 33 457 925.01 12 073 199.99 21 384 725.02
4.	Costs of using the road surface, maintaining repairs etc. of the motorway section built including – cars – lorries	−7 125 009.27 −1 829 264.52 −5 295 744.75
5.	Total—annual profits	811 797 066.09
6.	Total investment cost	737 500 000.00
7.	Investment return time (in years)	0.91

5 Conclusions

1. The results of the research show that in the areas through which the construction of modern road infrastructure is planned, new enterprises are often created, housing estates, stores, schools, etc. are built, which causes the development of a given area or region. One of such examples is the construction of beltways in the Warsaw Metropolitan Area, one of the beltways is still under construction, and modern investments are already being built along it. The second of the beltways is already put into use, which resulted in the expansion of housing estates, stores, and new enterprises were established, which thanks to this investment were located in areas very well communicated with, among others Warsaw.
2. The research carried out indicates that the area of impact of the investment is different. It depends on many factors, one of them is the connection of local roads to the main communication routes (expressways and motorways), and the other is the

density of population, the density of the network of other roads, the number of enterprises. Sometimes such an area of impact may reach several dozen kilometres from a modern transport route.

3. Currently, both in Poland and in the EU, transport is based on road transport, which is a system of economic circulation, the more modern transport infrastructure is, the higher quality, the more efficient the transport processes. As a result of the modernization of the transport network, we get more of its efficiency, which translates into greater possibilities of transport processes. However, it is only the range, its increase in accessibility to other areas creates opportunities for the development of a given region, area, which may also translate into going outside, reaching for new markets, i.e. local economic growth.
4. In Tables 1 and 2, the economic efficiency calculation for a given road section is presented. The time of reimbursement of costs incurred for the construction of a given section is extremely short, which means "congestion disaster". It can be said that such a road is already inefficient, it does not create opportunities for development and it must be modernized. In addition, the modernization of many sections, in particular those located in the Warsaw Metropolitan Area, will affect the introduction of sustainable development principles by increasing the flow of traffic, reducing exhaust emissions, noise, reducing accidents, deriving transit traffic from cities and built-up areas.

References

1. Jones, P.: The evolution of urban mobility: the interplay of academic and policy perspectives. IATSS Res. **38**(1), 7–13 (2014). https://doi.org/10.1016/j.iatssr.2014.06.001
2. Siegfried Rupprecht, Citizen and stakeholder involvement in the development and implementation of an SUMP—tools, benefits, and limitations, Perugia, 29–30 November 2012, POLIS Conference (2014). http://www.mobilityplans.eu/docs/SUMP_guidelines_web0.pdf, http://www.eltis.org/guidelines/what-sustainable-urban-mobility-plan
3. May, A.D.: Encouraging good practice in the development of Sustainable Urban Mobility Plans. Case Studies on Transport Policy **3**(1), 3–11 (2015). https://doi.org/10.1016/j.cstp.2014.09.001
4. Priester, R., Miramontes, M., Wulfhorst, G.: A generic code of urban mobility: how can cities drive future sustainable development? Transp. Res. Proc. **4**, 90–102 (2014). https://doi.org/10.1016/j.trpro.2014.11.008
5. Wefering, F., Rupprecht, S., Bührmann, S., Böhler-Baedeker, S.: Guidelines. Developing and Implementing a Sustainable Urban Mobility Plan. European Commission Directorate-General for Mobility and Transport, European Platform on Sustainable Urban Mobility Plans, Brussels (2013)
6. Research work: Analysis of the impact of the built road infrastructure on the level of economic activity in the surrounding territorial units, under the Operational Program Technical Assistance No. POPT 02.01.00-00-0021/15-00, titled: Support to the coordination institution Strategic Partnership Agreement in 2015–2016, implemented for the Ministry of Investment and Development, Warsaw (2017)
7. Research paper No. 6512/ITS, entitled: Shaping the transport system of the Warsaw Metropolitan Area. Motor Transport Institute, Warsaw (2015–2017)

8. Brdulak, J., Krysiuk, C.: Humanization of the communication network of the Metropolitan Area of Warsaw. Car transport No. 4/2016, Warszawa (2016)
9. Bryx, M.: Innovations in city management in Poland. SGH, Warsaw (2014)
10. Brdulak, J.: Problems programming infrastructure investments in transport. ITS, Warsaw (2016)
11. Brdulak, J., Pawlak, P., Krysiuk, C., Zakrzewski B.: The closing of the expressway and motorway network with a multiplier factor of the economic development of regions. Logistics 3/2014, Poznań (2014)
12. Krysiuk, C., Banak, M., Nowacki, G.: Problemy mobilności miast w Unii Europejskiej, Logistyka. Nr. 3 (2015)
13. Krysiuk, C.: Development of transport infrastructure in cities. Coaches. Technology, Operation, Transport Systems No. 9 (2016)
14. Nowacki, G., Mitraszewska, I., Kaminski, T.: Driving simulator analysis according to Directive 2003/59. In: Proceedings of the 7th International Conference Reliability and Statistics in Transportation and Communication, Relstat 2007, pp. 178–186 (2007)
15. Nowacki, G., Mitraszewska, I., Kaminski, T.: TachoScan PCNET Service digital tachograph equipment. In: Urban Transport Xiii, vol. 96, pp. 587–592 (2007)
16. Nowacki, G., Krysiuk, C., Kopczewski, R.: Dangerous goods transport problems in the European Union and Poland. Transnav-Int. J. Mar. Navig. Saf. Sea Transp. **10**(1), 143–150 (2016). https://doi.org/10.12716/1001.10.01.16
17. Warsaw Metropolitan Area in 2016. Central Statistical Office, Warsaw (2017)
18. Europe 2020: A strategy for smart, sustainable and inclusive growth, Brussels, 3 March 2010. COM (2010)
19. Cisowski, T., Szymanek, A.: Sustainable development of urban transport. Maintenance and Reliability, no. 1 (2006)

Factors Affecting the Efficiency Indicator of the Public Transport System—Case Study

Franciszek J. Restel(✉) and Łukasz Wolniewicz

Wroclaw University of Science and Technology,
27 Wybrzeze Wyspianskiego Street, Wroclaw, Poland
{franciszek.restel,lukasz.wolniewicz}@pwr.edu.pl

Abstract. In the paper, the problem of determining the crew efficiency indicator was analysed. The limits of the coefficient value were checked. In terms of the theoretical values the Wroclaw City case was studied. Differences between theoretical and empirical values were identified. Their genesis was found and guidance for constructing the timetable was provided. The designers constantly take measures to increase the value of this coefficient, for example by operating several lines by one vehicle. Increasing the length of communication lines - despite making them more susceptible to traffic disruptions—increases the achievable value of the efficiency index, as the length of the course does not affect the time required to perform duties on the loop. The highest values of the indicator are obtained on working days, when the planned transport is the highest. As the number of vehicle kilometres per day decreases, it becomes more difficult to achieve rational index values.

Keywords: Timetable · Efficiency indicator · Optimization · Public transport · Vehicle scheduling

1 Introduction

The value of the crew working time utilization index (also named: efficiency index) is one of the measures allowing to determine the rationality of prepared work schedule and it is an important indication for timetable constructors [1]. The level of this indicator has a crucial impact on the company profitability and the entire public transport network. It indicates what part of the drivers' work is devoted to driving. The operation of a communication network characterized by too low index value is very expensive and tends to strain the budget of local government units.

In order to adjust its transport offer to the needs of passengers, the public transport organizer [2] is forced to introduce frequent changes in the timetables. Therefore the value of the efficiency index is subject to significant fluctuations and changes after each modification of the public transport network.

Nowadays, knowledge on the rationalizing possibilities of the crew working time utilization index value is becoming a tool for increasing competitiveness between transport companies and allows for the development of the passenger transport services. Especially in urban areas and agglomerations. Moreover, implementing by the United Nations and the European Union of a policy of sustainable development aimed

I. Kabashkin et al. (Eds.): RelStat 2018, LNNS 68, pp. 57–68, 2019.
https://doi.org/10.1007/978-3-030-12450-2_6

at reducing the negative effects of human activity on the environment. Also saving raw materials and energy justifies the use of the least possible resources to carry out certain tasks (including transport operations).

The scheduling process is based on the experience of planners rather than on analysis and simulation. It is therefore reasonable to believe that there is a current problem with the determination of reasonable values for the drivers' working time factor in public transport. This should be the result of measurable factors and should contribute to the profitability of the whole transport network. On the other it should result in the expansion of the public transport offer.

Determination of the impact of various factors on the value of the efficiency index allows to use the indicator for system optimizing. System optimization on example of the air transport has been described in [3–5]. The crew working time utilization index should not be considered as the only parameter describing the rationality of timetables. Striving for an uncritical increase in the value of the indicator will have a negative impact on notes in other areas. I.e. the need may result to employ more drivers (in order to keep the law of Driver Working Time [6]), or the elimination of intermediate stops on routes.

2 Crew Working Time Utilization Index

The crew working time factor determines which part of the driver work is spent on driving the vehicle. This factor is therefore a measure of the effective working time. It can be calculated in relation to rolling stock, work time, brigade, line, group of lines or for the whole network [7]:

$$w_e = \frac{t_h}{t_e} = \frac{v_e}{v_h}[-], \tag{1}$$

where t_e—operating time $[h]$, which includes the times of operation: journey time between stops, stopping at intermediate stops, stops at stations and loops, technical journeys; t_h—commercial time $[h]$, consisting of: journey time between stops and stopping times at intermediate stops; v_e—service speed $\left[\frac{km}{h}\right]$ and v_h—commercial speed (communication speed v_k) $\left[\frac{km}{h}\right]$.

Formula (1) directly indicates that the value of the crew working time utilization coefficient we in public transport must be within the range of $0 \div 1$, however, achieving unity is difficult in practice, and often impossible. This is influenced by the factors classified into five categories:

1. Factors related to the assumptions of the work plan:
    - number of teams planned for the timetable [1],
    - changes vehicles between lines (this problem is labelled in [8–12]),
    - provision of breaks under the Act about Working Time of Drivers,
    - change of driver on the route.

2. Disturbance factors:
   - vehicle failure,
   - short-term detours not included in the working timetable,
   - incorrect driving technique of the driver,
   - service of passengers by the driver, e.g. sale of tickets,
   - delays due to congestion,
   - confusion of route by driver,
   - carrying out courses in teams intended for the operation of new rolling stock by older vehicles with lower parameters and characteristics,
   - weather conditions (road ice, leaves on tracks).
3. Factors rising from transport requirement and adaptation them timetable:
   - cyclicality of running (to maintain longer stopping times on the loops),
   - journey time between the start and end stops (based on technical speed and the road),
   - time of passenger exchange,
   - adapting journey times to different times of day,
   - the existence of impulse motion generators close to the loop,
   - need to differentiate capacity of the rolling stock used for the course.
4. Technical service factors:
   - time needed to prepare the vehicle for its course (e.g. exchange of relational tables),
   - technical journey times (journey between loops),
   - activities performed by rapid response teams (e.g. temporary cleaning on loops),
   - need to recharge the battery (in battery vehicles),
   - necessity to change the cab by the driver in two-way vehicles,
   - filling of the work (road) card.
5. Infrastructure related factors:
   - infrastructure failures,
   - capacity constraints (in tramway transport, the layout of the tracks and the type of loop, monorail sections),
   - stop by manually changing the junction,
   - lack of priority for mass transport in the traffic control system.

The influence of the factors on w_e index is determined intuitively, relying on design experience. The parameters affect the value of efficiency index. Some of them are difficult to predict (breakdowns, delays, technical service, driving technique, etc.), others are determined by calculations and mathematical-physical formulae (limitations resulting from Driver Working Time [6], running intervals, number of teams, etc.). Reliability methods are described in [13–16].

The Act on Driver Working Time minimizes breaks for public transport drivers whose route does not exceed 50 kilometres. On the basis of the abovementioned regulations, the maximum value of the analysed coefficient was obtained for different working times of crews. The results are summarized in Table 1.

Table 1. Maximum values of efficiency index is determined by the [6].

Daily driving time	Maximum index value $w_e[-]$
Under 6 h	1
More than 6 h less than 8 h	0.938
More than 8 h less than 10 h	0.925

For a daily driving time not exceeding 6 h, the driver is not entitled to a statutory break. When the driving time is between 6 and 8 h, he must have a break of 30 min. This may be divided into shorter periods, provided that one of the breaks lasts at least 15 min. If the driving time is between 8 and 10 h, the driver has a 45-min break. As in the case of driving time from 6 to 8 h, this break can be divided according to the aforementioned rule. It should be kept in mind, the driver during the break is not at the employer's disposal. Therefore, no service duties can be performed, i.e. completing the road card, change direction boards, change the cabin in two-way vehicles etc.

Also the running interval has a significant impact on limiting the crew working time utilization index value. The timetable periodicity results in longer stopping time at the end stops. Thus, decreasing the value of the discussed indicator.

The communication and operation speed depending on the time between courses; the duration of the courses was analysed in terms of the lowest possible number of brigades. The theoretical value of the indicator for typical running frequencies varies diametrically depending on the duration of the journey between the 1st and final stops.

Increasing the course frequency will result in lowering of the indicator. For example, for a 60-min interval and a 12-min course, this indicator reaches 0.4. It is not possible to increase this value when constructing a single line timetable. When the whole communication network is considered, the indicator can be increased, e.g. by introducing the operation of few lines by one brigade. In addition to the abovementioned changes of vehicles between lines, the indicator value can be increased by lengthening or shortening the route of a line.

A brigade is an organizational unit which consists of employees who operate one vehicle. It can be one shift or several shifts. The number of crew brigades needed to implement the assumed timetable is determined from the formula:

$$\lambda \geq \left\lceil \frac{\sum_{i=1}^{n} t_{k_i} + t_{s_i}}{\tau} \right\rceil [-], \tag{2}$$

where λ—number of brigades $[-]$; t_k—communication time between loops [min]; t_s—the length of service duties before the course [min]; i—number of brigade courses operated on the day $[-]$; τ—interval [min].

Explanation of t_{ki} and t_{si} is presented in [17, 18] where authors measure in graphical way the number of vehicles (trains) required to operate the line. Well-designed timetables need for implementation as few brigades as possible [19], so it is true that:

$$\left\lceil \frac{\sum_{i=1}^{n} t_{k_i} + t_{s_i}}{\tau} \right\rceil - \frac{\sum_{i=1}^{n} t_{k_i} + t_{s_i}}{\tau} \rightarrow \min. \tag{3}$$

Under the conditions of formulas (2) and (3), it is possible to achieve the highest possible value of the w_e index determined by the number of brigades. If calculations using formula (3) result in a number greater than or equal to one, the crew working time utilisation index may be increased by reducing the number of brigades.

3 Analysis of the Crew Working Time Utilisation Index for the Wroclaw City Case

In the previous chapter, the theoretical maximum values of the efficiency index were determined, taking into account the most important determinable restrictions. However, also other factors need to be taken into account for real systems. The analysis of transport tasks in the Wroclaw City transport network (bus and tram transport) shows how the value of the index develops in real conditions and how it changes depending on the assumed parameters. Data from public transport system in Wroclaw City source from [19] and [20].

The Wroclaw City communication network is characterised by different lengths of communication lines. There are relatively short courses, e.g. on the line no. 140, and also long courses, e.g. on the line no. 20.

For short lines, the value of the crew working time utilisation index depends on the type of day (working days, weekend and holidays). It varies from 0.66 to 0.76. With the increase of the route length, the value of the coefficient for all types of days is equal to 0.77. Regardless of the line length, the driver must have a standstill on the loop (the need to compensate for interference caused by a congestion, change of cabin, change of direction boards, filling of the work card). The time required to perform the above-mentioned service duties is constant regardless of the length of the course, so the percentage of time required to perform these duties during the course of work decreases with the increase of the driving time. This is illustrated by the graph in Fig. 1a. The analysis of the impact of the course average duration on the value of the performance indicator also leads to convergent conclusions, as illustrated in Fig. 1b.

The regression lines in Fig. 1a, b, and c base on the highest values of the correlation coefficient R^2 and allow for further analysis of changes in the crew working time utilisation index depending on the type of day. They illustrate the trend in the value of the performance indicator as a function of the length of the communication line and the duration of the course.

The value of crew working time utilisation index is directly proportional to the communication time and inversely proportional to the operation time, which consists of the communication time and the stopping time. As the waiting time for departure from the first stop increases, the value of the coefficient decreases. As the denominator in formula (1) increases. In the case of weekends, on the lines where the interval is the same during the week, it is necessary to extend the stoppage on the loop in order to

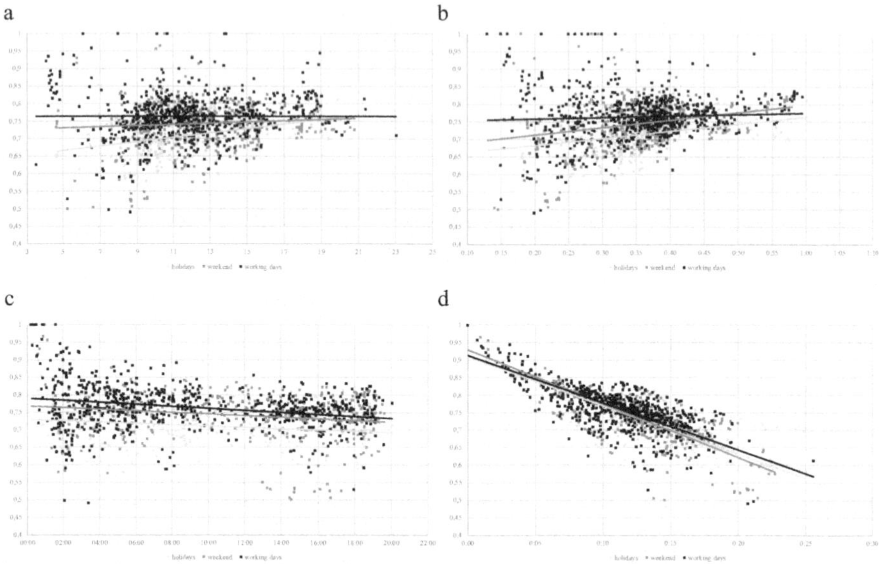

Fig. 1. Values of efficiency index of public transport system in Wroclaw City for: (a) line length; (b) course length; (c) crew work time; (d) average stopping time on the loop

keep the cycle. Thus, reducing the travel time between the first and the end stop. Figure 1d. shows the effect of the stopping time on the loop on the value of the crew working time utilisation index.

Crew working time utilisation index is strictly dependent on the provisions of the Act about Working Time of Drivers. As the crew working time increases, the w_e ratio in Wroclaw City slightly decreases and the discrepancy between the values obtained decreases. This is illustrated in Fig. 1c. For some of the short teams, one is achieved in the conditions of Wroclaw City, because the abovementioned regulations do not determine the need to take into account a break during work, when the driving period is shorter than 6 h. For very short teams there are no disturbances leading to the accumulation of delays (secondary delays) and thus there is no need to eliminate them.

The significant discrepancy between the values of the crew working time utilisation index for short and very short teams is influenced by the character of those teams. They are introduced in short periods of increased demand for public transport. Therefore, the ad hoc character determines the inability to standardise the we index value.

It should also be noted that mainly on working days there are few teams with a working time of 10–12 h. The reasons for this can be found in the statutory [6] limitation of the maximum driving time per driver to 10 h. The introduction of a two-person team with a total driving time of about 12 h is not economically justified.

For the indicator values shown above, the regression function parameters (Table 2) were estimated due to the basic regression definition:

Table 2. Regression parameters for the w_e indicator.

	Correlation between indicator w_e and average line length			Correlation between indicator w_e and average duration of course		
	Holidays	Weekend	Working days	Holidays	Weekend	Working days
Factor a	0.0064	0.0017	>0.0001	0.0020	0.0021	0.0004
Lower limit 95%	0.0042	0.0001	–0.0015	0.0011	0.0015	–0.0001
Upper limit 95%	0.0087	0.0034	0.0015	0.0030	0.0028	0.0009
Factor b	0.6330	0.7229	0.7649	0.6414	0.6699	0.7505
Lower limit b 95%	0.6050	0.7022	0.7471	0.6096	0.6472	0.7310
Upper limit b 95%	0.6610	0.7435	0.7827	0.6733	0.6926	0.7700
R^2	0.0834	0.0047	>0.0001	0.0523	0.0465	0.0029
r	–0.2887	–0.0683	–0.0001	–0.2288	–0.2155	–0.0534
	Correlation between indicator w_e and average brigade working time			Correlation between indicator w_e and average stopping time on the loop		
Factor a	>0.0001	>0.0001	>0.0001	–0.0139	–0.0154	–0.0135
Lower limit 95%	–0.0001	–0.0001	–0.0001	–0.0147	–0.0161	–0.0143
Upper limit 95%	>0.0001	>0.0001	>0.0001	–0.0132	–0.0147	–0.0127
Factor b	0.7286	0.7684	0.7907	0.9093	0.9297	0.9142
Lower limit b 95%	0.7143	0.7585	0.7824	0.8993	0.9206	0.9051
Upper limit b 95%	0.7429	0.7784	0.7990	0.9193	0.9388	0.9233
R^2	0.0215	0.0358	0.0633	0.6364	0.6725	0.5976
r	–0.1467	–0.1892	–0.2515	–0.7977	–0.8200	–0.7730

$$w_e = a \cdot x_i + b, \tag{4}$$

where: *a, b*—regression line parameters; x_i—chosen variable (average line length, average course duration, average brigade working time, average stopping time on the loop).

There are communication lines in Wroclaw City with low frequency of running, not exposed to the influence of traffic disruptions and operating small passenger streams. In such cases, taking into account the costs generated by the vehicle in motion (e.g. fuel, tyre and fluid consumption), it may sometimes be more economical to extend the duration of the stopping period rather than to increase the frequency.

It is noticeable that high values of the efficiency index are more difficult to achieve with the decrease in the frequency of running. This also applies to the tram and bus network of Wroclaw City. In order to perform the analysis, five groups of communication lines were distinguished (categorised depending on the course frequency on working days). The results obtained are summarised in Table 3.

Table 3. Values of efficiency index for different groups of public transport lines on working days for public transport system in Wroclaw commune.

Group of lines	Frequency of running		w_e
	In peak hours	In the interim hours	
I	min. 4 courses/hours		0.765
II	min. 3 courses/hours	min. 2 courses/hours	0.748
III	min. 2 courses/hours		0.729
IV	min. 1 courses/hours		0.677
V	running	no running	0.697

Table 4 shows the average values of the crew working time utilisation index depending on the type of day for the Wroclaw City public transport network. Lower we notes for holidays are determined by the relatively low frequency of running (trams every 30 min, bus lines in diluted mode, part of lines not running). Timetable designers in such situations use the vehicle changes between the communication lines to increase the value of the performance indicator.

Table 4. Average values of efficiency index for public transport system in Wroclaw commune.

Day type	Average value of an index w_e
Working day	0.765
Weekend	0.744
Holiday day	0.711

If buses and trams depart from stops more often, a higher value of the indicator can be achieved, as the loop times are shorter.

4 Influence of Travel Time Variability on Crew Efficiency Index

Operational research on course arrivals in Wroclaw it were performed. For the analysed period of time it was found, the arrival deviations from scheduled time can described by normal distributions N(3.6, 8.2)—for working days, N(2.4, 5.9)—for weekends and holydays. The sample size was more than 300 arrivals per type of day. The Chi squared goodness of fit test was performed. At significance level 0.05, there was no reason to reject the goodness of fit hypothesis.

For the matched distributions, the third quartiles of arrival deviation were calculated. Respectively 9.0 and 6.5 were obtained. Using the quartile values, the stopping times on the loops were recalculated. The resulting crew efficiency indicator has also been recalculated. Time was reduced for stops in the loop longer than 15 min to 15 min due to the Laws [2] and [6]. Other times were shortened to 4 min in order to perform service duties (for example, completing the road card, change direction boards, change the cabin in two-way vehicles) on the loop (Fig. 2).

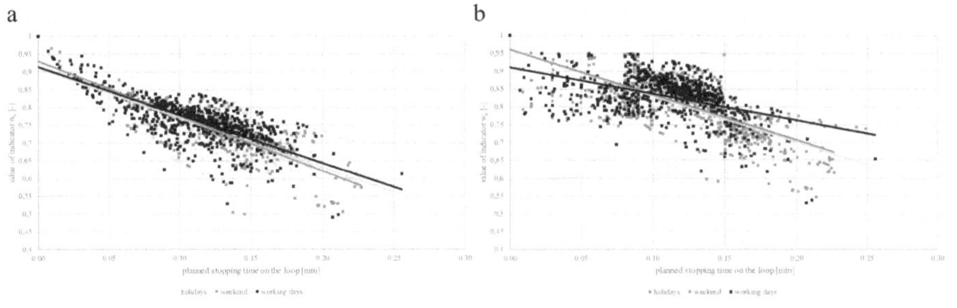

Fig. 2. Values of efficiency index reduced stopping time in the loop due to delays: (a) before recalculated; (b) after recalculated

Higher values of the coefficient were obtained. However, this is not a way of increasing it. This means that the timetable is impractical. In Wroclaw, a lot of delays in public transport are caused by traffic jams in peak hours. Analysis of critical situations and delays in transport are described in [21–28]. It is not easy to determine the correct journey times between stops. A factor calculated from the real journey times may be an indication for change times in timetable. A higher value of the real efficiency indicator compared to the value of the factor calculated from theoretical travel times means that too little time reserves have been planned in the timetable. In the case of a higher value of the real factor, the driver shall have fewer breaks in working time. When the actual value of the coefficient is lower than the planned value, this means driving ahead of time and taking longer breaks in working time. In Wroclaw driving according to the scheduled is 3 min. delay and 1 min. departure before time.

5 Conclusions

The theoretical values are overestimated due to the impossibility of taking into account non-measurable factors or some difficult to measure. There are many factors that have a significant impact. It is not defined by mathematical, physical and legislative rules or has not been assessed yet. These factors include: disruptions to the communication network and driver duties on the loops (requiring detailed chronometric research so popular on the railways). These factors affect the communication network in different ways.

The Wroclaw City experience shows that the global rational value of the we efficiency index for bus and tram transport should oscillate around 0.75 and increase as the frequency of courses increases.

The attempt made in Sect. 2 of this paper to estimate the maximum theoretically achievable values, adjusted by 5% (to take into account the necessity of delay compensation and other activities performed by drivers), after comparison with the global indication of the we coefficient for the Wroclaw City network, reveals that it is possible to increase the value of the quotient of transport and operating time (to the level of 0.8), and thus to contribute to the development of the transport offer without increasing the number of: vehicle hours, employees and vehicles.

The issue of the influence of factors determining the value of the crew working time utilisation index on the construction of the timetable was discussed. It was shown that the greatest influence on limiting the value of above mentioned indicator:

- tact of running on the line,
- Act about Working Time of Drivers,
- length of communication lines and time between loops,
- duties of driver (other than driving),
- time reserves to mitigate the negative effects of traffic disruptions.

The highest values of the indicator are obtained on working days, when the planned work transport—similar to the assumed frequency of journeys—is the highest. As the number of vehicle kilometres per day decreases, it becomes more difficult to achieve rational index values. This is largely due to the reduced frequency of running, which makes it difficult to select the optimum number of teams and forces the teams to extend their stoppages on the loops in order to keep the tact. The designers constantly take measures to increase the value of this coefficient, for example by operating several lines by one vehicle. Increasing the length of communication lines—despite making them more susceptible to traffic disruptions—increases the achievable value of the efficiency index, as the length of the course does not affect the time required to perform duties on the loop.

By comparing the maximum theoretical values of crew working time utilisation index possible to achieve with the values obtained practically, the authors proved that it is not possible to construct timetables solely on the basis of limitations resulting from legal regulations and mathematical accounts. They point out that it is necessary to take into account time reserves to offset the possible effects of traffic disruptions. They also point out that it is possible to increase the global value of the we coefficient for public transport in Wroclaw City.

References

1. Raff, S.: Routing and scheduling of vehicles and crews: the state of the art. Comput. Oper. Res. **10**(2), 63–67, 69–115, 117–147, 149–193 (1983)
2. Law of 16.12.2010 on the public transport, Warsaw 2010 (in Polish: Ustawa z dnia 16 grudnia 2010 r. o publicznym transporcie zbiorowym (Dz.U. nr 5 poz. 13), Warszawa 2010)
3. Kierzkowski, A., Kisiel, T.: A model of check-in system management to reduce the security checkpoint variability. Simul. Model. Pract. Theory **74** (2017). https://doi.org/10.1016/j.simpat.2017.03.002
4. Kierzkowski, A., Kisiel, T.: Simulation model of security control system functioning: a case study of the Wroclaw Airport terminal. J. Air Transp. Manag. **64**, Part B 173–185 (2017). https://doi.org/10.1016/j.jairtraman.2016.09.008

5. Kierzkowski, A.: Method for management of an airport security control system. Proc. Inst. Civil Eng.—Transp. **170**(4), 205–217 (2016). https://doi.org/10.1680/jtran.16.00036
6. Law of 16 April 2004 on the drivers working time, Warsaw 2004 (in Polish: Ustawa z dnia 16 kwietnia 2004 r. o czasie pracy kierowców (Dz.U. Nr 92 poz. 879), Warszawa 2004)
7. Madej, B., Pruciak, K., Madej, R.: Public transport. Rules of creating timetables, Academy of Transport and Entrepreneurship, Warsaw 2015 (in Polish: Madej, B., Pruciak, K., Madej, R., Publiczny transport miejski. Zasady tworzenia rozkładów jazdy, Akademia Transportu i Przedsiębiorczości, Warszawa 2015)
8. Ciancio, C., Laganà, D., Musmanno, R., Santoro, F.: An integrated algorithm for shift scheduling problems for local public transport companies. Omega **75**, 139–153 (2018)
9. Hassold, S., Ceder, A.: Public transport vehicle scheduling featuring multiple vehicle types. Transp. Res. Part B: Methodol. **67**, 129–143 (2014)
10. Beasley, J.E., Cao, B.: A tree search algorithm for the crew scheduling problem. Eur. J. Oper. Res. **94**, 517–526 (1996)
11. Shuia, X., Zuo, X., Chena, C., Smith, A.: A clonal selection algorithm for urban bus vehicle scheduling. Appl. Soft Comput. **36**, 36–44 (2015)
12. Lieshout, R., Muldera, J., Huisman, D.: The vehicle rescheduling problem with retiming. Comput. Oper. Res. **96**, 131–140 (2018)
13. Kierzkowski, A.: Model of reliability of security control operation at an airport. Tehnički Vjesnik—Technical Gazette **24**(2), 469–476 (2017). https://doi.org/10.17559/tv-20150812153802
14. Kierzkowski, A., Kisiel, T.: Modelling the passenger flow at an airport terminal to increase the safety level. In: International Conference on Military Technologies 2015, ICMT 2015, pp. 107–113 (2015). https://doi.org/10.1109/miltechs.2015.7153693
15. Kierzkowski, A., Kisiel, T.: Simulation model of logistic support for functioning of ground handling agent, taking into account a random time of aircrafts arrival. In: International Conference on Military Technologies 2015, ICMT 2015, pp. 115–120 (2015). https://doi.org/10.1109/miltechs.2015.7153694
16. Walkowiak, T., Mazurkiewicz, J.: Soft computing approach to discrete transport system management. Lecture Notes in Computer Science. Lecture Notes in Artificial Intelligence, vol. 6114, pp. 675–682 (2010). https://doi.org/10.1007/978-3-642-13232-2_83
17. Leibchen, C.: Periodic Timetable Optimization in Public Transport. Dissertation. Technische Universität Berlin, Berlin (2006)
18. Boyer, V., Ibarra-Rojas, O.J., Ríos-Solís, Y.A.: Vehicle and crew scheduling for flexible bus transportation systems. Transp. Res. Part B: Methodol. **112**, 216–229 (2018)
19. https://www.wroclaw.pl/rozklady-jazdy. Accessed 24 Oct 2017
20. iMPK application (operating systems: Amdroid and iOS). Accessed 17 Oct 2017
21. Larsen, R., Pranzo, M.: A framework for dynamic re-scheduling problems. Technical report 2012–01, Dip. Ingegneria dell'Informazione. University of Siena (2012)
22. Liebchen, C.: Computing delay resistant railway timetables. Computers and Operations (2010)
23. Ouelhadj, D., Petrovic, S.: A survey of dynamic scheduling in manufacturing systems. J. Sched. **12**(4) (2009)
24. Pender, B., et al.: Disruption recovery in passenger railways international survey. Transportation Research Record (2013)

25. Lusby, R.M., Larsen, J., Bull, S.: A survey on robustness in railway planning. Eur. J. Oper. Res. **266**, 1–15 (2018)
26. Schachtebeck, M., Schöbel, A.: To wait or not to wait and who goes first? Delay management with priority decisions. Transp. Sci. **44** (2010)
27. Walkowiak, T., Mazurkiewicz, J.: Analysis of critical situations in discrete transport systems. In: Proceedings of DepCoS—RELCOMEX 2009, Brunów, Poland, 30 June–02 July 2009, pp. 364–371. IEEE (2009). https://doi.org/10.1109/depcos-relcomex.2009.39
28. Kisiel, T., Zak, L., Valis, D.: Application of regression function—two areas for technical system operation assessment. CLC 2013: Carpathian Logistics Congress—Congress Proceedings, pp. 500–505 (2014). WOS:000363813400076

Directions of Development of Eastern Poland's Transport Infrastructure by the Example of the Lublin Province

Bartosz Zakrzewski[1](✉), Jacek Brdulak[2], Gabriel Nowacki[3], and Krzysztof Olejnik[4]

[1] Motor Transport Institute, Jagiellońska 80, 03-301 Warsaw, Poland
bartosz.zakrzewski@its.waw.pl
[2] SGH Warsaw School of Economics, al. Niepodległości 162, 02-554 Warsaw, Poland
jacek.brdulak@sgh.edu.pl
[3] Military University of Technology, Urbanowicza 2, 00-908 Warsaw, Poland
gabriel.nowacki@wat.edu.pl
[4] University of Technology, Generała Kazimierza Sosnkowskiego, 45-272 Opole, Poland
k.olejnik@po.opol.pl

Abstract. The paper refers to present the results of research carried out at the Motor Transport Institute in Warsaw on the direction of transport infrastructure development in the Eastern Poland and the EU, in the Lublin province, in the aspect of management and organization. The analysis carried out led to the forecasts how this development will affect the transport system of the region and the country. The article also presents the results of research on the economic calculation of postulated changes and transport investments being carried out in the Lublin province. Based on the economic calculation and the analysis carried out, the authors prepared appropriate conclusions and recommendations. We claim that competitiveness and development of the Lublin province will be greater with the development of modern road infrastructure, in the form of the A2 motorway along the section Siedlce—border of Poland in Kukuryki, as well as other expressways (S17, S19 or S12). The economic effect of these road investments will soon return and is associated with serving heavy freight traffic in transit through Poland.

Keywords: Logistical centres · Transport cooperation · European integration · Pan-European transport corridor no 2 · Transport infrastructure · Via Carpathia

1 Introduction

The aim of paper is to discuss the results of research carried out at MTI on the direction of transport infrastructure development in the Eastern Poland and the EU, in the Lublin province, in the aspect of management and organization.

The data refer to the European Union transport as follows [1]:

– transport sector accounts for 9% of EU gross value added (5% only for services),

I. Kabashkin et al. (Eds.): RelStat 2018, LNNS 68, pp. 69–79, 2019.
https://doi.org/10.1007/978-3-030-12450-2_7

- 20 million people employed—more than 9% of EU workforce,
- 13% of household consumption expenditure,
- machinery and transport equipment account for 40% of EU goods exports—transport services represent 17.3% of EU service exports,
- economic benefits: mobility needs of 500 million Europeans, flows of goods for 11 million companies, international trade. IMF: 1% GDP in infrastructure leads to 1.5% increase in GDP,
- infrastructure shapes mobility and help address negative externalities (fatalities of road safety; emissions, noise, congestion).

Main Policy Objectives of TEN-T [2] are to complete the EU internal market by ensuring seamless physical connections, creating missing links and removing bottlenecks, shape mobility by offering alternative routes on a more sustainable basis to operators and passengers—clean and efficient transport, boost economic growth by allowing greater transport and business opportunities and facilitate mobility and reduce congestion through traffic management systems and innovative technology.

The purpose of White Paper, published on 1 March 2017 [3] is to invite not only the European Union's Member States but also citizens and civil society at large for a lively and broad debate on how Europe shall develop by 2025.

Sustainable Urban Transport Planning (SUTP) for the European Union refers three dimensions of improvement [4–8]:

- helps the transport professionals to adopt a more strategic view on transport policy and to link urban transport planning with strategic economic, social, cultural and environmental objectives,
- increases the internal cooperation and capacity building,
- helps the cities to establish a new and better citizen approach.

The article presents the results of statutory research carried out by the Motor Transport Institute in the years 2008–2013 and the development of the Polish transport system [9] with particular emphasis on its Eastern border (thus the Eastern border of the EU) in the aspect of management and organization. The three-year ITS statutory research confirmed the desirability of establishing such a logistical centre. The preliminary economic calculation of such LC has become a supplement to the research report whose components were suggested and worked out by businessmen in the region [9].

In the EU and Polish policy, attention is paid to the fact that not only metropolises are important [10] and should develop, but also border regions [11]. The position of the Polish Foreign Office, despite political and economic turmoil in Ukraine, announces economic cooperation with Eastern countries, including Russia. This cooperation will take place through the Pan-European Transport Corridor No 2 running through Poland [11], including through the northern part [12] of the Lublin province. These are convenient places to carry out infrastructural transport building investments, e.g. the organization of logistical centres [12, 13].

The purpose of this article is to invoke arguments for the urgent need to expand road transport infrastructure in the Lublin province. The research objective will be achieved by answering the following questions:

1. What will be the investment expenditures for the construction of the four missing sections of the main highway roads, namely: the A2 motorway, the S12 Chełm bypass expressway, the S19 expressway Kraśnik–Janów Lubelski, and the section of the S17 Garwolin–Kurów?
2. How fast will the investment expenditures for these roads sections be paid back?
3. Is the construction of these sections profitable from the point of view of the development and modernization all of the Lublin province?

Studies carried out by Motor Transport Institute used such research methods as analysis, synthesis, deduction, induction and analogy. Empirical studies, such as diagnostic and field studies in the Lublin province, were also carried out.

The article presents the results of research on the directions of transport infrastructure development in the border region of Eastern Poland and the EU by the example of transport projects in the Lublin Voivodeship in terms of management and organization, as well as forecasting how this development will affect the transport system of the region and the country. Another goal was to carry out the economic calculation of postulated changes and transport (road) investments. The subject of the article is consistent with the "Main directions of ITS development in the years 2015–2020" approved by the Director of ITS on the 22 of December, 2014 in the areas of: economics of transport, organization, operation and efficiency of road transport.

2 Characteristics of the Researches

In the research carried out at the Institute, the analysis was carried out of directions and effects of transport infrastructure development in the border regions by selected examples in Poland and the EU, as well as analysis of directions and effects of transport infrastructure development in border regions, including the Lublin province. In the 2015–2016, the following activities were conducted:

1. Research and analysis of elements of the Polish transport system, including:
   - diagnosis of contemporary transport problems in Poland and the EU,
   - characteristics of the infrastructure of transport branches,
   - characteristics of regions, Euro-regions, and border regions in Poland and Europe, with particular emphasis on the Bug Euro-region,
   - the state of regional and transport policy of the EU and Poland.
2. Research and analysis of elements of the transport system of Poland's border regions, with particular emphasis on the Lublin province, such as:
   - border crossings,
   - Special Economic Zone (SEZ) and Free Customs Areas (FCA)—at a distance of 25 km from Biała Podlaska—the capital of Southern Podlasie—there is the country's largest "dry transhipment port" in Małaszewicze and the Free Customs Zone. It is possible to reload goods from rail to road transport and vice versa. In addition, the wide railway track from Belarus has been laid to there.

- logistical centres—where different transport branches are assumed to meet, they are infrastructural and organizational elements which, especially in Western Europe and in the United States where their network is already quite advanced, affect the ordering and increasing the efficiency of transport systems.

In 2017, the economic efficiency calculation of the investments was carried out (case studies) concerning the transport infrastructure network of the Lublin province, including:

- A2 motorway from Siedlce to the border in Kukuryki,
- sections of the S17 road from Warsaw to Lublin,
- sections of the S19 road, so-called Via Carpatia,
- Chełm bypass as part of the S12 route,
- a logistical centre in the area of Biała Podlaska and Małaszewicze,
- cargo airport—an important element of the transport infrastructure of the Lublin province is a post-military airfield in Biała Podlaska. Its reasonable development in the future provides the chance to develop passenger and cargo transport, initially on a small, and over time larger scale.

3 Economic Efficiency Calculation of the Investments in the Road Infrastructure in the Lublin Voivodeship (Province)

3.1 A2 Motorway

Within the Lublin Voivodeship there are two important sections of the planned future A2 motorway: the section from Siedlce to Biała Podlaska and the Biała Podlaska section of the state border in Kukuryki.

The section of Siedlce–Biała Podlaska is an important segment connecting Poland's capital, Warsaw with the country's Eastern border as part of the 2nd Pan-European Transport Corridor. The northern bypass of Międzyrzec Podlaski operates within DK No 2 (E30) route and the western bypass within DK 19 route, ultimately along the S19 expressway (Via Carpathia—Fig. 1). The plans are to build A2 motorway that is to run several kilometres north of the city.

The economic efficiency calculation was conducted of the Siedlce–Biała Podlaska motorway section as part of the study. Assuming the total investment cost on the Siedlce–Biała Podlaska motorway section in the amount of PLN 2,807,300,000.00, the length of the analyzed section of 67 km and annual profits resulting from the investment in the amount of PLN 132,534,508.51, the investment payback time arrived at indicates that the costs of the construction of the Siedlce–Biała Podlaska section of the motorway will be paid back within 21.18 years.

It should be noted that the A2 motorway in the Pan-European Transport Corridor No II will be used by freight road transport in cooperation with the railway container terminals, e.g. the Czech container terminal in Małaszewicze. According to estimates of

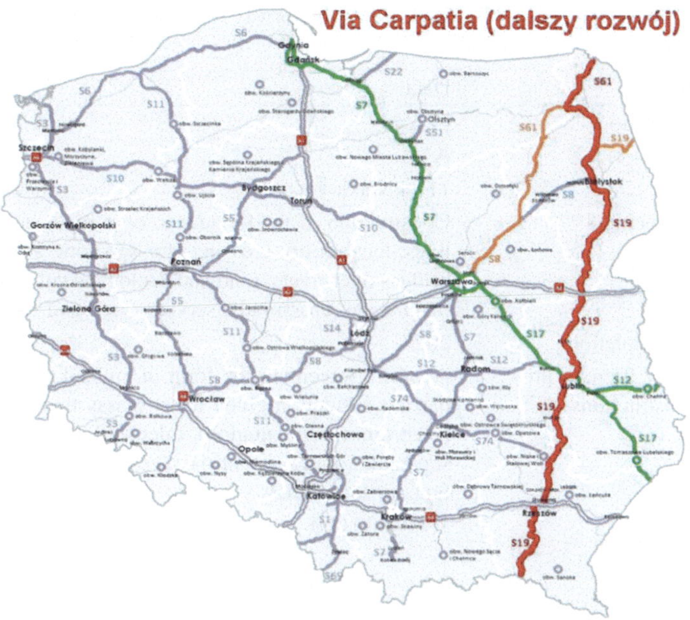

Fig. 1. Routing of the Via Carpathia and Via Baltica roads in Poland [13].

the administrative and customs services, the value of cargo transported by road in the A2 corridor is currently EUR 12 billion.

Assuming the total cost of the investment in the amount of PLN 1,759,800,000.00, the length of the analyzed section of 42 km and the annual profits resulting from the investment of PLN 46,032,317.64, the investment pay back time arrived at indicates that the costs of the construction of the Biała Podlaska—state border section will be paid back within 38.23 years.

In October 2017, an agreement was signed between Poczta Polska and China Post on distribution in Europe of small consignments from China, which will be imported by rail in containers. If 40 ha of wasteland at the post-military airfield in Biała Podlaska is purchased for PLN 60 million and a warehouse and distribution base will be created there, then building of the A2 motorway will be one of the main factors determining the success of this project. Table 1 presents technical data for the planned A2.

From the point of view of the authorities and residents of the Lublin Voivodeship, it is also necessary to restore and activate the agreement on the local border traffic with Belarus, including simplifying the issuing of visas to Poland so that Belarusians can come to our country to do business or shopping. Near our eastern border there is Brest–Litovsk with its 350,000 inhabitants. Such a large demographic potential located close to the northern part of the Lublin province is a large human and economic capital. Belarussians from Brest could do shopping in Poland and Polish companies could benefit from it.

Table 1. Technical data for the planned A2 motorway on the Biała Podlaska–Kukuryki section [14].

Specification	Data
Length of the section	42 km
Road class	A
Axle load	115 kN/axle
Engineering structures	Construction of road viaducts, over roads and railway lines, local utility crossings for servicing adjacent areas, bridges over rivers, road and ecological culverts under the main route, crossroads, service roads and exits
Deadline for completion	2028
Environmentally friendly pro-ecological constructions	Acoustic screens in the form of embankments and noise barriers insulation green belts, grassy trenches, retention reservoirs (sedimentation), inlet settling tanks and separators, independent passages for animals, reciprocal fencing for animals
Cost (in PLN)	1,759,800,000

3.2 S12 Expressway—Chełm Bypass

The proposed variant starts in Tytusin, bypassing the Stołpie from the south, then runs together with the variant I (from Parypse to Janiska), then turns south–east crossing the Uherka river and the Chełm-Włodawa railway line and runs to the "Srebrzyszcze" junction at the intersection with the county road 1828L. Variant II is recommended for further work on documentation.

The economic efficiency calculation of this investment was carried out: taking the total investment cost in the amount of PLN 712,300,000.00, the length of the analyzed section of 17 km and annual profits resulting from the investment of PLN 148,547,445.04, we receive a return on the investment, which indicates that the construction costs of Chełm bypass will be paid back in less than 5 years (4.79). It is a very good result, better than the results achieved in Western Europe. All the more necessary is to build appropriate roads to the border with Ukraine, which is, among the others, the Chełm bypass along the S12 route.

The Chełm ring road is very necessary and will quickly pay for itself, while being a supplement to the section of the S 12 Lublin–Dorohusk road in the eastern Poland transport system, including Piaski–Dorohusk investment construction, which will involve the building 62 km of the S 12 Piaski–Dorohusk expressway in the Lublin province. The signed Investment Program includes conducting the preparatory process for this task. The Chełm bypass will complement this important section of roads for the voivodship transport system. Table 2 describes technical data of Chelm bypass.

3.3 The S19 Expressway

Another very important element of the Lublin region's transport system is the transport route running through Lublin Voivodship called Via Carpathia, which runs along the

Table 2. Technical data of Chelm bypass [14].

Specification	Data
Length of the section	17 km
Road class	S
Axle load	115 kN/axle
Engineering structures	2 bridges, 8 viaducts over the roads, viaduct over the railway line, nodes: "Chełm West" (working name "Janów")—at the intersection with former DK12; "hełm North" (Horodyszcze)—at the intersection with DW 812; "Chełm Okszów" (Okszów)—at the intersection with DP 1823L; "Chełm East" (Srebrzyszcze)—at the intersection with DP 1828L. In addition, the service areas are planned
Deadline for completion	2024
Environmentally friendly pro-ecological constructions	As part of the investment, there will be, among the others: walkways for animals (large, medium and small), anti-noise screens storm water drains, pre-treatment devices and retention reservoirs
Cost (in PLN)	712,000,000

future S19 expressway (DK 19). Via Carpatia is the "life line" for all Eastern Poland. The route will represent a communication framework linking Lithuania, Belarus, Poland, Slovakia, Romania, Bulgaria and Greece. Almost all countries of Central and Eastern Europe are interested in its construction, and also those from outside the European Union. The interest in the accession to this project has been reported, among the others from Ukraine and Turkey, and recently the construction of roads connecting with this route, was also expressed by Serbia and Montenegro.

The construction of Kraśnik–Lasy Janowskie node section, as part of the Via Carpathia international road, is pan-European, international but also regional, important to the inhabitants of the southern part of the Lublin Voivodeship and the northern part of the Podkarpackie Voivodeship. The economic efficiency calculation for this investment was made as part of this study. Assuming the total investment cost in the amount of PLN 722,100,000.00, the length of analyzed section of 33 km and annual profits resulting from the implementation of this investment in the amount of PLN 59,874,552.56, we receive a return on the investment, which indicates that the cost of construction of the Kraśnik–Lasy Janowskie section will be paid back within 12.06 years. Table 3 presents technical data part of S19 expressway.

Assuming this type of calculation, it can be seen that the repayment of the value of entire construction project of the main motorway road from Kraśnik to the road junction in Lasy Janowskie will take place after 12.06 years. This investment will therefore pay for itself fairly quickly for Western European standards, which confirms its necessity and the growing implementation delay. In order for the profits to be larger, however, it is necessary to build the entire S19 route from the border with Belarus to Slovakia.

Table 3. Technical data part of S19 expressway Kraśnik–Janów Lubelski [14].

Specification	Data
Length of the section	33 km
Road class	S
Axle load	115 kN/axle
Engineering structures	A dual carriageway with two lanes, which in the future can be widened by the third lane. As part of the Kraśnik–Janów Lubelski project, there will be built: Szastarka and Modliborzyce nodes, Felinów service area (on both sides of the road), 7 viaducts along the expressway, 5 viaducts over the expressway and over other roads, 1 bridge
Deadline for completion	2018–2021
Environmentally friendly pro-ecological constructions	As part of the Janów Lubelski District task the following junctions will be built: Janów Lubelski North and Janów Lubelski South nodes, Janów Lubelski service area (on both sides of the road), 3 viaducts along the expressway, 1 viaduct over the expressway, 1 bridge. As part of built as well as 2 viaducts along the expressway, 1 viaduct over the expressway, 4 bridges
Cost (in PLN)	722,100,000

3.4 The S17 Expressway

One of the fragments of the S17 road analyzed in the study was the planned Garwolin–Kurów investment located entirely in the administrative area of the Lublin Voivodeship, administrative regions of: Ryki (Ryki commune, Ryki town, Ułęż) and Puławy (Żyrzyn and Końskowola commune). The beginning of the surveyed section is located in Niwa Babicka, on the border of the Mazowieckie and Lubelskie provinces, and its end is in Siedlce/Chrząchówek.

The location of the road that connects MA of W with Lublin will also enable the areas located in the investment impact zone to attract economic investors, be it in the area of tourism, industry or services, and will also allow better contact between smaller centres and larger ones related to job seeking, universities, tourism.

Assuming the total investment cost in the amount of PLN 879,600,000.00, the length of the analyzed section of 33.4 km and annual profits resulting from the investment in the amount of PLN 90,788,961.53, we receive a investment payback period which indicates that the cost of constructing the Garwolin (from border of Lublin province)–Kurów road section will be returned within 9.69 years. This is a very good economic result in the West European conditions, which indicates a serious traffic load on the analyzed road. Table 4 describes technical data for the section of the S17.

If the current trend of spontaneous development of road transport in Poland, and especially of individual motoring does not change, soon S17 will be loaded with traffic of 50 thousand vehicles per day. This would entitle to widen the route up to 6 lanes (in both directions).

Table 4. Technical data for the section of the S17 Garwolin–Kurów road [14].

Specification	Data
Length of the section	33.4 km
Road class	S
Axle load	115 kN/axle
Engineering structures	4 road junctions, road crossings (to the necessary extent) engineering structures, technical infrastructure, road technical equipment, drainage elements, access roads for local traffic service (construction of Service Areas (SA 1), including connections, target system: SA II (SA in the town of Niwa Babicka for the "to Lublin" direction, construction of the Expressway Maintenance Circuits (EMC), construction of road markings and traffic safety devices
Deadline for completion	2015–2019
Environmentally friendly pro-ecological constructions	Construction of 2 upper and 14 lower passages for large animals, 16 passages for large and medium animals, 15 culverts for amphibians and small mammals; construction of protective fences to reduce mortality of animals; construction of rainwater drainage devices, introduction of plantings of protective and insulating nature; construction of acoustic screens
Cost (in PLN)	879,600,000
Other remarks	The project is intended for co-financing with the participation of EU funds from OP IŚ 2014–2020 and from external and national funds at the disposal of GDDKiA

4 Conclusions

Competitiveness and development of the Lublin province will be greater with the development of modern road infrastructure, in the form of the A2 motorway along the section Siedlce–Biała Podlaska, as well as other expressways (S17, S19 or S12). The road infrastructure east of the Vistula River is an essential, though not the only, condition for limiting the risk of localization economic activity in the border regions of Eastern Poland, such as the Lublin Voivodeship. This is related to the wider problems of globalization and civilization development, as in the Lublin Voivodeship, on the section of the A2 motorway, two important communication corridors intersect: the Pan-European Transport Corridor II and the Via Carpathia international road along the S19 route.

Good road infrastructure in the Lublin Voivodeship, e.g. along the section of Siedlce–Biała Podlaska, will reduce the localization risk of companies in this region and help reverse migration trends: if it is possible to get to Warsaw within an hour, take care of the business there, and then return in one hour, it will not be necessary to live in a crowded capital city. It will be possible to live, for example, in Międzyrzec Podlaski, accessible from A2 motorway, or the S19 expressway, without losing time to "being stuck" in traffic jams. This situation will contribute to the fact that the Eastern regions of Poland will be more competitive, gaining the chance for faster civilization development.

We have received the following answers to the questions presented in the introduction:

1. capital expenditures for the construction of four roads sections presented in the article will be: planned A2 motorway on the Biała Podlaska–Kukuryki section PLN 1,759,800,000, Chelm bypass PLN 712,000,000, part of S19 expressway Kraśnik–Janów Lubelski PLN 722,100,000, the section of the S17 Garwolin–Kurów road PLN 879,600,000;
2. investment payback time will be:
   - of the Siedlce–Biała Podlaska section of the motorway A2—21.18 years,
   - of Chełm bypass (S12)—4.79 years,
   - of the Kraśnik–Lasy Janowskie section (S19)—12.06 years,
   - of the Garwolin (from border of Lublin province)–Kurów road section (S17)—9.69 years. These are a very good economic result in the West European conditions, which indicates a serious traffic load on the analysed roads;
3. The construction of the analysed road sections is extremely important and profitable from the point of view of the development and modernization of the Lubelskie Voivodeship, as it will stimulate the region's and the country's economic development, increase population mobility and reduce unemployment.

In addition to the extremely important construction investments in road infrastructure along the S19, S17 and S12 roads routes, a separate problem is (as already pointed out) extension of the A2 motorway to the Eastern border of the Polish state with Belarus. The economic effect of this investment is associated rather with serving heavy freight traffic in transit through Poland. The so-called "black hole" of the motorway in Central Europe needs to eliminated, as further to the East there continues a good road, the so-called Olympian. In the perspective of the coming years, the value of goods transported through Poland by road, in latitudinal directions, will reach 20 billion Euros annually. This can not be transported on one lane road along the current state road 2 East of Warsaw. Appropriate quality of road infrastructure must be created as a necessary condition, though not the only one, of further social and economic development of the Lublin Province and of Eastern Poland as a whole.

References

1. Meissonnier, S.: CEF Transport Results & future calls DG Mobility and Transport European Commission Ljubljana, 17 November 2016
2. Versini, F.: Delivering TEN-T Infrastructure Connecting Europe, Rome, 3 October 2017. https://www.espon.eu/sites/default/files/attachments/Frederic%20Versini.pdf

3. Łukasik, Z., Kuśmińska-Fijałkowska, A., Kozyra, J., Olszańska, S.: Evolution of costs in the activity of a transport company within the European Union. Ekonomicko-manazerskie spectrum **11**(2), 53–63 (2018)
4. Jones, P.: The evolution of urban mobility: the interplay of academic and policy perspectives. IATSS Res. **38**(1), 7–13 (2014). https://doi.org/10.1016/j.iatssr.2014.06.001
5. May, A.D.: Encouraging good practice in the development of Sustainable Urban Mobility Plans. Case Stud. Transp. Policy **3**(1), 3–11. https://doi.org/10.1016/j.cstp.2014.09.001
6. Priester, R., Miramontes, M., Wulfhorst, G.: A generic code of urban mobility: how can cities drive future sustainable development? Transp. Res. Procedia **4**, 90–102 (2016). https://doi.org/10.1016/j.trpro.2014.11.008
7. Sustainable Urban Transport Planning (SUTP), ELTIS, 29 August 2014. http://www.eltis.org/discover/case-studies/sustainable-urban-transport-planning-sutp
8. Wefering, F., Rupprecht, S., Bührmann, S., Böhler-Baedeker, S.: Guidelines. Developing and Implementing a Sustainable Urban Mobility Plan. Brussels: European Commission Directorate-General for Mobility and Transport; European Platform on Sustainable Urban Mobility Plans (2013)
9. Zakrzewski, B.: Investment efficiency ratio in a several-element, inter-industry logistics center in the South Podlasie region using the example of road transport. Report on the implementation of the work No 6029/ITS, ITS, Warszawa (2011)
10. Zakrzewski, B.: Infrastructural company in the development of the Euroregion Bug w: "Kwartalnik Nauk o Przedsiębiorstwie". Kolegium Nauk o Przedsiębiorstwie SGH 43 (2/2017), pp. 78–87 (2017)
11. Zakrzewski, B.: Logistics Centres of Eastern Poland in the Pan-European Transport Corridor No 2, "Problemy Transportu i Logistyki", Wyd. Naukowe Uniwersytetu Szczecińskiego, nr 2/2016 (34), pp. 215–224 (2016)
12. Zakrzewski, B.: Former military airfield in Biała Podlaska as a part of new Logistics Centre in Southern Podlasie. Secur. Dimens. **19**(12/2016), 116–134 (2016)
13. GDDKiA. https://www.gddkia.gov.pl/. Accessed 1–6 June 2017
14. Final report of MTI research projects from 2008 to 2013 for Ministry of Transport. The development of the Polish transport system with particular emphasis on its Eastern border in the aspect of management and organization. MTI, Warsaw (2014)

Power Plant Optimal Operation Time Approach in Best Energy Mixture Model with Simulation Analysis

Tatiana Endrjukaite[1(✉)], Alexander Dudko[2], Kunihiko Okano[3], and Hiromi Yamamoto[4]

[1] Transport and Telecommunication Institute, Riga, Latvia
tatianaendrjukaite@gmail.com
[2] Graduate School of Media and Governance, Keio University, Kanagawa, Japan
[3] Science and Technology, KEIO University, Yokohama, Japan
[4] Central Research Institute of Electric Power Industry, Kanagawa, Japan

Abstract. In this paper we proposed a new power plant optimal operation time approach in best energy mixture model. This approach takes into account detailed electricity consumption information and a set of parameters for a number of power plants, including temporal information and cost of the unit of every energy source. As an output the approach provides the capacity split for given power plants to meet the electricity demand. The approach is explained based on Latvia case example. Experimental results show that described approach can be successfully applied to any region or any county in the world individually, involving currently available as well as all potential energy sources, based on calculation and simulation of best energy mixture models.

Keywords: Best energy mix · Energy simulation · Renewable sources

1 Introduction

Energy is one of the most important resources required by modern human society. Our everyday life cannot be imagined without the use of electricity. Power consumption is continuously increasing globally. By 2035, global energy consumption is expected to increase by more than 50% from current levels. Countries all over the world depend on oil, coal, and natural gas to supply most of their energy needs. But, from a view point of energy security, reliance on fossil fuels presents a big problem because one day the world will run out of them or it will become too expensive to retrieve those that remain.

Another big problem is that fossil fuels cause air, water, and soil pollution, and emit greenhouse gases that contribute to the global warming. These problems makes the use of fossil fuels economically and ecologically inefficient. Efficient use of energy is an important step toward mitigating these effects.

The Great East Japan Earthquake that caused a nuclear accident at a power plant "Fukushima-1" in Japan [1, 2], prompted many countries to a parallel development and use of renewable energy sources (RES), the diversification of energy sources already in use, and active energy savings [3–5].

I. Kabashkin et al. (Eds.): RelStat 2018, LNNS 68, pp. 80–90, 2019.
https://doi.org/10.1007/978-3-030-12450-2_8

Renewables such as wind, biomass, hydropower, and solar offer clean alternatives to fossil fuels. They produce little or no pollution gases, and they will never run out. Nevertheless, excepting some RES like hydro and geothermal energies, most of them are unstable, not always available, and therefore very hard to maintain. Significantly important to find the ways of combining different sources of green energy into stable systems with high capacity and provide efficient energy management and distribution on countries levels.

This paper proposes a new stable best energy mix approach, which can be applied to the whole region or any county in the world individually, involving currently available as well as all possible energy sources. The uniqueness of this approach is that it calculates and simulates best energy mixture models including temporal information and cost of the unit of the chosen energy source.

2 Approach

2.1 Energy Sources

We choose all available energy sources for the region of our interest. It usually consists of coal, oil, gas, nuclear, solar, wind, and biomass. In the current approach only stable energy sources are used. To take into account unstable energy sources such as wind, solar and so on, they can be combined into a stable power sources, as it is presented in our previous researches on combined bio-wind system [15, 22].

2.2 Cost of Electricity Generation

As mentioned in our previous work, every energy source can be expressed in terms of cost of electricity (COE) as shown in (1) and (2), which is based on power station details and therefore can vary depending on a specific power station.

$$\mathrm{COE} = \frac{C_C \cdot (\mathrm{crf}) + C_{OM} + C_{SRC} + C_F}{P_e \cdot 8760 \cdot F_{av}}, \tag{1}$$

$$(\mathrm{crf}) = \left\{ \sum_{i=1}^{k-1} \frac{1}{(1+e)^i} \right\}^{-1} = \left[\frac{1}{e} \left\{ 1 - \frac{1}{(1+e)^{k-1}} \right\}^{-1} \right], \tag{2}$$

where: C_C—Capital cost (Construction cost); K—Operation life (years); E—Discount rate; C_{OM}—Operation & Maintenance cost of each year; C_{SRC}—Cost for periodic replacements of each year; C_F—Fuel cost of each year; F_{av}—Availability of plant (%); P_e—Net electric output (in kW); (crf)—Capital cost Recovery Factor; COE—Cost of electricity.

From (1) we can get a general overview, which is suitable only for rough comparison of energy sources and power stations, as it assumes full time operation time throughout the year. For the case when we need to combine various sources, we introduce a time-dependent equation for cost of electricity generation COEG(t) as shown in (3). It shows the price of 1 kWh generated by power station depending on how many hours the power station operates per year. It is essentially important when a power station is not necessary have to work a full year.

$$\mathrm{COEG(t)} = \frac{C_C \cdot (crf)}{P_e \cdot F_{av}} + \frac{(C_{OM} + C_{SRC} + C_F) \cdot t}{P_e \cdot F_{av} \cdot 8760}. \tag{3}$$

Figure 1 shows several examples of COEG(t) for different energy sources. It clearly explains, that some power stations might have higher construction cost, but due to lower operation and fuel costs become cheaper when operate for full year. However, Fig. 1 reveals the opposite cases as well. Some power stations might use costly fuel and might have high operation cost, but since construction cost is lower compared to other options, it becomes reasonable to use them for a short term operation.

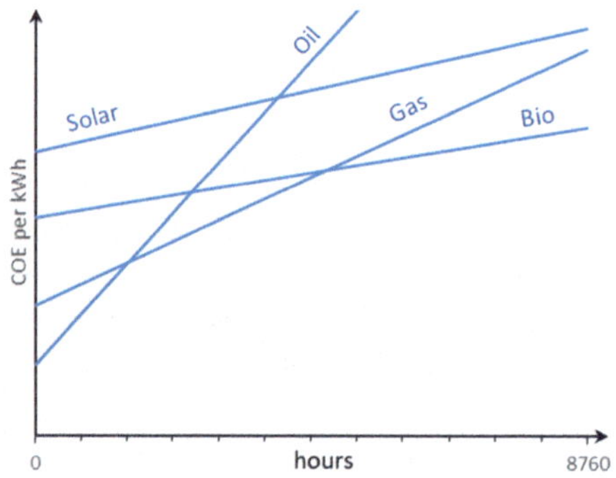

Fig. 1. Example of cost of electricity generation COEG(t) for different energy sources.

2.3 Optimal Operation Time

For every energy source we calculate cost of electricity generation COEG(t) over operation time and plot those into one chart (Fig. 2) We are interested in the minimal path, which consists of the straight lines between the crossing points. These points define the boundaries of the corresponding power sources usage amounts.

In the considered example, point t_1 shows the crossing point between Oil and Gas power stations. It means, that for the operation time less than t_1 it is economically more reasonable to use Oil power source than Gas power source. Point t_2 defines the crossing of Bio and Gas power stations, which means that Gas power station is more economically efficient if the operation time is less than t_2. For the operation time over t_2 point it is cheaper to use Bio power station. So, in this example, we determine that Oil source is used up to t_1, Gas up to t_2, and Bio up to t_3.

2.4 Energy Source Capacity

In this step, based on optimal operation times per each energy source we determine the overall capacity of power stations. This is done by means of energy consumption curve

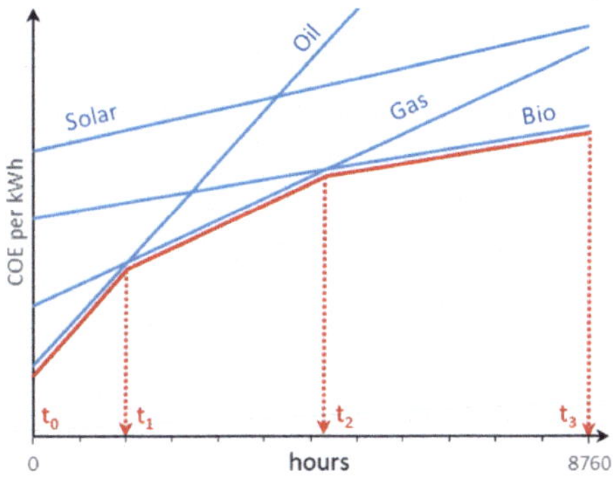

Fig. 2. Example of the optimal operation time of power stations in the given set of power stations.

for the region of interest. As displayed in Fig. 3, from the operation times t_1—t_3, we determine the cumulative power amounts for selected power sources. These values give us the required total power.

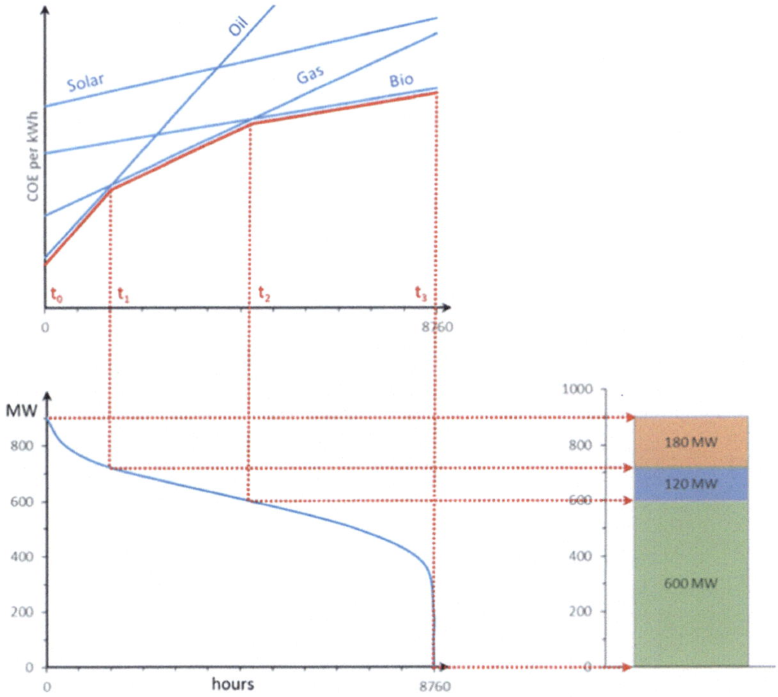

Fig. 3. Energy source capacity determination from COEG(t) and electricity consumption curve.

In this example, for the operation above t_2 we need the Bio power source. According to consumption curve it corresponds to 600 MW of capacity. Next is Gas power station, which should cover the consumption up to 720 MW by adding 120 MW by Gas power station. Oil power station adds last 180 MW to reach total peak of 900 MW in the considered example.

As a final step, we calculate how many power stations we need based on known capacity of the one power station of each type. For example, if we have Bio power station capacity equal to 50 MW, then to cover the required amount of 600 MW we need 12 Bio power stations. The same happens for Gas and Oil power stations to determine the exact number of power stations.

2.5 Simulation

In this step, we perform the model validation by simulating the power generation. Simulation can be done by using existing historical values of resources availability, such as wind speed by hour, power consumption by hour, and so on. Or otherwise it can use the generated time series values based on statistics and estimates.

From the simulation we get a detailed picture of the generated amounts per every power source. We can see that all demand is supplied, and we can calculate precise values of generated energy by time. Figure 4 shows a simplified model structure with an example of simulation output where electricity demand is supplied by multiple energy sources displayed in different colors. The model in this example utilizes 3 energy resources each having its own availability. The electricity demand for each power station is given from the bottom, and remaining uncovered demand is passed to the next power station above.

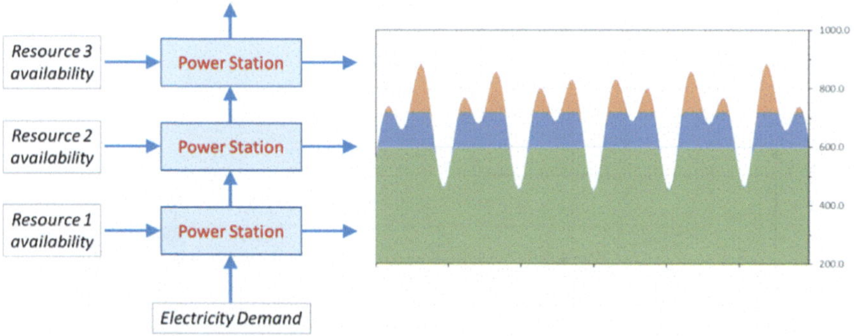

Fig. 4. Simulation results example.

3 Experiments

Our approach is explained based on Latvia case example, where we use existing gas power station with combination of introduced combined bio-wind energy system [15] to transform Latvia's current energy mixture model, which is depended on imported fossil fuels into locally produced energy from RES.

Latvia has significant potential for the expansion of RES, especially wind and biomass. Figure 5 represents share of total primary energy supply in Latvia for 2015 year. There are two most commonly used forms of renewable energy: biofuels/waste (34.6%) and hydro resources 3.9%. Wind energy is used to a less extent. Solar energy is used only in very small quantities of pilot projects. Natural gas accounted for 26.7% of gross inland energy consumption while the share of crude oil and petroleum products 33.4% is also significant. Nuclear-based electricity generation does not exist at all.

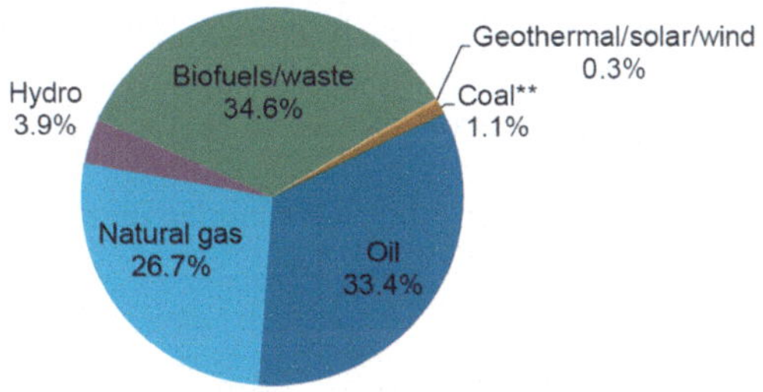

Fig. 5. Share of total primary energy supply for year 2015 in Latvia.

For electricity production for domestic consumption mainly the gas is used, which is exported from other countries. Latvia has to decrease energy dependence for electricity generation from natural gas and reduce impact of energy generation from fossil fuels for country's energy independence, ecology and global environment to meet Latvia's sustainable development goals.

3.1 Wind Energy

Despite the fact, that wind power stations in Latvia are not widely used, wind energy potential in this area is very high. The wind speed onshore is 6–7 m/s in average, but wind speed along west coastline is near 9 m/s, which is a very good value. An average load hours along coastline is up to 3000 h a year [11].

Grobina Farm Project. In this project there are 33 turbines combined into one power station with full capacity 19.8 MW located near west coastline of Latvia [12, 13]. Enercon E40/600 wind generators are used [18]. Each turbine has 600 kW power and

40 m diameter. Whole power station energy production is 45 GWh per year. Average work load is 2300 h per year, which gives its availability 26.3%, as in (4).

$$F_{av,w} = 2300h/8760h = 26.3\%. \quad (4)$$

Construction cost of the station is \$ 25 mln, and maintenance costs equal to \$ 0.5 mln that is calculated based on generally 1.5–2% of the original investment per year for wind turbines. Life span for this power station is considered to be equal to 20 years [13].

3.2 Biomass Energy

Bioenergy is expected to become one of the key energy resources in the future because bioenergy, if maintained adequately is renewable and free from net CO_2 emissions [14].

EcoGen Project. In the paper we will reference to 3.72 MW power stations [15]. Its energy production is 29.76 GWh per year by cultivating corn in 1593 ha area. The station can work up to 8000 h per year which gives its availability as follows:

$$F_{av,b} = 8000h/8760h = 91.3\%. \quad (5)$$

Construction cost of the station is \$ 5 mln, and maintenance costs \$ 1 mln for every year of its approximate 40 years of life span [15].

3.3 Operation Time and Capacity

In this sub-chapter, we describe comparison of Wind, Biomass, and Gas energy sources. The parameters of these power stations are presented in Table 1. These parameters are used for the approach calculation to determine optimal operation time and energy sources capacity for the considered energy sources.

Table 1. Power stations parameters.

	Wind	Biomass	Gas
C_C	50,000,000	10,000,000	595,980,000
k	40	40	40
e	0.03	0.03	0.03
C_{OM}	500,000	300,000	11,919,600
C_{SRC}	0	700,000	8,939,700
C_F	0	0	121,483,680
F_{av}	26.3%	91.3%	90.0%
P_e (kW)	19,800	3 720	400,000
(crf)	0.0438	0.0438	0.0438
COE	0.0590	0.0483	0.0534

Figure 6 presents the cost of electricity generation COEG(t) depending on the operation time throughout the year. From this picture we can see that Wind energy source becomes too expensive and its use is not efficient economically. The use of Biomass and Gas energy sources is reasonable as these two types of power stations have optimal operation time periods where electricity generation is efficient in terms of cost.

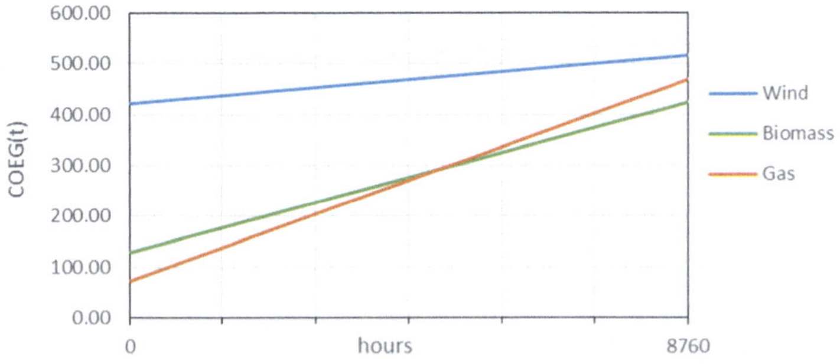

Fig. 6. Cost of electricity generation (COEG(t)).

Wind energy proves itself inefficient in Latvia as an independent standalone source of energy. However, it becomes more reasonable when used in a combined systems, such as described in [22]. Gas and Biomass energy sources lines have minimum values within the year bounds, and they cross at $t = 4898$ h. This makes both of these types of energy useful in the optimal energy mix.

According to consumption curve of Latvia, $t = 4898$ corresponds to 731 MW for Biomass power and 350 MW for Gas power. Figure 7 shows the amount of energy generated by the biomass equals to 88% of overall energy demand, and only remaining 12% of energy is covered by Gas energy.

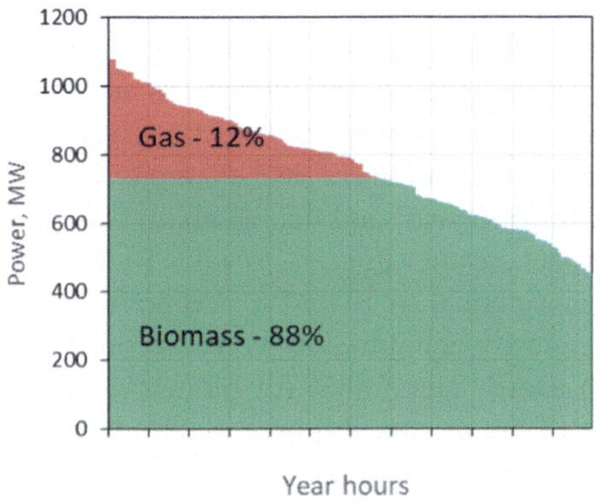

Fig. 7. Amounts of generated electricity.

3.4 Simulation

The energy mixture model in this experiment was simulated to validate the estimated energy share and visually show the energy split and distribution in time. The capacity of 731 MW power for biomass stations requires 197 biomass stations of 3.72 MW each, as considered above. One gas power station with 400 MW capacity is enough to meet the remaining 350 MW of power. Figure 8 shows a fragment of simulation output of generated electricity amounts for 1 week.

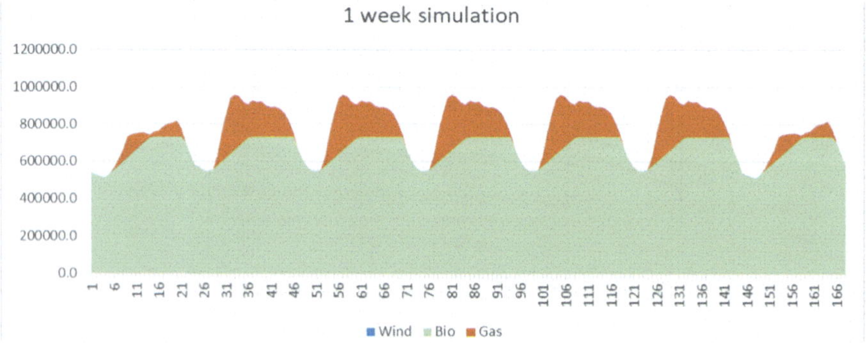

Fig. 8. Simulation experiment results for 1 week.

The simulation result for one year period shows that 86.4% of energy was generated by Biomass energy, and 13.6% was generated by Gas energy. These values are very close to estimated 88% versus 12% share.

4 Conclusions

This paper shows that power plant optimal operation time can be described and calculated by means of the presented approach. The approach takes into account electricity consumption curve, and a set of parameters for a number of power plants which exist or can be constructed in the region of interest. As an output the approach provides the capacity split for given power plants to meet the given electricity needs.

We explained this approach on Latvia case example, where we have taken existing electricity demand data for one year, and details about exiting wind, gas, and biomass power plants. The approach result suggests to take Biomass power plants as a base energy source with capacity equal to 731 MW and cover the remaining demand with Gas power plants with capacity equal to 350 MW. According to consumption curve of Latvia it corresponds to energy share 88% for biomass and 12% for gas energy.

Based on the results of the proposed power plant optimal operation time approach for Latvia case we performed simulation to get a visual picture of the generated electricity amounts and validate the approach estimated values. The simulation results in the experiment were very close to expected values which is fine as the exact energy share always depends on the consumption data.

Presented energy mix model in the experiment result for Latvia case is based on the renewable power generation which can help Latvia to meet its sustainable development goals through provision of access to clean, secure, reliable and affordable energy as well as to mitigate growing electricity consumption.

Acknowledgment. This work was financially supported by the specific support objective activity 1.1.1.2. "Post-doctoral Research Aid" (Project id. N. 1.1.1.2/16/I/001) of the Republic of Latvia, funded by the European Regional Development Fund. Tatiana Endrjukaite research project No. 1.1.1.2/VIAA/1/16/095 "Integrated Model for Energy Generation, Distribution and Management".

References

1. Leon, J.D., Jaffe, D., Kaspar, J., Knecht, A., Miller, M., Robertson, R., Schubert, A.: Arrival time and magnitude of airborne fission products from the Fukushima, Japan, reactor incident as measured in Seattle, WA, USA. J. Environ. Radioact. **102**, 1032–1038 (2011)
2. Kim, C.K., Byun, J.I., Chae, J.S., Choi, H.Y., Choi, S.W., Kim, D.J., Kim, Y.J., Lee, D.M., Park, W.J., Jim, S.A., Yun, J.Y.: Radiological impact in Korea following the Fukushima nuclear accident. J. Environ. Radioact. **111**, 70–82 (2012)
3. Butler, C., Parkhill, K.A., Pidgeon, N.F.: Nuclear power after Japan: the social dimensions. J. Sci. Policy Sustain. Dev. **53**(6), 3–14 (2011)
4. Ong, H.C., Mahlia, T.M.I., Masjuki, H.H.: A review on energy scenario and sustainable energy in Malaysia. J. Renew. Sustain. Energy Rev. **15**(1), 639–647 (2011)
5. Vivoda, V.: Japan's energy security predicament post-Fukushima. J. Energy Policy **46**, 135–143 (2012)
6. Denholm, P.: Improving the technical, environmental and social performance of wind energy systems using biomass-based energy storage. J. Renew. Energy **31**(9), 1355–1370 (2006)
7. Geographical Data—Key Indicators. Central Statistical Bureau Republic of Latvia, 5 October 2011. Accessed 17 May 2012
8. Latvia.eu. http://www.latvia.eu/news/latvia-third-renewable-energy-eu
9. Ministry of Economics of Republic of Latvia report for April (2014)
10. International Energy Agency, Latvia indicators report for 2015 year
11. Europe's onshore and offshore wind energy potential. Technical report No 6/2009. European Environment Agency (2009)
12. Latvian wind energy report. Development of Wind Energy in the Baltic Sea Region: Latvian Perspective (2004)
13. Wind turbines and wind farms database. Veju parks windfarm (Latvia) (2017)
14. Yamamoto, H.: Evaluation of bioenergy potential with a multi-regional global-land-use-and-energy model. J. Biomass Bioenergy **21**(3), 185–203 (2001)
15. Endrjukaite, T., Okano, K., Yamamoto, H., Okano, K.: Renewable energy introduction in latvia with best energy mixture model based on a concept of a combined power system of biomass and wind. In: Proceedings of Energy, Economy, and Environment Conference, Tokyo, January (2015)
16. Information Administration (EIA), Natural Gas Annual 2010, DOE/EIA-0131
17. Latvenergo AS portal (2017). https://www.latvenergo.lv/eng/about_us/generation/
18. Wind turbines. https://en.wind-turbine-models.com/turbines/68-enercon-e-40-6.44

19. World Bank Data (2017)
20. Review of RES perspective in Baltic countries till 2030. LITGRID (2015)
21. Latvian Environment, Geology and Meteorology Centre. https://www.meteo.lv/en/meteorologija-datu-meklesana/?nid=924
22. Endrjukaite, T., Dudko, A., Okano, K., Yamamoto, H.: Best energy mixture model and electricity generation simulation approach case study Latvia. In: Proceedings of 28th International Conference on Information Modelling and Knowledge Bases EJC2018, Riga, Latvia, June 2018
23. Gonzalez-Salazar, M.A., Kirsten, T., Prchlik, L.: Review of the operational flexibility and emissions of gas- and coal-fired power plants in a future with growing renewables. Renew. Sustain. Energy Rev. **82**, 1497–1513 (2018)

Decision-Making Process for Choosing Technology of Diesel Bus Conversion into Electric Bus

Kristine Malnaca[1,2(✉)], Mikhail Gorobetz[1,3], Irina Yatskiv (Jackiva)[2], and Aleksandr Korneyev[3]

[1] JSC Ferrus, 90 Gramzdas, Riga LV1029, Latvia
Malnaca.K@tsi.lv
[2] Transport and Telecommunication Institute, 1 Lomonosova, Riga LV1019, Latvia
[3] Riga Technical University, 1 Kalku, Riga LV1658, Latvia

Abstract. Many local public transport authorities and operators are in need of replacing ageing bus fleet with cleaner and more sustainable vehicles in order to meet standards, increase efficiency and reduce transport related emissions. There is a wide choice of technologies for urban bus operators in the market including electric buses but at the same time new vehicles beyond lower emission Euro VI diesel buses are still a challenge for transport operators due to high acquisition costs and lack of charging infrastructure. The alternative proposed is to convert used diesel city bus into electric bus. Decision-making process for choosing technology of the bus conversion requires thorough assessment of possible solutions from the technical, operational, and economic point of view under the given conditions and constraints. Within the framework of this research, mathematical models are developed for assessing the efficiency of an electric vehicle on the basis of various criteria which affect life cycle costs. The models include the definition of functional dependencies and dynamic performance equations of a diesel bus and a converted electric bus. The results help to choose the most suitable parameters of the traction motor's torque and power under the given conditions and determine the most suitable battery type and capacity for the selected bus route. Total Cost of Ownership model is utilized in the decision-making process to determine economic viability of a technological solution to convert a diesel bus into an electric bus. The proposed methodology is tested in the case study.

Keywords: Low-emission · Electric bus · Converted diesel bus · Economic analysis · Total cost of ownership · Energy consumption

1 Introduction

The European Commission's low carbon economy roadmap suggests that emissions from transport could be reduced to more than 60% below 1990 levels by 2050 [1]. One of behavioral changes identified that contributes to the de-carbonization of transport is the use of electric vehicles.

I. Kabashkin et al. (Eds.): RelStat 2018, LNNS 68, pp. 91–102, 2019.
https://doi.org/10.1007/978-3-030-12450-2_9

Following the European Commission's 2016 Strategy for Low Emission Mobility [3] many local public transport authorities and operators are in need of replacing ageing bus fleet with cleaner and more sustainable vehicles in order to meet standards, increase efficiency and reduce transport related emissions. There is a wide choice of cleaner fuel and engine technologies for urban bus operators in the market including electric buses but at the same time new vehicles beyond lower emission Euro VI diesel buses are still a challenge for public transport operators due to high acquisition costs of a new vehicle and lack of charging infrastructure. The alternative proposed is to convert used diesel city bus into electric bus which would significantly reduce the harmful impact of the used diesel bus on the environment and improve performance of the vehicle. Diesel bus (DB) conversion into electric bus (EB) can be realized using different technologies. In order to choose optimal solution of the bus conversion a thorough assessment of possible alternatives from the technical, operational, and economic point of view is necessary taking into account given conditions and constrains.

This paper explains the decision-making process for optimum technology selection of a diesel bus conversion into an electric bus where main technical and economic performance indicators are identified and assessment of alternatives is performed.

2 Methodology

2.1 The Decision-Making Process for Choosing the Optimum Technology

Before converting used DB into EB it is important to assess the technical, operational and economic performance of the existing vehicle by estimating the key characteristics of a motor (torque, speed, power), the basic physical characteristics of the movement (distance, speed, acceleration), and economic performance indicators. Economic efficiency in this study is assessed by calculating life-cycle cost of the bus which later can be used for comparison purposes among possible alternatives.

Once the values for key characteristics of the movement and the motor are obtained, possible configurations of EB can be defined. Alternatives are modelled, and it allows to assess characteristics of an electric motor. Values of the physical characteristics of the movement and the motor are important for selecting the battery type and capacity, which in turn has a great impact on the cost of the bus. The decision-making process for choosing the optimum technology of conversion is explained in Fig. 1.

Mathematical models and algorithms developed in this research are the following

- models of a diesel engine and an electric engine;
- model of an automatic gearbox and reduction gear;
- models of a generic bus, DB, and EB;
- model of a route for simulation;
- analytical models for DB and EB dynamics evaluation;
- algorithms for DB and EB motion simulation;
- algorithm for simulation of a bus movement at the predefined route;
- model and algorithm for the battery selection for EB.

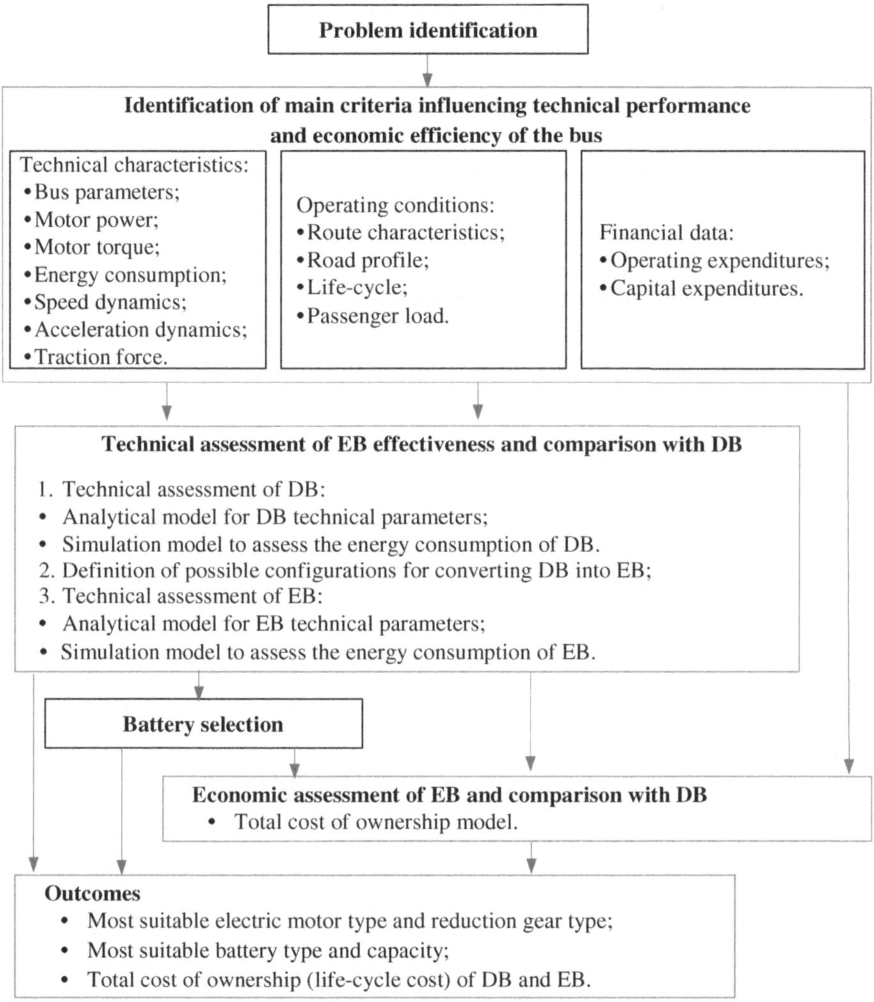

Fig. 1. Decision-making process for choosing technology of DB conversion into EB

The generic algorithm for the technical assessment has the following steps:

1. Use analytical model to evaluate technical parameters of DB: speed dynamics $v(t)$, acceleration dynamics $a(t)$, energy consumption $C(v)$, maximal slope $s(v)$, motor power $P(v)$, motor torque $T(v)$ and bus motion force $F(v)$.
2. Use simulation models of DB with different routes to assess the consumption of DB with different loads.
3. Define possible configurations as alternatives for conversion of DB into EB—reduction ratio of the rear axle, electric motor's nominal torque, nominal power, nominal rotation speed, battery types and capacity.

4. Use analytical model to evaluate conversion alternatives: speed dynamics $v(t)$, acceleration dynamics $a(t)$, energy consumption $C(v)$, maximal slope $s(v)$, motor power $P(v)$, motor torque $T(v)$ and bus motion force $F(v)$.
5. Use simulation models of EB with the same routes and the same set of different loads as for DB.
6. Use optimization model and algorithm for the battery selection.

2.2 The Main Models and Algorithms

The objective of **analytical models for DB** is to evaluate the dynamics of DB using the traction-speed parameter evaluation technique [5]. The method uses the DB model that inherits the generic bus model and includes models of the diesel engine and an automatic gearbox, and reduction gear.

Input parameters are: $TE_D(\omega) = \{t_1(\omega_1), \ldots, t_n(\omega_n)\}$—empiric values of the engine torque, $l \in \mathbb{R}$—length of the bus, m; $w \in \mathbb{R}$—width of the bus, m; $h \in \mathbb{R}$—length of the bus, m; $m_0 \in \mathbb{R}$—self mass of the bus, kg; $m_1 \in \mathbb{R}$—max load mass of the bus, kg; $w_t \in \mathbb{R}$—width of tire, m; $h_t \in \mathbb{R}$—relative height of tire; $r_t \in \mathbb{R}$—internal radius of tire, inch; $\alpha \in (0, 1) \in \mathbb{R}$—aerodynamics frontal resistance ratio, Ns2/m4; $\varphi \in (0, 1) \in \mathbb{R}$—road surface resistance ratio, m.

The outcomes of the analytical models are sets with momentary values of F_T—traction force of the bus; v velocity of the bus; F_f road friction force; F_a aerodynamic resistance force; $\mathbb{D}$ dynamic characteristic; ς_{max} maximal slope; a acceleration of the bus; t time travelled by the bus at the route stage; $P_D \in \mathbb{R}$—momentary value of the engine power; C consumed energy. These sets represent the dynamic of DB [4]: speed dynamics $v(t)$, acceleration dynamics $a(t)$, energy consumption $C(v)$, maximal slope $s(v)$, motor power $P(v)$, motor torque $T(v)$, and bus motion force $F(v)$.

The goal of the motion simulation model for DB is to obtain key performance indicators of DB on the defined route using the specific developed algorithm. The route model R is represented by following parameters: $n_R \in \mathbb{N}$—number of passenger stops; $l_R \in \mathbb{R}$—total length of the route; $t_R \in \mathbb{R}$—average time for movement between stops; $w_R \in \mathbb{R}$—waiting time for passengers embark/disembark; $v_R \in \mathbb{R}$—average motion speed for movement between stops; $s_R \in \mathbb{R}$—average distance between two stops, where

$$s_R = \frac{l_R}{n_R - 1}, \tag{1}$$

$$v_R = \frac{s_R}{t_R - w_R}. \tag{2}$$

Input parameters for simulation are dt—time step for simulation and $v_{\max}$—maximal speed to accelerate.

To obtain the comparable results of DB simulation with EB simulation the fuel consumption of DB is converted to the mechanical energy in kWh, taking into account the efficiency rates of the DB equipment: diesel motor, transmission, reduction gear etc. Thus, the main outcomes of the simulation for each route are $C \in \mathbb{R}$—consumed energy, kWh, $\theta_C \in \mathbb{R}$—relative consumed energy per distance unit, kWh/km and $\max(v) \in \mathbb{R}$—maximal velocity of the bus on the route, m/s.

It is obvious that various **different technical configurations of DB conversion into an EB** are possible. However, only two components are the most important for traction of the bus in general. These components are the motor and reduction gear. The simplified traction structure of DB is presented in Fig. 2. The blocks with the dashed outline in Fig. 2 are not parts of the traction system, but these components are necessary as power and control sources of the bus traction system. The different auxiliary systems have influence on the total energy consumption of the bus, but have no significant direct impact on the bus traction.

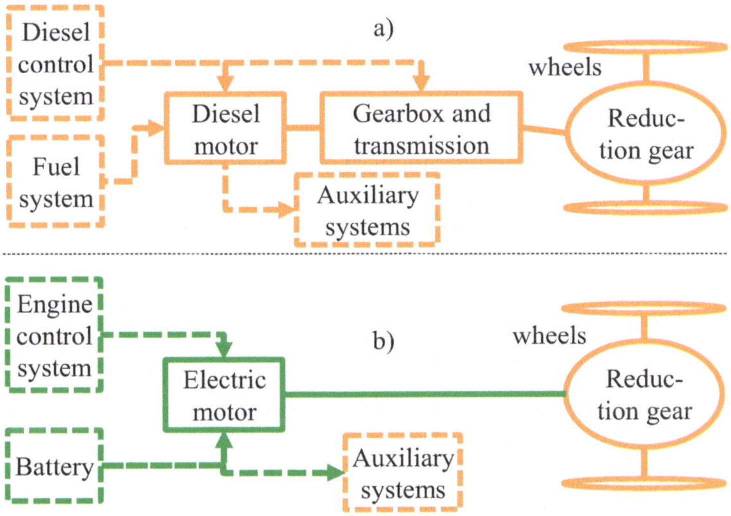

Fig. 2. Simplified traction structure of **a** a diesel bus and **b** an electric bus

As EB has no gearbox it is important to select the most suitable transmission rate of the reduction ratio for the stable and balanced motion of EB. Also, the power source (battery) should have the capacity, voltage and current characteristics to fit the bus. The control system is important too, but it requires the development of optimal control algorithms that is out of the scope of this paper.

Thus, the main outcomes of this step are different configurations (alternatives A) of the EB traction system represented by a Cartesian product sets of possible motors $M = (M_1, \ldots, M_k)$ and $R = (R_1, \ldots, R_p)$, i.e. $A = M \times R = (<M_1, R_1>, \ldots, <M_k, R_p>)$.

The objective of **analytical models for EB** is to evaluate the dynamics of each alternative of the EB configuration using the similar traction-speed parameter evaluation technique as for DB. The method uses the EB model that inherits the generic bus model and includes models of the electric engine and reduction gear. Input mechanical parameters are the same except $TE_D(\omega)$. But additionally for EB the following parameters are defined [8]: $P\max_E \in \mathbb{R}$—maximal power of the engine, kW;

$T\max_E \in \mathbb{R}$—maximal torque of the engine, N.m; $Pnom_E \in \mathbb{R}$—nominal power of the engine, kW; $Tnom_E \in \mathbb{R}$—nominal torque of the engine, N.m; $\eta_E \in \mathbb{R}$—efficiency rate of the engine; $Inom_E \in \mathbb{R}$—nominal current of the engine, A; $Unom_E \in \mathbb{R}$—nominal voltage of the engine, V; $\omega nom_E \in \mathbb{R}$—nominal rotation speed of the engine, rpm; $J_E \in \mathbb{R}$—inertia value of the engine, N.m.

The outcomes of these models represent speed dynamics $v_x(t)$, acceleration dynamics $a_x(t)$, energy consumption $C_x(v)$, maximal slope $s_x(v)$, motor power $P_x(v)$, motor torque $T_x(v)$ and bus motion force $F_x(v)$ of each EB configuration $x \in A$.

The goal of **the motion simulation for EB** is to obtain key performance indicators of each EB configuration $x \in A$ on defined routes using the specific developed algorithm. In order to obtain the comparable results the same routes are used as for the simulation of DB. Input parameters for simulation are dt—time step for simulation and $v_{\max}$—maximal speed to accelerate. Main outcomes of the simulation for each route and each configuration are $C \in \mathbb{R}$—consumed energy, kWh, $\theta_C \in \mathbb{R}$—relative consumed energy per distance unit, kWh/km and $\max(v) \in \mathbb{R}$—maximal velocity of the bus on the route, m/s.

The goal of **optimization model for battery selection** is to select the battery type and capacity with minimal costs under predefined constraints.

$$F = \psi(X, B, C_{AC}, \mathrm{A}) \rightarrow \min, \tag{3}$$

where F—total costs of the battery; X—set of input parameters; $B = (b_1, b_2, \ldots, b_n)$—set of n battery types; $\mathrm{A} \in (\alpha_1, \alpha_2, \ldots, \alpha_n) \in \{0, 1\} \in \mathbb{Z}$—selection of the battery type from the battery type set; $C_{AC} \in \mathbb{R}$—capacity of the battery, kWh.

A set of input parameters X consists of S_a—planned annual travelled distance by bus, km; c—consumption ratio of the bus, kWh/km; L—bus lifecycle, years; d—number of days per year of the bus exploitation, days; r—length of one route loop, km; U_c—charging voltage, V; U_d—discharging voltage, V; t_c—planned charging time after each loop, minutes.

For each battery type $b_i \in B$ the parameters $l_{AC} \in \mathbb{N}$—maximum number of full charge cycles; $I\max_{AC} \in \mathbb{R}$—relative maximal charge current of the battery per capacity, C; $\varpi_{AC} \in \mathbb{R}$—relative capacity of the battery per weight, Wh/kg; $c_{AC} \in \mathbb{R}$—relative costs of the battery per capacity, EUR/kWh are set for the selection.

The model for optimal battery selection is implemented in a Web environment and can be adjusted to the user needs.

Total Cost of Ownership (*TCO*) analysis is a method to assess life-cycle costs that include all costs of purchasing, operating, and maintaining the vehicle. The economic analysis model is prepared with the objective function to calculate *TCO* for DB and the converted EB. The comparison of the results allows to assess economic viability of the proposed technology of DB replacement with the EB for public transport services in the urban environment [6].

The *TCO* model is presented in detail in the paper by Malnaca, Yatskiv [6] and includes the vehicle costs, the charging infrastructure costs, and external costs:

$$TCO = CInv(bus) + CInv(charger) + CInv(grid) + C(O\&M) + I * C(ext), \quad (4)$$

where *CInv* (*bus*)—investment costs of a bus; *CInv* (*charger*)—investment costs of a charger; *CInv* (*grid*)—investment costs of a grid connection; *C*(*O&M*)—operating and maintenance costs for the vehicle and the charger; *C*(*ext*)—external (environmental) costs; and indicator *I*, that equal to 1 for DB, and to 0 for EB. Environmental costs are calculated as follows:

$$C(ext) = C(CO_2) + C(NO_X) + C(PM), \quad (5)$$

where $C(CO_2)$—costs of carbon dioxide emissions; $C(NO_X)$—costs of air pollution (nitrogen oxides); and *C*(*PM*)—cost of particulates.

For the comparison purposes *TCO* is expressed in Equivalent Annual Cost (EAC) as cost per kilometer (€/km), using a discount rate of 4% [2].

3 Case Study of Decision-Making Model Application

3.1 Model Initialization and Parametrization

The developed decision making steps are carried out for the conversion of the diesel bus Ambassador ALE 120-205/225 used in the mid-size city in Latvia as a public transport.

Technical assessment and economic analysis includes both mathematical and expert estimations regarding model development. A number of model variables requires expert evaluation based on specific location, situation, and mostly on availability of necessary infrastructure for the use of EB, whereas other items are included on constant bases.

Variables used in the models and their values are given in the Table 1. The variables are identified as the main indicators of technical performance and cost-effectiveness, which are the measures of outcome. All prices are given net of VAT.

Torque-rotation speed curves are obtained from the technical specification of the bus and the fragment is provided in the Table 2. The bus may be equipped with $ISBe^4$ 205 or $ISBe^4$ 225 diesel motor. For the simulation 2 routes are used with 6 different relative loads of the bus: 1—an abstract city route (25 stops/an average 1-minute time between stops); 2—a part of an intercity or the city route (70 km/h speed limit).

Different EB configurations are compared combining the sets of 6 motors and 3 reduction ratios and obtaining 18 alternatives. Parameter of energy consumption rate is compared with the same parameter of DB with $ISBe^4$ 205 and $ISBe^4$ 225 engines. Thus, influence of three different reduction ratios are tested: 4.88 ratio is the same gear as the DB has, 5.83 ratio is mostly used in similar EB, 8 ratio is mostly used in high torque EB. Six different most popular electric motors that are available in the market are simulated. Taking into account big amount of simulation and various criteria, it is assumed that EB is working at the nominal voltage in this study.

Table 1. Models' variables and their values.

Type	Variable	Value (DB)	Value (EB)
Technical	Self-mass	8900 kg	1.2 kWh/km
	Full mass	14,440 kg	
	Dimensions (length, width, height)	11,950 × 2500 × 2920 mm	
	Wheel and tire size	265/70/R19.5	
	Energy consumption	10.4 MJ/km (29 l/100 km)	
	Charging and discharging voltage	650 V	
	Drag coefficient	0.7	
	Aerodynamics frontal resistance rate	0.3	
	Transmission efficiency rate	0.9	
Operational	Useful life	10 years	7 years
	Length of one loop (distance travelled between charging)	34 km	
	Planned charging time after each loop	9 min	
	Road surface resistance rate	0.02	
Financial	Investment cost	200,000 €	196,700 €
	Energy price	1.00 €/L	0.11215 €/kWh
	Urea, oil	0.011 €/km	n/a
	Maintenance and repair	0.15 €/km	0.10 €/km
	Transport operating tax	0.002 €/km	n/a
Charging	Investment cost of charging infrastructure	n/a	150,000 €
Infrastructure	Charging infrastructure maintenance	n/a	1,000 €/year
Grid	Investment cost of grid connection	n/a	30,000 €
Connection	Transmission power maintenance	n/a	19.56 €/kW/year
	Electricity transmission tariff	n/a	0.02129 €/kWh

The estimated price of the converted EB is based on the remaining value of the 7-year old DB and the incremental cost of the conversion (the difference between the cost of battery, electric drive, other supplementary materials, and labor costs and the re-sell value of diesel engine and transmission system).

Table 2. The fragment of torque-rotation speed specification of DB.

ω_D, rpm	1000	1100	1200	1300	1400	1500	1600	1700	1800	1900	2000	…
TE_D—ISBe[4] 205, Nm	700	725	750	750	750	750	750	750	732	712	690	…
TE_D—ISBe[4] 225, Nm	700	775	850	850	850	850	850	850	840	828	790	…

In this study, the average travelled distance of the bus is assumed to be 200 km a day without returning to the depot. Fast charging infrastructure, also known as opportunity charging, with pantograph is selected in this analysis based on the results obtained from motion simulation model and battery selection model. As concluded by Malnaca, Yatskiv [6], smaller battery means less initial investment cost, less weight, and more room inside the bus for passengers.

The maintenance expenses of electric drive system are expected to be 30% less compared to combustion power transmission system because it requires less frequent service maintenance.

3.2 Modelling Results and Discussion

Average values of energy consumption rate of the diesel bus Ambassador ALE 120 is 1.8073 kWh/km for ISBe[4] 205 engine and 1.798 kWh/km for ISBe[4] 225 engine. Both models—analytical and simulation—provides similar results. Obtained criteria values are compared with different configurations of EB.

The average energy consumption of EB is estimated at 1.2 kWh/km in the conditions typical for the average mid-size city in a relatively flat area using the mathematical model which was created within the study to evaluate the effectiveness of EB. The efficiency of EB is higher for lower load values and lower for higher load values for the city route 1. The average energy efficiency is 31%.

The obtained simulation data and the results of acceleration dynamics analysis for each load allow selecting the most efficient configuration of EB.

Table 3 shows the comparison of average energy consumption of different configurations of EB and DB. The most efficient configurations of EB are motors 2300 Nm/200 kW/2350 rpm and 2600 Nm/250 kW/3500 rpm.

Analysis of the dynamics of EB allows comparing different reduction ratios of the rear axle for these motors. Reduction ratio 8.00 shows better maximal slope values, but in most cases it cannot achieve the maximum speed of 70 km/h within the maximal power, torque and rotation speed limitations. Also, acceleration values are higher and that may cause a discomfort for passengers. Reduction ratio 4.88 has higher energy consumption for the acceleration and lower maximal slope parameters. Reduction ratio 5.83 shows better parameters comparing to other reduction ratios, because it is able to achieve maximal speed of 70 km/h within the maximal power and the acceleration is

Table 3. Average energy consumption and its improvement by bus configurations.

Bus configuration (motor type, reduction gear ratio)	Average consumption rate EB, kWh/km	Average consumption rate DB, kWh/km	Improvement, %
1800/250/3200, 4.88	1.229	1.803	31.80
1800/250/3200, 5.83	1.232		31.66
1800/250/3200, 8.00	1.233		31.58
2100/200/3240, 4.88	1.232		31.67
2100/200/3240, 5.83	1.232		31.65
2100/200/3240, 8.00	1.234		31.55
2200/250/3500, 4.88	1.234		31.57
2200/250/3500, 5.83	1.234		31.54
2200/250/3500, 8.00	1.235		31.48
2300/200/2350, 4.88	*1.219*		*32.38*
2300/200/2350, 5.83	*1.220*		*32.30*
2300/200/2350, 8.00	*1.221*		*32.26*
2600/250/3500, 4.88	*1.220*		*32.31*
2600/250/3500, 5.83	*1.222*		*32.21*
2600/250/3500, 8.00	*1.223*		*32.14*
2700/250/3525, 4.88	1.235		31.51
2700/250/3525, 5.83	1.235		31.49
2700/250/3525, 8.00	1.237		31.41
Average:	1.229	1.803	31.81

faster than for 4.88. However, acceleration values are not so high and the change is smoother than for 8.00. The consumption is lower than for 4.88 and approximately the same as for 8.00. The recommended gear reduction ratio for the rear axle is 5.83. The recommended motors for the Ambassador EBs are provided in the Table 4.

Table 4. Recommended types of motor.

Motor type	Peak power (kW)	Cont. power (kW)	Peak torque (Nm)	Cont. torque (Nm)	Max operating speed (RPM)	Efficiency rate
2300/200/2350	200	150	2300	1230	2350	0.95
2600/250/3500	250	150	2600	969	3500	0.95

The recommended battery type is MpCO battery with 91 kWh capacity and 910 kg weight under the given constrains and operating conditions. The output of optimization model for battery selection is shown in the Table 5.

Table 5. Optimization results for battery selection.

Battery type	Capacity, kWh	Weight, kg	Cost, EUR	Max current, A
MpCO	91	910	70,707	2.99
LCO	91	758	81,900	2.99
LTO	91	1517	127,400	2.99
LFP	294	2262	147,000	0.93
NMC	588	4900	552,800	0.46

The main results of *TCO* analysis are shown in the Fig. 3. The calculated *TCO* of the diesel bus is 1.44 €/km, which include all costs related to owning, operation and maintenance of the bus and environmental costs 0.05 €/km. As concluded by Malnaca, Yatskiv [6], the price of the 7-years old DB converted into the EB is similar to the new DB, but because of shorter remaining lifetime (7 years), the investment costs comprise a bigger share in the *TCO* (39%). The total *TCO* of the electric bus is 1.40 €/km, which include costs of charging infrastructure (0.07 €/km) and grid connection costs (0.03 €/km).

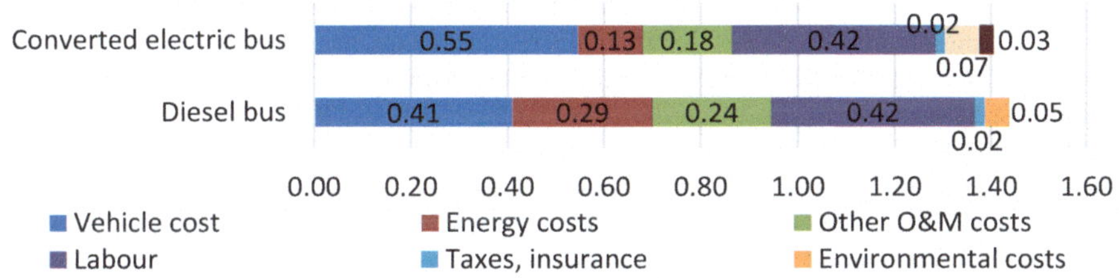

Fig. 3. *TCO* analysis results of DB in comparison with converted DB into EB (EUR) [6]

4 Conclusion

The results of the case study prove the workability of the developed decision-making model and it's applicability to the task of selecting technology for DB conversion into EB. The importance and necessity of each step of the process are confirmed by the case study. The provided methodology allows to assess technical, operational and economic performance of the existing DB and to find the optimum technological solution for DB conversion into EB which would have the same or even better values of technical and economic performance indicators.

The results of the models of the case study show that the EB which is converted from the Ambassador diesel bus may be by 31% more energy efficient. Average energy consumption rate for the analyzed EB is 1.229 kWh/km, however it may vary from 0.96 kWh/km to 1.54 kWh/km depending on the route type and the load.

The overall results of technical and economic assessment are in favour of used DB conversion into EB. The selected technical configuration of EB allows to achieve significant energy efficiency, reduce *O&M* costs of the vehicle, provides additional benefits to the environment, and extends the life of the used DB.

Acknowledgements. The paper is based on the research that has been conducted in the framework of the project No1.1.1.1/16/A/267 'Development of economically justified technology of conversion of the traditional diesel city bus into the environmentally friendly electrobus funded from the ERDF, and was financially supported by the ALLIANCE Project (GA no.: 692426) funded under European Union's Horizon 2020 research and innovation program'.

References

1. European Commission: Transport White Paper, Roadmap to a Single European Transport Area, Brussels (2011)
2. European Commission: Guide to Cost-Benefit Analysis of Investment Projects. Economic appraisal tool for Cohesion Policy 2014–2020. EC DG REGIO (2014)
3. European Strategy for Low-Emission Mobility. https://ec.europa.eu/transport/themes/stragies/news/2016-07-20-decarbonisation_en
4. Hibbeler, R.C.: Engineering Mechanics: Statics and Dynamics, 14th edn. Pearson, Prentice Hall, London, Upper Saddle River (2014)
5. Krivtsov, S.N., et al.: Measuring of traction and speed characteristics as well as of fuel economy of a car in road conditions. In: IOP Conference Series: Materials Science and Engineering, vol. 142, p. 012101 (2016)
6. Malnaca, K., Yatskiv (Jackiva), I.: Impact of critical variables on economic viability of converted diesel city bus into electric bus. In: Nathanail, E., Karakikes, I. (eds.) Data Analytics: Paving the Way to Sustainable Urban Mobility. CSUM 2018. Advances in Intelligent Systems and Computing, vol 879. Springer, Cham (2019)
7. Pihlatie, M., Kukkonen, S., Halmeaho, T., Karvonen, V., Nylund, N.-O.: Fully electric city buses—the viable option. In: IEEE International Electric Vehicle Conference, IEVC 2014, 17–19 December 2014, Florence, Italy (2014)
8. Terekhov, V.M.: Control Systems of the Electric Drives. Akademiya, Moscow (2008). (in Russian)

Statistics and Modelling in Transport Applications

Exploring Women Travel Behaviour in the Region of Žilina from Large Scale Mobility Survey

Ghadir Pourhashem(✉), Ľuboš Buzna, Tatiana Kováčiková, and Martin Hudák

ERAdiate Team, University of Žilina, Žilina, Slovakia
{ghadir.pourhashem, lubos.buzna, tatiana.kovacikova, martin.hudak}@erachair.uniza.sk

Abstract. Women and men often do not experience equal mobility opportunities. Increasingly, gender is being recognized to play a significant role in transport planning, particularly for addressing individual mobility needs in urban and rural areas. Previous studies have shown that transport policy and mobility planning are better suited to men's activities and mobility needs, while women experience the transport system differently than men in terms of accessibility, safety and security. To gain an in-depth understanding of women travel patterns in their daily activities, this paper investigates the factors that influence women's decisions between activity participation and choice of transport mode through analysis of mobility behaviour data gathered from 6000 households by Žilina self-governing region between 2013 and 2016. The results of this analysis provide valuable insights about the possible causal relation among socio-demographics, activity participation characteristics and daily travel mode choice of female and male travellers in the studied cities, which could be incorporated into more gender sensitive designs in mobility planning and travel demand modelling. Furthermore, the outcomes of this study can help decision makers, local and regional authorities and transport companies by raising awareness and providing support for (1) forming sustainable policies to overcome community mobility and accessibility challenges and (2) bringing more gender equality in the region of Žilina by changing traditional transport functionalities and providing more equitable provision of transport systems, particularly in public transport and active modes (non-motorised transport).

Keywords: Gender sensitive design · Mobility planning · Transport policy · Accessibility · Mode choice · Active mode

1 Introduction

Gender is being recognized to play a significant role in transport planning, particularly for addressing individual mobility needs in urban and rural areas since about 1980 [1–4]. In recent decades, many efforts toward understanding gender differences in mobility patterns have been accomplished in theory and practice. Investigating the relationships between changes in needs due to demographic and socio-economic

I. Kabashkin et al. (Eds.): RelStat 2018, LNNS 68, pp. 105–120, 2019.
https://doi.org/10.1007/978-3-030-12450-2_10

changes and spatio-temporal constraints proposed by Hägerstrand [5], was one of the most interesting aspects of identifying women's mode choice behaviour that has been researched in last decade.

A growing literature on travel behaviour indicates significant gender differences in mobility and travel patterns where women experience more trips than men in the developed world and in many developing countries [6]. However, women may be less mobile than men in some developing countries and their travel patterns are more complicated [7, 8]. Parameters like income, age, household size and structure, elder-child care responsibilities, ethnicity, employment status, degree of disability, location, class, and education can cause significant differences among women's mobility patterns. Moreover, some factors such as personal safety, security and quality of service are important concerns of women's mobility [9, 10]. Therefore, taking gender differences into account in travel behaviour it is important to identify women's transportation needs and mobility provisions [11–13].

Focus on differences in men's and women's travel patterns has been a characteristic feature in many discussions and the gender gap between developed and developing countries requires an acceleration of knowledge transfers between the two spheres [14]. A wide range of research has been done in developed and developing countries to report the gender difference impacts on the travel patterns where, gender differences in travel behaviour are relatively well-known in developed countries in comparing to the developing world [15–18]. Reference [19] well illustrates the facts that women have unequal access to means of transportation and to transport-related decision-making and that women's travel behaviour differs substantially from that of men. Some of the differences may be related to personal attitudes, temperaments and capabilities. Most of them, however, are directly related to male-female power relationships and gendered responsibilities for reproductive and caretaking tasks. Gender differences in travel behaviour are less marked in societies where women and men have equal access to the transport systems.

Works [20–22] explain important gender differences relating to perception of personal safety in public transport stations, car parks and other public places where women often change their travel behaviours due to fears of harassment and violence from other passengers. Moreover, gender issues in transportation has produced important results such as differences between women's and men's driving behaviour and safety outcomes, differences between women's and men's travel behaviours, e.g., mode choice, trip chaining, and vehicle design and safety features as they relate to women. In many western countries women are making more trips, and more frequently travel by car as drivers, also traveling longer distances [23]. Also, due to factors such as economic circumstances and age, subgroups of women meet notable differences in the travel patterns. Meanwhile, older women still do not drive as often as older men, and they stop driving earlier [24].

Reference [14] presented critical approach to generalization of men's and women's travel patterns, and to highlight the heterogeneity that exists within groups as well as similarities between traditional traveller categories. Researchers in Denmark, Finland, Norway and Sweden have problematized the stereotypes of gender and travel trajectories in these countries. The standard solution of equating travel behaviour (e.g. choice of transport mode) and travel needs (e.g. preferences) because of many reasons has come under criticism in their work [25, 26]. Promoting dialogue on constraints to women, providing technical assistance to governmental teams in designing and implementing policies and improvements of the institutional capacity of transport agencies in developing economies are considerable solutions. More inclusive urban access would also improve conditions for women and enable them to make choices according to their needs [27].

Despite the fact that many research studies have been accomplished to broaden the scope of gender dimension in transport planning and policies, there has been little attention to explore the influence of gender on activity participation and trip mode choice as a part of travel behaviour in Slovakia. Therefore, the current research aims to explore and understand gender differences in mobility patterns in the region of Žilina. The analysis of the relationships between type of activity participation and gender helps planers to better understand how men and women organise their daily time and how such organisation affects travel behaviours, particularly their daily travel mode. Besides, more knowledge about women mobility needs provides valuable intuition about the effectiveness of current transportation policy on changing women's activity-travel behaviour (travel pattern) and incorporating into more gender sensitive designs in mobility planning and travel demand modelling.

2 Data Collection and Sample Description

The data used in this paper were collected within the travel behaviour survey organised by the Žilina self-governing region between 2013 and 2016. Geographically, the survey covered the area of the Žilina region and it was conducted in two phases. The first phase was carried out in 2013 for districts of Žilina, Martin, Liptovský Mikuláš and Bytča. The second phase was executed in November 2016 for remaining districts of Čadca, Námestovo, Kysucké Nové Mesto, Turčianske Teplice, Ružomberok, Tvrdošín and Dolný Kubín.

The survey captured mobility behaviour of households for one workday in the mid of the week (Tuesday, Wednesday or Thursday). Participants were asked to fill in three questionnaires: household questionnaire, individual questionnaire and travel diary.

For survey, 6231 households (18,382 inhabitants) were approached to record the data about their trips. In diary questionnaire, people were asked to report their mobility behaviour (i.e. all out-of-home activities) across a one day along with their specifications including trip purpose, trip duration (departure and arrival time), destination (i.e. location of activity performance), mode of transportation, trip length and number of accompanying persons in car. Furthermore, the socio-demographic characteristics of households and of each member were inquired. After a cleaning and screening procedure, the sample was reduced to 17,907 persons among them 9048 (50.5%) females and 8859 (49.5%) males living in 5917 households. It should be noted that the survey was not intended to investigate participation of women in transport, however, the gender of participants seems to be well balanced over the districts of the Žilina region, age groups and social status (Fig. 1).

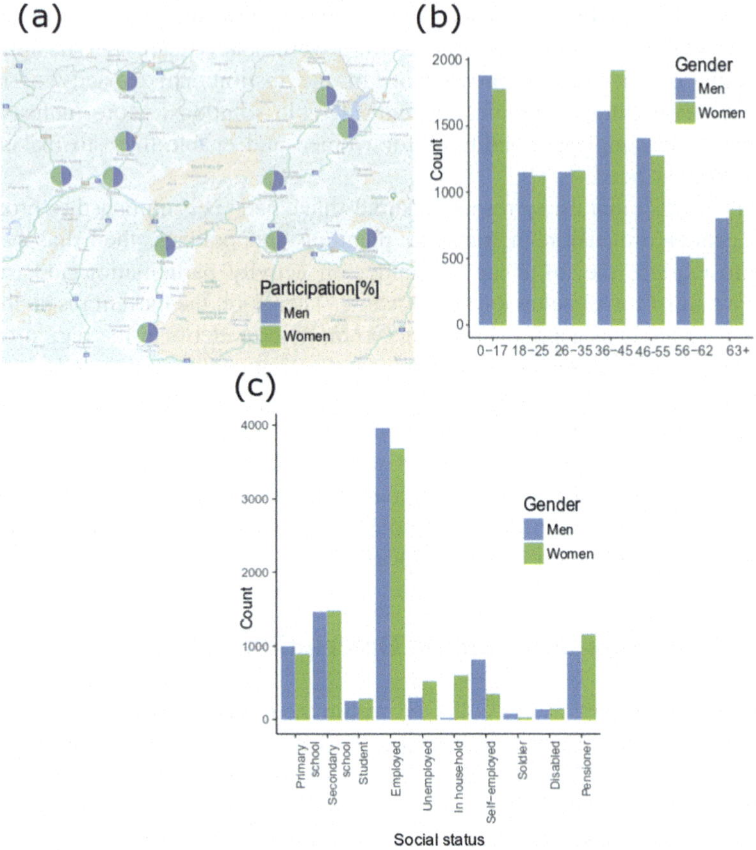

Fig. 1. Participation in the mobility behaviour survey by gender. **a** Distribution of participants' gender over the districts of the Žilina region. **b** Distribution of gender over age groups. **c** Distribution of gender as a function of social status.

3 Modelling Methodology

The mode choice is a complex decision-making process per se. To investigate the factors that influence women's decisions between activity participation and choice of transport mode, the Multinomial Logit (MNL) is applied. The MNL models have been applied widely in transportation research, particularly in travel choice modelling due to the fact that the formula for the choice probabilities takes a closed form and is readily interpretable. The MNL model is derived through the application of utility maximization concepts to a set of alternatives from which one, the alternative with maximum utility, is chosen [28].

The probability of selecting mode i by individual n is expressed as:

$$P_{ni} = \frac{e^{V_{ni}}}{\sum_{i=1}^{J} e^{V_{ni}}}, \quad (1)$$

where: P_{ni} is the probability that mode i is chosen by individual n; e is the Euler's number; V_{ni} is the estimate of the utility of mode i for individual n; J is the number of all available travel modes.

The perceived utility of mode i by individual n is derived as a linear function of explanatory variables (X_{ni}) as:

$$V_{ni} = \sum_{n} \beta_i * X_{ni}, \quad (2)$$

where β is a vector of parameters.

4 Results

4.1 Descriptive Data Analysis of the Gender Role in Transport Mode Choice

For out-of-home activity participation there are at least two activity type segmentations. The first group studies [29] categorise activities into two major groups: mandatory (work, work-related, school and school related activity) and non-mandatory (maintenance/ discretionary). The other group studies [30–32] consider three categories: subsistence (work or school), maintenance (such as child care, grocery shopping, banking, household work, and pickup/drop-off children) and discretionary (such as volunteer and religious activities as well as recreation and entertainment). Trip purpose classification adopted in this study is according to activities requiring travel, which are grouped in five categories: work (work and work-related activity), education (in school or university), shopping (grocery shopping and daily shopping), maintenance (e.g. medical care, banking, administrative services) and leisure (social and recreation activities).

The travel mode indicates the main transportation mode for the trip. Individuals for daily travel activities use different modes according to the household attributes (e.g. age, car ownership, income, etc.) and time of day. Hence, the travel mode reflects the most crucial decision that a traveller makes in terms of using a private car versus public transport and non-motorised or any other mode. Based on the reported data, individuals used eight transport modes for daily travel in the region of Žilina. With these data, we initially started analysing travel behaviour by comparing the number of trips as a function of the trip purpose, visualized in Fig. 2a. Here, we observe differences between behaviour of women and men. Women do more trips for leisure activities (visits, restaurants, cinema etc.) and for shopping activities, while men reported more trips for activities related to work. The number of trips for education (i.e. predominantly taken by youth generation) is very similar. Colours in Fig. 2a indicate the used mean of transport. Trips related to the education are very well balanced also in terms of the used mean of transport. Most often children get to schools on foot, by public transport or as car passengers. Thus, as expected, the gender transport inequalities emerge in the adult age. Independently on the trip purpose women tend to walk more and apart from commuting, it is for them the dominant mean of transport. For commuting the most often is used private car that is used much more often by men than women. Partly, this can be also attributed to the fact that within the population of adults that took part in the survey it is men who possess a driving license much more often. The only exception is age group from 56 to 62 (Fig. 2b).

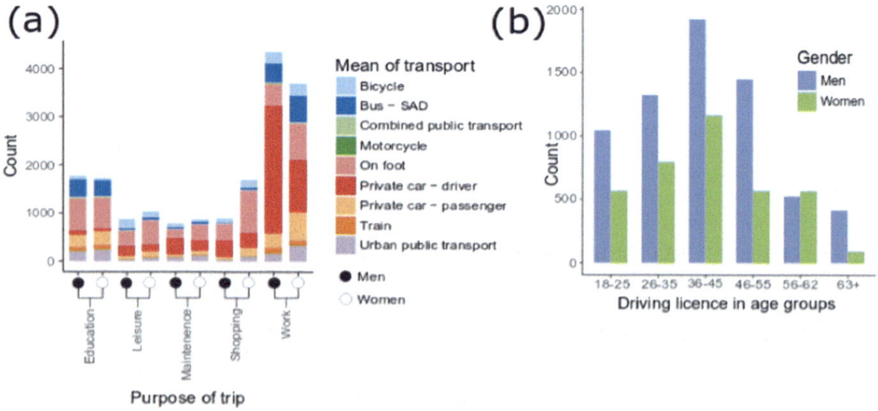

Fig. 2. **a** The number of trips by the purpose and the gender of travelers. Colours indicate used mean of transport. **b** Possession of the driving license among the participants of the survey by age groups.

To understand better the differences between how men and women use for commuting private cars and public transport, we show in Fig. 3 their usage as a function of age group and education level. Private cars are predominantly used by men independently of the age, while the higher level of education leads to lower level of gender inequality. When it comes to the usage of public transport, the situation is different. In young population (age up to 35 years), the usage of public transport is very well balanced and the bias toward women using public transport more, appears only in older population. This is potentially interesting finding, which should be noticed by public transport operators and by authorities to develop measures towards motivating more males in the age group of 30–40 not to change to individual transport, but to keep them using public transport. Similarly, the inequality decreases with the level of education, while in the group of participants with university education we observe good gender balance.

Apart from differences, descriptive analyses uncovered also many similarities in the travel behaviour of men and women. We have analysed origin-destination matrix of trips among districts of the Žilina region. As expected, the majority of trips are done within individual districts (short trips), but surprisingly the proportions of trips that

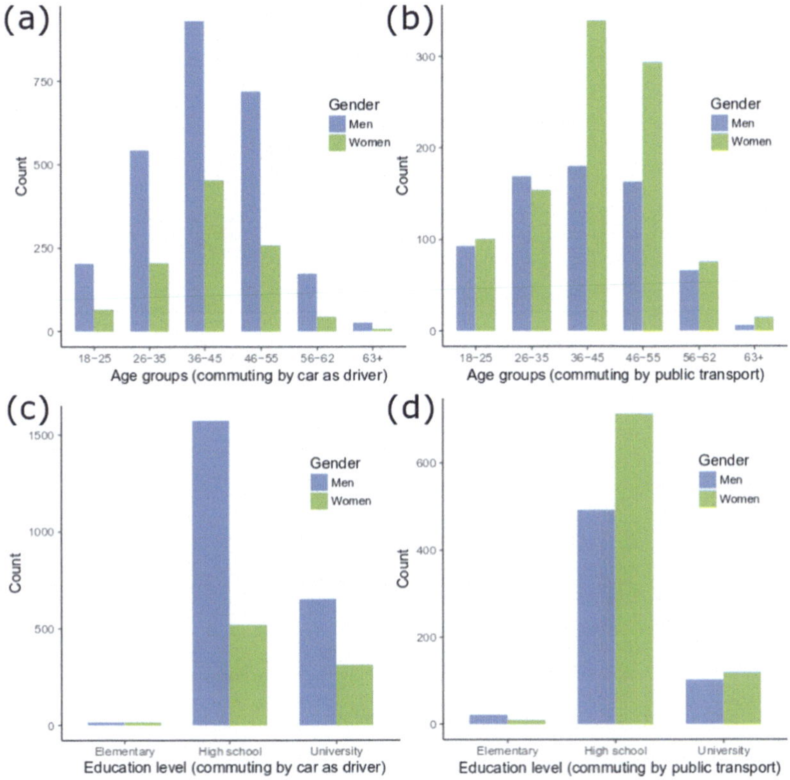

Fig. 3. Comparison of the participants driving to work a private car and using a public transport by gender. **a** Participants driving to work a private car by age groups. **b** Participants using for commuting a public transport by age groups. **c** Education level of participants driving to work a private car. **d** Education level of participants using for commuting a public transport.

originate and have a destination in different districts (long trips) are comparable for men and women. This can be also partly seen in Fig. 4a, where we displayed complementary cumulative distributions of travel distances of men and women.

From the distributions is clear that travelled distances are very similar. Furthermore, data in Fig. 4b imply that trips of mem and women are also similarly distributed over the time of day. Few more men start they trips early in the morning. Then, we observe a typical morning and afternoon peeks and also towards very late evening hours the number of trips by men and women is very similar. We checked more closely trips made by women after 10:00 p.m, finding that the most of them seem to be trips from work to home made by three-shifts production workers. The number of trips on weekdays is approximately twice the number of trips on weekend days (see Fig. 4c and d). Again, quite surprisingly, the numbers of trips reported in the survey by men and women are very similar.

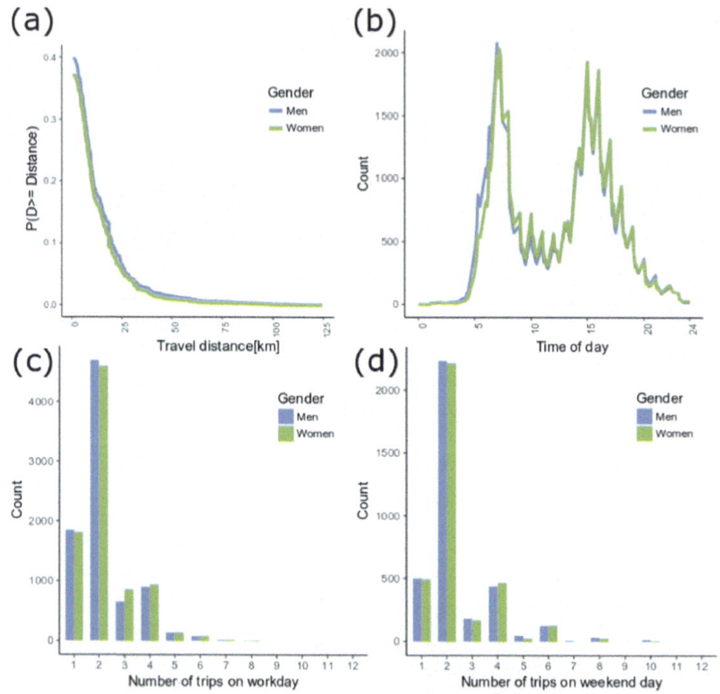

Fig. 4. Identified similarities in mobility behaviour of females and males in the Žilina region. **a** Complementary cumulative distribution of trip length in kilometres. **b** Number of travelers as a function of time of day. **c** Typical number of trips in a work day. **d** Average number of trips in weekends.

4.2 Analysis of Differences in the Travel Behaviour of Men and Women in the Žilina Region

Travel behaviour data of people over the age of 18 was analysed using MNL regression in SPSS23 that is appropriate when the outcome variables are categorical with more than two categories and the predictors are of any type. The modal choice model was estimated by using the maximum likelihood method. The results after modification according to final model represent that our model was fitted well (Table 1). The Pseudo R-Square indices in Multinomial Logistic regression is treated as R^2. Pseudo R-Square values can reach to a maximum of 1 and larger that indicate more of the variation is explained by the model.

Table 1. Pseudo R-Square.

Cox and Snell	0.587
Nagelkerke	0.607
McFadden	0.259

Estimation results of MNL is presented in Tables 2, 3 and 4. The estimated results reveal the significant effects of used variables on travel mode-choice which are discussed in following.

- ***Gender*** with regards to individuals' characteristics, gender has positive significant effect on women trips by public transport, private car as passenger, train, bicycle and walk. The multinomial logit for women relative to men is 1.047 unit higher for choosing private car as passenger given all other available travel modes in the model are held constant. Therefore, it can be interpreted women in region of Žilina are more likely than men to travel by car as passenger.
- ***Age*** respondents age as we expected was not significant predictor for choosing travel mode. According the results, age has negative impact on trips using intercity bus and foot. It means that young people more likely use less intercity bus for their daily trips compared with older people who mostly use private car as drivers (Tables 3 and 4).
- ***Driving license*** the results demonstrate that driving license availability has negative impacts on people trips by intercity bus, train, public transport, car as passenger,

Table 2. Multinomial modal choice model estimation in the Žilina region.

Travel mode[a]		B	Std. error	Wald	df	Sig.	Exp (B)	95% confidence interval for Exp(B)	
								Lower bound	Upper bound
1	2	3	4	5	6	7	8	9	10
Walk	Intercept	.997	.255	15.233	1	.000			
	Age	−.012	.003	16.838	1	.000	.988	.982	.994
	Travel distance	−.471	.019	618.060	1	.000	.624	.602	.648
	Social status	.105	.027	15.449	1	.000	1.111	1.054	1.170
	[Gender = 1]	.926	.075	153.745	1	.000	2.526	2.181	2.924
	[Gender = 2]	0	.	.	0	.	.	.	.
	[Trip purpose (work) = 1]	−.880	.172	26.097	1	.000	.415	.296	.581
	[Trip purpose (education) = 2]	.681	.314	4.709	1	.030	1.976	1.068	3.655
	[Trip purpose (shopping) = 3]	−.436	.171	6.513	1	.011	.646	.462	.904
	[Trip purpose (maintenance) = 4]	−.375	.140	7.160	1	.007	.687	.522	.905
	[Trip purpose (leisure) = 5]	0	.	.	0	.	.	.	.
	[Departure time (6:00–8:00) = 1]	3.565	.133	720.212	1	.000	35.332	27.233	45.839
	[Departure time (8:00–15:00) = 2]	0	.	.	0	.	.	.	.
	[Departure time (15:00–17:00) = 3]	−.253	.153	2.745	1	.098	.776	.575	1.047
	[Departure time (17:00–22:00) = 4]	−.261	.164	2.549	1	.110	.770	.559	1.061
	[Departure time (22:00–6:00) = 5]	−.737	.181	16.640	1	.000	.478	.336	.682
	[Driving license = 0]	−.538	.184	8.566	1	.003	.584	.407	.837
	[Driving license = 1]	0	.	.	0	.	.	.	.

(*continued*)

Table 2. (*continued*)

Travel mode[a]		B	Std. error	Wald	df	Sig.	Exp (B)	95% confidence interval for Exp(B)	
								Lower bound	Upper bound
1	2	3	4	5	6	7	8	9	10
Bicycle	Intercept	−.358	.364	.965	1	.326			
	Age	−.004	.004	.716	1	.397	.996	.988	1.005
	Travel distance	−.267	.019	196.241	1	.000	.766	.738	.795
	Social status	−.031	.040	.620	1	.431	.969	.896	1.048
	[Gender = 1]	.609	.110	30.637	1	.000	1.839	1.482	2.281
	[Gender = 2]	0	.	.	0	.	.	.	.
	[Trip purpose (work) = 1]	−.777	.246	9.992	1	.002	.460	.284	.744
	[Trip purpose (education) = 2]	−1.959	1.050	3.479	1	.062	.141	.018	1.105
	[Trip purpose (shopping) = 3]	−.335	.237	1.991	1	.158	.715	.449	1.139
	[Trip purpose (maintenance) = 4]	−.273	.199	1.892	1	.169	.761	.515	1.123
	[Trip purpose (leisure) = 5]	0	.	.	0	.	.	.	.
	[Departure time (6:00–8:00) = 1]	3.448	.158	475.505	1	.000	31.445	23.065	42.870
	[Departure time (8:00–15:00) = 2]	0	.	.	0	.	.	.	.
	[Departure time (15:00–17:00) = 3]	−.213	.220	.940	1	.332	.808	.525	1.243
	[Departure time (17:00–22:00) = 4]	−.366	.235	2.430	1	.119	.694	.438	1.099
	[Departure time (22:00–6:00) = 5]	−1.232	.273	20.408	1	.000	.292	.171	.498
	[Driving license = 0]	−.832	.270	9.502	1	.002	.435	.257	.739
	[Driving license = 1]	0	.	.	0	.	.	.	.

[a]The reference category is: Private car-driver.

bicycle and walk. For instance, multinomial log-odds of people who do not possess driving licenses to driving license holders is −1.379 unit higher for preferring intercity bus to private car given all available travel modes which means current users of intercity bus are more likely to prefer private car for daily trips in case of possession of driving license.

- ***Social status:*** the social status as a categorical variable was modelled for nine groups: 1. high-school student, 2. university student, 3. employee, 4. self-employed, 5. unemployed, 6. in household (homemaker), 7. pensioner, 8. soldier and 9. disabled. The estimated coefficient of social status was positive for selecting walk and motorcycle for performing out-of-home activities while it has significant negative effect on trips made by public transport, train and intercity bus. These results indicate that probability of using public transport, train and intercity bus was greater for students and employees than others (e.g. pensioners and in household).
- ***Trip purpose:*** the trip purpose associated to people's type of out-of-home activity participation evidentially has major impact on mode choice. Obtained results reveal that in the region of Žilina, work trip has negative impact on use of bicycle,

Table 3. Multinomial modal choice model estimation in the Žilina region (Cont.).

Travel mode[a]		B	Std. error	Wald	df	Sig.	Exp (B)	95% confidence interval for Exp(B)	
								Lower bound	Upper bound
1	2	3	4	5	6	7	8	9	10
Motorcycle	Intercept	–3.653	.968	14.230	1	.000			
	Age	–.008	.010	.557	1	.456	.992	.973	1.013
	Travel distance	–.063	.017	13.213	1	.000	.939	.907	.971
	Social status	.186	.088	4.458	1	.035	1.204	1.013	1.430
	[Gender = 1]	–.805	.317	6.446	1	.011	.447	.240	.832
	[Gender = 2]	0	.	.	0	.	.	.	.
	[Trip purpose (work) = 1]	–.990	.507	3.818	1	.051	.372	.138	1.003
	[Trip purpose (education) = 2]	.859	.856	1.007	1	.316	2.362	.441	12.654
	[Trip purpose (shopping) = 3]	–.661	.471	1.969	1	.161	.516	.205	1.300
	[Trip purpose (maintenance) = 4]	–1.083	.371	8.518	1	.004	.339	.164	.701
	[Trip purpose (leisure) = 5]	0	.	.	0	.	.	.	.
	[Departure time (6:00–8:00) = 1]	1.573	.401	15.407	1	.000	4.822	2.198	10.577
	[Departure time (8:00–15:00) = 2]	0	.	.	0	.	.	.	.
	[Departure time (15:00–17:00) = 3]	.094	.802	.014	1	.906	1.099	.228	5.292
	[Departure time (17:00–22:00) = 4]	.538	.184	8.566	1	.003	1.713	1.194	2.455
	[Departure time (22:00–6:00) = 5]	0	.	.	0	.	.	.	.
	[Driving license = 0]	–3.565	.133	720.212	1	.000	.028	.022	.037
	[Driving license = 1]	0	.	.	0	.	.	.	.
Private car - passenger	Intercept	–1.650	.295	31.276	1	.000			
	Age	–.004	.004	1.465	1	.226	.996	.989	1.003
	Travel distance	.005	.003	3.079	1	.079	1.005	.999	1.010
	Social status	–.034	.031	1.198	1	.274	.966	.909	1.027
	[Gender = 1]	1.407	.087	263.433	1	.000	4.084	3.446	4.840
	[Gender = 2]	0	.	.	0	.	.	.	.
	[Trip purpose (work) = 1]	–.017	.217	.006	1	.939	.984	.642	1.506
	[Trip purpose (education) = 2]	1.040	.340	9.338	1	.002	2.830	1.452	5.517
	[Trip purpose (shopping) = 3]	–.435	.242	3.213	1	.073	.647	.403	1.041
	[Trip purpose (maintenance) = 4]	.150	.190	.625	1	.429	1.162	.801	1.686
	[Trip purpose (leisure) = 5]	0	.	.	0	.	.	.	.
	[Departure time (6:00–8:00) = 1]	3.194	.138	536.492	1	.000	24.395	18.617	31.967
	[Departure time (8:00–15:00) = 2]	0	.	.	0	.	.	.	.
	[Departure time (15:00–17:00) = 3]	–.592	.140	17.946	1	.000	.553	.421	.728
	[Departure time (17:00–22:00) = 4]	–.845	.157	28.827	1	.000	.430	.316	.585
	[Departure time (22:00–6:00) = 5]	–.644	.175	13.524	1	.000	.525	.373	.740
	[Driving license = 0]	–.716	.185	14.936	1	.000	.489	.340	.703
	[Driving license = 1]	0	.	.	0	.	.	.	.

(*continued*)

Table 3. (*continued*)

Travel mode[a]		B	Std. error	Wald	df	Sig.	Exp (B)	95% confidence interval for Exp(B)	
								Lower bound	Upper bound
1	2	3	4	5	6	7	8	9	10
Public transport	Intercept	−.058	.340	.029	1	.865			
	Age	.007	.004	2.953	1	.086	1.007	.999	1.016
	Travel distance	−.315	.020	258.831	1	.000	.730	.702	.758
	Social Status	−.213	.040	28.462	1	.000	.808	.747	.874
	[Gender = 1]	.664	.103	41.528	1	.000	1.942	1.587	2.376
	[Gender = 2]	0	.	.	0	.	.	.	.
	[Trip purpose (work) = 1]	−.221	.246	.806	1	.369	.802	.495	1.299
	[Trip purpose (education) = 2]	1.983	.371	28.564	1	.000	7.266	3.511	15.036
	[Trip purpose (shopping) = 3]	−.545	.261	4.368	1	.037	.580	.348	.967
	[Trip purpose (maintenance) = 4]	.153	.206	.551	1	.458	1.165	.778	1.746
	[Trip purpose (leisure) = 5]	0	.	.	0	.	.	.	.
	[Departure time (6:00–8:00) = 1]	3.828	.151	641.600	1	.000	45.970	34.185	61.818
	[Departure time (8:00–15:00) = 2]	0	.	.	0	.	.	.	.
	[Departure time (15:00–17:00) = 3]	−.981	.184	28.406	1	.000	.375	.262	.538
	[Departure time (17:00–22:00) = 4]	−.769	.199	14.968	1	.000	.464	.314	.684
	[Departure time (22:00–6:00) = 5]	−1.202	.227	28.042	1	.000	.301	.193	.469
	[Driving license = 0]	−1.024	.230	19.864	1	.000	.359	.229	.563
	[Driving license = 1]	0	.	.	0	.	.	.	.

[a]The reference category is: Private car-driver.

motorcycle and walk that demonstrate people are more likely use private car than other travel modes for work trips. In contrast education trips was identified to be positively significant for choosing train in long distance trips, private car as passenger and walk in short distance trips to go to school or university.

- ***Travel distance:*** the travel distance was identified to be negatively significant on choice of walk, bicycle and motorcycle among other alternatives. It obviously indicates with increasing travel distance people has less tendency to walk or use bicycle for daily trips in urban areas.
- ***Departure time:*** according to results, departure time substantially affects individuals' mode choice. As can be seen in Tables (2, 3 and 4), in the evenings people tend to select less public transport and intercity bus and train that reflects the high tendency to use private car for trips after 5 p.m. Results also show the early morning departure time positively affects the use of public transport, bicycle and walking to the destination.

Table 4. Multinomial modal choice model estimation in the Žilina region (Cont.).

Travel mode[a]		B	Std. Error	Wald	df	Sig.	Exp (B)	95% Confidence Interval for Exp(B)	
								Lower Bound	Upper Bound
1	2	3	4	5	6	7	8	9	10
Train	Intercept	−.915	.432	4.493	1	.034			
	Age	−.006	.005	1.347	1	.246	.994	.984	1.004
	Travel distance	.018	.003	32.705	1	.000	1.018	1.012	1.024
	Social status	−.347	.055	39.249	1	.000	.707	.634	.788
	[Gender = 1]	.813	.118	47.892	1	.000	2.256	1.792	2.840
	[Gender = 2]	0	.	.	0	.	.	.	.
	[Trip purpose (work) = 1]	−.178	.353	.256	1	.613	.837	.419	1.671
	[Trip purpose (education) = 2]	1.931	.431	20.036	1	.000	6.898	2.961	16.068
	[Trip purpose (shopping) = 3]	−.633	.445	2.020	1	.155	.531	.222	1.271
	[Trip purpose (maintenance) = 4]	.402	.322	1.562	1	.211	1.495	.796	2.807
	[Trip purpose (leisure) = 5]	0	.	.	0	.	.	.	.
	[Departure time (6:00–8:00) = 1]	3.459	.161	460.068	1	.000	31.786	23.172	43.602
	[Departure time (8:00–15:00) = 2]	0	.	.	0	.	.	.	.
	[Departure time (15:00–17:00) = 3]	−1.291	.184	49.100	1	.000	.275	.192	.395
	[Departure time (17:00–22:00) = 4]	−1.365	.200	46.567	1	.000	.255	.173	.378
	[Departure time (22:00–6:00) = 5]	−1.333	.225	35.181	1	.000	.264	.170	.410
	[Driving license = 0]	−1.288	.238	29.220	1	.000	.276	.173	.440
	[Driving license = 1]	0	.	.	0	.	.	.	.
Intercity Bus-SAD	Intercept	−.509	.303	2.819	1	.093			
	Age	−.011	.003	10.433	1	.001	.989	.982	.996
	Travel distance	.019	.002	73.231	1	.000	1.019	1.015	1.023
	Social status	−.233	.033	49.195	1	.000	.792	.742	.845
	[Gender = 1]	.807	.081	100.487	1	.000	2.241	1.914	2.625
	[Gender = 2]	0	.	.	0	.	.	.	.
	[Trip purpose (work) = 1]	.284	.247	1.321	1	.250	1.328	.819	2.154
	[Trip purpose (education) = 2]	1.665	.334	24.763	1	.000	5.283	2.743	10.177
	[Trip purpose (shopping) = 3]	−.591	.306	3.736	1	.053	.554	.304	1.008
	[Trip purpose (maintenance) = 4]	.718	.228	9.926	1	.002	2.050	1.312	3.203
	[Trip purpose (leisure) = 5]	0	.	.	0	.	.	.	.
	[Departure time (6:00–8:00) = 1]	3.511	.136	669.727	1	.000	33.498	25.676	43.704
	[Departure time (8:00–15:00) = 2]	0	.	.	0	.	.	.	.
	[Departure time (15:00–17:00) = 3]	−1.132	.127	79.890	1	.000	.322	.251	.413
	[Departure time (17:00–22:00) = 4]	−.885	.136	42.611	1	.000	.413	.316	.538
	[Departure time (22:00–6:00) = 5]	−1.477	.163	82.405	1	.000	.228	.166	.314
	[Driving license = 0]	−1.379	.172	64.256	1	.000	.252	.180	.353
	[Driving license = 1]	0	.	.	0	.	.	.	.

[a]The reference category is: Private car-driver.

5 Conclusions

This paper focused on investigation of factors that influence women's choice of transport mode for daily trips in the region of Žilina. A Multinomial Logistic model (MNL) was applied for estimating effects of selected variables on individual's modal choice using mobility behaviour data gathered from 6000 households by Žilina self-governing region between 2013 and 2016.

Findings from this study show that women without driving license choose public transport and intercity bus more than men for their daily education trips when trip starts between 6:00 to 8:00 a.m. Furthermore, the results indicate the influence of departure time on women use of car as passenger. It means that females who don't have driving license and make their education trip as a first trip of day in the morning prefer to use car as passenger to their destination. In addition, modelling findings prove with increasing travel distance for education trips, women use more likely train among available transport modes than men. According to outcome of this study, we found the significant negative effect of evening departure on use of public transport by females. This is probably due to the low-frequency of public transport (30-min headway) services after 5:00 p.m and women's perception of security, safety and risk of violence in public transport.

Understanding effective factors in people decision-making process about travel and activity participation could help planners, policy makers and authorities to implement more gender sensitive designs in urban mobility plans and to prepare inclusive transport policies in the region of Žilina. These findings also have important implications for the Žilina self-governing development agency (RAZSK) for providing better environment for walking and cycling in terms of being safe and guide the public transport service providers that are interested in attracting more users.

In this research, the lack of required and reliable data enforced us to limit the scope of our analysis and we were not able to analyse the interaction among household members for joint-activity participation and travel behaviour, particularly the modal choice. In recent years several elements have been changed: the increasing women's participation in the labour force, the decreasing importance of the traditional family nucleus, individuals' type and frequency of activity, and travel behaviour. To better evaluate how women, choose the travel mode, the results should be extended by focusing on assessing women's experiences and specific transport requirements, by the mean of an exploration of their own perception of their mobility constraints and needs, and of innovative solutions to meet these needs. Women's perception and attitudes about ICT and new business model's solution that already exists in the mobility systems and that women already experienced should be explored to understand, if these types of solutions could improve their mobility or to what extent it could constitute new constraints and inequality sources. Further research is required to evaluate the effects of adopting inclusive transport policies and long-term investments for employing new technologies changing people attitudes and travel behaviour in the direction of sustainable transport management.

Acknowledgments. Authors are grateful to Development Agency of the Žilina Self-governing Region for providing the mobility behaviour survey data conducted in Žilina region and Daniel Gašpierik for pre-processing the data and for preparing visualizations used in Sect. 4.1. Ľuboš Buzna has been partially supported by the research grants VEGA 1/0463/16 "Economically efficient charging infrastructure deployment for electric vehicles in smart cities and communities" and APVV-15-0179 "Reliability and emergency systems on infrastructure with uncertain functionality of critical elements". This work was facilitated by the FP 7 project ERAdiate-Enhancing Research and Innovation dimension of the University of Zilina for Intelligent Transport Systems – Grant agreement no. 621386. Authors also declare that the procedures followed in publishing this paper were in compliance with the ethical and personal data protection standards of EU GDPR.

References

1. Rosenbloom, S.: Editorial: the need for study of women's travel issues. Transportation **7**(4), 347–350 (1978)
2. Giuliano, G.: Public transportation and the travel needs of women. Traffic Q. **33**(4), 607–615 (1979)
3. Giuliano, G.: Getting there: women and transportation. In: Zimmermann, J. (ed.) The Technological Woman. Interfacing with Tomorrow, pp. 102–112. Praeger, New York (1983)
4. Scheiner, J., Holz-Rau, Ch.: Gendered travel mode choice: a focus on car deficient households. J. Transp. Geogr. **24**, 250–261 (2012)
5. Hägerstrand, T.: What about people in regional science? In: Ninth European Congress of the Regional Science Association, vol. 24, pp. 7–21 (1970)
6. Law, R.: Beyond women and transport: towards new geographies of gender and daily mobility. Prog. Hum. Geogr. **23**(4), 567–588 (1999)
7. Pucher, J., Renne, J.L.: Transp. Q. **57**(3), 49–77 (2003)
8. Dobbs, L.: Wedded to the car: women, employment and the importance of private transport. Transp. Policy **12**(3), 266–278 (2005)
9. Li, H., Guensler, R.L., Ogle, J.: Comparing women's and men's morning commute trip chaining in Atlanta, Georgia, by using instrumented vehicle activity data. In: Conference on Research on Women's Issues in Transportation. Transportation Research Board, Chicago (2004)
10. Whitzman, C.: Building Inclusive Cities: Women's Safety and the Right to the City, vol. xv, 216 p. Routledge, Abingdon (2013). [8] p. of col. plates
11. Dupont, A., Krakutovski, Z.: Travel time and distance regarding gender patterns in the paris region: past trends and forecasts to 2030. In: Conference on Research on Women's Issues in Transportation. Transportation Research Board Conference, Chicago (2009)
12. Meyer, M.D.: Research on Women's Issues in Transportation, Report of a Conference: Conference Overview and Plenary Papers, in Conference on Research on Women's Issues in Transportation, vol. 1. Transportation Research Board Conference, Chicago (2004)
13. Kalter, M.-J.O., Harms, L., Jorritsma, P.: Changing travel patterns of women in the Netherlands. In: Conference on Research on Women's Issues in Transportation. Transportation Research Board Conference, Chicago (2009)
14. Levin, L., Faith-Ell, Ch.: Methods and tools for gender mainstreaming in Swedish transport planning. In: The 5th International Conference on Women's Issues in Transportation Proceedings, Paris, France, pp. 215–223 (2014)

15. Duchène, C.: Gender and Transport: Discussion Paper 11, International Transport Forum (2011)
16. Peters, D.: Breadwinners, homemakers and beasts of burden: a gender perspective on transport and mobility. In: Habitat Debate, vol. 4, no. 2. UNCHS, Nairobi (1998)
17. Nobis, C., Lenz, B.: Gender differences in travel patterns: role of employment status and household structure. In: Conference on Research on Women's Issues in Transportation. Transportation Research Board, Chicago (2004)
18. Babinard, J., Scott, K.: What do existing household surveys tell us about gender and transportation in developing countries? In: Conference on Research on Women's Issues in Transportation. Transportation Research Board Conference, Chicago (2009)
19. Peters, D.: Gender issues in transport – applying an integrative perspective, keynote presentation for commission C4 "Urban Mobility Management. In: Metropolis Network Conference in Toronto, Canada, 15 June 2006
20. Loukaitou-Sideris, A., Fink, C.: Addressing women's fear of victimization. Transportation settings a survey of US transit agencies. Urban Aff. Rev. **44**(4), 554–587 (2009)
21. Lahiri-Dutt, K., Ahmad, N.: Considering gender in social impact assessment. In: Vanclay, F., Esteves, A.M. (eds.) New Directions in Social Impact Assessment. Conceptual and Methodological Advances, pp. 117–137. Edward Elgar, Cheltenham (2012)
22. Markovich, J., Lucas, K.: The Social and Distributional Impacts of Transport: A Literature Review. Transport Studies Unit School of Geography and the Environment, Oxford (2011)
23. Rosenbloom, S.: Understanding women's and men's travel patterns: the research challenge research on women's issues in transportation. In: Report of a Conference, Conference Overview and Plenary Papers, Chicago, 18–20 November, vol. 1, pp. 7–28. Transportation Research Board, TRB, Washington (2004)
24. Hjorthol, R., Levin, L., Sirén, A.: Mobility in different generations of older persons: the development of daily travel in different cohorts in Denmark. Norway and Sweden. J. Transp. Geogr. **18**(5), 624–633 (2010)
25. Christensen, H.R., Poulsen, H., Oldrup, H., Malthesen, T., Hvidt, M., Holmen, M.: Gender Mainstreaming European Transport Research and Policies. Building the Knowledge Base and Mapping Good Practices. TRANSGEN, Köpenhamn (2007)
26. Cresswell, T., Uteng, T.P. (eds.): Gendered Mobilities. Ashgate, Aldershot (2008)
27. Thynell, M.: The quest for gender-sensitive and inclusive transport policies in growing Asian cities. Transp. Policy Soc. Incl. **4**(3), 72–82 (2016). ISSN 2183-2803
28. Koppelman, F.S., Sethi, V.: Closed - form discrete-choice models. In: Hensher, D.A., Button, K.J. (eds.) Hand Book Transport Modeling, pp. 211–227. Elsevier Science Ltd., Amsterdam (2000)
29. Atlanta Regional Commission: Implementing activity-based models in Virginia. VTM Research Paper 09-01 (2009)
30. Bhat, C.R., Koppelman, F.S.: A conceptual framework of individual activity program generation. Transp. Res. Part A **27**(6), 433–446 (1993)
31. Bhat, C.R., Misra, R.: Discretionary activity time allocation of individuals between in-home and out-of-home and between weekdays and weekends. Transportation **26**(2), 193–229 (1999)
32. Bowman, J.L., Ben-Akiva, M.E.: Activity-based disaggregate travel demand model system with activity schedules. Transp. Res. Part A **35**, 1–28 (2000)

Assessment of Inventory Indicators for Nomenclature Groups with Rare Demand

Valery Lukinskiy(✉), Vladislav Lukinskiy, and Anna Strimovskaya

National Research University Higher School of Economics (HSE), St. Petersburg, Russia
{lukinskiy, vladas27, astrim26}@mail.ru

Abstract. Rare demand is an important part of the inventory management, nevertheless there are no any appropriate analytical descriptions or numerical examples of it except separate papers where considered the possibility to describe rare demand with Poisson distribution. The divergence between forecasts and actual data could be explained by the following reasons: the first one—extreme values in preforecasting period, as well as not significant 'length' of the analyzed time period; the second one—taking data for the forecast from the period of conduction certain actions (sales, promo, etc.), consideration of these actions might be done with combined forecasting methods. The paper describes the approach to assessment of inventory consumption for rare demand based on Poisson distribution. Besides, the paper contains the numerical examples and analysis of the results.

Keywords: Supply chains · Inventory classification · Forecasting · Poisson distribution

1 Introduction

Logistics usually becomes involved in forecasting in terms of how much should be ordered from its suppliers (through purchasing), and how much of finished product should be transported or held in each market that the organization serves. Small disruptions, if are not taken into consideration, could cause huge damages to the entire supply chain even if it performs well in cost reduction or benefits improvement [1]. As Amazon.com well knows, inventory is one of the most expensive assets of many companies, representing as much as 50% of total invested capital. Although retailers assume a position of risk on a wide variety of products, their position on any one product is not deep. Risk is spread across more than 30,000 stock keeping units (SKUs) in a typical supermarket. A mass retailer offering general merchandise and food often exceeds 50,000 SKUs. Faced with this width of inventory, retailers attempt to reduce risk by pressing manufacturers and wholesalers to assume greater and greater inventory responsibility. The four competition factors are time, cost, flexibility and quality to ensure customer satisfaction [2]. The dilemma for every manager, whether they are a provider, producer or vendor, is not only to obtain a service rate of 100% to satisfy all their clients, but also to have the lowest possible storage cost. Each penny lost in variable costs inevitably entails huge additional expenses along with reduced

I. Kabashkin et al. (Eds.): RelStat 2018, LNNS 68, pp. 121–129, 2019.
https://doi.org/10.1007/978-3-030-12450-2_11

efficiency. Market competitiveness is simply not compatible with over-purchasing and throwing away an extra product [3]. Most companies have to accumulate some level of inventory. There are good reasons for such accumulation. For example, during periods when demand has decreased, a company may continue to buy from some vendors to maintain the relationship and/or to keep employees by producing to inventory [4]. As the desired service level rises, it takes a disproportionate investment in inventory to achieve small incremental improvements in availability. However, if it is possible to find alternative service strategies for servicing customers, say, for example, by speeding up the flow of information about customer requirements and by using faster modes of transport, then the same level of service can be achieved with less inventory. Variety reduction can make substantial savings in inventory by standardising and rationalising the range of materials, parts and consumables kept in stock [5]. The irregularity of removing quantities from a definite picking location in a warehouse very often results in the stock out in the particular picking address. Managing demand is challenging because of the difficulty in forecasting future consumer requirements accurately [6]. Inventory used to absorb uneven rates of demand or supply, which businesses often face, is referred to as anticipation inventory. The primary lever to reduce anticipation inventory is simply to match demand rate with production rate. Secondary levers can be used to even out customer demand in one of the following ways:

1. Add new products with different demand cycles so that a peak in the demand for one product compensates for the seasonal low for another.
2. Provide off-season promotional campaigns.
3. Offer seasonal pricing plans.

Extrapolation of historical data is the most common approach when forecasting demand in connection with inventory control. To determine a suitable technique, we need to have some idea of how to model the stochastic demand. In principle, we should try to determine the model from analysis of historical data. In practice this is very seldom done. With many thousands of items, this initial work does not seem to be worth the effort in many situations. In other situations there are not enough historical data. A model for the demand structure is instead determined intuitively. In general, the assumptions are very simple. Items with low demand are referred to as *slow-moving items* and typically have a high coefficient of variation, whereas items with high demand are referred to as *fast-moving items* and typically have a low coefficient of variation. When demand for items is intermittent, because of low volume over all and a high degree of uncertainty as to when and at what level demand will occur, the time series is said to be "lumpy", or "irregular". This pattern is often found in products that are phasing in or out of the product line, demanded by relatively few customers, divided among many stocking locations so that demand at each location is low, or derived from the demand for other items. Such demand patterns are particularly difficult to forecast using the more popular techniques. However, because such items may represent as much as 50 percent of the products a firm handles, they represent a special demand forecasting problem for the logistician.

Thus, literature analysis has showed that the authors pay attention to different questions, related to inventory management. Herewith, it is worth mentioning that rare demand is an important part of the inventory management theory. Nevertheless there are no any appropriate analytical descriptions or numerical examples of it.

2 Methods of Grouping and Inventory Classification

Enlargement of stored goods at the warehouses of different levels has required the development of methods related to systematization and grouping of multinomenclature inventory. In the papers devoted to logistics and supply chain management there are mentioned about 10 methods, some of them are presented in Table 1.

Table 1. Analysis, systematization and grouping methods of multinomenclature inventory.

The method title	Characteristics of resulting classification feature	Comment (percentage correlation)
ABC—analysis	According to sales volume, storage costs, purchasing, etc. According to correlation price-quality	*A* (80% according to sales volume; 20% according to quantity) *B* (15% according to sales volume; 30% according to quantity) *C* (5% according to sales volume; 50% according to quantity)
XYZ—analysis	According to consumption nature (demand) According to accuracy of the forecast	*X*—stable consumption ($v < 10\%$)* *Y*—accuracy of the forecast is restricted $10 \leq v < 25\%$) *Z*—significant fluctuations ($v > 25\%$); forecast is not accurate
LMN—analysis	According to physical volume of goods unit	*L*—small *M*—medium *N*—large
FMR—analysis	According to customer requests According to "choice frequency" for order fulfillment	*F*—most frequent *M*—less frequent *R*—rare request

*v—variation coefficient (recommended in some research papers)

From the analysis of Table 1, it is followed that, firstly, for some analysis methods the characteristics of classification features given in literature sources differ considerably; secondly, percentage correlations reflecting application to a certain nomenclature group depend on chosen by the author dividing criterion and group forming way; thirdly, data volume and its nature is not considered (quantity, dynamics, etc.); fourthly, it is worth highlighting that all the analysis methods could be divided on two groups: the first group supposes division according to a certain feature of the nomenclature among the whole aggregate (for example, *ABC* or *FMR*); the object of the second group analysis is presented by each nomenclature in separation, and index reflects demand dynamics (for example, *XYZ*-analysis).

Given in Table 1 methods of analysis and nomenclature assignments to different groups allow, in their turn, to systemize the choice process of calculation methods for inventory indexes and appropriate strategies of inventory management. However, in spite of a long discussion, the goods assignment problem on different groups, in our mind, is still not completed.

For example, in the paper [7], it is considered that strict optimization should be done for *A* group (so 10–15% of all nomenclature); for *B* group it is possible to use simple methods; for *C* group—the simplest "two hopper" strategy, herewith a big safety stock is possible. The same point of view is presented by Bowersox and Closs [8], they suppose that, firstly, for *A* group it is better to use accurate methods of requirements planning; secondly, for groups *B* and *C*—reactive inventory management methods, that almost do not require "data collection and processing", because «reactive system allows to deliver products by smaller lots»; thirdly, for 85–90% of nomenclature groups there are used simple methods, or this assessment is not conducted at all.

Given in the papers [7, 8] ideas that nomenclature group *A* should be "strictly optimized" and for each position "there should be used accurate forecasting methods" have recommended nature, because cost criterion of *ABC*-analysis do not consider consumption dynamics and, due to different reasons, for some positions the forecast could not be done.

That is why, collection and data systematization about real perspectives of requirement, as well as the possibility of accurate forecast calculations for different nomenclature groups, related to rare events, are of a current interest.

3 Development of Calculation Methodology for Inventory Indexes Related to the Group of Rare Demand

In Fig. 1, there are given the results of weekly nomenclature consumption during a year. Obviously, zero consumption is observed for most of the weeks during a year (30 weeks), while maximum inventory consumption meaning, which is equal to four positions, is observed once a year; to three positions is twice a year and so on.

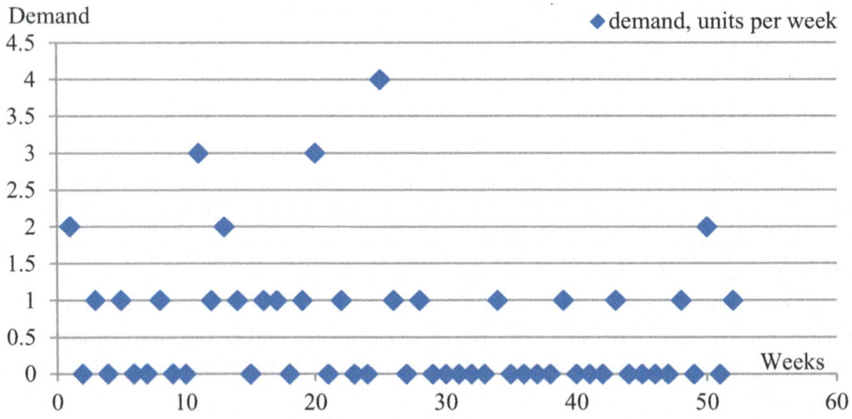

Fig. 1. Demand for nomenclature position (units per week) during a year.

Performed literature analysis has showed that for parameters assessment of such processes should be used distribution functions of discrete random variables: Poisson, binominal, geometrical and others.

Considered [9], that if t—time between occurring of rare events happening with average intensity of λ, than t has an exponential distribution.

It is worth highlighting that exponential distribution is connected with Poisson distribution, which sometimes is referred as distribution of rare events. It means that if intervals between the orders for consumed stock units have exponential distribution, than the amount of orders during a certain period Δt is distributed according to Poisson law with parameter $a = \lambda$.

It is known, that density of Poisson distribution is written as the follows:

$$p(k) = \frac{a^k}{k!} e^{-a}, \tag{1}$$

where a—average amount of the orders during a time period Δt; k—order amount, $k = 0, 1, 2, \ldots, N$.

Remind that the parameter of Poisson distribution a is equal to mathematic expected value (the first moment), and the standard deviation (the second central moment) $\sigma = \sqrt{a}$.

In order to calculate the parameter a for different time periods Δt there is used the formula:

$$a = \lambda \cdot \Delta t, \tag{2}$$

where λ—intensity of order flow.

The function of probability distribution for inventory availability is determined by the formula:

$$P(k) = \sum_{k=0}^{N} p(k) = \sum_{k=0}^{N} \frac{a^k}{k!} e^{-a}, \tag{3}$$

In analogy with accepted terminology of inventory management, let us consider parameter a of Poisson distribution as a current stock S_{cs}. Than safety stock S_{ss} is determined by the formula:

$$S_{ss} = S_{\max} - a = S_{\max} - S_{cs}, \tag{4}$$

where $S_{\max}$—the maximum inventory or total stock size, related to a certain (or chosen) probability $P(k)$.

4 The Numerical Example

Example 1. In Table 2 there is initial data on weekly inventory consumption during two months. As a result of statistic processing there have been get average meanings of

Table 2. Initial data on inventory consumption for three types of product, units.

Product type	Weeks								λ_i, unit/week
	1	2	3	4	5	6	7	8	
1	0	1	0	1	0	0	2	0	0.5
2	2	1	1	0	1	0	1	2	1.0
3	1	4	2	1	1	3	2	2	2.0

order intensity λ (for the time period Δt, which is equal to one week), which allows in their turn to calculate appropriate meanings of parameter a_i for three types of product using formula (2).

Probabilistic estimates of total stock size, including safety and current stocks, will be conducted for the second product type (a = 1.0). Based on Table 3 for order periodicity equal to one week, average stock size is 1 unit, where maximum or total stock size should be 4 units (for probability P_k = 0.996) or 5 units (for probability P_k = 0.999). For order periodicity equal to two weeks (Δt = 2), accordingly average stock size is 2 units, and total stock size is 5 units (P_k = 0.980) or 7 units (P_k = 0.996).

Table 3. Probability of inventory availability (according to Poisson distribution).

Inventory level	Average order amount			
	a = 0.5	a = 1.0	a = 2.0	a = 4.0
0	0.606	0.368	0.135	0.018
1	0.910	0.736	0.405	0.091
2	0.986	0.920	0.676	0.237
3	0.998	0.981	0.856	0.432
4		0.996	0.946	0.627
5		0.999	0.980	0.783
6			0.992	0.887
7			0.996	0.947
8				0.977
9				0.980

Note: the values in the table are rounded off

Example 2. Consider indexes of inventory consumption for nomenclature given in Fig. 1 for different periods of delivery.

In Table 4 there are given the results of statistic data processing about inventory consumption:

- Empirical frequency $n_e(k)$ and distribution density $p_e(k)$;
- Theoretical distribution densities $p_T(k)$, calculated with the formula (1) for a = 0.614;

Table 4. Results of statistic data processing of inventory consumption.

Inventory size, k	$n_e(k)$	$p_e(k)$	$k \cdot p_e(k)$	$p_T(k)$	$P_T(k)$	$n_T(k)$	$\frac{(n_Э(k) - n_T(k))^2}{n_T(k)}$
0	30	0.577	0.000	0.541	0.541	28.132	0.124
1	16	0.308	0.308	0.332	0.873	17.264	0.093
2	3	0.058	0.116	0.102	0.975	5.304	1.001
3	2*	0.038	0.114	0.021	0.996	1.092*	2.46
4	1*	0.019	0.076	0.004	1	0.156*	–
Total	52	1	$a = 0.614$				$\chi^2 = 3.678$

Note: meaning are often summarized when calculating criterion χ^2

- Theoretical distribution function $P_T(k)$, formula (3);
- Theoretical frequencies $n_T(k)$.

Checking with Chi-square goodness of fit has showed that hypothesis on probability of approximation of empiric sampling by Poisson distribution is not contradictory.

Consider probabilities of inventory availability for different delivery period: week, month, quarter (Table 5). Based on the analysis of Table 5 there occurs ability to design options of inventory management strategies with fixed periods, aliquot to delivery weeks amount (Table 6).

Table 5. Probability of inventory availability of goods for different delivery periods.

Inventory size	Poisson distribution parameter		
	Week (a = 0.615)	Month (a = 2.46)	Quarter (a = 7.40)
0	0.5406	0.0854	0.0006
1	0.8731	0.2956	0.0050
2	0.9754	0.5541	0.0214
3	0.9963	0.7661	0.0615
4	–	0.8965	0.1359
6	–	0.9869	0.3814
8	–	–	0.6573
10			0.8470
12			0.9343

Table 6. Options of inventory design for different delivery periods.

Inventory type	Week	Month	Quarter
	a = 0.615	a = 2.46	a = 7.40
Current stock (average)	1	2	7
Safety stock	2	4	5
Total (probability of inventory availability P)	3 (0.9963)	6 (0.9869)	12 (0.9343)

Thus, for weekly delivery let us assume that average stock size is equal to one unit, total (maximum) one is equal to three units (probability of inventory availability is $P = 0.9963$), safety stock—two units. At the same time, depending on costs related to deficit, there are several alternative options possible, when total stock size will be two units ($P = 0.9754$) or one unit ($P = 0.8731$). It is worth mentioning that if total order size is equal to one unit, than division for current and safety stocks is not considered.

For deliveries once a quarter the forecast of current stock is $a = 7$ (8) units, maximum amount $Q_{\max} = 12$ (for $P = 0.9342$), safety stock $Q_s = 5$ units.

Numerical examples of inventory consumption indexes on the basis of Poisson distribution for different nomenclature related to rare events have showed that satisfied results were in 60% of cases. In 40% of cases probabilistic forecasts of inventory consumption have been claimed as not reliable. The divergence between forecasts and actual data could be explained by the following reasons: the first one—extreme meanings in pre forecasting period, which leads to mistaken assessment of parameter a, as well as not significant 'length' of the analyzed time period; the second one—taking data for the forecast from the period of conduction certain actions (sales, promo, etc.), consideration of these actions might be done with combined forecasting methods (for example, apart from extrapolation ones, using experts forecasting methods).

5 Conclusions

The approach proposed, in our opinion, allows to cover significant amount of nomenclature positions related to the group of rare events; increase assessment accuracy of inventory management indexes, i.e. current (average), safety and total stocks; improve customer service by high probability of deficit absence; optimize costs; decrease illiquid stocks; that will finally lead to the increase of supply chain efficiency.

Application of the approach designed to assessment of inventory indexes for goods groups related to rare events allow to increase efficiency of decision making process for enterprise managers in purchasing (resource management), manufacturing management and distribution logistics system management (channels of supply chains).

The subsequent research of rare demand should be focused on the following directions:

- Development of combined forecasting methods, envisaging apart from extrapolation, experts methods as well;
- Designing and forming of new methods for inventory classification, in particular multi criteria assessment;
- Generalization and refinement of classic models and methods of inventory management, in particular defining economic order quantity (EOQ) and its modifications (restrictions, discounts, multinomenclature), inventory management strategies in multi-level supply chains, special calculation models (one cycle task, goods with restricted store period, etc.).

References

1. Kersten, W., Blecker, T.: Managing Risks in Supply Chains. How to Build Reliable Collaboration in Logistics. Erich Schmidt Verlag, Berlin (2006)
2. Trojahn, S.: Logistics strategies for resource supply chains. Transp. Telecommun. **19**(3), 244–252 (2018). https://doi.org/10.2478/ttj-2018-0021
3. Jackson, I., Tolujevs, J., Reggelin, T.: The combination of discrete-event simulation and genetic algorithm for solving the stochastic multi-product inventory optimization problem. Transp. Telecommun. **19**(3), 233–243 (2018). https://doi.org/10.2478/ttj-2018-0020
4. Coyle, J., Bardi, E., Langley, C.: The Management of Business Logistics: A Supply Chain Perspective, 7th edn. South-Western Thomson Learning, Boston (2003)
5. Lysons, K., Farrington, B.: Purchasing and Supply Chain Management, 7th edn. Pearson Education Limited, Harlow (2006)
6. Wisner, J., Tan, K., Leong, G.: Supply Chain Management: A Balanced Approach, 3rd edn. Cengage Learning, South-Western, Boston (2012)
7. Ryzhikov, U.: Theory of Queues and Supply Management. Piter, Saint-Petersburg (2001). (in Russian)
8. Bowersox, D.J., Closs, D.J.: Logistical Management. The Integrated Supply Chain Process. McGraw-Hill Companies Inc., New York (1996)
9. Ventsel, E.S., Ovcharov L.A.: Applied Problems of Probability Theory. Radio and Communications, Moscow (1983). (in Russian)

Multi-agent Approach to Optimization of Tariffs for the Air Navigation Service

Igor Bessmertny[1(✉)] and Nikolai Sukhikh[2]

[1] Faculty of software engineering and computer science, ITMO University, 49, Kronverksky pr., Saint Petersburg, Russia
bia@cs.ifmo.ru

[2] Saint Petersburg State University of Civil Aviation, 38, Pilotov str., Saint Petersburg, Russia

Abstract. The significant part of air fares is air navigation service charge. Air navigation service providers have opportunity as cover the air traffic control expenses as get a geo-spatial rent abusing their monopolistic nature. Establishing fair air traffic control tariffs needs a model to play different scenarios of tariff changes. An approved method for such a complex task is multi-agent approach. We propose a set of two types of agents: agents of air companies build air routes for all flights minimizing total expenses, agents of air navigation service providers update tariffs trying to maximize income. Based on just two figures reflecting change the tariff and income from air traffic control charge the agent can find a rational tariff in changing environment. Experimental research of this approach on a small air traffic network demonstrated that it is possible to assess effects of different agent's behavior, as "greedy" when the agent tries to raise the tariff to get maximal spatial rent, as "humble" when it decreases the tariff in order to apply more flights to it's air control zone.

Keywords: Air navigation tariffs · Multi-agent · Graph search

1 Problem Statement

1.1 An Introduction

The air traffic control (ATC) tariffs should as reimburse expenses of the air navigation service operation as get a spatial rent because of the geographical position. So, a reasonable motivation exists to establish the maximal level of tariffs unless the transit air traffic participants prefer to bypass this air traffic zone. Air navigation service providers (ANSP) often to use the monopolistic nature of such service and state unfair ATC tariffs. IATA coordinates efforts to promote effective dialogue between airlines and ANSPs. This activity needs an instrument to define fair ATC pricing. Fair pricing means avoidance of inconsistent methodologies and discriminatory practices, transparency in cost information, etc. [1].

Having the statistics of operation on previous periods of time we can easily calculate different scenarios for varying air traffic tariffs and predict the carriers' behavior from the point of view of our ANSP. Coincidently, ANSPs in other countries could

I. Kabashkin et al. (Eds.): RelStat 2018, LNNS 68, pp. 130–137, 2019.
https://doi.org/10.1007/978-3-030-12450-2_12

change their tariffs as well, so our forecast will be not correct and we get not the expected earnings. To solve this problem we need to model the behavior of many ANSPs where each one eager to maximize the spatial rent. Thus, the goal of this research is enhancing the long-term strategic vision of the airlines and their partners [1].

1.2 Formal Task Statement

The formal task statement is to be outlined as follows. Let say we have a finite indirected graph $G = (V, E, Z)$, where V—set of nodes, E—set of edges, Z—ATC zones. Edge $e_{ij} = \left(f_{ij}, a_{ij}\right)$ links nodes V_i and V_j, where f_{ij}—fuel cost, \$, a_{ij}—ATC cost for the flight from V_i to V_j or from V_j to V_i. ATC cost $a_{ij} = t_{ij}d_{ij}$, where t_{ij}—ATC tariff, \$/km, d_{ij}—distance between V_i and V_j, k Each $t_{ij} \in Z_r$, r—ANSP. Each air carrier k fulfils a set of flights $\{F_k\} = \left\{V_i, V_j\right\}$. For the simplicity, we mean f_{ij} and t_{ij} are equal for all aircraft types. Each k flight builds a chain of several edges $s_k = \left\{e_{ix}, \ldots, e_{yj}\right\}$ so the operation cost of k flight

$$C_k = \sum_{x,y \in s_k} \left(f_{xy} + a_{xy}\right). \tag{1}$$

Each air carrier has a goal to minimize sum of operation costs

$$S_k = \min \sum_k n_k C_k, \tag{2}$$

where n_k—number of flights during the period of time. Each r ANSP aims to maximize the income for their service

$$I_r = \max \sum_{i,j \in Z_r} n_{rij} a_{ij}, \tag{3}$$

where n_{rij}—number of flights through the edge e_{ij}, belonging to zone r.

These two goals are contradictory, so the optimal solution is a subject of compromises reached between air carriers and ANSPs. Hence, if one tries to raise the tariffs then not only air companies would avoid this ATC zone but other ANSPs could change their tariffs. The existing solutions like Dijkstra's algorithm [2], Bellman-Ford algorithm [3], Floyd-Warshall algorithm [4], are focused just on search for the shortest paths on the graph where its parameters are unchangeable. An approved solution for such complex tasks is a multi-agent approach that widely used at air traffic control modeling [5, 6].

2 A Proposed Approach to ATC Tariffs Optimization

2.1 Multi-agent Model

The formal task statement described in the Sect. 1.2 could be used for maximizing income of ANSP but in reality we will face the problem of insufficiency of data. Some data related to operational activity of air companies could be unavailable. Beyond this, the model is too simple. Particularly, if we try to state the tariff using formulas (1)–(3) we will see that there is no limit for raising the tariff because this model does not take into account the elasticity of demand: the significant growth of average ATC tariff will decrease the demand for air carriages and vice versa. At last, there could exist some non-economical reasons for change the ATC tariff policy.

Agent approach is a universal way to solve the problems described above. Multi-agent systems prove their efficiency at real-time control of transportation networks [7], power transmission [8], production processes [9], GSM tariff choosing [10], et al. where deregulation allows decrease operational costs or provide customers better service because of better reflecting the market environment. Moreover, agent approach allow act using incomplete data [11].

Let an agent of the air company k be responsible for air routes using the data from the open sources: flights $\{F_k\}$, fuel costs $\{f_{ij}\}$, air treks $\{e_{ij}\}$ and ATC zones tariffs $\{t_{ij}\}$ for each edge e_{ij}. The set of such agents could build all routes. At first glance, the routing task could be solved using well known methods like Dijkstra algorithm as shown by Rebezova [12]. In fact, formal algorithms are too sensitive for minor change of parameters: the algorithm re-route the flight even if the economy is infinitesimal. We program the agent to range the possible routes and to choose the most economical one with probability 0.8 and the second one with probability 0.2. This approach allows fuzzify the solutions to reflect incompleteness of the model.

Another type of agents is to be responsible for ATC tariff for the ATC zone. Using the same data this agent is able calculate the income for the ATC service of its zone. Then, the agent should try to increase income changing the tariff of its zone. The only information available to the agent is change of income after the last change. There are several policies the agent could choose depending on combination of changes. We have approved two agent's behavior: "greedy" and "humble" as shown in the Table 1.

Table 1. Agent actions depending on income dynamics.

Tariff	Air service income	Agent behavior	Tariff to change
Up	Up	Any	Up
Down	Down	Any	Up
Down	Up	Greedy	Up
		Humble	Down
Up	Down	Greedy	Up
		Humble	Down

The greedy behavior supposes raising tariff in non-obvious cases as the humble behavior will decrease. Each zone can choose as greedy as humble behavior for tariff change. To take account to demand elasticity we assess market share instead of absolute values of income.

2.2 Experimental Research

A network of air treks for the experiments is shown on the Fig. 1. There are 12 nodes and four air traffic control zones.

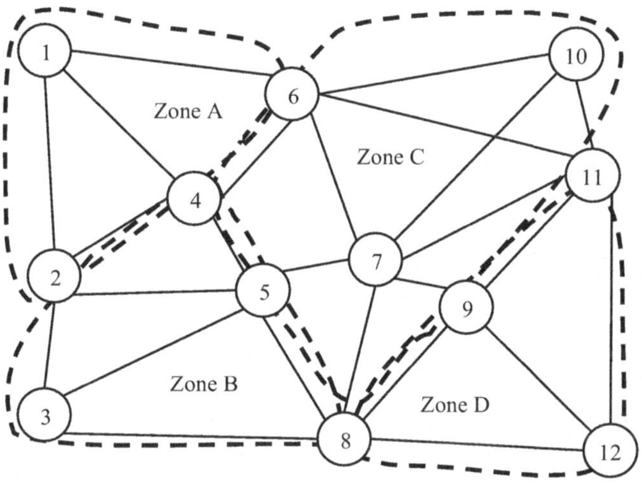

Fig. 1. The topology of an air traffic control zones.

Distances between nodes are shown in the Table 2. For the simplicity, the fuel cost and the ATC charge assumed to be independent on the model of aircraft.

Flights fulfilled by air companies join nodes 1–8, 1–11, 1–12, 2–11, 3–10, 3–11, 4–9, 4–11, 4–12, 5–10, 5–12, 6–8, 6–12, 8–10 so the zone C is a transit one for almost all flights and this zone could be bypassed via zones B and D. This makes reasonable to distinguish zone C from others and to model the behavior of its ANSP.

Each iteration of the experiment contains the steps listed below:

1. Agents of air companies build routes for all flights using current values of ATC tariffs.
2. Calculation of income for each ANSP.
3. Agents of ANSPs update ATC tariffs according their model of behavior.

Figure 2 contains the results of experiment where the agent of zone C is "humble" as the others are "greedy". The fuel cost assumed to $200/100 km and the initial tariff for all ATC zones is $100/100 km.

Table 2. Distances between the nodes.

Node	1	2	3	4	5	6	7	8	9	10	11	12
1		600		500		600						
2	600		300	400	550							
3		300			600			650				
4	500	400				350		650				
5		550	600				300	400				
6	600			350			400			700	450	
7					300	400		450	250		550	
8			650	650			450		350			600
9							250	350		350		
10						700			350		300	
11						450	550			300		700
12								600			700	

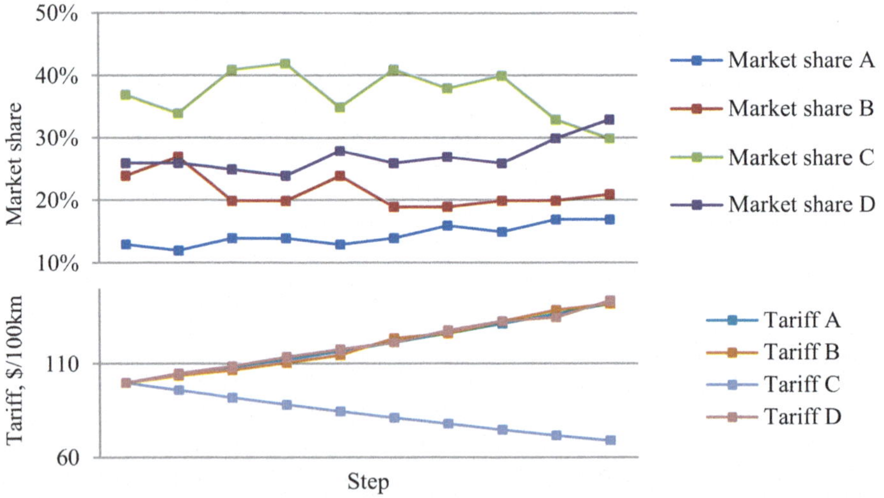

Fig. 2. Market share change when zone C uses "humble" behavior while other zones—"greedy".

As C could be a transit ATC zone for the most flights the reason of "humble" behavior is to raise income appealing all the flights from the adjacent zones due to low tariffs. Other zones try to increase tariffs to compensate leakage of transit flights. The results shown on the Fig. 2 demonstrate that flow-in of flights to the zone C just cover sinking the tariffs and vice versa, raising the tariffs by other zones just compensate the loss of some flights. This model does not reflect a possible overload of air treks in case of appealing to many flights by the low ATC tariff.

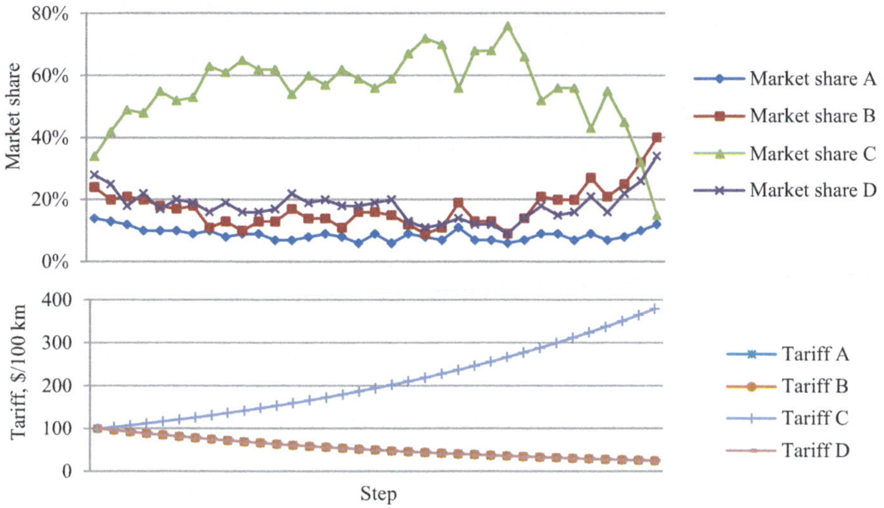

Fig. 3. Market share change when zone C uses "greedy" behavior while other zones—"humble".

Another experiment represented on Fig. 3. There are "greedy" behavior of the zone C agent and "humble" for others. The initial tariff of all ATC zones is equal to \$100/100 km. Then, the zone C tries to raise its income increasing the ATC rate. In fact, we can see that primarily the market share of zone C raises. It is caused by a simple reason: bypass of the zone C needs to much additional fuel. Then, the market share of zone C decreases dramatically because dropping down the ATC tariffs by other ANSPs covers the raising of fuel cost.

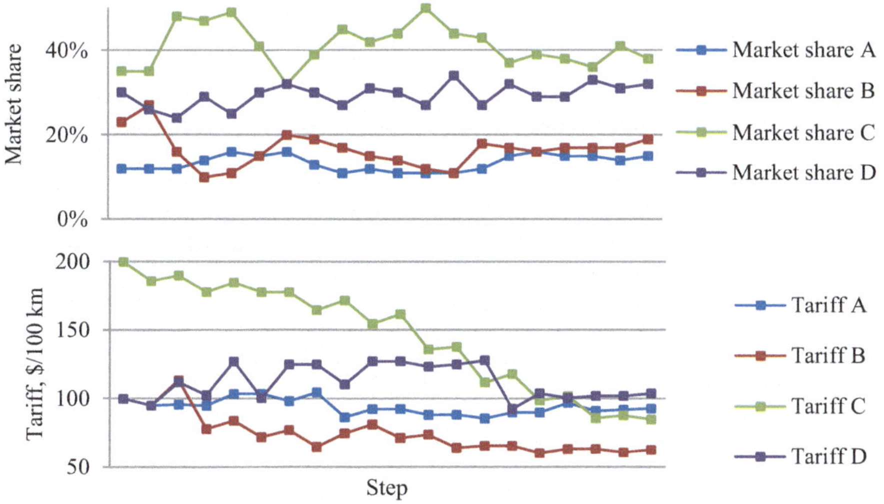

Fig. 4. Tariff balancing process when primarily the tariff for zone C was overstated.

The last experiment has a goal to assess how agents cope with unbalanced tariffs. Let state the ATC tariff for the zone C twice higher then ones for other zones and the same (humble) behavior for all agents. Figure 4 demonstrates that not only the tariff for zone C decreases but the tariff for zone B drop down as well to keep the market share.

3 Conclusion

The experiments described above show that in contrary to analytical solutions or classical algorithms the agent approach allows find solution applying diverse strategies and using incomplete data. Despite the limitedness of the model ignoring break even points for some flights or for entire air carriers it could reflect the market share holding by ANSPs. More realistic multi-agent model taking into the account market elasticity and restriction of air treks throughput could be used to support negotiation between air carriers and air navigation service providers to establish fair air traffic control tariffs.

References

1. IATA Homepage. http://www.iata.org/whatwedo/ops-infra/air-traffic-management/Pages/air-traffic-control-charges.aspx. Accessed 29 June 2018
2. Cormen, T.H., Leiserson, C.E., Rivest, R.L., Stein, C.: Section 24.3: Dijkstra's algorithm. In: Introduction to Algorithms, 2nd ed., pp. 595–601. MIT Press and McGraw–Hill, Cambridge (2001)
3. Bellman, R.: On a routing problem. Q. Appl. Math. **16**, 87–90 (1958)
4. Cormen, T.H., Leiserson, C.E., Rivest, R.L.: Section 26.2 The Floyd–Warshall algorithm. In: Introduction to Algorithms, 1st ed., pp. 558–565. MIT Press and McGraw-Hill, Cambridge (1990)
5. Agogino, A.K., Tumer, K.: Auton: a multiagent approach to managing air traffic flow. Auton. Agent Multi-Agent Syst. **24**(1), 1–25 (2010)
6. Lancelot, F., et al.: Human-in-the-loop multi-agent approach for airport taxiing operations. In: Trends in Practical Applications of Agents, Multi-Agent Systems and Sustainability. Advances in Intelligent Systems and Computing (2015)
7. Adler, J.L., Satapathy, G., Manikonda, V., Bowles, B., Blue, V.J.: A multi-agent approach to cooperative traffic management and route guidance. Transp. Res. Part B: Methodol. **39**(4), 297–318 (2005)
8. Yen, J., Yan, Y., Contreras, J., Ma, P-Ch., Wu, F.F.: Multi-agent approach to the planning of power transmission expansion. Decis. Support Syst. **28**(3), 279–290 (2000)
9. Lažanský, J., Ŝtepánková, O., Mařík, V., Pěchouček, M.: Application of the multiagent approach in production planning and modelling. Eng. Appl. Artif. Intell. **14**(3), 369–376 (2001)
10. Kaya, B.: An intelligent method for optimization of tariffs in GSM networks. In: IEEE/ACM International Conference on Advances in Social Networks Analysis and Mining (ASONAM), San Francisco, CA, pp. 917–920 (2016)

11. Torreño, A., Onaindia, E., Sapena, Ó.: An approach to multi-agent planning with incomplete information. In: 20th European Conference of Artificial Intelligence, ECAI 2012, vol. 242, pp. 762–767 (2012)
12. Rebezova, M.: Repayment problem at a settlement of debts. Comput. Model. New Technol. **17**(2), 7–14 (2013)

Exploring the Potential of Social Media Content for Detecting Transport-Related Activities

Dmitry Pavlyuk[1](✉), Maria Karatsoli[2], and Eftihia Nathanail[2]

[1] Transport and Telecommunication Institute, Riga, Latvia
Dmitry.Pavlyuk@tsi.lv
[2] Department of Civil Engineering, University of Thessaly, Volos, Greece
{makarats, enath}@uth.gr

Abstract. The wide spread of social media encourages the users to share more often their activities as well as their location, leading to a rapid growth of the data volume. Current research retrieves this user-generated content on social media platforms in an effort to convert them into powerful tools, enabling transport related data collection. In this paper data from Twitter are retrieved and processed to explore their potential for providing transport related data. The main objective is to investigate the reliability of the transport related content retrieved from tweets and the transferability of analytics methods to other cities and languages. The research data set includes thousands of tweets collected in three cities: Minneapolis-Saint Paul twin cities (USA), Riga (Latvia), and Volos (Greece) in May–June 2018. Selection of the research areas is owed to substantially different environments in terms of population, language and transport infrastructure. The collected data were classified into five classes: general transport-related information, real-time information, complain, advice/question, unrelated to transport. Based on the obtained results, a cross comparison was made about efficiency of Twitter as a social media source of transport-related information in different urban environments.

Keywords: Text mining · Twitter · Big data · Classification models · Location-based data

1 Introduction

The increase of traffic volumes and public transport use in modern cities makes important the optimization of the existing mobility systems. The continuous collection of traffic data is necessary for decision making, prediction of changes in a transport system as well as objective analysis of commuters' expectations and needs [1].

Social media data is highly available for researchers and public, and this fact creates a potential for using them as a supporting tool for traffic-related decisions [2].

In the current research, data from Twitter are explored for transport purposes. Twitter is a micro blogging site that has experienced increasing popularity among social media users mainly due to the restriction of 140 characters on tweets that makes the retrieved information easily consumed [3]. Twitter users have the tendency to share

I. Kabashkin et al. (Eds.): RelStat 2018, LNNS 68, pp. 138–149, 2019.
https://doi.org/10.1007/978-3-030-12450-2_13

information regarding traffic conditions and incidents as well as their opinion/ complains about transport services [4–6].

This paper investigates the potential of transport related content retrieved from Twitter and the transferability of analytics methods to other cities and languages. The paper is structured as follows. Section 2 is a brief overview of previous transport studies based on Twitter data use. In the following section the methodology used for data collection, preliminary grading, data pre-processing and classification is described. Empirical results are included in Sect. 4. Lastly, discussion and conclusions are given in Sect. 5.

2 State of the Art

In recent years several studies have analysed the use of social media for transport related purposes. This analysis opens up potential for various applications including accident – incident detection, transport planning and decision making, human mobility analysis and travel behaviour [7]. Social media users use often their accounts on platforms (such as Twitter) to share traffic information. That information that stem from Twitter users who can be drivers, pedestrians or passengers, correspond to larger coverage of traffic conditions as compared to point sensors [8]. In fact, due to their nature, it is believed that Twitter data can complement traditional collection methods [3, 9]. However, tweets contain usually misspellings or grammatical errors ending up to a percentage of 60% of useful content [10]. Another drawback of Twitter use is the limitations on the amount of data that can be retrieved in real time [4].

Studies related to incident detection have shown that incidents can be spotted earlier in regular users' social media accounts compared to official transport sources, making traffic related tweets important for real time incident detection [11]. Analytically, D'Andrea et al. [12] classified tweets with accuracy 95.75%, based on traffic event occurrence in order to present a real-time traffic event system. Gutierrez et al. [13] detected incident relevant tweets by using a classifier to identify traffic event evolution. Wanichayapong et al. [14] developed a tool that provides traffic report after having classified tweets using Natural Language Processing (NLP). Following this approach, Gal-Tzur et al. [15] investigated the correlation between tweets and traffic congestion, while Gu et al. [16] detected five major incident types employing Naive Bayes classification. The work of Mai and Hranac [17] showed that there is a correlation between accidents mentioned on tweets and incident records.

Previous studies highlighted the potential of social media as a source of information for transport stakeholders to better understand users' needs [18]. Collins et al. [19] used Twitter keywords to determine users' dissatisfaction in specific metro lines. Zhang et al. [20] used 500,000 geo-located tweets to explore the potential in transport management improvement. Pathak et al. [21] investigated how transport authorities manage user generated content for calculation of traffic flow indicators.

Regarding use of Twitter data for mobility analysis, Krumm et al. [22] estimated travel behaviour with urban routing patterns. Hawelka et al. [23] correlated tweet locations with socioeconomic characteristics of population. Jurdak et al. [24] found that tweets can be used in a similar way as mobile phone tracks. The use of Twitter in

European transport networks was investigated by Lenormand et al. [25], comparing the daily traffic reports with tweets.

3 Methodology

The research in this paper addresses the problem of detecting of transport-related activities in social media sources. Social media is acknowledged as a valuable source of information in recent literature, but utility of obtained information is highly dependent on intensity of social media activities in the specified area. The key objective of this research is to investigate the reliability of the transport related content retrieved from tweets and the transferability of findings analytics methods to smaller other cities and other languages. We selected Twitter as a social media source for research experiments due to its international popularity and well developed automated access to source date (using Twitter API). Note that although this research is focused on social media source, we consider Twitter data only as a supplement to classical urban traffic data sources.

The research methodology is common for studies, based on social media data. The main methodological stages of the research are (Fig. 1):

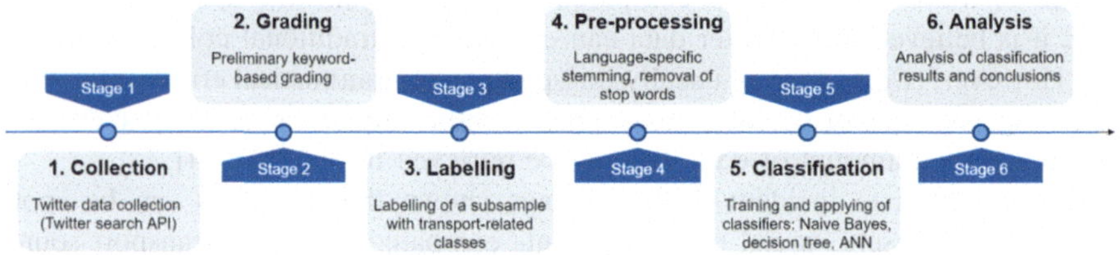

Fig. 1. Research stages.

1. Twitter data collection.
2. Preliminary keyword-based grading of tweets.
3. Manual labelling of tweets.
4. Text pre-processing.
5. Training the classifier algorithm.
6. Analysis of classification results and discovering language and area specifics.

3.1 Data Collection

Data collection is implemented using the standard Twitter API and scripts, developed by the authors. Each tweet is considered as a separate document and collected information includes the following properties:

- Tweet submission date and time,
- Message text,
- Author's nickname,
- Geo-reference (if available).

Tweets were separately collected for 3 analysed research areas for the same period of time.

3.2 Preliminary Grading

Collected tweets were preliminary graded on the base of a predefined keyword list. This grading allowed initial identification of potentially transport-related tweets in the research corpuses for their further analysis. Following the results of Kuflik et al. [1], the authors composed lists of transport-related keywords in 4 languages (English, Latvian, Russian, and Greek). Each keyword was associated with a grade from 0 (not related to transport) to 5 (highly related to transport). The complete lists of keywords are available from the authors by request. Further each document was graded according to the sum of its words.

3.3 Transport Domain Classes' Identification

The authors used their expert opinion for manual labelling of a defined set of documents (500 or 1000 tweets depending on the corpus volume). The set of classes for labelling includes:

- General transport-related information: includes all the tweets with general information about public transport, traffic, activities or opinion of the users regarding transport issues.
- Real-time transport-related information: class of tweets that inform about real time transport issues such as accidents, lane closures or public transport lines out of order.
- Transport-related complain: this category includes tweets of users complaining about transport services or driving behaviours of other road users.
- Transport-related advice/question: tweets advising/asking users regarding a place visit, a route choice, travel speed or public transport lines/schedule.
- Unrelated to transport: this category of tweets includes information irrelevant to transport. In some cases transport related words are used metaphorically.

3.4 Twitter Data Pre-processing

The data pre-processing was implemented for translation of text data into machine-readable format. This procedure includes usual text mining techniques, namely:

- Text stemming reduces all words to the base form, usually by direct exclusion of word's root. As the research corpuses are multilingual, we applied different stemming algorithms for 4 languages: Porter's algorithm for English, Kreslin's algorithm for Latvian, Dolamic's algorithm for Russian, and Ntais's algorithm for Greek.
- Removal of stop words (language-specific), punctuation, numbers and other symbols, uninformative in the context of this research (links, emoji, etc.).
- Document-term matrix representation of the corpuses. We applied term frequency-based matrix, considering tf-idf (term frequency—inverse document frequency) as

an alternative. The resulting size of the document-term matrix was reduced by removing sparse terms (with sparsity less than 95%).

3.5 Classification

Automated classification of tweets to transport-related classes is one of the most important practical challenges [1, 26]. We applied three different classification techniques to ensure that our results are not classifier-specific:

- Naive Bayes classifier (unigram),
- Decision tree (recursive partitioning),
- Artificial neural network (feed-forward network with one hidden layer).

Classifiers were separately examined for 4-class and 2-class cases.

Evaluation of classifiers was implemented with classical Accuracy (ACC) and Cohen's kappa metrics. All metrics of each classifier was cross-validated for 10 resampling iterations with a training sample size of 75% of the labelled tweets.

4 Empirical Results

4.1 Data Description

In the current research, tweets of three cities: Minneapolis-Saint Paul twin cities (USA), Riga (Latvia), and Volos (Greece) were collected during the period 22nd May–1st June 2018. Selection of cities in the research sample corresponds to the main question of the research—opportunity to utilise methodologies, successfully applied in monolingual English-speaking agglomerations, for analysis of social media data in smaller English-speaking (Minneapolis-Saint Paul), non-English speaking (Volos) and multilingual (Riga) environments. Further selection is explained by the authors' language proficiency.

The first study area is Minneapolis/Saint Paul. With 3,600,618 residents (Table 1), Minneapolis/Saint Paul is a unique geographic region in east central Minnesota with two urban central business districts. The region has a multi-modal network which includes a Metro Transit bus fleet with over 100 routes, two Light Rail lines, Commuter and Intercity Rail service, Bus Rapid Transit (BRT) on city arterials, bicycle trails and car share services [27]. According to Texas A&M University Transportation Institute the examined region is among the least congested major metropolitan areas in the U.S [28].

Riga is the second study area, in which the population amounts up to 641,423. According to Riga's municipality (2015) the road network consists of 1778 streets with total length 1173 km. The public transport system includes trams, trolleybuses and buses. The particular problems of Riga's transportation system are the narrow street network in the urban core and the crossing of the river Daugava, leading to a congested city core due to dense traffic.

The third study area Volos is located in the centre of Greece and is the sixth largest city of the country with 144 thousand inhabitants. Roads in the city centre are

Table 1. Population and Twitter statistics of research areas.

Area	Country	Population	Population density	Collected tweets	Tweets per week per thousand inhabitants
Minneapolis–Saint Paul (MSP)	USA	3,600,618[a]	214*	372,077	120
Riga	Latvia	641,423[b]	2000**	35,206	55
Volos	Greece	144,449[c]	370***	6851	47

[a] Data source: US Census Bureau (2017);
[b] Data source: Central Statistical Bureau of Latvia (2017);
[c] Data source: Hellenic Statistical Authority (2011).

organised as a grid and in the top season serve intensive vehicle traffic. The lack of parking areas and the illegal parking of private or UFT vehicles are the main reasons that lead to congestions in the city centre.

Filtering by the coordinates, bounding boxes were set in the three study areas obtaining 372,077 tweets in MSP, 35,206 tweets in Riga and 6851 tweets in Volos in a time period from 22nd May to 1st June 2018.

The most used language on MSP's tweets is English (88.7%) and the rest of them are in other languages that are not further analyzed. Regarding Riga's tweets, 36.4% of them are in Latvian, 27% in English and 19.8% in Russian. The majority of tweets collected in Volos city are in Greek language (87.6%) and only 4.5% of them in English (Fig. 2).

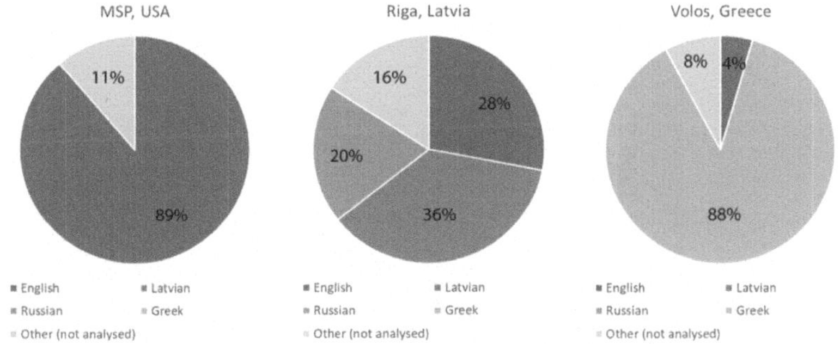

Fig. 2. Twitter language distributions.

In the following table (Table 2) the tweets are categorized based on the language used along with the grade per document. The number of tweets for the analyzed classes based on the manual labelling is also included (Fig. 3).

4.2 Preliminary Grading Results

The keyword-based grading allowed preliminary selection of transport-related tweets and provided initial results on potential of social media as a source of transport-related

Table 2. Twitter corpuses' descriptive statistics.

Corpus	Number of documents	Number of terms	Grade per document	Labelled documents				
				Complain	General information	Real-time information	Unrelated	Total
MSP (en)	330,082	428,478	0.995	87	379	60	474	1000
Riga (en)	9846	25,795	1.096	9	255	4	232	500
Riga (lv)	12,814	42,361	0.615	31	167	0	382	580
Riga (ru)	6963	19,138	0.341	16	37	3	444	500
Volos (el)	6000	21,607	0.460	5	89	2	404	500
Volos (en)	306	1580	0.605	0	2	0	304	306

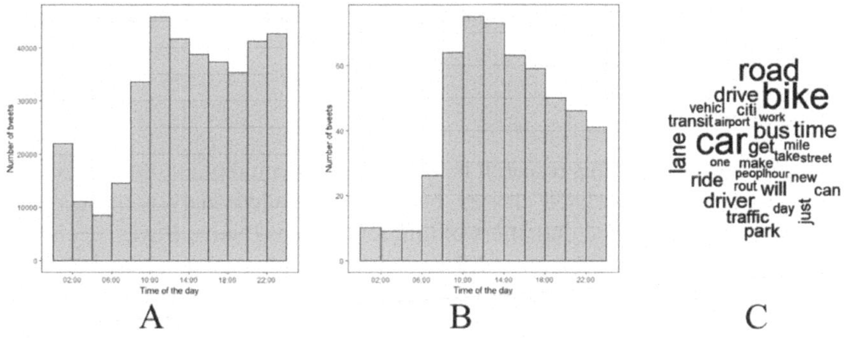

Fig. 3. Daily patterns of Twitter activity. A—total number of tweets; B—number of transport-related tweets; C—word cloud of transport-related tweets.

data. The keyword lists are developed to be very similar for all languages, so we expect the grading results to be comparable.

First, we identified a percentage of graded tweets (with at least one transport-related word) for different research areas. This percentage reached 15% for Volos, 22% for Riga, and 27% for MSP (all differences are statistically significant). There are two area-specific characteristics, analysed in this research, that can be used to explain this difference—Twitter popularity (intensity on tweets is higher for Riga and MSP, Table 1) and language distribution (Fig. 2). To test the hypothesis about language specifics we considered intensity of graded tweets for Volos and Riga for different languages (average grade per document values are presented in Table 2). Values support the hypothesis about higher potential of tweets in English as a source of transport-related information: for Riga the average grade per document for tweets in English is higher (1.096) than for tweets in Latvian and Russian (0.615 and 0.341) and comparable with this value for tweets in English in MSP (0.995). For Volos the results are similar—average grade for tweets in English is higher (0.605) than in Greek (0.460). Obtained average grades are lower for Volos, which leads us to a preliminary conclusion of lower Twitter potential for areas with smaller population.

4.3 Transport-Related Domains

The collected tweets of each city were manually classified into the four classes (complain, general information, real-time information, unrelated to transport) that were described in Sect. 3.3. The class transport-related advice/question was merged to general information due to the limited number of related tweets.

It is worth noting that more than the half of the MSP's (52.6%) and Riga's (en-53.6%) tweets that were manually analysed are transport related. The higher average grades shown in Sect. 4.2 are matching these high percentages of transport related tweets. However, low percentages of the analysed tweets of Volos (0.65% en & 19.2% gr) and Riga (ru-11.2%) are related to transport, matching again the lower average grades shown above. The limited number of collected tweets of Volos city as well as the low average grades are responsible for the high number of unrelated to transport tweets, concluding again that the potential of Twitter data use is higher to areas with larger population such as MSP.

4.4 Classification

The classification stage of this research is focused on potential of automated association of a tweet with transport-related classes, but not on detailed analysis of classification techniques. Thus, we applied classifiers of different groups (Naive Bayes, decision tree, and artificial neural networks) for different number of classes (2 and 4 classes). Application of different classification techniques allows making the conclusions more reliable and technique-independent. Classification accuracy is presented in Table 3.

Table 3. Classification accuracy statistics.

Corpus	Classifier	Binomial classification (2 classes)		Multinomial classification (4 classes)	
		Accuracy	Kappa	Accuracy	Kappa
MSP (en)	Naïve Bayes	0.622	0.257	0.626	0.265
	Decision tree	0.613	0.248	0.609	0.238
	ANN	0.653	0.309	0.658	0.320
Riga (en)	Naïve Bayes	0.806	0.617	0.750	0.523
	Decision tree	0.792	0.590	0.780	0.576
	ANN	0.790	0.585	0.806	0.624
Riga (lv)	Naïve Bayes	0.729	0.338	0.698	0.158
	Decision tree	0.730	0.300	0.736	0.308
	ANN	0.745	0.372	0.723	0.316
Riga (ru)	Naïve Bayes	0.886	0.000	0.888	0.000
	Decision tree	0.878	0.106	0.884	0.031
	ANN	0.886	0.262	0.890	0.184
Volos (el)	Naïve Bayes	0.808	0.000	0.812	0.000
	Decision tree	0.798	0.115	0.802	0.137
	ANN	0.808	0.020	0.800	0.137

First, we note a general consistency of results for different classifiers, so we pretend that the results are not classifier-specific. Second, accuracies are higher for 2-class classification, than for 4-class. This fact is expected and matches findings main by other authors—separating of related and unrelated tweets is easier, than separating different classes within transport-related tweets (general information, real-time information and complains). Third, overall classification accuracy is low (the average Cohan's kappa is 0.275), so the potential of automated classification of tweets is very limited. The only corpus with significantly better results includes English tweets for Riga (the average Cohan's kappa is 0.597), and this fact supports our hypothesis about higher potential of tweets in English (even as a foreign language) as a source of transport-related information.

5 Discussion-Conclusions

The explosive growth of social media use and the amount of the publicly shared information has led to huge volumes of available data. In this study, tweets in three cities of different population, language and transport infrastructure were examined. The city of MSP produced higher Twitter activity compared to Riga and Volos in the examined period, even though the number of geo-referenced tweets in the three cities was limited.

The collected tweets were preliminary graded based on a keyword list and potential transport-related tweets were identified. According to Fig. 3 which illustrates the daily patterns of Twitter activity, between 00:00–06:00 the number of transport-related tweets is low. MSP reached the highest percentage (27%) of graded tweets (with at least one transport-related word) compared to Riga (22%) and Volos (15%). The popularity of Twitter in Riga and MSP as well as the language distribution (Fig. 2) can explain this difference. The average grade per document for English tweets of the three cities is higher than for tweets in other used languages, supporting the hypothesis about higher potential of tweets in English as a source of transport-related information.

The most frequent domain classes of messages (general transport-related information, real-time information, complain, advice/question, unrelated to transport) were identified. The class transport-related advice/question was merged to general information due to the limited number of related tweets. A training sample was prepared for classification, which includes tweets labeled by experts as related to one of identified classes. The majority of MSP's and Riga's (en) tweets that were manually analysed are transport related, while the percentages of transport related tweets of Volos (el & en) and Riga (ru) are low. In general, the overall intensity of transport-related tweets in the three samples is low and the most useful transport-related tweets are shared by official bodies (Total Traffic & Weather Network) and automated volunteered sources (Waze, etc.). The limited number of transport related tweets in the city of Volos and the low average grades prove that the utility of retrieved information depends on the intensity of social media use in the study area as well as on the tweets' language. This observation leads to the conclusion that areas with smaller population have a lower Twitter potential.

Three different classification techniques were used to the research sample in order to ensure that the results are not classifier-specific and were separately examined for the aforementioned 4-class cases and 2-class cases (transport related and transport unrelated). Analytically, the results of classification were explored in terms of classification precision and specific attributes of discovered classes. A general consistency of results was noted for the three classifiers, showing that the results are not classifier-specific, while accuracies are higher for 2-class classification than for 4-class. This finding matches the research results of other authors that a classification of related and unrelated tweets is easier compared to a classification of transport related tweets into different classes. The overall classification is relatively weak but significant (Table 1), concluding that the potential of automated classification of tweets for transport-related information is very limited. Possible reasons of weak classification results could be related to short length of tweets, ambiguity of transport-related terms, and a lower popularity of Twitter as an information channel in the research environments (comparatively to large US and UK cities).

Although this research is based on collected qualitative data of social media, we consider that Twitter data can only be used to enrich datasets of quantitative data sources. Future work includes the comparison of data from social media with a classical traffic data source (loop detectors) and investigation of how data from Twitter can supplement other datasets.

Authors' Contribution All authors conceived of the presented idea and contributed to the final manuscript. D.P. collected and executed an empirical analysis of social media data, including data pre-processing and classification. M.K. reviewed the existing scientific and research developments, labelling manually the tweets, analysed the social media data. E.N supervised the findings of this work.

Acknowledgements. The first author was financially supported by the specific support objective activity 1.1.1.2. "Post-doctoral Research Aid" (Project id. N. 1.1.1.2/16/I/001) of the Republic of Latvia, funded by the European Regional Development Fund. Dmitry Pavlyuk's research project No. 1.1.1.2/VIAA/1/16/112 "Spatiotemporal urban traffic modelling using big data".

References

1. Kuflik, T., Minkov, E., Nocera, S., Grant-Muller, S., Gal-Tzur, A., Shoor, I.: Automating a framework to extract and analyse transport related social media content: the potential and the challenges. Transp. Res. Part C: Emerg. Technol. **77**, 275–291 (2017)
2. Zheng, X., Chen, W., Wang, P., Shen, D., Chen, S., Wang, X., Zhang, Q., Yang, L.: Big data for social transportation. IEEE Trans. Intell. Transp. Syst. **17**, 620–630 (2016)
3. Kokkinogenis, Z., Filguieras, J., Carvalho, S., Sarmento, L., Rossetti, R.J.F.: Mobility network evaluation in the user perspective: real-time sensing of traffic information in Twitter messages. In: Advances in Artificial Transportation Systems and Simulation, pp. 219–234. Elsevier (2015)
4. Salas, A., Georgakis, P., Petalas, Y.: Incident detection using data from social media. Presented at the 2017 IEEE 20th International Conference on Intelligent Transportation Systems (ITSC), October 2017
5. Rusitschka, S., Curry, E.: Big data in the energy and transport sectors. In: Cavanillas, J.M., Curry, E., Wahlster, W. (eds.) New Horizons for a Data-Driven Economy, pp. 225–244. Springer International Publishing, Cham (2016)
6. van Oort, N., Brands, T., de Romph, E.: Short-term prediction of ridership on public transport with smart card data. Transp. Res. Rec.: J. Transp. Res. Board **2535**, 105–111 (2015)
7. Steiger, E., Resch, B., de Albuquerque, J.P., Zipf, A.: Mining and correlating traffic events from human sensor observations with official transport data using self-organizing-maps. Transp. Res. Part C: Emerg. Technol. **73**, 91–104 (2016)
8. Wang, S., Zhang, X., Cao, J., He, L., Stenneth, L., Yu, P.S., Li, Z., Huang, Z.: Computing urban traffic congestions by incorporating sparse GPS probe data and social media data. ACM Trans. Inf. Syst. **35**, 1–30 (2017)
9. Zhang, Z., He, Q., Gao, J., Ni, M.: A deep learning approach for detecting traffic accidents from social media data. Transp. Res. Part C: Emerg. Technol. **86**, 580–596 (2018)
10. Pearanalytics: Twitter Study. San Antonio, USA (2010)

11. Sakaki, T., Okazaki, M., Matsuo, Y.: Earthquake shakes Twitter users: real-time event detection by social sensors. Presented at the (2010)
12. D'Andrea, E., Ducange, P., Lazzerini, B., Marcelloni, F.: Real-time detection of traffic from Twitter stream analysis. IEEE Trans. Intell. Transp. Syst. **16**, 2269–2283 (2015)
13. Gutiérrez, C., Figuerias, P., Oliveira, P., Costa, R., Jardim-Goncalves, R.: Twitter mining for traffic events detection. In: 2015 Science and Information Conference (SAI), pp. 371–378 (2015)
14. Wanichayapong, N., Pruthipunyaskul, W., Pattara-Atikom, W., Chaovalit, P.: Social-based traffic information extraction and classification. In: 2011 11th International Conference on ITS Telecommunications, pp. 107–112 (2011)
15. Gal-Tzur, A., Grant-Muller, S.M., Kuflik, T., Minkov, E., Nocera, S., Shoor, I.: The potential of social media in delivering transport policy goals. Transp. Policy **32**, 115–123 (2014)
16. Gu, Y., Qian, Z.S., Chen, F.: From Twitter to detector: real-time traffic incident detection using social media data. Transp. Res. Part C: Emerg. Technol. **67**, 321–342 (2016)
17. Mai, E., Hranac, R.: Twitter interactions as a data source for transportation incidents. In: TRB 92nd Annual Meeting Compendium of Papers, p. 11. Washington DC, United States (2013)
18. Cottrill, C.D., Derrible, S.: Leveraging big data for the development of transport sustainability indicators. J. Urban Technol. **22**, 45–64 (2015)
19. Collins, C., Hasan, S., Ukkusuri, S.: A novel transit rider satisfaction metric: rider sentiments measured from online social media data. J. Public Transp. **16**, 21–45 (2013)
20. Zhang, Z., Ni, M., He, Q., Gao, J.: Mining Transportation Information from Social Media for Planned and Unplanned Events. Final Report. Transportation Informatics Tier I University Transportation Center, New York (2016)
21. Pathak, A., Patra, B.K., Chakraborty, A., Agarwal, A.: A city traffic dashboard using social network data. Presented at the (2015)
22. Krumm, J., Caruana, R., Counts, S.: Learning Likely Locations. In: Carberry, S., Weibelzahl, S., Micarelli, A., Semeraro, G. (eds.) User Modeling, Adaptation, and Personalization, pp. 64–76. Springer, Heidelberg (2013)
23. Hawelka, B., Sitko, I., Beinat, E., Sobolevsky, S., Kazakopoulos, P., Ratti, C.: Geo-located Twitter as proxy for global mobility patterns. Cartogr. Geogr. Inf. Sci. **41**, 260–271 (2014)
24. Jurdak, R., Zhao, K., Liu, J., AbouJaoude, M., Cameron, M., Newth, D.: Understanding human mobility from twitter. PLoS ONE **10**, e0131469 (2015)
25. Lenormand, M., Tugores, A., Colet, P., Ramasco, J.J.: Tweets on the road. PLoS ONE **9**, e105407 (2014)
26. Khan, S.M., Ngo, L.B., Morris, E.A., Dey, K., Zhou, Y.: Social media data in transportation. In: Chowdhury, M.A., Apon, A. (eds.) Data analytics for intelligent transportation systems, pp. 265–282. Elsevier, Amsterdam (2017)
27. U.S. Department of Transportation: Beyond Traffic: the Smart City Challenge. Minneapolis, USA (2016)
28. Texas A&M University Transportation Institute: Annual Urban Mobility Report (2014)

Simulation Model of Check-in Management Regarding the IATA Level of Service Standards

Artur Kierzkowski, Tomasz Kisiel(✉), and Maria Pawlak

Transportation and Hydraulic Systems, Faculty of Mechanical Engineering, Wroclaw University of Science and Technology, W. Wyspianskiego 27, 50-370 Wroclaw, Poland
{artur.kierzkowski, tomasz.kisiel, maria.pawlak}@pwr.edu.pl

Abstract. The article presents a simulation model serving for the planning of resources (number of desks) that are necessary for the execution of the check-in process at the airport terminal. The elaborated model may be applied at any airport using conventional desks for the check-in process. We present a stochastic model that presents the degree of schedule adjustment for the fulfillment of the level of service assumptions according to the IATA standards as output data. A combination of quantitative and qualitative methods in the check-in process modeling, taking into consideration a stochastic character of the process, has not been fully considered yet. A simulation model uses the Monte-Carlo method, basing on the universal characteristics of passenger reports to the check-in with a division of the traditional, low-cost, and charter passengers. The simulation model obtains information on the degree of schedule adjustment for check-in desks to the C level of the service category according to the IATA standards. The application of the simulation model may be applied in the management of the check-in process in order to avoid an overestimation or underestimation of the service work desks schedules.

Keywords: Airport · Check-in · Simulation model

1 Introduction

The airport as a punctual element of the critical infrastructure of the air transportation system fulfills a crucial role in terms of the reliable functioning of the air transportation process. The tasks related to the exploitation of critical infrastructure are mainly intended to prevent and minimize interferences in the functioning of the system. The airport is a place on the premises of which the main subsystems of ramp handling of the aircrafts are located as well as the check-ins of passengers arriving and departing from a given airport. There is a series of measures enabling the assessment of the functioning of logistics and transportation systems that are presented in each domain of transportation [1]. These measures mostly include operational indicators (e.g. performance of the system) [2–4]. Globally, the issues related to the reliability of transportation

I. Kabashkin et al. (Eds.): RelStat 2018, LNNS 68, pp. 150–158, 2019.
https://doi.org/10.1007/978-3-030-12450-2_14

systems have been addressed many times [5–8]. Models of evaluation of the functioning of complex exploitation transportation systems were presented in [9–13].

Taking into consideration the purpose of an airport, one of the most important measures is a timely execution of the aircraft handling processes. Therefore, it is crucial to ensure that the turnaround of the aircraft falls within the assumed time interval. An appropriate tactical management of flight timetables and execution of the resources of processes should also enable the minimization of results of the original delays of the aircrafts at a given airport.

The performance of the systems is currently very often related to information concerning the passengers' level of service. Assuming some time ranges, in which the passenger should be handled in order to minimize the delay, it is also possible to control the system in order to ensure the highest possible comfort for the passenger. This article presents a practical simulation model that enables the choice of resources for the execution of the check-in process at the airport terminal in order to ensure an appropriate level of service. We present a quantitative and qualitative approach that enables the classification of adjustment of the schedule to the optimal class of the level of service, and through quantitative indicators, to subsequently present the level of overestimation or underestimation of the system.

In Sect. 2, we present an overview of the methods applied in check-in modeling as well as the service quality indicators. In Sect. 3, we describe the structure of the model whose exemplary validation is presented in Sect. 4. At the end (in Sect. 5), we present the discussion and conclusion.

2 Modeling of the Check-in Process Regarding the Level of Service

The issues of the dynamic management of check-in processes at airports have been addressed many times. It was proposed in a study [14] to manage the check-in system in such a way as to balance the operating expenses and the waiting time in line.

A knowledge-based simulation system to predict resource requirements at an international airport used by a check-in desk allocation system was proposed in a study [15]. The authors take into consideration the minimization of the use of resources with the fulfillment of conditions determined by the level of service at the same time.

The same assumptions were also adopted in a study [16]. It takes into account the level of service with the minimization of the use of resources at the same time. The authors indicated that the process is easy to manage if the desk capacity is constant. The authors presented two exemplary alternative fixed plans with variable capacity.

The analyses presented in the references indicate that an appropriate optimization level may be achieved by a choice of an appropriate time interval between subsequent changes in schedule. Thirty-minute long intervals applied in a study [17] enabled the authors to achieve higher profit than in the case of another study [16]. In a different study [16], one-hour long intervals were assumed. Another study [15] offers a possibility to adjust the time intervals between the subsequent ones, starting from 15-min long intervals.

The above-mentioned papers make reference to the level of the service concept and comprise analyses that have been conducted in such a manner as to meet the assumed requirements. However, there are no papers that make reference directly to standards for the level of service that have been standardized by the IATA. The simulation model presented in the further part of the present paper has been elaborated for this purpose.

3 Structure of the Simulation Model

Each passenger departing from a given airport must be subjected to a passenger check-in process. One of the sub-processes of a passenger check-in is the ticket and baggage check. The ticket and baggage check may be conducted with the use of an IT system (departure control system—DCS) or manually. The manual check-in is usually used as an alternative version in case of DCS system damage. The ticket and baggage check with the use of DCS is conducted in a traditional way at check-in desks or through alternative methods enabling the execution of the process by the passengers all by themselves (self service). The passengers using self-service methods may register their baggage at specially dedicated points (baggage drop off). In reality, we can also find mixed methods in which the passengers have a possibility to use self-service methods, but it is not obligatory and they can check-in at the airport terminal.

The ticket and baggage check-in, at check-in desks, is conducted with the use of different strategies. A common system for one's own air transport operations means that, for a given carrier, there is an appropriate number of desks for the check-in carried out for all the flights of such carrier. Due to an unfavorable effect of mixing passenger streams with different time restraints (different departure time of the aircraft) in appropriate time intervals, dedicated desks for a flight in question may be appointed. Only the passengers of the indicated flight may be checked-in at a given desk. This strategy enables you to minimize the probability of a delay of the aircraft check-in or delays of passengers for a given flight. The dedicated method is primarily used for charter flights in which a high intensity of passengers' reports with a large amount of baggage occurs.

The simulation model assumes a probability to carry out analyses in a dedicated or common system. For this purpose, the software user must enter the flight number, number of passengers, and desks' numbers at which the passenger may be handled as input data. Through an indication of the same desks for different flight numbers, the check-in will be conducted in the common system for all flights indicated for them. For flights numbers to which exclusive desks have been assigned within a given time interval, the check-in is conducted in the dedicated system.

Based on the conducted analyses in the actual system, it has been observed that the reports stream varies considerably depending on the nature of transport that is carried out. The probability density functions (PDF) for the reports of passengers to the system have been indicated on this basis. It is possible to assign flight numbers to low-cost, traditional, and charter groups.

The moment of reporting the passenger for the low-cost carrier (t_{rep}^{lc}) is determined on this basis (1). The moment of reporting the passenger for the traditional carrier (t_{rep}^{tr})

is determined on this basis (2). The moment of reporting the passenger for the charter carrier (t^{ch}_{rep}) is determined on this basis (3). On a similar basis, the PDF for the times of passenger services at the desks have been determined. For the low-cost group, the service duration time at the check-in desk is compatible with (4). For the traditional group, the service duration time at the check-in desk is compatible with (5), whereas for the charter group, the service duration time is determined on this basis (6).

A simplification of the actual system has been adopted through an assumption that the passengers report themselves to the system individually and they are handled as a single item.

$$f\left(t^{lc}_{rep}\right) = \frac{4}{104.8} \cdot \left(\frac{t^{lc}_{rep}}{104.8}\right)^{3} \cdot exp\left(-\left(\frac{t^{lc}_{rep}}{104.8}\right)^{4}\right), \tag{1}$$

$$f\left(t^{tr}_{rep}\right) = \frac{3.8}{92.9} \cdot \left(\frac{t^{tr}_{rep}}{92.9}\right)^{2.8} \cdot exp\left(-\left(\frac{t^{tr}_{rep}}{92.9}\right)^{3.8}\right), \tag{2}$$

$$f\left(t^{ch}_{rep}\right) = \frac{8}{131.2} \cdot \left(\frac{t^{ch}_{rep}}{131.2}\right)^{7} \cdot exp\left(-\left(\frac{t^{ch}_{rep}}{131.2}\right)^{8}\right), \tag{3}$$

$$f\left(t^{lc}_{ser}\right) = \frac{1.57}{1.3} \cdot \left(\frac{t^{lc}_{ser}}{1.3}\right)^{0.57} \cdot exp\left(-\left(\frac{t^{lc}_{ser}}{1.3}\right)^{1.57}\right), \tag{4}$$

$$f\left(t^{tr}_{ser}\right) = \frac{exp\left(-\frac{1}{2} \cdot \left(\frac{ln\left(t^{tr}_{ser}\right) - 1.18}{0.68}\right)^{2}\right)}{t^{tr}_{ser} \cdot 0.68 \cdot \sqrt{2 \cdot \pi}}, \tag{5}$$

$$f\left(t^{ch}_{ser}\right) = \frac{exp\left(-\frac{1}{2} \cdot \left(\frac{ln\left(t^{ch}_{ser}\right) - 1.64}{0.39}\right)^{2}\right)}{t^{ch}_{ser} \cdot 0.39 \cdot \sqrt{2 \cdot \pi}}. \tag{6}$$

The simulation model functions on the basis of a basic queuing model in which any number (assumed by the user) of service channels occurs (see Fig. 1).

Each desk is equipped with its own queue, and the passengers choose the entry to the queue from a set of check-in desks dedicated to their flight number according to the lowest number of waiting passengers. The passengers are checked-in according to the FIFO (First In First Out) strategy. The execution of the process for a single passenger is presented in Fig. 2.

The simulation experiment is carried out many times on the basis of the Monte-Carlo method that enables you to obtain the value of the expected output data. It has been assumed in the simulation model that it is important to determinate whether the assumed schedule allows for obtaining the C category in the aspect of the level of

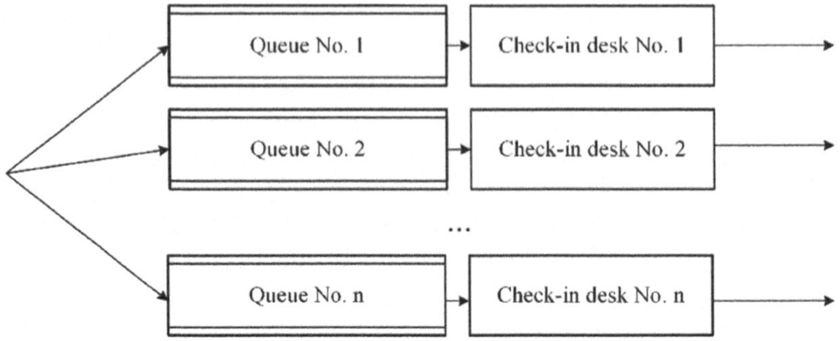

Fig. 1. Structure of the check-in system.

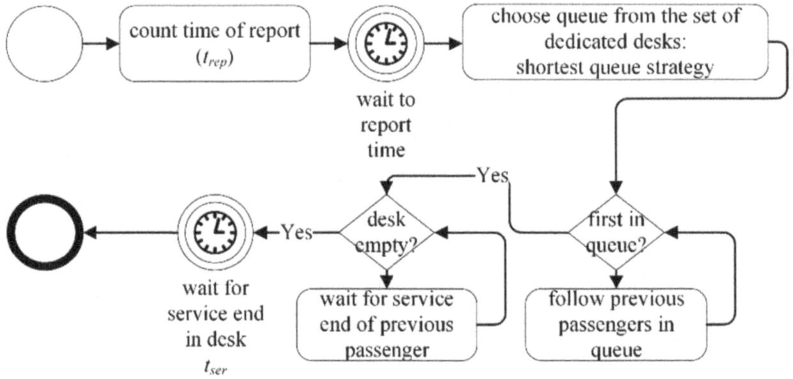

Fig. 2. Algorithm of the execution of a simulation for a single passenger.

service according to the IATA [18] standards. It has been assumed on this basis that the system may be in three states during the execution of the service process. The state in which the system finds itself is determined on the basis of the actual time t_q [min] of waiting of the passengers in queue according to (7).

$$S = \begin{cases} S_{overestimated}\ for t_q < 10 \\ S_{optimal}\ for t_q \in \langle 10.20\rangle \\ S_{suboptimal}\ for t_q > 20 \end{cases}. \tag{7}$$

On the basis of formula (7), it is possible to elaborate the output data in the form of:

- percentage of occurrence of particular states,
- occurrence of particular states in the function of time (with an indication of the amount of assigned resources, the time of process execution).

The model enables you to carry out the analysis on three different scales:

- Macroscopic—indicators presented for the entire stream of handled passengers,

- Mesoscopic—indicators presented with a division of the nature of the transportation or of the chosen carrier,
- Microscopic—indicators presented with a division according to the flight number.

4 Validation and Verification of the Model

A validation and verification of the proposed method have been carried out on an actual system. Below, we present an exemplary analysis for a given day executed at the Wrocław Airport. The analysis is presented in mesoscopic scale for a given carrier. The performed simulation gave the results shown in Figs. 3 and 4.

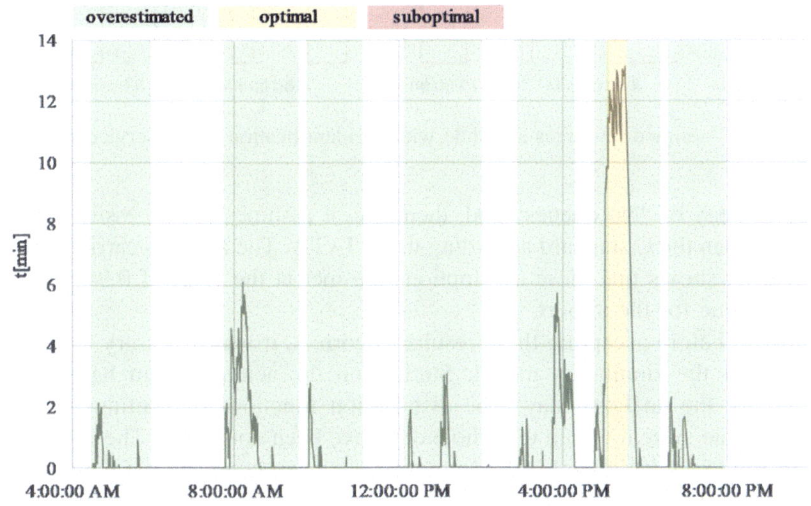

Fig. 3. Expected time in queue with an indication of service states.

Figure 3 shows an expected value of the time in queue to the check-in desks of a chosen carrier from the low-cost group. The areas of the system remaining in a given state are also indicated on the chart. A similar comparison is presented in Fig. 4, but the areas of the system remaining in a given state was compared to the number of desks dedicated for service provided by the chosen carrier. The simulation model also determined the number values of probability of remaining in particular states. For the obtained simulation, they are in accordance with (8)

$$P_S = \begin{cases} P_{S_{overestimated}} = 0.94 \\ P_{S_{optimal}} = 0.06 \\ P_{S_{suboptimal}} = 0.00 \end{cases} . \quad (8)$$

It can be concluded on the basis of the obtained results that the airport correctly adjusted the desks functioning schedule. The Wrocław Airport implements the

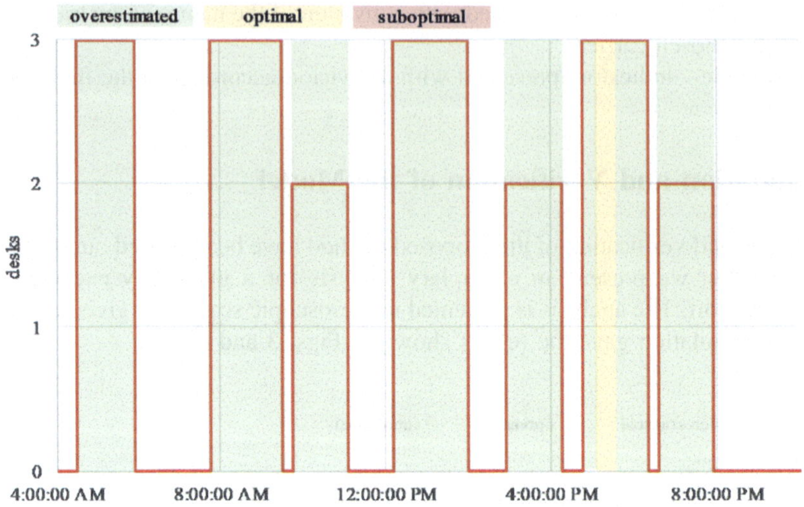

Fig. 4. Assumed resources schedule with an identification of the service states.

assumed strategy of "No queues" and, therefore, it assumes the necessity to ensure a higher class than the C standard according to the IATA. The analysis carried out for the chosen carrier shows that these assumptions are met at the level of 0.94 which is an acceptable value for the airport.

However, before accepting these results as correct, it was necessary to carry out a verification of the simulation model. Studies on the actual system have also been conducted for the analyzed day. The distribution functions of waiting times in the passenger queue system within the whole day have been compared. The comparison is shown in Fig. 5. A Kolmogorov-Smirnov test has been conducted at the significance level of 0.05. The obtained statistic value was lower than the critical value and, therefore, it could be concluded that the model generates correct results.

5 Summary

The present paper presents a model whose purpose is to support the scheduling processes of the technical resources for the execution of the check-in processes at an airport terminal. The main purpose of the model is to enable the management of the process in order to ensure the required level of service class according to the IATA standards.

The model may be applied for systems using conventional check-in desks. This method is applied in most European airports as the main method or as an auxiliary method for persons who do not want to use the self-service methods.

The proposed simulation model assumed the starting of the ticket and baggage check-in within a time frame of 40–120 min before the planned departure. The further development stage will consist of enabling the execution of any configuration of the

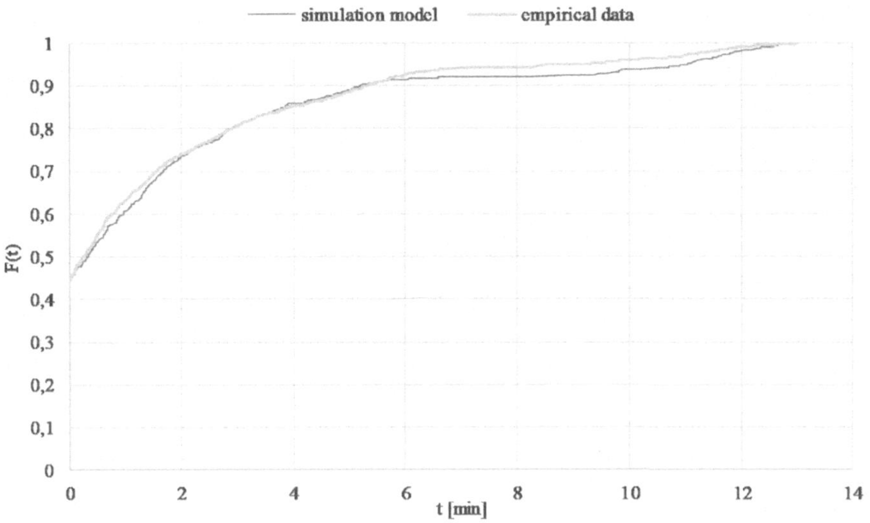

Fig. 5. CDF functions for the queue waiting times obtained from the actual system and the simulation model.

system by the user, thanks to which the model will be more universal and will be more suitable for a larger group of recipients. Another application may be a verification of the service quality class of the airport according to the IATA standards.

References

1. Walkowiak, T., Mazurkiewicz, J.: Analysis of critical situations in discrete transport systems. In: Proceedings of DepCoS—RELCOMEX 2009, Brunów, Poland, 30 June–02 July, pp. 364–371. IEEE (2009). https://doi.org/10.1109/DepCoS-RELCOMEX.2009.39
2. Nowakowski, T., Tubis, A., Werbińska-Wojciechowska, S.: Vulnerability assessment of passenger transport system—case study, Risk, reliability and safety: innovating theory and practice. In: Walls, L., Revie, M., Bedford, T. (eds.) proceedings of the 26th European Safety and Reliability Conference, ESREL 2016, Glasgow, Scotland, 25–29 September 2016, pp. 2254–2259. CRC Press/Balkema, Boca Raton (2016)
3. Milewski, R., Smal, T.: The use of multidimensional comparative analysis in the choice of means of transport for the army. In: Safety and Reliability of Complex Engineered Systems —Proceedings of the 25th European Safety and Reliability Conference, ESREL 2015, pp. 87–92 (2015)
4. Mazurkiewicz, J., Walkowiak, T.: Discrete transportation systems quality performance analysis by critical states detection. In: Zamojski, W., Mazurkiewicz, J., Sugier, J., Walkowiak, T., Kacprzyk, J. (eds.) New Results in Dependability and Computer Systems. Advances in Intelligent Systems and Computing, vol. 224, pp. 307–314. Springer, Heidelberg (2013). https://doi.org/10.1007/978-3-319-00945-2_27
5. Ryczyński, J., Smal, T.: Proposition of a model for risk assessment in the transport of the oversized loads in the army. In Proceedings of the International Conference on Military Technologies (ICMT), 31 May–2 June 2017, Brno, Czech Republic, pp. 166–170 (2017)

6. Restel, F., Wolniewicz, Ł.: Tramway reliability and safety influencing factors. Proc. Eng. **187**, 477–482 (2017). https://doi.org/10.1016/j.proeng.2017.04.403
7. Friedrich, J., Restel, F., Wolniewicz, Ł.: Railway operation schedule evaluation with respect to the system robustness. In: Zamojski, W., Mazurkiewicz, J., Sugier, J., Walkowiak, T., Kacprzyk, J. (eds.) Contemporary Complex Systems and Their Dependability. DepCoS-RELCOMEX 2018. Advances in Intelligent Systems and Computing, vol. 761, pp. 195–208. Springer, Cham (2019). https://doi.org/10.1007/978-3-319-91446-6_19
8. Vintr, Z., Valis, D.: A Tool for Decision making in k-out-of-n system maintenance. mech. Aerosp. Eng. PTS **1–7**, 110–116 (2012). https://doi.org/10.4028/www.scientific.net/AMM.110-116.5257
9. Tubis, A., Werbińska-Wojciechowska, S.: The scope of the collected data for a holistic risk assessment performance in the road freight transport companies. In: Zamojski, W., Mazurkiewicz, J., Sugier, J., Walkowiak, T., Kacprzyk, J. (eds.) Advances in Dependability Engineering of Complex Systems. DepCoS-RELCOMEX 2017. Advances in Intelligent Systems and Computing, vol. 582, pp. 450–463. Springer, Cham (2018)
10. Restel, F.J., Tubis, A., Wolniewicz, Ł.: Investigation of small-consequence undesirable events in terms of railway risk assessment. In: Zamojski, W., Mazurkiewicz, J., Sugier, J., Walkowiak, T., Kacprzyk, J. (eds.) Contemporary Complex Systems and Their Dependability. DepCoS-RELCOMEX 2018. Advances in Intelligent Systems and Computing, vol. 761, pp. 384–399. Springer, Cham (2019). https://doi.org/10.1007/978-3-319-91446-6_36
11. Milewski, R., Smal, T.: Decision making scenarios in military transport processes in refers to military security. Part 1—approach to logistics and transport costs. In: Proceedings of 21st International Scientific Conference. Transport Means 2017. Part II. Juodkrante, Lithuania, September 20–22, pp. 356–361 (2017)
12. Milewski, R., Smal, T.: Decision making scenarios in military transport processes in refers to military security. Part 2—approach to decision making in military transport. In: Proceedings of 21st International Scientific Conference. Transport Means 2017. Part II. Juodkrante, Lithuania, 20–22 September, pp. 362–367 (2017)
13. Kwasniowski, S., Zajac, M., Zajac, P.: Telematic problems of unmanned vehicles positioning at container terminals and warehouses. In: Transport Systems Telematics, pp. 391–399. Springer, Heidelberg (2010)
14. Bruno, G., Genovese, A.: A mathematical model for the optimization of the airport check-n service problem. Electron. Notes Discrete Math. **36**, 703–710 (2010)
15. Chun, H.W., Mak, R.W.T.: Intelligent resource simulation for an airport check-in counter allocation system. IEEE Trans. Syst. Man Cybern. Appl. Rev. **29**(3), 325–335 (1999)
16. Van Dijk, M., van der Sluis, E.: Check-in computation and optimization by simulation and IP in combination. Eur. J. Oper. Res. **171**, 1152–1168 (2006)
17. Araujo, G.E., Repolho, H.M.: Optimizing the airport check-in counter allocation problem. J. Transp. Lit. **9**(4), 15–19 (2015)
18. IATA: Airport Development Reference Manual, 10th edition, 2nd release, Montreal (2014)

Reduction of Dimensionality of Feature Vectors in Subject Classification of Text Documents

Tomasz Walkowiak(✉), Szymon Datko, and Henryk Maciejewski

Faculty of Electronics, Wrocław University of Science and Technology, Janiszewskiego St. 11/17, 50-372 Wrocław, Poland
{tomasz.walkowiak, szymon.datko, henryk.maciejewski}@pwr.edu.pl

Abstract. Within a paper we investigate the influence of dimensionality reduction of feature vector (PCA and random projection) on the results of subject classification of text documents in Polish. Two state of the art methods of text representation have been applied, i.e. bag of words consisting of the most frequent 1000 lemmatized nouns and the emerging fastText based on the word embedding dimensionality equal to 100. The methods have been evaluated on four corpora in Polish (two different data sets, divided into training and testing set in different proportions). Results show that PCA gives better accuracy in all analyzed cases. In case of fastText, a significant reduction of up to 10 times is possible without loss of quality regardless the quality of corpus. To analyse this phenomenon, we have performed a set of experiments in which we train fastText with different values of word embedding dimensionality (from 2 to recommended 100). Experiments show that even the dimension 3–10 (depending on the quality of data) allows to achieve very good accuracy.

Keywords: Text mining · Polish · Subject classification · Dimensionality reduction · PCA · Random projection · Bag of words · FastText

1 Introduction

Subject classification of text documents is a task of automatic assigning a given document into one of defined groups. It is based on a usage of natural language processing (NLP) methods [13] and algorithms of machine learning [6]. The documents are represented by high-dimensional feature vectors which are then used for training classifiers, such as SVM [6] or multilayer perceptron [6], given a collection of documents with known categories (class labels).

Within this paper we investigate the influence of dimensionality reduction of feature vector on the results of subject classification of text documents in Polish [14, 17–19]. Two state of the art methods of text representation are used, i.e. the bag of words [5] consisting of the most frequent 1000 lemmatized nouns [19] and the emerging fastText [17] based on the word embedding technique [10].

The paper is organized as follows. In Sect. 2, we discuss the text representation methods providing the technicalities: bag of words and the fastText method. In Sect. 3

I. Kabashkin et al. (Eds.): RelStat 2018, LNNS 68, pp. 159–167, 2019.
https://doi.org/10.1007/978-3-030-12450-2_15

we describe two methods of dimensionality reduction: principal components analysis (PCA) and random projection. Then, the used in experiments corpora are presented. Finally, in Sect. 5, the performed experiments and achieved results are discussed.

2 Text Representation Methods

2.1 Introduction

Features can refer to different levels of the language analysis or text forms. They should reveal properties of text that are characteristic of the semantic content. The most common vector representation of texts is the bag of words [5] (BoW) discussed by us in the next section. Recently the word embedding [10] methods gained large popularity. They mostly act as a dictionary that maps each word into a continuous multidimensional vector space. Among many word embedding methods, the most popular for text classification with known groups is the fastText algorithm [9] (described in Sect. 2.2). Both methods assume that text can be represented as an unordered collection of words.

2.2 Bag of Words

The bag of words features consist of numbers of word occurrences in each text [5]. BoW features forms a sparse feature matrix due to a fact that not all words appear in each text. A number of modifications of this method are available. The most common includes filtering out the most common and the rarest words in the whole corpora and lemmatization of words (by grouping words under lemmas). The literature commonly suggests weighting the raw counts of word occurrences, e.g. in relation to the total length of each document and uniqueness of a word for a single category [7, 16].

For experiments reported in this paper, we have followed the schema proposed in [19], which was selected among many BoW modifications as the most effective for the subject classification of texts in Polish. It is based on using the most frequent 1000 of lemmatized nouns. To found nouns and their lemmas in text we have used a morphosyntactic tagger designed for Polish language (WCRFT) [15]. The feature generation for BoW was performed using CLARIN-PL infrastructure [13, 20].

As suggested in [18], we have not used any dedicated weighting schema only standard weighting required by classification algorithm was applied (by removing the mean and scaling to a unit variance).

2.3 FastText

The fastText [9] is a recently proposed method that utilizes word embedding approach, i.e. each word in training corpora is mapped to a vector of fixed dimension containing real numbers. The document feature vector is calculated as an averaged of word embedding's from text (so called doc-to-vec [10]). Therefore, fastText (and most of word embedding) features forms a dense feature matrix. The main concept of fastText, in supervised version, compared with other word-to-vec methods, is to build the word embedding map based on the training corpora used for classification and in simultaneous generation of word representation and classifier (linear soft-max [4]) learning process. The learning algorithm and its implementation is highly effective [9]. Following, the default value from fastText and results of work reported in [17], the dimension of word embedding and hence the dimension of feature vectors is equal to 100 in our experiments.

3 Dimensionality Reduction

In this work we compare two methods of dimensionality reduction: principal components analysis (PCA) and random projection (RP). Given a data set X_{nxd} composed of n observations in d-dimensional feature space (1000 for BoW and 100 for fastText), both methods project the observations onto k-dimensional subspace ($k < d$). The methods differ in how the projection matrix R_{dxk} is constructed. In the PCA method, columns of the projection matrix are taken as eigenvectors of the covariance matrix of the features (of the columns of the original matrix X), where the eigenvectors are ordered by descending value of the corresponding eigenvalues. The resulting k-dimensional subspace is orthogonal, with the most variance in data captured along the first axis (principal dimension), the second most along the second axis, etc. The drawback of this method is high computational cost.

Random projection method constructs the projection matrix much simpler. The method is motivated by the Johnson-Lindenstrauss lemma [8] which states that a dataset of n observations can be mapped onto $k \sim O\left(\frac{\log n}{eps}\right)$ dimensional subspace, with distortion in Euclidean distance between the observations not bigger than $(1 \pm \text{eps})$. Interestingly, the minimum required dimensionality k depends (logarithmically) on the number of observations n, but does not depend on the original dimensionality d.

As shown, e.g. in [3], such mapping can be constructed by randomly selecting the subset of original dimensions, which motivates the commonly used procedure to construct the projection matrix R_{dxk} as a random matrix from the Gaussian distribution (GRP). Achlioptas proposed even a simple procedure to construct the projection matrix as a sparse matrix using integer arithmetic's [1], which in numerous evaluation studies e.g. [2] was found as equally successful as the Gaussian procedure.

4 Data Set

To test the influence of dimensionality reduction on the classification results we have used two corpora. The first corpus (Press) [19] consists of Polish press news that were assigned by press agency to five subject categories. All the subject groups are very well separable from each other and contains reasonably large number of members. There are 6564 documents in total, which gives an average of ca. 1300 documents per class.

The second corpus (Wiki) [11, 12] consists of articles extracted from the Polish Wikipedia. It consists of 34 subject categories. It includes 9837 articles in total, which translates into about 300 articles per class.

We have randomly divided each data sets into training and testing sets in two proportions as presented in Table 1 in the third column. Different sizes of training and testing set were selected to check if results of dimensionality reduction depend on the training set size. Next, for each two proposed methods of text representation (Sect. 2) we have trained multilayer perceptron (MLP) using with Broyden—Fletcher—Goldfarb—Shanno (BFGS) nonlinear optimization learning algorithm. The classification results on testing set for all corpora are presented in Table 1 in last two columns. Since the MLP incorporates randomness during the learning process, the presented results are the averages of achieved among 20 runs.

Table 1. Classification results.

Corpus	Division schema	Number of per-class training examples	Accuracy for BoW	Accuracy fastText
Press	A	1042	0.9364	0.9814
	B	51	0.8625	0.8762
Wiki	C	276	0.8219	0.8545
	D	85	0.7600	0.7806

5 Experiments and Results

We have applied principal components analysis (PCA) and Gaussian random projection (GRP) to feature vectors for eight data sets (see Table 1). The results for BoW features and Press corpus with 1042 (A) and 51 (B) examples per class and for different values of k-dimensional subspace are presented in Fig. 1. We present the error defined as the difference between accuracy achieved for no dimension reduction (absolute values are presented in Table 1 in the fourth column) and the accuracy achieved on data after dimension reduction. The results for the BoW features and Wiki corpus with 276 (C) and 85 (D) examples per class and for different values of projected subspace dimension are presented in Fig. 2.

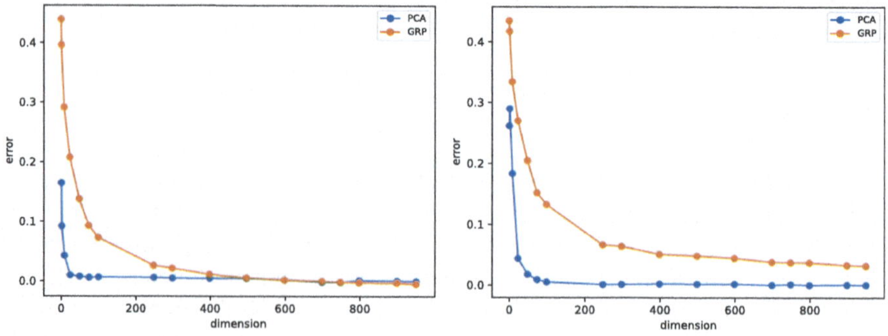

A B

Fig. 1. Results of dimension reduction for BoW and Press corpus for different number of per-class training examples (A 1042, B 51).

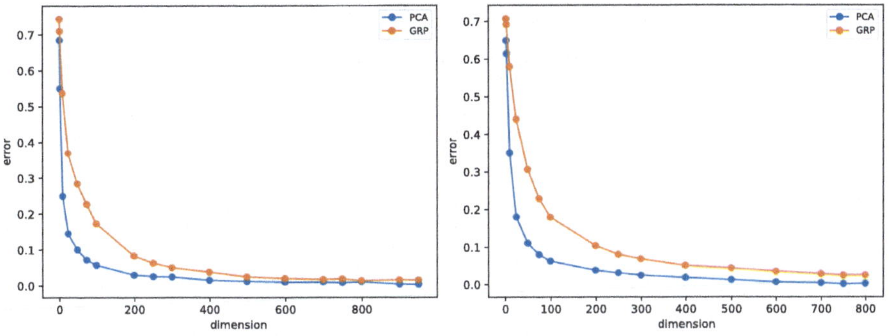

C D

Fig. 2. Results of dimension reduction for BoW and Wiki corpora for different number of per-class training examples (C 276, D 85).

The results for fastText features (with original dimension equal to 100) are presented in Figs. 3 and 4. The classification was done using multilayer perceptron as in case of BoW features, not by a linear soft-max (the classifier built in fastText algorithm).

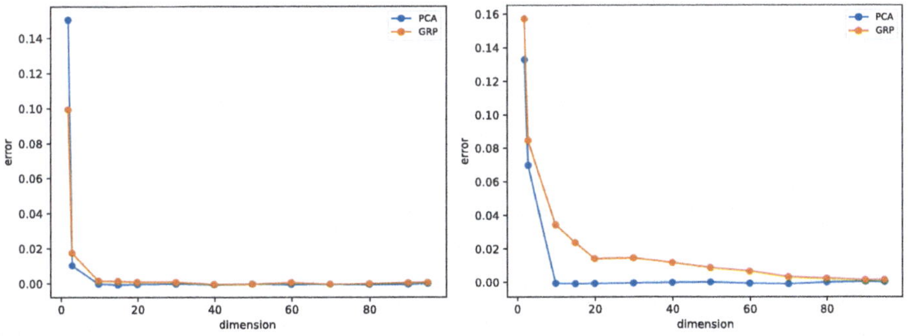

A B

Fig. 3. Results of dimension reduction for fastText features and Press corpora.

We can notice that for almost all analyzed cases PCA gives better accuracy (smaller error). Even for some values of projected subspace dimension, achieved accuracy is slightly better than for original dimensionality. Results show that used features vectors are redundant. In case of BoW features the level of reduction noticeable depends on the quality of data. The largest reduction is possible for data giving the best accuracy (see PCA results in Fig. 1A), it lowers when the quality of data lowers (see PCA results in Fig. 2D data set).

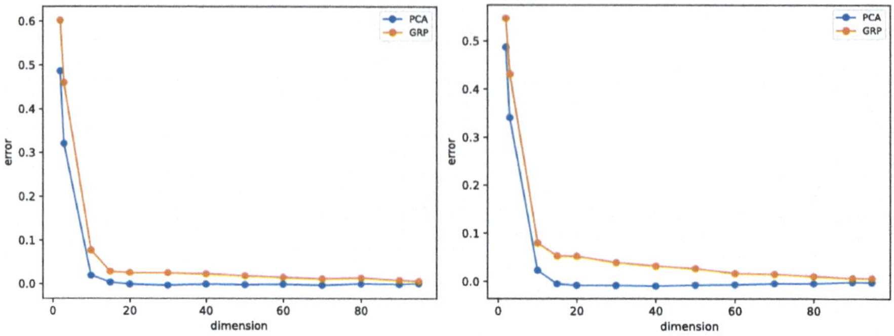

C D

Fig. 4. Results of dimension reduction for fastText and Wiki corpora.

In case of fastText features, a significant reduction of up to 10 times is possible without loss of quality regardless the quality of data. It raises the question why not build fastText models with smaller dimensions than recommended 100?

To answer this question, we have performed a set of experiments in which we train the fastText with different values of word embedding dimensionality (from 2 to 100) and calculate the accuracy on the test set using the multilayer perceptron as a classifier. Results (difference between the achieved accuracy and the accuracy on the default model with 100 dimensionality) are presented in Figs. 5 and 6. Similarly, to PCA results, the quality of dimension reduction depends on quality of data set. For Press corpus we can decrease the dimension to 3–4 with very small decrease of achieved accuracy. Whereas for Wiki corpora the limit is 10 dimensions.

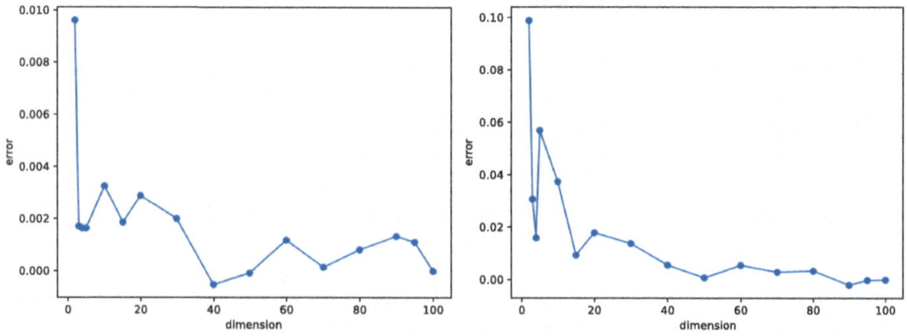

A B

Fig. 5. Change in accuracy (error) for fastText features with different values of word embedding dimensionality for Press corpus.

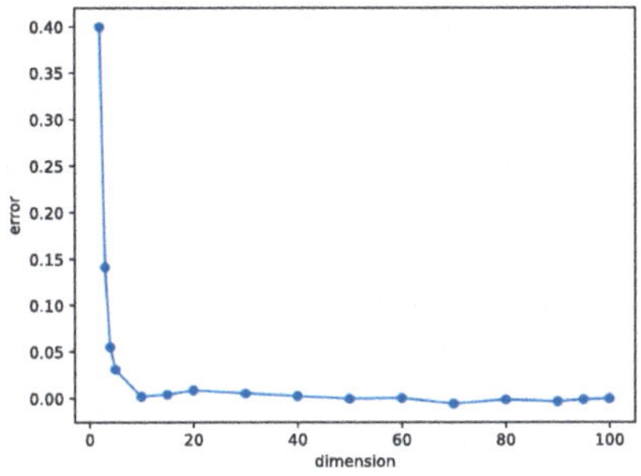

C

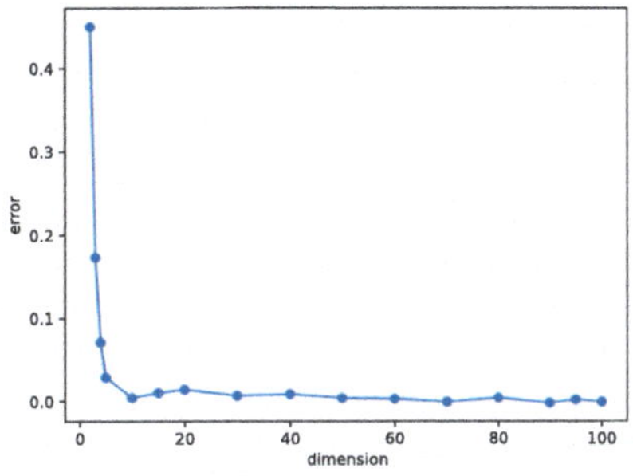

D

Fig. 6. Change in accuracy (error) for fastText features with different values of word embedding dimensionality for Wiki corpus.

Moreover, the results for Press corpus for 3 dimensions are very interesting since they do not differ a lot from the best one and such small dimension allows to use the techniques for data visualisation.

6 Conclusion

This paper presented a comparison of two approaches to reduce the dimensionality of feature vectors based on linear projection of feature matrix on lower dimensional space, i.e. principal component analysis and random projection. The methods have been evaluated on four corpora in Polish (two different data sets, each divided into training and testing set in a different way) using two different methods of feature generation: bag of words and the emerging fastText. Results show that PCA gives better accuracy than random projection in all analyzed cases. In case of fastText, a significant reduction of up to 10 times is possible without a loss of quality regardless the quality of corpus. To analyse this phenomenon, we have performed a set of experiments in which we train fastText with different values of word embedding dimensionality (from 2 to recommended 100). Experiments show that even the dimension 3–10 (depending on the quality of data) is sufficient to achieve very good accuracy.

In future work we plan to investigate the influence of dimensionality reduction in open-set classification.

Acknowledgements. This work was sponsored by National Science Centre, Poland (grant 2016/21/B/ST6/02159).

References

1. Achlioptas, D.: Database-friendly random projections. In: Proceedings of the Twentieth ACM SIGMOD-SIGACT-SIGART Symposium on Principles of Database Systems, pp. 274–281 (2001)

2. Bingham, E., Mannila, H.: Random projection in dimensionality reduction: applications to image and text data. In: Proceedings of the Seventh ACM SIGKDD International Conference on Knowledge Discovery and Data Mining, pp. 245–250 (2001)
3. Dasgupta, S., Gupta, A.: An elementary proof of the Johnson-Lindenstrauss lemma. Int. Comput. Sci. Inst. Tech. Rep. **22**(1), 1–5 (1999)
4. Goodman, J.: Classes for fast maximum entropy training. In: Proceedings of 2001 IEEE International Conference on Acoustics, Speech, and Signal Processing (Cat. No. 01CH37221), vol. 1, pp. 561–564 (2001)
5. Harris, Z.: Distributional structure. Word **10**, 146–162 (1954)
6. Hastie, T.J., Tibshirani, R.J., Friedman, J.H.: The elements of statistical learning: data mining, inference, and prediction. Springer Series in Statistics. Springer, New York (2009). Autres impressions: 2011 (corr.), 2013 (7e corr.)
7. Joachims, T.: A probabilistic analysis of the Rocchio algorithm with TFIDF for text categorization. In: Fisher, D.H. (ed.) Proceedings of ICML-97, 14th International Conference on Machine Learning, pp. 143–151. Morgan Kaufmann Publishers, San Francisco (1997). http://citeseerx.ist.psu.edu/viewdoc/summary?doi=10.1.1.21.7950
8. Johnson, W.B., Lindenstrauss, J.: Extensions of Lipschitz mappings into a Hilbert space. Contemp. Math. **26**, 189–206 (1984)
9. Joulin, A., Grave, E., Bojanowski, P., Mikolov, T.: Bag of tricks for efficient text classification. In: Proceedings of the 15th Conference of the European Chapter of the Association for Computational Linguistics: Volume 2, Short Papers, pp. 427–431. Association for Computational Linguistics (2017). http://aclweb.org/anthology/E17-2068
10. Mikolov, T., Sutskever, I., Chen, K., Corrado, G.S., Dean, J.: Distributed representations of words and phrases and their compositionality. In: Advances in Neural Information Processing Systems, pp. 3111–3119 (2013)
11. Młynarczyk, K., Piasecki, M.: Wiki test—34 categories (2015). http://hdl.handle.net/11321/217. CLARIN-PL digital repository
12. Młynarczyk, K., Piasecki, M.: Wiki train—34 categories (2015). http://hdl.handle.net/11321/222. CLARIN-PL digital repository
13. Piasecki, M.: User-driven language technology infrastructure – the case of CLARIN-PL. In: Proceedings of the Ninth Language Technologies Conference, Ljubljana, Slovenia (2014). http://nl.ijs.si/isjt14/proceedings/isjt2014_01.pdf
14. Piskorski, J., Sydow, M.: Experiments on classification of polish newspaper. Arch. Control Sci. **15**, 613–625 (2005)
15. Radziszewski, A.: A tiered CRF tagger for Polish. In: Intelligent Tools for Building a Scientific Information Platform, Studies in Computational Intelligence, vol. 467, pp. 215–230. Springer (2013)
16. Salton, G., Buckley, C.: Term-weighting approaches in automatic text retrieval. Inf. Process. Manag. **24**(5), 513–523 (1988)
17. Walkowiak T., Datko S., Maciejewski H.: Bag-of-words, bag-of-topics and word-to-vec based subject classification of text documents in Polish - a comparative study. In: DepCoS-RELCOMEX 2018. Advances in Intelligent Systems and Computing, vol. 761. Springer, Cham (2019)

18. Walkowiak, T., Datko, S., Maciejewski, H.: Feature extraction in subject classification of text documents in polish. In: Artificial Intelligence and Soft Computing. Springer International Publishing, Cham (2018)
19. Walkowiak, T., Malak, P.: Polish texts topic classification evaluation. In: Proceedings of the 10th International Conference on Agents and Artificial Intelligence-Volume 2: ICAART, pp. 515–522. INSTICC, SciTePress (2018)
20. Walkowiak, T.: Language processing modelling notation – orchestration of NLP microservices. In: Zamojski, W., Mazurkiewicz, J., Sugier, J., Walkowiak, T., Kacprzyk, J. (eds.) Advances in Dependability Engineering of Complex Systems, pp. 464–473. Springer International Publishing, Cham (2018)

Optimization of the Regional Placement of the Mobile Division Bases of the Commercial Drone Company by Means of the Combinatory Method

Nikolajs Sulima(✉)

Tecnova S.A., Villa La Jolla 17, A Suneion de Bellu, 325 m,
San Hose, Costa Rica
nikolajs.sulima@gmail.com

Abstract. The problem of optimizing of the number and geographical location of mobile units of a commercial company of unmanned aircraft vehicles (drones) services, which provide aerial observation and/or aerial mapping services to the relevant customers and facilities on the territory of the region assigned to them. The efficiency criteria and initial conditions are determined. The optimization task is assigned as a Boolean programming task. When choosing the initial conditions, the potentially possible locations of the mobile units are analysed taking into account the geography of the region, the specifics of the weather conditions and of the facilities for the air service delivery. The combinatory method is proposed for solving the optimization task. The difference between the proposed method for solving the given problem and the previously published linear programming method is that, firstly, there are no restrictions on the number of customers assigned to each mobile unit base, and secondly, the method gives optimal, and not suboptimal solutions. The method was applied for optimization of the number of mobile units bases, their locations and attaching customers to them for a real Costa Rican company.

Keywords: Mobile unit · Unmanned aircraft vehicle services · Basing task · Combinatory method · Optimization

1 Introduction

The problem of basing of a commercial drone's company's (DCo) mobile units (MU) that provides aerial observation and/or aerial mapping services using unmanned aircraft vehicles (UAV) in a particular country or region is examined here.

Depending on the principles of organization of work, DCo can include:

- geographically distributed MU, each of which serves customers' orders on the territories of the region assigned to them;
- the central MU based in the location of the head office of the DCo and serving the customers of the whole region (i.e., this is a particular case of territorial distribution, when only one MU is used in the DCo).

I. Kabashkin et al. (Eds.): RelStat 2018, LNNS 68, pp. 168–177, 2019.
https://doi.org/10.1007/978-3-030-12450-2_16

- the main customers of UAV-services (UAVS) of comparable cost, including both aerial observation and post-processing of the observation results (production of photos, booklets, videos, etc.) for DCo are:
- owners of real estate (hotels, warehouses, private houses, land, etc.);
- organizers of video shooting of events (weddings, commemorative events, local sports competitions, holidays, etc.);
- organizers of excursions and ecological tourism.
- another group of UAVS customers who require only video surveillance services for the purposes of monitoring and controlling of the state of objects/processes and do not require fixation of their condition consists of:
- farms for assessing the condition of soil, lands, crops;
- construction companies for remote control over the progress of work as well as its quality and compliance with the plan;
- environmental organizations for environmental assessment;
- local authorities for monitoring of emergencies and other situations in the local infrastructure, etc.

The classically considered task belongs to the group of typical problems of production plant location, for example, of industrial and agricultural enterprises, warehouses, bank branches, supermarket chains, etc. [1–3]. In the field of commercial use of UAV in the field under consideration, a similar problem has not been raised in the scientific literature before and has not been solved. In the field of commercial use of UAV under consideration, a similar problem has not been raised in the scientific literature before and has not been solved. For the first time the version of the formulation and solution of a similar problem as a task of linear programming, has been presented in [4]. However, the solution of the problem by this method is bound with imposing significant limitations on its dimensionality in order to obtain a result in acceptable computer time. In addition, the method of solving of the problem developed in this work in the general case gives suboptimal but not optimal solutions.

The purpose of this work is to optimize the choice of the number and specific locations of MU bases for a particular Costa Rican Drone Company that provides UAVS considering its customers' location in the region. To achieve the purpose in the work, the following tasks are set:

- to propose a two-stage combinatory method for solving the optimization problem formulated in [4], taking into account the distinctive features and specific factors of the UAVS rendering, which would substantially reduce the dimension of the problem;
- to solve the optimization problem of MU basing under the given initial conditions for a specific Costa Rican firm that provides UAVS for its customers geographically distributed throughout the country.

2 Development of a Two-Stage Combinatory Method for Solving the Optimization Problem of MU Basing with Considering the Regional Specifics of Rendering

2.1 Problem Statement and the Combinatory Method of Its Solving

To achieve this goal, we use the statement of the problem carried out by the author in [4], formulated as a Boolean programming task:

to minimize the function

$$\min F(x,z) = \sum_{i=1}^{n} f_i x_i + \sum_{i=1}^{n}\sum_{j=1}^{m} c_{i,j} z_{i,j} \tag{1}$$

on Boolean variables x_i, $z_{i,j}$ under constraints:

$$\sum_{i=1}^{n} x_i \le k, \tag{2}$$

$$x_i \le 1,\ i = 1,\ldots,n, \tag{3}$$

$$\sum_{i=1}^{n} z_{i,j} = 1,\ j = 1,\ldots,m, \tag{4}$$

$$\sum_{j=1}^{m} z_{i,j} \le s x_i,\ i = 1,\ldots,n. \tag{5}$$

The following notation keys are accepted in the assigned task:

i is the number of possible locations of MU, the total quantity of which is $n = |I|$;

j is the number of places of UAVS which is equal to the number of the served customers UAVS($m = |J|$);

f_i is the cost of organizing and maintaining of a MU base in the location $i \in I$, provided for the given period of time;

$c_{i,j}$ are costs of fixing of the j-th customer to the i-th MU base ($j \in J$), determined mainly by the costs of transportation to the place of UAVS delivery;

k is the maximum possible number of potential MU base sites;

s is the maximum number of UAVS customers assigned to the same MU base;

x_i, z_{ij} are Boolean variables;

$x_i = 1$ if the base opens in the i-th site and 0—otherwise; $x = (x_1, \ldots, x_n)$;

$z_{i,,j} = 1$ if the base in the i-th site serves the j-th client site and 0—otherwise; $z = (z_{i,,j})_{n,m}$.

The required variables in the task are:

k^* is the optimal number of bases

The assigned task of minimizing of the criterion (1) under the constraints (2–5) determines that the costs of a DCo depend mainly on the number and specific places of the MDU, and on the territorial location of the clients' facilities being serviced, but the income is constant. On the whole, the solution of the task in the formulation (1–5) should ensure the minimum costs of the DCo when servicing customers with a given number of UAVS in certain areas of the region.

The efficiency of the choice of the MU base locations is significantly influenced by the costs of the MU facilities and the post-processing means of the results of aerial mapping of objects and processes, UAV application factors and characteristics.

In the composition of DCo, there can be distinguished:

- special UAV means for commercial purposes with appropriate attachments;
- universal means of providing—means of remote delivery of UAV to the places of use (mainly cars, although motorcycles, boats, etc. also can be used), stationary drone chargers, navigators, etc.;
- crews (pilots, operators, drivers of specially equipped cars).

The distinctive features of the task under consideration and its initial conditions are the following factors:

- the need to transport drones to the service delivery places due to the limited range of commercial UAVs;
- the significant dependence of the process of the UAV service delivery on force majeure circumstances (weather conditions, natural disasters, transport infrastructure, etc.);
- taking into account of the fact that the approach of MDU bases to the place of service delivery potentially increases the number of orders in the region concerned;
- limitation of the period of service delivery by the duration of daylight hours;
- exclusion of territories subject to regulatory and legislative restrictions, such as regime-protected areas, from potential locations for the service delivery and, therefore, from the potential deployment of MDU.

Among the basic parameters that do not affect the choice of drones for DCo, since the problem is solved in the conditions of their identity for all possible versions of the MU basing, there can be distinguished:

- distance and time of flight without refuelling (additional charge);
- load capacity, taking into account the payload;
- distance and reliability of the control channel;
- multifunctionality of application;
- cost indicators.

For effective customer service, the DCo should conduct processing of survey data, and for this purpose, it needs a powerful enough computer, specialized operating systems for UAV programming and controlling, software application and trained specialists in shooting direction and material editing.

The basis of a MU usually consists of a crew, a UAV with appropriate attachments and a specially equipped car. In the minimal version, the crew includes only one person performing the functions of a pilot, as well as a drone operator and a car driver.

2.2 Description of the Algorithm for Calculating of the MU Base Locations Using the Combinatory Method

Let us assume that there is only one MU on one base. The more the number of MUs will be, the more will be the cost of ensuring their functioning, but the less time will be required for servicing UAs for the customers of the corresponding area, and the more UAVS can be serviced. And vice versa, the less the number of MUs will be, the lower will be the costs, but the time losses for the transportation of drones to the place of UAVS delivery for the customers will increase and, accordingly, the number of UAVS delivered will reduce.

In the described task, there are $(2n + m + 1)$ constraints and $n\ (m + 1)$ variables. For small countries, the dimension of the task is low ($n < 10$, $m < 20$), which allows using exact combinatory methods of solving it [5–7].

The combinatory solution method is applicable to this formulation of the task, since the number n of possible base locations and the number k of potential MU bases are relatively small.

The application of the combinatory method is proposed in two stages:

- the first stage—reduction of the number of potential locations (k);
- at the second stage—the problem solving by scanning of the remaining options.

The reduction of the number k is proposed to be carried out in the expert way taking into account the climatic features of the regions of the country that affect the average time of providing UAVS (average flying time)—$T_{f,mid}$.

In the second stage of solving the task, all combinations of base distributions in all possible places can be viewed. Generation of all the combinations I of object (base) distributions k over n possible MU base locations is effected as follows.

First, each allocation of MU bases is described by an n-dimensional Boolean vector. The vector components correspond to the possible locations of a MU. If the component is equal to 1, then the base MU is placed in this site, and if it is equal to zero, then no. Each such n-dimensional vector contains k units—according to the number of MU bases being placed. The number of such different vectors, i.e. the number of different options for placing MU bases is equal to $n! / (k!(n - k)!)$. Creating of the list of all possible combinations begins with an n-dimensional Boolean vector in which the units occupy k low-order classes of the vector. Further, *one* is added to the current Boolean vector. If after that the number of *ones* in the new vector is equal to k, the vector determines a new combination and is memorized. The described procedure is completed when all the required combinations represented by the set I are obtained.

After formation of all the possible options of base placing, there takes place transition to the second stage—to the choosing of the allocation variant that would ensure the optimal value of the goal function (1). To do this, all the allocations from I are analysed successively. For a fixed allocation, all the customers are methodically examined. For each client $j \in J$, the base I, for which the expenses $c_{i,j}$ are minimal, is detected. The sum of the expenses $\{f_i\}$ for organizing the planned bases I is added to the sum of these costs for all the customers. As a result, we get the value of the goal function (1) for the considered variant of the MU base allocation. The minimum value among all accommodation options determines the optimal solution of the task under consideration. This value determines both the optimal number of MU base sites and assignment of customers to the bases. The mathematical programming language Mathcad was used to solve the problem.

3 Specification of the Conditions and the Initial Data of the Assigned Task Using the Example of the MU Basing by a Costa Rican DCo

Let us consider an example of solution of the task in the formulation (1) under the initial conditions (2–5) with respect to the choice of the MU locating sites of a Costa Rican DCo, for which we adopt the following expert assumptions and averages.

A. Orders for MDS services are concentrated in 7 provinces, major cities, national parks (NP), and reserves.

The following areas account for the largest order volume in the country:

1. Alajuela Province—the cities of Alajuela and Quesada, the Arenal Volcano, and Irazú Volcano and Poás Volcano National Parks (NP);
2. Cartago Province—the cities of Cartago, Paraíso, the Irazú Volcano NP;
3. Guanacaste Province—the city of Liberia, the little town of Samara, the Santa Rosa, Juan Castro, Palo Verde, Rincon de La Vieja Volcano NP;
4. Heredia Province—eredia, "the city of flowers";
5. Limón Province—the city of Limón, the Tortugero NP;
6. Puntarenas Province—the city of Puntarenas, the Manuel Antonio, Corcovado, Carrara NP;
7. San José Province—the city of San José.

In the provinces of *Limón*, Cartago, Heredia and San José the majority of orders also tend to be oriented on urban targets in big same-named cities or about an hour's drive from them. *In* Guanacaste, orders are also usually placed by guided and eco tour operators. And in the Puntarenas province, covering the major part of the west coast, accounts for orders are of various types.

B. Based on the data from the Synoptic charts of Costa Rica [8, 9], it is calculated that the daily flight time for the order fulfillment, which significantly affects the revenues, and therefore the spending limits.

- Let us evaluate the average duration of the potential UAVS flight time on the basis of statistical data for different periods of the year for the most contrasting regions of the country: the west and the east coasts.

On the west coast, we select:

- the coast of the province of Guanacaste and the north of the province of Puntarenas (G&P);
- the part of the coast in the province of Puntarenas in the vicinity of the city of Jaco (MP);

– the southern part of the province of Puntarenas in the area of the cities of Golfito and Puerto Jimenez (ZP), including the Corcovado National Park.

On the east coast, we select the region of the city of Puerto-Limón in the province of Limón (PL).

The results of the assessment are shown in Fig. 1.

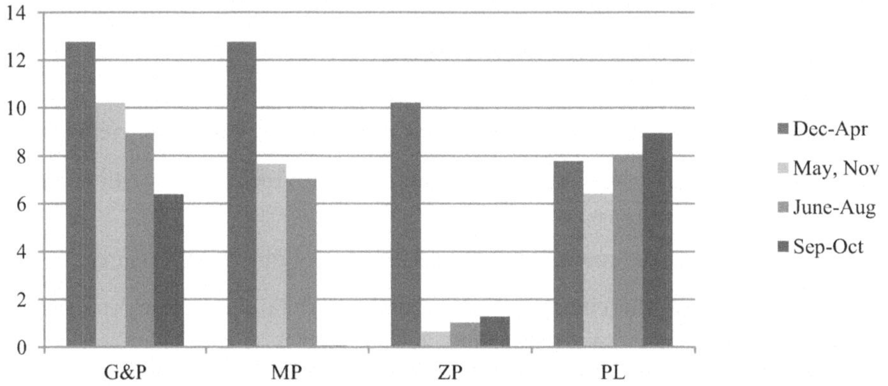

Fig. 1. Examples of the average duration of $T_{f,\ mid}$ (in hours) for the selected regions and periods of the year.

The average annual values of T_f, mid for the selected regions (Fig. 1) are:

- region G&P: $T_{f,mid} = 10.1$ h;
- region ZP: $T_{f,mid} = 4.61$ h;
- region MP: $T_{f,mid} = 8.35$ h;
- region PL: $T_{f,mid} = 7.37$ h.

Obviously, the most favourable regions for MU basing are G&P and MP, and the least favourable region is ZP. Note that the fulfilment of orders for UAVS in the centre of the province of Limón, if it is possible under the terms of the customers, is advisable to be planned for the period June–October, when in the other regions the period of the weather conditions is the worst. The number of potential sites expertly selected is $n = 9$. Their supposed locations are: the neighborhoods of San Jose, Cartago, Alajuela, Limón, Jaco, Uvita, Golfilo, Cobano and Tilaran.

C. The expenses f_i are represented by the vector:

$$f = (5 \quad 4 \quad 5 \quad 4 \quad 5 \quad 4 \quad 3 \quad 5 \quad 4).$$

Note that he order in this vector is not related to the order of enumeration of the assumed base locations from the item C.

D. The number of potential customers in the places (closed customer service areas) $m = 15$. Limitations on the number of customers assigned to any MU base are not imposed.
E. The numerical values (given) of expenses $c_{i,j}$ are presented in Table 1.

Table 1. Expenses $\{c_{i,j}\}$.

$j \setminus i$	1	2	3	4	5	6	7	8	9
1	8.6	8.0	2.0	1.0	4.2	4.2	2.2	1.7	0.9
2	2.4	6.3	3.6	0.3	6.0	7.8	4.8	0.7	0.8
3	0.5	5.9	2.9	2.6	4.1	1.5	4.1	9.6	3.8
4	1.7	1.9	1.0	2.3	6.2	0.9	8.4	4.5	5.2
5	2.2	2.3	7.1	8.6	0.0	4.7	4.3	1.8	1.4
6	7.9	2.4	6.7	0.5	3.9	3.3	7.3	4.7	0.7
7	0.8	4.1	0.1	0.6	3.2	6.1	0.8	2.5	9.1
8	1.4	2.7	6.0	0.9	8.5	2.8	8.5	2.8	7.6
9	5.6	3.8	2.9	9.6	2.0	4.4	3.6	8.2	1.0
10	6.3	2.1	7.5	8.8	4.7	2.2	0.9	6.3	6.8
11	4.5	6.0	9.1	4.0	4.3	5.2	1.9	0.1	8.2
12	3.3	3.6	6.1	6.6	4.2	0.1	9.1	2.9	2.7
13	2.8	9.8	7.8	9.5	5.8	7.7	3.1	8.6	8.7
14	8.7	9.0	5.4	0.0	3.4	9.7	3.4	7.1	9.7
15	8.7	9.5	3.9	0.5	9.5	9.0	2.7	9.2	2.2

Expenses $c_{i,j}$ in Table 1 include also losses due to impossibility of UAVS delivery because of weather conditions or other force majeure circumstances at selected locations of the MU bases.

The calculated values of the criterion $F(x,z)$ as a function of k under the initial conditions described above are shown in Fig. 2.

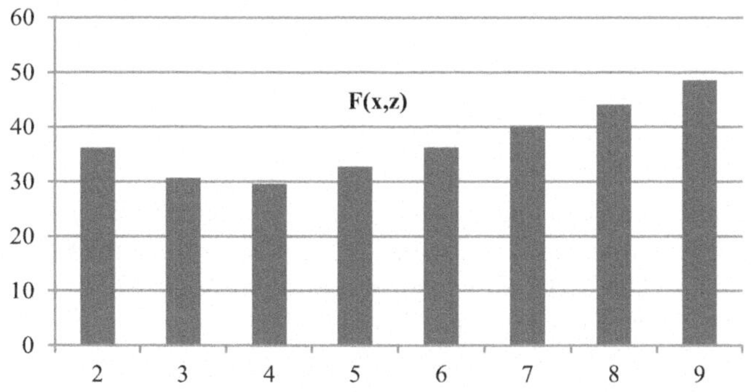

Fig. 2. The graph of dependency *of* $F(x,z)$ on k.

In accordance with the Fig. 1, the optimal number of MU bases is 4. These bases, in accordance with the all conditions, are located in the areas of the settlements Alajuela, Tilaran, Limón and Jaco, marked with triangles on the map Fig. 3.

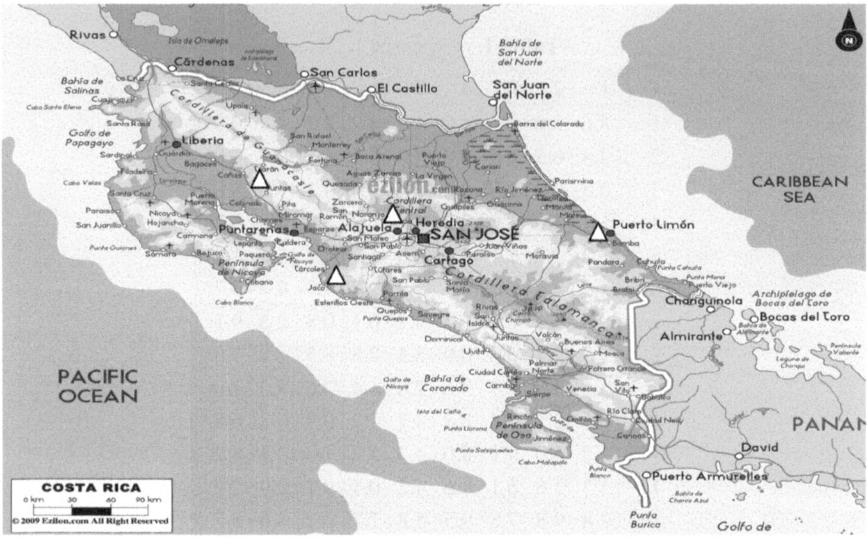

Fig. 3. The map of optimal locate of MU-base.

In this case, the expenses will be minimal and equal to 29.5 (Table 2). Optimal assigning of customers to the bases with $k = 4, 5$ is shown in Table 2.

Table 2. Optimal assigning of customers j to the bases I with $k = 4$ and $k = 5$.

j	1	2	3	4	5	6	7	8	9	10	11	12	13	14	15	F
$i, k = 4$	3	3	5	5	8	3	3	3	8	6	6	5	6	3	3	29.5
$i, k = 5$	8	3	5	5	8	3	3	3	8	6	7	5	6	3	3	32.7

4 Conclusion

A task of optimal basing of the mobile units of a company providing commercial UAVS is formulated. The combinatory method is used to solve the task applied to the selected conditions. The success of its application is associated with such additional conditions as the integer solution and absence of restrictions on the number of customers assigned to any MU base. Numerical calculation using the example of the Costa Rican company illustrates the suggested approach to the task solving.

References

1. Krarup, J., Pruzan, P.M.: The simple plant location problem: survey and synthesis. Eur. J. Oper. Res. **12**, 36–81 (1983)
2. Khumawala, B.M.: An efficient branch-bound algorithm for the warehouse location problem. Manag. Sci. **18**, 718–731 (1972)
3. Cornuejols, G., Fisher, M.L., Nemhauser, G.L.: Location of bank accounts to optimize float. Manag. Sci. **22**, 789–810 (1977)
4. Sulima, N., Stecenko, I.: The Task of Optimizing Regionsl Deployment of Commercial Drone Bases. In: DANĚ – TEORIE A PRAXE. Sborníkreferátůaabstraktů XX. ročníkumezinárodníodborné conference. Brno. Akademie Sting, o.p.s. 158–167 (2017)
5. Mirchandani, P.B., Francis, R.L.: Discrete Location Theory. Wiley, Hoboken (1990)
6. Bruadli, R.A.: Introductory Combinatorics, 3rd edn. Prentice Hall, New Jersey (1999)
7. Nocedal, J., Wright, S.: Numerical Optimization, 2nd (edn.) Springer, Heidelberg
8. LNCS Homepage (2006). http://www.encyclopedia.com/places/latin-america-and-caribbean/costa-rican-political-geography/costa-rica#CLIMATE. Accessed 03 Mar 2018
9. Electronic source. https://weatherspark.com/averages/32645/San-Jose-Alajuela-Costa-Rica. Accessed 03 Mar 2018

Data Preparation Framework Development for Markov-Modulated Linear Regression Analysis

Irina Jackiva (Yatskiv)(✉) and Nadezda Spiridovska

Transport and Telecommunication Institute,
1 Lomonosova Str., 1019 Riga, Latvia
{Jackiva.I, Spiridovska.N}@tsi.lv

Abstract. The modern data collection and analysis methods enhance reasonable and effective data-driven decision making in transport planning and have significantly expanded the scope of potential data applications. But if the goal is to use data in transport models for evaluating and predicting, the quality of data becomes crucial.

This research is focused on data pre-processing issues such as data understanding, data exploring, and data transformation as an important part of data analysis life cycle rather than transformation models themselves. These phases involve many different tasks and many of the data preparation activities are routine, tedious, and time consuming.

In order to resolve this problem the data preparation framework for Markov-modulated linear regression model, considering the limitations and assumptions, was developed. This kind of regression model can be used for public transport passenger flow prediction or other transport planning tasks and suggests that the model parameters vary randomly in accordance with the external environment. The developed framework is applied on data concerning Riga tram route trip validation captured by e-ticket system "E-talons" and the Latvian Environment, Geology and Meteorology Centre database. R software is used in conjunction with a set of libraries.

Keywords: Data analysis · Public transport passenger flow · Regression · Data preparation and exploring

1 Introduction

Transport models are popular and effective tools for analyzing a variety of problems associated with complex processes in transportation systems and transport planning. Nowadays as noted in [1] transformed technologies (Internet of Things connected and automated vehicles, etc.) give additional opportunities for transportation system analysis and Big Data has the potential to be transformational technology to change transport planning. It provides many opportunities, for instance, to enhance public transport efficiency and improve the quality of service. Big Data is a powerful "tool" for knowledgeable decision making regarding city transportation [2]: for planning the public transport (PT) and understanding travelers' journey patterns, informing transport

I. Kabashkin et al. (Eds.): RelStat 2018, LNNS 68, pp. 178–190, 2019.
https://doi.org/10.1007/978-3-030-12450-2_17

operators on how different social groups use the public transport system, providing information on routes and schedules to citizens based on real-time information and intelligent predictions on passenger loads, etc. [3–5].

Each year number of applications to look for opportunities and challenges in using Big Data are increased dramatically in transport field. The directions of Big Data applications in transport are as follows [6]: Intelligent Transport Systems; Open Data; Automated driving; and Smart mobility. Trends in Big Data and how they can and should inform plans for the evolution of ITS technologies and practices are discussed in [7]. One of the above mentioned public transport trends concerns the Dublin case, where Big Data application is aimed at optimizing traffic flow of public transport vehicles. The system estimates speed, time, and informs users on their mobile phones, advising them on the alternative itineraries and based on Data mining algorithms: Max Likelihood and Kernel Density Estimation [9]. Great number Big Data applications for understanding the passengers' travel behavior are being developed [10–12].

Big Data has huge potential: faster, cheaper data, but at the same time there are some drawbacks such as non-representativeness and unreliability. The data analysis results heavily rely on the quality of preliminary collected data. And if the goal is to connect data to transport models for evaluating and predicting these drawbacks become crucial. Data processing techniques, applied before modeling stage, can substantially improve the overall quality of the obtained patterns or the time required for the actual analysis. It is the process that involves many different tasks and which cannot be fully automated. Data preparation process is time consuming and laborious. It has been estimated that data preparation accounts for 60–80% of the time spent on a data mining project [13].

In this research we will not focus primarily on data modelling algorithms, the present study is focused on data preprocessing issues. It is critical to fit the models to data and transform data into knowledge. For this purpose we need to look into the nature of data and understand it. In the present study, the important stage of data analysis life cycle—large datasets understanding and preparing—is investigated through in-depth analysis and as a result methodology is prepared. The methodology is applied to concrete data sets and R software was used in conjunction with a set of libraries.

The outline of the paper is as follows: the next Section explains the model and assumptions that are crucial for regression based on Markov-modulated case. Section 3 proposes methodology of data preparing and exploring and in Sect. 4 the methodology applied to the case study of the passenger flow prediction in Riga city for particular tram line. Finally, the paper discusses the findings and recommendations for future applications.

2 Markov-Modulated Linear Regression Model Assumptions

Markov-modulated linear regression (MMLR) suggested in [14] and theoretical findings were supported by large amount simulation data (67,500 observations). The model operates in the external environment $J(.)$, which has final state space

$S = \{s_j, j = 1, \ldots, m\}$, for the fixed state $s_j \in S$, $j = 1, \ldots, m$, and generalized linear regression model in matrix notation as follows

$$Y(t) = \begin{pmatrix} Y_1(t_1) \\ Y_2(t_2) \\ \cdots \\ Y_n(t_n) \end{pmatrix} = \begin{pmatrix} \overrightarrow{t_1} \otimes x_1 \\ \overrightarrow{t_2} \otimes x_2 \\ \cdots \\ \overrightarrow{t_n} \otimes x_n \end{pmatrix} vec\,\beta + diag(\sqrt{t_1}, \sqrt{t_2}, \ldots, \sqrt{t_n})Z, \tag{1}$$

where n is the number of observations; $Y_i(t)$ are time-additive scale responses $(Y_i(0) = 0)$; the $1 \times m$ vector $\overrightarrow{t_i} = (t_{i,1}, \ldots, t_{i,m})$, which component $t_{i,j}$ means a sojourn time for response Y_i in the state j (note that $t_i = t_{i,1} + \ldots + t_{i,m}$); $\otimes$ is Kronecker product; $(x_{i,1}, x_{i,2}, \ldots, x_{i,k})$ is $1 \times k$ vector of regressors; the $k \times m$ matrix $\beta = (\beta_1, \ldots, \beta_m) = (\beta_{v,j})$ of unknown parameters; *vec* operator *vec*(A) of matrix A, the n-dimensional diagonal matrix *diag(v)* with the vector v on the main diagonal; $Z = (Z_i)$ is the $n \times 1$ vector, where $Z_i(t)$ is Brown motion scale disturbance.

The whole trajectory of the environment $J(.)$ is unknown and the estimated conditional average sojourn time is used instead of unknown sojourn times $T_{i,j}$ in the state s_j. All necessary formulas for the calculation of the conditional average sojourn time that allow to get estimations of unknown parameters can be found in [14].

Assumptions of MMLR can be divided into 3 groups defined by:

1. The classical regression assumptions: the disturbances Z_i are independently, identically normally distributed with mean zero and variance σ^2, the $n \times k$ matrix $X = (x_{i,v}) = (x_i^T)^T$ has rank $r(X) = k$, so $(X^T X)^{-1}$ exists.
2. The model formulation: $Y_i(t)$ are time-additive scale responses $(Y_i(0) = 0)$.
3. The external environment that is a random one and is described by a continuous-time Markov chain $J(t)$, $t > 0$, with the finite state set S. $\lambda_{i,j}$ is the known transition rate from state s_i to state s_j, and $\Lambda_i = \sum_{j \neq i} \lambda_{i,j}$. A continuous-time stochastic process $\{X(t) : t \geq 0\}$ is called a continuous-time Markov chain if it has the Markov property. The Markov property is a "forgetting property" suggesting memorylessness in the distribution of time a continuous-time Markov chain spends in any state. In other words, if T_j denotes the sojourn (holding) time in state j, for $j \neq i$, then T_j is exponentially distributed with transition rate λ_j and T_j is independent of T_i.

MMLR model was applied to the data of coaches delay time (on the particular route for the period from 2012 to 2017) provided by Riga Coach Terminal [15]. Despite a large amount of initial data (for example, 5414 records in the sample for 2012) and due to data preparation phase and the assumptions of the model (time-additive response variable), the number of observations has significantly decreased. The application of the model showed low accuracy of the prediction and it was concluded that it wasn't related to the incorrectness of the proposed model, rather it was due to the low quality of the classical regression model as a tool for describing the delays of coaches: supposedly the day of the week wasn't the only factor determining the delay time. One more reason was connected to the quality and amount of data: sparse and repeating data that could cause unreliable results. The next approbation of the model was carried out on the data of trip validations provided by Rigas Satiksme [16]. Initial data set of time period (4.5 months at 2017) contained n = 1,048,001 trip validations, but after data transformation and aggregation the number of observations has significantly decreased and prediction didn't show convincing results for the same reasons.

The authors concluded that it is crucial for the application of the Markov-modulated linear regression model to have a large amount of high quality data. And due to routine and time-consuming procedure of data preparing it is critical to build a data preparation framework that would also include model assumptions testing.

3 Methodology of Data Preparing and Exploring

The process of data preparing is illustrated in Fig. 1 and consists of next stages.

Problem identification. Identify response variable, external environment and internal variables considering model assumptions. Main output of the phase: defined number of states of the external environment ($m = 2, 3, \ldots$).

Data acquisition and understanding. Use two kinds of data sources—primary and secondary. Initial data review to understand the data, and discover initial insights; evaluate the quality of the data.

Data pre-processing: Cleaning. Fill in missing values, smooth noisy data, identify/remove outliers, and resolve inconsistencies.

Data pre-processing: Structuring. Write or choose algorithms based on the criteria of different usage requirements with a purpose to automate the process of data classification, categorization or aggregation partially or fully. Depending on the data processing environment, use different technics to bring the data into the required format (macros or pivot tables in Excel, scripts in *R*, etc.).

Data pre-processing: Exploring. Apply the descriptive statistics: measures of central tendency (mean, median, mode), measures of data spread (quartiles, interquartile range, variance, etc.) to describe the basic features of the data, to fit the distribution. Prepare data for the aggregation: consider different ways of aggregation.

Data pre-processing: Merging. (Non-obligatory) If the data on the external environment is presented in a small amount or the behaviour of the environment differs at different stages, use inferential statistics to test the possibility of data merging or aggregation. The research questions were: is it possible to combine some months to describe the behavior of the external environment? Are transition intensities the same for different months? If the homogeneity hypothesis is not rejected, the data for regression is also merged according to the principle chosen for external environment data.

Markov property testing. Use inferential statistics for Markov property testing: exponential distribution of sojourn times fitting in each state) and hypothesis about no significant correlation between sojourn time T_j and T_i, ($j \neq i$). If null hypothesis is rejected return to data merging or acquisition phase. Also use other technics, for example, use a convolution of exponential densities for a parametrical estimation [16].

Data pre-processing. Aggregation. Based on the model assumption about time-additive responses ($Y_i(0) = 0$), aggregate dependent variable values according to decision, made previously.

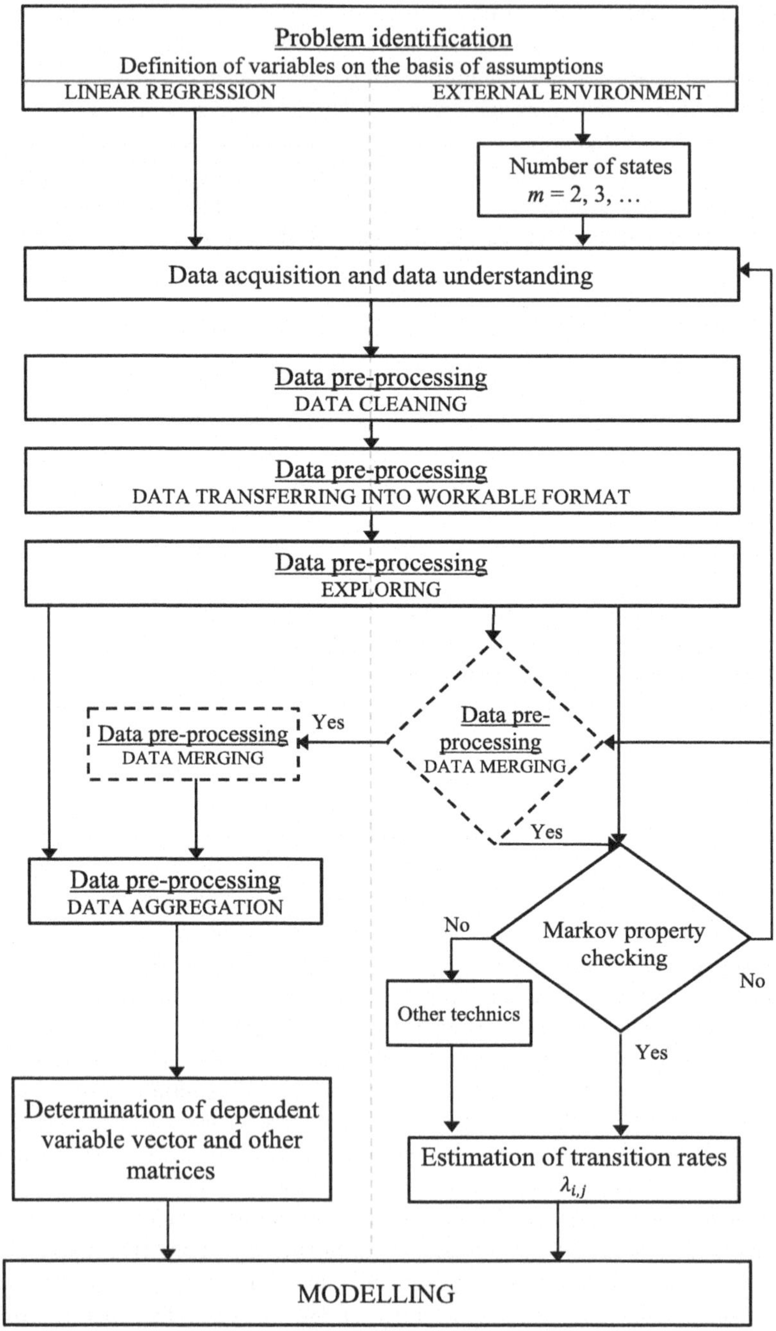

Fig. 1. Data exploring and preparing framework.

Estimation of transition rates $\lambda_{i,j}$. Estimate unknown parameters according to parameter estimation theory.

Determination of dependent variable vector and other matrices. For each realization i the following data should be available:

- total observation time $t_i = T_{i,1} + \ldots + T_{i,m}$;
- initial $J_{i,0}$ state of Markov chain $J(.)$;
- vector of independent variables $x_i = \left(x_{i,1}, x_{i,2}, .., x_{i,k}\right)$;
- the response $Y_i = Y_i(t_i)$.

A methodology is outlined here to develop, implement and monitor quality of data using available data exploring tools.

4 Case Study: Results and Discussion

The model developed for passenger volume prediction on one of the longest tram (No 11) routes (Ausekļa street/city centre to Mežaparks/city border) in the Riga city is considered in [15]. This route consists of 23 stops and has different schedule for working days and weekends (direction "forth"):

1. Mon–Fri: 43 runs from 4:00 till 10:00 a.m.; 23 runs from 11:00 a.m. till 14:00 p.m; 38 runs from 15:00 till 19:00 p.m and 19 runs from 20:00 till 24:00 p.m.
2. Sat, Sun and public holidays: 33 runs from 4:00 till 10:00 a.m.; 31 runs from 11:00 a.m. till 14:00 p.m; 35 runs from 15:00 till 19:00 p.m and 21 runs from 20:00 till 24:00 p.m.

In Latvia, the e-tickets are used in Riga region for PT fare validation. When entering PT, all passengers have to apply an e-ticket "E-talons" to an electronic validator.

The predicted variable is passenger volume (per day or per some intervals of the day, depending on scenario) and as regressors are chosen: day of the week (dummy variable) with "Monday" as key variable; direction (forth and back, dummy variable), "back" as key variable, autoregressor (optionally).

Weather conditions in the Riga city were chosen as an external environment. It was assumed that there are 2 states of the external environment: "No precipitation" (dry) and "Precipitation" (wet) as alternating states ($m = 2$).

Data acquisition and understanding. Trip validation data was provided by Riga Urban PT Operator "Rīgas Satiksme" through the company "Rīgas karte", which was established to develop, manage and provide modern technological solutions by integrating them with the Riga PT system including buses, trams and trolleybuses. The smartcard data for the half-year 2017 was extracted for a specific tram route, so each smartcard transaction record contained information on date, respective time stamp, destination (forth/back) and type of e-ticket (the system uses different types).

The data about weather conditions was obtained from the Latvian Environment, Geology and Meteorology Centre (LEGMC) database (www.meteo.lv) and included meteorological observations at 33 stations (1 station per 1500 km^2), having fixed location over the territory of Latvia. For this case data from the station named “Riga” was obtained. All available data that measured in categorical scale (‘0’–‘9’) was downloaded from LEGMC website in an excel format.

Data with measurement points of 3 h was selected and categorised on 2 above mentioned states of the external environment:

- 0, 1, 2—“No precipitation” or “dry” weather conditions (1st state);
- (3–9)—“Precipitation” or “wet” weather conditions (2nd state).

Monthly data for Riga city is available from 2006 till current moment. For this case study data was processed from 2006 till 2017 inclusively.

Both initial data sets were not appropriate for further usage and analyses and required pre-processing.

Data pre-processing. Cleaning. All initial data files about weather conditions for this study had the same format and didn’t contain missing or shifted data or extra time intervals, which would need to be processed additionally.

Data pre-processing. Structuring.

1. *Data on weather conditions.* To estimate the parameters of the external environment, the critical point was to obtain data on the sojourn times in each state (m =2). Since the data was acquired in coded format, it was necessary to transfer data into workable format. The Visual Basic code was written to divide the available codes (0–9) into 2 groups (states) and to calculate the duration of a sojourn time in each state (multiplying the number of consecutive values in each group by 3 h). Thus, data to estimate the distribution of the sojourn time in each state was obtained (Fig. 2).

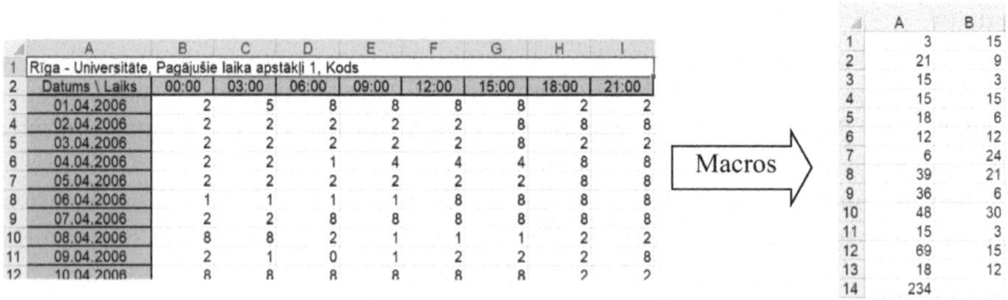

Fig. 2. Results of transferring initial data into workable format by means of macros in excel.

Totally 144 separate excel sheets (12 years multiplied by 12 months) were processed and combined into one tidy data file.

2. *Data on trip validations.* Initial data set of time period (4.5 months at 2017) contained n = 1,048,001 trip validations. For calculation the total number of validations per day the Excel built-in functions and the pivot tables were used.

Data pre-processing. Exploring. It was necessary to draw conclusions from both data sets for further data pre-processing. Only some of the data exploring results are reflected here as an illustration.

1. *Data on weather conditions.* The main goal of descriptive analysis implementation for this data set was the initial diagnostics for homogeneity.

Table 1 reflects the main characteristics (mean, trimmed mean, sum, max etc.) of the sojourn time in each state ("No precipitation" and "Precipitation") for some observed months. Data showed that there was difference between mean values in sojourn times, especially between September and November, but still significance of this difference should be tested.

Table 1. Descriptive statistics of the sojourn time in each state for each observed month.

Variable	Valid N	Mean	Trimmed mean	Sum	Max	Std. Dev.
Sept/NoPr[a]	206	30.83010	23.87097	6351	255	44.99133
Sept/Prec[b]	199	11.56281	9.98883	2301	81	10.93529
Oct/NoPr	227	25.98238	18.76098	5898	306	42.10364
Oct/Prec	221	13.83258	12.06030	3057	135	13.73066
Nov/NoPr	261	17.91954	14.66809	4677	159	22.22121
Nov/Prec	257	15.42023	13.58442	3963	114	14.30289

[a] – NoPr – "No Precipitation";
[b] – Prec – "Precipitation".

The next step involved the construction of histograms and estimation of distribution. Histograms showed that Markov property obviously should be checked, inasmuch as even visually not all histograms fit to exponential distribution (see Fig. 3).

2. *Data on trip validations.* Used data set: n_{tot} = 1,048,001 (trip validations); from 2017-08-16 till 2017-12-31; two directions: forth (from the city centre): $n_{forth} = 530,422$ and back (to the city centre): $n_{back} = 517,579$. Figure 4 shows that the average number of trips is about the same for the autumn months and this information could be used later in aggregation phase.

Data pre-processing. Merging. Test it is possible to combine some months to describe the behavior of the external environment, for example autumn months. Two-sample nonparametrical tests were used for homogeneity analysis at the significance level of 0.05: Mann-Whitney U test and Kolmogorov-Smirnov two-sample test with null hypothesis H_0: the sojourn time in particular state ("dry" or "wet") has non-significant difference for Month#1 (M1) and Month#2 (M2) (see some of the results in Table 2).

It was decided for one of the scenarios to combine months: August, September and October since their average values of sojourn time in each state differ insignificantly.

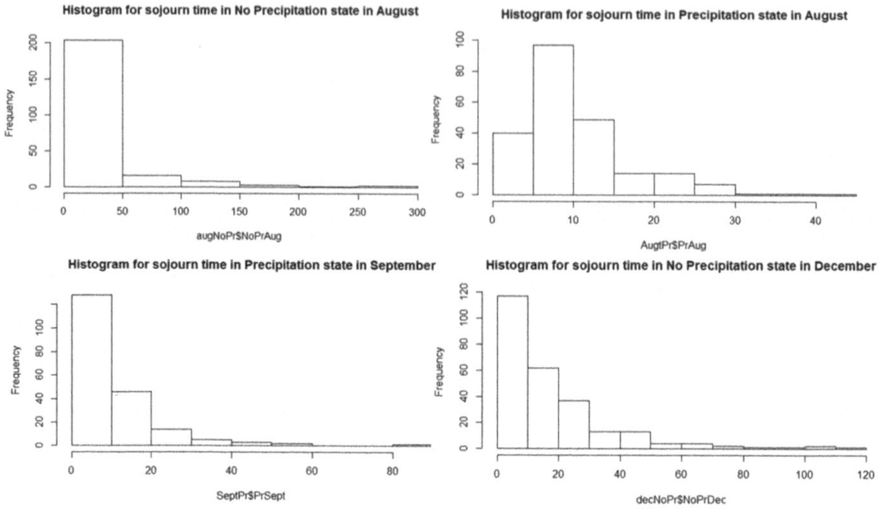

Fig. 3. Histograms for sojourn time.

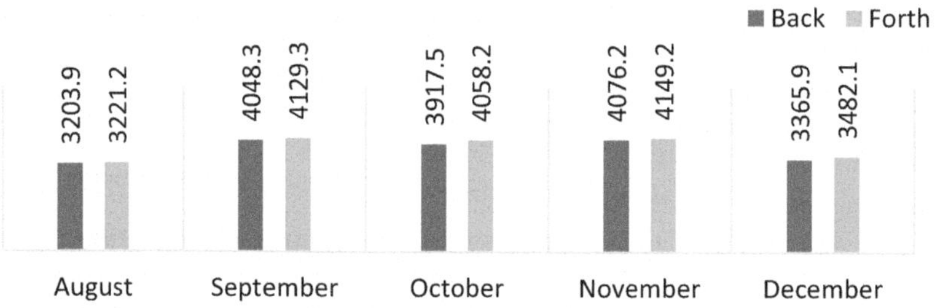

Fig. 4. Average number of trip validations distributed by months.

Table 2. Two-sample hypothesis testing results.

H_0: State, Month#1 vs Month#2		Mann-Whitney test Z/p-value	Kolmogorov-Smirnov/p-value	Decision about H_0
M08 vs M09	"No prec"	−0.28062/0.7790	0.0659/p > .10	Accept
	"Precipitation"	−0.5996/0.5488	−0.0622/p > .10	Accept
M08 vs M10	"No prec"	−0.9301/0.3524	−0.0423/p > .10	Accept
	"Precipitation"	1.7336/0.0829	0.0965/p > .10	Accept
M11 vs M12	"No prec"	−0.8627/0.3883	−0.0683/p > .10	Accept
	"Precipitation"	−0.2837/0.7766	−0.0444/p > .10	Accept
M10 vs M11	"No prec"	1.9542/0.0507	0.1039/p > .10	Reject
	"Precipitation"	−1.6225/0.1047	−0.0825/p > .10	Accept
M9 vs M11	"No prec"	2.3259/0.0200	0.1227/p < .10	Reject
	"Precipitation"	−3.7672/0.0002	−0.1532/p < .025	Reject

Markov property checking. Two criteria were selected to check Markov property of the external environment: distribution of the sojourn time in state (is supposed to be exponential) and independence of the observations' pairs (memoryless property).

Distribution fitting. It was necessary to test hypothesis about the sojourn time distribution in each state. The null hypothesis in general can be stated as follows: the sojourn time in particular state ("No precipitation" or "Precipitation") has an exponential distribution: H_0: $F_{emp}(x_1, \ldots, x_n) = F_{exp}(x_1, \ldots, x_n)$. Nonparametric Chi-squared goodness-of-fit test was used for it. Some of the results for different combinations of parameters at the significance level 0.05 are shown in the Table 3.

Table 3. Distribution fitting results.

H_0: State, Month	Chi-Square test/p-value	Decision about H_0
"No prec", August	24.32154/0.00007	Reject
"Precipitation", August	79.13903/0.00000	Reject
"No prec", October	15.43725/0.00148	Reject
"Precipitation", October	2.79030/0.59351	Accept
"No prec", December	5.00020/0.41586	Accept
"Precipitation", December	6.95228/0.13843	Accept

At this stage, if negative results were obtained, it was possible to return to the data acquisition stage and (1) rethink the environment states, or (2) try to extract new information from the original data, or (3) try to obtain additional data. Even though the results of distribution fitting obtained were partially negative, it was decided that the data is relevant for describing the external environment in the context of this task. It does not seem to be a problem, since some investigations prove using an approximation of arbitrary nonnegative density by a convolution of exponential densities [16].

Testing of independence. Correlation analysis indicated that there is no linear dependence between the observation pairs "No precipitation—Precipitation" for each month and combinations of months. Some of the results are presented in Table 4.

Table 4. Correlation analysis results.

Variable I	Variable II	Correlation coefficient	Significance at 0.05 level
NoPrAug	PrAug	0.01982	Non-significant
NoPrSept	PrSept	0.05627	Non-significant
NoPrDec	PrDec	−0.01014	Non-significant
NoPrAugSeptOct	PrAugSeptOct	−0.01192	Non-significant

[a] – NoPr – "No Precipitation";
[b] – Pr – "Precipitation".

Data pre-processing. Aggregation. Due the model assumption about time-additive responses, one of the vectors related to the initial data should be a total observation time vector. Observations can be aggregated for any period of time: for year, for month, for day, for hour, etc. The most obvious solution would be to aggregate the data by day, but in order to expand the sample it was decided to split day into four-time intervals based on some considerations.

Distribution of trip validation times during the day was explored (Wednesday as middle day of the week, Fig. 5) and then to make time intervals more or less uniform it was decided to choose four intervals: (5:00–10:00); (10:00–16:00), (16:00–19:00) and (19:00–00:00). Trip validations were summed for each interval of each day.

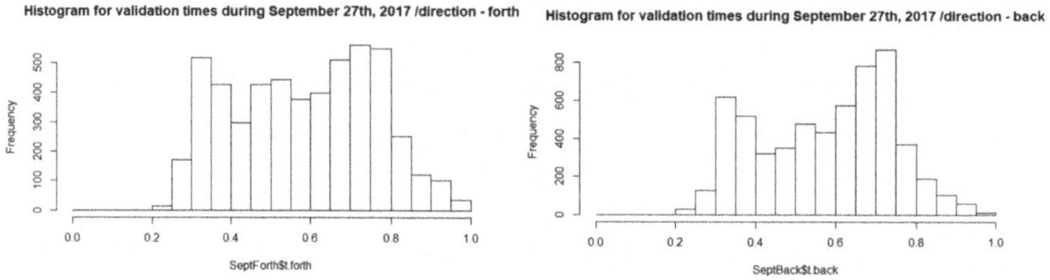

Fig. 5. Distribution of converted trip validation times during the day.

Estimation of transition rates $\lambda_{i,j}$. The transition intensities from state i to state j are calculated as reciprocals to the mean sojourn time in each state. For example, for combined months results are reflected in Table 5.

Table 5. The transition intensities from state i to state j.

Combined Aug Sept Oct	Mean	$\lambda_{i,j}$
No precipitation (i = 1)	28.16942	$\lambda_{1,2} = 0.035499$
Precipitation (i = 2)	12.02795	$\lambda_{2,1} = 0.08314$

Determination of dependent variable vector and other matrices. The most scrupulous part of data processing was the formation of matrices that were saved in separate files with .xlsx and .csv extensions. One of the examples is shown in Fig. 6, from the left to the right: matrix of independent variables, total observation time, initial $J_{i,0}$ state of Markov chain $J(.)$ and the response $Y_i = Y_i(t_i)$ vector (totally 543 observations in each matrix/vector).

5 Conclusions

Each year number of notable Big Data applications in transport field increases dramatically and end users that should have positive impacts from Big Data analytics in the transport sector: government, business and citizens. At the same time, the data preparation stage is still routine and often time consuming. The research object in this study is data pre-processing: understanding, exploring, and transformation for special

	A	B	C	D	E	F	G	H	I
1	1	0	0	1	0	0	0	0	787
2	1	0	0	1	0	0	0	1	783
3	1	0	0	1	0	0	0	0	1620
4	1	0	0	1	0	0	0	1	1533
5	1	0	0	1	0	0	0	0	998
6	1	0	0	1	0	0	0	1	1057
7	1	0	0	1	0	0	0	0	485
8	1	0	0	0	1	0	0	1	339
9	1	0	0	0	1	0	0	0	862
10	1	0	0	0	1	0	0	1	845
11	1	0	0	0	1	0	0	0	2018
12	1	0	0	0	1	0	0	1	1652
13	1	0	0	0	1	0	0	0	893
14	1	0	0	0	1	0	0	1	1024
15	1	0	0	0	1	0	0	0	472
16	1	0	0	0	0	0	0	1	445
17	1	0	0	0	0	0	0	0	811
18	1	0	0	0	0	0	0	1	801
19	1	0	0	0	0	0	0	0	1494

	A
1	5
2	6
3	6
4	3
5	3
6	5
7	5
8	5
9	5
10	6
11	6
12	3
13	3
14	5
15	5
16	5
17	5
18	6
19	6

	A
1	0
2	0
3	0
4	1
5	1
6	1
7	1
8	0
9	0
10	0
11	0
12	0
13	0
14	1
15	1
16	1
17	1
18	1
19	1

	A
1	783
2	1620
3	1533
4	998
5	1057
6	485
7	339
8	862
9	845
10	2018
11	1652
12	893
13	1024
14	472
15	445
16	811
17	801
18	1494
19	1545

Fig. 6. Final data sets for modelling.

kind Markov-modulated linear regression model applied for traffic flow prediction taking into account external environment.

The framework for data exploring and preparing, considering the limitations and assumptions of formulated regression model was developed in this study and demonstrated through particular case: preparing data for passenger numbers on longest Riga tram route prediction. Authors discussed the problems of the real data sets (data captured by e-ticket system “E-talons” and the LEGMC database about weather conditions) and suggested possible decisions realized in automatized framework on the basis of *R* software.

Acknowledgement. This work was financially supported by the specific support objective activity 1.1.1.2. “Post-doctoral Research Aid” (Project id. N. 1.1.1.2/16/I/001) of the Republic of Latvia, funded by the European Regional Development Fund. Nadezda Spiridovska research project No. 1.1.1.2/VIAA/1/16/075 “Non-traditional regression models in transport modelling”.

References

1. TRB’s Transportation Research E-Circular 208: Transformational Technologies in Transportation: State of the Activities
2. van Oort, N., Cats, O.: Improving public transport decision making, planning, and operations by using Big Data, September 2015
3. Big data and transport (2016). http://internatinsportforum.org. Accessed 29 May 2016
4. Big Data’s Implications for Transportation Operations: An Exploration. White Paper—19 December 2014 FHWA-JPO-14-157. www.its.dot.gov/index.htm
5. International Transport forum (OECD) Big Data and Transport Understanding and assessing options (2015). http://www.internationaltransportforum.org/pub/pdf/15CPB_BigData.pdf
6. Leveraging Big Data for Managing Transport Operations, May 2018
7. McGurrin, M.: Big Data and ITS. Noblis, Inc. (2013)
8. de Regt, K., Cats, O., van Oort, N., van Lint, H.: Investigation potential transit ridership by fusing smartcard and GSM Data. Transp. Res. Rec. J. Transp. Res. Board **2652** (2017). https://doi.org/10.3141/2652-06

9. Bouillet, E., Gasparini, L., Verscheure O.: Towards a real time public transport awareness system: case study in Dublin. In: Proceedings of the 19th International Conference on Multimedea 2011, Scottsdale, AZ, USA (2011). https://doi.org/10.1145/2072298.2072463
10. Lathia, N., Capra, L.: How smart is your smartcard? Measuring travel behaviours, perceptions, and incentives. In: Proceedings of the 13th International Conference on Ubiquitous Computing, UbiComp 2011, pp. 291–300. ACM, New York (2011)
11. Lathia, N., Quercia, D., Crowcroft, J.: The hidden image of the city: sensing community well-being from urban mobility. In: Proceedings of the 10th International Conference on Pervasive Computing, Pervasive 2012, pp. 91–98. Springer, Heidelberg (2012)
12. Ma, X.-L., Wu, Y.-J., Wang, Y.-H., Chen, F., Liu, J.-F.: Mining smart card data for transit riders' travel patterns. Transp. Res. Part C: Emerg. Technol. **36**, 1–12 (2013)
13. Xu, G., Zong, Y., Yang, Z.: Applied Data Mining, 1st edn. (2013)
14. Andronov, A., Spiridovska, N.: Markov-modulated linear regression. In: Proceedings' Book: International conference on Statistical Models and Methods for Reliability and Survival Analysis and Their Validation (S2MRSA), Bordeaux, France, pp. 24–28 (2012)
15. Spiridovska, N.: A quasi-alternating Markov-modulated linear regression: model implementation using data about coaches' delay time. Int. J. Circ. Syst. Sig. Process. **12**, 617–628 (2018)
16. Spiridovska, N., Yatskiv (Jackiva), I.: Public transport passenger flow analysis and prediction using alternating Markov-modulated linear regression. In: 29th European Conference on Operational Research (Euro 2018) Handbook, p. 208 (2018)
17. Andronov, A., Spiridovska, N., Santalova, D.: On a parametrical estimation for a convolution of exponential densities. In: Book of Abstracts: The 27th Nordic Conference in Mathematical Statistics (Nordstat 2018), Tartu, Estonia, p. 10 (2018)

Natural Language Processing Knowledge Network Approach for Interactive Highlighting and Summary

Alexander Dudko[1(✉)], Tatiana Endrjukaite[2], and Yasushi Kiyoki[1]

[1] Graduate School of Media and Governance, Keio University, Yokohama, Kanagawa, Japan
aleksandrsdudko@gmail.com

[2] Transport and Telecommunication Institute, Riga, Latvia
tatianaendrjukaite@gmail.com

Abstract. This paper describes a new approach of data retrieval from free text documents in medical domain. Proposed approach creates the document summary and highlights most important keywords in the text by means of document ontology, as an internal representation of the document. This document ontology is a graph with concepts, relations between them, and concept points as a metric of relevance. By means of points the approach performs ambiguity resolution, selects most relevant concepts to display in the summary, and votes for keywords highlighting in the text. Moreover, interactive highlighting reveals additional information from document ontology as reader interacts with the keywords and summary in the document. The described approach helps to speed up analysis and decision-making processes by means of providing aggregated summary for a document and highlighting most meaningful parts of the text.

Keywords: Information retrieval · Text mining · Summary generation · Interactive highlighting

1 Introduction

In the paper we present a novel approach of text mining and data retrieval from free text. Medical documents are used to demonstrate the new text mining approach for summary generation and most important keywords highlighting in the original texts. The approach starts with identification of medical keywords within the document text and then aggregates them into a document ontology in a form of graph. The document ontology is used to produce the summary in a form of a short list of concepts ranged by relevance.

Moreover, the document ontology is used to highlight the most meaningful keywords in the original document. Keywords highlighting provides the explanation of the concepts in the summary and shows the most important places in the text, which made the most significant contribution to the generated summary. Finally, document ontology is used to retrieve all relations between keywords and summary concepts to provide valuable and descriptive information for interactive use.

I. Kabashkin et al. (Eds.): RelStat 2018, LNNS 68, pp. 191–200, 2019.
https://doi.org/10.1007/978-3-030-12450-2_18

In this paper document is a piece of text in medical domain, which describes patient's visit to a doctor. Such document includes a section of standard general information such as patient's name, age, gender and so on, as well as a free-text part, where the doctor describes a patient's complaints, examination results, and other notes related to the patient health. Usually we can easily process the standard information section since it is already organized in a form of predefined key-value pairs. However, the most valuable information is always written in the free-text part which cannot be processed that straightforward and require text data mining approaches for information retrieval.

A particular application of medical document processing in this work is to enable rapid browsing of large patients' documents collections by looking at generated summaries and highlighted text pieces of automatically found most representative parts within the text. This is especially actual for people who review, analyze and track patients' health status from medical records, such as in medical organizations, health care centers, insurance companies, and so on.

The proposed approach does not try to suggest a diagnosis or make any other kind of decision automatically. It shows what concepts are written about in the document, and which keywords explain these concepts. The described approach is general enough, so it can be applied to perform the analysis of any kind of text and in any domain. However, to achieve good results the approach requires a well-designed ontology of the target domain.

2 Related Work

Automatic data summarization is part of machine learning and data mining, where the main idea is to find a subset of data which contains the "information" of the entire set. There are two general approaches to automatic summarization: extraction and abstraction. Today researches are mostly focused on extractive methods [6, 7].

Biomedical text mining refers to text mining applied to texts and literature of the biomedical and molecular biology domain. It is a rather recent research field on the edge of natural language processing and biomedical informatics. Doms and Schroeder in [4] introduced GoPubMed web server, which allows to explore search results with a hierarchically structured vocabulary for molecular biology.

Jenssen et al. demonstrated applications of text mining to research in biomedical text mining. They have carried automated extraction of explicit and implicit biomedical knowledge from publicly available gene and text databases to create a gene-to-gene co-citation network. The associations between genes have been annotated by linking genes to terms from the medical subject heading index and terms from the gene ontology database [5]. Sarker et al. in [16], utilised an annotated corpus that is specialized for the task of evidence-based summarization of text. The approach focuses on the use of corpus-based statistics, and domain-specific lexical knowledge for the identification of summary contents. They also apply a target-sentence-specific summarization technique that reduces the problem of under fitting that persists in generic summarization models.

Molla et al. describe a corpus designed for the development and testing of text processing tools for evidence based medicine, in particular for tasks related to the extraction and summarization of answers and corresponding evidence related to a clinical query. It has been used for the original summarization task for which it was designed, as well as for other related tasks such as the appraisal of clinical evidence and the clustering of the results [17].

Moreover, there are a lot of researches on multi-document summarization. It is an automatic procedure aimed at extraction of information from multiple texts written about the same topic. Resulting summary report allows individual users, such as professional information consumers, to quickly familiarize themselves with information contained in a large cluster of documents [8, 10, 11].

3 Approach

Proposed approach performs medical document analysis, generates summary and highlights important keywords in the document by using natural language processing (NLP) approaches and medical domain ontologies.

This approach is based on points and voting in the ontology graph. Initial document is a natural free text, which is written by a human or in other words by a doctor, and describes the symptoms and treatment of the patient. This document is processed into a descriptor in the form of a graph, that represents information about the document under analysis by using identified keywords from the given medical ontology.

As a result, an output document is created with a meaningful generated summary, which is based on important keywords that summarize the document and which are highlighted along with the original text. It aims to support quick reading of medical documents and make it easier to understand long pieces of texts without the need of reading them completely.

Figure 1 shows the document processing approach for automatic summary generation and keywords highlighting in the document. White boxes represent data, which is an output of one step and is an input for another step. Blue boxes represent the steps themselves. The arrows show how the data flows between the processing steps to generate the output document.

3.1 Keywords Recognition and Related Concepts Identification

The keywords recognition in the text of the given document is performed by means of typical natural language processing approach which includes sentence boundary detection, tokenization, part-of-speech tagging, etc. To process documents and to identify keywords in the text of the document related to medical domain we use Apache cTAKES [12], a publicly available open source natural language processing system for extraction of information from electronic medical record clinical free-text.

Related concepts identification is performed by querying medical ontologies such as SNOMED [13] and MeSH [14] with SPARQL query similar to the one shown below.

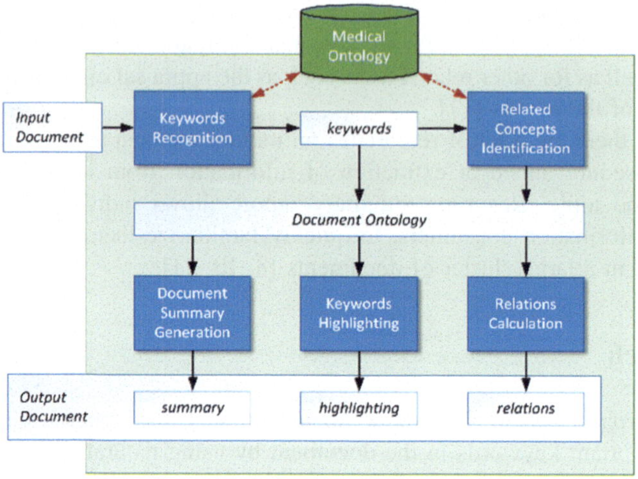

Fig. 1. Medical documents processing approach high-level architecture and steps.

```
PREFIX mesh: http://id.nlm.nih.gov/mesh/
PREFIX meshv: <http://id.nlm.nih.gov/mesh/vocab#>
PREFIX rdfs: <http://www.w3.org/2000/01/rdf-schema#>

SELECT ?child ?child_label ?parent ?parent_label
WHERE {
  ?x rdfs:label "Stomach"@en.
  ?x meshv:broaderDescriptor* ?child.
  ?child meshv:broaderDescriptor ?parent.
  ?child rdfs:label ?child_label.
  ?parent rdfs:label ?parent_label.
}
```

3.2 Document Ontology

Document ontology includes information about all recognized keywords in the document, as well as all related concepts identified in the previous step. Document ontology creation also includes the measure of relevance for every concept which is calculated through assignment of points to concepts. [3, 8].

Every identified keyword in the document text gets 1 point. If a keyword is repeated multiple times in the text, every occurrence adds 1 point. There could be one or more concepts corresponding to a keyword in the text. The points of the identified keyword are propagating to the corresponding concepts and then to its parents, every time dividing value equally between all parent concepts.

The equation for points calculation is shown in (1) and (2).

$$U(p) = \sum_{c \in C(p)} u(p, c), \tag{1}$$

$$u(p, c) = \frac{U(c)}{|P(c)|}, \tag{2}$$

where $U(p)$ is number of points assigned to node p, $u(p, c)$ is number of points propagated from node c to p, $C(p)$ is a set of children for node p, $P(c)$ is a set of parents for node c.

When a concept has only one parent concept, parent concept gets the same number of points as the current concept. When a concept has more than one parent concept, then the points are divided equally between all parent concepts. If a concept is a parent for multiple child concepts, points coming from all child concepts are added together.

This points propagation step is repeated recursively on every level up until the top-level of concepts is reached. The parent-child relation guarantees that the graph will be unidirectional with no cycles.

Figure 2 represents an example how points are propagated from bottom nodes up. Numbers in the nodes show the assigned points. Numbers at the edges display what amount of points is given from a concept to its parent concepts. The graph of concepts with assigned points and the document keywords makes up the document ontology.

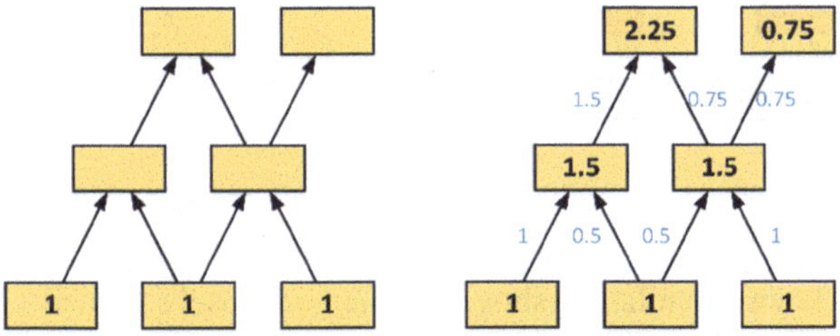

Fig. 2. Points propagation from keywords to top-level concepts.

3.3 Document Summary

Document summary is generated from the document ontology. Top levels in the ontology hierarchy are the most suitable and meaningful for the summary, because they represent high level concepts, which aggregate more specific concepts from the underlying layers. However, it can be controlled to what level of detail it is required to go in the summary. Either the summary is kept short but less informative, or the summary is made longer, and it provides detailed information about the document. It is very important to keep in mind that in the case of fully detailed summary it may become as long as the original text, but it is still valuable because it represents information in the structured form.

3.4 Relations Calculation and Interactive Highlighting

To highlight important meaningful keywords in the document, we use keywords voting approach which makes the default selection of most important parts in the document. The approach, as described in [3], identifies which keywords have to be highlighted in the text based on votes for every keyword, which reached the summary of the document.

Moreover, in this work we propose interactive highlighting which reveals additional information from document ontology as reader interacts with the keywords and summary in the document. Such interactive highlighting gives explanations to the reader why specific items appear in the document summary, and how keywords in the text relate to each other.

Summary concept item to keywords relation is calculated by top-to-bottom walk in the document ontology as presented in Fig. 3a. And the opposite relation from keyword to summary items is determined by a bottom-to-top walk through the document ontology, as presented in Fig. 3b.

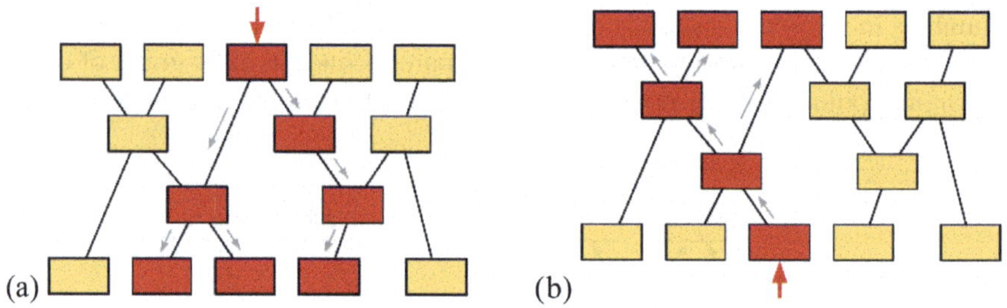

Fig. 3. Relations calculation between summary concepts and keywords.

Keyword to keyword relation shows similar or closely related concepts directly mentioned in the document. In a simple case, this type of relation is calculated as a distance in the document ontology graph as the minimum number of edges, which has to be passed to reach from one keyword to another. Figure 4 shows an example of distances calculation for a single selected keyword.

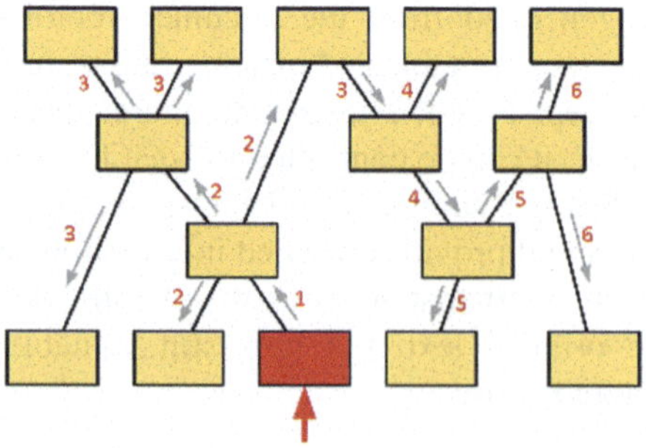

Fig. 4. Example of keyword to keyword relation calculation.

However, for a better result edges in the graph can be valued with different weights. For example, concept to concept relation gets distance equal to 1, keyword to concept gets distance equal to 0.5, and top-level concept to ultimate concept "thing" gets distance equal to 1.5. In this case, keywords pointing to the same concept would become closer to each other, but keywords from different top-level concepts will be less close.

4 Experiments and Discussions

In this experiment the following input document text was taken:

> The patient said that sometimes the blood pressure is too high. Feels pain in the chest. The previous doctor said that patient has had heart trouble such as an abnormal electrocardiogram. Currently the patient suffers from frequent cramps in his legs and has difficulty breathing. The cholesterol level is high. Need to test the patient on abdominal aortic aneurysm and critical aortic stenosis.

Keywords recognition step found 18 keywords in the specified text. The identified keywords are following:

> pain, aortic aneurysm, aortic, legs, blood, stenosis, heart, aortic, heart trouble, abdominal aortic aneurysm, aneurysm, aortic stenosis, cholesterol, electrocardiogram, blood pressure, cramps, abdominal, chest.

Document ontology has been constructed based on these keywords as described in the approach. In total, the ontology includes 18 keywords and 72 related concepts, where top-level concepts are 17 of them. A list of the top-level concepts is shown in the Table 1.

Summary concepts to keywords relations and keywords to summary concepts relations can be both described at once in a form of a table as shown in Fig. 5. As expected, most concepts which appear higher in the summary are supported by multiple keywords from the document text. A good example we can see for "Cardiovascular Diseases" concept which is supported by keywords "heart trouble", "abdominal aortic aneurysm", "aortic aneurysm", "aneurysm", "aortic stenosis".

On the other side, when a keyword has relations to multiple concepts, it means ambiguity and we should give priority to concepts which have greater value of points to resolve the uncertainty. For example, keyword "pain" which appears in the text has very general meaning. It could refer to a wide variety of concepts by itself. After considering points for ambiguity resolution, the most likely concept for this keyword becomes "Pathological Conditions, Signs and Symptoms", but concepts like "Psychological Phenomena and Processes" become very unlikely for this document text.

Keyword to keyword relations are presented in Fig. 6. We can see that the keyword "heart trouble" is very close to keywords "aortic stenosis" and "aneurism", but keyword "electrocardiogram" has a close relation to "blood pressure". These relations show a good example of revealed meaningful information that supports reading and understanding the document.

Table 1. Top-level concepts with points.

Concept description	Points
Cardiovascular diseases	5.0
Cardiovascular system	3.0
Body regions	3.0
Diagnosis	1.5
Pathological conditions, signs and symptoms	1.5
Polycyclic compounds	0.75
Nervous system diseases	0.75
Fluids and secretions	0.5
Hemic and immune systems	0.5
Circulatory and respiratory physiological phenomena	0.5
Lipids	0.25
Musculoskeletal diseases	0.25
Musculoskeletal and neural physiological phenomena	0.25
Therapeutics	0.0625
Natural science disciplines	0.0625
Behavioral disciplines and activities	0.0625
Psychological phenomena and processes	0.0625

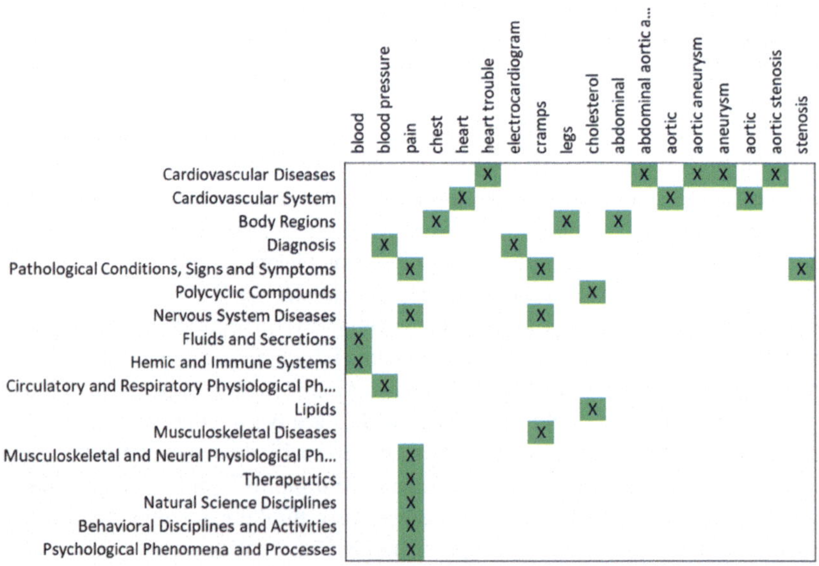

Fig. 5. Relations between keywords and summary concepts.

	blood	blood pressure	pain	chest	heart	heart trouble	electrocardiogram	cramps	legs	cholesterol	abdominal	abdominal aortic a...	aortic	aortic aneurysm	aneurysm	aortic	aortic stenosis	stenosis
blood	0	9	8	8	7	7	9	9	9	9	8	10	9	9	8	9	9	8
blood pressure	9	0	10	10	9	9	6	11	11	11	10	12	11	11	10	11	11	10
pain	8	10	0	9	8	8	10	4	10	10	9	11	10	10	9	10	10	6
chest	8	10	9	0	8	8	10	10	6	10	3	11	10	10	9	10	10	9
heart	7	9	8	8	0	7	9	9	9	9	8	10	5	9	8	5	9	8
heart trouble	7	9	8	8	7	0	9	9	9	9	8	6	9	5	4	9	3	8
electrocardiogram	9	6	10	10	9	9	0	11	11	11	10	12	11	11	10	11	11	10
cramps	9	11	4	10	9	9	11	0	11	11	10	12	11	11	10	11	11	7
legs	9	11	10	6	9	9	11	11	0	11	6	12	11	11	10	11	11	10
cholesterol	9	11	10	10	9	9	11	11	11	0	10	12	11	11	10	11	11	10
abdominal	8	10	9	3	8	8	10	10	6	10	0	11	10	10	9	10	10	9
abdominal aortic aneurysm	10	12	11	11	10	6	12	12	12	12	11	0	12	2	3	12	8	11
aortic	9	11	10	10	5	9	11	11	11	11	10	12	0	11	10	1	11	10
aortic aneurysm	9	11	10	10	9	5	11	11	11	11	10	2	11	0	2	11	7	10
aneurysm	8	10	9	9	8	4	10	10	10	10	9	3	10	2	0	10	6	9
aortic	9	11	10	10	5	9	11	11	11	11	10	12	1	11	10	0	11	10
aortic stenosis	9	11	10	10	9	3	11	11	11	11	10	8	11	7	6	11	0	10
stenosis	8	10	6	9	8	8	10	7	10	10	9	11	10	10	9	10	10	0

Fig. 6. Identified keyword to keyword relations.

5 Conclusions

In this paper we described a new approach of data retrieval from free text documents in medical domain. Proposed approach generates the document summary and highlights most important keywords in the text as a result, which is determined based on natural language processing and document ontology in a form of graph. To achieve this result we process the document to find the medical-related keywords, match them with known medical concepts and build hierarchy from related concepts to generate document ontology. This document ontology is a graph with concepts in the nodes and assigned points that show relevance. The document ontology is used to generate a summary as a list of most related top-level concepts, and to perform highlighting in the original text to emphasize most important parts in the document.

The proposed approach has an important feature of ambiguity resolution, which is a very important factor in text mining. This feature is achieved by means of concept points in the document ontology. Ambiguity resolution is explained on an example in this work.

Besides the direct representation of identified information in the summary and static highlighting, this work proposes a way of relations information extraction to provide interactive behaviour support. Such interaction can give explanations regarding summary items, and can display related information within the document text through interactive highlighting.

The paper presents experimental results which were explained and discussed in detail. The results show that automatic summary generation and keywords highlighting can be successfully performed by the proposed method to achieve meaningful and highly relevant results. Moreover, identified knowledge about relations between keywords-to-keywords and keywords-to-concepts provides valuable information for useful explanations and additional knowledge revealing in interactive use.

References

1. Wan, X., Zhang, J.: CTSUM: extracting more certain summaries for news articles. In: Proceedings of the 37th International ACM SIGIR Conference on Research and Development in Information Retrieval, Queensland, Australia, 6–11 July 2014
2. Ren, P., Chen, Z., Ren, Z., Wei, F., Ma, J., Rijke M.: Leveraging contextual sentence relations for extractive summarization using a neural attention model. In: Proceedings of the 40th International ACM SIGIR, Tokyo, Japan, 7–11 August 2017
3. Dudko, A., Endrjukaite, T., Kiyoki, Y.: An information retrieval approach for text mining of medical records based on graph descriptor. In: Proceedings of the International Conference on Information Modelling and Knowledge Bases (EJC 2018), Riga, Latvia (2018)
4. Doms, A., Schroeder, M.: GoPubMed: exploring PubMed with the Gene Ontology. Nucleic Acids Res. **33**, 783–786 (2005)
5. Jenssen, T.K., et al.: A literature network of human genes for high-throughput analysis of gene expression. Nat. Genet. **28**(1), 21–28 (2001)
6. Mihalcea, R., Tarau, P.: TextRank: bringing order into text. In: Proceedings of the International Conference on Empirical Methods on Natural Language Processing (EMNLP), Barcelona, Spain (2004)
7. Yatsko, V.A., Starikov, M.S., Butakov, A.V.: Automatic genre recognition and adaptive text summarization. Autom. Doc. Math. Ling. **44**(3), 111–120 (2010)
8. Mittal, S., Mittal, A.: Versatile question answering systems: seeing in synthesis. Intell. Inf. Database Syst. **5**(2), 119–142 (2011)
9. Dudko, A., Endrjukaite, T., Kiyoki, Y.: Medical documents processing for summary generation and keywords highlighting based on natural language processing and ontology graph descriptor approach. In: Proceedings of the 19th International Conference on Information Integration and Web-based Applications & Services (iiWAS 2017), Salzburg, Austria (2017)
10. Lin, H., Bilmes, J.: Learning mixtures of submodular shells with application to document summarization. In: Proceedings of the Conference on Uncertainty in Artificial Intelligence, Catalina Island, US, 14–18 August 2012
11. "Simplish—Simplification and Summarization Tool". The Goodwill Consortium. Accessed Oct 2018
12. Apache cTAKES: Clinical Text Analysis and Knowledge Extraction System. http://ctakes.apache.org/
13. SNOMED CT. http://www.snomed.org/
14. Medical Subject Headings. https://www.nlm.nih.gov/mesh/
15. History and Physical Examination Examples. http://www.clinicaladvisor.com
16. Sarker, A., Moll'a, D., Paris, C.: Query-oriented evidence extraction to support evidence-based medicine practice. J. Biomed. Informatics **59**, 169–184 (2016)
17. Molla, D., Santiago-Martınez, M.E., Sarker, A., Paris, C.: A corpus for research in text processing for evidence based medicine. Lang. Resour. Eval. **50**, 1–23 (2015)
18. National Drug File—Reference Terminology. https://bioportal.bioontology.org/ontologies/NDFRT

Intelligent Transport Systems

Use Cases and Introductory Analysis of the Dataset Collected Within the Large Network of Public Charging Stations

Milan Straka(✉) and Ľuboš Buzna

University of Žilina, Univerzitná 8215/1, Žilina, Slovakia
milan.straka@fri.uniza.sk

Abstract. The recent rise of electric vehicles (EV) brings social and technological changes in the transportation and energy sectors, including the massive deployment of charging stations. To provide effective decision support for the operators of charging stations, we are exploring possibilities for exploiting the available dataset and present results of preliminary data analyses. Our dataset contains over 32 million meter-readings from charging of plug-in electric vehicles (PEV) on more than 1700 charging stations, located in the Netherlands. Based on the discussions with experts and the available literature, three main application areas were identified: forecasting of demanded energy, identification of customer segments and characterization of suitable locations for charging stations. As an example of a use case, we forecast the consumption of the electric energy on charging stations in the COROP region of Utrecht. Two kinds of SARIMAX model together with three kinds of training-forecasting procedure are used with various exogenous predictors to identify, which combination provides the best long-term forecasts.

Keywords: Data analysis · Electric mobility · Energy forecasting · Vehicle charging

1 Introduction

Electric mobility became quickly popular as a research topic over the last years, what is also reflected by a large number of studies appearing in the scientific literature. First, the lack of field data formed an obstacle and researches studied a potential to change conventional vehicles to electric indirectly. An approach, where GPS traces of vehicles collected in two Italian cities Modena and Firenze were used to extract the travel behaviour and hence to estimate the expected demand for charging vehicles, has been used in [1]. A Similar approach was applied in [2]. Data from 1.6 million mobile phones collected in Japan were used to extract travel trips. Clustering methods were used to analyse the data and four classes of travel behaviour were identified. The EV adoption potential for the resulting classes was analysed by evaluating the feasibility of trips while considering battery capacity and charging opportunities. Another work adopting this approach is the reference [3]. GPS traces of 76 vehicles collected by the University of Winnipeg were analysed and combined with simulation models to evaluate the energy consumption while assuming charging at home, at work and near to

I. Kabashkin et al. (Eds.): RelStat 2018, LNNS 68, pp. 203–213, 2019.
https://doi.org/10.1007/978-3-030-12450-2_19

shopping malls. A stochastic method, capturing the relationship between the departure, arrival and travel time more accurately than other existing methods, was proposed. Such approaches can be valuable when searching for suitable positions of charging stations or when estimating the adoption potential of EVs, however, they may not capture the charging behaviour of EV drivers in a realistic way.

Studies based on the operational data from charging stations started to appear in the scientific literature only very recently. In [4], data from mobile phones were combined with the data from charging stations to link travel patterns of EV drivers with the charging behaviour. By shifting the time when the vehicles are charged within the time window that is given by the travel behaviour, the potential to shave the demand peak was explored. Considering the electricity tariffs, the monetary gains to incentivize the adoption of recommendations were calculated. The potential to coordinate the charging load of EVs in order to flatten the load curve was analysed in [5]. Clustering methods were applied to data from charging stations to identify types of EVs and characteristics of charging sessions of users in each cluster were analysed. Considering the flexibility in charging an optimization approach was developed to evaluate the load flattening scenarios. It was concluded that near workplace charging should be targeted to fill the afternoon valley and the near home charging sessions should be targeted to fill night valley. Results of the Western Australian Electric Vehicle Trial (2010–2012) were evaluated in [6]. Data from 11 EVs and 23 charging stations were collected and evaluated. The study found that most charging is related to the business and home locations, while public charging stations were used for 33% of charging events only. Most of the trips are short and on average vehicles are connected to charging stations when the battery has more than 50% of the charge remaining. Interestingly, authors concluded that their results suggest that parking spaces are in higher demand than charging facilities. Initial results of analyses of the actual usage patterns of the public charging infrastructure in the city of Amsterdam, based on more than 109,000 charging sessions, were presented in [7]. The time series describing the evolution of the number of charging stations, the number of charge sessions, and consumed energy (in kWh) were descriptively analysed, concluding that the system is intensively growing. Utilization of charging stations was compared across different city districts, while finding that utilization improves over time in all cases. As possible influence factor were identified, the learning process of EV owners and the increasing popularity of Car2Go sharing system, which leads to the increasing number of EVs that are charged at public charging stations. In [8], authors argued that in order to make solid predictions of the future use of charging stations it is important to account for significant differences in the behaviour among user types. Therefore, they analysed charging patterns of different user types characterized by the timing, charging amount and location preferences. Initial set of six user types and the set of eleven measurable user characteristics was created based on the discussion with the policymakers. Then, they compared differences in connection patterns, distribution of start times, distribution of consumed energy per sessions and other characteristics of identified user groups. The ability of the stochastic logistic regression approach to make one day ahead profile prediction of the energy consumption on single charging stations was tested in [9]. Time series modelling was applied to aggregated charging stations load in [10]. Factors influencing the success of public charging stations were tested in [11]. Paper revisits four hypotheses used in the past by the city of Hague to place new public charging stations and validates them by analysing the available data, concluding that the location of the charging stations plays a central role for their usage. Models to predict the need for public charge points were developed in [12]. Factors such as high average gross monthly income and high number of cars seem to be the most significant indicators of the need for public charging infrastructure. Effects of local incentives, regulations and spatial structures related to the public charging infrastructure were evaluated in [13]. A set of key performance indicators (KPI) was used to compare the situation in five Dutch cities. The KPIs involve the amount of kWh charged, the number of unique users, the occupancy, the number of active charging stations, etc. The results of analysis indicate that the number of profitable charging stations is still very low, and cities should not only be focused on building new infrastructure but should improve the utilization of already established stations. Another useful source of information on charging behaviour are surveys. Psychological dynamics underlying the behaviour of EV drivers was examined in [14]. Another survey-based approach to examine characteristics of Dutch EV users was presented in [15].

This paper is organized as follow: Sect. 2 presents the datasets and the use cases that are suitable for further research. Section 3 introduces some preliminary results of time series analyses. To conclude, we summarize our main findings in Sect. 4.

2 Description of the Dataset and the Use Cases

2.1 ElaadNL Dataset

The dataset provided by the Dutch company ElaadNL covers the time period from January 2012 to March 2016, contains information about more than 1700 stations placed all over the Netherlands and 27,000 users identified by Radio-frequency identification (RFID) cards. The dataset consists of two tables, the first one is *Meter-readings*, containing 32 million rows. Each row is a measurement taken every 15 min when the vehicle is plugged into a station. Columns are the timestamp, RFID card, transaction ID and charged power. The second table maintains information about individual charging *Transactions*. The Transactions table contains attributes charging station id, RFID card, charging station GPS coordinates, socket number, charged power, transaction start time, transaction end time, connection time and idle time (i.e. the length of the time interval when an EV is connected but not charging).

2.2 Proposed Research Problems to Be Addressed by Analyzing Operation Data from a Network of Charging Stations

After discussions with representatives of Dutch company ElaadNL and Slovak company Voltia, that are developing and deploying the charging infrastructure, three topics were identified to be addressed by the data analysis methods.

(A) Identification of suitable candidate locations for charging stations. Improperly selected locations or the insufficient number of charging stations can negatively affect the number of drivers interested in electric vehicles [16]. Therefore, it is important to provide effective decision support to operators and public authorities to enable the more efficient process of charging infrastructure deployment. In the literature, there is currently a lack of statistical models that could be used to identify suitable candidate locations for charging stations. Building such models requires identification of attributes potentially affecting the success of charging stations. Equally important is to find a suitable set of explanatory variables that describe the success of charging stations such as for example occupancy, idle time, the average count of transactions, the average energy consumption, etc. Charging stations in the ElaadNL dataset are heterogeneous w.r.t. these characteristics (Fig. 1), e.g. the number of transactions on charging stations ranges from less than 10 to more than 1000. Expected problems that need to be properly handled are multicollinearity among predictors and low explanatory and predictive power of models. These problems can be handled by using tolerant methods such as variable selections methods, e.g. Lasso and Elastic net [17] and by developing models to find specific clusters of charging stations (e.g. stations predominantly used for home charging, work charging or opportunistic charging).

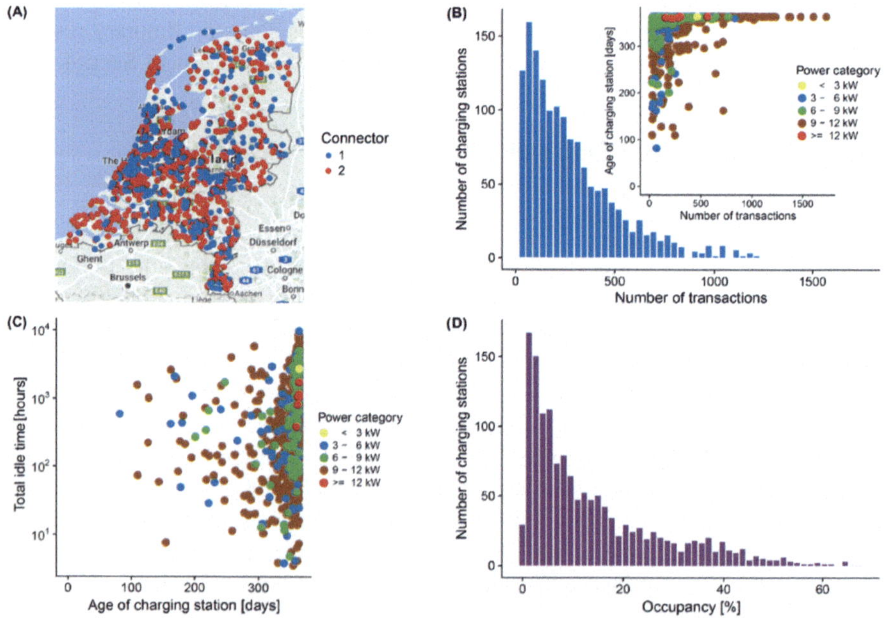

Fig. 1. Selected characteristics of charging stations in the ElaadNL dataset observed within the period of 03/2015–03/2016 when the number of charging stations was relatively stable. **A** Geographical positions of charging stations with one or two connectors; **B** Histogram of the number of transactions. The inset is a scatter plot of the number of transactions vs. the age of the charging station colored according to the power category; **C** Scatter plot of the accumulated idle time as a function of the age of the charging stations, colored according to the power categories; **D** Histogram of the occupancy.

(B) Forecasting applications. Charging station operators need to ensure supply of electricity for charging stations. Thus, the ability to predict electricity consumption can help them in using renewable energy sources or buy electricity for better prices. Long and short-term forecasting of the electric load is a very well-studied problem in the scientific literature [10, 18, 19]. If the load from several charging stations is aggregated, it has a character of time series (Fig. 2) and thus methods for the time series analyses can be applied for predictions. Other possible approaches are autoregressive conditional Markov chain-based models, neural network models, classification approaches and wavelet models [10]. In the literature, there are only a few works focusing on forecasting of the load from charging station and we did not find any work applying autoregressive forecasts and evaluating the importance of exogenous variables that could improve the accuracy of predictions. Similar to the load forecasting application is the prediction of the occupancy of individual charging stations. Such predictive models could be used to guide EV drivers when choosing charging stations.

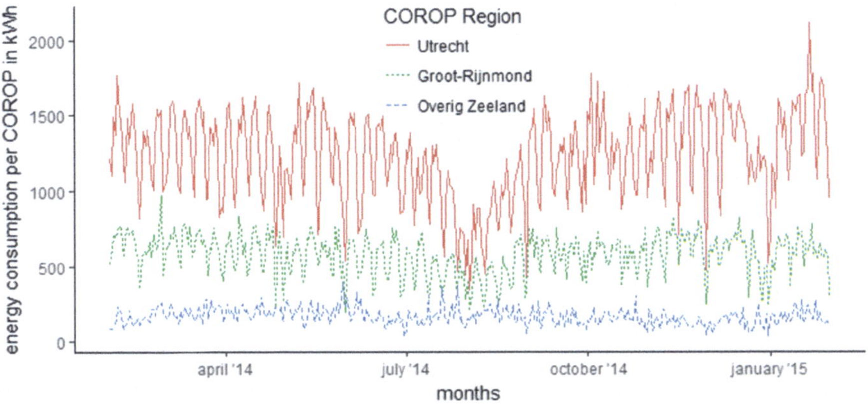

Fig. 2. The plot of aggregated daily energy consumption on charging stations in selected Dutch COROP regions from March 3rd, 2014 till March 1st, 2015 (364 days). (Color figure online)

(C) Identification of customer segments. The goal is to look for different attributes and their combinations that would enable to identify customer segments (important classes of customer that behave in a similar way). In two or three dimensions it is possible to inspect patterns in data by visualization techniques (Fig. 3) and in general they can be identified by using various unsupervised clustering methods (k-means, self-organizing maps, dbscan algorithm [20]). Identified clusters can be afterward interpreted and described by their properties. From the business perspective, it is also interesting to predict how clusters are going to evolve in time, what could be helpful in targeting the business strategies and provision of services more efficiently.

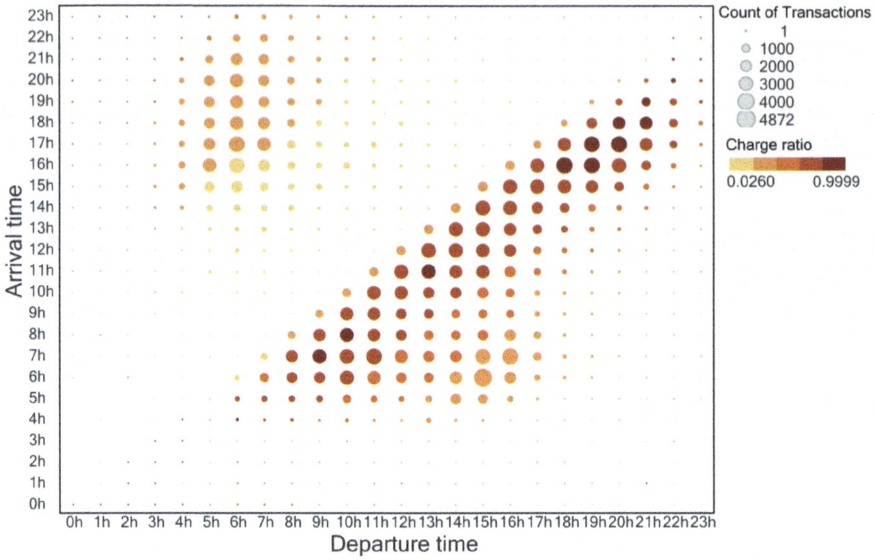

Fig. 3. The number of charging transactions on weekdays categorized by the arrival and departure times (rounded to hours). The size of circles is proportional to the number of transactions that occurred at given hours and the color indicates the charge ratio (charge time divided by connection time).

3 Energy Consumption Analysis

To forecast energy consumption for a single charging station is difficult task and typically it leads to poor accuracy of predictions [9]. As a test case we consider the aggregated load time series of the Utrecht COROP region extracted from the ElaadNL dataset, because the data is the most complete and consequently the energy load profile is also reasonably stable. Analyses of aggregated data from charging stations with ARIMA models have been already done in [10], however, without considering exogenous variables. In this section, we investigate how exogenous variables can enhance the predictions.

By inspecting the energy consumption time series of the Utrecht region, we found apparent seasonal patterns. To identify seasonal patterns on a short time scale autocorrelation function (ACF) and partial autocorrelation function (PACF) were used, showing very strong weekly seasonality. Average energy charged during weekends is much lower (by 30% on average) than in weekdays, hence, we added a predictor to distinguish them. Moreover, weekends are similar to bank holidays and bridge days [21], therefore a dummy predictor was defined also for them. Energy consumption often exhibits double or higher seasonality patterns [18]. By eyeballing the data, a larger drop-down on summer holidays and towards the end of December was identified, which occurs for all 4 years of the data, therefore we considered the energy consumption lagged 364 days (the same day of the week approximately one year ago) as the exogenous predictor. By investigating the relationship between users and energy consumption, we found out that 300 users (super users) with the largest consumption among all 15,000 that have charged their EVs in Utrecht region are responsible for 50% of total energy consumption. The count of super users charging on a given day has the correlation coefficient of 0.91 with charged power. The count of super users lagged 364 days is more significant for predicting consumed energy than energy consumed lagged 364 days (confirmed by linear regression). This might be explained by more regular patterns in the count of super users compared to energy consumed by all users. Thus, rather than the energy consumption lagged 364 days, we used as exogenous predictor the count of super users lagged 364 days.

Let the historical data be denoted by $y_1, \ldots, y_T$, where T is the number of observations and $\hat{y}_t$ be the estimate of y_t value, where t is a time index. Based on the preliminary investigations, to include seasonality and exogenous variables, the SARIMAX model will be used. SARIMAX is an extension ARIMA *(p, d, q)* model, which can be formulated as follows

$$y'_t = \phi_1 y'_{t-1} + \phi_2 y'_{t-2} + \cdots + \phi_p y'_{t-p} + \theta_1 \epsilon_{t-1} + \theta_2 \epsilon_{t-2} + \cdots + \theta_q \epsilon_{t-q} + \epsilon_t,$$

where $\phi_1, \ldots, \phi_p$ are the parameters associated with lagged variables (*AR*(*p*) part of the model), $y'_t = y_t - y_{t-1}$ are the first order (i.e. $d = 1$) differences (in this paper the differences of higher order are not considered), $\theta_1, \ldots, \theta_q$ are the parameters associated with errors $\epsilon_{t-l} = y_{t-l} - \hat{y}_{t-l}, \mathrm{l} \in \{1, \ldots, q\}$ of q previous models (*MA*—part of the model) and ϵ_t is the white noise. This model can be extended to a SARIMA (*p*, *d*, *q*) (*P*, *D*, *Q*)$_m$ model adding seasonal component, whose seasonal parameters *P*, *D* and *Q* have analogous meaning as (*p*, *d*, *q*), but they are applied to observed values lagged *m* steps. By adding exogenous variables to the SARIMA model, we obtain a SARIMAX model. As a model error measurement, we will use Mean Absolute Percentage Error (MAPE)

$$\text{MAPE} = \frac{100\%}{|N|} \sum_{t=1}^{|N|} \frac{y_t - \hat{y}_t}{y_t},$$

where N is the set of estimated values.

As another potential improvement the *Alternating model*, composed of two SARIMAX models (one for weekends and one for weekdays), was devised. To account for the information about the weekdays in the weekend part of *Alternating model*, we have added the average consumption over a weekday before the weekend as an exogenous variable to the model. The use of exogenous predictors is summarized in Table 1.

As a basic training set for models, we used energy consumption in the Utrecht region dating from March 3rd, 2014 till March 1st, 2015 (364 days) as depicted in Fig. 2 with the red line. To each model three types of training-forecasting procedures depicted in Fig. 4 were applied. "Static procedure" uses the same model for all forecasts. "Fixed origin" uses fixed start and updates the model in time. "Rolling origin" procedure shifts the origin of the training set while maintaining the training set length constant.

Table 1. Exogenous predictors used in the SARIMAX models.

	Single model	Alternating model	
		Weekdays	Weekend
Super users 364 lag	✓	✓	✓
Holidays	✓	✓	–
Fourier terms	–	2	1
Weekend dummy	✓	–	–
Weekday average	–	–	✓

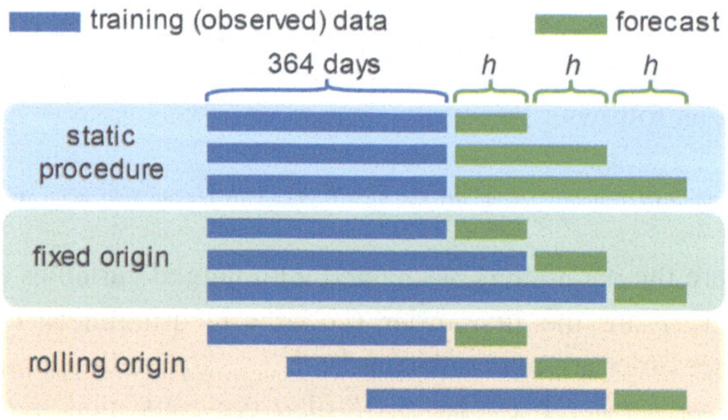

Fig. 4. Illustration of applied training-forecasting procedures. Forecast horizon is denoted by h.

3.1 Results of Numerical Experiments

We have evaluated all three types of modelling approaches by evaluating their forecasting ability of the energy consumption up to 280 days ahead. For forecasting we used three different forecast horizons $h \in \{7, 14, 28\}$ days, i.e. we do 280/7 = 40, 280/14 = 20, 280/28 = 10 forecasts, respectively. We calculated mean values of MAPE for each forecast horizon and listed them in Table 2. We used R software and auto.arima function from forecast library to perform the forecasts. The function does the automatic search for the parameter value of Box-Cox stationarisation [22] and the SARIMAX model parameter values.

Table 2. Values of MAPE for the proposed models and the used forecast horizon values.

Forecast horizon	Single model			Alternating model		
	Static procedure (%)	Fixed origin (%)	Rolling origin (%)	Static procedure (%)	Fixed origin (%)	Rolling origin (%)
7 days	16.35	12.75	12.58	14.06	10.54	10.46
14 days	16.35	12.68	12.66	14.06	10.42	10.36
28 days	16.35	13.56	13.87	14.06	10.51	10.50

As the first experiment, we tested the impact of exogenous predictors on forecasts of basic 7 days horizon static procedure while using the single model. In this case, adding all exogenous predictors to the model lowered MAPE from 37.35 to 16.35%.

The results are strongly affected by the used training-forecasting procedure and less affected by the forecast horizon (Table 2). As expected, extending the training set by using the most recently observed values is improving the precision of predictions. Please note that static procedure provides the same accuracy independently on the forecast horizon. This is a trivial effect of using only one model. There is no notable difference between fixed origin and rolling origin procedures. Alternating models outperform single model by more than 2%. We also have tested, whether the forecast errors of the models tend to change over time. Applying linear regression to this dependency and by examining the p-value of the slope coefficient, we found with both models that static procedure has a significant positive slope parameter (p-value <0.005), meaning that the error is increasing over time. But the trend of energy consumption is also increasing in time. Forecast error in remaining two procedures does not exhibit any significant trend.

4 Conclusions

We have overviewed articles in the field of electric vehicles charging, aiming at data analysis approaches. Available dataset and three use cases that arose from discussions with experts were described. First use case concerns the identification of suitable candidates for charging station locations using regression methods. The second use case is forecasting of charging station demand. As a third use case, we identified analyses of customer segments using clustering algorithms, with the consecutive aim to analyse and predict the evolution of these segments. Furthermore, we addressed the problem of forecasting energy consumed by charging stations in the Utrecht COROP region. We used single SARIMAX model and added exogenous variables as the lagged count of super users, bank holidays, bridge days and weekend dummy variables leading to large improvement in the precision of the forecasts. Moreover, we created an alternating model, which uses two SARIMAX models (one for weekdays and one for the weekend) and includes exogenous variables as well. Three training-forecasting procedures were applied, and combined with two proposed models, and evaluated on three different forecast lengths. The best results were obtained from procedures learning in time, while alternating model reached MAPE about 10%. This level of accuracy, while taking into account the high variability of data, might be considered as already sufficient for long-term prediction applications.

As future steps, we envisage inclusion of weather data and the use of machine learning models as Deep Neural Networks, Support Vector Regression and Gradient Boosted Decision trees. Moreover, the clustering of charging stations based on charging behaviour could be used to improve the accuracy of results.

Acknowledgements. This work was supported by the research grants VEGA 1/0463/16 "Economically efficient charging infrastructure deployment for electric vehicles in smart cities and communities", APVV-15-0179 "Reliability and emergency systems on infrastructure with uncertain functionality of critical elements" and it was facilitated by the FP 7 project ERAdiate [621386].

References

1. De Gennaro, M., Paffumi, E., Martini, G.: Customer-driven design of the recharge infrastructure and Vehicle-to-Grid in urban areas: a large-scale application for electric vehicles deployment. Energy **82**, 194–311 (2015)
2. Zhang, H., Song, X., Xia, Y., Yuan, T., Fan, Z., Shibasaki, R., Liang, Y.: Battery electric vehicles in Japan: human mobile behavior based adoption potential analysis and policy target response. Appl. Energy **220**, 527–535 (2018)
3. Ashtari, A., Biberau, E., Shahidinejad, S., Molinski, T.: PEV charging profile prediction and analysis based on vehicle usage data. IEEE Trans. Smart Grids **3**(1), 341–350 (2012)
4. Xu, Y., Çolak, S., Kara, E.C., Moura, S.J., González, M.C.: Planning for electric vehicle needs by coupling charging profiles with urban mobility. Nat. Energy **3**, 484–493 (2018)
5. Sadeghianpourhamani, N., Refa, N., Strobbe, M., Develder, C.: Quantitative analysis of electric vehicle flexibility: a data-driven approach. Electr. Power Energy Syst. **95**, 451–462 (2018)
6. Speidl, S., Bräunl, T.: Driving and charging patterns of electric vehicles for energy usage. Renew. Sustain. Energy Rev. **40**, 97–110 (2014)
7. van den Hoed, R., Helmus, R.J., de Vries, R., Bardok, D.: Data analysis on the public charge infrastructure. In: EVS27 Symposium, Barcelona (2013)
8. Helmus, R.J., van den Hoed, R.: Unraveling user type characteristics: towards a taxonomy for charging infrastructure. World Electr. Veh. J. **7**(4), 589–604 (2015)
9. Bikcora, C., Refa, N., Verheijen, L., Weiland, S.: Prediction of availability and charging rate at charging stations for electric vehicles. In: Probabilistic Methods Applied to Power Systems (PMAPS), pp. 1–6 (2016)
10. Louie, H.M.: Time-series modeling of aggregated electric vehicle charging station load. Electr. Power Compon. Syst. **45**, 1498–1511 (2017)
11. Montfort, K., Kooi, M., Poel, G., van den Hoed, R.: Which factors influence the success of public charging stations of electric vehicles? In: 6th Hybrid and Electric Vehicles Conference, London (2016)
12. Montfort, K., Poel, G., Visser, J., van den Hoed, R.: Prediction of necessary public charging infrastructure of electric vehicles. In: 6th Hybrid and Electric Vehicles Conference, London (2016)
13. Wolbertus, R., van den Hoed, R., Maase, S.: Benchmarking charging infrastructure utilization. In: EVS 29 Symposium, Quebec, pp. 754–771 (2016)
14. Franke, T., Krems, J.: Understanding charging behavior of electric vehicles users. Transp. Res. Part F **21**, 75–89 (2013)
15. Hoekstra, A., Refa, N.: Characteristics of Dutch EV drivers, In: EVS30 Symposium, Stuttgart, pp. 1–13 (2017)
16. Sandén, B., Wallgren, P.: System Perspectives on Electromobility. Chalmers, Goteborg (2017)
17. Hastie, T., Tibshirani, J., Friedman, J.: The Elements of Statistical Learning, 2nd edn. Springer, Heidelberg (2017)

18. Dannecker, L.: Energy Time Series Forecasting: Efficient and Accurate Forecasting of Evolving Time Series from the Energy Domain. Springer, Heidelberg (2015)
19. Weron, R.: Electricity price forecasting: a review of the state-of-the-art with a look into the future. Int. J. Forecast. **30**(4), 1030–1081 (2014)
20. Han, J., Kamber, M., Pei, J.: Data Mining: Concepts and Techniques, 3rd edn. Morgan Kaufmann, Burlington (2012)
21. OfficeHolidays: Office Holidays in Netherlands. https://www.officeholidays.com/countries/netherlands/2015.php. Accessed 20 May 2015
22. Hyndman, R.J., Athanasopoulos, G.: Forecasting: Principles and Practice. OTexts. https://otexts.org/fpp2/. Accessed 25 May 2018

Influence of Constructive Materials of Road Cover on Magnetic Field Dispersion of Wireless Power Transmission Systems

Rodions Saltanovs[1], Aivars Rubenis[1], and Alexander Krainyukov[2(✉)]

[1] SIA "TransfoElectric", Akmenu 47, Ogre, Ogres n. LV-5001, Latvia
[2] Transportation and Telecommunication Institute, Lomonosova Str. 1, Riga LV-1019, Latvia
krainukovs.a@tsi.lv

Abstract. The paper is focusing on the research of the influence of road construction materials on magnetic field dispersion and transmission efficiency. During the research samples of the road surfaces were created using most widely used road surface construction materials in Latvia and the test bench was built in accordance with SAE J2954. The influence of asphalt and concrete road surface materials on the magnetic field of the transmitting coil of the WPT system at a frequency of 85 kHz was investigated. It was determined that efficiency dropped with NaCl solution on concrete road surface.

Keywords: Road cover · Magnetic field · Transmission efficiency · Concrete · Asphalt

1 Introduction

Currently, the technologies wireless power transmission (WPT) for charging the electric vehicle batteries are given sufficient attention [1, 2]. Charging the electric vehicles on the move seems very perspective which is confirmed by the standards of the leading world countries [3].

Many studies and design developments have been performed to improve the efficiency of WPT systems using resonantly inductive energy transfer technology [4], for example, using different coil designs, applying various amplification and conversion schemes on the transmitting and receiving side, using a variety of electronic circuits for frequency and load regulation [5–7].

For electric vehicle dynamic charge in motion the transmitting module of the system must be built into the road cover to ensure the safety of traffic, as a high-speed highway cover must be smooth and all components of wireless charge system must be hidden under it. Most WPT system energy transfer efficiency studies [6, 7] were performed in laboratory or in-doors just with an air gap between the transmitting and receiving coils, and only occasionally asphalt or concrete, without taking into account existing road surface material features. In real life the WPT transmitter has to be embedded under the road surface, not to be an obstacle for the traffic. The road surface materials may affect the WPT magnetic field, and hence the WPT. The purpose of the

I. Kabashkin et al. (Eds.): RelStat 2018, LNNS 68, pp. 214–223, 2019.
https://doi.org/10.1007/978-3-030-12450-2_20

this research was to determine influence of most commonly used road surface construction materials in Latvia on the efficiency of electric vehicle WPT system.

2 Modelling the Road Surfaces

All roadways are subdivided into roadways with rigid pavement and roadways with flexible pavement on the bending resistance [8, 9]. Figure 1 shows the roadway structure with flexible pavement. It is seen that this roadway is created by three structural elements: a road surfacing, road foundation and subgrade [8].

The flexible pavements consists of two upper structural elements: a road surfacing and road foundation (Fig. 1). The roadway surfacing layers (wearing course, binder course and base course) are created from different kinds of asphalt or concrete, and materials, stabilized with cement, bitumen, lime and etc. The roadway foundation layers (road base and subbase) are created from crushed stone, slag, gravel and other weakly granular materials.

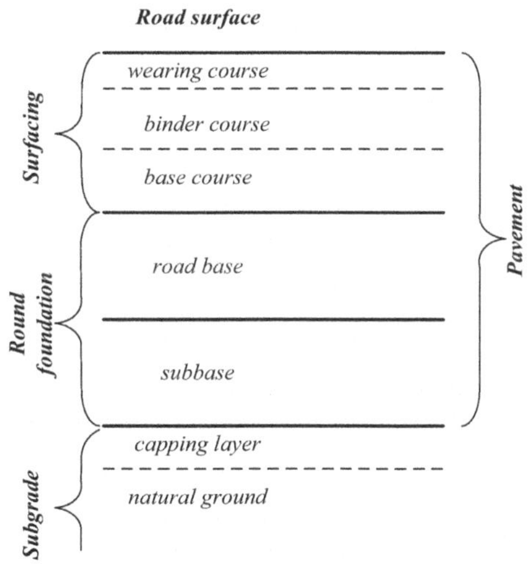

Fig. 1. Roadway structure with flexible pavement.

Road surfacing—the top, most durable layer of cover directly carrying the load of vehicles. The roadway surfacing may be a single layer, double-layer or triple-layer. In the first case the surfacing consists of the wear layer, in the second case the surfacing consists of the wear layer and the lower layer of the surfacing, and in the third case the surfacing consists of the wear layer, the cohesive layer (the top layer of the road surfacing) and the lower layer of surfacing. The top layer of surfacing due to a flat surface provides the necessary transport and roadway performance. The top layer of the roadway surfacing directly exposed to the wheels of vehicles and weather conditions, so it is of strong stone materials with the use of binders. The assumption was used that the transmitting coils of WPT system are placed below the road base in a subbase layer of road foundation (Fig. 1).

In accordance with [8, 9] 4 samples of road covers were created for carrying out our investigations. Parameters of samples of the road covers are in Table 1. Figure 2 shows the appearance of one of the samples of road covers.

These samples fully corresponded to all types of car roads in Latvia in terms of structure and materials of the upper layers of road covers. The first and fourth samples of road covers correspond to internal roads in Latvia. The second sample of the road surface corresponds to ordinary roads, and the third sample corresponds to high traffic roads in Latvia. The horizontal dimensions of the samples were 1200 mm and 800 mm, and the thickness of the samples corresponded to the road type.

3 The Test Bench for the Research Influence of Constructive Materials of Road Cover on Magnetic Field Dispersion

In order to conduct the research, a test bench was created consisting of an electric vehicle, a system of electric vehicle wireless charge with power 3 kW, an trestle of non-magnetic material, a system of coils orientation and an automatic data logger of magnetic field with mechanic dispersion (Fig. 3a). The road surface samples can place between an electric vehicle supplied with the receiver of wireless charge system and a transmitter coil of WPT system (Fig. 3a).

The test bench was built in accordance with SAE J2954 [10]. The magnetic field sensor is based on the MLX 91205HB chip [11] and provides measurements in the range from 0 to 25 mT. The linear manipulator provides movement of the magnetic field sensor in three coordinates with the following values of movement along the axes: $x = 0.8$ m, $y = 0.8$ m and $z = 1.6$ m. The mechanized platform provides the location of the source of the magnetic field under electric vehicle with the following values of displacement along the axes: $x = 0.8$ m, $y = 0.6$ m and $z = 0.2$ m. The block diagram of control of the linear manipulator and the mechanized platform is shown in Fig. 4.

A specialized software application was used to control the linear manipulator and the mechanized platform.

4 Results of Research of Influence Road Cover on the Efficiency WPT System

In the course of research, we determined the influence on the efficiency of the distance between the transmitting and receiving coils of the system and the road cover. The transmitting coil was placed on a mechanized platform, and a sample of the road cover was placed on the transmitting antenna. The transmitter of the system operated in the mode of maintaining a stable current in the transmitting coil, the effective value of the transmitting coil current was equal to 20 amperes. Both the resonant frequency of the transmitting circuit and the operating frequency of the transmitter were 85 kHz.

The magnetic field sensor was moved by a linear manipulator over the road surface sample with a step of 50 mm in horizontal coordinates and with a step of 5 mm in vertical coordinate. At these points, the magnetic field values were recorded. The

Table 1. The parameters of road cover samples.

No	Wearing course	Thickness, mm	Binder course	Thickness, mm	Base course	Thickness, mm	Road base	Thickness, mm
1	Hot asphalt mark AC 8	25	Hot asphalt solids mark AC 16	40	–	–	Crushed stone	30
2	Hot asphalt mark AC 11	40	Hot asphalt solids mark AC 16	65	–	–	Crushed stone	30
3	Hot asphalt mark SMA 11	40	Hot asphalt bindings AC 16	60	Hot asphalt substrate mark AC 22	80	Crushed stone	30
4	Concrete bricks	60	–	–	–	–	Sand leveling layer	30

Fig. 2. The road cover samples: **a**—all samples; **b**—samples No. 1.

Fig. 3. The test bench for the research influence of constructive materials of road cover on magnetic field dispersion: **a**—general view; **b**—the platform under an electric vehicle.

recorded magnetic field values were used to distribute the intensity of the magnetic field over the road cover sample.

DD inductance coils (Fig. 5a) used on transmitting side of WPT system for creation of the magnetic field. Figure 5b, c and d show the intensity distribution of the magnetic field of the transmitting coil along one of the trans-spatial coordinates above the sample, where the color indicates the intensity of the magnetic field: The blue color corresponds to the minimum levels of the magnetic field, and the dark red color to the maximum levels of the magnetic field close to 20 mT. The measured values of the magnetic field intensity of the transmitting coil were used for determining the power at the output of the receiving circuit and determining the efficiency as the ratio of the

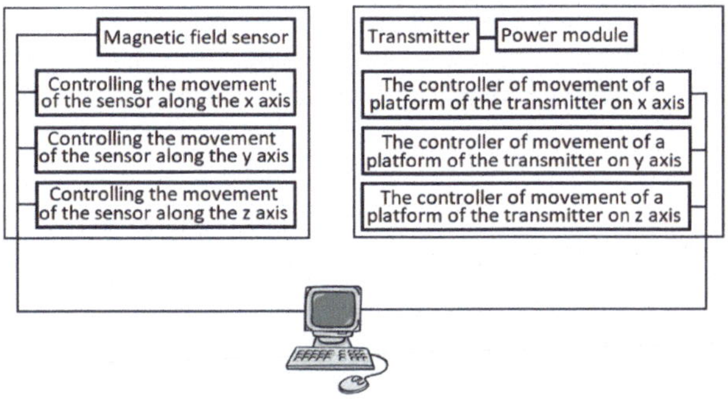

Fig. 4. The block diagram of control of the linear manipulator and the mechanized platform.

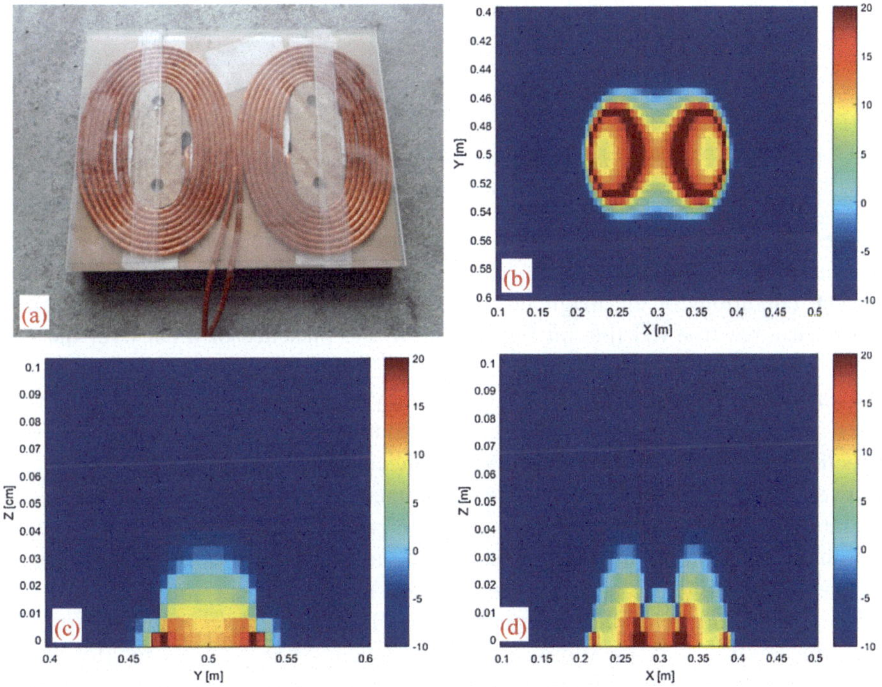

Fig. 5. DD inductance coil construction **a** and the intensity distribution of the magnetic field of this inductance coil in three planes: **b**—in the XY plane at Z = 0 m; **c**—in the YZ plane at X = 0.25 m; **d**—in the XZ plane at Y = 0.5 m.

power at the output of the receiving antenna to the power at the input of the transmitting circuit of the WPT system without taking into account losses in the generating and rectifying circuits of WPT system.

To determine the effect of the road covers on the efficiency of the system, the distribution of the magnetic field over the upper surface of each road cover samples was determined. The measurements were made using each road cover specimen for three conditions: a dry sample, a wet sample, and a sample in a saline solution.

Figure 6 shows the change in WPT efficiency with increasing distance.

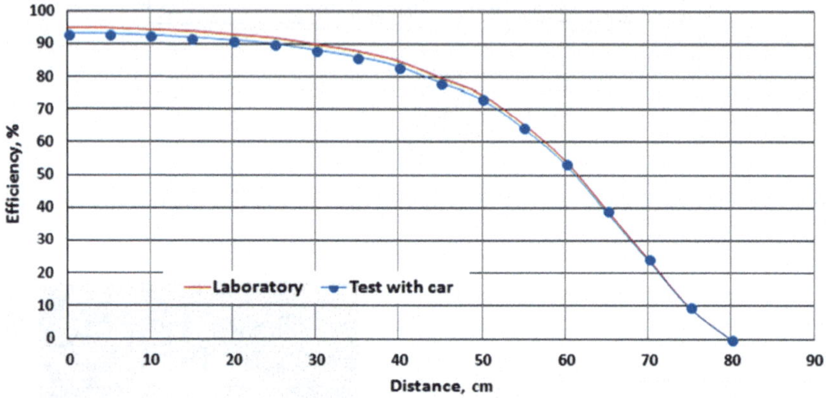

Fig. 6. Dependence of efficiency on the distance from the center of the coil along the Y—axis.

These dependences were obtained in the absence of a road cover sample over the transmitting antenna WPT system. These dependences were obtained for two measurement conditions of the magnetic field distribution of the transmitting coil: in the absence of a car (the red line on Fig. 6) and in the presence of a car over the transmitting coil (the blue line on Fig. 6).

It can be seen that maximum WPT efficiency is ensured at the minimum distance values. The presence of a car above the transmitting coil reduces the intensity of the magnetic field, and as a result, the WPT efficiency also decreases for all distances. However, the decrease in efficiency does not exceed 3%.

The effect of asphalt on WPT efficiency is shown in Fig. 7. These dependences were obtained by measuring the intensity of the magnetic field of the transmitting coil over the third sample of the road cover (Table 1).

Dry asphalt and wet asphalt do not have any effect on the level of the magnetic field of the transmitting coil, therefore, the WPT efficiency dependences for dry and wet asphalt coincide (Fig. 7a) and are similar to the dependences shown in Fig. 6. However, the presence of a salt solution with a small concentration on the surface of the road sample reduces the magnetic field of the transmitting coil and this is manifested in a decrease of WPT efficiency to 5% at initial distances (green line on Fig. 7b). When using the first and second asphalt samples (Table 1), WPT efficiency researches gave the same dependencies as in Fig. 7.

Figure 8 shows the effect of concrete (fourth sample in the Table 1) on WPT efficiency. Concrete also does not affect the magnetic field of the transmitting coil when it is dry or wet (Fig. 8a).

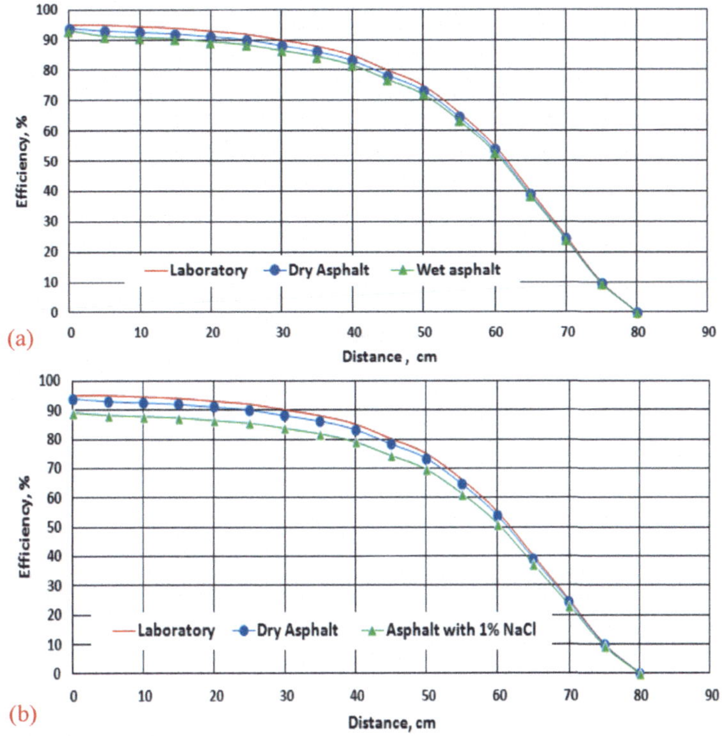

Fig. 7. The effect of asphalt on WPT efficiency: **a**—a wet asphalt; **b**—asphalt with saline.

Therefore, the WPT efficiency dependences in Fig. 8a are the same as for the asphalt road cover (Fig. 7a). At the same time, a saline solution with an insignificant concentration significantly weakens the magnetic field of the transmitting coil. This weakening is manifested in a decrease in efficiency of almost 20% (green line on Fig. 8b).

5 Conclusions

Analysis of research results leads to the following conclusions:

- testing road cover materials most used in Latvia the difference between laboratory and tested materials was insignificant for asphalt and dry concrete road surfaces—losses are less than from misalignment of transmitting and receiving coils;
- the efficiency dropped with NaCl solution on concrete road surface;
- the water containing defrosting agent as sand layer, saturated with 1% NaCl solution absorb 20% of magnetic stream;

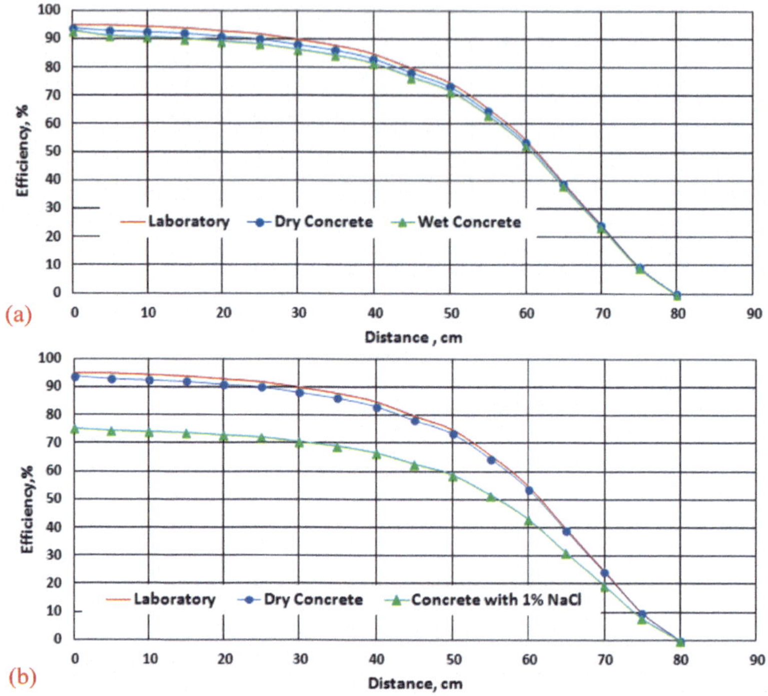

Fig. 8. The effect of concrete on WPT efficiency: **a**—a concrete; **b**—concrete with saline.

- the WPT application should be limited to private roads not treated with NaCl for melting snow;
- further research is needed to see how the road surface would be physically affected with built-in WPT charger within road cover.

Acknowledgment. This paper has been published within the research project "Research on development of pre-production series of transmitting module for wireless charging of electric vehicles" The project is co-financed by European Regional Development Fund under dedicated support initiative, project identification number 1.2.1.1/16/A/008, and the activities are carried out in cooperation with "Competency center of advanced engineering system, transport and energy".

References

1. Covic, G., Boys, J.: Modern trends in inductive power transfer for transportation applications. IEEE J. Emerg. Sel. Top. Power Electron. **1**(1), 38–48 (2013)
2. Li, S., Mi, C.: Wireless power transfer for electric vehicle applications. IEEE J. Emerg. Sel. Top. Power Electron. **3**(1), 4–17 (2015)

3. Pereirinha, P.G., Trovão, J.P.: Standardization in Electric Vehicles. http://www.uc.pt/en/efs/research/EESEVS/f/XIICLEEE_1844_PereirinhaTrovao.pdf. Accessed 18 Sept 2018
4. Grajski, K.A., Tseng, R., Wheatley, C.: Loosely-coupled wireless power transfer: physics, circuits, standards. IEEE, Qualcomm Incorporated (2012)
5. Shin, J.: Design and implementation of shaped magnetic-resonance based wireless power transfer system for roadway-powered moving electric vehicles. IEEE Trans. Ind. Electron. **61**(3), 1179–1192 (2014)
6. Imura, T., Okabe, H., Hori, Y.: Basic experimental study on helical antennas of wireless power transfer for electric vehicles by using magnetic resonant couplings. In: Proceedings of IEEE Vehicle Power and Propulsion Conference, September 2009, pp. 936–940 (2009)
7. Hui, S.Y.R., Zhong, W., Lee, C.K.: A critical review of recent progress in mid-range wireless power transfer. IEEE Trans. Power Electron. **29**(9), 4500–4511 (2014)
8. Ceļu specifikācijas 2016 ar 2017. gada 29. maijā VAS "Latvijas Valsts ceļi" Tehniskajā komisijā apstiprinātajiem grozījumiem. http://lvceli.lv/lat/sadarbibas_partneriem/normativie_dokumenti/autocelu_specifikacijas. Accessed 12 Oct 2018
9. VAS: Latvijas Valsts ceļi. Ceļa segu tipveida konstrukciju katalogs (Metodiskie norādījumi autoceļu segu projektēšanai latvijas apstākļiem). Rīga. 88 lp., 01 September 2018
10. SAE International Publishes J2954™ Recommended Practice Enabling Wireless Charging to 11 kW. WiTricity. http://witricity.com/sae-international-publishes-j2954-recommended-practice-enabling-wireless-charging-11kw/. Accessed 01 Sept 2018
11. MLX 91205HB. Melexis. https://www.melexis.com/-/media/files/documents/datasheets/mlx91205-datasheet-melexis.pdf. Accessed 13 Aug 2018

The Performance Wall of Large Parallel Computing Systems

János Végh(✉), József Vásárhelyi, and Dániel Drótos

University of Miskolc, 3515 Miskolc, Hungary
j.vegh@uni-miskolc.hu

Abstract. Since the beginning of using parallelized computing units it is known that the actual performance is less than the possible nominal performance: the unproductive part of the computing performance remains "dark". It is also known that the amount of dark performance strongly depends on the number of parallelly working processing units, so its role must be crucial for supercomputers, where in the coming exa-scale models millions of processors are utilized, as well as for the exa-scale applications they are running, like brain simulation and Earth simulation. Although the effects affecting parallel performance are known from the beginning, their relative weights have been considerably changed with the development of the field and strongly depend on the type of application. For large computer systems the "dark performance" represents a new major obstacle, in addition the former ones, like "heat wall", "memory wall", "dark silicon", etc. The careful reconsideration discovers that in contrast with the general belief, supercomputer performance *has* an upper limit, and reaching that limit explains some strange and mysterious events, like canceling projects immediately before their target date or that the special-purpose brain simulator cannot outperform the many-thread simulator running on a general-purpose supercomputer.

Keywords: Parallel computing · Efficiency · Performance · Limitations · Supercomputer · Brain simulator

1 Introduction

In computing today supercomputers with performance above 1 Eflop/s [1] are on the horizon, and no change in their behavior is expected: computer experts see no upper bound on supercomputer performance [2].

To achieve higher performance, however, gets day by day harder. The main reason of this experience is that quietly some bounds are approached or reached, exceeding of which are prohibited by physical laws: signals cannot propagate with speed exceeding speed of the light, the always smaller electronic components show up quantal behavior though they are in conventional computers, the core temperature approached melting point of the material and the age of "dark silicon" era arrived [3]. Since a single processor cannot any more achieve higher processing performance [4], computer systems are prepared that contain several processors working in parallel, and their resulting (apparent) performance considerably exceeds the performance of a single

I. Kabashkin et al. (Eds.): RelStat 2018, LNNS 68, pp. 224–237, 2019.
https://doi.org/10.1007/978-3-030-12450-2_21

processor. Today even our personal computers comprise several cores (essentially full-value processors), cloud services are offered for enterprises, and special-purpose supercomputers are built for solving extremely large-scale scientific/technical tasks.

Those latter ones comprise hundreds of thousands of processors (even, the biggest one [5] ten million). The continuous race between manufacturers and owner institutions forces to target building ever-bigger supercomputers. One meets mysteries and experiences the prohibiting effect of the "dark performance" [6–8].

At the same time, Europe Union's Action plan [9] states that "*With a differentiated strategy and sufficient investment and political will, Europe can be a global player in HPC*". During the preparations detailed reports are prepared, like [10]. These studies scrutinize different aspects of the exa-scale (Eflop/s, performing 10^{18} floating operations in a second) supercomputer, from energy consumption to computing algorithms, the only exception being *whether supercomputer of that size can be built at all*, i.e. whether some laws of nature or computing science forbid building it. Such a calculation, however, can be quickly carried out, with surprising result: such a limit is inherent for the present sequential-parallel systems and it is already reached in the existing supercomputers.

2 Amdahl' Law

The most commonly known and cited limitation on parallelization speedup, the so called Amdahl's law [11], is based on considering that some parts (P_i) of a code can be parallelized, some (S_i) must remain sequential. In practice "*Amdahls Law is one of the few, fundamental laws of computing*" [12].

Amdahl only wanted to draw the attention to that when putting together several single processors, and using Single Processor Approach (SPA), the achievable speed gain due to using large-scale computing capabilities has a theoretical upper bound. He also mentioned that the data housekeeping (non-payload calculations) causes some overhead, and that *the nature of that overhead appears to be sequential, independently of its origin*. Actually, *Amdahl's law is valid for any partly parallelizable activity (including computing unrelated ones) and the non-parallelizable fragment shall be given as the ratio of the time spent with non-parallelizable activity (independently of their origin) to the total time*.

Amdahl speaks about "the fraction of the computational load"; algorithmic reasons like "computations required may be dependent on the states of the variables at each point"; architectural aspects like "may be strongly dependent on sweeping through the array along different axes on succeeding passes" as well as "physical problems" like "propagation rates of different physical effects may be quite different". His point of view is valid also today: one has to consider the load of the complex HW/SW system [13–15].

3 A Model for Deriving Performance Gain

Although Amdahl's classic model [16] is pretty outdated due to the increased complexity of the computing systems, its idea can serve as a good starting point. With separating the different contributions according to their origin and interpreting them properly, the experienced new issues can be properly explained.

3.1 Deriving the Achievable Performance Gain

One of the key results of Amdahl (also confirmed theoretically by [17]) is that the non parallelizable (apparently sequential) fragment of the computing load causes an upper bound on the performance gain (speedup). If α fragment of the computational load can be parallelized on k processors, then the speedup S will be given as

$$S^{-1} = (1 - \alpha) + \alpha/k. \tag{1}$$

The resulting performance gain (or maximum speedup), *provided that* α *does not depend on* k, cannot exceed value

$$G = (1 - \alpha)^{-1} \tag{2}$$

and the efficiency (the specific speedup)

$$E = S/k = (k(1 - \alpha) + \alpha)^{-1} \tag{3}$$

(for more details of the calculation see [15]). A practical difficulty with applying this result to supercomputers is that the value of α is not *ab ovo* known, and the complexity and diversity of supercomputers make hopeless to derive some value for G from a technical, parameterized model.

One can create, however, a theoretical (in some sense naive) model, which qualitatively describes the behavior of parallelized many-processor systems. In the second phase, utilizing the reliable data on supercomputers [18] one can estimate parameters for that simple model. Since in that database values of E (calculated as $R_{\max}/R_{peak}$) are provided, an empirical value

$$\alpha_{Eff} = (Ek - 1)/E/(k - 1) \tag{4}$$

and so the achievable performance gain (see Eq. (2)) for the individual supercomputers can be derived. Although (mainly because of the lack of dedicated measurements) these estimations are only good to get the order of magnitude of the performance gain, its behavior (mainly its dependence on the number of processors) explains qualitatively some mysteries in connection with supercomputing, furthermore provides hopes to (more) accurately describe supercomputer performance and in this way to understand the behavior of supercomputer applications.

3.2 Our Model of Parallel Execution

To understand the operation of computing systems working in parallel, one needs to extend Amdahl's original (rather than that of the successors') model in such a way, that non-parallelizable (i.e. apparently sequential) part comprises contributions from HW, OS, SW and propagation delay PD, and also some access time is needed for the operation. The contributions of model component XX to α_{eff} will be denoted by α_{eff}^{XX} in the followings. Notice that the different contributions have only one common feature: *they all consume time*. The extended Amdahl's model is shown in Fig. 1. The vertical scale displays the actual activity for processing units shown on the horizontal scale. Notice that our model assumes no interaction between processes running on the parallelized systems in addition to the absolutely necessary minimum: starting and terminating the otherwise independent processes, which take parameters at the beginning and return results at the end. It can, however, be trivially extended to the more general case when processes must share some resource (like a database or physical device). Concurrent objects have inherent sequentiality [19], and synchronization and communication among those objects [20] considerably increase the non-parallelizable fraction, so in the case of extremely large number of processors special attention must be devoted to their role on the efficiency of the application.

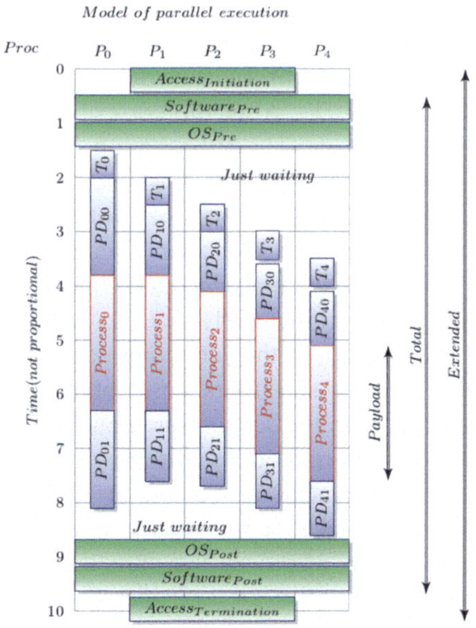

Fig. 1. The extended Amdahl's model (somewhat idealistic).

Let us notice that all contributions have their role during measurement: contributions due to SW, HW and OS cannot be separated, though dedicated measurements can reveal their role, at least approximately. The relative weights of the different contributions are very different for the different parallelized systems, so in every single parallelization case a careful analysis is required.

Access time. Initiating and terminating the parallel processing is usually made from within the same computer, except when one can only access the parallelized computer system from another computer. This access time is independent from the parallelized system, and *one must properly correct for the access time when derives timing data for the parallelized system.* Amdahl's law is valid only for properly selected computing system.

Execution time. The execution time *Total* covers all processing on the parallelized system. All applications, running on a parallelized system, must make some non-parallelizable activity at least before beginning and after terminating parallelizable activity. This SW activity represents what was assumed by Amdahl as the total sequential fraction. In more general terms, it is a kind of synchronization and communication that affects drastically the parallelism of the system [20].

Role of OS. All applications must use OS services and some HW facilities to initiate themselves as well as to access other processors. Because operating system works in a different (supervisor) mode, a considerable amount of time is required for switching context. Notice that processors (in SPA) must be handled one-by-one. This activity causes the contributions from SW and OS to increase linearly as the number of processors increases, so handling them properly is crucial for high number of processors.

OS initiates only accessing the processors, after that HW works partly in parallel with the next action of the OS and with other actions initiating accessing other processors. This period is denoted by T_x. After the corresponding signals are generated, they must reach the target processor. The Propagation Delays are denoted by PD_{xn}, corresponding to actions delivering input data and result, respectively.

The contributions critical for large processor numbers. Notice that non-payload processing comprises two contributions of variable length. The first one is OS contribution loop iteration overhead, that simply increases linearly with loop count, and the second one is propagation delay overhead PD that increases with the physical size of the supercomputer (and the quality of the interconnection). These two contributions may be different for different processing units, so combining short ones from the first overhead class with long ones from the second one reduces the overall overhead time, see Fig. 1. From the figure one can find out the meaning of the introduced metric α_{eff}: it is simply the ratio of *Payload* time (the processors are utilized to do actual work) to *Total* execution time on the parallel system, exactly as in the classic Amdahl model. Notice that from the point of view of processing units there is no difference *why* they cannot do *Payload* work: all activities are considered as contributions to the non-parallelizable fraction. Notice also that using *Extended* time in place of *Total* falsifies characteristics of the parameters of the parallelized system: it adds foreign contribution (the time consumed by systems *other* than the parallelized system).

It is also clear from the figure that the origin of the "dark performance" is that *in a considerable fraction of the time the processors are just waiting for something* and the amount of non-payload time depends strongly on the actual conditions.

3.3 Amdahl's Law for Supercomputers

Figure 2 shows how the theoretically achievable performance gain (the upper bound for the architectural solution) has developed during the history of supercomputers. The data show a semi-logarithmic behavior, akin to Moore's law. Recall, however, that when calculating these values, the efficiency values R_{Max}/R_{Peak} was used, so any effect of Moore's law (like clock frequency, component density, absolute performance, etc.) was eliminated. Since all data have been derived utilizing the standard benchmark HPL, and that code has long been standardized, and similarly, only minor improvements happened in the implementation of OS, the impressing improvement shall mainly be attributed to the improvement of parallelization HW. Though there is a considerable scatter in the data, they obviously fit a semi-logarithmic line (although with some signs of saturation in the past few years, also akin to Moore's Law). It looks like that *Amdahl's law appears in this form for supercomputers.*

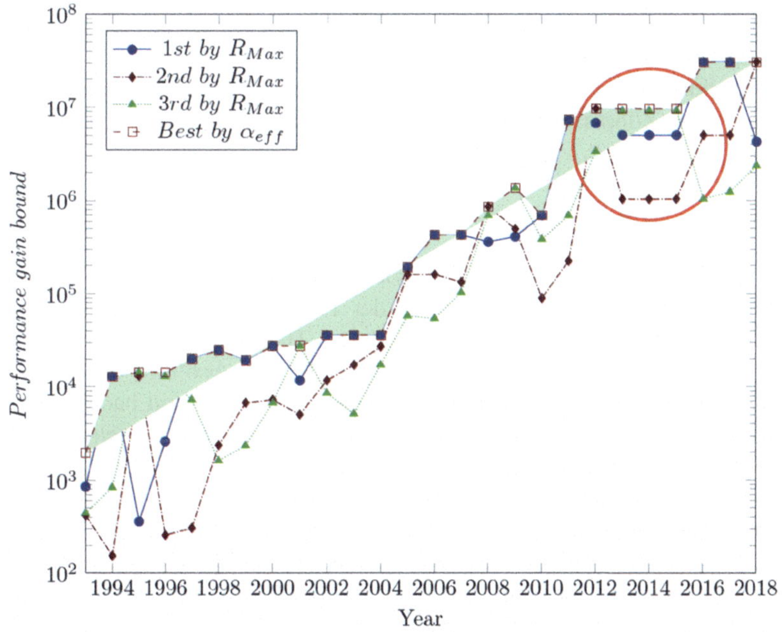

Fig. 2. Performance gain of supercomputers in function of the time during their quarter of century history. Data derived from the TOP500 database, measured using the benchmark HPL. The marks show the actual (measured) performance gains for the first three (by R_{Max}) supercomputer and that for the best by α_{eff}.

Let us notice that (see the marks in the big circle) the only place where some kind of stalling occurred, is the 3–4 years immediately before year 2016. The performance gain bound for the actual first 3 supercomputers did not change for 3–4 years, which is the time of a complete generation change in the field of supercomputing. Knowing "the exponential laws of computing growth" [21] that growth is described by a logistic curve rather than an infinite exponential, one might think that performance bound has approached the point where the saturation feature of the logistic curve manifests.

4 A Model for Deriving Performance Bound

4.1 The Inherent Bound

It is crucial to understand that the present technology, using clock driven processors, inherently has an upper bound. Suppose that a supercomputer application can be started and stopped in one clock period. There is no way to distribute these two actions to parallelly working computers. Since this is a fixed amount of time, the absolute parallelization efficiency depends on the total measurement time. Some concrete values are known from benchmarking supercomputer Taihulight [22]: the 13,298 s benchmark runtime on the 1.45 GHz processors means 2 × 1013 clock periods. (In the followings 1.0 GHz frequency will be used in the calculations.) This means that a supercomputer that spends no time with controlling its available processors, cannot exceed performance gain 1013. Considering a presently typical 10 Gflop/s = 1010 flop/s single-processor performance, it also means that the absolute dream limit is around 1023 flop/s = 100 Zflop/s, and present supercomputers have million(s) times less performance.

4.2 The Practical Limitations

In practice, however, there are some serious other limitations. Some of the supercomputer processors are in a distance of order 100 m from each other, so the signal round trip time is cca. 10^{-6} s, or 10^3 clock cycles. That is, a typical supercomputer of that size cannot provide performance gain above 10^{10}, or absolute performance above 100 Eflop/s. The signal should deliver information bits, there and back, reducing this bound further. At this point the high-speed interconnections cannot help a lot, because unfortunately this is not the dominating contribution.

The program must set up itself, initialize its internal structures, prepare for the calculation. Similarly, at the very end it must collect all results from the helper processors, and terminate orderly. This SW activity is surely sequential, and depends strongly on the executed program. It can hardly avoid utilizing OS services for this.

Mainly because of the required context change, about 2×10^4 clock cycles [23] shall be used for that goal, which consideration alone reduces the achievable performance gain to 10^9: when utilizing OS, the absolute limit shall be reduced to 10 Eflop/s. In addition, the processors must be addressed individually, so the time spent in the OS increases linearly with the number of the processors. At relatively low number of processors (say a few thousands) this contribution is not noticeable (compared to that of the context switching), at extremely large number of processors (say several millions) it dominates.

4.3 The Efficiency of Supercomputers

As discussed above and depicted in Fig. 2, the achievable performance gain bound is somewhere in the range 10^6–10^7, and the efficiency of running exa-scale applications can also be calculated. When a supercomputer will be built from processors with single-processor performance 10 Gflop/s, a supercomputer with nominal performance 1 Eflop/s will need 10^8 processors. Even if the optimistic performance bound 10^7 is assumed, the efficiency will be less than 10%, even when measured with the benchmark HPL. Recall that parallelization efficiency is worse by at least 100 times when measured using benchmark HPCG, the efficiency cannot exceed 0.1%. (Recall that *Taihulight* produced efficiency 0.3%, with using 10^7 processors "only" [18].)

4.4 Ways of Enhancing Efficiency of Parallelization

As shown in Fig. 2, in 2016 *Taihulight* took over the "pole position", both when ranked by R_{Max} and by $(1 - \alpha_{eff})$. In 2018, it is ranked as first by $(1 - \alpha_{eff})$ but only second by $R_{\max}$. *Summit* took over the pole position thanks to its high single-processor performance. However, *Summit* uses clustering (i.e. reduces the loop count by addressing cluster members rather than single processors), and *Taihulight* does not. Despite this, the $(1 - \alpha_{eff})$ value of *Taihulight* is about ten times better than that of *Summit*. The solution is hidden in their home-grown processor [24]: they make a kind of clustering at the other end, inside the processor. The much better $(1 - \alpha_{eff})$ value is caused by two more tricks. They use a drastically different computing paradigm: the cores can transfer data directly from core to core and the computing units are working in kernel mode. This also underpins our conclusion that *reducing cycle count, context switching time and amount of inter-process communication all contributes to achieving better* $(1 - \alpha_{eff})$ *value.*

Increasing the single processor performance, however, has its limitations when those processors are utilized in supercomputer environment. A counter-example is the case of "AI Bridging Cloud Infrastructure" supercomputer: although its single-processor performance is outstanding, utilizing GPU for acceleration results in tragically low efficiency: less than 3% of its processors could be used in the competition. The reason is that copying data between the main memory and GPU memory represents a kind of inter-thread communication and strongly increases the value of $(1 - \alpha_{eff})$.

4.5 The Role of the Benchmarking and the Application

The benchmarks, utilized to derive numerical performance parameters for supercomputers, are specialized programs, which run in the HW/OS environment provided by supercomputer under test. One can use benchmarks for different goals. Two typical fields of utilization: to describe the environment supercomputer application runs in, and to guess how quickly and efficiently an application will run on a given supercomputer.

As it is obvious from our model, the (apparently) sequential fraction $(1 - \alpha_{eff})$ cannot distinguish between the (at least apparently) sequential processing time contributions of different origin, even the SW (including OS) and HW contributions cannot be separated. Similarly, it cannot be taken for sure that those contributions sum up linearly. Different benchmarks provide different SW contributions to the non-parallelizable fraction of the execution time (resulting in different efficiencies and ranking [25]), so comparing results (and especially establishing ranking!) derived using different benchmarks shall be carried out with maximum care. Since the efficiency depends heavily on the number of cores, different configurations shall be compared using the same benchmark and same number of processors (or same R_{Peak}).

If the goal is to characterize the supercomputer's HW + OS system itself, a benchmark program should distort HW + OS contribution as little as possible, i.e. the SW contribution must be much lower than the HW + OS contribution. In the case of supercomputers, benchmark HPL is used for this goal since the beginning of the supercomputer age. The mathematical behavior of HPL enables to minimize SW contribution, i.e. *HPL delivers the possible best estimation HW + OS contribution for* α_{eff}.

If the goal is to estimate the expectable behavior of an application, the benchmark program should imitate the structure and behavior of the application. In the case of supercomputers, a couple of years ago the benchmark HPCG has been introduced for this goal, since "*HPCG is designed to exercise computational and data access patterns that more closely match a different and broad set of important applications, and to give incentive to computer system designers to invest in capabilities that will have impact on the collective performance of these applications*" [8, 26]. However, its utilization can be misleading: the ranking is only valid for the HPCG application, and only utilizing that number of processors. According to our model, in the case of using the HPCG benchmark, the SW contribution dominates, i.e. *the benchmark HPCG delivers the best possible estimation for* a_{eff}^{SW} *for this class of supercomputer applications*.

Figure 3 depicts how payload performances of two TOP10 supercomputers depend on SW contribution (including the benchmarks). As shown, the payload performance of real-life supercomputer applications cannot exceed even the 1 Pflop/s.

4.6 The Role of the Software

As discussed above, in parallelization efficiency SW contributions to α_{eff} cannot be separated from the contributions of HW and OS. It is experienced, however, that the efficiency of a supercomputer changes 2–3 orders of magnitude when using benchmark HPCG instead of HPL and even ranking of supercomputers is drastically different when utilizing another benchmark. The reason is clear: the instruction mix (and so: the performance) defines the behavior of the complex computing system (that involves also the SW) rather than HW alone. As benchmarking with HPCG shows, the SW contribution of real-life programs dominates the efficiency of program on a given computer. As shown in Fig. 4, in benchmarking the same supercomputer with different (benchmark) applications, the software contribution dominates, i.e. the measuring device (rather than the device under test) defines the apparent performance.

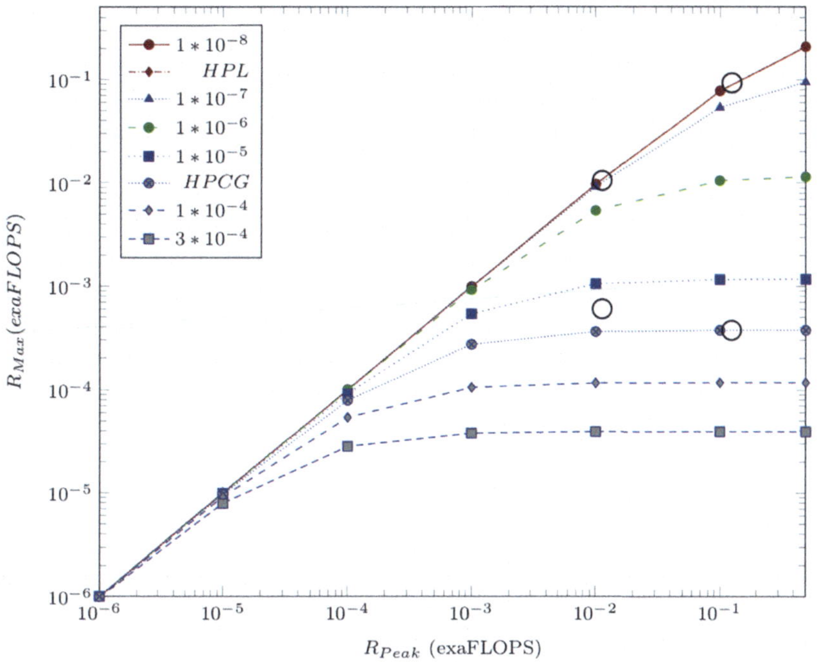

Fig. 3. R_{Max} performance in function of peak performance R_{Peak}, at different $(1 - \alpha_{eff})$ values. Bubbles display measured values when using HPL and HPCG benchmarks, for Taihulight and *K* supercomputer, respectively.

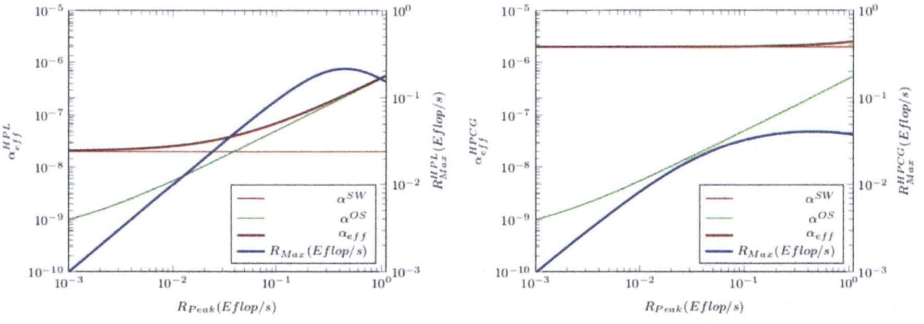

Fig. 4. Contributions α_{eff}^{X} to α_{eff} and max payload performance R_{Max} of a fictive super-computer (P = 1 Gflop/s @ 1 GHz), imitating the behavior of benchmarks HPL and HPCG. The α_{eff} values refer to the left scale, R_{Max} to the right scale.

Even if a new computer principle solves the "too many processors" syndrome, the best performance gain that can be expected for real-life supercomputer applications is in the range $10^4 \ldots 10^5$; an extremely wrong efficiency for utilizing $10^7 \ldots 10^8$ processors.

4.7 The Case of Brain Simulation

The case of brain simulation provides an excellent example how decisive the operating principle of the HW/SW system can be. If one makes an analysis based on the model, as was done for the case of supercomputers, the reasons will be obvious: in this special case two more contributions can become dominating.

The parallel processes in the biological neurons are simulated by solving differential equations in the sequential-parallel computing systems, which needs considerable time. The neurons can start with the next calculation only when the previous "grid time" finished, and in this way a hidden "grid time" clock frequency is introduced. Considering the nature of the biological processes, a 1 ms "grid time" (integration time) is used. Repeating with this clock period the analysis given above, and considering that the "benchmark time" is 1000 times shorter, one can see that the dream limit of performance gain in this case cannot be better than 10^4. This explains why even the special-purpose hardware simulator cannot provide better performance.

There is, however, an even worse issue. Several (about 100) neurons share the computing and memory resources of a core, and they all must have their own thread. The thread handling consumes a horrible amount of time: as [11] estimates, this time grows up to 10%, i.e. the value of $(1 - \alpha_{eff})$ is 10^{-1}. Such low parallel performance has not yet been produced since the beginning of the supercomputer age. [27] proved that the present reasoning is correct: changing the way of the handling of the threads (reducing the neuron handling overhead due to context switches) the efficiency changed by a factor of 10. Compare these values to the parameters of the diagram lines shown in Fig. 4: the future Artificial Intelligence applications (brain simulation) will run with performance (efficiency) much worse than today the HPCG class applications do. As formulated in [27]: "*the algorithms creating instances of model neurons and their connections scale well for networks of ten thousand neurons, but do not show the same speedup for networks of millions of neurons*". *That is, using the present computing model not only the simulation of the complete brain is not achievable, but also some of its largest subsystems* [28].

4.8 Where the Less is More

Figure 4 depicts some α_{eff} contributions, calculated as discussed above, in the function of the nominal performance, assuming the shown operating parameters of a fictive supercomputer; together with the corresponding R_{Max} diagram. As shown, the linearity of its dependence on R_{Peak} breaks down around the current (as of fall 2018) top performance 0.1 Eflop/s. As discussed (and depicted in Fig. 4), (at least) the looping delay of the sequential contributions depends on the number of processors. In the case of the benchmark HPL the contribution due to the operation of OS is at a few percent of the Eflop/s dream value, the value of α_{eff} starts to rise steeply, and correspondingly breaks down R_{Max}. In this way adding more processors to the system will decrease the payload performance. In the case of benchmark HPCG the case is similar, but—since the SW contribution is by orders of magnitude higher—the decrease is less definite near to the 1 Eflop/s dream value.

4.9 "Experimental Proofs" of the Existence of the Performance Limit

As shown in Fig. 4, after some point *increasing* the nominal performance (the number of processors) results in *decreasing* payload performance. Although this behavior of the parallelized processors seems to be shocking, there are indirect and direct proofs of the statement. Aurora failed ("retooled"); Summit uses only 4300 nodes out of the available 4600; Gyokou collapsed and withdrawn in a half year (it could only use 12% of its total number of processors); the new AI Bridging Cloud Infrastructure, (because of the cloud infrastructure) can participate in the TOP500 competition with using only 11 K out of its 400 K processors [18].

There is, however, a direct measurement that demonstrates that the breakdown of the computing performance is reality. In [27] the authors measured the execution time of their fixed length brain simulation in the function of the nominal computing performance. In their Fig. 7 they show that *with increasing the number of the processors (or threads as they use it) the execution time started to rise*, in accordance with the theoretical expectations shown in Fig. 4.

5 Conclusions

It is crucial to understand that the computing, using the present digital electronic technology and thinking in the 70 years old Single Processor Approach, has some inherent performance limitations. Those limitations are almost not remarkable in most of the cases, including parallel computing systems built from several processors. The huge systems (comprising several thousands of processors), however, start to show up different behavior, a "too many processors" syndrome: a considerable part of computing performance remains "dark" (unproductive). The paper reconsidered the validity of the Amdahl's law, and demonstrated that many of the formerly neglected terms shall be considered again (and even they can dominate) in extremely large parallelized computing systems. The technical implementation and the computing paradigm together form a strong upper limit for the supercomputer performance and the applications running on them [6]. The experienced low efficiency of some applications is of principle rather than of technical nature. For more see [15].

Acknowledgements. Project no. 125547 has been implemented with the support provided from the National Research, Development and Innovation Fund of Hungary, financed under the K funding scheme [29].

References

1. Dongarra, J.: The Global Race for Exascale High Performance Computing (2017). http://ec.europa.eu/newsroom/document.cfm?doc_id=45647
2. Bourzac, K.: Streching supercomputers to the limit. Nature **551**, 554–556 (2017)
3. Esmaeilzadeh, H.: Approximate acceleration: a path through the era of dark silicon and big data. In: Proceedings of the 2015 International Conference on Compilers, Architecture and Synthesis for Embedded Systems, CASES 2015, pp. 31–32

4. Fuller, S.H., Millett, L.I.: Computing performance: game over or next level? Computer **44**, 31–38 (2011)
5. Fu, H., Liao, J., Yang, J., Wang, L., Song, Z., Huang, X., Yang, C., Xue, W., Liu, F., Qiao, F., Zhao, W., Yin, X., Hou, C., Zhang, C., Ge, W., Zhang, J., Wang, Y., Zhou, C., Yang, G.: The Sunway TaihuLight supercomputer: system and applications. Sci. China Inf. Sci. **59**(7), 1–16 (2016)
6. Végh, J.: Limitations of performance of exascale applications and supercomputers they are running on. ArXiv e-prints; Submitted to Special Issue of IEEE J. Parallel Distrib. Comput. (2018)
7. Top500.org: Retooled Aurora Supercomputer will be Americas First Exascale System (2017). https://www.top500.org/news/retooled-aurora-supercomputer-will-be-americas-first-exascale-system/
8. Inside HPC: Is Aurora Morphing into an Exascale AI Supercomputer? (2017). https://insidehpc.com/2017/06/told-aurora-morphing-novel-architecture-ai-supercomputer/
9. European Commission: Implementation of the Action Plan for the European High-Performance Computing strategy (2016). http://ec.europa.eu/newsroom/dae/document.cfm?docid=15269
10. US DOE: The Opportunities and Challenges of Exascale Computing (2010). https://science.energy.gov/~/media/ascr/ascac/pdf/reports/Exascale_subcommittee_report.pdf
11. van Albada, S.J., Rowley, A.G., Senk, J., Hopkins, M., Schmidt, M., Stokes, A.B., et al.: Performance comparison of the digital neuromorphic hardware SpiNNaker and the neural network simulation software NEST for a full-scale cortical microcircuit model. Front. Neurosci. **12**, 291 (2018). https://doi.org/10.3389/fnins.2018.00291
12. Paul, J.M., Meyer, B.H.: Amdahl's law revisited for single chip systems. Int. J. Parallel Prog. **35**(2), 101–123 (2007)
13. Amdahl, G. M.: Validity of the single processor approach to achieving large scale computing capabilities. In: Proceedings of the April 18-20 1967 spring joint computer conference on - AFIPS 67 (Spring) AFIPS 67 (Spring) (1967)
14. www.supercomputersnotes.info
15. Végh, J., Molnár, P.: How to measure perfectness of parallelization in hardware/software systems. In: 18th International Carpathian Control Conference on ICCC, pp. 394–399 (2017)
16. Amdahl, G.M.: Validity of the single processor approach to achieving large-scale computing capabilities. In: AFIPS Conference Proceedings, vol. 30, pp. 483–485 (1967)
17. Karp, A.H., Flatt, H.P.: Measuring parallel processor performance. Commun. ACM **33**(5), 539–543 (1990)
18. TOP500: November 2017 List of Supercomputers (2017). https://www.top500.org/lists/2017/11/
19. Ellen, F., Hendler, D., Shavit, N.: On the inherent sequentiality of concurrent objects. SIAM J. Comput. **43**(3), 519536 (2012). https://doi.org/10.1137/08072646X
20. Yavits, L., Morad, A., Ginosar, R.: The effect of communication and synchronization on Amdahl's law in multicore systems. Parallel Comput. **40**(1), 1–16 (2014)
21. Denning, P.J., Lewis, T.: Exponential laws of computing growth. Commun. ACM 54–65 (2017)
22. Dongarra, J.: Report on the Sunway Taihulight system. Technical report UT-EECS-16-742 (2016)
23. Tsafrir, D.: The context-switch overhead inflicted by hardware interrupts (and the enigma of do-nothing loops). In: Proceedings of the 2007 Workshop on Experimental Computer Science, ExpCS 2007, p. 3. ACM, New York (2007). http://doi.acm.org/10.1145/1281700.1281704

24. Zheng, Fang, et al.: Cooperative computing techniques for a deeply fused and heterogeneous many-core processor architecture. J. Comput. Sci. Technol. **30**(1), 145–162 (2015)
25. IEEE Spectrum: Two Different Top500 Supercomputing Benchmarks Show Two Different Top Supercomputers (2017). https://spectrum.ieee.org/tech-talk/computing/hardware/two-different-top500-supercomputing-benchmarks-show-two-different-top-supercomputers
26. www.studentclustercompetition.us
27. Ippen, T., Eppler, J.M., Plesser, H.E., Diesmann, M.: Constructing neuronal network models in massively parallel environments. Front. Neuroinform. **11**, 30 (2017). https://doi.org/10.3389/fninf.2017.00030
28. doaj.org
29. Végh, J.: Introducing the explicitly many-processor approach. Parallel Comput. (2018)
30. Dettmers, T.: The Brain vs Deep Learning Part I: Computational Complexity or Why the Singularity Is Nowhere Near (2015). http://timdettmers.com/2015/07/27/brain-vs-deep-learning-singularity/
31. HPCG Benchmark (2016). http://www.hpcg-benchmark.org/
32. Markov, I.: Limits on fundamental limits to computation. Nature **512**(7513), 147–154 (2014)
33. Top500.org: DOE Withholds Details of First Exascale Supercomputer (2018). https://www.top500.org/news/doe-witholds-details-of-first-exascale-supercomputer-even-as-it-solicits-researchers-to-apply-for-early-access/
34. Végh, J.: Renewing computing paradigms for more efficient parallelization of single-threads (chap. 13). In: Advances in Parallel Computing, pp. 305–330. IOS Press (2018)

Telematics

Availability of Applications in Container-Based Cloud PaaS Architecture

Igor Kabashkin(✉)

Transport and Telecommunication Institute, Lomonosova 1, Riga 1019, Latvia
kiv@tsi.lv

Abstract. In the paper the reliability of container-based cloud PaaS architecture is discussed. In practice, when organizing container-based access to user applications, it is of interest to determine the availability of a separate dedicated container with a user application. In this case if the failure of another service for another user has come, for this particular user this failure from the position of reliability is not critical. In the article the availability of a dedicated user container in the system with container-based PaaS architecture is studied. The mathematical model of the reliability of a separate dedicated container with a user application in the real conditions of operation is developed. Expressions for availability of application operation in dedicated user container are developed.

Keywords: Reliability · Container · Container-based architecture · Software failures · Cloud computing

1 Introduction

Platform as a Service (PaaS) allows users to focus on running applications instead of managing infrastructure. With PaaS the virtual infrastructure is provided and users can deploy any application on it, including web applications such as nginx and ghost; custom services and more [1]. This approach provides total independence from the underlying infrastructure allowing you to run your applications anywhere, without investing an effort in maintaining the infrastructure such as updates, security, hardware failures and such.

Containers are one of the building blocks of PaaS, and in a certain manner are enablers of it. The typical architecture for container virtualisation is shown at the Fig. 1 [1]. Containers allow users easily to spawn applications on any virtual or physical infrastructure, integrating all its dependencies such as code, runtime, system tools and system libraries. Since these components are embedded, the software can be deployed transparently, notwithstanding the environment where they are running. Applications can run as a single unit in a single cloud infrastructure or run separately in multiple cloud infrastructures. PaaS provides a late binding mechanism that allows applications to communicate with each other, regardless of where the application runs [2].

I. Kabashkin et al. (Eds.): RelStat 2018, LNNS 68, pp. 241–248, 2019.
https://doi.org/10.1007/978-3-030-12450-2_22

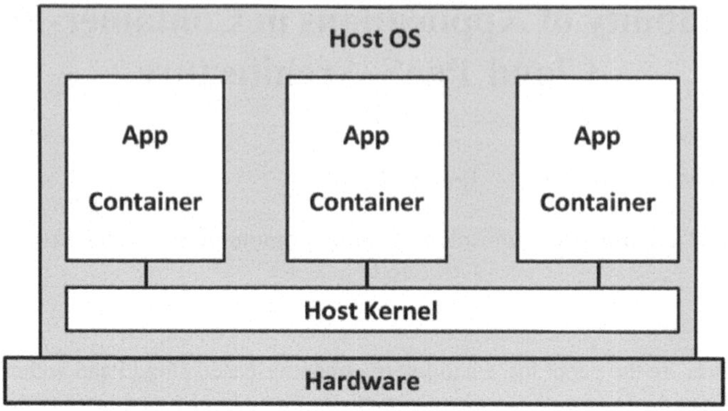

Fig. 1. Architecture for container virtualisation.

From an operator's view, reliability would provide a good target for establishing reliability services besides infrastructure services. This would be an important indicator in private or hybrid cloud because the operator may provide integration services to the business' reliable requirements. Most cloud providers' benchmarks would be on availability, which is the ratio of operational time compared with all-time [3].

From a user's view, reliability would be even more important for a services renting on cloud. Even if the performance is broken for only a few hours, the damage from this violation can reach thousands of dollars per minute [4].

However, approaches to the analysis of container availability from the provider of cloud services and from the user's side are significantly different. For the provider it is important to ensure the reliability of the entire architecture, which ensures the availability of all services and applications. For the user in terms of reliability, only the availability of the used application is important. At the same time, if the failure of another service at another container has come, for this particular user this failure from the position of reliability is not critical.

In the paper availability of applications in container-based PaaS cloud architecture from user point of view is discussed.

The content of this paper is organized as follows. In Sect. 2 some important publications are reviewed. In Sect. 3 the main definitions and assumptions are presented and a model of reliability for container-based cloud architecture is proposed. In Sect. 4 the conclusions are presented.

2 Related Works

Containers are an effective tool for the transformation of the load into several blocks, which greatly facilitates the movement of software applications between cloud architectures.

The article [5] discusses the suitability of containerization for cloud technologies, the principles of virtualization and the main technical requirements for their architecture. Particular emphasis in the work is placed on the problems of orchestration in PaaS systems.

Technical solutions should also determine economic efficiency, which should ensure a relatively quick return on investment. In [6] the Maslow's Hierarchy of Cloud was proposed as hierarchy of needs for the cloud by users. These are basically the expectation dimensions that are imposed on users from the supplier side of cloud infrastructures. Among rating of these needs the reliability is the main factor of influence.

Virtual systems have the same reliability issues as natural systems. However, their use in cloud computing creates additional risks, which are discussed in [7].

The article [8] discusses the problems of reliability and security of cloud computing using virtualization as a key technology, and proposes measures to improve the availability and security of services for these architectures.

Despite the fact that cloud structures operate in a dynamic environment, task scheduling algorithms are mostly static without taking into account reliability issues. This problem is solved in article [9], in which a new linear programming algorithm is proposed for dynamic computations in distributed cloud architectures.

A joint solution to the problems of reliability and energy efficiency of cloud computing is considered in [10]. Issues of improving the reliability of cloud structures based on fog architectures oriented to distributed networks are considered in [11].

Cloud technologies provide service simultaneously to many users. In [12], a model was proposed in which the total available resources with certain reliability are provided to users on the basis of flexible agreements.

If it is possible to process requests in a private or public cloud, [13] proposed a cloud architecture that takes into account reliability issues in the resource provision policy.

According to traditional methods, fault tolerance is determined during the development phase of systems. In [14], an approach was proposed to ensure the resiliency of cloud structures based on the provision of third party services using the concept of resiliency as a service.

In practice, when organizing container-based access to user applications, it is of interest to determine the availability of a separate dedicated container with a user application. In this case the user is not interested in the availability of all other containers and applications provided for other users. In this article, we study the availability of a dedicated user container (DUC) in the system with container-based PaaS architecture.

3 Model Formulation and Solution

The following symbols have been used in the paper:

λ—Failure Rate of container; μ—Repair Rate of container; k—Number of containers; A—Container Availability, U—Container Unavailability, $U = 1 - A$, l—Number of repair bodies.

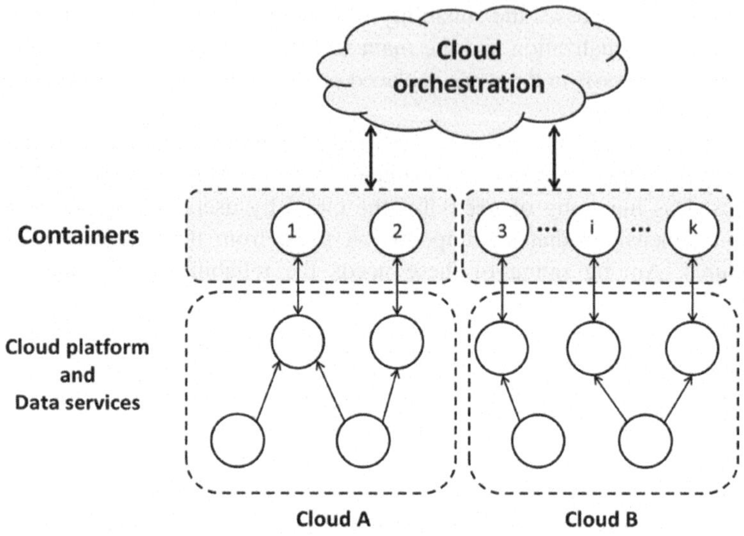

Fig. 2. Container-based platform as a service architecture.

Let us study a PaaS architecture (Fig. 2) [4] with k containers.

For real system with $l = 1$ the behaviour of the examined system is described by the Markov Chain state transition diagram (Fig. 3), where: H_i—state with i failed containers, but dedicated user container is fault-free; H_{if}—state with i failed containers unused by dedicated user and one failed DUC.

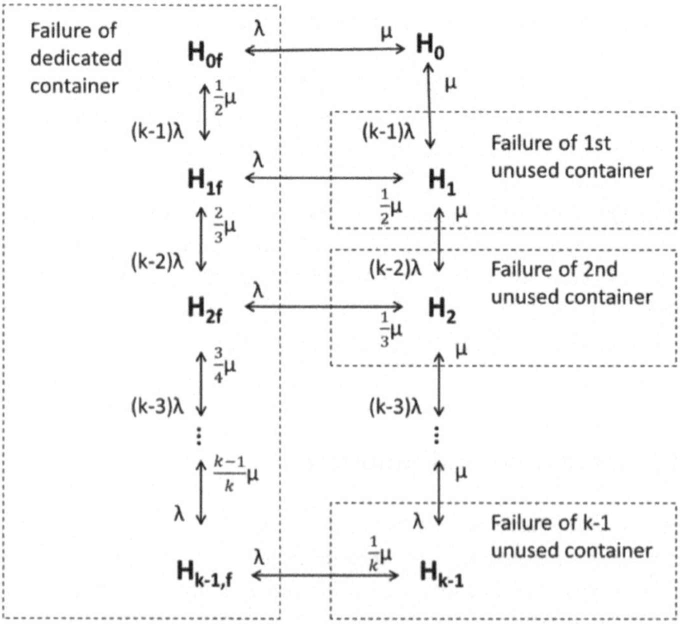

Fig. 3. Markov chain state transition diagram.

On the base of this diagram the system of Chapman–Kolmogorov's equations can be writing in accordance with the general rules [15].

$$
\begin{gathered}
p'_0(t) = -k\lambda p_0(t) + \mu p_1(t) + \mu p_{0f}(t), \\
p'_1(t) = (k-1)\lambda p_0(t) - [(k-1)\lambda + \mu]p_1(t) + \mu p_2(t) + \tfrac{1}{2}\mu p_{1f}(t), \\
\ldots \\
p'_{k-1}(t) = \lambda p_{k-2}(t) - (\lambda + \mu)p_{k-1}(t) + \tfrac{1}{k}\mu p_{k-1,f}(t), \\
p'_{0f}(t) = \lambda p_0(t) - [(k-1)\lambda + \mu]p_{0f}(t) + \tfrac{1}{2}\mu p_{1f}(t), \\
p'_{1f}(t) = \lambda p_1(t) + (k-1)\lambda p_{0f}(t) - [(k-2)\lambda + \mu]p_{1f}(t) + \tfrac{2}{3}\mu p_{2f}(t), \\
\ldots \\
p'_{k-1}(t) = \lambda p_{k-1}(t) + \lambda p_{k-2,f}(t) - \mu p_{k-1,f}(t).
\end{gathered}
$$

The solution of this system of equations gives the expression for availability of selected container:

$$A = 1 - \sum_{\forall i,f} P_{if} = \frac{a_1}{a_1 + a_2}, \tag{1}$$

where

$$
\begin{aligned}
a_1 &= (k-1)! \sum_{i=0}^{k-1} \frac{\gamma^i}{(k-i-1)!}, \\
a_2 &= (k-1)! \sum_{i=0}^{k-1} \frac{(i+1)\gamma^{i+1}}{(k-i-1)!}, \\
\gamma &= \frac{\lambda}{\mu}.
\end{aligned}
$$

3.1 Numerical Example

Let us study the reliability of dedicated user container with the help of the unavailability parameter $U = 1 - A$, where A is determined in accordance with the expression (1).

One of the most popular container programs is Docker [16]. A quarter of companies, applying Docker, use at least 18 containers simultaneously (Fig. 4) [17].

At the Fig. 5 the unavailability parameter U is shown as function of number k containers in the PaaS architecture with different reliability parameters of container itself $\gamma = \lambda/\mu$ for typical number containers per host.

Analysis of the curves at Fig. 5 shows that unavailability of dedicated user container in PaaS architecture increases with increasing number of containers and decreases with increasing of reliability of containers.

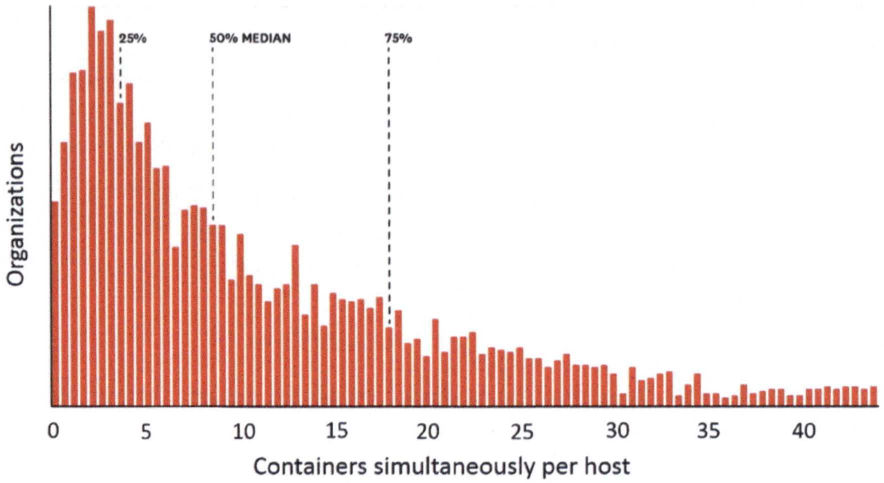

Fig. 4. Container density per host.

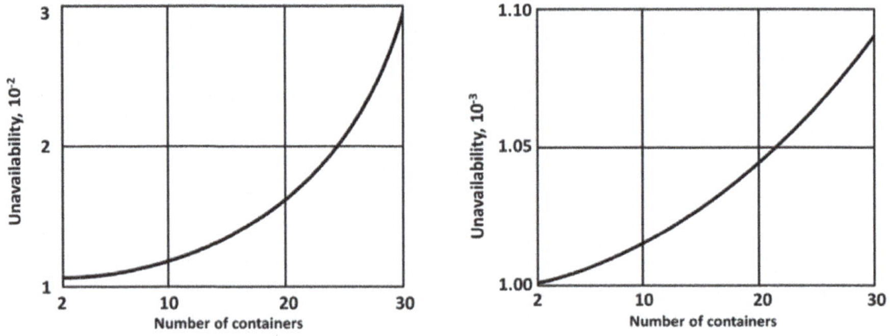

(a) (b)

Fig. 5. The unavailability parameter for **(a)** $\gamma = 10^{-2}$ and **(b)** $\gamma = 10^{-3}$.

4 Conclusions

Containerization is one of the efficient methods to increase productivity of cloud technology. As a result the increase in container density is a main trend in virtualization. In such situation, it is interesting to analyse the effect of the number of containers on the availability of applications that use them.

In the paper the mathematical model of the reliability of a separate dedicated container with a user application in the real conditions of operation is developed. This model was used to determine the availability of a dedicated user container in the container-based PaaS architecture. Expressions for availability of application operation in dedicated user container are developed.

For typical number containers per host and typical reliability parameters of container itself it is shown that unavailability of dedicated user container in PaaS architecture increases with increasing number of containers and decreases with increasing of reliability of containers itself.

Reducing the availability of applications and increase the density in the host to a large number of containers required transition from manual to automatic monitoring of containers in virtual systems.

References

1. Practical Guide to Platform-as-a-Service. Version 1.0. Cloud Standards Customer Council (2015)
2. Managing Multi-Cloud Micro-Services with CloudFoundry APIs. https://cloudfoundryideas.wordpress.com/2015/01/12/multi-cloud-micro-services-using-cloudfoundry/
3. Liu, Y.: Using reliability to measure your cloud (Part two). IBM, Cloud computing news, 12 May 2014. https://www.ibm.com/blogs/cloud-computing/2014/05/using-reliability-measure-cloud-part-two/
4. Linthicum, D.: Moving to autonomous and self-migrating containers for cloud applications. IEEE Cloud Comput. **3**(6), 6–9 (2016)
5. Pahl, C., Lee, B.: Containers and clusters for edge cloud architectures—a technology review. In: 3rd International Conference on Future Internet of Things and Cloud (FiCloud-2015), Rome, Italy, 24–26 August 2015 (2015)
6. Abbassi, P.: Maslow's Hierarchy of Cloud. Giant Swarm, 20 March 2017 (2017). https://blog.giantswarm.io/maslowhierarchyofcloud/
7. Bauer, E., Adams, R.: Reliability and Availability of Cloud Computing. Wiley-IEEE Press (2012)
8. Sabahi, F.: Cloud computing reliability, availability and serviceability (RAS): issues and challenges. Int. J. Adv. ICT Emerg. Reg. **4**, 12–23 (2011)
9. Kumar, S., Balasubramanie, P.: Dynamic scheduling for cloud reliability using transportation problem. J. Comput. Sci. **8**(10), 1615–1626 (2012)
10. Sharma, Y., Javadi, B., Si, W.: On the reliability and energy efficiency in cloud computing. In: Proceedings of the 13th Australasian Symposium on Parallel and Distributed Computing (AusPDC 2015), Sydney, Australia, 27–30 January 2015, pp. 111–114 (2015)
11. Madsen, H., Burtschy, B., Albeanu, G., Popentiu-Vladicescu, F.: Reliability in the utility computing era: towards reliable Fog computing. In: Proceedings of the 20th International Conference on Systems, Signals and Image Processing (IWSSIP), 7–9 July 2013, Bucharest, Romania, pp. 43–46 (2013)
12. Limrungsi, N., Zhao, J., Xiang, Y., Lan, T., Howie Huang, H., Subramaniam, S.: Providing reliability as an elastic service in cloud computing. In: Proceedings of the IEEE International Conference on Communica-tions (ICC), 10–15 June 2012, Ottawa, Canada, pp. 1–6 (2012)
13. Javadi, B., Abawajy, J., Buyya, R.: Failure-aware resource provisioning for hybrid cloud infrastructure. J. Parallel Distrib. Comput. **72**(10), 1318–1331 (2012)
14. Jhawar, R., Piuri, V., Santambrogio, M.: Fault tolerance management in cloud computing: a system-level perspective. IEEE Syst. J. **7**(2), 288–297 (2013)

15. Rubino, G., Sericola, B.: Markov Chains and Dependability Theory, p. 284. Cambridge University Press, Cambridge (2014)
16. Build, Manage and Secure Your Apps Anywhere. The Dev to Ops Choice for Container Platforms, 28 August 2018 (2018). https://www.docker.com/
17. 8 surprising facts about real Docker adoption, 28 August 2018 (2018). https://www.datadoghq.com/docker-adoption/

Increasing the Efficiency of the Wireless Charging System for Mobile Devices that Support Qi Standard

Aleksandr Krivchenkov[1(✉)] and Rodion Saltanovs[2]

[1] Transportation and Telecommunication Institute, Lomonosova str. 1, Riga LV-1019, Latvia
krivchenkovs.a@tsi.lv
[2] TransfoElectric SIA, Akmenu 47, Ogre, Ogres n. LV-5001, Latvia

Abstract. The charging systems for mobile devices that support Wireless Power Transfer technology are under consideration. Most popular wireless charging standards are used and on this basis effective prototype is realized. Power Transfer Efficiency of the two-stage prototype is no worse than 60–70% at distances of 15–20 cm. The measurements of emitted magnetic fields are presented.

Keywords: Wireless charging · Power transfer efficiency

1 Introduction

There is an active development of wireless power transfer (WPT) technologies for battery charging systems. In a number of widespread models of phones, smart watches and similar devices the corresponding circuits are built-in. Wireless charging, based on the Qi standard, received more than 80 smartphones, and this list is regularly updated.

The best parameters for power transfer in charging systems are provided when the carrier is a magnetic field. For the Qi standard, for example, some technical solutions are presented in Fig. 1.

The main purpose of this work is to develop a prototype of the charger, which should have the following features: compatibility with the standards; effective charging also through non-metallic surfaces; the permissible level of electromagnetic interference (EMI).

2 Overview of WPT for Charging Batteries

The technology used has the most significant effect on the parameters and design of the charger. For energy transfer, a magnetic field in the IPT (Inductive power transfer), an electric field and an electromagnetic field (EM) can be used. The analysis of WPT fixes 3 technologies (Table 1) for these applications.

IC suggests a "strong" coupling of magnetic flux interacting circuits. The basic principle of energy transfer is changing the magnetic induction of the transmission coil.

I. Kabashkin et al. (Eds.): RelStat 2018, LNNS 68, pp. 249–258, 2019.
https://doi.org/10.1007/978-3-030-12450-2_23

Fig. 1. Qi1001—built; Qi2001—the receiver; Qivolino—built Qi1001 in the Table [1]

Table 1. Comparison of different technologies of wireless charging

Technology	Advantages	Limitations	Published power transfer efficiency	Applications
Inductive coupling (IC)	Safe use, simple implementation	Short distances, thermal effects, requires alignment transmitter and receiver	From a few mm to several cms. From 5.81% to 57.2% in range 16.2–508 kHz	Mobile electronics, household appliances, RFID tags, non-contact smart cards
Magnetic resonance coupling (MRC)	No need to align devices, charging multiple devices at the same time. High PTE without line of sight	Limited charging distance. Complex implementation	From a few cms up to several m. 90% to more than 30%, for the distance from 0.75 to 2.25 m	Mobile electronics, household devices, televisions and desktop computers, electric vehicles
Electromagnetic radiation (EM)	Long charging distance	It is not safe at high RF power. Low charging efficiency. The need for line of sight	From several m to several km. 0.4%, 18.2% more than 50% at −40 dBm, −20 dBm, −5 dBm of input power	RFID devices, wireless sensors, implanted devices, LED

Power Transfer Efficiency (PTE) in such a system depends on the coupling factor k and the quality factor Q of the circuits. The method provides high PTE, but at short distances. Usually, this distance is less than the diameter of the circuit. This method does not involve the leakage of the magnetic flux of the transmitting circuits.

MRC is based on the generation and transfer of energy between two resonant circuits through an alternating magnetic field. This method involves working at a resonant frequency, which for both circuits should be the same for better efficiency.

EM allows the transfer of energy over sufficiently long distances. Isotropic power transmission is provided in the frequency range 10 MHz to 6 GHz for the EIRP in the 1–3 W range. RF power should be −17 dBm and the RF–DC conversion efficiency in the receiver reach 80%. Note that EM technology has significant "drawback" for charging applications of mobile devices. The power transferred to the load is estimated as $P_L = 0.5 \cdot 10^{-0.5} = 0.16$ mW (Table 1), which leads to significant charging times.

Manufacturers support several wireless charging "standards".

Specification Qi (English pronunciation "chee") is developed and maintained by the Wireless Power Consortium (WPC) from the 2008 year. The Qi specifications are publicly available, but the latest version is only available to WPC members. Qi devices correspond to a model describing both WPT and the informational exchange. Principal and important characteristics of Qi are as follows:

- IC is used, and the distance between the transmitter and the receiver is up to 40 mm;
- two categories for the power of receivers: low power up to 5 W in the frequency range 110–205 kHz and average power up to 120 W in the range 80–300 kHz;
- strong coupling between the transmitter and receiver coils and 3 possible constructive ways of "alignment";
- in-band communication; signalling is a set of messages; the standard implementation of the communication controller and the control device in integrated circuits.

Specification PMA is named after the alliance Power Matters Alliance. Since 2015 the alliance PMA has merged with the alliance A4WP (Alliance for Wireless Power). The new alliance AirFuel was formed. A brief consideration of the PMA leads to conclusions that specifications at the "physical" level are close to the Qi (IC technology, 277–357 kHz range) however, it was most likely intended to introduce more complex information processes accompanying charging processes.

Specification A4WP is accompanied by the alliance A4WP. The standard supported by this alliance was called Rezence, and its specifications are likely replaced by new specifications [2]. The important characteristics of the A4WP are as follows:

- MRC technology; the distance can be several meters according to some data;
- due to the use of MRC, foreign objects can be placed in the working space of the charger without adversely affecting them;
- power transfer at 6.78 MHz (one of the ISM range); signalling is in the 2.4 GHz range (also ISM range);
- power transmitting unit (PTU) of the charger can be in one of the 6 states and power receiving unit (PRU) of the charged device can be in one of the 5 states;
- PTU and PRU are divided into classes and categories;
- BLE (Bluetooth Low Energy) is used for wireless data transmission.

The structure supporting the charging network based on A4WP and currently existing in the form of prototypes is described in [3].

Qi, PMA, A4WP implementation in one device become popular for some solutions. The simplest solution is the implementation of PRU that can receive energy from transmitters operating under different standards. Taking into account that Qi and PMA operate in the frequency range 100–300 kHz, and A4WP at 6.78 MHz, the simplest solution is the use of 2 different receiving coils [4]. The presence of the influence of the frequency of the transmitting circuit on the receiving circuit of a "not its" standard leads to losses in the transmitted power. Tests of the effectiveness give that PTE is not more than 60% for Qi/PMA/A4WP PRU.

The transmitting circuit can only support one standard at the same time, with the corresponding capabilities of the receiving part. In this case remains the open question of information interaction between the PTU and the PRU. The solution with PTU supporting 3 standards is tested experimentally [5]. The total efficiency in this prototype does not exceed 65%. In fact, these prototypes show the possibility of extensive use of the wireless charging network even in the absence of a single unified standard.

From the point of view of the WPC consortium, a good assessment of the effectiveness of WPT is the evaluation of transmitted power, which can be compared with wired charging systems. For TI (Texas Instruments) system based on the Qi power transfer efficiency is shown in the Fig. 2. The efficiency of power transmission from the DC source to the charging unit is 0.95 • 0.89 • 0.89 = 0.75 for a distance 3.7 mm.

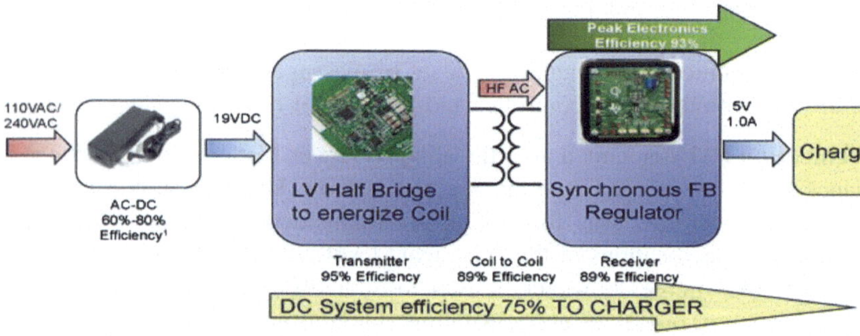

Fig. 2. The power transfer efficiency of the TI wireless Qi charging device

For WPT the RLT (reflected load theory) models, which uses LCR equivalent circuits and Kirchhoff laws, are used. Such RLT models for 2, 3 and 4 contours are presented in the Fig. 3. Some values, which correspond to the experimental ones, are in Table 2.

The following designations are accepted:

$$Q = \omega \frac{L}{R} = \frac{1}{\sqrt{LC}} \frac{L}{R} = \frac{1}{R} \sqrt{\frac{L}{C}} \tag{1}$$

– quality factor of the LCR circuit; $\omega = \frac{1}{\sqrt{LC}}$—resonance frequency of the LCR; $k_{xy} = \frac{M}{\sqrt{L_x L_y}}$—coupling coefficient of the circuits, where M is the mutual induction.

In the Table, the power transfer efficiency (PTE) parameter is indicated by the η symbol. This parameter for the 2-circuits model follows from the law of conservation of energy: $P_S = P_2 + P_3 + P_L$, where P_S—the power produced by the energy source; P_L—power allocated to the load; P_2 and P_3 are the powers allocated in circuits 2 and 3.

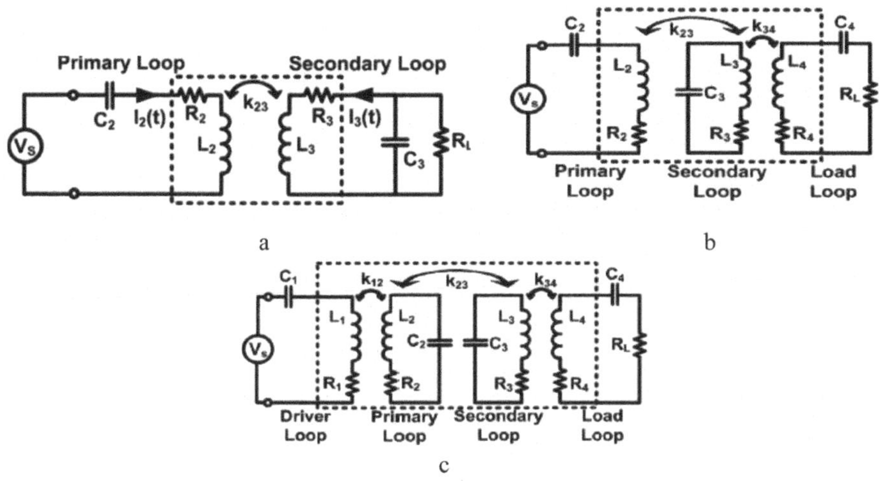

Fig. 3. RLT models for coupled **a** 2 circuits; **b** 3 circuits; **c** 4 circuits in WPT system

Table 2. Specifications of 2, 3, 4 circuits WPT systems for calculations and measurements

Parameters	Symbols	2-coil	3-coil	4-coil
Inductance (μH)	L_1/L_4	–	0.57	0.57
Coil outer diameter (cm)	D_1/D_4	–	5.2	5.2
Quality factor	Q_1/Q_4	–	183	183
Inductance (μH)	L_2/L_3	0.9	0.9	0.9
Coil outer diameter (cm)	D_2/D_3	16.8	16.8	16.8
Quality factor	Q_2/Q_3	255	255	255
L_2 and L_3 distance (cm)	d_{23}	30	30	30
L_2–L_3 mutual coupling	k_{23}	$6.6\ 10^{-3}$	$6.6\ 10^{-3}$	$6.6\ 10^{-3}$
L_1–L_2, L_3–L_4 mutual coupling	k_{12}/k_{34}	0.09	0.09	0.09
PTE calculated (%)	η	1.1	31.4	31.3
PTE measured (%)	η	1.37	29.7	27.9
Nominal load (Ω)	R_L	100	100	100
Power carrier frequency (MHz)	f_0	13.56	13.56	13.56

$$\eta = \frac{P_L}{P_S} = \frac{k_{23}^2 Q_2 Q_{3L}}{1 + k_{23}^2 Q_2 Q_{3L}} \cdot \frac{Q_{3L}}{Q_L} \tag{2}$$

– PTE. $Q_L = \frac{R_L}{\omega L_3}$ a load quality factor and $Q_{3L} = \frac{Q_3 Q_L}{Q_3 + Q_L}$. PTE dependence on the distances between circuit 2 and 3 for 2, 3, and 4 contour models, and comparison with experimental data are presented in the Fig. 4.

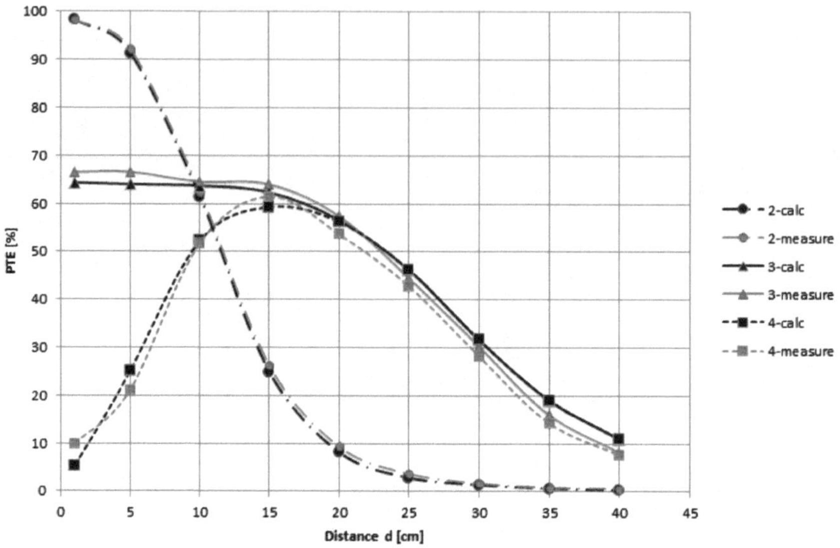

Fig. 4. Comparison between the calculated and measured value of PTE, depending on the distance between circuits d_{23} (variants with 2, 3 and 4 circuits)

2 contour system, in which the power source is connected directly to circuit 2, and the load to circuit 3, is effective at short distances between the circuits; so for a distance of 5 cm, the PTE is of the order of 90%, which agrees well with 89% from the TI measurements.

3 and 4 contour systems have a lower maximum efficiency (60–65%), but a more significant range of distances (at a distance of 10–25 cm efficiency is more than 50%); this is achieved by connecting the load through the 4th circuit, which reduces the effect of load on the quality factor of the receiving circuit 3.

3 contour system has an advantage over 4 at a small distance, since connecting a power source directly to the transmission coil reduces its Q-factor and the resonance effect (at a distance of about 15 cm) on the transfer efficiency.

3 An Effective Solution for the WPT System

It is proposed a two-stage functional scheme of the WPT system (Fig. 5). As a load for the first stage different secondary transmitters are used. The output of the rectifier and further the stabilizer is provided with a rectified voltage of 5 V. The transmitting contour (dimensions 15 × 24 cm) has 7 turns and a 1 cm thick ferrite core; the receiving contour has similar parameters. The resonance of the transmission loop is provided by a digital phase-locked loop system; the regulation of the transmitted power is provided by tuning the frequency down from the resonance frequency of the circuit; by providing resonance in the transmission loop, the PTE achieved at rated power is highly dependent (decreases with increasing or decreasing) on the load resistance. The

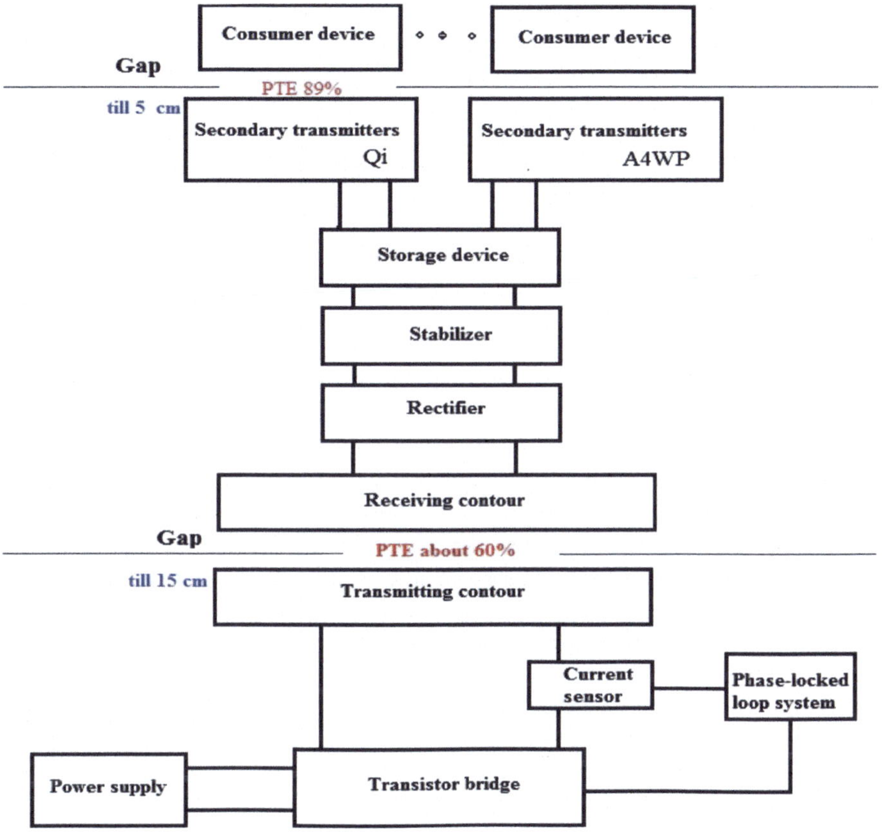

Fig. 5. Functional scheme of a two-stage charging device with a PTE of about 53% through 2 gaps for a distance up to 20 cm for Qi and A4WP devices

nominal transmit power must be 100 W; the maximum, which is determined by the system linearity, is limited by heating and is near 400 W. The distance for devices supporting the Qi charging for a two-stage system reaches not less than 20 cm. The charging is also performed in the presence of separate surfaces.

4 Analysis of the Magnetic Fields Generated by the System

To determine the main characteristics of magnetic fields in the WPT system, we consider the fundamental situation that arises when using the IPT (inductive power transfer) technology. In Fig. 6 the interacting circuits and the resulting fluxes of magnetic induction are presented.

In the transmitting coil T_x, a current is generated due to a voltage source V_s. This current generates (in accordance with the Ampere law) a magnetic flux, which can be determined through the vector of magnetic induction B. If the magnetic flux changes

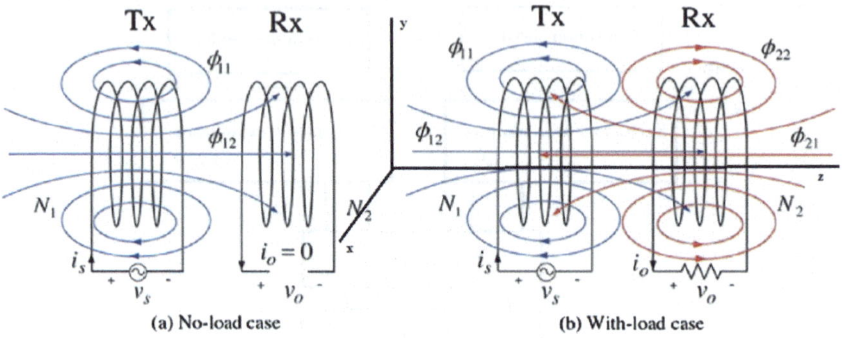

Fig. 6. Magnetic flows of coupled coils. **a** the load is switched off; **b** the load is connected and current flows in the receiving coil [6]

with time, in accordance with the Faraday law part of the flux Φ_{12} passing through receiving coil R_x, generates i_0. This current, in turn, generates streams Φ_{21} affecting the current in T_x, and Φ_{22}, which for the IPT is a "leakage stream", as it participates in energy dissipation in the surrounding space and reduces the efficiency of the WPT.

Exactly the same role (leakage) is played by the flow Φ_{11} in the contour T_x. Due to the low frequency and with a system size much smaller than the wavelength, the electric induction D is zero.

The Fig. 7 shows the pattern of the field of two contours calculated in simulation system [7]. A series of measurements were carried out. The WPT system corresponded to the two-circuit model. The plane of the right contour corresponded to the XY plane; the z-axis coincided with the axis of the contours. Measurements were made in different

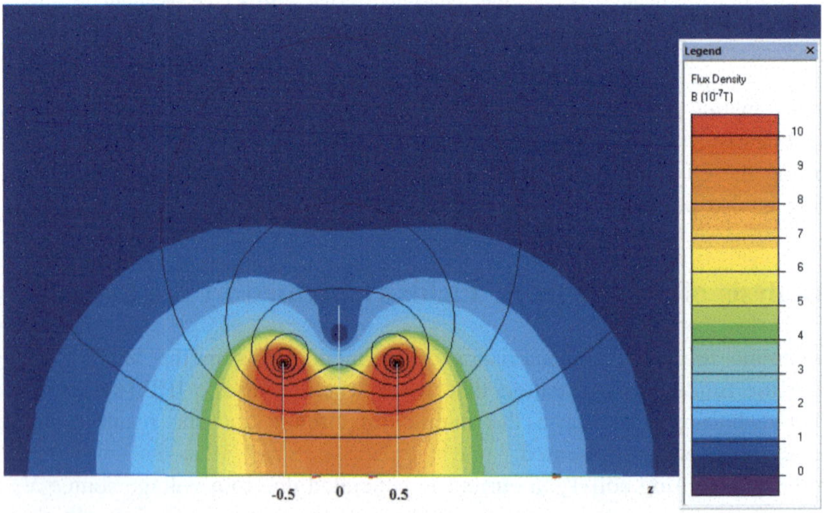

Fig. 7. Induction B of two contours, z = 0 in the middle between the contours

planes, separated from the right contour by a distance 0–200 mm. The simulation and measurement data are presented in Fig. 8. Here the allowed limits of magnetic field H are also indicated.

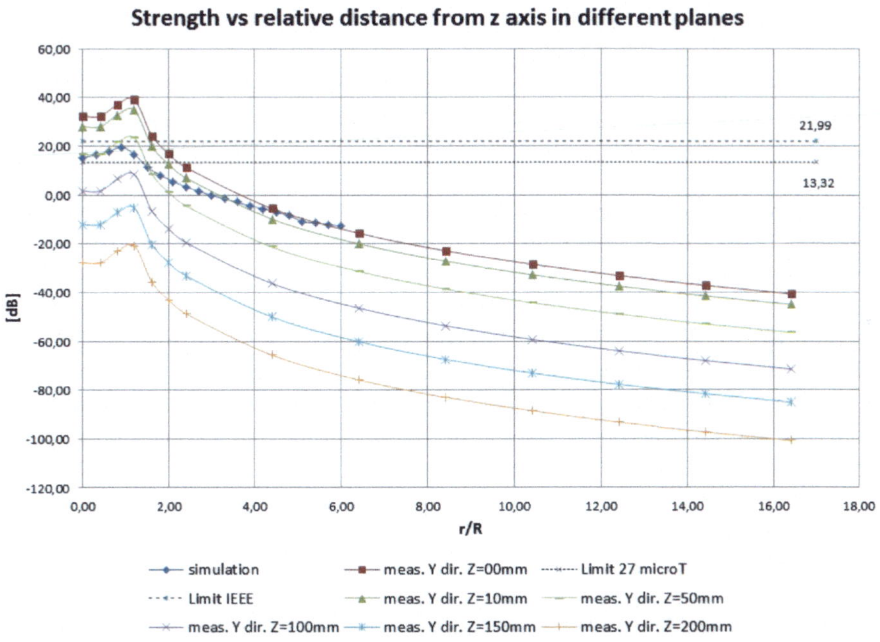

Fig. 8. Changes in the field in different planes at a distance of 0–200 mm from the contour

The systems using IPT do not generate an electric field, and the generated magnetic field must satisfy the conditions formulated in some recommendations. In 10 Hz–100 kHz range, the recommendations of the ICNIRP and IEEE organizations are quite different, and the maximum (but allowable) values for the H in this frequency range are 652–158 A/m or 28–22 dB (units in which measurement data are presented). And it is true for limits in [8].

If we accept the recommendation that the effective value (RMS) of the magnetic induction B should not exceed 27 μT (2 times smaller than the magnetic field of Earth, estimated as 53 μT), and take into account the relationship between B and H in the absence of magnetic materials (in the air) $1\,T = 4\pi \cdot 10^{-7}\frac{A}{m}$, we obtain a limit for the value of the magnetic field strength equal to 21.5 A/m. It corresponds to the level of 13.32 dB for the presentation.

After the measurements, we come to the conclusion that condition is not satisfied in full for all values of r/R of the prototype WPT in the plane coinciding with the plane of the coil. At a distance of about 50 mm from the plane of the circuit, the magnetic field strength is within the requirements and rapidly decreases with further distance.

5 Conclusions

For charging systems the most acceptable technologies are Inductive coupling (IC) and Magnetic resonance coupling (MRC).

Two standards are leading for mobile devices: Qi and A4WP. With some advantage of the A4WP before Qi over the working distance, the processes of management are much more complicated in A4WP.

The most significant characteristics for effective charging systems for mobile devices are power transfer efficiency (PTE); distance more than Qi (4.7 cm in the standard); permissible values of electromagnetic interference (EMI); embedding in surfaces.

End-to-end PTE of the prototype system is no worse than 60–89% in a gap. Sum of distances is of 15–20 cm. The total transmitted power is not less than 20 W.

Magnetic field measurement data was obtained from the prototype at distances of 0–200 mm from its working surface. In the 10 Hz–100 kHz range, the recommendations of IEEE for H intensity are 650–160 A/m, and it is more liberal than in the ICNIRP. This condition is not satisfied in full in the prototype in the plane coinciding with the plane of the contour (immediate proximity). In this connection, the conditions for carrying out the experiments should be clarified if they are carried out in order to prove compliance with the EMI requirements.

References

1. Wireless charging solutions. http://www.qinside.biz/en/solutions. Accessed July 2018
2. AirFuel Wireless Power Transfer System Baseline System Specification v1.3: AirFuel. https://www.airfuel.org/build-with-airfuel/specs/. Accessed July 2018
3. Lu, X., Wang, P., Niyato, D., Kim, D., Han, Z.: Wireless charging technologies: fundamentals, standards, and network applications. IEEE Commun. Surv. Tutor. **18**(2), 1413–1452 (2016)
4. Hwang, J.: An all-in-one (Qi, PMA and A4WP) 2.5 W fully integrated wireless battery charger IC for wearable applications. In: IEEE International Solid-State Circuits Conference (ISSCC) (2016)
5. De Rooij, M., Zhang, Y.: A 10 W multi-mode capable wireless power amplifier for mobile devices. In: PCIM Asia 2016. IEEE (2016)
6. Chun, R., Chris, M.: Wireless Power Transfer for Electric Vehicles and Mobile Devices. Wiley, Hoboken (2017)
7. QuickField: Tera Analysis. http://www.teraanalysis.com/products.htm. Accessed July 2018
8. IEEE C95: ANSI/IEEE C95.1-2005—IEEE Standard for Safety Levels with Respect to Human Exposure to Radio Frequency Electromagnetic Fields, 3 kHz to 300 GHz. IEEE Standards Associations (2014)

A System of Data Processing as Two-Phase Queueing System

Iakov Dalinger(✉)

Saint Petersburg State University of Civil Aviation,
Saint Petersburg 196210, Russia
iakovdalinger@gmail.com

Abstract. Modern telecommunication and data processing systems perform the processing of calls or data packages in several stages. These processes are simulated and investigated by using tandem queues or networks. The following case of two stages is considered in the paper. There exist N independent Poisson flows of elementary messages. The intensity of the i-th flow equals $\lambda_i > 0$. Based on the incoming elementary messages, a complex message is formed according to the following rules. (1) A complex message does not contain more than $h_i \geq 1$ elementary messages of the i-th flow. (2) The generation of a complex message ends if the number of elementary messages in some flow reaches the maximal value. The complex message that has been generated is sent to a one-line queueing system. Such a system has not been previously discussed in the literature. Different indices of the described process are calculated, such as the distribution of elementary message generation time, the distribution of the queue length for complex messages, etc. The obtained results conform to airport control systems.

Keywords: Poisson flow · Queueing system · Complex message · Elementary message

1 Introduction

Modern telecommunication and data processing systems perform the processing of calls or data packages in several stages. These processes are simulated and investigated by the use of tandem queues or networks [6, 7]. We will consider the following case.

There exist N independent Poisson flows of elementary messages [1, 5, 9]. The intensity of the i-th flow equals $\lambda_i > 0$. Based on the incoming elementary messages, a complex message is formed according to the following rules. (1) A complex message does not contain more than $h_i \geq 1$ elementary messages of the i-th flow. (2) The generation of a complex message ends if the number of elementary messages in some flow reaches the maximal value.

The complex message that has been generated is sent to a one-line queueing system. The processing time for one complex message does not depend on the composition of the complex message and has exponential distribution with an intensity μ. The discipline of the queue is FIFO [9], the length of the queue is unrestricted.

I. Kabashkin et al. (Eds.): RelStat 2018, LNNS 68, pp. 259–264, 2019.
https://doi.org/10.1007/978-3-030-12450-2_24

Note that such a system has not been previously discussed in the literature. Only the case of one input flow has been considered.

Different indices of the described queueing system are calculated, such as the distribution of complex message generation time, the distribution of the queue length for complex messages, etc. The obtained results conform to airport control systems.

The peculiarity of this paper consists in applying an embedded Markov chain on arrival moments of complex messages. It allows for designing a simple numerical procedure for obtaining results. The paper is organized as follows. The analysis of the service of elementary messages is presented in Sect. 2. Section 3 is devoted to the flow of complex messages. A numerical example will be considered in Sect. 4. The paper ends with a conclusion and some open problems discussion.

2 The Distribution of Complex Message Generation Time

Let us consider a flow with number i and a random time when the h_i-th elementary message is received. This time equals a sum of h_i independent identically distributed exponential random variables having a mean $1/\lambda_i$. It is known [2, 4, 8, 10] that such a sum has Erlang distribution [6, 7] with a density

$$f_i(t) = \frac{1}{(h_i - 1)!}\lambda_i(\lambda_i t)^{h_i - 1}\exp(-\lambda_i t),\ t \geq 0. \tag{1}$$

The corresponding cumulative distribution function is as follows:

$$F_i(t) = 1 - \sum_{n=0}^{h_j - 1}\frac{1}{n!}(\lambda_j t)^n \exp(-\lambda_i t),\ t \geq 0.$$

The probability that all other flows do not reach their own maximal numbers of arrivals is as follows:

$$Pr_i(t) = \prod_{j \neq i}\sum_{n=0}^{h_j - 1}\frac{1}{n!}(\lambda_j t)^n \exp(-\lambda_i t),\ t \geq 0. \tag{2}$$

Therefore, a density function of complex message generation time is calculated by the following equation:

$$f(t) = \sum_{i=1}^{n} f_i(t) Pr_i(t),\ t \geq 0. \tag{3}$$

3 An Analysis of Complex Messages Service

In this case, we have the queueing system $GI/M/1/\infty$ [9]. Let us consider an embedded Markov chain at the moments when complex messages arrive [2–4]. Let $p_{i,\ j}$ be the probability of the one-step transition from i to j for the number of complex messages in the system. Obviously, for $i = 0, 1, \ldots, j = 0, 1, \ldots, i + 1$

$$\begin{gathered} p_{i,j} = \int_0^\infty f(t) \frac{1}{(j-i-1)!}(\mu t)^{j-i-1} \exp(-\mu t),\, j = 1, \ldots, i+1, \\ p_{i,0} = 1 - \sum_{j=1}^{i+1} p_{i,j}. \end{gathered} \tag{4}$$

Note that the numerical use of these equations requires that the infinite limit of integration should be replaced with some finite number Up.

The stationary probability PSt_j to see j message at the arrival time satisfies the equation system

$$PSt_j = \left[\begin{array}{l} \sum_{i=0}^{\infty} PSt_i\, p_{i,0},\ j = 0, \\ \sum_{i=j-1}^{\infty} PSt_i\, p_{i,j},\ j = 1, 2, \ldots \end{array} \right. \tag{5}$$

The solution of this equation system, satisfied to the normalization condition, yields the needful probabilities.

The distribution $\{PSt_j\}$ allows calculating such numerical Indices of service efficiently as the expectation and the variance of the number of complex messages at the arrival time.

A cumulative distribution function $Gw(t)$ of a waiting time is as follows. The probability that a waiting absents $Gw(t) = PSt_0$, therefore

$$\begin{gathered} Gw(t) = PSt_0 + \sum_{j=1}^{\infty} PSt_i \left(1 - \sum_{i=0}^{j-1} \frac{1}{i!}(\mu t)^i \exp(-\mu t)\right) = \\ = 1 - \exp(-\mu t) \sum_{i=0}^{\infty} \frac{1}{i!}(\mu t)^i \sum_{j=i+1}^{\infty} PSt_j = \\ = 1 - \exp(-\mu t) \sum_{i=0}^{\infty} \frac{1}{i!}(\mu t)^i \left(1 - \sum_{j=0}^{i} PSt_j\right) = \exp(-\mu t) \sum_{i=0}^{\infty} \frac{1}{i!}(\mu t)^i \sum_{j=0}^{i} PSt_j,\ t \geq 0. \end{gathered} \tag{6}$$

To understand this equation, note that there is no waiting time if the system is empty. The corresponding probability is PSt_0. If an incoming message finds j messages in the system, then the waiting time has an Erlang distribution with the cumulative distribution function $F_i(t)$. This produces Eq. (6).

4 Numerical Example

Our example contains the following initial data. There are five elementary flows. The intensity of the flows $\{\lambda_i\}$ and the maximal number of elementary messages in the complex message $\{h_i\}$ are determined by vectors

$$\lambda = (\lambda_1, \ldots, \lambda_N)^T = (2, 3, 4, 5, 2)^T, \quad h = (h_1, \ldots, h_N)^T = (4, 6, 7, 8, 8)^T.$$

The intensity of the exponential distribution of service time $\mu = 1.3$.

Figure 1 shows the density function (3) for complex message generation time. The average generation time equals 0.966, the variance equals 1.815, and the mean-root-square error equals 1.347.

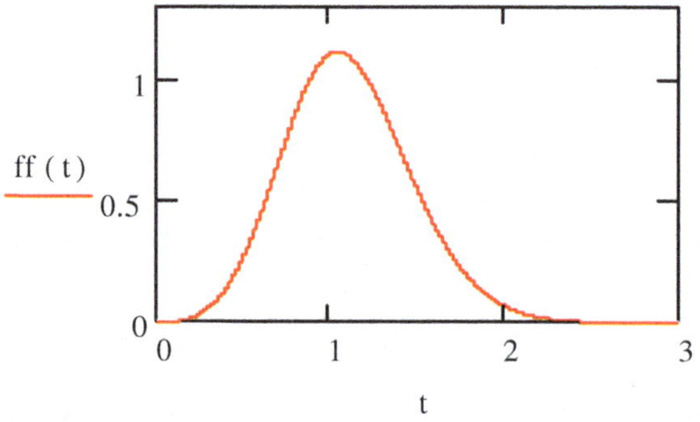

Fig. 1. The grafic of density function (3).

The probabilities (4) of the transition from i to j for the number of complex messages are presented by matrix P in Table 1. Note that the infinite limit of inte-

Table 1. Matrix of transition probabilities $P = (p_{i,j})$.

i\j	0	1	2	3	4	5	6	7	8	9
0	0.739	0.261								
1	0.415	0.325	0.261							
2	0.188	0.227	0.325	0.261						
3	0.072	0.116	0.227	0.325	0.261					
4	0.024	0.048	0.116	0.227	0.325	0.261				
5	0.007	0.017	0.048	0.116	0.227	0.325	0.261			
6	0.002	0.005	0.017	0.048	0.116	0.227	0.325	0.261		
7	0.000	0.002	0.005	0.017	0.048	0.116	0.227	0.325	0.261	
8	0.000	0.000	0.002	0.005	0.017	0.048	0.116	0.227	0.325	0.261
9	0.000	0.000	0.000	0.002	0.005	0.017	0.048	0.116	0.227	0.325

gration in (4) has been replaced with $Up = 5$.

The stationary probabilities $\{PSt_j\}$ to see j complex messages at the arrival time are presented in Table 2. The average number of complex messages equals 1.114. Figure 2 shows how this average number *Average* (5, μ) depends on service intensity μ. Here $Up = 5$ is the upper limit of the integral in Eqs. (4). A graph for the variance *Variance* (5, μ) is presented in Fig. 3.

Table 2. Stationary probabilities $\{PSt_j\}$.

j	0	1	2	3	4	5	6	7	8	9
PSt_j	0.507	0.250	0.123	0.061	0.030	0.015	0.007	0.004	0.002	0.001

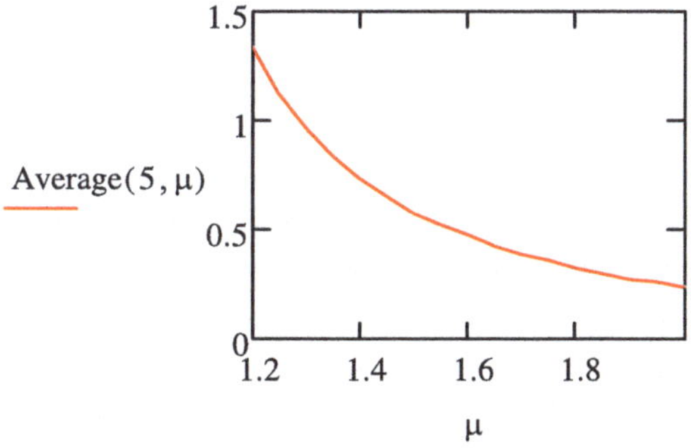

Fig. 2. The average number of complex messages as function of service intensity μ.

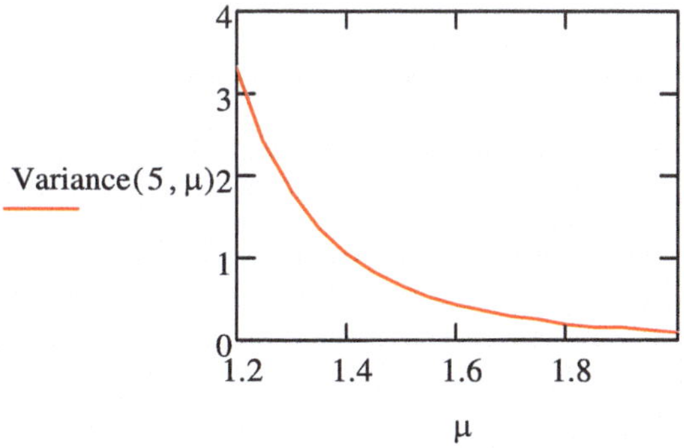

Fig. 3. The variance of the number of complex messages as function of service intensity μ.

Figure 4 contains the cumulative distribution function $Gw(t)$ of the waiting time for a complex message. Here $\mu = 1.3$, $Up = 5$ and 10 means that $PSt_j = 0$ for $j > 10$ in Eq. (6).

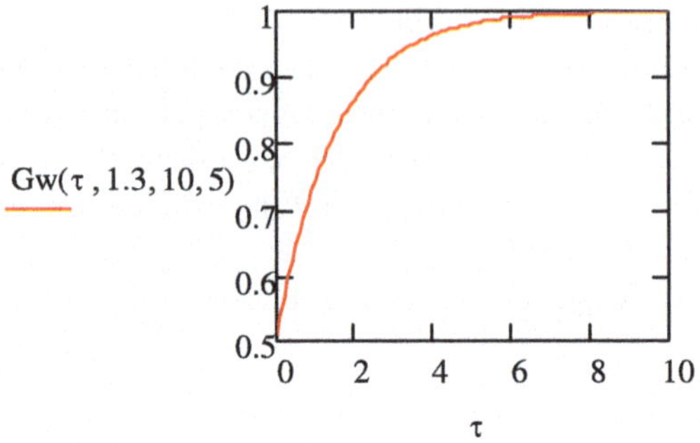

Fig. 4. The cumulative distribution function of the waiting time.

5 Conclusion

The two-phase queueing system with many input Poisson flows has been investigated. The model under consideration can be generalized in different ways. We plan to consider cases with many servers and non-exponential service time. The obtained results conform to airport control systems.

References

1. Bocharov, P.P., D'Apice, C., Pechinkin, A.V., Solerno, S.: Queueing Theory. VSP, Utrecht-Boston (2004)
2. Feller, W.: An Introduction to Probability Theory and Its Applications, vol. 1. Wiley, New York (1970)
3. Gnedenko, B.V., Kovalenko, I.N.: Introduction to Queueing Theory. Nauka Publisher, Moscow (1986). (in Russian)
4. Kijima, M.: Markov Processes for Stochastic Modeling. Chapman & Hall, London (1997)
5. Kleinrock, L.: Queueing Systems: Theory, vol. 1. Wiley, New York (1975)
6. Klimenok, V., Kim, C.S., Dudin, A.: Tandem queueing system with different types of customers. In: Al-Begain, K., Balsamo, S., Fiems, D., Marin, A. (eds.) ASMTA 2011. LNCS, vol. 6751, pp. 99–112. Springer, Berlin (2011)
7. Moiseev, A., Nazarov, A.A.: Tandem of infinite-server queues with Markovian arrival process. In: Vishnevskiy, V., Kozyrev, D. (eds.) DCCN 2015. CCIS, vol. 601, pp. 323–333. Springer, Switzerland (2016)
8. Ross, S.M.: Applied Probability Models with Optimization Applications. Dover Publications Inc., New York (1961)
9. Saaty, T.: Elements of Queueing Theory. McGraw-Hill Inc., New York (1961)
10. Sleeper, A.: Six Sigma Distribution Modeling. McGraw-Hill Inc., New York (2006)

Significant Simulation Parameters for RESTART/LRE Method in Teletraffic Systems of SDN

Elena Ivanova[1(✉)], Teodor Iliev[1], Grigor Mihaylov[1], Ventsislav Keseev[1], and Ivaylo Stoyanov[2]

[1] Department of Telecommunication, University of Ruse, 7017 Ruse, Bulgaria
{epivanova, tiliev, gmihaylov, vkeseev}@uni-ruse.bg
[2] Department of Electrical Power Engineering, University of Ruse, Ruse, Bulgaria
stoyanov@uni-ruse.bg

Abstract. The last years the telecommunication sector constantly evolve. Estimating parameters as the probability of a rare event has applications in reliability, telecommunications, insurance, and several other areas. The rare events such as blocking, overflows and losses in new generation networks, especially in systems with very low probability, are one of the major estimation problems for ordinary simulation methods. Speed-up simulation is the most convenient method for estimation of rare events. The main purpose of the research is to determine and compare the significant parameters for speed-up simulation of SDN. The simulations opportunities represent a chance to reduce the digital divide and to ensure SDN with minimum cost of time and financials. The simulation of teletraffic models allows the modelling of resource management issues, admission of system users with different parameters.

Keywords: RESTART algorithm · Limited Relative Error · Rare events · SDN Software Defined Networking · Teletraffic models · Pareto distribution · Blocking probability

1 Introduction

The last years the telecommunication sector constantly evolve. The new vision of the network is about unlocking the every one of potential markets and profitable revenue growth by engaging the new technologies in the smart home, small and big business, all smart device and multi-user infrastructure of broadband network. Estimating parameters as the probability of a rare event has applications in reliability, telecommunications, insurance, and several other areas. For complicated models, this could be calculated done in principle by the Monte Carlo simulation, but when the event of interest is rare [1], straightforward simulation would in most cases require an excessive number of runs for the rare event to happen frequently enough so that the estimator is meaningful.

Simulations represent a good opportunity to reduce the digital divide and to ensure the new networks based on SDN. Many universities and researchers are currently

I. Kabashkin et al. (Eds.): RelStat 2018, LNNS 68, pp. 265–273, 2019.
https://doi.org/10.1007/978-3-030-12450-2_25

involved in simulation modeling to provide better results. The implementation of simulation program is made with .NET. The teletraffic models with Pareto and Interrupted Poisson Process arrival distributions, which are versatile, possessing a pseudo-memoryless property which makes the solution of many GE-type queuing systems and networks analytically tractable [2]. The new communication systems are interested in the modeling and performance evaluation of single communication links, but in the modeling of more devices in the whole network. Typically, the performance of the whole network containing many links working in parallel and interfering with each other is of primary interest.

This type of simulation allows the modeling of resource management issues, admission of system users with different priority, interaction and network loading by several types of traffic sources, etc. The aim of such modeling is network capacity evaluation and probabilities of failure. Combination of RESTART and LRE (Limited Relative Error) is described by Carmelita Görg, Garvels, Cerou [3–5]. The LRE measures the complementary distribution function of the queue occupancy and performs the Run Time Control (RTC) of the simulation. The LRE performs the Run Time Control with two conditions: the Large Sample Conditions and the Relative Error Condition [3]. The algorithm is described in our previous work [6, 7].

In order to be able to compare the project measures with known results, the part of the project is study different types of models of queues. Evaluation of blocking probability for different teletraffic systems is simulated [6, 7]. RESTART/LRE is implemented the in Ptolemy, a powerful, object-oriented simulator in .NET. The multifactor regression analysis is based on initial simulation results, which are manipulated with program Statistica v10 [6, 8]. The purpose of the analysis is to explore the relationship among the variables of algorithm RETART/LRE and parameters of the teletraffic model for SDN and defining the significant simulation parameters for those systems.

2 Trends of SDN

The enormous spread of internet connected smart phones are changing people's communications with others and their relationships with information. User's ability to access data immediately through applications and web browsers is creating a new culture of real-time information seekers and problem solvers. Recently the number of mobile users equipped with wireless devices capable of video streaming on the go has increased immensely. Recent survey of CISCO shows that mobile data traffic in 2017 was reachless times the amount of entire global internet traffic in 2000 and the mobile video traffic constitutes more than 50% of the entire mobile traffic. If this traffic goes on increasing, network based problems such as delay, packet loss, time-outs etc. may occur. The user gets distracted for the longer delay times, there are some researches which indicate that longer delays lead to consumer frustration and a negative attitude toward the product or service displayed. The ability to transmit video and support real-time multimedia applications is considered important in mobile networks. In order to provide service with high capacity and good quality performance analysis of mobile networks comes into account.

Last year's delivering enormous digital video content with enhanced Quality of Experience *QoE* to the end user over wireless networks has become a challenging issue for mobile service providers. It is important to know whether the user is satisfied with mobile streaming service and there is a need to assess the *QoE*. *QoE* depends on Quality of Delivery *QoD* of packets, i.e., *QoE* links user reactions to delivery problems.

Market and technologies promise that:

- Ultra Fast Access technology bring high bandwidth without the need to rewire;
- Internet of Things promise automation and intelligent remote managements;
- NFV insert the concept of low-cost networked compute and storage;
- SDN enable these functions and user control to create responsive experience;
- Rapid development of software oriented solutions via open source;
- The higher bandwidth is appropriate with new generation of 5 G.

NFV and SDN are complementary and mutually dependent technologies (see Fig. 1).

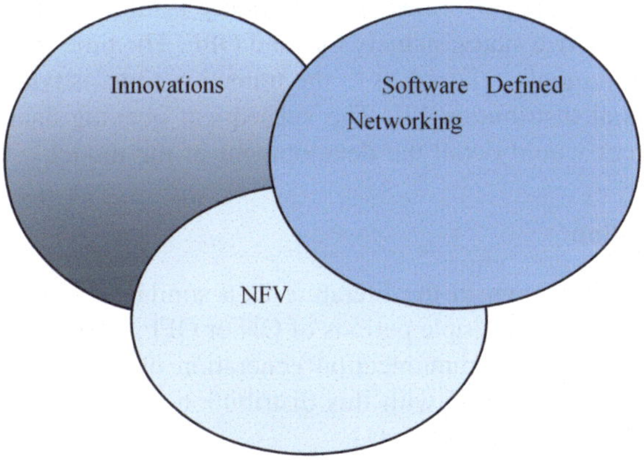

Fig. 1. NFV and SDN.

Software-defined networking (SDN), network functions virtualization (NFV), and network virtualization (NV) are all complementary approaches [9]. They each offer a new way to design deploy and manage the network and its services: SDN separates the network's control (brains) and forwarding (muscle) planes and provides a centralized view of the distributed network for more efficient orchestration and automation of network services; and NFV focuses on optimizing the network services themselves.

NFV decouples the network functions, such as DNS, caching, etc., from proprietary hardware appliances, so they can run in software to accelerate service innovation and provisioning, particularly within service provider environments. NFV ensures the network can integrate with and support the demands of virtualized architectures, particularly those with multi-tenancy requirements

The definition of the basic requirements and characteristics of this type of system sets the traffic parameters, which have to be described with heavy distributions. Determining the traffic systems and their simulation parameters, we can determine the significant parameters for reduction of the blocking events. And this cannot happen without generating a huge amount of traffic without speed-up simulation. On the other hand, the definition of significant parameters could greatly facilitate the introduction of these systems.

3 Teletraffic Models for SDN

The heavy traffic models must be simulated, because the various expectations must be fulfilled on very high level [10]. The type of investigated queueing systems is with different arrival distributions such as Pareto, Geo, Poisson, Erlang and etc.

The design and development of an ON-OFF model relies on an accurate description of traffic entities from link level to application level. The model is generally used, when it is necessary to capture the scaling behaviors of network traffic. For instance, analysis of the structure of IP traffic is performed predominantly using ON-OFF models. The ON-OFF model uses only two states, namely ON and OFF. The time spent between the ON and OFF states, commonly referred to as the transition time, is typically expected to follow an exponential distribution [7]. The subsequent queuing analysis of multiplexed ON-OFF sources would detail the development of the model.

3.1 Pareto Distribution

The Pareto distribution is known in the literature as a similar or long-term character describing network traffic with multiple periods of ON or OFF, where it is assumed that there is traffic or not. New telecommunication generation of networks such as SDN based on IP traffic can be described with this distribution. The Pareto distribution is described with the PDF and CDF set by (1).

$$\begin{aligned} F(x) &= 1 - \left(\frac{b}{x}\right)^{a}, \\ G(x) &= \frac{ab^{a}}{x^{a+1}}, \end{aligned} \tag{1}$$

where $x \geq b$, the parameters a and b are positive. They are the shape and location parameters, respectively. The Pareto distribution is applied to model self-similar arrival in packet traffic. Other important characteristics of the model are that the Pareto distribution has infinite variance, when $b \geq 2$ and achieves infinite mean, when $b \leq 1$.

The parameter a defines shape of queue or its index. If the parameter is assigned with very low value, then the probability is very high, combined with extremely high values of x.

The mean value of mathematical expectation of Pareto distribution is described with (2).

$$E(x) = \frac{ab}{a-1}. \tag{2}$$

Obtaining analytical values of probability parameters for system evaluation is not applicable in this case, which necessitates the simulation studies of such teletraffic systems. The responders of the ON/OFF model can illustrate future energy-saving systems.

3.2 The Interrupted Poisson Process

The Interrupted Poisson Process is yet another two state process. The network channel is one of the two states, ON or OFF. In a discrete time IPP or Interrupted Bernoulli Process, a packet arrives in each of the time slots of the ON state, following a Bernoulli distribution. Though the IPP model is similar to the ON-OFF model, there is a slight variation that differentiates the two models. The difference is that in case of the IPP model, there is no traffic or in other words, no packets arrive during the OFF state.

Teletraffic systems for SDN could be described with other heavy traffic models such as geometrical, Erlang, Weibull, Markov, Markov Modulated Fluid Models and Embedded Markov models.

4 Simulation Results of Teletraffic Models for SDN

Telecommunication tools for simulations are various, such as OPNET, NS2, other MATLAB, java implementations. The reasons of implementing a new tool are many, but we create new simulation module in teletraffic aspect. The program for Simulation of Rare events in teletraffic systems is made with the .NET Framework. The program product .NET is used because a large amount of generated data has to be processed efficiently. Standard simulators generate a pseudo random stream of information. The generated information that is needed for this study should be modelled by changing the specified parameters so that the RESTART method is applied. Simulations represent a good opportunity to reduce the digital divide and to ensure the new trend of SDN networks. The implementation of simulation program is made with .NET.

The simulation of the three main types of queuing systems is made of the AMD Athlon(tm) II X2 250 Processor 3.00 GHz and 3 GB RAM, running 64-bits operation system is Windows 10 Professional.

First simulations are made with M / M / 1 / N queuing systems with the .Net tools, because it was necessary to compare analytical and simulation results, to prove and validate the new results. The algorithm and results are described before [6].

After validation of results, the next step of determine significant parameters is calculating of correlational matrix for the teletraffic systems with Pareto distribution of arrival rate.

Correlational matrix for system Pareto / Geo / 1 / N is calculated, after the important proof that the parameters of the system and the algorithm are without linear dependence. The values of these parameters don't have such a linear correlation, because the correlation index is not bigger then 0.9 (Table 1).

Table 1. Correlational matrix providing nonlinear dependence between parameters in teletraffic system Pareto / Geo / 1 / N.

	n	L	N	RE	μ	FIFO/LIFO
n	1	–0.246	–0.327	0	0	0
L	–0.246	1	0.107	0.302	–0.171	–0.02
N	–0.327	0.107	1	0	0	0
RE	0	0.302	0	1	0	0
μ	0	–0.171	0	0	1	0
FIFO/LIFO	0	0.02	0	0	0	1
PB	–0.411	–0.156	–0.017	0.212	0.034	–0.105

The next step of validation of significant parameters of the RESTART/LRE algorithm for such a system is multifunctional regression analysis. The analysis defines which parameters will have significant influence of final results. The simulation is made for 180 cases with a large variation of the variables. The studied parameters vary in ranges from 10,000 to 1,000,000 for n number of observations, service distribution parameters μ from 0.1 to 0.9, and the number of service places N from 90 to 150.

The correlation matrix of the system and algorithm parameters for the different cases determines the following correlations affecting the simulation probability of blocking in the teletraffic system Pareto / Geo / 1 / N in descending order:

- The highest correlation coefficient is 0.41 the influence of n of PB;
- The relative error correlates with 0.212;
- The levels of splitting L is 0.156;
- The μ correlates with 0.034.

The remaining correlation coefficients are shown in Table 1. In conclusion for the correlation coefficients of the investigated parameters of those teletraffic system and the algorithm are within the admissible range of values demonstrating the lack of linear correlation between them, from which follows that multiple regression analysis can be applied to this system and algorithm.

After the statistical calculations, the parameters of the experiments are set, with Table 2 showing the obtained simulation probabilities for a teletraffic system with $N = 90$ service places compared to the parameter n = 10 000 and n = 100 000. The intensity of the geometric distribution varies widely ranging from 0.1 to 0.9 with no significant impact, the many cases in the system with $N = 90$ are considered for greater completeness. The levels of splitting L of the algorithm are 2 or 3, with very low variance. As with previous systems like M / M / 1 / N, [X] / M / 1 / N and M / [X] / 1 / N, the service discipline has a negligible impact [6].

Table 2. Simulation results with $N = 90$ places for service in the system, the relative error is 10%, the received blocking events.

L	RE	μ	FIFO/LIFO	n	P_{BO}
1	2	3	4	5	6
3	5	0.9	FIFO	10 000	1.0E−12
	10	0.9	FIFO	10 000	7.29E−10
	5	0.8	FIFO	10 000	7.29E−10
	10	0.8	LIFO	10 000	7.29E−10
4	5	0.6	FIFO	10 000	3.43E−09
	10	0.6	LIFO	10 000	1.728E−9
	5	0.3	FIFO	10 000	1E−13
	10	0.3	LIFO	10 000	1.33E−10
	5	0.2	FIFO	10 000	3.43E−10
	10	0.2	LIFO	10 000	3.37E−13
1	2	3	4	5	6
3	5	0.9	FIFO	100 000	2.704E−12
	10	0.9	FIFO	100 000	1.4884E−11
	5	0.8	FIFO	100 000	3.136E−12
	10	0.8	LIFO	100 000	1.5376E−11
	5	0.6	FIFO	100 000	3.964E−12
	10	0.6	LIFO	100 000	1.664E−11

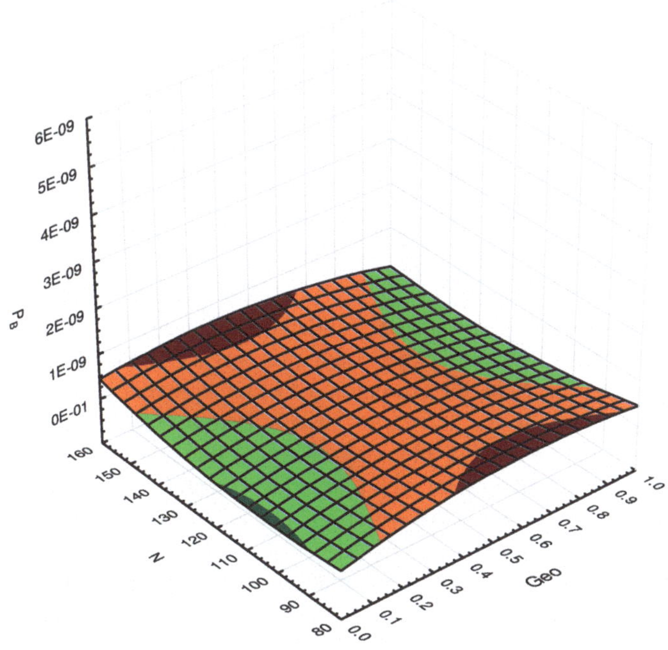

Fig. 2. Relation of PB with n and service rate, received with the algorithm RESTART/LRE.

The graphical relationship between the simulation probability of blocking, N and the service rate μ is illustrated in Fig. 2, where N is more significant parameter then service distribution.

The priorities in the modelled system are dynamic; the service rates of G are with values 0.2, 0.3, 0.4, 0.44, 0.48 and 0.5. The number of examples are 10 000, 30 000, 50 000 and 100 000. The G service rate is with values from 0.2 to 0.5. Table 2 shows simulation results with $N = 60$ places for service in the system, the relative error is 10%, the received blocking events. The results of teletraffic system with $N = 60$ and

Table 3. Simulation results with $N = 120$ and 150 places for service in the system, the relative error is 10%, the received blocking events.

N	μ	n	P_{BO}
120	0.9	100 000	6.724E−12
	0.8	100 000	1.744E−12
	0.7	100 000	1.256E−11
	0.6	100 000	6.889E−12
	0.5	100 000	5.929E−12
	0.4	100 000	4.448E−11
150	0.9	10 000	2.16E−10
	0.8	10 000	1.43E−10
	0.7	10 000	2.16E−10
	0.6	10 000	1.534E−10
	0.5	10 000	5.12E−10

$N = 90$ are described in Tables 2 and 3.

The need of very low level of blocking events in new generations networks is very important part of the process of standardization.

5 Conclusion

In this paper we have reviewed, presented, validated and discussed the scenarios for teletraffic models for SDN. Secondly, this paper presents an implementation of teletraffic models developed in .NET for estimation of blocking events.

The simulation results of the teletraffic system Pareto / Geo / 1 / N are based primarily on the statistical results for blocking events of the analyzes. The simulation results are received for queuing system with Pareto arrivals and non Markovian service rate is implemented with geometrical distribution. The service is organized with FIFO and a finite buffer size N.

Significant parameters for estimating simulation results of RESTART/LRE algorithm for teletraffic system Pareto / Geo / 1 / N are:

- The parameters of finite buffer size N;

- The number of observations *n*.

The future work is the study for the blocking probability with RESTART method for with c servers and finite capacity *N*. In this direction will focus on developing a multi-serving system. Introducing a heavy arrival and service distribution combined with a multi-service system will make simulation to unexpected results.

Acknowledgements. The study was supported by contract of University of Ruse “Angel Kanchev”, No BG05M2OP001-2.009-0011-C01, “Support for the development of human resources for research and innovation at the University of Ruse “Angel Kanchev”. The project is funded with support from the Operational Program “Science and Education for Smart Growth 2014–2020” financed by the European Social Fund of the European Union.

The present document has been produced with the financial assistance of the European Social Fund under Operational Programme “Human Resources Development”. The contents of this document are the sole responsibility of “Angel Kanchev” University of Ruse and can under no circumstances be regarded as reflecting the position of the European Union or the Ministry of Education and Science of Republic of Bulgaria:

Project № BG05M2OP001-2.009-0011-C01 “Support for the development of human resources for research and innovation at the University of Ruse “Angel Kanchev”.

References

1. Bucklew, J.: Introduction to Rare Event Simulation. Springer Series in Statistics, 268 p. Springer, New York (2004)
2. Balabanova, I., Georgiev, G., Kogias, P., Sadinov, S.: Selection of plan of experiment by statistical analysis of the parameters of teletraffic model with voice services. J. Eng. Sci. Technol. Rev. **9**(6), 76–81 (2016)
3. Cerou, F., Guyader, A.: Adaptive multilevel splitting for rare event analysis. Research report, INRIA, (RR-5710) (2000)
4. Garvels, M., Kroese, D.: A comparison of RESTART implementations. In: Proceedings of the 30th 1998 Winter Simulation Conference, December 1998, Washington, DC, pp. 601–609. IEEE Press, Washington (1998)
5. Garvels, M.: The splitting method in rare event simulation. Ph.D. thesis, 175 p. University of Twente (2000)
6. Ivanova E.: Modelling of Traffic in Broadband Convergence Networks, 156 p. Ph.D. thesis, University of Ruse Press, Ruse (2013)
7. Radev, D.: Teletraffic engineering, 146 p. University of Ruse Press, Ruse (2012)
8. Mitkov, A.: Theory of experiments, 226 p. University of Ruse Press, Ruse (2010)
9. Nguyen T., Le, H., Zinner, T.: A context-aware traffic engineering model for software-defined networks. In: Proceedings of the International Conference on Nature of Computation and Communication, ICTCC 2014: Nature of Computation and Communication. November 2014, Ho Chi Minh, Vietnam, pp. 73–82 (2014)
10. Iversen, V.: Teletraffic engineering and network planning, 382 p. Technical University of Denmark (2015)

Intelligent Methods in Digital Forensics: State of the Art

Aleksandr Krivchenkov, Boriss Misnevs(✉), and Dmitry Pavlyuk

Transport and Telecommunication Institute, 1 Lomonosova, LV1019 Riga, Latvia
{krivcenkovs.a,misnevs.b,dmitry.pavlyuk}@tsi.lv

Abstract. This paper contains a review of modern intelligent methods' applications for digital forensics. One of the main problems of digital forensics, which was investigated by authors, is related to the fact that a huge volume of data needs to be analysed for evidence of crime. The primary aim of this work is to improve this challenging forensic process through application of intelligent methods for analysis of digital evidences. The desired outcome of this work is to encourage advancing these methods in a forensic science discipline.

Keywords: Network forensics · Machine learning · Data mining · Intrusion detection

1 Introduction

Nowadays cyber-attacks, executed using malware software agents, are widely utilised to cause damage by means of computers. Different types of attacks could break a critical infrastructure, potentially causing significant harm to people [1].

Implications and scopes of cybercrimes have rapidly expanded during last decades and thus greater research attention is focused on apprehending and prosecuting of offenders. New technologies and legislation could be developed to facilitate the investigation of digital criminal activities. Investigating human misuse of computers creates new puzzles and technical challenges, particularly when offenders attempt to conceal incriminating evidence and their activities on computer systems and networks. One of the main problems of digital forensics is related to the fact that a huge volume of data needs to be analysed for evidence of crime.

Methods of digital forensic analysis are classified by the application area as End-point Forensics, Network Forensics, Mobile Forensics, IoT Forensics, Email and Social Media Forensics [1]. Procedures of all mentioned areas include capturing data, data storage, and data analysis and thus meet well-known big data challenges—volume, velocity, variety, and others.

This work is a feasibility study of methods of information security and big data analyses for solving digital forensic problems, in particular on intrusion detection and timeline reconstruction. Timeline reconstruction is focused on identification of a limited number of investigator-friendly high-level events on the base millions of low-level events (networks packets, registry updates, file modifications, etc.).

I. Kabashkin et al. (Eds.): RelStat 2018, LNNS 68, pp. 274–284, 2019.
https://doi.org/10.1007/978-3-030-12450-2_26

2 Information Security and Big Data Standards

The framework for the research is defined by the set of Information security and Big Data standards (Drafts) [2], presented in Table 1.

The analysis allows drawing a conclusion that the approaches developed in the field of digital forensics appeared in international standards related to Information security. However, the concepts from the Big Data area do not pay much attention to the fact that the processes of digital forensics are the active supplier of problems in this area.

Table 1. Information security and Big Data standards.

Standard	Short description
1	2
ISO 22320:2011 Societal security—Emergency management—Requirements for incident response	Will be replaced by ISO/FDIS 22320. This International Standard specifies minimum requirements for effective incident response and provides the basics for command and control, operational information, coordination and cooperation within an incident response organization. It establishes requirements for operational information for incident response which specifies processes, systems of work, data capture and management in order to produce timely, relevant and accurate information [3]
ISO/IEC 27031:2011 Information technology—Security techniques—Guidelines for information and communication technology readiness for business continuity	This International Standard describes the concepts and principles of information and communication technology (ICT) readiness for business continuity, and provides a framework of methods and processes to identify and specify all aspects (such as performance criteria, design, and implementation) for improving an organization's It also enables an organization to measure performance parameters that correlate to its IRBC in a consistent and recognized manner [4]
ISO/IEC 27035-1:2016 Information technology—Security techniques—Information security incident management—Part 1: Principles of incident management	This International Standard presents basic concepts and phases of information security incident management, and how to improve incident management. This part combines these concepts with principles in a structured approach to detecting, reporting, assessing, and responding to incidents, and applying lessons learnt [5]

(*continued*)

Table 1. (*continued*)

Standard	Short description
ISO/IEC 27035-2 Information technology—Security techniques—Information security incident management—Part 2: Guidelines to plan and prepare for incident response	This standard describes how to plan and prepare for incident response. This part covers the "Plan and Prepare" and "Lessons Learnt" phases of the model presented in ISO/IEC 27035-1. The standard also intends to inform decision-makers that need to determine the reliability of digital evidence presented to them. It is applicable to organizations needing to protect, analyse and present potential digital evidence. It is relevant to policy-making bodies that create and evaluate procedures relating to digital evidence, often as part of a larger body of evidence [6]
1	2
ISO/IEC 27037:2012 Information technology—Security techniques—Guidelines for identification, collection, acquisition and preservation of digital evidence	This International Standard provides guidelines for specific activities in handling potential digital evidence; these processes are: identification, collection, acquisition and preservation of potential digital evidence. This International Standard also provides general guidelines for the collection of non-digital evidence that may be helpful in the analysis stage of the potential digital evidence [7]
ISO/IEC 27043:2015 Information technology—Security techniques—Incident investigation principles and processes	This standard provides guidelines that encapsulate idealized models for common investigation processes across various investigation scenarios. This includes processes from pre-incident preparation up to and including returning evidence for storage or dissemination, as well as general advice and caveats on processes and appropriate identification, collection, acquisition, preservation, analysis, interpretation, and presentation of evidence [8]
ISO/IEC 27050-1:2016 Information technology—Security techniques—Electronic discovery—Part 1: Overview and concepts	This document provides an overview of electronic discovery and describes related terminology, concepts, and processes that are intended to be leveraged by other parts of ISO/IEC 27050. Electronic discovery is the process of discovering pertinent Electronically Stored Information (ESI) or data by one or more parties involved in an investigation or litigation, or similar proceeding. This document also identifies other relevant

(*continued*)

Table 1. (*continued*)

Standard	Short description
	standards and how they relate to, and interact with, electronic discovery activities [9]
ISO/IEC 27050-3:2017 Information technology—Security techniques—Electronic discovery—Part 3: Code of practice for electronic discovery	This document provides requirements and guidance on activities in electronic discovery, including, but not limited to, identification, preservation, collection, processing, review, analysis and production of electronically stored information (ESI). In addition, this document specifies relevant measures that span the lifecycle of the ESI from its initial creation through to final disposition [10]
ISO/IEC DIS 20546 Information technology—Big data—Overview and vocabulary	The big data paradigm is a rapidly changing field with rapidly changing technologies. This document provides definitions and vocabulary needed to promote improved communication and understanding of this area. This document provides a conceptual overview of the field of big data, its relationship to other technical areas and standards efforts, and the concepts ascribed to big data that are not new to big data [11]
ISO/IEC TR 20547-2:2018 Information technology—Big data reference architecture—Part 2: Use cases and derived requirements	This document is focuses on forming a community of interest from industry, academia, and government, with the goal of developing a consensus list of big data technical considerations across all stakeholders. This included gathering and understanding various examples of use cases from diversified areas (i.e., application domains) [12]

3 Digital Forensics from a Practical Point of View

There are two points of view for the digital forensics:

- **Scientific** [13]—the use of scientifically derived and proven methods toward the preservation, collection, validation, identification, analysis, interpretation, documentation and presentation of digital evidence derived from digital sources for the purpose of facilitating or furthering the reconstruction of events found to be criminal or helping to anticipate unauthorized actions are shown to be disruptive to planned operations.
- **Practical** [1]—the collection of methods and tools that can provide the proper steps to react when a cyber-incident occurs; can provide capture of evidence that is useful for potential legal action; will be able to explain what happened by hunting for the digital footprint left by a malicious party.

During the last years, significant attention is paid to searching breaches in security and investigations of many kinds of incidents. Digital forensic methods are involved in the process of responding to a data breach. Their role is mainly technical and lays in providing the details around the breach. It is important to point out that an incident response is different to digital forensics; however, many incident response plans include using forensics to understand what occurred, proving theories about the incident, or preparing for potential future legal action. We look at some details required to accomplish incident response and forensic tasks. Digital forensics is multi-staged process starting from identification of digital media from a scene (possible criminal) as potential evidence to the stage, where it is presented as evidence by an expert witness in a court of law. The sequence of activities is illustrated at a high level in Fig. 1.

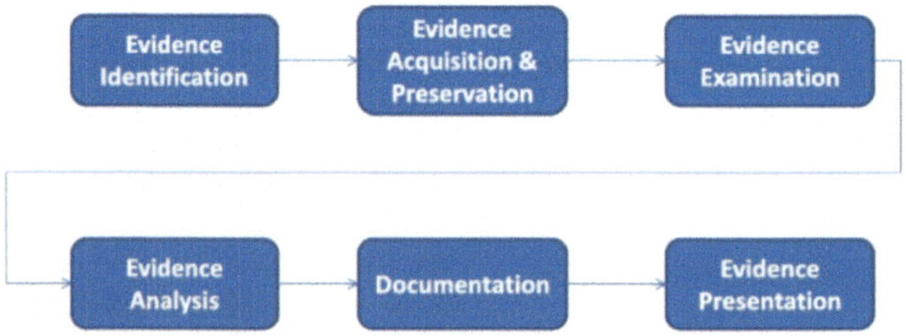

Fig. 1. Digital forensic multi-staged high-level process [13].

From a practical point of view, it is important what an investigator is doing during the entire life cycle of the forensic investigation process. The initial questions that you need to be answered are of a very high level and simple. Next, it is important to open a case properly, leveraging the forensic case management technology. There are several main investigation steps: initial response, data collection, search and seizure, chain of custody, and reporting. A step for closing the case is also needed as an evaluation of performance to improve digital forensic practices. Digital forensic performance is recommended to estimate from the point of view of maturity [1, 14] and the IEEE looks at things like assessing and measuring digital forensic capabilities, people, processes, tools, knowledge base, a repository of procedures, skill profiles, and training. Different types of media sources, used for collecting information, create different fields or types of digital forensics.

The Endpoint system forensics. There are so many different types of endpoints connected to our networks today. From a security standpoint, dealing with all these endpoints is a big challenge. The most important endpoint data sources are:

- file systems for different types of operation systems (OS),
- special OS database like MS Windows registry,
- slack space and corrupt clusters on external hard drives,
- alternate data streams that could be used to hide malicious files or other messages inside the file record of an innocent file,

- MAC OS artefacts locations,
- Log files (system event log, security event log, application event log, etc.).

Theoretically, this is possible to identify the endpoint behaviour of interest like anomalies in user account activity and geographical irregularities.

The IoT forensics. The Internet of Things has introduced many new types of endpoints on corporate networks. The majority of these devices come from manufacturers that are not in the business of creating secure devices as it is mentioned in [1].

The Network Forensics. In this field of forensics, the data sources include network protocols, security tools that monitor networks, various types of network-based attacks, and the footprint they leave behind. Developed network forensic solutions could be classified to inline and passive. Inline (or real-time) systems act directly on traffic (dropping some packets for an example). The passive systems cannot act on live data without triggering another tool because they work with a copy of real traffic.

Figure 2 represents intrusion detection system (IDS), which is passive, versus an intrusion prevention system (IPS), which is inline. IDS is working with a big amount of data, but IPS does this but in real-time.

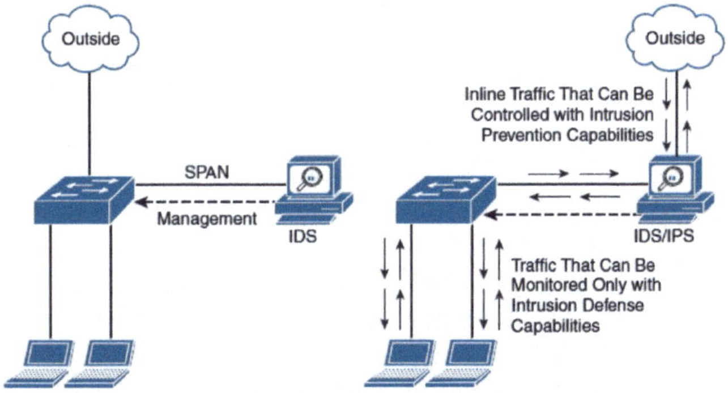

Fig. 2. Passive system (left, SPAN—Switched Port Analyser or ports mirroring) and Inline or real-time system (right) [13].

Some special features have also ***The Mobile Devices Forensics*** and ***The Email and Social Media Forensics***, but they also deal with a big amount of analysed data and need for automation tools. Some quick reference of forensic tools for digital data investigations is provided by Muniz and Lakhani [1].

4 Intelligent Methods Applied to Digital Forensics

The term "intelligent methods" refers to the ability of a computer to learn a specific task from data [15] and include methods of data mining, machine learning, soft computing and traditional artificial intelligence. We conventionally use this term for approaches to automated solving of digital forensic problems. One of the main goals of the intelligent methods can be stated as intrusion identification (for IDS) or for preventing intrusion in real-time (for IPS). There are two main intelligent approaches in support of digital forensics: rule-based and anomaly-based.

Rule-based methods are designed to detect known intrusions using the database of their signatures—low-level computer or network features. Rules can be implemented in a predefined standardised form (e.g. Snort's rules) or as independent scripts. The common application of intelligent methods in this area is related to construction of new rules or optimisation of enabled set of rules. Unified storage of rules and corresponding security problems is another emerging problem for this approach. Generally, rules are weakly structured and thus relational databases are not appropriate for storing them. A modern approach is based on development of ontologies, structured concepts and categories in a domain that show their properties and the relations between them. Recently several researches were focused on development of ontologies for intrusion detection and timeline events [16, 17].

Anomaly-based methods use the reference system behaviour for unsupervised or semi-supervised identification of potential intrusions as deviations from the normal behaviour. In terms of intelligent methods, anomaly detection can be considered as a classical clustering and outlier detection problems. Note that a detected anomaly is not necessary an intrusion evidence and can be related to unusual but proper user behaviour, so this approach normally suffer from a high false alarm rate. Number of pure anomaly-based methods in literature is very limited—many methods combine anomaly detection with rule-based recognition to reduce the false alarm rate. Another potential barrier for application of the anomaly-based approach is related to its weak human-oriented reasoning, which is highly required for presenting evidences of a crime in a court.

The second-stage problem of intrusion identification is an automated classification of discovered intrusion into classes of a different nature and level of danger. Classical set of classes includes attacks (DoS, probe, scan) and minority attacks (user-to-root (U2R) and remote-to-local (R2L)). The problem of classification is another forensics area for extensive application of intelligent methods.

Summarizing the problems stated above, we identify three main areas for application of intelligent methods: (1) rule extraction, (2) anomaly detection, and (3) intrusion classification.

Several recent studies present extensive literature reviews on application of intelligent methods for digital forensics. Bhuyan et al. [18] provided a review and classification of intelligent anomaly-detection methods and empirical evidences of their applications for automated network forensics. Buczak and Guven [19] presented an excellent literature survey, focused on application of data mining techniques for intrusion detection, and discussed application areas and computational complexity of different intelligent methods. Shah and Issac [20] provided a review of performance of popular IDS and related rule-based methods. Umer et al. [21] also analysed applications of intelligent methods for flow-based intrusion detection and formulated emerging methodological challenges in this area. Intelligent methods of almost all mentioned classes can be applied for different problems: rule extraction, anomaly detection and intrusion classification. A literature review, conducted by the authors, revealed a list of most widely used methods. Table 2 presents references to recent studies, classified by applied intelligent methods and by solved digital forensic problems.

Table 2. Application of intelligent methods for intrusion detection tasks.

	Rule extraction	Anomaly detection	Intrusion classification
Artificial neural networks (feed forward and deep neural networks)	Fisch et al. [22]	Shenfield et al. [23]	Akashdeep et al. [24]
Association rule learning (breadth-first, frequent pattern, sequential pattern mining)	Brahmi et al. [25]	Jie et al. [26]	Tajbakhsh et al. [27]
Decision tree learning (iterative dichotomiser, multivariate adaptive regression splines)	Kim et al. [28]	Khammassi and Krichen [29]	Aziz et al. [30]
Probabilistic graphical models (Bayesian networks, hidden Markov models)	Jemili et al. [31]	Devarakonda et al. [32]	Kruegel et al. [33]
Classical clustering methods (*k*-means, hierarchical, self-organising maps)	Blowers and Williams [34]	Roshan et al. [35]	–
Classical classifiers (naive Bayes, support vector machines)	Panda and Patra [36]	Wang et al. [37]	Mukherjee and Sharma [38]
Ensemble learning (boosting trees, random forests)	Zhang et al. [39]	Chebrolu et al. [40]	Aburomman and Reaz [41]
Evolutionary algorithms (genetic algorithms, particle swarms)	Hajisalem and Babaie [42]	–	–

The list of studies, presented in Table 2, should be considered as only typical recent examples of corresponding classes. Each class represents a wide direction of research, covered by many empirical studies.

In addition, we note a growing academic attention to a feature selection problem in intrusion detection. A large volume of heterogeneous data with multiple characteristics lead to a classical problem of machine learning—feature selection and extraction. Recently many researchers [24, 29, 37] addressed the feature selection problem and proved its empirical importance for intrusion detection problems.

5 Conclusions

According to the reviewed literature, there are three most important emerging tasks related to IDS/IPS and requiring artificial intelligence:

- Construction or selection of the optimal set of rules.
- Detection of anomalies (as a sign of an attack).
- Classification of attacks.

The numerous scientific sources analysed in this study allow us to state that there is a real opportunity to improve this challenging forensic process through computer automation by application of intelligent methods for analysis of digital evidences.

The main recommended methods of Artificial Intelligence are artificial neural networks, association rule learning, decision tree learning, probabilistic graphical models, classical clustering methods and classifiers, ensemble learning, and evolutionary algorithms.

Authors' Contribution. A.K. reviewed existing works in digital forensics from a practical point of view. B.M. performed a review of standards and implemented general editing of the paper. D.P. executed a survey on intelligent methods applied for digital forensics. All authors conceived of the presented idea and contributed to the final manuscript.

References

1. Muniz, J., Lakhani, A.: Investigating the cyber breach: the digital forensics guide for the network engineer. Pearson Education Inc., Indianapolis (2018)
2. Kao, D.-Y.: Exploring the cybercrime investigation framework of ATM Heist from ISO/IEC 27043:2015. Presented at the 19th International Conference on Advanced Communication Technology (ICACT), Bongpyeong, South Korea (2017)
3. International Organization for Standardization: ISO 22320:2011 Societal security—Emergency management—Requirements for incident response (2011)
4. International Organization for Standardization: ISO/IEC 27031:2011 Information technology—Security techniques—Guidelines for information and communication technology readiness for business continuity (2011)
5. International Organization for Standardization: ISO/IEC 27035-1:2016, Information technology—Security techniques—Information security incident management—Part 1: principles of incident management (2016)
6. International Organization for Standardization: ISO/IEC 27035-2:2016 Information technology—Security techniques—Information security incident management—Part 2: guidelines to plan and prepare for incident response (2016)
7. International Organization for Standardization: ISO/IEC 27037:2012 Information technology—Security techniques—Guidelines for identification, collection, acquisition and preservation of digital evidence (2012)
8. International Organization for Standardization: ISO/IEC 27043:2015 Information technology—Security techniques—Incident investigation principles and processes (2015)
9. International Organization for Standardization: ISO/IEC 27050-1:2016 Information technology—Security techniques—Electronic discovery—Part 1: overview and concepts (2016)
10. International Organization for Standardization: ISO/IEC 27050-3:2017 Information technology—Security techniques—Electronic discovery—Part 3: code of practice for electronic discovery (2017)
11. International Organization for Standardization: ISO/IEC DIS 20546 Information technology—Big data—Overview and vocabulary
12. International Organization for Standardization: ISO/IEC TR 20547-2:2018 Information technology—Big data reference architecture—Part 2: use cases and derived requirements (2018)

13. Raghavan, S.: Digital forensic research: current state of the art. CSI Trans. ICT **1**, 91–114 (2013)
14. Al Hanaei, E.H., Rashid, A.: DF-C2M2: a capability maturity model for digital forensics organisations. Presented at the IEEE Security and Privacy Workshops, USA, May 2014
15. Russell, S.J., Norvig, P., Davis, E.: Artificial Intelligence: A Modern Approach. Prentice Hall, Upper Saddle River (2009)
16. Chabot, Y., Bertaux, A., Nicolle, C., Kechadi, T.: An ontology-based approach for the reconstruction and analysis of digital incidents timelines. Dig. Invest. **15**, 83–100 (2015)
17. Turnbull, B., Randhawa, S.: Automated event and social network extraction from digital evidence sources with ontological mapping. Dig. Invest. **13**, 94–106 (2015)
18. Bhuyan, M.H., Bhattacharyya, D.K., Kalita, J.K.: Network anomaly detection: methods, systems and tools. IEEE Commun. Surv. Tutor. **16**, 303–336 (2014)
19. Buczak, A.L., Guven, E.: A survey of data mining and machine learning methods for cyber security intrusion detection. IEEE Commun. Surv. Tutor. **18**, 1153–1176 (2016)
20. Shah, S.A.R., Issac, B.: Performance comparison of intrusion detection systems and application of machine learning to Snort system. Future Gener. Comput. Syst. **80**, 157–170 (2018)
21. Umer, M.F., Sher, M., Bi, Y.: Flow-based intrusion detection: techniques and challenges. Comput. Secur. **70**, 238–254 (2017)
22. Fisch, D., Hofmann, A., Sick, B.: On the versatility of radial basis function neural networks: a case study in the field of intrusion detection. Inf. Sci. **180**, 2421–2439 (2010)
23. Shenfield, A., Day, D., Ayesh, A.: Intelligent intrusion detection systems using artificial neural networks. ICT Express **4**, 95–99 (2018)
24. Akashdeep, Manzoor, I., Kumar, N.: A feature reduced intrusion detection system using ANN classifier. Expert Syst. Appl. **88**, 249–257 (2017)
25. Brahmi, H., Brahmi, I., Ben Yahia, S.: OMC-IDS: at the cross-roads of OLAP mining and intrusion detection. In: Tan, P.-N., Chawla, S., Ho, C.K., Bailey, J. (eds.) Advances in Knowledge Discovery and Data Mining, pp. 13–24. Springer, Heidelberg (2012)
26. Jie, X., Wang, H., Fei, M., Du, D., Sun, Q., Yang, T.C.: Anomaly behavior detection and reliability assessment of control systems based on association rules. Int. J. Crit. Infrastruct. Prot. **22**, 90–99 (2018)
27. Tajbakhsh, A., Rahmati, M., Mirzaei, A.: Intrusion detection using fuzzy association rules. Appl. Soft Comput. **9**, 462–469 (2009)
28. Kim, G., Lee, S., Kim, S.: A novel hybrid intrusion detection method integrating anomaly detection with misuse detection. Expert Syst. Appl. **41**, 1690–1700 (2014)
29. Khammassi, C., Krichen, S.: A GA-LR wrapper approach for feature selection in network intrusion detection. Comput. Secur. **70**, 255–277 (2017)
30. Aziz, A.S.A., Hanafi, S.E.-O., Hassanien, A.E.: Comparison of classification techniques applied for network intrusion detection and classification. J. Appl. Logic **24**, 109–118 (2017)
31. Jemili, F., Zaghdoud, M., Ahmed, M.B.: A framework for an adaptive intrusion detection system using Bayesian network. Presented at the 2017 IEEE Intelligence and Security Informatics, New Brunswick, NJ, USA, May 2007
32. Devarakonda, N., Pamidi, S., Kumari, V.V., Govardhan, A.: Intrusion detection system using Bayesian network and hidden Markov model. Procedia Technol. **4**, 506–514 (2012)
33. Kruegel, C., Mutz, D., Robertson, W., Valeur, F.: Bayesian event classification for intrusion detection. Presented at the 19th Annual Computer Security Applications Conference (2003)
34. Blowers, M., Williams, J.: Machine learning applied to cyber operations. In: Pino, R.E. (ed.) Network Science and Cybersecurity, pp. 155–175. Springer, New York (2014)

35. Roshan, S., Miche, Y., Akusok, A., Lendasse, A.: Adaptive and online network intrusion detection system using clustering and Extreme Learning Machines. J. Franklin Instit. **355**, 1752–1779 (2018)
36. Panda, M., Patra, M.R.: Network intrusion detection using Naïve Bayes. Int. J. Comput. Sci. Netw. Secur. **7**, 258–264 (2007)
37. Wang, H., Gu, J., Wang, S.: An effective intrusion detection framework based on SVM with feature augmentation. Knowl.-Based Syst. **136**, 130–139 (2017)
38. Mukherjee, S., Sharma, N.: Intrusion detection using Naive Bayes classifier with feature reduction. Procedia Technol. **4**, 119–128 (2012)
39. Zhang, J., Zulkernine, M., Haque, A.: Random-forests-based network intrusion detection systems. IEEE Trans. Syst. Man Cybern. Part C (Appl. Rev.) **38**, 649–659 (2008)
40. Chebrolu, S., Abraham, A., Thomas, J.P.: Feature deduction and ensemble design of intrusion detection systems. Comput. Secur. **24**, 295–307 (2005)
41. Aburomman, A.A., Reaz, M.B.I.: A survey of intrusion detection systems based on ensemble and hybrid classifiers. Comput. Secur. **65**, 135–152 (2017)
42. Hajisalem, V., Babaie, S.: A hybrid intrusion detection system based on ABC-AFS algorithm for misuse and anomaly detection. Comput. Netw. **136**, 37–50 (2018)

Using Impact-Oscillatory Loading and Nanotechnologies for Improving Mechanical Properties of Two-Phase Titanium Alloy VT23

Mykola Chausov[1], Pavlo Maruschak[2(✉)], and Olegas Prentkovskis[3]

[1] National University of Life and Environmental Sciences of Ukraine, Heroiv Oborony St, 15, 03041 Kyiv, Ukraine

[2] Ternopil National Ivan Pul'uj Technical University, Ruska 56 str., 46001 Ternopil, Ukraine
Maruschak@tu.edu.te.ua

[3] Vilnius Gediminas Technical University, Plytinės g. 27, LT-10105 Vilnius, Lithuania
olegas.prentkovskis@vgtu.lt

Abstract. New methods for enhancing the plasticity and strengthening the surface layers of materials for aircraft industry are proposed. The modes of dynamic non-equilibrium processes (DNP) are grounded, which under the symmetric impact-oscillatory loading can significantly improve the plastic properties of two-phase high-strength titanium alloys. Of particular note is that processing was carried out at room temperature. The combined application of impact-oscillatory loading and nanotechnology has enhanced the strength and hardness of the surface layers of titanium alloy VT23. The results obtained can be used to improve the reliability of materials of aviation transport, as well as spacecraft.

Keywords: Modern transport systems · Non-equilibrium processes · Fracture

1 Introduction

Stringent requirements are placed on the quality of parts and components of modern transport systems, as well as their materials. Therefore, the improvement of materials of aviation transport is impossible without the development of new methods for modifying their properties [1]. Relatively new materials used in aircraft industry include titanium alloys, which are successfully used in aviation transport due to their low density, relatively high strength, and resistance to corrosion media [2].

High machinability, good weldability by all types of welding, low sensitivity to thermal strengthening account for their wide application in the manufacture of aircrafts. At the same time, approaches for enhancing the plasticity of titanium alloys with the preservation of their high strength, or modifying their surface layers are being developed actively. It is known that the modification of thin surface layers allows preserving the ductility and crack resistance of materials [3, 4]. At the same time, the properties of

I. Kabashkin et al. (Eds.): RelStat 2018, LNNS 68, pp. 285–292, 2019.
https://doi.org/10.1007/978-3-030-12450-2_27

the modified material are affected both by the modification process and the initial structure of the material, which requires a comprehensive investigation into the influence of these technological factors.

The authors were the first to show that structural instability occurs during the impulse introduction of energy into materials of various classes (steel, titanium alloys, aluminum alloys, armco-iron, copper) under impact-oscillatory loading with a load frequency of 1–2 kHz. As a result, with subsequent loading, the plastic properties of materials change significantly either for the better or for the worth. In this case, the strength properties of materials practically correspond to the original state. The consequences of this physical process associated with the self-organization of a structure in dynamic non-equilibrium processes (DNP) depend on many factors, including the degree of initial static deformation, in which the materials are subjected to additional impulse force loads, the intensity of additional impulse loads, the number of successive additional impulse loads, changes in temperature modes of loading after the realization of DNP, the initial structure of materials, etc. [5]. It is established by experiment that optimal modes of impact-oscillatory loading can be grounded for virtually any material, ensuring the maximum increase in their plastic properties compared with the initial state. One of the manifestations of major structural changes in the material under impact-oscillatory loading is the occurrence of microextrusions on the specimen surface in the form of "ridges" ("mountains") due to the formation of low density dissipative structures [6, 7]. This indicates changes in the structure and mechanical properties not only in the volume of materials, but primarily in the surface layers. The obtained effect was used by the authors to develop an effective method for strengthening and nanostructuring the surface of materials, and obtaining uniformly controlled nanostructures of the surface layer with the enhanced mechanical strength [8].

New experimental results on the effects of changing the initial mechanical properties of two-phase high-strength titanium alloy VT23 ($\sigma_{us} \geq 1150$ MPa, $\delta = 15\%$), which are manifested under impact-oscillatory loading with the simultaneous use of nanotechnologies, are summarized in this research.

2 Methods of Mechanical and Physical Research

The method of impact-oscillatory loading is realized using the ZD-100Pu modified hydraulic setup for static tests and described in detail in [9, 10]. The main idea of the proposed method is to subject the material to a high-speed stretching, with the high frequency vibration process (several kilohertz) that corresponds to the own frequency of the test machine superimposed. Mechanical tests were performed on specimens (Fig. 1) from sheet industrial titanium alloy VT23 with the thickness of 3 mm. The strain measurement base was 16 mm.

X-ray examination was used to estimate the percentage ratio of α and β phases in the VT23 alloy in the initial state. X-ray examination was performed in monochromatic CuKα radiation on the DRON-UM1 diffractometer. This technique is described in detail in [11].

The chemical composition of the VT23 alloy is given in Table 1.

The microstructure of the VT23 alloy in the initial state is shown in Fig. 2.

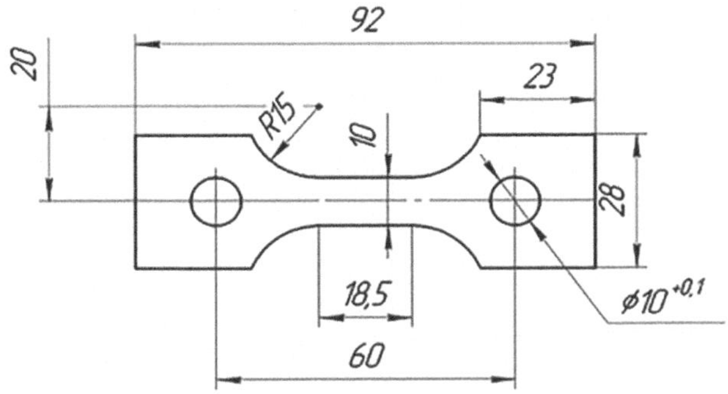

Fig. 1. Test specimen.

Table 1. Chemical composition of the VT23 alloy.

Fe	Cr	Mo	V	Ti	Al
0.5–0.8	1.0–1.4	1.8–2.5	4.3–5.0	86.0–89.3	4.4–6.3

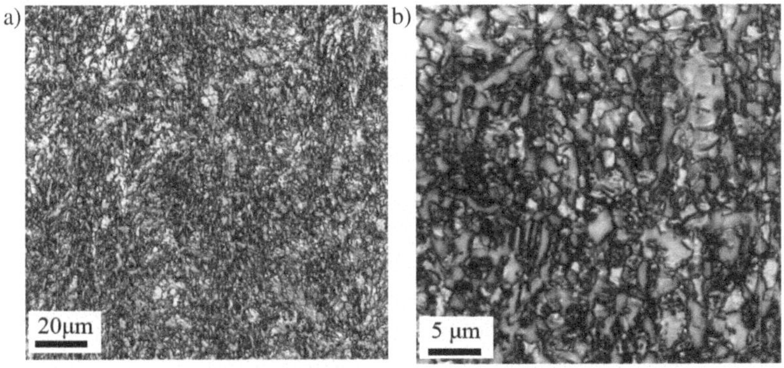

Fig. 2. Microstructure of the VT23 alloy: **a**—x410; **b**—x1660.

To obtain colloidal solutions of materials and the possibility to preserve the physical and chemical activity of their particles, the method for electroerosive dispersion of the material granules in water and deionized water was used [12]. Deionized water, which has a conductivity of $\sim 10^{-6}$ to $3 \times 10^{-6}\ \Omega^{-1}$/cm, is a convenient medium for obtaining aqueous nanosolutions of carbon.

The X-ray photoelectron (XPS) spectroscopy method was used to investigate the surface layers of specimens from the VT23 alloy after the realization of the impact-oscillatory loading. The XPS spectra were measured in the UHV-Analysis-System chamber at a residual pressure of less than 8×10^{-8} Pa. The XPS spectra were investigated using a source of X-ray MgKα-radiation (E = 1253.6 eV) and recorded at a constant delaying potential of 30 eV.

3 Results and Discussion of Experimental Findings

The results of the diffractometric study have shown that in the investigated titanium alloy VT23, the β-phase occupies 43.5% by weight, and the α-phase −56.5% by weight. Mechanical tests were performed in two stages. In the first stage, a number of specimens were subjected successively to the static loading to a growing deformation level of 0.025–0.4%, and next the specimens were subjected to the additional impulse loading within a narrow range of 120 ± 5 kN. Sudden increases in deformation under the impulse introduction of energy ε_{imp} were chosen and justified as the parameter characterizing the intensity of the impulse introduction of energy into the materials. The choice of ε_{imp} as the parameter that describes the intensity of the impulse introduction of energy into materials greatly simplifies the test procedure. Using this parameter, we controlled the modes of impact-oscillatory loading, which can be worked out on hydraulic test machines of different rigidity. This eliminates the need for complex calculations to determine the transfer of a certain force directly on the specimen depending on the total impulse applied to the mechanical system.

The results of the mechanical tests with the symmetric impact-oscillatory loading revealed the modes, in which, during the impulse introduction of energy into the VT23 alloy, the plasticity of the alloys is significantly improved with repeated static stretching. The analysis of the experimental results obtained showed that the optimal value for the VT23 alloy is ε_{imp} = 3.5–4.0% (Fig. 3). In this case, an increase in the plastic deformation of the alloy is obtained, as compared to the initial state, by 30–35%.

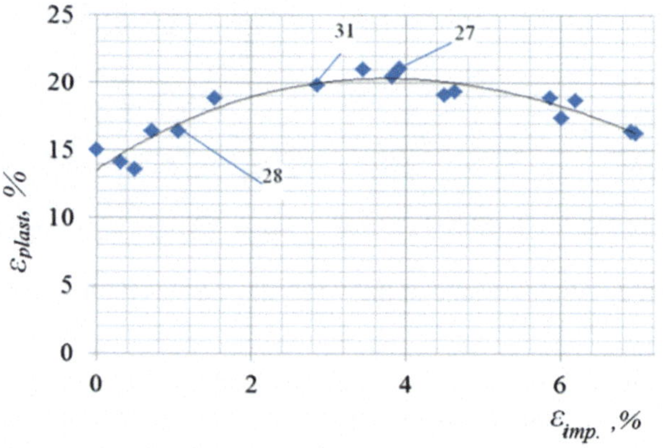

Fig. 3. Curve showing dependence of plastic deformation that occurs at repeated static stretching of specimens from titanium alloy VT23 on sudden increases in deformation of the alloy under impulse introduction of energy (the numbers of specimens, which were subsequently used for metal-physical studies, are indicated on the curve).

Using this optimal influence of the intensity of the impulse introduction of energy into the VT23 titanium alloy, the strengthening of the surface layers of the VT23 alloy was performed on the second batch of specimens using colloidal solutions of tungsten nanoparticles, tungsten carbide and carbon.

The technology of this method is as follows. Prior to the application of the additional impulse loading, the surfaces of specimens from the VT23 alloy are wetted with the relevant colloidal solutions of nanoparticles of materials, and nanoparticles are precipitated from solutions onto the surface of the VT23 alloy by drying.

The nanoparticles located on the surface of the investigated titanium alloy are "embossed" into the surface as a result of its preliminary wetting by a given colloidal solution in a dynamic non-balanced process caused by impact-oscillatory loading, which leads to significant structural changes in the surface layer with the formation of a controlled nanostructured surface layer. The controlled structure of the material surface is obtained by using nanoparticles of the appropriate size in a colloidal solution, which provides for the appropriate concentration of the solution. After the application of the additional impulse loading, the specimens were completely unloaded, wiped off with alcohol, and the hardness of the surface layer was measured in the working area and on the heads of specimens N 44, 47, 57, 58 at a working load of 10 kg on the VIKERS-10 hardness meter.

The number of injections for each study area of the surface was not less than 30. Figure 4 gives some experimental results obtained for the colloidal solution of W (Fig. 4a), colloidal solution of W–C (Fig. 4b), and colloidal solution of carbon (Fig. 4c, d), respectively. The analysis of the results obtained shows that the colloidal solution of W can increase the microhardness of the surface layer of the titanium alloy VT23 by 8.5–9.0%, the colloidal solution of W–C—by 18.5–19.0%, the colloidal solution of carbon—by up to 8.5%. It is clear that by using other colloidal solutions of the optimum concentration, it is possible to achieve even greater strengthening of the surface layers of titanium alloys under impulse introduction of energy. In addition, experiments were carried out on the application of colloidal solutions onto the polished surfaces of specimens from the VT23 alloy. On the pre-polished specimens, the effect of increasing the microhardness of the surface layers by using colloidal solutions is less pronounced. For example, the use of the same colloidal solution of carbon for the unpolished and polished surface of the specimen leads to a decrease in the strength of the surface layer by the value from 8.5% to 4.4% (see Fig. 4c). This can be explained as follows: nanoparticles of the colloidal solution of carbon are "embossed" into the surface layers of the VT23 alloy easier, because there is already the initial roughness of the surface, onto which microectrusions are superimposed due to the formation of low density dissipative structures of the alloy. To explain the mechanical effects of strengthening the surface layers of the investigated titanium alloy by using various colloidal solutions, detailed physical examinations of one of the unpolished specimens were performed using the X-ray photoelectron (XPS) spectroscopy method. In this case, a specimen with a colloidal solution of carbon was selected (the effect of increasing the strength of the surface layers compared with the initial state was 8.4%).

Investigation into the specimen by X-ray photoelectron (XPS) spectroscopy was performed before and after the cleaning of its surface with Ar^+ ions. The analysis of the energy position of the C1s-spectra revealed that before cleaning, the position of the 1s-line corresponds to the clean carbon EC1s = 284.6 eV, whereas after the argon cleansing, this line shifted by 1.2 eV towards lower binding energy, and its peak corresponds to energy EC1s = 285.8 eV (Fig. 5).

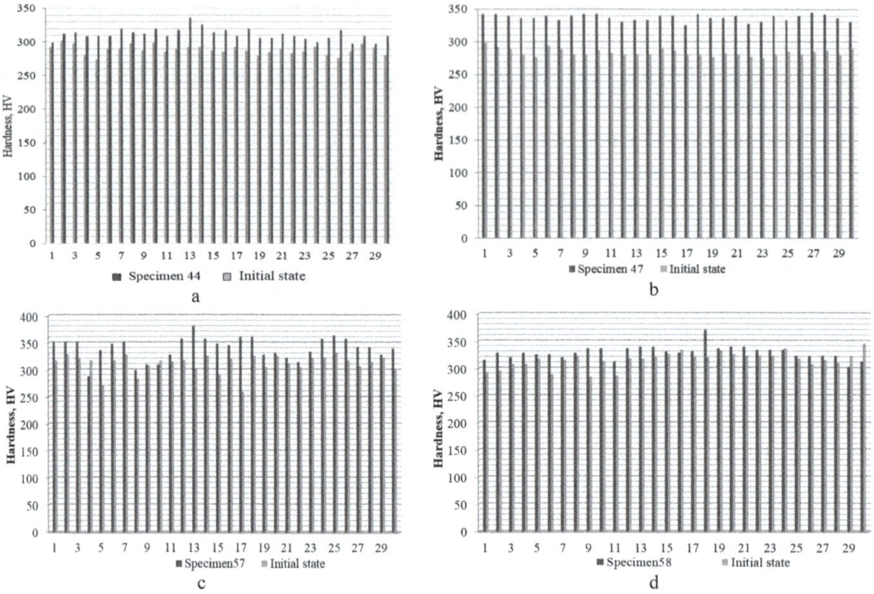

Fig. 4. Increase in the hardness (HV) of the surface layer of specimens from the VT23 alloy treated with various colloidal solutions: **a** colloidal solution of W; **b** colloidal solution of W–C; **c** colloidal solution of carbon, unpolished surface; **d** colloidal solution of carbon, polished surface (the horizontal axis shows the number of indentations).

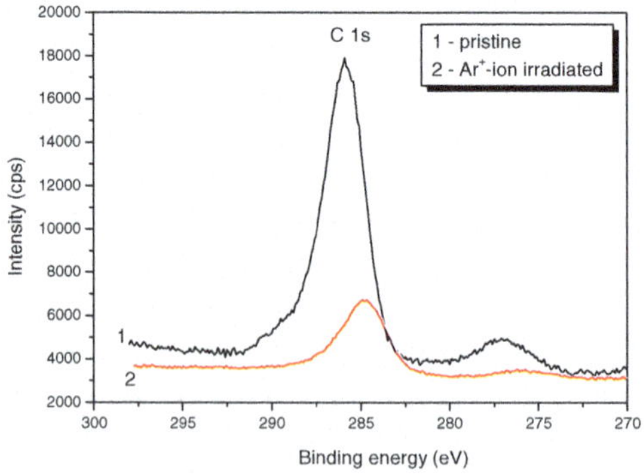

Fig. 5. XPS-spectrum of internal C1s-electrons of the investigated specimen before **1** and after **2** cleaning of its surface with Ar^{+} ions.

This indicates the joining of electrons by carbon and the reduction of its positive charge. Since the electron affinity in carbon is greater than in titanium, and much lower than in oxygen, the electron transfer is possible only from titanium. Consequently, such transfer is possible only with the formation of Ti-C bonds. In addition, after cleaning, $Ti2p_{3/2}$- and $Ti2p_{1/2}$-lines shifted toward lower binding energies by 0.2 eV and 0.4 eV, respectively (Fig. 6). This is due to the fact that carbon takes fewer electrons than oxygen. However, small shifts of the titanium lines indicate that there are more Ti-O bonds than Ti-C bonds. Thus, there is a mixture of titanium oxide and titanium carbide, or titanium oxycarbide of type $TiO_{2-x}C_x$ on the specimen surface after cleaning.

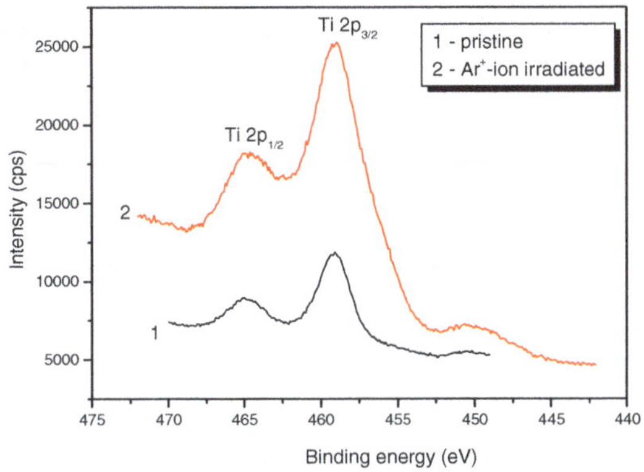

Fig. 6. XPS-spectrum of internal Ti2p-electrons of the investigated specimen before (1) and after (2) cleaning of its surface with Ar^+ ions.

Thus, the proposed simple technological methods for the modification of mechanical properties of two-phase titanium alloys using impact-oscillatory loading and nanotechnologies are highly effective for increasing their initial plastic deformation and strengthening their surface layers.

4 Conclusions

On the example of tests of sheet titanium alloy VT23, the positive effect of impact-oscillatory loading on the improvement of plastic deformation of two-phase high-strength titanium alloy was confirmed. It is shown that when the deflections of plastic deformation reach the values of ε_{imp} = 3.0–4.0% in the process of impulse introduction of energy into the VT23 alloy, the plastic deformation of the alloy increases by 35.0–40.0% with repeated static stretching. A new method for strengthening the surface layer of titanium alloys due to the impulse introduction of energy and the use of various colloidal solutions of materials has been tested. It has been shown experimentally that, due to the impulse introduction of energy and the use of colloidal solutions of materials, one can not only significantly increase the initial plastic deformation of sheet high-strength two-phase titanium alloy VT23, but also increase the strength of its surface layers. Thus, in particular, when using a colloidal solution of W, it is possible to increase the microhardness of the surface layer of the titanium alloy VT23 by 8.5–9.0%, the colloidal solution of W–C—by 18.5–19.0% and the colloidal solution of carbon—by up to 8.5%.

The method of X-ray photoelectron (XPS) spectroscopy has established that the strengthening of the surface layers of the titanium alloy VT23 due to impact-oscillatory loading and the use of carbon nanosolution is associated with the formation of a mixture of titanium oxide and titanium carbide, or titanium oxycarbide of type $TiO_{2-x}C_x$ on the surface. Thus, the fact that titanium carbide Ti-C was formed due to impact-oscillatory loading and the use of carbon nanosolution at room temperature was recorded for the first time.

References

1. Boyer, R.R.: An overview on the use of titanium in the aerospace industry. Mater. Sci. Eng. A **213**(1–2), 103–114 (1996)
2. Uhlmann, E., Kersting, R., Klein, T.B., Cruz, M.F., Borille, A.V.: Additive manufacturing of titanium alloy for aircraft components. Procedia CIRP **35**, 55–60 (2015)
3. Panin, V.E., Sergeev, V.P., Panin, A.V., Pochivalov, Y.I.: Nanostructuring of surface layers and production of nanostructured coatings as an effective method of strengthening modern structural and tool materials. Phys. Met. Metallogr. **104**(6), 627–636 (2007)
4. Panin, S.V., Vlasov, I.V., Sergeev, V.P., Maruschak, P.O., Sunder, R., Ovechkin, B.B.: Fatigue life improvement of 12Cr1MoV steel by irradiation with Zr+ ion beam. Int. J. of Fatigue **76**, 3–10 (2015)
5. Chausov, M., Pylypenko, A., et al.: Influence of dynamic non-equilibrium processes on strength and plasticity on materials of transportation systems. Transport **33**(2), 580–590 (2018)
6. Chausov, M.G., Maruschak, P.O., Pylypenko, A.P., Prenkovskis, O., Hutsaylyuk, V.: Mechanical properties of the sheet two-phase-high-strength titanium alloy VT23 due to impulse introduction of energy. Procedia Eng. **187**, 414–419 (2017)
7. Chausov, M., Zasimchuk, E., Hutsaylyuk, V., Volyanskaya, E.: Research methods of properties of dissipative structures formed while the impulse energy input into the material. J. Ternopil Nat. Tech. Univ. Part 2, 92–97 (2011). (Special edition)
8. Chausov, M.G., Kaplunenko, V.G., Kosinov, M.V., Porokhnyuk, K.M.: Patent of Ukraine N 98493. Method of the modification of mechanical properties of materials, Decl. 29.12.2009; Publ. 25.05.2012, Bull. N 10
9. Chausov, N.G., Voityuk, D.G., Pilipenko, A.P., Kuz'menko, A.M.: Setup for testing materials with plotting complete stress–strain diagrams. Strength Mater. **36**(5), 532–537 (2004)
10. Chausov, N.G., Pilipenko, A.P.: Influence of dynamic overloading on fracture kinetics of metals at the final stages of deformation. Mechanika **48**, 13–18 (2004)
11. Chausov, M., Maruschak, P., Prentkovskis, O., Karpets M.: Risks associated with the use of high-strength titanium alloys in transportation systems. In: Kabashkin, I., Yatskiv, I., Prentkovskis, O. (eds.) Reliability and Statistics in Transportation and Communication. Lecture Notes in Networks and Systems, vol 36, pp. 213–222. Springer (2018)
12. Azharonok, V.V., Lopat'ko, K.G., Aftandilyants, E.G.: Influence of acoustic and electromagnetic actions on the properties of aqueous nanoparticle dispersions used as tempering liquids for dental cement. J. Eng. Phys. Thermophys. **89**(3), 702–713 (2016)

Employment of SIC MOSFETS and GaN-Transistors for Wireless Power Transmission Systems

Rodion Saltanovs[1] and Alexander Krainyukov[2(✉)]

[1] SIA "ElGoo Tech", Peldu iela 7, 3002 Jelgava, Latvia
[2] Transportation and Telecommunication Institute, Lomonosova str. 1, 1019 Riga, Latvia
krainukovs.a@tsi.lv

Abstract. The paper is focusing on to the research of the possibility of using of SiC MOSFETs and GaN-transistors in wireless power transmission systems. To carry research eGaN EPC2034 transistors and SiC transistors C3M0065090J have been used, since these transistors have good frequency properties and small cases. The power inverter hybrid modules were created with chosen transistors. The frequency properties of modules of power inverters were researched when silicone oil was used for direct liquid cooling by of modules. The efficiency assessments is presented for the created modules.

Keywords: Power transistor · Wireless power transmission · Power inverter · Power inverter efficiency · Liquid cooling

1 Introduction

The development and improvement of electric vehicles (EVs) and hybrid electric vehicles (HEVs) is a modern trend of the auto industry. EVs have such advantages as reducing fossil fuel consumption and emissions from the tailpipe. The problems of using EV are manifested in a short range of driving and long time of accumulators charging.

These problems can be solved by dynamically charging EVs accumulators in motion using wireless power transmission (WPT) systems, in which resonantly inductive energy transfer technology is realized [1, 2]. In this case, such WPT systems should be characterized by a high energy transfer coefficient. The effectiveness of the WPT system depends on the design features and technical characteristics of the functional elements of the system, among which the transmitting side power inverters are the most important devices [3, 4].

The WPT systems inverters must have a high conversion efficiency provided that their operating frequency is increased, their sizes is reduced and they have high reliability, since the WPT systems transmitting modules are built in the road cover. Parameters of power electronic inverters are determined mainly by parameters of power semiconductor devices, employmented to design inverters. Currently, power electronic inverters for systems are mainly created on the basis of the insulated gate bipolar

I. Kabashkin et al. (Eds.): RelStat 2018, LNNS 68, pp. 293–301, 2019.
https://doi.org/10.1007/978-3-030-12450-2_28

transistors (IGBT) and control drivers for IGBTs [5]. IGBTs provide a sufficient level of power, but they are low-frequency, besides they have large dimensions and require a stable temperature regime. These features of the IGBTs determine the dimensions of the power electronic inverters and complicate their use in WPT systems.

Silicon carbide (SIC) and gallium nitride (GaN) field-effect transistors are characterized small mass, small dimensions, greater efficiency, increased operating frequency and improved reliability [6, 7] compared to silicon IGBTs. Small dimensions of these power semiconductor devices allow use the modular principle of designing power electronic inverters to get the required power output. In this case of power inverters, the modular construction requires the use of cooling systems of the entire power module, rather than individual power transistors. This research was purposed at design of power inverters modules with use GaN-transistors and SiC MOSFETs and determining the possibility of using them for WPT system.

2 Power Inverter Hybrid Modules with use GaN-transistors and SiC MOSFETs

To create power inverters modules we chose C3M0065090J [8] transistors of new family SIC MOSFETs and eGaN-transistors EPC2034 [9]. Selected transistors has the such technical characteristics which allow to provide the required parameters of power inverters modules.

Cree Power is developer of SiC transistors C3M0065090J. C3M0065090J transistor is silicon carbide silicon transistor of the third generation. The following factors explain choice of C3M0065090J transistors:

- C3M0065090J transistor has parameters whose values are several times better than the same parameters of the best silicon power MOSFETs, namely: switching period —16 ns, the peak drain current is 90 A, the gate charge is 30 nCl and open channel resistance is 65 mΩ.
- C3M0065090J transistors are placed in the D2PAK7 package. The D2PAK7 package provide oscillatory processes reduce and switching losses reduce, since the “leakage distance” between the source and drain is a rather large and equal 7 mm.
- C3M0065090J transistors frequency characteristics provide to increase the operating frequency of power devices and by this to reduce the transformers dimensions and significantly to reduce heat generation, and, accordingly, to reduce the used radiators dimensions.
- C3M0065090J transistors can be applied in voltage inverters, electric drives, power factor correctors, voltage converters, induction heating, welding equipment, as well as in wind power and solar power instead of silicon IGBTs.

Company “Efficient Power Conversion” (EPC) is developer of eGaN EPC2034 transistor. The EPC2034 transistor belong the eGaN series transistors, which are creating by Enhancement Mode Gallium Nitride Power Transistor Technology. The eGaN series transistors outperform the traditional power MOS-FET transistors in a number of parameters. Advantages of eGaN-transistor EPC2034:

- eGaN-transistor EPC2034 main parameters: the open channel resistance is 45 mΩ, the maximum drain-source voltage is 60 V, the maxi-mum gate voltage is 6V, the output capacitance is 90 pF, the peak drain current is 24 A, the gate charge is 1.15 nCl and the input capacitance—115 pF.
- eGaN-transistor EPC2034 is created for operation in the temperature range from −40 to +150 °C at a low thermal resistance of the transistor crystal-case is equal 6.5 °C/W.
- eGaN-transistor EPC2034 BGA-case has the miniaturized sizes.
- The specialized drivers are manufacturing to use eGaN-transistors EPC2034. The specialized drivers have the form of microcircuits with case sizes equals several mm. Driver operating temperature range is the same as range the temperature range of eGaN-transistors EPC2034.

The case features of chose transistors determined the methods of mounting of transistors on the printed circuit board. The surface mounting method was used to mount the D2PAK7 case of the C3M0065090J transistor on board of the hybrid power module. The EPC2034 transistor case contains solder balls instead of the pins, this feature allowed to save space on the module board.

A contact one-sided micro-welding technology was used to mount the chosen transistor cases, drivers and other components to conductive pads of a printed circuit board. Power inverter hybrid modules with use SiC transistors C3M0065090J and eGaN-transistors EPC2034 are showed Fig. 1(a).

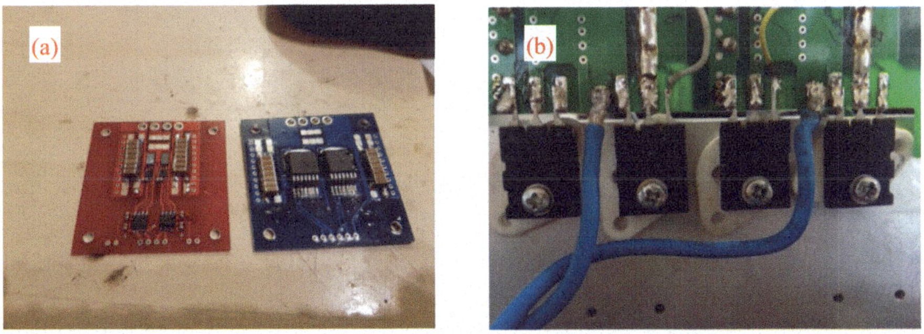

Fig. 1. Power inverters: **a** hybrid modules with eGaN-transistors EPC2034 (on the left) and SiC—transistors C3M0065090J (on the right); **b** IGBTs of full H—bridge.

The half H—bridge was mounted using selected transistors and discrete components on both the module boards. The created micro modules are printed circuit boards the sizes of which are 12 × 10 cm. For comparison, the IGBTs of full H—bridge are shown on Fig. 1b. It can be seen that the transistors are large size and mounted on a large-sized metal radiator.

3 The Test Bench for the Investigation of the Power Inverters Hybrid Modules

The test bench was designed and created to investigation the parameters of the created power inverters hybrid modules. The test bench can be represented by two functional diagrams: electrical (Fig. 2) and hydraulic (Fig. 3).

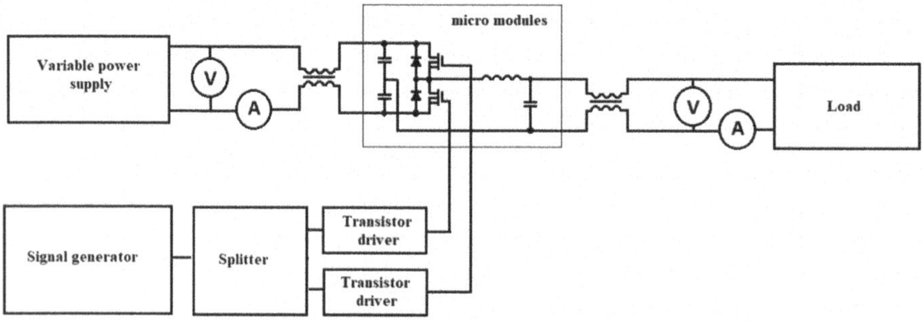

Fig. 2. The functional electrical diagram of the test bench.

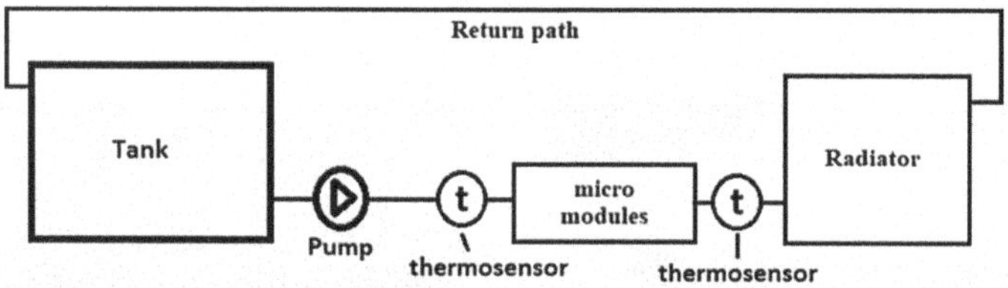

Fig. 3. The functional hydraulic diagram of the test bench.

Such a test bench construction allows to investigate power inverters hybrid modules under various cooling and operating conditions.

The electric test bench includes: variable power supply with reservoir capacitor unit, the generator of control impulses, module of power inverter (half H—bridge) and its load. Variable power apply contains reservoir capacitors, which are a powerful source of current when the inverter transistors are in the open state, and the discharge current of the capacitors flows through the inverter load. The half H—bridge transistors are opening by the control impulses, which is formed by the transistors drivers and are given to the transistors gates.

The signal generator is the master generator of the bridge inverter control circuit. The signal generator generates rectangular pulses with adjustable repetition frequency and duration. The pulses magnitude is equal to 5 V.

The signal splitter form a meander signal from a input periodic sequence of rectangular pulses. This pulses are generated by the master oscillator, in which the pulse repetition frequency is adjustable. Transistor drivers provide the following : increase of power control pulse, galvanic isolation on all control circuits and current protection of power transistors. To control the bridge power inverter on eGaN-transistors EPC2034, the transistor drivers were developed using a microcircuit UCC27611 [10]. The microcircuit UCC27611 is a single-channel, high-speed, gate driver optimized for 5-V drive, specifically designed for efficient control GaN FETs. To control the bridge power inverter on SIC MOSFETs C3M0065090J the transistor drivers were developed on base a microcircuit IR21271 [11]. The microcircuit IR21271 is fast one-channel driver with current monitoring and this feature explains its choice for creating a transistor driver.

The created power inverters modules have small sizes, but in order to use them in WPT systems, highly efficient cooling of modules is necessary. Currently, electronic power devices are cooled by various methods. However, the maximum cooling efficiency is provided by direct liquid cooling technology, which involves immersing the electronic device in the cooling fluid. This is explained by the fact that with this technology, thermal energy is removed directly from the surface of power electronic components and devices [12]. However, to implement the technology of direct liquid cooling, special dielectric liquids and ensuring their circulation are necessary. The hydraulic part of the experimental installation provides such cooling of created power inverters modules.

The hydraulic part of the test bench (Fig. 3) consists of a storage tank, a feed pump, a special case with a board, a radiator and connecting pipes. Silicone oil [13] was chosen for direct liquid cooling.

The power inverters hybrid modules were placed in specially designed cases (Fig. 4) through which the silicone oil flowed to perform direct liquid cooling of modules.

Fig. 4. Cases of power inverters modules for their direct liquid silicone oil cooling.

To define the temperature of silicon oil flow the incoming and outcoming of cases, two temperature sensors (Fig. 3) were used, which were mounted on the printed circiut boards of modules. Temperature sensors made it possible to know the temperature of the transistors on the printed circuit boards, and calculate the temperature gradient of the silicone oil flow in the case.

4 The Investigation of the Power Inverters Hybrid Modules

In the course of investigation, we determined frequency dependencies of the output power and the efficiency of power inverters hybrid modules when they are cooled silicone oil by direct liquid cooling. Investigation was carried out under the following conditions:

- the voltage of variable power supply (Fig. 2) was equal 300 V;
- the master oscillator frequency was changed in range from 100 kHz to 300 kHz, that corresponding range of the operating frequencies of WPT systems;
- the temperature of the modules was about 30° C, that provided by a direct liquid cooling system. Obtained frequency dependencies of the output power and the efficiency of power inverters shown in Fig. 5. All output power and efficiency values are average values of the input and output powers of power inverters measured for each frequency value in the course of 10 experiments. The measurement results had a relative standard deviation of less than 2%.

(a)

(b)

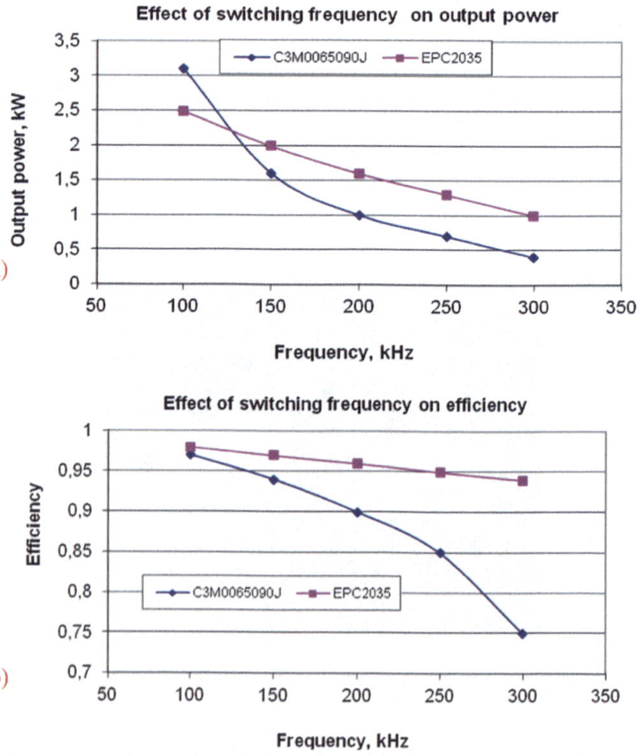

Fig. 5. The output power **a** and the efficiency **b** frequency dependencies of of power inverters with direct liquid cooling.

If the switching frequency of inverter increases, the output power decreases for both created modules of power inverters (Fig. 5a). For the inverter module created with SiC MOSFETs C3M0065090J the output power exceeds 3 kW, if the switching frequency equal 100 kHz. However, the output power value less 2.5 kW for the inverter module with eGaN transistors EPC2034 at the same frequency. Output power of power module with SiC MOSFETs C3M0065090J significant decreases if switching frequency increase. When switching frequency is 200 kHz the output power is 1 kW. The output power drop is smaller with increasing switching frequency for inverter module with eGaN transistors EPC2034. EPC2034 transistors are open-core transistors, and The liquid cooling is more efficient removes heat from EPC2034 transitors. Only the electrical characteristics of the EPC2034 transistors determine the power decrease of inverter module with eGaN transistors EPC2034.

The efficiency of inverter module with SiC MOSFETs C3M0065090J significant decreases with the increase in switching frequency: the efficiency equals about 0.97 if switching frequency equals 100 kHz, hence the power loss equal 100 watts. For this module the efficiency decreases to 0.1, hence the power loss equal 300 watts, if switching frequency will equals 200 kHz. The efficiency of inverter module with eGaN-transistors EPC2034 decreases to a lesser extend with the increase in switching frequency: the efficiency equals about 0.98, hence the output power equals 2.4 kW and the power loss equal 50 watts, if switching frequency will equals 100 kHz. However, this modules with eGaN-transistors EPC2034 has the high efficiency, which exceeds 0.9 even if switching frequency will equals 300 kHz.

Figures 6 and 7 illustrate the influence of the operating frequency on the shape of the output pulses of half-bridges of power modules. The vertical division price—50 V/div. It can be seen that the shape of the pulses is close to rectangular at the outputs of both half-bridge power modules, if the switching frequency equals 100 and 200 kHz (Figs. 6a, b, 7a and b). When the switching frequency equals 300 kHz. The SiC MOSFET C3M0065090J do not have time to fully become open until the moment when the blocking voltage is applied to the gates of the transistors, therefore the output voltage of the bridge is similar to the harmonic voltage (Fig. 7c). The output voltage of the bridge, performed on EPC2034 transistor, remains close in form to the meander (Fig. 6c). This explains the decrease of the efficiency and output power of inverter modules with SiC MOSFETs C3M0065090J.

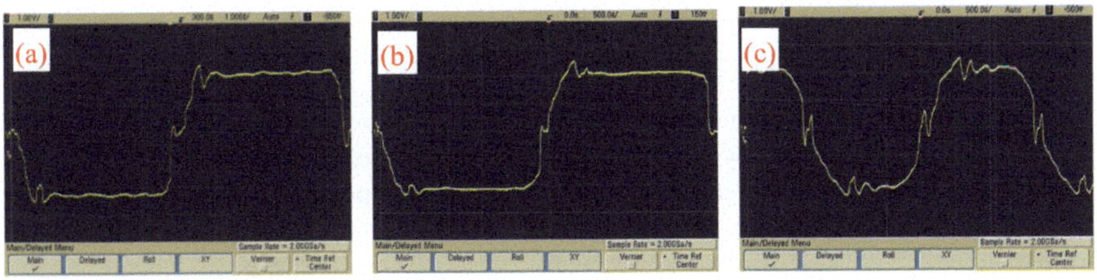

Fig. 6. Time diagrams of output pulses of half-bridge power modules on eGaN-transistors EPC2034 for different operating frequency: **a** 100 kHz; **b** 200 kHz and **c** 300 kHz.

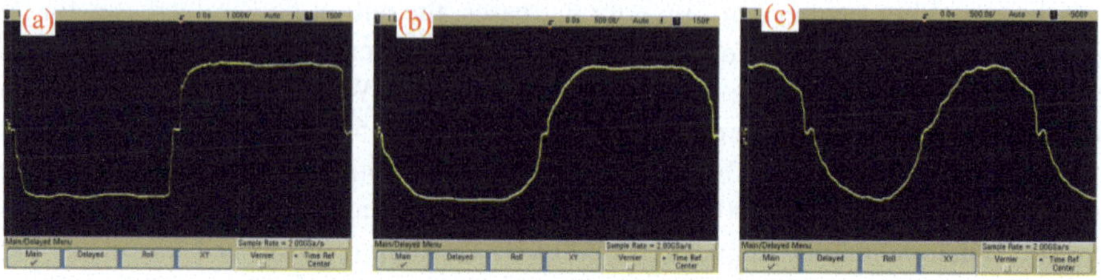

Fig. 7. Time diagrams of output pulses of half-bridge power modules SiC MOSFETs C3M0065090J for three operating frequency: **a** 100 kHz; **b** 200 kHz and **c** 300 kHz.

Despite a output power decrease of the created inverterers modules with an increase in the operating frequency, the output power levels of the modules remain high enough for wireless power transfer and dynamic battery charging. However, such levels of output power and high efficiency of power modules were achieved using the hydraulic system of direct cooling of the modules with silicone oil. The use of such a hydraulic system in WPT systems is impossible, since the cooling system increases the size and reduces the efficiency of the transmitting part of WPT. To cool the power inverter modules in JE systems, it is necessary to use 3M™ Fluorinert™ specialized thermostatic fluids designed for cooling electronic equipment [14], since pumps and cooling radiators are not required when using them.

5 Conclusions

This work main results:

- if the inverter module with eGaN-transistors EPC2034 is used, the efficiency is greater than 90% if the operating frequency is in the range from 100 to 300 kHz;
- if the inverter module with C3M0065090J transistors is used, the efficiency decreased to 75% in the same frequency range;
- GAN—transistors and SIC MOSFETS have a prospect of use in WPT systems, since created of on their basis inverters hybryde modules provide the necessary power levels in the frequency range defined for WPT;
- the direct liquid cooling of the inverter modules ensures their high efficiency, but for cooling the inverter modules in the systems, it is necessary to use specialized thermostatic fluids 3M™ Fluorinert™.

Acknowledgements. This paper has been published within the research project "Smart Engineering, Transport and Energy Competence Centre" project number: 1.2.1.1/16/A/008. "Development of hybrid circuit technology for high frequency power conversion modules using micro assembly".

References

1. Shin, J.: Design and implementation of shaped magnetic resonance based wireless power transfer system for roadway-powered moving electric vehicles. IEEE Trans. Ind. Electron. **61**(3), 1179–1192 (2014)
2. Imura, T., Okabe, H., Hori, Y.: Basic experimental study on helical antennas of wireless power transfer for electric vehicles by using magnetic resonant couplings. In: 2009 Proceedings of IEEE Vehicle Power Propulsion Conference, pp. 936–940 (2009)
3. Covic, G., Boys, J.: Modern trends in inductive power transfer for transportation applications. IEEE J. Emerg. Sel. Top. Power Electron. **1**(1), 38–48 (2013)
4. Li, S., Mi, C.: Wireless power transfer for electric vehicle applications. IEEE J. Emerg. Sel. Top. Power Electron. **3**(1), 4–17 (2015)
5. Wintrich, A., Nicolai, U., et al.: Application Manual Power Semiconductors, 466 p. SEMIKRON International, Ilmenau (2011)
6. Biela, J., Schweizer, M., Waffler, S.: SiC versus Si–evaluation of potentials for performance improvement of inverter and DC–DC converter systems by SiC power semiconductors. IEEE Trans. Ind. Electron. **58**(7), 2872–2882 (2011)
7. Everts, J., Dasz, J., Van den Keybusx, J., et al.: GaN-based power transistors for future power electronic converters. In: PIEEE Benelux Young Researchers Symposium (YRS2010). https://lirias.kuleuven.be/handle/123456789/268624. Accessed 12 Sept 2018
8. C3M0065090J. Silicon Carbide Power MOSFET. https://www.wolfspeed.com/media/downloads/145/C3M0065090J.pdf. Accessed 18 Sept 2018
9. EPC2034—Enhancement Mode Power Transistor (2013). https://epc-co.com/epc/Portals/0/epc/documents/datasheets/EPC2034_datasheet.pdf. Accessed 1 Sept 2018
10. UCC27611. Texas Instruments. http://www.ti.com/lit/ds/symlink/ucc27611.pdf. Accessed 11 Sept 2018
11. IR21271—Infineon. https://www.infineon.com/cms/en/product/power/gate-driver-ics/ir21271/. Accessed 15 Sept 2018
12. Advanced Cooling for Power Electronics. https://www.electronics-cooling.com/2017/07/advanced-cooling-power-electronics/. Accessed 2 Aug 2018
13. Boukhanouf, R., Haddad, A.: A CFD analysis of an electronics cooling enclosure for application in telecommunication systems. Appl. Therm. Eng. **30**(16), 2426–2434 (2010)
14. Thermostatic fluids 3M™. http://multimedia.3m.com/mws/media/-782409O/3mtm-thermal-management-fluids.pdf?fn=ThermalMgmtFluid.pdf. Accessed 12 Aug 2018

Off-the-Shelf Convolution Neural Networks Low-Light Object Detection Comparison

Przemysław Śliwiński[1(✉)], Jacek Mazurkiewicz[1], Jarosław Sugier[1], Tomasz Walkowiak[1], and Krzysztof Helt[2]

[1] Wrocław University of Science and Technology, Wybrzeże Wyspiańskiego 27, 50-370 Wrocław, Poland
przemyslaw.sliwinski@pwr.edu.pl
[2] Teleste sp. z o.o, Szybowcowa 31, 54-130 Wrocław, Poland

Abstract. In the note, we examine several off-the-shelf object detecting convolutional neural networks in simulated low-light conditions, typical for surveillance and automotive vehicle applications. We propose the test procedure and its present its formal background. The preliminary results suggest that such ready-to-use networks are viable tools in the mentioned applications.

Keywords: Object recognition · Low-light imagery · Poisson noise · Bernoulli trial · Off-the-shelf convolutional neural networks · Test procedure

1 Introduction

The article describes the test routine, which mimic the actual scenario, where a company (be it a start-up, or an established enterprise) want to verify to what extent the open source off-the-shelf neural network-based (usually of convolutional type, aka as CNNs) object detection tools are applicable in applications involving:

- Surveillance or
- Autonomous vehicles (drones and/or cars).

The common issue in these applications is a lack of sufficient scene illumination, as it is assumed that they need to reliably work from "dusk to dawn" as well. This results in a low signal-to-noise ratio, SNR. In turn, the available CNN-based tools are trained with the help of well-light scenes and objects, that is, in conditions where SNR is high. The recent tragic accident in Arizona, where allegedly (according to the recent findings of the National Transportation Safety Board, NTSB, which investigates the accident, [11]) the "hardware" detected the trespasser but the "software" falsely ignored her presence, testifies to such a need.

2 A Probabilistic Model of Light

The low SNR of low-light scenes is an inevitable consequence of light's random nature. We will shortly recall its model based on Poisson and Bernoulli distributions; see [9, Ch. 1]. The model will eventually reveal why this randomness can be neglected

I. Kabashkin et al. (Eds.): RelStat 2018, LNNS 68, pp. 302–310, 2019.
https://doi.org/10.1007/978-3-030-12450-2_29

when the amount of light is sufficient. The probability that a number of photons, X, generated in a unit time by the source of Poisson distribution is equal to $k = 0, 1, 2, \ldots$, is given by a well-known formula

$$P(X = k|\kappa) = \frac{k^{\kappa}}{k!}e^{-\kappa}, \tag{1}$$

where κ (called the rate), is the Poisson distribution parameter (and, in our context, can be interpreted as the light intensity). The light sources, which are well modelled by the Poisson law, are:

- Sun and incandescent bulbs,
- Discharge lamps,
- Lasers and light emitting diodes (LEDs).

Once generated, the photons can be:

- Reflected (by mirrors) and/or refracted (in lenses),
- Scattered (by matte objects),
- Filtered (by translucent/color objects), and eventually,
- Captured (by a sensor).

The above phenomena can, in turn, be modeled as a Bernoulli (binomial) trial experiment with the appropriate success probabilities η_r, η_s, η_f, and η_c. The usefulness of the combined Poisson-Bernoulli model stems from the following binomial selection theorem:

Theorem 1. *Binomial selection of a Poisson process yields a Poisson process, and the rate λ of the output of the selection process is the rate κ of the input times the success probability η (see* [1])

$$\lambda = \kappa \times \eta. \tag{2}$$

So, the resulting distribution of the number of photons collected by the sensors, regardless of the way the traveled from the lights source to the sensor, is the Poisson one with the rate parameter

$$\lambda = \kappa \times \eta r \times \eta s \times \eta f \times \eta c. \tag{3}$$

Recall now that, the rate λ is, simultaneously, equal to the mean value, EX, and to the variance, var X, of the collected photon number X. This implies that the SNR, defined as

$$\text{SNR }(X) = \frac{EX}{\sqrt{varX}} = \sqrt{\lambda} \tag{4}$$

grows rather slowly with growing light intensity, and explains the low SNR of the images captured in low-light conditions (i.e. when values of *Xs*, representing the single pixels, are small).

Remark 2. Using the equivalent 'signal + noise' representation of *X*:

$$X = \lambda + Z, \text{ where } Z = X - \lambda, \tag{5}$$

where $EZ = 0$ and $var\ Z = \lambda$, one can observe that, if the value X is stored in n bits, then approximately then $n/2$ of least significant bits contain the noise (sometimes, in this context, Z is referred to as the shot noise).

Remark 3. In the report we assume that in low-light conditions the shot-noise is a dominant factor, and neglect the other disturbances, like dark current, thermal, read, and cross-talk noises; cf. [9] (Fig. 1).

Fig. 1. Template images.

The experiment setup described and used in this report exploits the following:

Conclusion 4. The close to realistic low-light versions of a well-lighted scene image **should not be** computed by simple re-scaling the image intensity by a factor n, say. In such a case, the SNR of the resulting image would have decreased by n rather than by $\sqrt{n}$ (as the variance of X/n decreases by n^2 with respect to the original X). Instead, they should be generated using scaled values of each template image pixels as the rate parameters of Poisson random variables.

3 Surveillance and Autonomous Vehicle Imaging Specifics

Object detection in both surveillance and autonomous vehicle domains remains a difficult (and open) problem. First, there is an increasing pressure from companies, which (fueled by the impressive neural network-based classifiers effectiveness and potential) are working to implement "intelligent" solutions in their devices. Second, there are still significant obstacles of twofold origin. Objective:

- Insufficient and/or uneven light on a scene,
- Potentially fast moving objects, quick interactions between them (occlusions, collisions),
- Low-SNR conditions, lack of formal justification of their effectiveness. and (subsequently).

And subjective:

- Lack of standardized (law enforced) tests verifying reliability of the proposed tools,
- Human-factor present during the learning process (heuristics applied to network architectures, ad hoc network modifications, rather arbitrary, hand-crafted, re/pre-training methods).

Remark 5. At present, the CNNs, while based, in principle, on bio-cybernetic archetypes (like a human vision system; see e.g. [7, 8]), are, in fact the black-box model devices. This means that we do not know how they actually detect and classify objects, and therefore, we do not know when they work and when (and why) they fail.

The currently accepted solution to this conundrum bases on comprehensive learning and testing of the CNNs at hand in order to obtain the smallest available test error rate. This approach requires both huge and adequate training and testing image sets. We believe that the proposed test will help to solve the testing part problem.

4 De-Noising Preprocessing

The tests were performed on the raw images and their version de-noised by select filters.

Anscombe Transform. Many de-noising algorithms are designed (and shown to be effective or optimal) under the assumption that the noise is of Gaussian distribution. This is clearly not the case in our problem. Nevertheless, we can use the following standard technique, the *Anscombe* transform, which turns the image pixels X being random variables of Poisson distribution into the new ones

$$A = 2\sqrt{X + \frac{3}{8}}, \tag{6}$$

which have approximately Gaussian distribution with a unit variance (i.e. var $A = 1 + O(1/\lambda^2)$; see e.g. [5]. The last property that is, the known and constant variance– recall that the original random variable X has an unknown variance λ equal to its mean (i.e. the pixel intensity) is a very convenient feature of the transform, especially because the aforementioned de-noising algorithms usually assume that the noise variance is known. Clearly, after de-noising, the inverse Anscombe transform can be applied (here, in its approximate form; see [5] for the exact one):

$$X = (A/2)^2 - \frac{1}{8}. \tag{7}$$

Convolution Filters. This is a basic and one of the first de-noising technique. The original signal A is *convolved* with some filter function F

$$W(n) = (A * F)(n) = \sum\nolimits_m A(n-m) \cdot F(m). \tag{8}$$

F can be a rectangular, triangular or Gaussian density function or any interpolation function, like e.g. the Keys one; see [4]. In the 2D image (Fig. 2) case, the filters are usually the products of its 1D versions (Fig. 1)

$$F(n,m) = F(n) \cdot F(m). \tag{9}$$

Fig. 2. A simulated low-light (EV = –7) counterpart of a template image and its filtered versions (clock-wise: convolution, wavelet smoothing and bilateral filters)

Remark 6. Convolution operations are – as their name implies – performed by the input layer in CNNs (and in the subsequent ones in their deep versions). Nevertheless, they are rather oriented at edge (feature) detection and, because of the computational overhead, are rather local (i.e. encompass regions of 3×3 or 5×5 pixels) and therefore are not effective noise reduction filters.

Bilateral Filtering. One of the most important decision in convolution-based denoising is to set the filter width. Too large one results in over-smoothed images (lacking edges and other features) while too small does not effectively remove the noise. The *bilateral filter* is an adaptive version of the convolution one whose width decreases in the image parts where it varies. In the simplest case, the filtering is governed by the convolution-like formula

$$W(n) = A(n-m)A(F(n) - F(m)) \cdot F(m). \tag{10}$$

Remark 7. Bilateral filtering is an adaptive (nonlinear with respect to the image) technique and requires careful tuning. That also means, that the settings developed for one image can be useless for others. It also lack effective (with performance comparable to the basic convolution filters) FFT-based implementations.

Wavelet Thresholding and Smoothing. *Wavelet thresholding* combines computational effectiveness of convolution filters with adaptiveness of the bilateral ones. Moreover, it was shown that it is an optimal technique if the noise is Gaussian of known variance; [6]. The algorithm has three steps:

- Wavelet transformation of the image

$$W(n) = WT(A(n)), \tag{11}$$

- Coefficient thresholding (here, the soft-thresholding was used)

$$W_T(n) = \max(|W(n)| - T, 0) \cdot \mathrm{sgn}(W(n)), \tag{12}$$

- Inverse wavelet transformation

$$A_T(n) = IWT(W_T(n)). \tag{13}$$

Remark 8. It was demonstrated that the soft-thresholding algorithm is equivalent to the LASSO algorithm; see e.g. [3] (Fig. 3).

The wavelet *smoothing* filter exploits the multi-resolution representation of the wavelet transform of the image and (by removing the detail coefficients) results in smoothed low-resolution-like version of the image.

Median Filter. Observing that in the extremely low-light conditions (*e.g.* for EV = −7) the image noise resembles the ‘salt and pepper’ pattern, we also implemented the median filter in the pre-processing phase (Fig. 4).

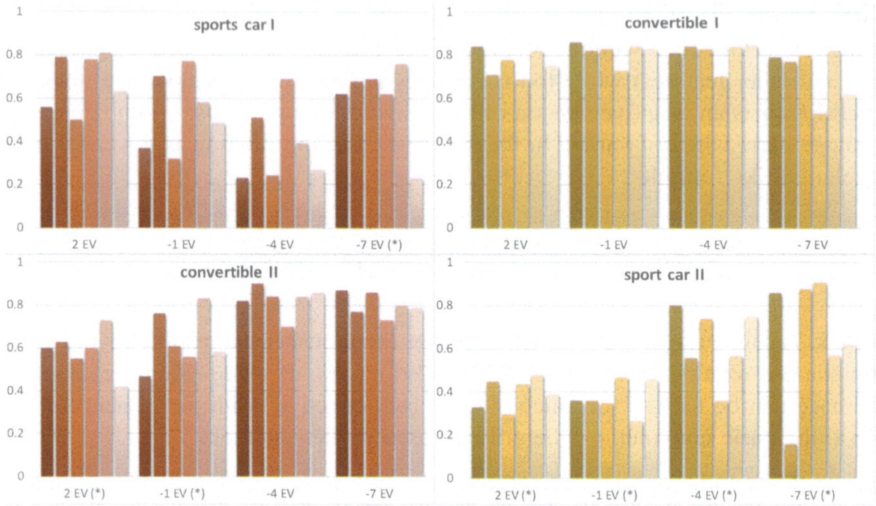

Fig. 3. Classification results for the raw and filtered images. EV = 2, −1, −4 and −7 (columns for each value of EV correspond to raw image, and to convolution, bilateral, wavelet thresholding, wavelet smoothing and median filters, respectively)

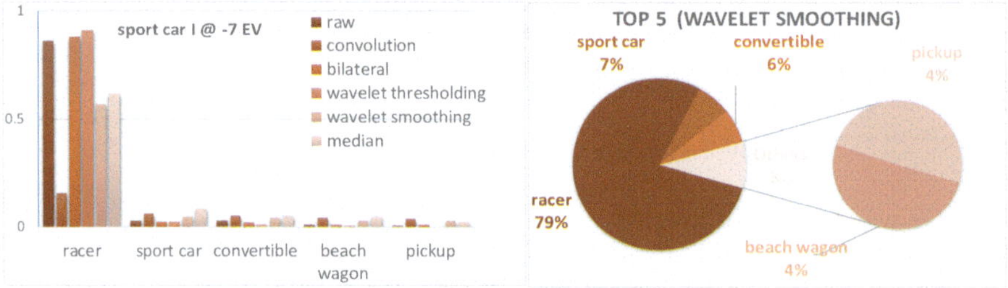

Fig. 4. Exemplary classification results for EV = −7

5 Experimental Setting

The reported tests are in their initial phase and the full account of conclusions and recommendations is not accomplished yet. Therefore, in order to make the experiment both concise and illustrative, we have selected a few template images obtained in daylight conditions, for which the tested networks successfully detected the object. Then, according to Conclusion 3, the low-light versions of the images were generated. (The networks were not re-trained for the low-light new images.).

Convolution filters used Keys' (bicubic) function, bilateral filters were based on Gaussian one, while wavelet algorithms exploited 9/7 biorthogonal wavelets; cf. e.g. [2, 6, 10]. The 'inceptionresnetv2' neural network was pre-selected for the tests; [12]. Other networks, including 'inceptionv3', 'resnet50', 'resnet101', 'squeezenet', 'googlenet', 'alexnet', 'vgg16', 'vgg19' and 'bvlc_ reference_caffenet' appeared to perform much poorer in the (simulated) low-light conditions.

5.1 Implementation Remarks

It was assumed that the template images are taken from the cameras (that is, they already are in the JPG format). JPG offers 8-bit color depth while the sensors have usually the full-well capacity of 16 bits (i.e. c.a. 60 ke − or more, see e.g. [13]). That means, in particular, that the images generated from such templates behave (according to the Poisson model) as if they are obtained with the exposure setting decreased by 8 eV.[1]

Exposure value and camera exposure settings are related by the following formulas

$$EV = log_2 \frac{LS}{K} \text{ and } EV = log_2 \frac{N^2}{t} \tag{14}$$

respectively, where L is a scene luminance (calculated over the measurement area) in cdm^{-2}, S is the ISO speed (e.g. 25, 50, 100, 200, etc.), K is the standardized luminance calibration constant (e.g. 12.5 cdm^{-2}), N is the relative (with respect to the lens focal length) aperture (*f*-number), and t is the exposure time (in seconds).[2]

Example 8. Lenses of relative aperture close to 1 (e.g. 1.2) are in the common use however, in automotive and surveillance applications, in order to avoid motion blur, the shutter times are usually of order 1/100 or 1/200 or shorter. The typical value of exposure value for a daily light scene is 12 ÷ 16 eV, while for a night one it scores from −6 ÷ 6 eV. Assume now that the unit amplification (that is, the case, when the pixel's ADC converter output value corresponds directly to the number of electrons captured by a single pixel) is obtained for ISO 800. Then we need, for a camera with a lens with $N = 1$, the shutter speed $t = 1/200$, and at ISO 800, the EV of the scene need to be as high as $\log_2 (1 \cdot 800/200) = 2$ eV.

The template images used in experiments were captured in daily-light conditions, with 10 eV. Hence, when converted to the template JPG files, they correspond to the images obtained from a real camera with the scene luminance close to this 2 eV.

6 Conclusions

The presented, preliminary findings, suggest that the off-the-shelf CNNs are:

- Applicable in general in surveillance/autonomous vehicle applications. Moreover,
- The simple convolution- and wavelet-based smoothing algorithms increases the probability of proper classification,

[1] We ignore here the fact that the resulting 8-bit depth in JPG files is obtained from the raw 16-bit ones through a non-linear ("log-gamma") transformation.

[2] Note that the camera exposure setting definition of EV ignores units.

- The other (non-linear) filtering algorithms do not offer significant advantages (and, being dependent on some handcrafted parameters, can in fact make object recognition less accurate).

The available computational resources in cameras can be used to pre-process images (smoothing and down- sampling, in particular) prior to their transmission to the object recognition module ("edge computing approach"). Alternatively, one can consider incorporating into the CNN an additional input layer dedicated to image pre-filtering only.

Acknowledgements. This work was supported by the Polish National Centre for Research and Development (NCBR) within the Innovative Economy Operational Programme, grant No. POIR.01.01.01-00-0235/17 as part of the European Regional Development Fund (ERDF).

References

1. Barrett, H.H., Myers, K.J.: Foundations of Image Science. Wiley Series in Pure and Applied Optics. Wiley, Hoboken (2003)
2. Daubechies, I.: Ten Lectures on Wavelets. SIAM Edition, Philadelphia (1992)
3. James, G., Witten, D., Hastie, T., Tibshirani, R.: An Introduction to Statistical Learning, vol. 112. Springer, Heidelberg (2013)
4. Keys, R.: Cubic convolution interpolation for digital image processing. IEEE Trans. Acoust. Speech Sign. Proces. **29**(6), 1153–1160 (1981)
5. Makitalo, M., Foi, A.: Optimal inversion of the Anscombe transformation in low-count Poisson image denoising. IEEE Trans. Image Proces. **20**(1), 99–109 (2011)
6. Mallat, S.G.: A Wavelet Tour of Signal Processing: The Sparse Way. Academic Press/Elsevier, Burlington (2009)
7. Poggio, T., Serre, T.: Models of visual cortex. Scholarpedia **8**(4), 3516 (2013). https://doi.org/10.4249/scholarpedia.3516,revision%23149958
8. Schmidhuber, J.: Deep learning. Scholarpedia **10**(11), 32832 (2015). https://doi.org/10.4249/scholarpedia.32832,revision%23152272
9. Seitz, P., Theuwissen, A.J.: Single-photon imaging, Springer Series in Optical Sciences, vol. 160. Springer (2011)
10. Taubman, D., Marcellin, M.: JPEG2000. Image Compression Fundamentals, Standards and Practice, The Kluwer International Series in Engineering and Computer Science, vol. 642. Kluwer Academic Publishers (2002)
11. Preliminary Report Released for Crash Involving Pedestrian, Uber Technologies, Inc., Test Vehicle, NTSB News Release. https://ntsb.gov/news/press-releases/Pages/NR20180524.Aspx. Accessed 14 Aug 2018
12. Szegedy, C., Ioffe, S., Vanhoucke, V., Alemi, A.: Inception-v4, Inception-ResNet and the impact of residual connections on learning, arXiv, https://arxiv.org/abs/1602.07261v2. Accessed 14 Aug 2018
13. Koifman, V.: Image sensors world. http://image-sensors-world.blogspot.com. Accessed 14 Aug 2018

Performance Evaluation of Event-Driven Software Applied in Monitoring Systems

Jarosław Sugier[1], Tomasz Walkowiak[1(✉)], Jacek Mazurkiewicz[1], Przemysław Śliwiński[1], and Krzysztof Helt[2]

[1] Faculty of Electronics, Wrocław University of Science and Technology, Janiszewskiego St. 11/17, 50-372 Wrocław, Poland
{jaroslaw.sugier, tomasz.walkowiak, jacek.mazurkiewicz, przemyslaw.sliwinski}@pwr.edu.pl
[2] Teleste Sp. z o.o, Szybowcowa 31, 54-130 Wrocław, Poland
krzysztof.helt@teleste.com

Abstract. Real-time monitoring systems now often use complex event processing software and in this paper we analyze performance of such an approach in a simple case study. After introducing briefly the main concepts of event processing networks we define the case of detecting anomalies in a stream of two events which, in correct operation, should appear in pairs keeping precise temporal relation. For such a stream we propose two event processing networks for detection of specific anomalies which are then implemented on the PROTON software platform and tested as RESTful web applications. The performance tests include two scenarios for generation of the input events and show the average system response time as a function of the increasing load (the stream of incoming events). The results identify limitations of the tested implementation which are related to deficiencies of massive multi-threaded processing in the Java environment.

Keywords: Event processing · Monitoring system · Web application · Performance analysis

1 Introduction

Operation of various contemporary real-time information systems is often based on Detect-Decide-Respond working scheme: in a stream of events which are coming from the environment, functionality which the system must provide can be defined as a detection of some specific temporal and semantic patterns within the event stream, followed by an evaluation of their various characteristics and generation of appropriate reactions as a result of their classification. Such a style of processing can be found e.g. in security or safety monitoring, active diagnostics, predictive processing or business process management. In these cases it can be advantageous to ground the construction of the whole system on the event-driven processing paradigm.

The performance [1] of event processing system is an important aspect in monitoring systems. It can be assessed in three aspects: their capability to provide responses

I. Kabashkin et al. (Eds.): RelStat 2018, LNNS 68, pp. 311–319, 2019.
https://doi.org/10.1007/978-3-030-12450-2_30

in the desired timespan, the capability to respond correctly (without errors), and the ability to handle large, cumulated workloads [2].

Within this paper the authors examine the performance of software implementing the event processing (namely the Proactive Technology Online—PROTON [3]) deployed as a web server.

The paper is structured as follows. Firstly, an event-driven data processing paradigm is analyzed. It is followed by a description of the case study. Finally, performance analysis results are presented and discussed.

2 Basics of Event-Driven Data Processing

2.1 Organization of the Event Processing System

The general structure of a typical Complex Event Processing (CEP) system is a network of some basic components which communicate with themselves by sending event messages (Fig. 1). In this introduction, we will describe organization of such systems using a generic perspective presented by the textbook [4]. The primary source of *raw events* (i.e. the basic events which enter the system) are external *event producers* which can be as simple as real-world physical sensors or as complex as other information systems capable of communicating their output in the form of events. The essential part of the processing is done within an *Event Processing Network* (EPN) which is a composition of *Event Processing Agents* (EPAs). Each agent is capable of reading events on its inputs, analyzing them according to some specific processing scheme and generating *derived events* on its outputs. The derived events can be forwarded to other agents within the network for further processing or can be sent to *event consumers* as the result of system operation.

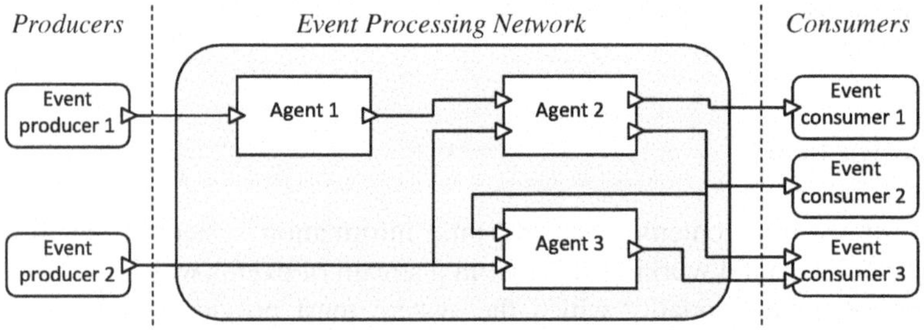

Fig. 1. Basic components of an event processing system.

It should be noted that efficient computer implementation of such a CEP system can be in general a very complex task which is accomplished with application of numerous hardware/software resources and communication technologies. The EPN grid can be implemented on a single computer server or can be divided into partitions dispersed over LAN or WAN networks. Efficient distribution of events, proper synchronization between the agents and realization of their concurrent operation in such a network is a crucial challenge which determines effectiveness of the whole system.

2.2 An Event Processing Agent and Its Contexts

The Event Processing Agents are the essential components which encapsulate actual operations executed on events and the CEP model defines their various types. In general an agent of any type performs a 3 step processing: filtering (selection of the events from the input streams), matching (finding patterns among the accepted events which will trigger generation of the output) and derivation (forming the output events and evaluating their attributes using the results of the matching step). This generic scheme results in various functionalities (operators) of the EPAs which range from simple event filtering based on attribute values (a stateless operation), through transformations (e.g. aggregating a set of input events and composing a derived event from their characteristics), up to different pattern detection schemes which analyze temporal and/or semantical relations between the events.

A *context* is a fundamental CEP concept (although not directly visualized in the EPN schemas) which is used, in the most general sense, for specification of different assignment strategies between event instances and EPA agents. Formally, a context specifies a set of conditions which group related events into *context partitions* so that they can be processed by a certain agent. In particular, a *temporal context* selects a set of events which occur within some time window and thus defines a period when an agent is active, while a *segmentation context* groups events with the same attribute value, e.g. UserID, so that each partition contains events referencing—for example—the same user. A *composite context* is a construction which joins several contexts yielding a Cartesian product of their partitions: for example selection of events from a specific time period (a temporal context) is joined with a segmentation context which further splits the selected events into groups referring to different users. An agent should operate on each partition concurrently and independently.

Special mechanisms are provided for specification of time intervals, i.e. the methods for *opening* and *closing* temporal contexts. Such a context is started by an *initiator* which can be an event or an absolute time moment, and it is ended by a *terminator*—an event or an absolute or relative time moment. By these means dynamic reactions to various event occurrences can be modelled and the agents can detect and adaptively react to their patterns.

3 The Case Study

3.1 The Task: Detecting Anomalies in a Stream of Correlated Events

In this paper we will analyze a stream of two raw events: `A` and `B` where the expected correct behavior is when each occurrence of the `A` event is followed by the `B` within some constant time-out window $\mathtt{T}_{AB}$ (Fig. 2). This is a common elementary inter-event dependence found for example in security systems when an initial user login (event `A`) is required to be followed by some extra authentication e.g. with an SMS password (event B), or in industrial control systems when an issued actuator command must be followed by a sensor confirmation of a part reaching its destination position, etc. An additional requirement is to consider an ObjectID attribute stored in each A and B event which would identify a user or object generating the event—the events may refer to multiple objects and the system must be able to distinguish them and to monitor their behavior individually and independently.

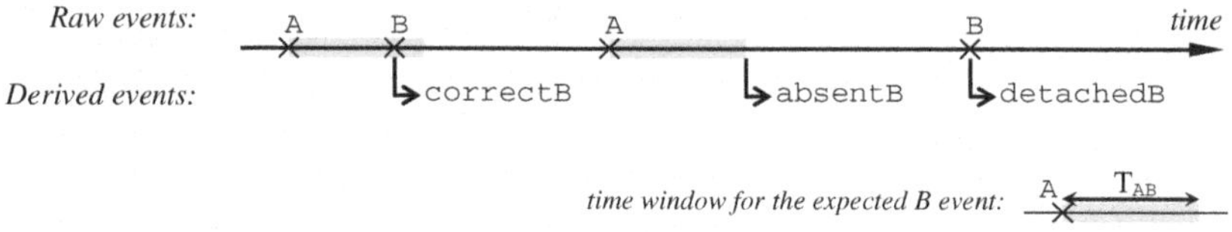

Fig. 2. The stream of two events with the expected behavior and the two alerts.

In such an stream we want to detect and report two alerts as presented in Fig. 2:

1. absentB—an event generated at the end of the time-out period after A if no occurrence of B has been registered;
2. detachedB—an event signaling reception of B which was not properly preceded by an occurrence of A.

Moreover, in the case 2. an auxiliary internal derived event correctB will be used for signaling proper occurrence of B within a T_{AB} window after some preceding event A.

Even in such a simple stream other abnormal conditions could be identified (e.g. multiple A events before a B one) but efficiency tests in this paper will be limited to EPNs detecting the two above alerts.

3.2 The Proactive Technology Online (PROTON) Engine

The Proactive Technology Online (PROTON, [3]) is a scalable, integrated platform to support the development and deployment of CEP systems and will be used in this paper as the implementation environment for the above test cases of event processing. Developed at IBM Haifa Research Lab as a component of the open-source FIWARE solution platform [5], PROTON follows closely the philosophy and methodology of active event processing presented in the canonical textbook by Opher Etzion and Peter Niblett [4]. Being consistently built from ground up as a dedicated CEP environment, it is not attached to any legacy programming or database management language like many other solutions.

The PROTON server is written entirely in Java. After loading a specification of the EPN network from a JSON definition file it can communicate input/output events through either local files or a RESTful Web interface [6] so that it can seamlessly integrate with other FIWARE solutions. The JSON definition file can be prepared with a dedicated Web interface where the user specify event types, EPA agents, producers/consumers and all other aspects of system organization.

3.3 The PROTON Implementations of the Two Event Processing Networks

The first task—detection of an absence of a B event within a T_{AB} timeout—can be accomplished by a (very simple) EPN no. 1 presented in the upper part of Fig. 3. The network consist of only one agent with the "Absence" operator—it generates the derived event if its input event(s) (B in this particular case) has(have) *not* occurred within a time window when the agent was active.

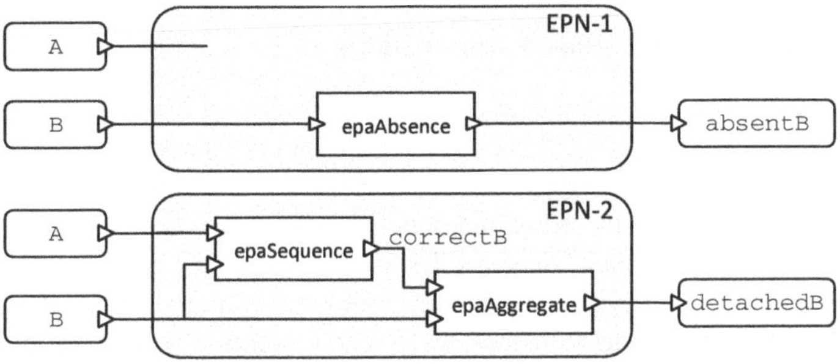

Fig. 3. The two EPN configurations detecting absent and detached events B.

Operation of the agent is trivial in this aspect but more attention must be paid to definition of its contexts. First, the temporal context has the event A as its only initiator so each time this event is received a new widow is opened (hence events A, even if seem to be ignored in the EPN scheme as they are not routed to any agent, do take important part in the solution). The context terminator is defined as a relative time with the value of the T_{AB} parameter—after this period the context is closed and the absence of event B is validated. If it is confirmed the derived event absentB is emitted—which fulfills our detection task.

The last aspect of the solution is discrimination of the ObjectID attributes. For this purpose a segmentation context must be defined which includes references to this attribute in both events A and B. Effectively, the absence EPA operates with a composite context which consists of the temporal and the segmentation ones.

The second task—detection of detached events—requires somewhat more complex network with two agents which are presented in the lower part of Fig. 3. Here the sequence agent (epaSequence) is waiting for the proper order A–B of its input events and its temporal context is defined with the A event as the only initiator, i.e. detection of the sequence starts every time the A event arrives. Sequence agents can be defined with so called immediate evaluation policy (i.e. the sequence can be detected and reported immediately after receiving its last event instead of at the end of the context window) and this mode must be used in our case so that the correctB event is emitted instantly after arrival of B. If not, a relative interval T_{AB} closes the window like in the EPN-1 case so the agent will not consider detached B events as the proper ones. This relative interval is the only terminator of the temporal context assigned to epaSequence agent.

The aggregate agent is activated either by `B` or `correctB` events and operates as a "temporal AND gate": it counts occurrences of both input events in a small time window and generates the `detachedB` alert only if the counter for the `correctB` events is zero.

As in the case of the EPN-1 scheme, both agents operate with composite contexts with the segmentation part defined on the `ObjectID` attribute in all the events.

4 Performance Evaluation

4.1 Performance of Web Based Applications

The concept of performance evaluation of web based application is generally well known and also implemented in many commercial and open source software tools. It is important to notice that results depends a lot on how the traffic is generated [7]. The simplest approach, adopted by most of the benchmarking software, is bombarding the system with a stream of requests modeled by a Poison process. As it is discussed in [7], more information on the system behavior could be gained by a different schema: keeping the number of clients constant. Each client sends a request to the server and next waits for the server response. The performance of typical web application is presented in Fig. 4 [2]. The application operates in two ranges: in the normal utilization range, until it reaches the maximum number of clients that it can handle correctly (roughly 300 clients in Fig. 4). The exact values of the average service response time and the number of clients depends on task complexity, CPU performance and type of used web server. But the shape is constant [7] for CPU consuming tasks (e.g. the ones doing some calculations).

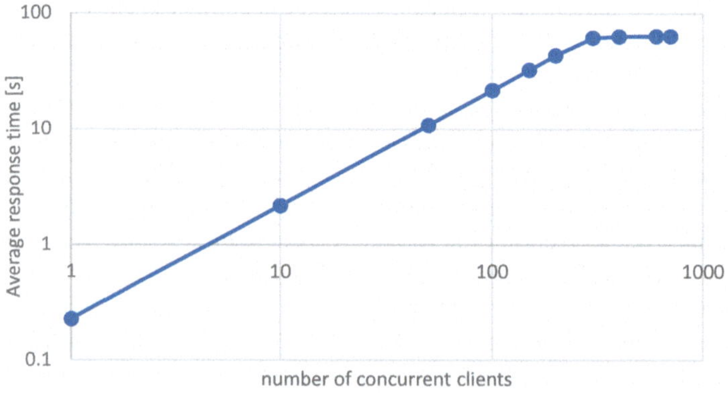

Fig. 4. Average service response when interacting with various number of concurrent clients.

4.2 Concurrent Web Clients

We have tested the PROTON instance deployed on a Tomcat server by bombarding the system by different number of concurrent users. For each number of users the system was tested for 120 s and next the web application was restarted.

At first the EPN-1 (see Subsect. 3.3) was tested with T_{AB} timeout set to 2 s. Two scenarios were used. In the first scenario (A) we model a security breaking situation, i.e. the EPN receives only events of type `A`. Each event comes with different `ObjectID` attribute (see Subsect. 3.1). In the second scenario (B), we model the correct situation, i.e. we generate a stream of pairs of events: event `A` followed by event `B` after 0.5 s, both with the same `ObjectID` attribute. The results for different number of concurrent users are presented in Fig. 5.

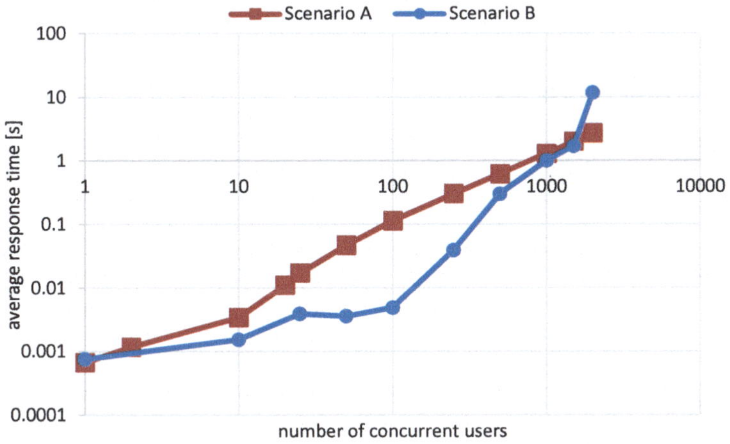

Fig. 5. Average service response time for EPN-1.

We can notice that the shape of performance curve differs from that one in Fig. 4. It looks like the server is operating in normal utilization rate. Deeper analysis of application behavior showed that at 20 concurrent users the PROTON starts to delay the detection of alarms. The alarms were detected in a time much larger than the T_{AB} timeout. The detection delay was consequently increasing, achieving even 40 min (for the A scenario).

Next, we tested the EPN-2. Now, in scenario A, only events `B` are generated. The results are presented in Fig. 6. Like in the previous case, after more than 20 concurrent users PROTON starts to delay the detection of alarms (for both scenarios).

Analysis of Tomcat server resources consumption showed that for 40 concurrent users more than 800 threads were used. It could be the main reason of the system behavior (very long delays of alarms).

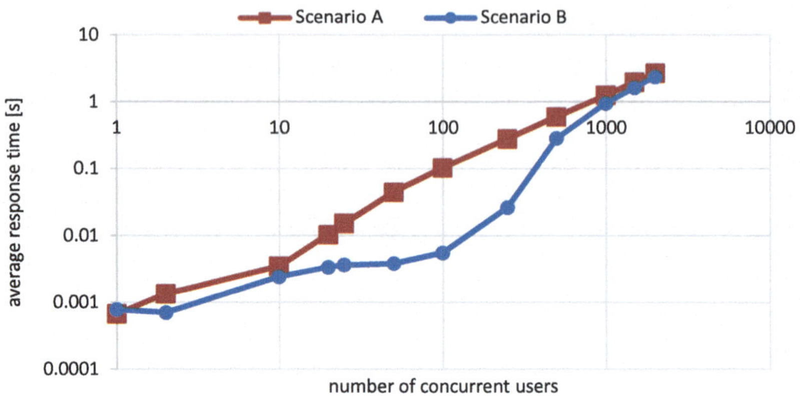

Fig. 6. Average service response for the EPN-2.

4.3 Central Broker Schema

The previous experiment assumed that request of each user comes in a different TCP/IP connection. In real applications the CEP receives the events from some kind of a central broker. In such case, all events come from the same IP. To speed up the processing and save resources a connection pool [8] is very often used. Therefore, we have modelled such a schema. Results for the EPN-1 for both scenarios are presented in Fig. 7.

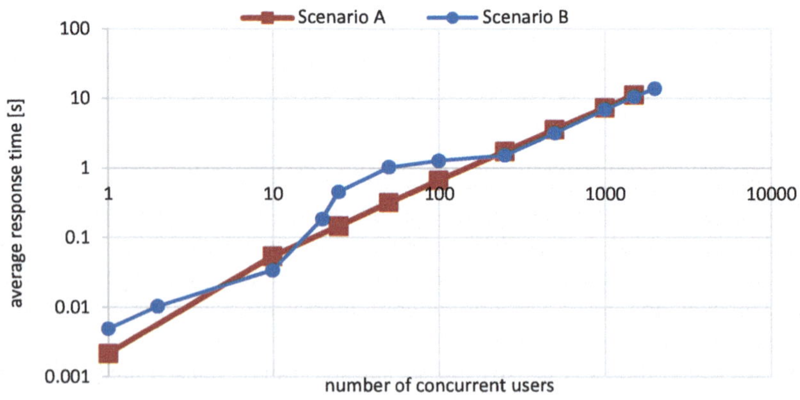

Fig. 7. Average service response time for the EPN-1 tested with a central broker.

We can notice that for scenario A the performance curve has the same shape as in Fig. 4, but nor for scenario B. Deeper analysis showed the detection of alarms in scenario A was correct, however in scenario B at 50 concurrent users the application hangs up, and it was unable to detect any alarms until its restart.

5 Conclusions

We examined the performance of the PROTON software implementing the event processing deployed in the Tomcat web server. Results showed that the software is not capable of processing of more than 20–50 concurrent users due to hanging-up or loosing real time alarm detection capability. This in an unacceptable behavior for most of security systems. Probably, the reason of such behavior is the method of implementation of event-driven data processing paradigm. PROTON, implemented in Java, uses large numbers of threads, whereas Java standard threads are reasonably heavy [9].

Acknowledgements. This work was supported by the Polish National Centre for Research and Development (NCBR) within the Innovative Economy Operational Programme grant No. POIR.01.01.01-00-0235/17 as a part of the European Regional Development Fund (ERDF).

References

1. Miller, L.C.: Application Performance Management for Dummies, Riverbed Special Edition edn. Wiley, Hoboken (2013)
2. Caban, D., Walkowiak, T.: Prediction of the performance of web based systems. In: Zamojski, W., Sugier, J. (eds.) Dependability Problems of Complex Information Systems, pp. 1–18. Springer, Cham (2015)
3. IBM Research Lab: Proactive Technology Online User Guide Version 5.4.1. Haifa (2016). https://github.com/ishkin/Proton/blob/master/documentation/ProtonUserGuidev5.4.1.pdf. Accessed July 2018
4. Opher, E., Niblett, P.: Event Processing in Action. Manning Publications Co., Stamford (2011)
5. FIWARE: The Open Source Platform for Our Smart Digital Future. https://www.fiware.org/. Accessed July 2018
6. Complex Event Processing Open RESTful API Specification. http://forge.fiware.org/plugins/mediawiki/wiki/fiware/index.php/Complex_Event_Processing_Open_RESTful_API_Specification. Accessed July 2018
7. Walkowiak, T.: Behavior of Web servers in stress tests. Adv. Intell. Syst. Comput. **286**, 467–476 (2014)
8. Wischik, D., Handley, M., Braun, M.B.: The resource pooling principle. SIGCOMM Comput. Commun. Rev. **38**(5), 47–52 (2008)
9. Luo, Y., John L.K.: Workload Characterization of Multithreaded Java Servers. In: Proceedings of 2001 IEEE International Symposium on Performance Analysis of Systems and Software (ISPASS), pp. 128–136 (2001)

Checkers Player Next Move Aided System

Agata Filar and Jacek Mazurkiewicz(✉)

Faculty of Electronics, Department of Computer Engineering, Wrocław University of Science and Technology, Wybrzeze Wyspianskiego St. 27, 50-370 Wroclaw, Poland
Jacek.Mazurkiewicz@pwr.edu.pl

Abstract. Main aim of the paper was to create checkers player move aided system. For this purpose two different approaches were used. First one was well known searching algorithm—Negamax and second one was Reinforcement Learning algorithm—SARSA. For the purpose of making experiments on algorithms' performance special environment was created. It was checkers program which main goal was to give its user possibility of launching the game between two different agents. One of its constraints was also to make creating and adding new agent easy for the future use in other research.

Keywords: Checkers · Negamax · SARSA · Softcomputing · Agent system

1 Introduction

Checkers is very popular game known by almost everyone and loved by many. Its origins dates back to ancient times and shows that people's taste in board games did not change that much since then. Nowadays it is still favored form of entertainment and even sport discipline for more demanding players. Really interesting fact about this seemingly simple game is that it is a very complex problem which many scientists have been trying to solve over the years. Rules of checkers can be explained in less than few minutes and implementation of the game itself can be as easy as an exercise for programming classes. What can be so hard in this game that makes solving it so difficult? It is its uniqueness that comes from numerous combinations that can occur during the game. Most of traditional approaches that tries to find perfect solution are to computationally expensive while other ones cannot provide the optimal solution and will not be able to play unimpeachable games [1, 2].

Some years ago one type of checkers was announced to be weakly solved. This is American Checkers solved by Chinook [2] program developed by scientists from the University of Alberta. Chinook combines different approaches. Its algorithm includes a library of opening moves, a deep search algorithm and 'end-game' database [3]. Since then after over a dozen years of research scientists proven that solving the game of checkers is possible. Computer is able to win at checkers playing with master players but it still did not outdated the need of further research on this problem. That is because solving the game itself is not the only reason to try.

What is really important and useful in the game of checkers is the fact that it provides a great environment for research on techniques that can be later used for real life

I. Kabashkin et al. (Eds.): RelStat 2018, LNNS 68, pp. 320–330, 2019.
https://doi.org/10.1007/978-3-030-12450-2_31

problems. Creating and testing agents operating in well-known and fully defined abstract environment where the correctness of decisions can be easily evaluated can bring a lot of new conclusions and ideas. Playing checkers is definitely less expensive and easier than simulating most of real life problems but a lot of mechanisms stays similar [3]. Games like checkers can be useful in the process of developing new valuable techniques and exploring already known ones. Starting from something simple can develop into something complicated and useful. While Chinook program is not scalable and playing checkers is probably the only thing it is good for there are multiple AI techniques like Reinforcement Learning that are able to generalize problems [4]. They are waiting to be further explored. This paper should be treated as an attempt to explore such techniques using the problem of making decision about move in checkers game as a starting point.

Following chapter tries to bring the topic of checkers closer. Rules of American Checkers that is variant of the game used for this research were explained in details. Then in Chap. 3 implementation of environment was described in details. All information about what is the input for developed playing agents and how state of environment is encoded was also provided in this chapter. Later—Chap. 4—two previously mentioned approaches were introduced. First one that was Negamax algorithm with its alpha beta pruning variation and used evaluation function and the second one—SARSA algorithm. After this presentation of approaches research results were presented with comparison of algorithms and methods. After the sections with results summary and conclusions from whole project were taken and ideas for future development were given.

2 Game of Checkers

The game of checkers is one of the most valued board games in the world. There is no one person who did not heard about this game. For centuries it was played by kings, learned and ordinary people. Big fans of the game were for example Charles Darwin, Lev Tolstoi, Benjamin Franklin, Napoleon Bonaparte and Frederic Chopin. It is even a subject of poem named 'Checkers' by Adam Mickiewicz. It is hard to be surprised with those facts because simplicity of rules and creation of checkers that comes in pair with complexity of the game itself makes it enjoyable entertainment and brain training for the ones that look for something more demanding. Roots of checkers date back to ancient times and they are not losing popularity until these days [5].

Variant of checkers that was used during the research presented in this paper is already mentioned American Checkers called also Straight Checkers or English Draughts. The game is played on 8×8 checkerboard. Both players have 12 pieces. In starting position all pieces are put in three first lines on both sides of the checkerboard. Black moves first and then players alternates their moves. Rules of the game:

- pieces can move only diagonally forward to not occupied dark places,
- while making simple (not capturing) move a piece may move only one square,
- while capturing a piece jumps over opponent's piece and lands in straight diagonal line two squares further from starting position,

- in a single jump only one piece can be captured but multiple jumps are allowed,
- captured piece gets removed from the checkerboard,
- capturing opponent's pieces is mandatory and breaking this rule is illegal,
- when more than one capture move is possible player is free to select,
- when a piece gets to the last row of checkerboard it gets promoted to be a king,
- kings are limited to move diagonally but they can move and catch backward,
- jumps performed by kings can be combined in several directions on the same turn.

Player is perceived to win the game when opponent cannot make any move. It can be caused by blocking of pieces that left on the checkerboard or capturing all of opponent's pieces. This type of checkers was chosen as it is perceived less complicated than International Checkers. Number of possible positions in this variant is 500, 995, 484, 682, 338, 672, 639. It has game-tree complexity of approximately 10^{40} while for chess is estimated to have between 10^{43} and 10^{50} legal positions [6, 7].

3 Checkers System

The main idea was to create the system that implements a game of American Checkers according to the rules that were introduced in the previous chapter. This system have had to give a user possibility of making different kinds of agents playing against each other in order to collect statistics about the game.

The word 'agent' does not have a standardized definition. The word should be understood as an entity that is acting in the environment and is characterized by features:

- communication—it is communicating with the environment in some way,
- perception—it monitors the changes that happens in the environment,
- autonomy—to make decisions about future actions to win the game.

Two agents of different types (or two instances of the same type with different parameters) can be chosen to play with each other. Whole program is prepared in the way that provides possibility of adding new algorithms and ideas for agent's 'brain' (Fig. 1) In order to do this new class of agent has to be created. Such class gets all needed information about actual state of environment and returns chosen action. During the research five kinds of such agents were created: Human Agent, Random Agent, Negamax Agent, SARSA Agent and SARSA(λ) Agent. Every one of them was based on individual algorithm and can get slightly different input but all of them provides the same output—an action to be taken [8]:

1. *Human Agent*—is just a simple interface between human player and the game. Human Agent can use a console or graphical interface to see the actual state of the checkerboard and opponents actions. While using console interface in the beginning of every turn Human Agent gets a list of possible moves. Then it provides chosen action as number of move provided with keyboard or by dragging a piece from start to destination position on the checkerboard.
2. *Random Agent*—is a pseudo-random algorithm. It does not get full information about state of environment, because it does not need one. All it gets as an input is a

list of actions that it is allowed to perform in given state. It returns one of the moves chosen pseudo-randomly.

3. *Negamax Agent*—uses searching technique. Exact algorithm that was used here is Negamax with alpha beta pruning. It gets full information about state of environment—how many pieces are left on the checkerboard, what are their types, colors and positions. It chooses the best possible action in a process of maximizing an evaluation function for a few future moves.
4. *SARSA Agent*—an algorithm used in reinforcement learning area of machine learning. It also uses full information about environment. First during the process of learning it plays with other agent and maps pairs of environment state and taken action with a reward calculated by evaluation function. Later learned SARSA can use this mapping to play with other agents.
5. *SARSA(λ) Agent*—a variant of SARSA algorithm that works very similar basing on the same input.

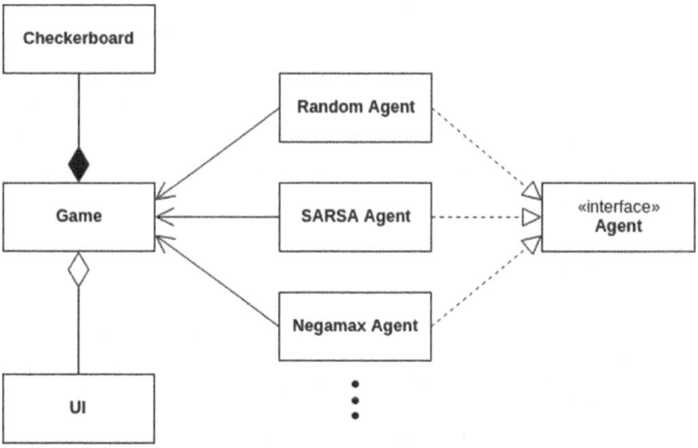

Fig. 1. Simplified class diagram for checkers system.

Full information about checkerboard is encoded using a structure containing vectors of decimal numbers describing its actual state. Using such numbers following information are stored: places of all black and white pieces and kings together with possible moves and jumps for active player. This way of encoding the checkerboard may look incomprehensible but it becomes pretty obvious when decimal numbers are converted into binary (Fig. 2). The idea behind checkers system was to give a user possibility of performing different kinds of experiments when launching games between agents. Two algorithms that were used in this paper: SARSA and Negamax agents [9, 10].

Scenarios of experiments for both of them looked different. Typical scenario for Negamax agent was first launching the game against Random agent with the use of graphical interface. Stepping through the game validity of agent's moves was checked

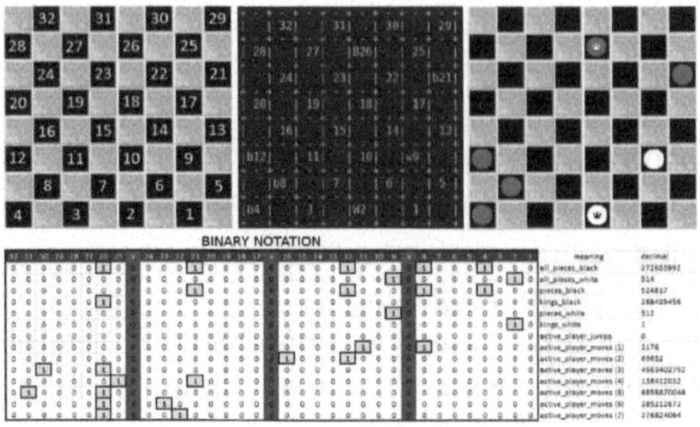

Fig. 2. Example of checkerboard structure.

and how it performs 'on first sight'. Later higher number of games was lunched one after another in order to provide some statistics. When basing on the outputs parameters of the agent were set to the best performing ones then agent was ready to compete with human. Scenario typically used for SARSA agent was more complex, because of the nature of this algorithm. First learning process had to be performed. For this purpose most commonly Negamax algorithm was used to play against SARSA. After launching 20,000 games one after another statistics were gathered and 'brain' of learned SARSA agent was serialized to the file. After that smaller amounts of games were launched for learned agent and other algorithms like random or other variants of Negamax to see if it is not over-fitted. Finally it played with human agent [11, 12].

4 Searching Algorithms

The problem that is tried to be solved in this paper is making decision about choosing the best possible move during the game of checkers. It can be treated as a search problem. From the space of possible moves the action that will make player closer to winning the game should be always taken. While for the human player thinking about many future moves and taking into consideration different scenarios played by opponent can be hard (especially for not experienced ones) intuition tells us that computer should be perfect for this job and perform much better. Also there cannot be better way of choosing the best possible move than checking every single scenario. Searching algorithms brings us this possibility but unfortunately computational complexity of this task for game of checkers is very high. What can be used here for searching algorithm advantage is the fact that to win with beginner player taking into consideration next six moves should be enough. The constraints that were taken when preparing searching algorithm to be used for playing with human opponent and during learning process of other agents are that time of computation should not take more than 3 min while searching depth of 7 moves. It was important to make it possible to play with a human. For this purpose reducing the size of searching tree was implemented. That is why the algorithm that was chosen for this purpose was Negamax with alpha-beta pruning.

4.1 Negamax Algorithm with Alpha Beta Pruning

Negamax search algorithm is a variant of Minmax search. This algorithm can be used for games between two opponents. The game also has to rely on zero-sum [13] property. Zero-sum is a concept taken from game theory. It represents a situation where for one players gain of the opponent is exactly balance by player's loss and vice versa player's loss is balanced by opponents gain. When we add gains of both players and subtract their losses than they will sum to zero. Other example of such games can be Tic-tac-toe or Chess. Negamax algorithm operates on binary trees. Every node of the tree is a state of the environment (situation on the checkerboard) and transitions to the child nodes represents actions that can be taken to get to the particular state. Negamax algorithm tries to find the node score value for the active player that starts from the root node. In the algorithm the root node gets score calculated by value function for its child nodes. Next move to play is represented by the child node that got the best score and set the ultimate score for root node. Negamax returns heuristic score from the point of view of current player. It always calculates maximum score for every node first but then for nodes corresponding to passive player are taken as negation of the original score. That is why we can call active player a maximizing player and passive player a minimizing player. While pure Negamax is full search algorithm that needs a lot of computation there are some possible optimization. The one that was used in the paper was alpha-beta pruning [8, 14] that by 'cutting' some parts of the tree can reduce its parts that potentially will not bring a good solution. Alpha and beta in this algorithm represents two bounds for child nodes values—lower bound and upper bound. During the initialization they equal to lowest and highest possible value and later are updated with new calculated values. When Negamax encounters a child node value outside an alpha/beta range, it cuts off a part of the tree and there is no need for further extrapolation of this part.

4.2 Evaluation Function

Evaluation function was used for calculating value of every node. It is very important as 'intelligence' of the algorithm depends on this one. In the project different version of the function were used but all of them were combinations of 6 factors. The idea of this function was inspired by other researchers works and also by books for checkers players and own ideas. It is also worth mentioning that this function could not be too complicated to make agent move fast. The six factors that were used [15].

Piece score—calculates the difference between number of pieces and kings left on the checkerboard that belongs to the active player and all pieces and kings of passive player. Kings are treated three times as valuable as pieces, because they give the player more possibilities for moving and catching.

Position score—not only amount of pieces but also their position on checkerboard is very important. Pieces standing in the edges of the board most of the time are more valuable than those standing in the center. Popular hint for beginner players says that it is good to leave first row of pieces not moved for as long as it is possible. It is also important to get to the last row because it gives a player a new promoted king. Positions in center are said to be dangerous as pieces can be caught from different sides.

Advancement—tells about in how advanced stadium is active player's game. It gives one point for every active player's piece standing in 5th or 6th row. It subtracts one point for every opponent's piece standing in 3rd or 4th row.

Mobility—lack of possible move is the loose conditions in the game so it is very important to always create situations where player can move or jump in the future actions. Mobility gives one point for every place on the checkerboard where active player can move its piece. While calculating mobility possibility of player's piece being caught after such move is not take into consideration.

Threat—is a mobility with one special condition—place where player's piece can be moved threatens the opponent with catching his piece in following move. In other words threat gives one point for every square to witch player can move and in next move will have possibility to capture opponent's piece.

Scare is an opposite of threat. It subtracts one point for every square if on the next move active player's piece occupying it could be captured by opponent without an exchange.

4.3 SARSA and SARSA(λ) Algorithm

SARSA is on-policy algorithm. It takes new actions following some policy π based on actually estimated value function Q_t. Its name seems to be very appropriate as it gives its meaning and describes its following steps. Operations that are performed by SARSA were shown in Fig. 3. At the time t environment state is described as state s_t agent chooses action at following its policy. Depending on what action was taken it gets a reward r_t and the environment transforms into its next state s_{t+1}. Afterwards agent like it did before chooses another action a_{t+1}. This is the origin of the abbreviation SARSA —s_t, a_t, r_t, s_{t+1}, a_{t+1}.

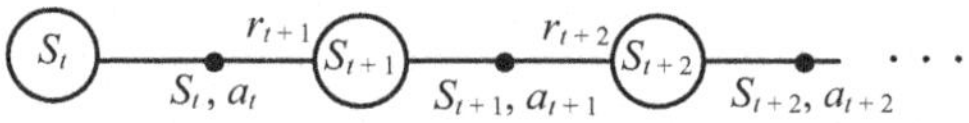

Fig. 3. Sequence of states and state-action pairs in single SARSA episode [16].

As it was mentioned before behaviour of SARSA agent should follow some policy. Policy tells about what action should be chosen in particular state—it maps states of environment with actions to be taken. Main goal of the agent is to find the best possible policy π and follow this policy to maximize results. In order to do this agent learns action-reward function $Q^{\pi}(s, a)$. This function says what reward we predict to get when we start from state s and perform action a following policy π. If we managed to find optimal function Q^* than it means that we had also found optimal policy according to the equation $\pi^*(s) = argmax_x\ Q^*(s, a)$. First Q values for all possible pairs of states s and actions a are initialized arbitrarily. Then for each episode first state is initialized (for checkers game it would be starting position with all pieces standing on the checkerboard) and first action is chosen. Then during the whole game stepping through next turns previously chosen action a is performed. Now agent observes how the environment have changed s' an gets reward r for taken action. Then it chooses next action a' that will be taken in next turn. Function Q gets updated basing on all gathered information about rewards. Finally at the end of the single step future action a' and state s' becomes actual ones to be taken in next turn. There are two things worth mentioning about implementation of SARSA agent for problem stated in this paper. First one is about starting step of the algorithm that cannot be taken. Initializing $Q(s, a)$ arbitrarily for every possible state and action is not possible without sufficient hardware. That is why this step was skipped and $Q(s, a)$ was initialized only for state-action pairs that occurred during the game. Second thing that is worth of mentioning is that actions are taken using ϱ-greedy policy that most of the time chooses the most maximizing move but depending on ϱ value it sometimes chooses randomly. This mechanism is important because thanks to new previously not visited states of the environment can be explored [11, 12]. SARSA(λ) is a variant of SARSA algorithm that was invented to increase its efficiency. In order to do this eligibility traces were added. Those eligibility traces are a temporary record of visited states and taken actions. Using eligibility traces SARSA estimates Q value basing not only on one step but also using all state-action pairs that led to the actual state. What is significant issue for eligibility traces depends on set of parameters—γ λ, where γ is a discount rate and λ is importance factor. Importance is decayed taking following steps starting from the most recent and ending at the oldest [16].

5 Results

First test was launched to see how Negamax performs with depth parameter $d = 6$ and basic evaluation function that was taking into consideration only piece score and position score $F = PS \times 2^{20} + PO \times 2^{14}$, where PS—piece score, PO—position score. Piece score was chosen as more important component, because catching all opponents pieces is the clue of the game. Putting them in good positions can help but it is not mandatory to win. To check how it works 100 games were launched. As it was expected for this algorithm it was able to win all the games with random agent. This observation opened a path for further experiments. In the next one evaluation function was changed for more complex while depth parameter was left the same. As expected this time also 100% of games was successful for Negamax agent. For both 'simple' and 'advanced' Negamax mean of overall score gained after every game was calculated in order to choose better evaluation function. This final score function was equivalent of PS component of Negamax evaluation function that was created to sum up situation on the checkerboard after the finished game. For first version of the algorithm calculated mean was 13.12 points and for the second one 12.4 points. In order to choose better performing Negamax algorithm two described versions were launched to play with each other 100 games. The problem in this idea was the fact that because of the nature of searching algorithms all game scenarios always looked exactly the same. That is why to both playing agents some portion of randomness was implemented. With 5% probability both agents were choosing the next move randomly not basing on searching the tree. Outcomes of this experiment shown that Negamax with more complex evaluation function version was able to win 64% of games. As better performing the second agent was chosen for further research. Last verification of performance of the second algorithm was facing human. Beginner human player played 20 games with this agent. With such opponent Negamax agent was able to win 5 games what gave it 25% of effectiveness. This result was perceived as a good start for further research. Process of learning SARSA agent started with launching both agents to play for over a dozen thousands of episodes. In this way the best parameters for SARSA were chosen. The best achieved parameters was following set: $\varrho = 0.1$, $\alpha = 0.2$, $\gamma = 0.9$. Outcomes from the learning process were presented in Fig. 4. It is visible that the process of learning took place but in the end it achieved only around 35% of effectiveness and it did not seem to be able to improve during longer process of learning. In order to make learning faster and see if it can achieve better results SARSA(λ) variation of basic algorithm was launched to play with the same Negamax agent. Also basic SARSA parameters were not changed and discount rate λ was set to 0.9. Comparison of SARSA and SARSA(λ) was shown in Fig. 4. It definitely improved time of learning but unfortunately did not influenced effectiveness of algorithm.

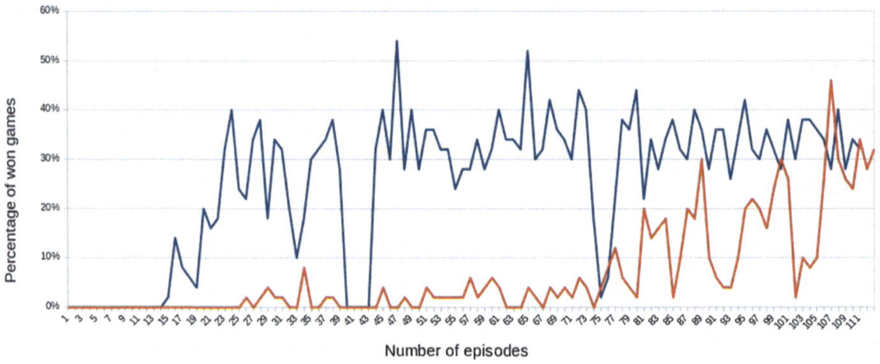

Fig. 4. SARSA—red and SARSA(λ)—blue after 5650 games, percentage by every 50 games.

Next we significantly reduced possible number of states. After the game just like before average score for every 100 games was calculated together with percentage of won games. Results are presented in Fig. 5. It seems to be very satisfying outcome.

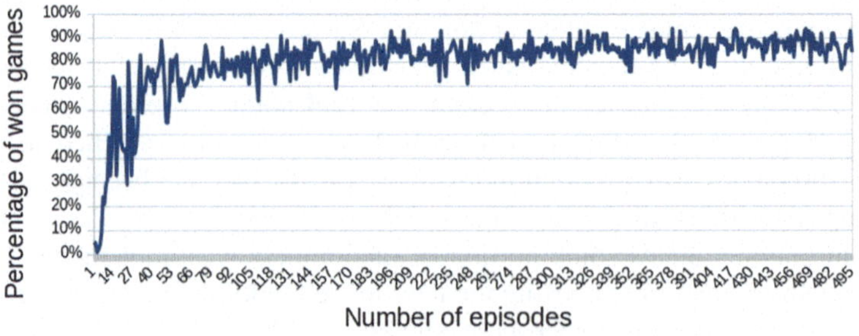

Fig. 5. SARSA learning with Negamax, 50,000 games—percentage by every 100 games.

During learning process SARSA agent was winning about 85% of games. Also average of 5 points (in the late phase of the process) says that most of the time not even one of black pieces was taken (5 point is equivalent of 2 pieces and 1 king). Final test for SARSA agent was playing with human player. It managed to win only one game. What was the main agent's problem was the fact that still it was easy to find not known by SARSA state and use it for advantage. Playing 'on time' until the moment when agent started to make visible mistakes was very effective strategy. In such situation SARSA was playing like random agent that for unknown states made it worse than Negamax playing with human.

6 Conclusions

Agent based on searching algorithm with good evaluation function got to the level a bit below an amateur player. Tested SARSA and SARSA(λ) reinforcement learning agents that were implemented also did not brought satisfactory results. After seemingly successful process of learning during which they showed ability of beating other no-human agents they still were almost useless during the game with human. Even if reinforcement learning used for this complex problem did not brought revolutionary results it proved to be very interesting and promising method. The agent could learn in foreign environment during the process of unsupervised learning. That could make it useful for solving different problems. Also using possible skill for generalization of such agent is very promising. A good idea would be to test mixing Negamax with SARSA. Thanks to this agent could learn new moves and know exactly how to behave most efficiently in particular situations using SARSA and deal with unknown situations using searching algorithm. Also some new techniques could be added. Some heuristic for searching also could be implemented and tested. Experiments showed that it could be interesting field for more thorough studying.

References

1. Schaeffer, J.: Man Versus Machine: The Silicon Graphics World Checkers Championship. Department of Computing Science, University of Alberta (1992)
2. Schaeffer, J.: Checkers is solved. Am. Assoc. Adv. Sci. **317**, 1518–1522 (2007)
3. Oxland, K.: Gameplay and Design. Pearson Education, London (2004)
4. Kimble, G.A.: Learning Theory. Encyclopaedia Britannica (1998)
5. Wilkinson, J.G.: Ancient Egyptians. Kegan Paul Limited, London (2005)
6. Fortman, R.L.: Duke vs Lowder. Personal Computing (1978)
7. Walker, C.C.: What is the Origin of Draughts? Checkers Magazine (1988)
8. Schaeffer, J.: The history heuristic and alpha-beta search enhancements in practice. IEEE Trans. Pattern Anal. Mach. Intell. **11**(11), 1203–1212 (1989)
9. Levy, D.N.L., Beal, D.F.: Heuristic Programming in Artificial Intelligence: The First Computer Olympiad. Ellis Horwood, Chichester (1989)
10. Levy, D.N.L., Beal, D.F.: Heuristic Programming in Artificial Intelligence: The Second Computer Olympiad. Ellis Horwood, Chichester (1991)

11. Samuel, A.L.: Some studies in machine learning using the game of checkers. IBM J. Res. Dev. **3**(3), 210–229 (1959)
12. Samuel, A.L.: Some studies in machine learning using the game of checkers: recent progress. IBM J. Res. Dev. **11**(6), 601–617 (1967)
13. Peterson, M.: An Introduction to Decision Theory. Cambridge University Press, Cambridge (2009)
14. Joshi, P.: Artificial Intelligence with Python. Packt Publishing, Birmingham (2017)
15. Gollapudi, S.: Practical Machine Learning. Packt Publishing, Birmingham (2016)
16. Sutton, R.S., Barto, A.G.: Reinforcement Learning: An Introduction. The MIT Press, Cambridge (2017)

Required Depth of Electricity Price Forecasting in the Problem of Optimum Planning of Manufacturing Process Based on Energy Storage System (ESS)

Aleksandr Krivchenkov(✉), Alexander Grakovski, and Ilya Balmages

Transport and Telecommunication Institute, 1 Lomonosova Str., 1019 Riga, Latvia
avg@tsi.lv

Abstract. The issue of appropriated range of electricity market price forecasting in the problem of enterprise's electricity consumption expenses optimisation is considered. Various kinds of battery energy storage systems (BESS) for small enterprises, as well as control algorithms for energy consumption expenditures, cost reduction, are considered. The estimation of accuracy of hourly electricity price time series forecasting with the artificial neural networks (ANN) algorithm is made with step-by-step increasing depth of the forecast. It is shown that for the optimal control of electricity consumption forecast makes sense no more than for about of few hours (or tens hours) ahead only depending on the forecast errors.

Keywords: Energy storage system · Battery · Electricity consumption control · Time series forecasting · Artificial neural network

1 Introduction

The problem of enterprise's electricity consumption expenses optimisation became topical in the early 2000s due to the growth of electricity consumption in industry, especially for small and medium enterprises (SMEs), introducing automated production lines against the background of a gradual annual increase in the price of electricity [1].

From the other hand, the creation of regional exchange structures, for example, the *Nord Pool Spot* market in the Baltic Sea Region [2], uniting producers and suppliers of electricity, has allowed consumers to have a dynamic electricity tariff with an hourly change in the price of 1 kWh. The hourly prices for consumers of a certain region are becoming known in advance for the next day, which enable enterprises to plan electricity consumption in advance [3].

If an enterprise has the ability to plan the load freely during the day, the easiest way to optimize electricity costs is to disconnect part of the equipment at the hour of the highest price (Fig. 1a), but it usually leads to the transfer of work to night hours and increasing of other costs. Most popular saving method is the use of the Energy Storage System (ESS) presented in Fig. 1b. The use of additional sources of 'green' energy (solar, wind, hydro and other) increases the possibility of charging the battery and from

I. Kabashkin et al. (Eds.): RelStat 2018, LNNS 68, pp. 331–342, 2019.
https://doi.org/10.1007/978-3-030-12450-2_32

grid consumption [4], as well as, with the relevant agreements with the supplier, allows the return (sell) part of the electricity to the grid. For the SMEs, the Battery Energy Storage System (BESS) architecture, including only the battery and AC\DC converter-inverter (Fig. 1c), often is most suitable.

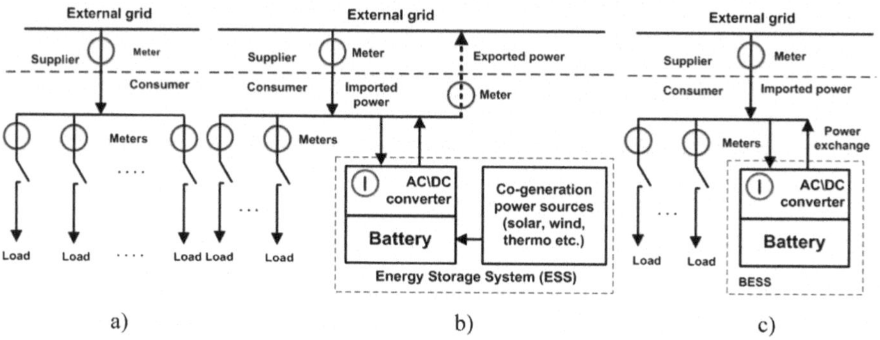

Fig. 1. Traditional architectures for the enterprise's consumption expenses optimisation: **a** load manipulation without ESS; **b** full system ESS together with additional energy sources, and **c** battery based energy storage system (BESS).

Formulation of the optimal control basic problem for BESS (Fig. 2) is that: on base of a known or forecasted hourly prices for the next day (Fig. 2a), and the planned or forecasted of hourly electricity consumption by the enterprise (Fig. 2b), try to determine the lowest-cost time slots for charging the battery and the largest-expensive time slots for discharging the battery, reducing consumption from the grid (Fig. 2c). It is necessary to take into account a number of limitations: charging power consumed from the grid should not exceed the limits established for the enterprise; the operating conditions of the batteries (permissible level of charge and discharge, temperature, charging and discharging current, number of charge-discharge cycles per day, etc.) should be performed so as to ensure the maximum lifetime of the battery.

On the other hand, it is obvious that an increase of the optimal planning period enhances the economic effect of BESS because allows us to forecast the maximum spending slots and reduce them if the forecast is valid for several days, weeks or months. Therefore, the great importance has the quality and depth of forecasting the price and the consumption. This issue is considered as the main goal of proposed study.

2 Battery Energy Saving System Management Algorithms

Charging and discharging the ESS batteries (Fig. 1c) is monitored. The Imported power is purchased in the wholesale electricity market and is supplied through the external grid. For the owner of local power grid, there is a principal possibility to return excess electricity to the external grid but it is not expected in this discussion.

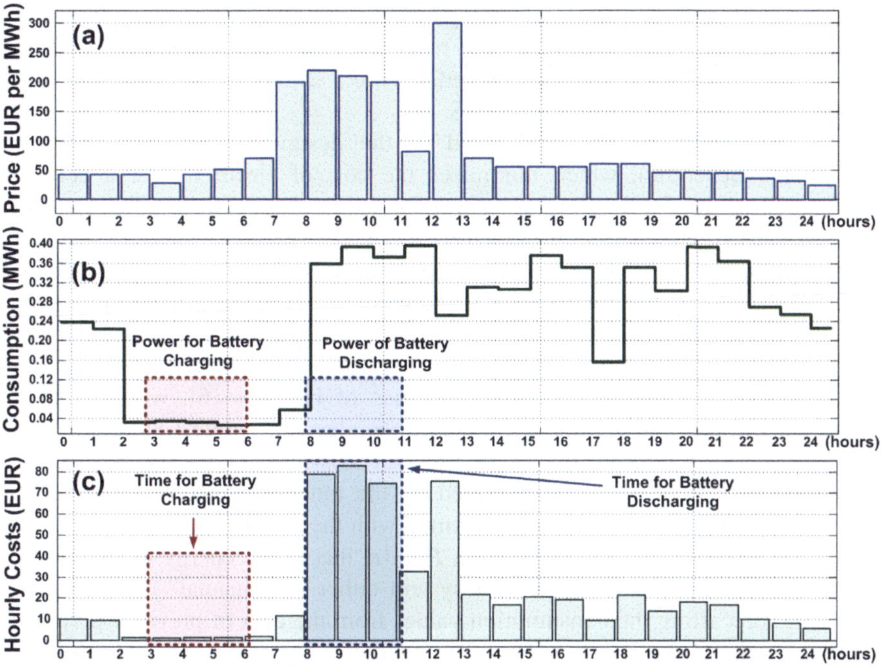

Fig. 2. The optimal scheduling for BESS: **a** hourly prices (real data on 19/05/2014 from *Nord Pool Spot* [3]); **b** hourly consumption of typical SME working in two shifts, and **c** electricity costs (need to be minimized).

With known (or forecasted) electricity prices, it is possible to reduce the cost of energy by reorganizing the production process. This can be achieved by transferring energy-intensive processes to those times of the day for which the prices will be minimal. This approach conflicts with traditional methods of production organization and is not considered here.

The advantage of the owners of ESS, even in the absence of alternative sources, is the possibility of obtaining lower bills due to the changes in tariffs for energy as a

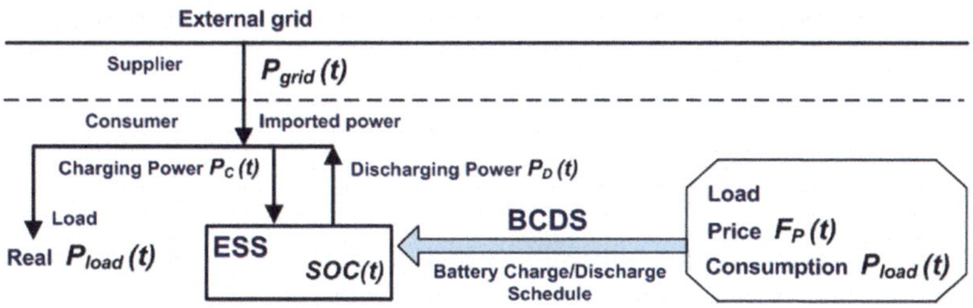

Fig. 3. The main components considered for the management of ESS.

function of time. Savings are achieved through the use of electricity accumulated by ESS batteries at those intervals when the price reaches its maximum. When the price of electricity is low, the ESS batteries are charged. Figure 3 shows the main components that are taken into account managing the ESS.

The task of optimal control is reduced to the design of a BCDS (Battery Charge/Discharge Schedule), which minimizes the cost of electricity for a certain period of time. The basis for drawing up such a plan is known or forecasted prices and the plan or forecast of the energy consumed by the load.

The problem of optimal planning and management of ESS was solved for various systems. In [4] a planning and control strategy based on the predictive control model is presented. In [5] two BCDS algorithms are given. According to the authors, these algorithms are the simplest and most effective. In a number of works, attempts are made to take more detailed account the properties of batteries [6], which further complicates the planning algorithm. A common planning and control algorithm for ESS with a proven level of efficiency at the moment most likely does not exist.

The data set of Price (Fig. 3) is described by the function $Fp(t)$ is the price of electricity as a function of time. This set is formed with the help of stock data [3]. The data set of Load is described by the function $P_{load}(t)$ that is the energy of load consumption. It is formed in the management system either by "manual" consumption planning, or by forecasting the consumption values from the data of previous periods.

The time period T for BCDS plan consists of intervals of a fixed length. If such time interval is 1 h, then $t = 1, 2, \ldots, T$ and $T = 24$, the time period is 1 day (24 h). Thus, the planning algorithm assumes discrete time intervals and operates with some average parameter values at each of the intervals. We introduce the function $P_{grid}(t)$—the energy consumption from the external grid. The cost of energy for the period T is determined by the following expression:

$$C_T = \sum_{t=1}^{T} (P_{grid}(t) \cdot Fp(t)). \tag{1}$$

In the case where ESS is not used at all, obviously, the following equality holds:

$$P_{grid}(t) = P_{load}(t). \tag{2}$$

The use of ESS makes it possible to reduce the consumption of $P_{grid}(t)$ at intervals when $Fp(t)$ tariffs are large. However, for this purpose it is necessary to charge the battery at an earlier time. The decrease in value of cost over the interval t will be achieved if the following conditions are met:

$$\begin{cases} P_{load}(t) \cdot (Fp(t) - Fp(t_c)) \geq 0, \\ T > t > t_c. \end{cases} \tag{3}$$

Here, the interval t_c is the time interval in which the batteries are charged. At each interval, conditions are formulated that depend on the properties of the inverter and the batteries. They can vary depending on the specific equipment.

The following conditions may be applied:

- the power of the charge and, consequently, the energy saved on one interval, does not exceed a certain maximum:

$$0 \le P_C(t) \le P_C\text{max}, \tag{4}$$

- the discharge power does not exceed a certain maximum:

$$0 \le P_D(t) \le P_D\text{max}, \tag{5}$$

- SOC (state of charge): the battery status should be in a certain range (for an example, 20–80%):

$$SOC\text{min} \le SOC(t) \le \text{SOCmax}. \tag{6}$$

There are also the requirements for compliance with a number of other requirements, which depend on the equipment of a particular system.

Thus, the problem of optimization of management is reduced to the selection of such a BCDS plan that would maximize the sum over t in the time period T of expression (3) and under the action of conditions (4)–(6). It will ensure the minimization of the cost of energy (1).

The computational complexity of the algorithm for determining the minimum cost in the "brute force" realization is proportional to 3^T (for a day time period of 3^{24}). The implementation of planning is illustrated by an example in Table 1. As a research methodology in this case we have taken the real prices, typical consumption and compare profits for two planning algorithms.

The Table 1 shows the real prices of Nord Pool Spot on 19.11.2013. The average price for this day was 0.0412 €/kWh, and consumption is typical for a small enterprise without energy-intensive processes [5] (about 21 kWh per day). In the process of providing a BCDS plan, the load was redistributed using ESS. Two BCDS plans were calculated, and the algorithms had the following features:

- in both cases, only one Charge/Discharge cycle was allowed per day; this condition can significantly increase battery life; for both plans BCSD1 and BCDS2 the charge was carried out at the time interval 01:00–02:00;
- for the BCDS1 plan (described in [5]), the discharge of batteries to the load was assumed at those times for which "large" prices were observed (at hours 16–18, 23); for the BCDS2 plan, the algorithm selected the time slots for which the product of the price and consumption was "large" (these are slots 9, 21–23);
- for BCDS1 it was assumed that the maximum power consumption from the grid was not exceeded (2.993 kW $\le P(2) \le P\text{max}$) and the charging power is permissible (1.891 kW $\le P_C(t) \le P_C\text{max}$); for the BCDS2 plan, these requirements are significantly higher: (8.8 kW $\le P(2) \le P\text{max}$) and (7.7 kW $\le Pc(2) \le Pc\text{max}$), otherwise this plan must be considered as invalid.

Table 1. Data and calculation of the battery charge/discharge schedule (BCDS).

Hour #	Prices €/kWh	Load kWh	Cost €	BCDS1 kWh	BCDS1 Cost €	BCDS1 Profit €	BCDS2 kWh	BCDS2 Cost €	BCDS2 Profit €
1	0.033	1.482	0.05	1.482	0.05		1.482	0.05	
2	0.0328	1.102	0.04	2.993	0.10	−0.06	8.776	0.29	−0.25
3	0.0329	0.730	0.02	0.730	0.02		0.730	0.02	
4	0.0337	0.465	0.02	0.465	0.02		0.465	0.02	
5	0.0357	0.156	0.01	0.156	0.01		0.156	0.01	
6	0.0451	0.108	0.00	0.108	0.00		0.108	0.00	
7	0.0443	0.106	0.00	0.106	0.00		0.106	0.00	
8	0.0439	0.290	0.01	0.290	0.01		0.290	0.01	
9	0.0395	1.901	0.08	1.901	0.08		0.000	0.00	0.08
10	0.039	1.497	0.06	1.497	0.06		1.497	0.06	
11	0.039	1.274	0.05	1.274	0.05		1.274	0.05	
12	0.0395	0.770	0.03	0.770	0.03		0.770	0.03	
13	0.0419	0.464	0.02	0.464	0.02		0.464	0.02	
14	0.0427	0.299	0.01	0.299	0.01		0.299	0.01	
15	0.0461	0.135	0.01	0.135	0.01		0.135	0.01	
16	0.0478	0.143	0.01	0.143	0.01	0.01	0.143	0.01	
17	0.0477	0.150	0.01	0.000	0.00	0.01	0.150	0.01	
18	0.0425	0.279	0.01	0.000	0.00	0.01	0.279	0.01	
19	0.0392	1.189	0.05	0.000	0.00		1.189	0.05	
20	0.0385	1.620	0.06	1.620	0.06		1.620	0.06	
21	0.038	2.261	0.09	2.261	0.09		0.000	0.00	0.09
22	0.0429	2.194	0.09	2.194	0.09		0.000	0.00	0.09
23	0.0687	1.319	0.09	0.000	0.00	0.09	0.000	0.00	0.09

(continued)

Table 1. *(continued)*

Hour #	Prices €/kWh	Load kWh	Cost €	BCDS1 kWh	BCDS1 Cost €	BCDS1 Profit €	BCDS2 kWh	BCDS2 Cost €	BCDS2 Profit €
24	0.0348	1.327	0.05	1.327	0.05		1.327	0.05	
Sum:	–	21.262	0.86	21.262	0.80	0.05	21.262	0.76	0.09
Average	0.0412	0.8859	–	–	–	–	–	–	–
Cost of energy without ESS €:			0.86		0.86			0.86	
Minimal possible cost with ESS:			0.70		0.80			0.76	
Possible decrease in cost:			−18.6%		−6.4%			−11.0%	

- It should be noted that for a corresponding "variability" of prices, there is some maximum possible reduction in energy costs for the corresponding period of time using ESS. In this fact lies a simple physical sense: the minimum is reached when the battery has a large capacity and can be charged when the price is minimal, and then it is discharged to the load at other times. For our example, the maximum possible cost reduction was 18.6%. The BCDS1 plan cuts costs by 6.4% and the BCDS2 plan by 11%. In this sense, Plan 2 is preferable to Plan 1 but there is no Plan with cost reduction more than 18.6%.

3 Artificial Neural Network for Electricity Market Price Forecasting

The time series is a sequence of the values obtained at the marked instants of time. Each time series is one of the implementations from an infinite set created by a random process, which is affected by many factors [7]. There are many methods for forecasting the values of time series. According to the literature review in [8], the forecasting can be divided into two categories: statistical methods and artificial intelligence methods.

The statistical category includes stochastic time series, general exponential smoothing, multiple linear regression, support vector regression, state space, etc. The artificial intelligence includes fuzzy inference, expert system, artificial neural network (ANN), etc. However, none of these forecasting methods can be distinguished as a method with the greatest accuracy of forecasting.

Consider forecasting using ANN. Differing from the traditional time-series methods that may not be able to capture the nonlinear pattern in data, ANN establishes non-linear dependencies between future and actual process values. ANN adapts to the surrounding environment, (however, does not always provide robustness). Under adverse conditions, the ANN performance degrades slightly. Because of their uniformity of analysis and design, it can be used in different applications [9]. For time series forecast don't needs to find a trend, periodic and random components, and external influences. In the training process, ANN should "identify" them and take into account.

Disadvantages of neural networks are the inability to see intermediate computations, the complexity of the choice of architecture and the learning algorithm [10]. In paper [11], a literature review of the ANN techniques for forecasting:

- Multilayer Feed-forward ANN used in causal forecasting. The information flow comes from the input level to the output level.
- Recurrent ANN is a system with the loop where the output signals are fed back to the input.
- Time delay ANN integrates time delay lines.

In our work, we use the nonlinear autoregressive exogenous model of ANN which is a combination of all above ANN and is successfully applied in time series forecasting.

The learning algorithm of Levenberg-Marquardt is chosen. It is a kind of Newton's method and is intended to minimize functions, which are the sums of the squares of other nonlinear functions. The advantage of the Levenberg-Marquardt's algorithm is fast convergence. Disadvantage—the need for a large amount of memory for storing all data. A neural network trained by the Levenberg-Marquardt's algorithm can be less time-consuming and get the least forecasting error [12]. The number of neurons in the hidden layer was 15. As activation function, in the hidden layer is the sigmoid function and in the output layer is linear. The activation function must have two important properties: (1) nonlinearity (signal, passing through each neuron, can get more and more complex functional dependence); (2) limited range of admissible values for an unbounded range of definition. The values of the output signal should always remain within the specified limits.

Randomly (on base of uniformly distribution) the time for training of ANN is chosen, which can be in the range from 14 to 28 days. Three attempts are made to train the network, based on which the forecast is made 3 times, and the results of the three forecasts calculate the average, that passes through the "smoothing" filter.

Let's define the depth of the forecast. Producing the forecast (in our case the day ahead), in first time for training set, the actual data is used. Moving along the time axis and already having the forecasted data of past forecasts, we use it instead of actual data. Thus, over time, the data on which we make the forecast is less and less dependent on the actual data, and more on the forecasted data. After 4 weeks, the training set will be fully built on the forecast.

Let's forecast prices of electricity for a time period for 2–2.5 months in advance. Our goal is to study how much the forecast which is based on the forecasted data differs from actual values and how it's change over time. The results of the experiment will give an estimate of the depth of electricity price forecast. To do this, define the accumulated error (mean absolute percentage error, MAPE) in expression (7):

$$MAPE(N) = \frac{1}{N}\sum_{i=1}^{N}\frac{|x_i - \hat{x}_i|}{x_i} \cdot 100\%. \tag{7}$$

As a research methodology we have taken here the numerical experiments on hourly electricity price forecasting by ANN in Matlab using the real data of prices and weather conditions. Since the time series describing the price of electricity is not stationary, the experiment was performed not on one interval of time, but on different ones. In order to make sure of the repeatability of the algorithm, at each selected time interval, the experiment was staged several times. In repeated experiments on the same times, the results were similar.

4 Results of Price Forecast Experiments

As the result of implemented forecasting algorithm, at different time intervals, the trend of the accumulated error behavior was similar (Figs. 4c and 5c). Accumulated error rapidly increases during the first four days. Then within a month, there is a smooth rise (during which there may be local drops). Perhaps this is due to the fact that this is the

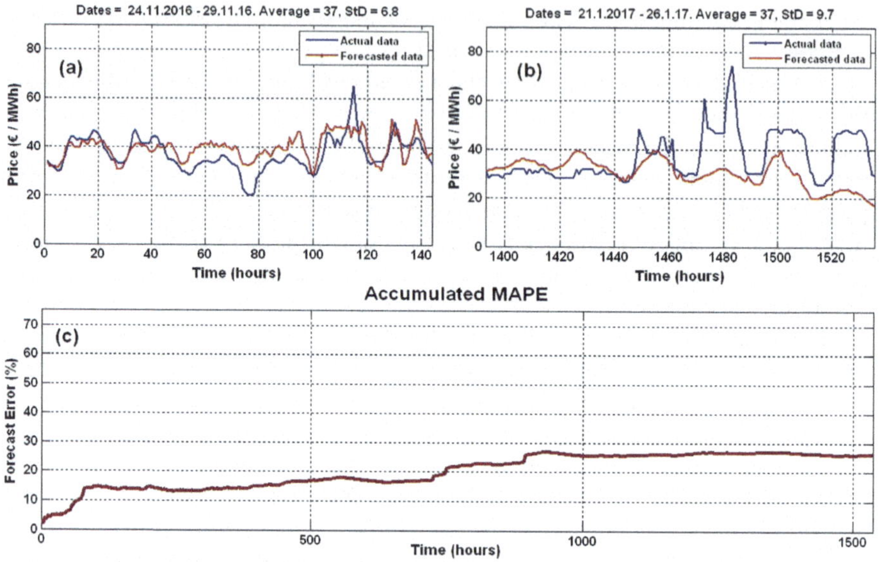

Fig. 4. Comparison of actual data and forecasted data for the case with the *smallest* accumulated error: **a** first 6 days from the beginning of forecasting; **b** sequence of 6 days of forecasting for 59–64 days from the beginning of the observation, and **c** overall accumulated error (64 days from 24.11.2016 to 26.1.2017).

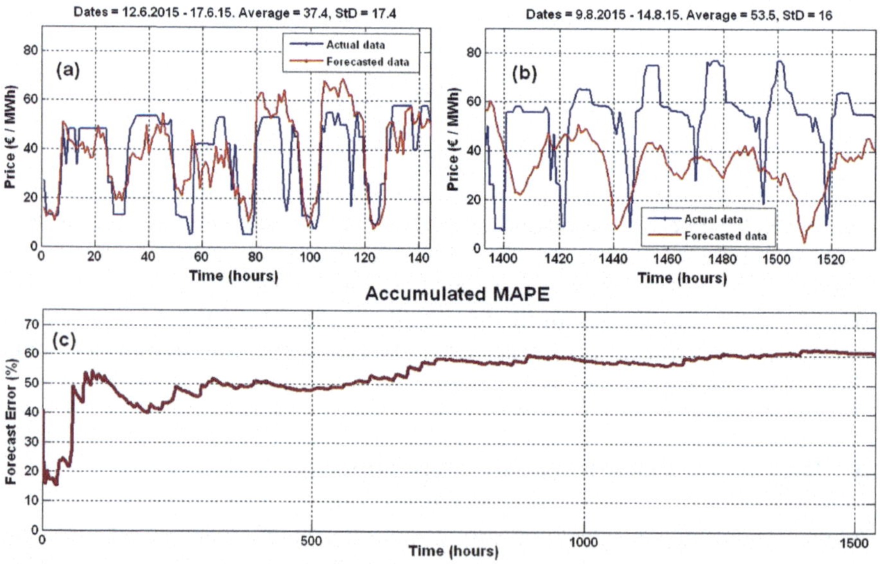

Fig. 5. Comparison of actual data and forecasted data for the case with the *biggest* accumulated error: **a** first 6 days from the beginning of forecasting; **b** sequence of 6 days of forecasting for 59–64 days from the beginning of the observation, and **c** overall accumulated error (64 days from 12.06.2015 to 14.08.2015).

time when the entire training set is based on forecasted data only. After that, it "stabilizes" at some level.

This is a general trend, but the values of the accumulated error were significantly different between the time sections. Two sections were selected for demonstrating the results: Fig. 4(a–b) showing the best results (64 days from 24.11.2016 to 26.01.2017) and the worst result on Fig. 5(a–b) (64 days from 12.06.2015 to 14.08.2015) were investigated. Such a difference is due to the non-stationary behavior of data.

On the actual data with the smallest accumulated error, average of values = 35 (€/MWh) and standard deviation = 9 (€/MWh). In the actual data with the biggest accumulated error, average of a values = 44 (€/MWh) and standard deviation = 20 (€/MWh).

Forecasted data in these cases has in the same range of values. Accordingly, when the actual data and the forecasted data begin to divaricated, especially if the so-called 'anti-phase' is formed (Fig. 5b), in the case where the spread (standard deviation) is greater, and significant error is observed.

5 Conclusions

The analysis of theoretic and experimental results of the problem of BESS optimum control on base of price forecasting shows that:

- The use of the Battery Energy Storage System (BESS) for SMEs theoretically reduces the company's energy costs by an average of 7–18%, depending on the nature of the production, the parameters of the BESS, and primarily on the volatility of the hourly price of electricity.
- Increasing the period of optimal planning on several days in advance cannot produce the desired result due to a significant increase in errors in the Artificial Neuron Network (ANN) method for predicting the hourly price (or load consumption) of electricity. There are conditionally "good" and "bad" periods of time where the errors multiply differ in magnitude. The explanation for this fact is that the "bad" sets are characterized by a large standard deviation (or large volatility).
- At periods of several tens of hours, the error almost linearly increases. At longer periods, there are significant differences in the forecast with a real price. Applicability of the ANN method for short-term forecasts is limited to a range from several hours till a few days when the error does not exceed several percent.
- Optimal control strategy based on a 24-h planning period with established hourly prices [3] remains the most preferable since forecast errors reduce the economy, and if the synchronism of real and predicted price fluctuations (so-called 'anti-phase' effect) is lost, on the contrary, lead to additional costs.

Acknowledgements. This research was granted by ERDF funding, project "Optimum planning of an energy-intensive manufacturing process and optimisation of its energy consumption depending on changes in the market price (2017–2019)", No 1.1.1.1/16/A/280 (Subcontract No L-s-2017/12-9).

References

1. Payne, J.: Survey of the international evidence on the causal relationships between energy consumption and growth. J. Econ. Stud. **37**(1), 53–59 (2010)
2. Varfolomejeva, R., Gavrilovs, A., Iļjina, I.: The regulation possibility of energy-intensive enterprises according to the market price change. In: 2017 IEEE International Conference on Environment and Electrical Engineering and 2017 IEEE Industrial and Commercial Power Systems Europe, Italy, Milan, 6–9 June 2017, pp. 1118–1123. IEEE, Piscataway (2017)
3. Nord Pool Spot market data. Electricity hourly prices. https://www.nordpoolgroup.com/Market-data1/Dayahead/Area-Prices/ALL1/Hourly/?. Accessed 13 Aug 2018
4. Xu, Y., Xie, L., Singh, C.: Optimal scheduling and operation of load aggregators with electric energy storage facing price and demand uncertainties. In: North American Power Symposium (NAPS), pp. 1–7 (2011)
5. Lebedev, D., Rosin, A.: Modelling of electricity spot price and load. In: Proceedings of 55th International Scientific Conference on Power and Electrical Engineering of Riga Technical University (RTUCON), pp. 222–226. IEEE (2014)
6. Barnes, A., Balda, J., Geurin, S., Escobar-Mejía, A.: Optimal battery chemistry, capacity selection, charge/discharge schedule, and lifetime of energy storage under time-of-use pricing. In: Proceedings of Innovative Smart Grid Technologies (ISGT Europe), 2nd IEEE PES International Conference and Exhibition, pp. 1–7 (2011)
7. Box, G.E.P., Jenkins, G.M.: Time Series Analysis, Forecasting and Control (1976)
8. Yang, J.: Power System Short-term load forecasting. Thesis for Ph.D. degree. Germany, Darmstadt, Elektrotechnik und Informationstechnik der Technischen Universität, 139 p. (2006)
9. Haykin, S.: Neural Networks and Learning Machines. Prentice Hall, Upper Saddle River, 938 p. (2009)
10. Mazengia, D.H.: Forecasting Spot Electricity Market Prices Using Time Series Models: Thesis for the Degree of Master of Science in Electric Power Engineering. Gothenburg, Chalmers University of Technology, 89 p. (2008)
11. Mitrea, C.A., Lee, C.K.M., Wu, Z.: A comparison between neural networks and traditional forecasting methods: a case study. Int. J. Eng. Bus. Manag. **1**(2), 19–24 (2009)
12. Hagan, M.T.: Neural Network Design, 2nd ed., PWS, MA, 1012 p. (1996)

Smart Solutions for Supply Chain Management

Operational Simulation-Based Decision Support in Intralogistics Using Short-Term Forecasts

Marcel Müller(✉), Tobias Reggelin(✉), and Stephan Schmidt(✉)

Otto von Guericke University Magdeburg, Universitätsplatz 2, 39106 Magdeburg, Germany
{marcell.mueller, tobias.reggelin, stephan.schmidt}@ovgu.de

Abstract. Simulation models are still often only part for decision support in the planning area. For short-term decisions at the operational level, there have been good fundamentals since the 1990s, but still relatively few implementations, especially in the logistics sector. Our approach is to use real-time data to provide short-term forecasts, by using a simulation model that provides required information. Due to current hardware and a well-chosen degree of abstraction of the model, real-time decision support ("real-time" means in this context: fast enough to support the decision) is possible. This paper presents a concept of a procedure model for the realization of such operational simulation-based decision support, applied to the picking area of an industrial laundry. The operational use of the simulation model is part of the project "Laundry Order Consolidation System (LOCSys)", which aims to improve the picking & storing processes in the clean area of an industrial laundry through automation.

Keywords: Real-time simulation · Real-time decision making · Virtual commissioning · Industrial laundry

1 Introduction

Today's decision-makers have to analyse rapidly complex problems in the intralogistics to make the right decision. There are not many tools to support this process but usually many data is available. A concept to use this data, in an easy way to optimize the own decision based on forecasts, would deliver a great benefit for the decision-makers.

In planning of processes in intralogistics, the use of material flow simulation models is quite common. The benefit of such simulation models does not necessarily end in the planning process. Modern simulation approaches and the current rapid hardware enable simulation runs in an extremely short time, fast enough to deliver decision support for operational decisions. According to Rogers and Gordon [9] "real-time" decision support implies a reaction time of the support tool that is smaller than the time until the decision has to be made.

Many simulation software already offer numerous integrated optimization tools: Beside classical optimization algorithms, this can also be artificial neural networks and

I. Kabashkin et al. (Eds.): RelStat 2018, LNNS 68, pp. 345–352, 2019.
https://doi.org/10.1007/978-3-030-12450-2_33

genetic algorithms for an artificial intelligence approach. Furthermore, many programs offer also a wide range of interfaces to connect the model with databases or a programmable logic controller (PLC).

For our approach, we develop a simulation model, which provides valid short-term forecasts by using real-time data to validate different picking order sequences. The optimization tools can directly use these data for changing various control parameters to optimize the target parameters or alternatively an extern algorithm uses the simulation results to create new (optimized) control parameters. Renewed simulation runs enable the validation of this optimization. The decision-maker can use the highly accurate results for the decision in his specific situation. The idea of real-time simulation is not new and appeared beside the idea of real-time decision making to control manufacturing systems [7, 9]. Since the 90s researchers already applicate real-time simulation for example to "assign due dates on logistics-manufacturing networks" [10] or in combination with artificial neural networks to control flows in sewerage networks in real-time [6]. Despite the progress in this area, research continues to focus on other use cases and new concepts nowadays.

We applicate our approach initially in the area of industrial laundries. These companies are under a high cost pressure. Although the turnover of the laundry industry as a whole is increasing, market concentration is taking place in Germany. With increasing costs due to rising minimum wages, some laundries are trying to tap new potential with innovative solutions. For example, the use of simulation models helped to find optimization potentials and to improve the processes [1]. Real-time simulation models go one-step further and provide meaningful results in a timely and therefore more useful way. However, there are not many applications of real-time simulation for industrial laundries. At least one approach describes the use of a real-time algorithm that optimizes the sequence in order to save resources with a predictive control [8]. A specific simulation-based approach is missing in this context.

As part of the current project "LOCSys—Laundry Order Consolidation System", a group of researchers and small- and medium-sized enterprises develop an automatic picking solution for industrial laundries. In addition to the classic simulation model for the planning of this solution, the picking solution communicates also to an operational simulation model, which is used during operation to map an emulation and to make short-term forecasts. This paper presents the conceptual model for this real-time connection of the simulation model.

2 Laundry Logistics

2.1 Closed-Loop Supply Chain

Industrial laundries are, in contrast to the classical manufacturing industry, characterized by a material cycle, which is the basis for the business model. There are many similar terms to describe laundry cycles because the topic of sustainability has a strong influence describing cycles in the economy. We use a closed-loop supply chain to describe the laundry cycle in the best way. Figure 1 shows a schematic representation of the laundry cycle. The circular structure consists of the relationship between customer and industrial laundry. The figure also shows clearly that the circuit is not completely closed. Purchases of laundry items or worn out laundry items vary the amount of items in the circulation.

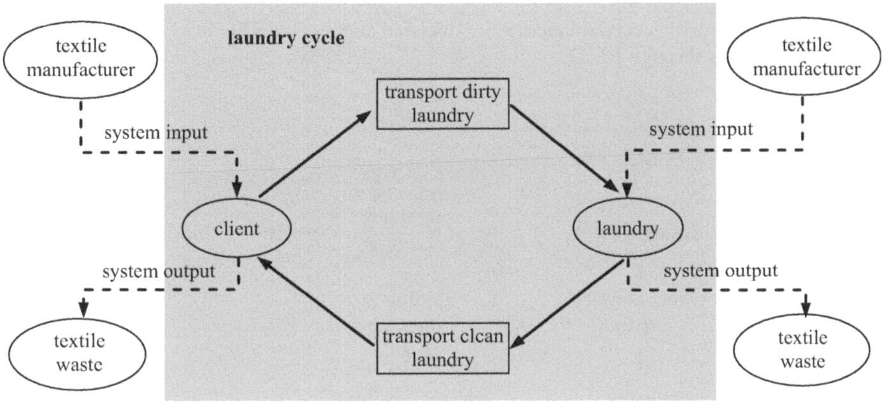

Fig. 1. Schematic representation of the laundry cycle.

The focus of our application is on industrial laundries, as there is no influence on the customer. The consideration of transport logistics between customer and laundry is an interesting field, but the conventional and new approaches for typical transport logistics problems are applicable here.

2.2 Processes in Industrial Laundries and RFID

The difficult handling of the laundry items decisively influences the logistical processes of industrial laundries. Although large areas, in particular in the washing process, have already been automated, the workers are still processing many operations manually, especially between some process steps and above all in the picking area. Therefore, there are current efforts to automate the picking area. This automation solution is based on real-time data in order to react in an optimal way to every current situation. An important prerequisite for this is the use of radio-frequency identification (RFID). Industrial laundries are increasingly using this technology, for example to obtain information about the loss of laundry in the laundry cycle. The use of this technology makes it possible to process data, with the help of further data from the merchandise management system, in order to make a real-time decision.

Since the considered system is the industrial laundry, it would make sense to collect data using RFID in the incoming goods department to identify the incoming laundry items and to have as much time as possible to make a decision. Figure 2 shows the processes in an industrial laundry and puts the simulation model in a context. There will not only be an identification point in the goods receipt, but also in the picking area. However, the latter will probably just make fast calculations and smaller variant comparisons to make a real-time decision.

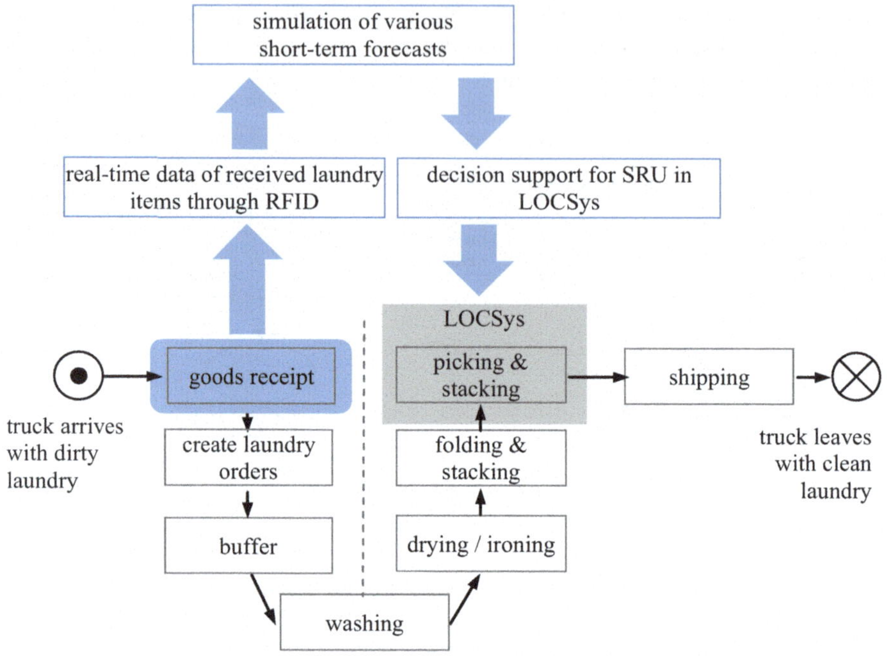

Fig. 2. Typical processes in an industrial laundry and early concept of the core processes for operational decision support of LOCSys.

The picking solution "LOCSys" consists currently of one rack row and a mounted on rails storage and retrieval unit.

3 Problem Description and Real-Time Decision Making

In our application, we face the complex challenge of combining a typical scheduling issue with a space allocation and stacking problem. Due to this, a decision making based on real-time data is essential, because several questions must be answered based on this data:

- Which is the next laundry stack to transport (storage, relocate or retrieve)?
- Where should the laundry stack be stored or relocated?
- Should it put the laundry stack on another laundry stack?

Inaccurate data would quickly lead to incorrect or non-optimal answers to these questions. The incoming material flow in industrial laundries is very variable in its quantity and composition. A simple use of historical data to estimate the overall state of the system supposedly known is thus not possible.

The optimization of sequence of different picking orders using certain limited resources is a typical scheduling problem. Scheduling problems are not new [2] and neither are the real-time approaches to solve them: with the emergence and spread of computers and networks in production and logistics, researchers were already looking for real-time solutions to optimize these problems with these new technologies in the 1990s [7, 9]. According to Rogers and Gorden there are three different approaches for solving scheduling problems: the chosen simulation-based one, "O.R. tools" and "A.I. concepts" [9]. Many approaches in the literature focus on applications in manufacturing systems and less on typical logistics solution as a warehouse.

Our simulation-based approach is also called "Simulation-based Real-time Decision-Making (SRDM)" [3] or "On-Line" simulation [4, 9, 12]. Looking less at the decision-making component and focusing more on testing the real system, our approach can also be described as a virtual commissioning [11], which leads also to typical results of an emulation [5]. The intended picking algorithm in our project could be considered as an A.I.-approach but will not be discussed further in this paper as it is still in development and not yet completely clear. We can already state at this point that the combination of the two approaches (simulation and A.I.) should be able to overcome the three-part problem.

4 Concept

4.1 Simulation Model, Picking Algorithm and PLC

To optimize order picking and warehousing, an efficient data exchange between simulation model, picking algorithm, PLC and the real logistics solution is required. Figure 3 shows the conceptual data exchange model at a high aggregation level. The central point of exchange are various data tables, which can be located on a normal computer with network connection in the enterprise. The merchandise management system or the corresponding part of an ERP software provides data about the current state of the overall system at regular intervals (e.g. every hour/day). This allows the system to know exactly which and approximately where a laundry item is currently located in the industrial laundry. Customer orders that result in picking orders can either be transmitted via the ERP system, separately in another software or directly by the customer. The actual real-time data comes from the real automated picking solution. This can be primarily the position of the SRU, as well as data on the instantaneous speed of the SRU and possibly faults that occur. With this data, the picking algorithm can specify the (approximately) optimized sequence of the picking orders. For this purpose, the algorithm uses not only its normal decision patterns but also the simulation model, as long as the real system allows sufficient simulation runs in time. The algorithm can conclude this period of time from the open picking orders and the current items of laundry in the picking area.

The simulation model needs the real-time data from the data tables for initialization in order to establish the current state of the real system. Already completed transport orders have been noted in the data tables and give conclusions about the current occupancy of the warehouse, despite the lack of sensor technology at this point. The current position of the SRU is also transferred after a processed transport order and is thus at least partially given in real time, if the SRU has no current transport order.

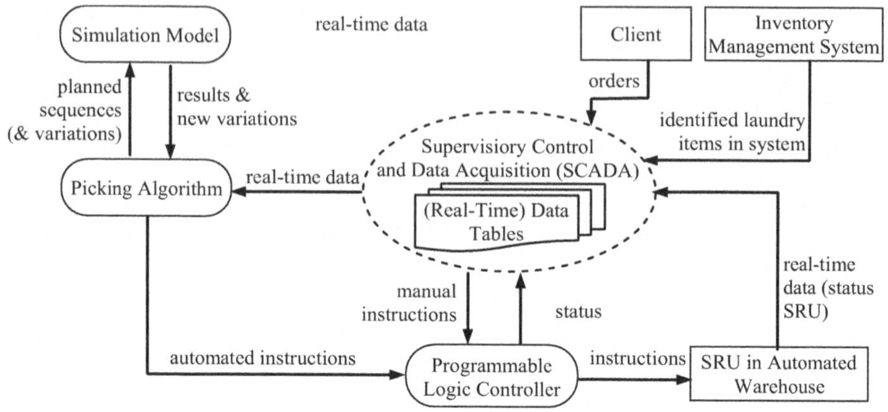

Fig. 3. Conceptual data exchange model between simulation, algorithm, PLC and real system.

After simulation, the picking algorithm receives the results. This can also result in possible new variations of the sequence that the algorithm did not sent before. The picking algorithm then forwards the sequence of the picking orders to the PLC as separate transport requests. The PLC then translates the transport requests into corresponding control commands. However, manual control commands can also be issued via the user interface of the SCADA and allow the change of automated created sequence or to create new transport orders.

4.2 Data Tables

Table 1 gives an overview of the planned entities for the central exchange of data. In addition to the mentioned attributes, there are relationships between the entities that

Table 1. Overview of the data tables.

Entity	Primary key	Further attributes
SRU	ID	Maximum speed; acceleration
Laundry item	ID	Length, width, height, location object, type of laundry, weight, position in stack
Laundry stack	ID	Location object, x-position, y-height
Shelf level	ID	Length
Transport order	ID/order number	Type of transport (store, relocate, retrieve), creation time stamp, completion time stamp, target object, target x-position
Picking order	ID	Date of delivery, creation time stamp, completion time stamp
Client	Client number	Name

involve further data fields.

A central component of data exchange will be the table with transport requests. Already assigned transport orders must be remembered in order to be taken into account when assigning new transport orders. This is always the case when a new item of laundry arrives via the conveyor belt to the picking area. In addition, the timestamps enable a statistical evaluation of the assigned transport requests. Table 2 shows some example data fields from the transport order table. The records of the transport orders have beside the shown columns also data fields for "stack height", "target stack", "transport mode" and might get more in the future. In this example transport order 9 and 10 are finished yet and transport order 11 would be taken next by a storage and retrieval unit.

Table 2. Example of typical transport orders and some of their attributes.

Transport order number	Creation date	Completion date	Transported stack	Shelf layer	x-position
9	5:24.6667	5:53.3417	stack:3	layer:1	6.556
10	5:25.4333	6:20.4126	stack:4	layer:3	5.514
11	5:51.8333		stack:11	layer:2	0.539

5 Conclusion and Outlook

The project is currently still in the planning phase and the approaches to real-time decision support by means of simulation-based forecasts are to be implemented next year. It has already become apparent that a database as a middleware is required for smooth communication between the simulation model, the picking algorithm and the PLC. The coordination of the data fields and transfer protocols is an important prerequisite.

The paper has given a brief overview of the most important aspects and their state of the art. In addition, it also presented a first data exchange model, which shows how the future interaction should work.

Acknowledgements. As part of thc "Zentrales Innovationsprogramm Mittelstand" (ZIM; central innovation program), the Federal Ministry for Economic Affairs and Energy of Germany launched a research and development project to find an innovative solution for automating the picking area in industrial laundries. The project is called "LOCSys—Laundry Order Consolidation System". Part of LOCSys is the use of an operational simulation model to provide short-term forecasts and support the decision-making of the automated picking system.

We are very thankful for the financial support of the ministry, which allows us to research for new picking solutions in the laundry industry.

References

1. Brandau, A., Weigert, D., Tojulew, J.: Applications of simulation to improve operations. In: Simulation in Production and Logistics, Germany, pp. 289–298 (2015)
2. Conway, R.W., Maxwell, W.L., Miller, L.W.: Theory of Scheduling. Addison-Wesley Publishing Company, Reading (1967)
3. Dalal, M., Groel, B., Prieditis, A.: Real-time decision making using simulation. In: Proceedings of the 2003 Winter Simulation Conference, USA (2003)
4. Davis, W.J.: On-line simulation: need and evolving research requirements. In: Handbook of Simulation: Principles, Methodology, Advances, Applications, and Practice. Wiley (1998)
5. Hofmann, W., Langer, S., Lang, S., Reggelin, T.: Integrating virtual commissioning based on high level emulation into logistics education. Procedia Eng. **178**, 24–32 (2016)
6. Hunter, A., Hare, G., Brown, K.: Genetic design of real-time neural network controllers Neural Comput. Appl. **6**(1), 12–18 (1997)
7. McConell, P.G., Medeiros, D.: Real-time simulation for decision support in continuous flow manufacturing systems. In: Proceedings of the 1992 Winter Simulation Conference, USA (1992)
8. Peitz, S., Gräler, M., Henke, C., Hessel-von Molo, M., Dellnitz, M., Trächtler, A.: Multiobjective model predictive control of an industrial laundry. In: 3rd International Conference on System-Integrated Intelligence: New Challenges for Product and Production Engineering, Germany (2016)
9. Rogers, P., Gordon, R.J.: Simulation for real-time decision making in manufacturing systems. In: Proceedings of the 1993 Winter Simulation Conference, USA (1993)
10. Ruiz-Torres, A.J., Nakatani, K.: Application of real-time simulation to assign due dates on logistic-manufacturing networks. In: Proceedings of the 30th Conference on Winter Simulation, USA (1998)
11. VDI/VDE-Gesellschaft Mess- und Automatisierungstechnik (GMA): Virtual commissioning model types and glossary. In: VDI/VDE 3693, Beuth Verlag GmbH, Germany (2016)
12. Wu, S.-Y.D., Wysk, R.A.: An application of discrete-event simulation to on-line control and scheduling in flexible manufacturing. Int. J. Prod. Res. **27**(9), 1603–1623 (1989)

Choosing the Localisation of Loading Points for the Cargo Bicycles System in the Krakow Old Town

Vitalii Naumov(✉) and Jakub Starczewski

Cracow University of Technology, str. Warszawska 24, 31155 Kraków, Poland
vnaumov@pk.edu.pl

Abstract. Loading points are indispensable elements in the systems of goods delivery by cargo bikes. The choice of the loading points location is a basic issue to be solved when designing cargo distribution systems. The paper proposes an approach to justifying the location of loading points based on computer simulations of the delivery process with the use of cargo bikes. In the developed mathematical model, we present consignees as vertices of the transport network, while the graph edges reflect the sections of the road network. We propose to quantify requests for transport services with numerical parameters, among which the most significant are the size of the consignment, its dimensions, the time interval between requests, and the delivery distance. The developed mathematical model was applied for choosing the loading point localisation for the system of cargo deliveries in the Krakow Old Town: demand parameters for transport of goods in the Old Town area were presented as random variables, the existing road network model was formed, and the system simulations for the alternative locations of a loading point were provided. The simulation results allowed us to substantiate such a loading point location which is characterised by minimal value of the total transport work.

Keywords: Cargo bicycles · Loading point · Localisation problem

1 Introduction

Effective freight deliveries within cities is a highly problematic issue, which generates relatively high costs in comparing to costs of the entire delivery process. Neglect of this issue by transport companies and researchers cause that its practical aspect meets with numerous physical, administrative and organisational limitations. Furthermore, last mile deliveries are usually characterised by a large diversity of cargos in terms of natural susceptibility followed by low utilisation of transport capacity [1]. In such situation, cargo distribution systems based on cargo bikes could be considered as the alternative to traditional deliveries by automobile transport.

The basis for the operation of the mentioned systems is the use of various types of cargo bikes and bicycle carts [2] equipped with a cargo basket. There are many technical solutions of such vehicles, among others: two-, three-, four-wheeled vehicles, with a load box at the front and/or at the rear, with front or rear axle drive. The wide range of these vehicles allows adjusting system parameters to certain city conditions

I. Kabashkin et al. (Eds.): RelStat 2018, LNNS 68, pp. 353–362, 2019.
https://doi.org/10.1007/978-3-030-12450-2_34

and to the given demand for transport services. Nevertheless, the transport of loads by bikes has many limitations, one of which is delivery distance. Its maximum value depends on the system's environment and the distribution of target points. According to [3], the distance of delivery with the use of cargo bicycles can vary between 1 and 166 km and 90% of deliveries are completed within the range of up to 75 km.

Therefore, when servicing big city areas, it is necessary to use a loading point (one or several) in order to shorten the distance of delivery by bicycle. Examples of such points are a trailer or a semi-trailer, a cargo container unit used as a mobile point, the dedicated transhipment bay, a "parcel locker" or a storage space having the character of a permanent loading point [4]. In the case of mobile variants, it is mandatory to specify a rational location of loading points. The specialised literature describes many ways to determine the loading point localisation, one of the most popular is the method of gravity centre. Despite its simplicity, this solution works well just for a single loading point [5]. For more complex delivery systems, it is necessary to use more advanced algorithms that take into account the volume of cargo, the cost of transport and the number of recipients. The approach proposed in [5] attempts to determine the location using heuristic methods that consider these factors. However, existing approaches usually do not take into account features of the technological process of transport, such as the used routing method, scheduling procedures, etc.

The goal of this paper is to present an approach to choose the location of a loading point for a cargo bicycle system considering routing procedures.

The paper has the following structure: the mathematical model of the cargo bikes delivery system is briefly described in the second part; the third part contains the description of software implementing the proposed model; the fourth section introduces a case study of choosing the loading point location for cargo bikes delivery system in the Krakow Old Town; the last part offers brief conclusions.

2 Formalisation of the System of Goods Delivery by Bikes

The process of goods delivery by bicycles is implemented within a logistics system of a city. As basic elements of this system, the following subsystems should be listed:

- transport network where the process of goods delivery takes place;
- demand for the supply of goods reflecting the customers' needs for the delivery of cargoes;
- a subsystem serving the demand for the delivery of goods which includes means of transport (cargo bicycles) and loading points.

Thus, the mathematical model $\mathbf{M}_{SD}$ of the delivery system in a general form could be presented as the set of three basic elements:

$$\mathrm{M}_{SD} = \langle \mathbf{\Omega}, \mathbf{D}, \mathbf{\Phi} \rangle, \tag{1}$$

where $\mathbf{\Omega}$ is the transport network model; $\mathbf{D}$ is the model of demand for transport services; $\mathbf{\Phi}$ is the model of the servicing subsystem.

2.1 Transport Network Model

The most commonly used approach to modelling transport networks is the use of mathematical structures based on graph models. Let us formalise the transport network as a pair of subsets—the nodes and the edges:

$$\mathbf{\Omega} = \langle \{\eta_i\}, \{\lambda_j\} \rangle, \ \forall \eta_i \in \mathbf{N}, \ \forall \lambda_j \in \mathbf{\Lambda}, \tag{2}$$

where η_i is the *i*-th node being the element of the all network nodes set $\mathbf{N}$; λ_j is the *j*-th edge being the element of the all network edges set $\mathbf{\Lambda}$.

In the mathematical model, as the nodes of the transport network, we present customers (for which the goods are delivered), while the edges reflect the relevant sections of the road network connecting the recipients. The loading points, from which the cargoes are delivered by bicycles, can be also represented as the network vertices.

The basic characteristics of the node are its geographical coordinates (e.g. GPS coordinates) and lists of ingoing and outgoing edges:

$$\eta = \langle x, y, \boldsymbol{\lambda}_{in}, \boldsymbol{\lambda}_{out} \rangle, \tag{3}$$

where x and y are the coordinates characterizing the node location; $\boldsymbol{\lambda}_{in}$ and $\boldsymbol{\lambda}_{out}$ are the sets of ingoing and outgoing edges for the node: $\boldsymbol{\lambda}_{in} \subset \mathbf{\Lambda}$, $\boldsymbol{\lambda}_{out} \subset \mathbf{\Lambda}$.

The necessary parameters describing the edge are its weight and end vertices:

$$\lambda = \langle w, \eta_{out}, \eta_{in} \rangle. \tag{4}$$

Where w is the edge weight (e.g.: length of the road section, the travel time, the transport costs, etc.); η_{out} and η_{in} are the beginning and the ending nodes of the edge: $\eta_{out} \in \mathbf{N}$, $\eta_{out} \in \mathbf{N}$.

It should be noted that in a more advanced version of the network model, the set of node and edge parameters could be extended depending on the problem being solved.

2.2 Model of Demand for Goods Delivery

The basic unit shaping demand is a request for the delivery of goods, understood as the customer's need for services, supported by his purchasing ability and presented on the market in order to be satisfied. A set of requests for the services of a given company forms the demand for the services of a transport company. Each request could be quantified on the basis of a set of numerical parameters.

In the general form, the demand model can be presented as an ordered set:

$$\mathbf{D} = \{\rho_1, \rho_2, \ldots, \rho_N\}, \tag{5}$$

where ρ_i is the *i*-th request in the flow: $\rho_i \prec \rho_{i+1}$ if $t_i \leq t_{i+1}$, t_i is the moment of appearance of the *i*-th request; N is the number of requests in a flow.

A single request for the cargo delivery by bikes is characterised by means of the following parameters:

$$\rho = \langle \eta_o, \eta_d, \zeta, \omega, \langle \theta_l, \theta_w, \theta_h \rangle \rangle, \tag{6}$$

where η_o and η_d are the transport network nodes defining the localisation of a sender and a recipient: $\eta_o \in \mathbf{N}$, $\eta_d \in \mathbf{N}$; ζ is the time interval between the moments of appearance of the given and previous requests [min.]; ω is the consignment weight [kg]; $\langle \theta_l, \theta_w, \theta_h \rangle$ are dimensions of the load unit – its length, width and height [cm].

For a flow of requests with a finite number of elements, the division of requests by sender and recipient can be presented in the form of a travel matrix $\mathbf{\Delta}$. An element δ_{ij} of such a matrix reflects the number of requests for which the sender is located in the node η_i, while the recipient—in the node η_j.

For a single request, its numerical parameters are deterministic characteristics, but for the requests flow, these parameters are random variables. It follows that the demand model for cargo transport can be presented as a set of random variables characterizing the numerical parameters of requests for cargo deliveries:

$$\mathbf{D} = \left\langle \mathbf{\Delta}, \tilde{\zeta}, \tilde{\omega}, \left\langle \tilde{\theta}_l, \tilde{\theta}_w, \tilde{\theta}_h \right\rangle \right\rangle, \tag{7}$$

where $\tilde{x}$ is a random variable characterising a numerical parameter x of demand.

Given that the set of ordered requests for the cargo deliveries characterise the demand, the task of demand modelling can be presented as the task of generating numerical parameters of the requests in a flow [6]. Implementation of the demand model as a flow of N elements is a set of requests in the form (5), which numerical parameters are values from samples, generated for the respective random variables, and the locations of senders and recipients are defined by the matrix $\mathbf{\Delta}$.

2.3 Model of the Servicing System

Elements of the servicing system are the fleet of transport means (cargo bicycles) and a set of loading points:

$$\mathbf{\Phi} = \langle \{b_i\}, \{p_j\} \rangle, \; b_i \in \mathbf{B}, \; p_j \in \mathbf{P}, \tag{8}$$

where b_i is the i-th cargo bicycle being the element of the vehicles fleet $\mathbf{B}$; p_j is the j-th loading point being the element of all the loading points set $\mathbf{P}$.

The main parameters of cargo bikes as elements of the servicing system are their load capacity, dimensions of the cargo space and technical speed:

$$b = \langle q_b, v_b, \langle l_b, w_b, h_b \rangle \rangle, \tag{9}$$

where q_b is the load capacity of a cargo bicycle [kg]; v_b is an average technical speed of a cargo bicycle [km/h]; l_b, w_b, and h_b are dimensions of the bicycle cargo space—its length, width, and height, respectively [cm].

Loading points within the frame of the proposed model are short-term (temporary) storage points to which cargo is transported by other means of transport (usually—by delivery vans) in order to deliver them to the final recipient. The basic characteristics of loading points are their location and capacity:

$$p = \langle \eta_p, q_p \rangle, \quad (10)$$

where η_p is the node of the transport network where the loading point is located, $\eta_p \in \mathbf{N}$; q_p is the loading point capacity (possible maximum amount of cargo that can be stored at the loading point) [t or m^3].

2.4 Defining the Routes for Delivery of Cargos by Bikes

During the simulation of the system of goods transport by freight bicycles, it is necessary to determine the traces of delivery routes. Assuming that transport operators undertake activities related to shaping rational (or optimal) transport routes, in this simulation model we propose to define the routing procedure.

As input data in the routing methods, the matrix of the shortest distances between the vertices of the transport network is used. In order to estimate such a matrix, any of the known methods of searching for the shortest paths (e.g. the Dijkstra algorithm or the Floyd-Warshall algorithm [7]) should be used. For the generated flow of requests, the task of shaping the cargo delivery routes is solved as the traveling salesman problem (TSP), where the location of the load sender (parameter η_o) is defined as the location of the loading point, and locations of the recipients (parameter η_d)—as the location of the respective nodes of the transport network. Any known heuristic methods can be used for solving the TSP (e.g.: Clarke-Wright method, simulated annealing method, methods based on genetic algorithms ant colony optimisation methods, etc. [8]). The result of the routing procedure is a set of delivery routes, being the ordered sets of vertices of the transport network. The first and last element in the set of nods shaping the route are the vertices defining the locations of the loading points.

The basic characteristics of goods delivery routes, calculated on the basis of known parameters of the mathematical model, are the route length, the total weight of transported cargo, and the performed transport work.

3 Software Implementation of the Cargo Delivery System

In order to implement simulation models of the systems of goods transport by cargo bikes, a library of basic classes was created. This library has been developed in the Python programming language, and it is available in an open source in the repository https://github.com/naumovvs/cargo-bikes-system.git.

The developed library contains the following basic classes, on the grounds of which the elements of the cargo bicycle transport system are created:

- *Net* class is used to implement the proposed mathematical model of the transport network; this class contains lists of *Node* and *Link* elements defining the network configuration and the list of *Consignment* objects implementing the model of demand for goods delivery; the servicing system is implemented within a given class as the lists of *LoadPoint* and *CargoBike* elements;
- *Node* and *Link* classes allow implementing the network nodes and edges;

- *Consignment* class is used to model requests for the delivery of goods; the sender and the consignee locations are defined within the given class as references to the appropriate *Node* objects of the transport network model;
- *Route* class allows implementing the program model of the delivery route; due to the fact that the delivery route model could be defined only within a specific transport network, this class contains an object of the *Net* type being a reference to the network program model; the shape of the route is defined as a list (ordered collection) of elements of the *Consignment* type; numerical characteristics of the route (transport work, route length, consignment weight) are calculated by methods implemented as properties of this class;
- *CargoBike* class is a program model of a freight bicycle as a transport mean used in the process of handling requests for the goods delivery;
- *LoadPoint* class is used to develop program models of loading points; the point location is defined by the reference to the object of the *Node* type, which represents the respective vertex of the transport network.

The *Net* class contains a method for generating demand as a flow of requests for the given random variables defining parameters of requests, as well as a method for calculating the shortest distance matrix based on the Floyd-Warshall algorithm and a method for forming delivery routes implementing the Clarke-Wright algorithm.

4 Case Study: Simulation Experiment for Choosing the Loading Point Location in the Krakow Old Town

The Old Town district of Krakow (Poland) is the area with limited traffic: only 3 streets of this district are not closed for automobile transport. In the same time, Krakow Old Town is the main touristic attraction of the region, where hundreds of restaurants, shops, and other touristic objects are located (localisations of the 730 commercial objects is shown in Fig. 1). As far as the district is closed for heavy vehicles, the supply of the commercial objects in the district is allowed in the morning from 7 a.m. to 10 a.m. under condition that only light good vehicles with the carrying capacity not exceeding 1.5 tons deliver the goods. Such restrictions lead to the increase of logistics expenses due to additional storage costs and the lack of the ability to implement on-time deliveries. The use of cargo bicycles is a solution that allows reducing logistics costs in this situation, while preserving the tourist appeal of the district and low level of environmental pollution.

The location of a loading point for the bikes delivery system in the Old Town of Krakow should be chosen from the set of alternative variants, which must satisfy the conditions of automobile and bicycle transport accessibility and take into account the architectural features of historic buildings in the district. The set of such alternative locations of a loading point is presented in Table 1.

Using the developed software, the transport network model, which considers roads available for bicycles, was developed for the Krakow Old Town district. Initial data for the network model were obtained with the use of Google Maps API. As the nodes of the graph representing the network model, commercial objects located in the Krakow

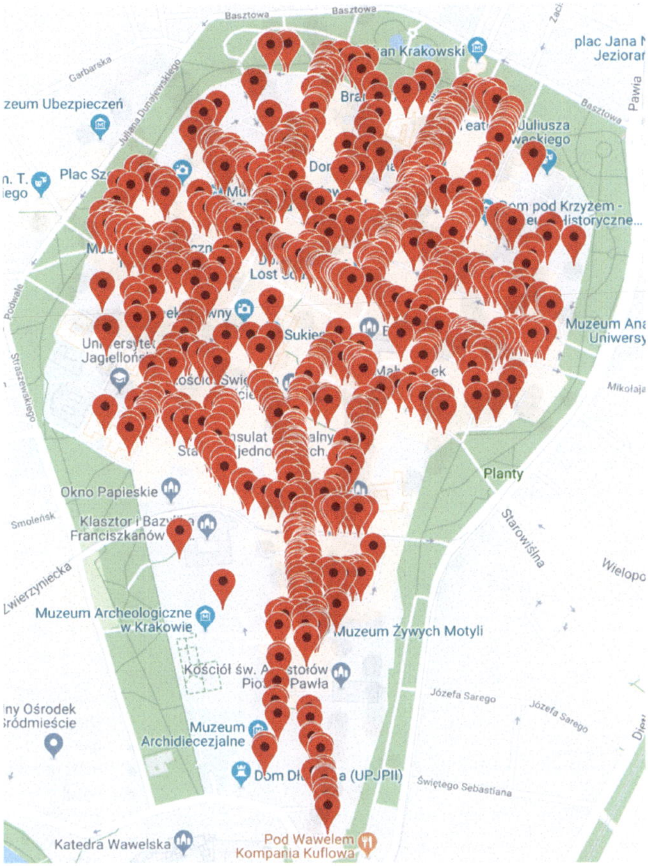

Fig. 1. Localisations of clients in the Krakow Old Town.

Table 1. Characteristics of the alternative locations of a loading point.

Variant of location	Latitude	Longitude	Address of location
A	50.064231	19.935140	Szczepański sq.
B	50.063645	19.942277	Holy Spirit sq.
C	50.060262	19.941684	Sienna str.
D	50.061151	19.932600	Gołębia str.
E	50.055391	19.937715	Napoleon's Great Army sq.
F	50.058972	19.936962	All Saints sq.

Old Town (including hotels, shops, restaurants, cafés, and bookstores) and the set of alternative locations of a loading point were considered. The graph of the obtained network model is presented in Fig. 2.

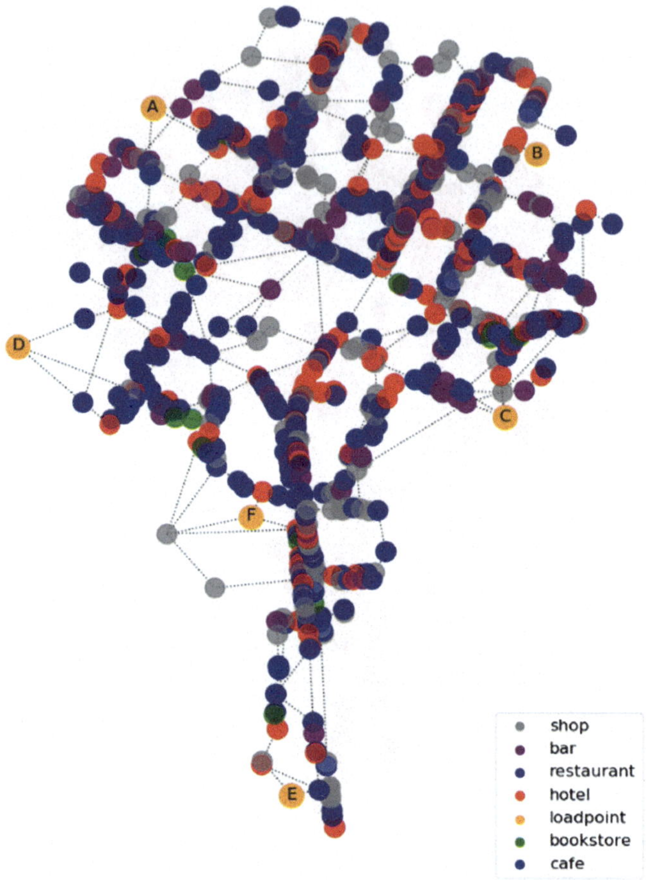

Fig. 2. The transport network model of the Krakow Old Town.

In order to choose the best localisation of the loading point, Monte-Carlo simulations were performed for the developed model. For each variant of location, 300 runs of the following procedure were completed: the requests for goods delivery were generated for the set of the clients, the delivery routes were formed for the generated requests, and the total transport work was calculated for the obtained delivery routes. While the demand simulation, the probability of the request appearance was taken equal to 0.1 for each client and the random variable of the consignments weight was accepted as the normally distributed variable with the average value equal to 30 kg and standard deviation equal to 5 kg. As the result, the samples characterising the random variable of the total transport work were obtained for the considered localisations of a loading point (distributions of the total transport work are shown in Fig. 3).

More detailed characteristics of the random variables representing the total transport work for the considered locations are given in Table 2. To confirm that the obtained samples are big enough to make statistically significant conclusions, the sufficient number of observations was calculated for the level of significance equal to

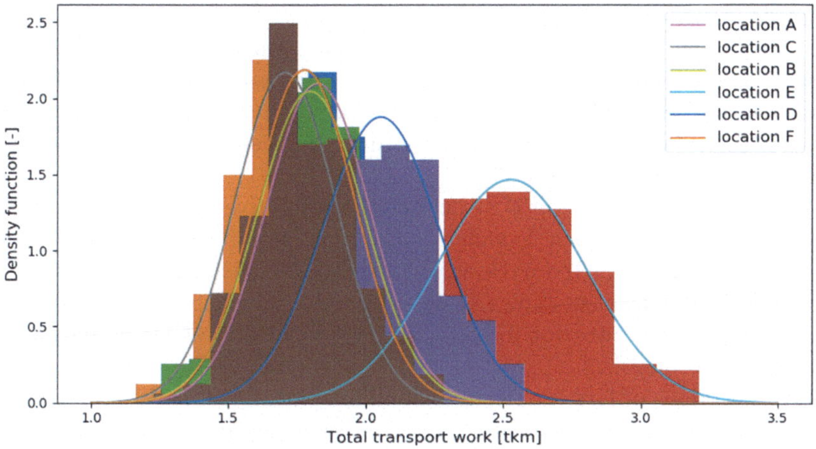

Fig. 3. Distribution of the total transport work for the considered loading point localisations.

0.05 assuming that the random variables representing the total transport work have a normal distribution. As it follows from the data presented in Table 2 (the number of performed observations is bigger than its sufficient number), the obtained results could be considered as statistically significant with the confidence probability equal to 0.95.

Table 2. Results of computer simulations for the considered loading point localisations.

Variant of location	Mean value of total transport work [tkm]	Variance of total transport work [tkm^2]	Sufficient number of observations [obs.]
A	1.827	0.0362	12
B	1.801	0.0380	13
C	1.708	0.0337	12
D	2.056	0.0451	11
E	2.526	0.0741	12
F	1.779	0.0332	11

On the grounds of the obtained results of simulations, the localisation C for the loading point should be chosen, as far as it's characterised by the smallest mean value of the total transport work.

5 Conclusions

The proposed approach to the modelling of goods deliveries by cargo bicycles allows considering the stochastic nature of the demand for freight transport and the used technology of the delivery process. The numerical parameters described in the mathematical model are the main characteristics, but the proposed model can be extended to be adapted to other issues. Unlike the existing approaches, the proposed method allows taking into account the use of rational delivery routes, which increases the adequacy of the results and the validity of the choice for the loading point location. Since the main tool in choosing the loading point location is the delivery process simulation, the proposed approach can be expanded by including other technological procedures in the program model, in addition to routing.

With the use of the developed model and the respective software, the problem of choosing the localisation for a loading point in the cargo bikes delivery system was solved for the Krakow Old Town. However, it should be noted that the obtained results are preliminary; the detailed studies of demand and its simulation in the frame of the proposed model are needed for the real-life implementation of the system of goods delivery by cargo bicycles.

As the directions of further research, the development of the advanced procedures of demand simulations and the implementation of algorithms of such technological operations as vehicles scheduling and stacking of load units in a bicycle cart should be mentioned.

References

1. Iwan, S.: Wdrażanie dobrych praktyk w obszarze transportu dostawczego w miastach. Wydawnictwo Naukowe Akademii Morskiej w Szczecinie, Szczecin (2013)
2. Allen, J., Thorne, G., Browne, M.: BESTUFS Good Practice Guide on Urban Freight Transport (2007)
3. Gruber, J., Kihm, A., Lenz, B.: A new vehicle for urban freight? An ex-ante evaluation of electric cargo bikes in courier services. Res. Trans. Bus. Manage. **11**, 53–62 (2014)
4. Chiffi, C.: Cyclelogistics micro hubs and pick-up points. In: Cycle Logistic Conference, San Sebastian (2015)
5. Iwan, S.: Zarządzanie dostawami ostatniego kilometra realizowanymi z wykorzystaniem rowerów towarowych. Prace Naukowe Uniwersytetu Ekonomicznego we Wrocławiu, vol. 383, pp. 867–880 (2015)
6. Naumov, V., Kholeva, O.: Studying demand for freight forwarding services in Ukraine on the base of logistics portals data. Procedia Eng. **187**, 317–323 (2017)
7. Simchi-Levi, D., Chen, X., Bramel, J.: The Logic of Logistics: Theory, Algorithms, and Applications for Logistics Management. Springer, New York (2014)
8. Labadie, N., Prins, C., Prodhon, C.: Metaheuristics for Vehicle Routing Problems. Wiley, London (2016)

The Role of Reverse Logistics in the Transition to a Circular Economy

Irina Makarova[1], Ksenia Shubenkova[1(✉)], Anton Pashkevich[2], and Vladimir Shepelev[3]

[1] Kazan Federal University, Syuyumbike pr-t, 10a, 423822 Naberezhnye Chelny, Russia
kamivm@mail.ru, ksenia.shubenkova@gmail.com
[2] Tallinn University of Technology, Ehitajate tee 5, 419086 Tallinn, Estonia
[3] South Ural State University, Lenin pr-t 76, 454080 Chelyabinsk, Russia

Abstract. In the transition to a circular economy, there is a need of new logistic approach, reverse logistics. Organization of spare parts delivery in the PC "KAMAZ" in transition to circular economy is considered. We have developed the concept of the Decision Support System to predict, plan and organize new spare parts delivery and faulty spare parts return delivery. The algorithm of spare parts delivery including reverse deliveries for remanufacturing is proposed. Decision Support System based on this algorithm consists of Data Storage module, OLAP-cube, spare parts failure statistics analysis module and simulation modules. Mathematical model to plan spare parts delivery is presented.

Keywords: Circular economy · Reverse logistics · System of branded automotive service · Spare parts logistics · Decision Support System

1 Introduction

In today's world, the basic model for the production of goods is linear: natural resources are extracted and turned into products that are used for a limited time. While some of these products are recycled, most eventually end up in the world's landfills. This model is becoming increasingly unsustainable and expensive for businesses. Demand for goods and services coupled with increasing resource scarcity and price volatility is causing companies to move from the linear model to a circular economy (CE) [1], which has a restorative and closed character [2] and is based on minimizing the consumption of primary raw materials and reducing waste burial (Fig. 1). Beyond the cost, there is increasing awareness of the impact of the linear model on the well-being of the planet, and a growing sense of urgency as we see the ultimate ramifications on society and the environment. CE is currently a popular concept promoted by the EU, by several national governments, including China, Japan, UK, France, Canada, the Netherlands, Sweden and Finland, and by many businesses around the world. However, the scientific and research content of the CE concept is superficial and unorganized. So, European Commission adopted an ambitious CE Package, which includes

I. Kabashkin et al. (Eds.): RelStat 2018, LNNS 68, pp. 363–373, 2019.
https://doi.org/10.1007/978-3-030-12450-2_35

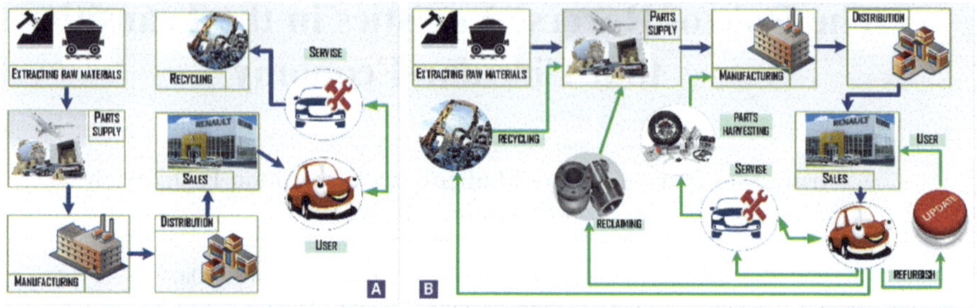

Fig. 1. **a** Linear economy model; **b** Circular economy (Source: own figure).

measures that will help stimulate Europe's transition towards a CE, boost global competitiveness, foster sustainable economic growth and generate new jobs.

The Circular Economy Package consists of an EU Action Plan for the CE that establishes a concrete and ambitious programme of action, with measures covering the whole cycle: from production and consumption to waste management and the market for secondary raw materials and a revised legislative proposal on waste. The EU is confident that the proposed actions will contribute to "closing the loop" of product lifecycles through greater recycling and re-use, and bring benefits for both the environment and the economy. According to a report by the World Economic Forum, a shift away from a Linear Economy to a Circular Economy by 2025 could generate an estimated $1 trillion annually in economic value globally, create more than 100,000 new jobs, and prevent 100 million tons of waste within the next five years.

The basis of the CE is formed by closed-loop supply chains, for which it is necessary to apply fundamentally new logistics approaches. One of such approaches is a reverse logistics (RL) and implies the movement of material flow from the consumption point to the original place of production [3].

2 Logistics in a Circular Economy

2.1 Circular Economy's Problems and Examples of Solutions

It is easy see how reverse logistics fits in a Circular Economy. Bringing products back for their next use is a central focus of the reverse logistics profession. But, the reverse logistics industry will need to adapt a completely new mindset to move to a CE.

Businesses also need to examine their own supply chain and operations to identify areas where they can make improvement. Each product in a company's supply chain must be analyzed based on its unique characteristics from raw material to end-of-life. Circular planning must then be used to determine whether reclaimed products and resources should be transported back to a central hub facility or dealt with on a local level. Increasingly, more companies are looking for help in implementing viable, sustainable solutions across their operations. There are a number of barriers to widespread adoption; from the geographic dispersion of supply chains; to the complexity of materials and deconstructing products. Digital and technology innovations are providing companies with the opportunity to overcome such barriers [2]. Machine-to-machine and data analytics enable companies to match the supply and demand for underused assets and products. "The cloud", in combination with mobile and social media, can dematerialize products or even entire industries. And 3D printing creates opportunities for manufacturing inputs that are biodegradable [4].

For a long time, reverse logistics has been seen only as logistics going in the "wrong direction". But that should no longer be the case. Today return flows are becoming the norm rather than the exception. In the last 15 years, more than 260 Extended Producer Responsibility (EPR) policies, which drive return flows, have been adopted worldwide. In France, one out of every five tons of material flowing through the economy is waste (and therefore, return flow), and the importance of reverse logistics continues to grow as the transition towards a circular economy accelerates.

However, the cost of reverse flows is usually high, while comparably, the residual value of goods is usually low. Collection of goods is often expensive due to geographic dispersion. Yet the cost challenge, like most other hurdles, can be overcome.

To reach sufficient scale and build effective and efficient reverse logistics, companies need to consolidate their return flows by collaborating along the incumbent value chain, and adjacent or cascaded activities. Return flows are usually easier to consolidate across companies than forward flows because they are not subject to the same timing and confidentiality constraints. Some service providers offer to aggregate return flows within industry sectors. In the textile industry, CoremanNet, a subsidiary of Bosch Group, has set up a dedicated logistics network and associated information system to manage the return flows in the automotive remanufacturing industry.

Because users of end-of-life or end-of-use products trigger reverse flows, they need to be included in the reverse logistics. To get their products back, companies can incentivize their users to return them. For example, Caterpillar links used engine cores to a deposit and a discount system to maximize the capture of used components into their remanufacturing operations.

2.2 Automotive Spare Parts' Logistic Systems

Modern trends in the automotive industry show that to ensure the competitiveness of the business it is necessary to have a developed logistic system. In case of automotive spare parts delivery, the decision on spare parts repair or remanufacturing have to be based on the failures statistics. If the spare part cannot be repaired in the branded service autocenter (BSA), it have to be delivered to the manufacturer for disassembly, replacement of components where necessary and assembly of a product to bring it back into as-good-as-new condition. In this case, one of the most urgent tasks is to build spare parts' reverse supply chains with account of a lot of factors influencing on delivery efficiency, and, accordingly, on the whole manufacturing operation. In most studies [5–7], the efficiency of supply chains is assessed mainly in terms of minimizing transportation costs and delivery time, and the Vehicle Routing Problem takes an important place [8] in a lot of researches, because of a wide range of factors that need to be considered when routing and scheduling: speed [9], the weighted load [10], traffic load on the route [11], and negative impact on the environment [12], etc.

When planning spare parts delivery, reasonable transportation mode selection is the priority task. Authors of the research [13] have considered such alternatives as rail, truck and air mode of transport, but this study does not consider the possible options for intermodal transportation. At the same time, globalization lengthens supply chains so that companies tend to expand their production patterns offshore or source from more distant locations [14], the use of intermodal transport can provide cost effective solutions for long distance transport needs of the supply chains for the movement of raw materials, semi-finished or finished products, as well as for reverse logistics. Intermodal transportation can ensure the benefits of each mode of transport used in the supply chain. However, in this case companies face such problems as the lack of coordination at the various intermodal transfer points, causing delays. One of the most promising areas for solving this problem is the development of intelligent systems, that has became possible in the era of economics digitalization.

3 Organization of Spare Parts Delivery in the PC "KAMAZ"

3.1 Logistic System of the PC "KAMAZ"

PC "KAMAZ" is a major manufacturer of trucks with a high level of production localization and is the basis of transport security in Russia. The company dominates the Russian truck market and plays a significant role in export markets. PC "KAMAZ" has a developed system of corporate service, which goals are to create the effective sales and distribution network of the genuine spare parts in the regions, to minimize the deficit of spare parts and provide the fastest delivery to the clients. The spare parts are shipped directly from the PC "KAMAZ" plant stocks in different quantities and different transportation modes depending on the type and location of dealer's warehouse. On the map (Fig. 2a), the plant-manufacturer PC "KAMAZ", as well as some points of its spare parts' sales and automobiles' service are marked.

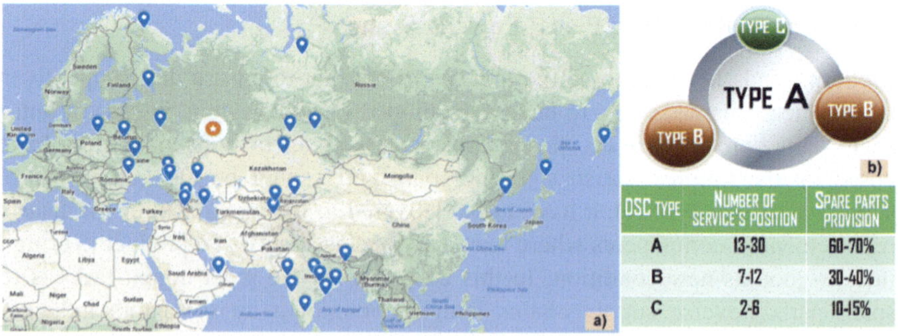

DSC TYPE	NUMBER OF SERVICE'S POSITION	SPARE PARTS PROVISION
A	13-30	60-70%
B	7-12	30-40%
C	2-6	10-15%

Fig. 2. **a** Location of some authorized dealers of PC "KAMAZ"; **b** Types of DSCs.

When creating a Dealer and Service Center (DSC) for large trucks in the regions, the following scheme is often used: a large DSC of A type is in the center of the "bush" and there are DSC s of type B or C (depending on the location) on the periphery of the "bush" (Fig. 2b). The small DSCs (B and C types) usually have a small storage space, so the spare parts are usually stored in A-types DSCs and delivered only when there is a need. In A-type DSCs, there are usually 2 parts of warehouse: for storing spare parts for sale (it is focused on seasonal needs) and spare parts for servicing (it is formed depending on the type-age structure of the vehicle fleet). Since transportation of automotive spare parts is a complex process, that is influenced by a lot of different factors, decisions are often made under conditions of incomplete information. To identify all significant factors, the complete, relevant and adequate information, as well as the application of tools and methods of its processing and analysis are needed. Multicriteria analysis methods, OLAP-technologies, simulation, elements of situational management have to be used to make the final managerial decision.

Since any error in supply chain management can lead to financial, time and other losses, methods of risk analysis and management have to be used. Managerial decisions in complex systems, must be comprehensive, consistent and scientifically based. Therefore, it is proposed to create a Decision Support System (DSS) consisting of modules, each of which will perform its function using all of the above methods.

The first module is a specially created data storage (DS), that is formed for integrating, actualizing and coordinating of the operative data from heterogeneous sources in order to arrive at a single, non-controversial view of the object as a whole. DS contains the information collected from several operative databases of an on-line transaction processing systems (OLTP). The intelligent heart of the developed DSS is an OLAP cube, which collects, stores and formalizes the parameters of Dealer and Service Network (DSN) (Fig. 3). To implement the OLAP cube, the hybrid option has been chosen. Hybrid OLAP (HOLAP) is a combination of ROLAP's (Relational OLAP) high scalability and MOLAP's (Multidimensional OLAP) fast computation. HOLAP servers allow storing the large data volumes of detailed information. Reporting

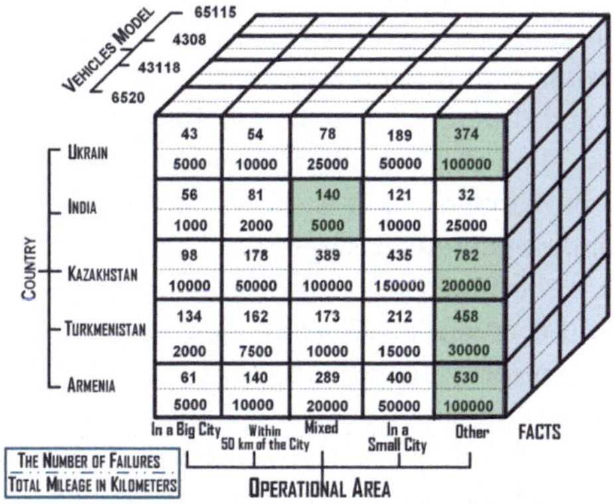

Fig. 3. One of OLAP-cube's variants (Source: own figure).

Services (SSRS) allow creating reports for a large number of data sources.

This service is completely integrated with the SQL Server tools and components [15], that helped to create the software, which allows outputting data on all possible measurements of an OLAP-cube for the intelligent analysis.

3.2 Algorithms and Models on Which DSS Is Based

In transition to the CE, there are three types of the faulty spare parts that have to be returned for remanufacturing:

- For repair. They are sent to the warehouse after recovery and in the future can be used for vehicles' maintenance.
- For remanufacturing. They are stored separately, and if it is a DSC of B or C type, these items are sent to the larger warehouse of A-type DSC for storage. When there is enough items to form the delivery lot, they are sent to the manufacturer by the return trips of vehicles, thus excluding empty runs.
- For recycling—can be stored and transported in an extruded form to save space.

The most important in the organization of customer service is the warranty period. Therefore, it is important that during this period the customer does not have problems with the vehicle. If buyer of the vehicle detects failure during the warranty period, he addresses to the nearest BSA. The employee prepares a reclamation report, according to which the repair is carried out. DSS records the flow of service requests, the analysis of which allows building the laws of time distribution between receipts of requests, as well as the time of the vehicles' stay at the repair station. These distribution laws are compared with similar laws of the previous period, which allows determining the state of the system and adjusting managerial process. If the time between requests has increased, this may indicate a drop in services demand. The increase in the time of the vehicles' stay at the repair station can be caused by the lack of spare parts, as well as inefficient organization of the technological process. Spare parts management is carried out by analyzing data from reclamation reports, according to which the most appropriate law for failures distribution is being built and an optimal supply plan is calculated. To improve the planning efficiency of the spare parts delivery structure and time, it should be taken into account that different aggregates, units and systems of a truck have different useful lives and different levels of reliability which, in turn, depend on numerous stochastic factors. The vehicle's failure occurs at the time point Tfail that can be predicted with a certain probability. As it is shown in [16], the vehicle's failure rate $\lambda(t)$ can be divided into three operational stages, and the mechanisms of spare parts supply on each stage have to be functionally different.

Since the warranty period is crucial in securing the customer loyalty, a high-quality service during this period is a priority. To ensure uninterrupted service at this stage, we have developed the mechanism for calculating the qualitative and quantitative structure of the warranty spare parts kit (WSPK) to be forwarded together with the next vehicles' kit. The WSPKs are needed for timely replacement of spare parts, if failure happens during warranty period; their contents are specific for each region and formed using special techniques allowing DSCs to eliminate the greatest number of failures in the sold group of trucks likely to occur during the warranty period. To predict possible customer requests during the after-warranty period, we have developed method based on the study of the probability of failure of different components depending on the mileage and age of the vehicle. This method helps to form the spare parts kits' structure, to calculate the date of the each part's replacement, and to create the delivery schedule. Algorithm, on which the created DSS is based, is in the Fig. 4.

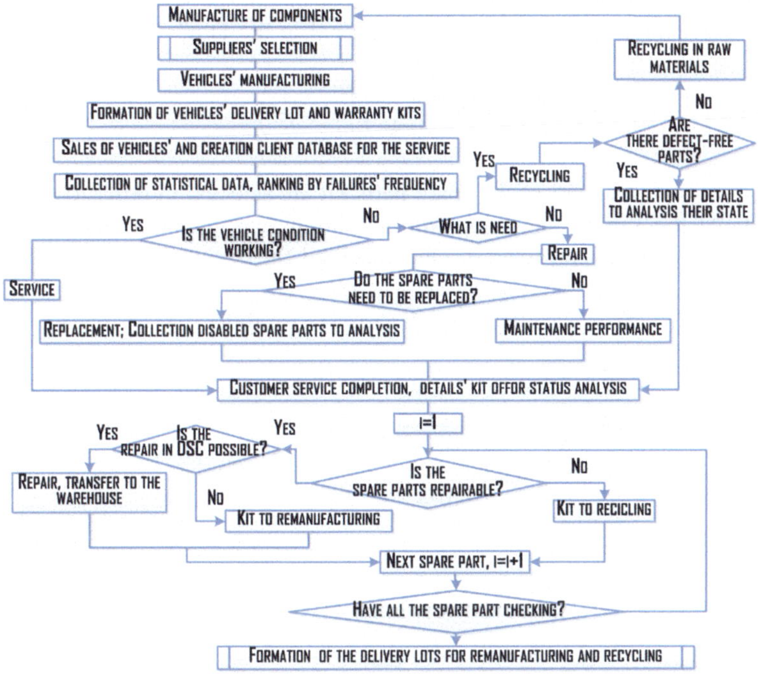

Fig. 4. Algorithm of spare parts delivery including reverse deliveries for remanufacturing (Source: own figure).

The problem's mathematical formulation consists in balancing between the costs to urgent delivery of spare parts required but not available in the warehouse and the cost of storing surpluses in DSN warehouses [17]. We have supposed that the spare parts kit has this volume: $q(i, j)$ ($i = 1, \ldots, N$), ($j = 1, \ldots, M$). It must be distributed from the logistic management center in M foreign regions, where N is the total number of nomenclature units in spare parts kits. Deliveries are carried out by sets of the i-th name of a spare part, which can be considered the aggregated nomenclature.

The reserves' level at the logistic warehouses is reduced in accordance with λ_{ij} which is the failure rate of the i-th details, aggregate or system in the j-th operational region. Delivery takes place either on the deliveries date τ, which depends on the distance between the j-th operational region to logistic management center, or by reducing the inventory level k_{ij} in DSN reserve to the critical values S_{ij} of the i-th component. In this case, the kit's size is equal to q_i, and the costs should aim to minimize the g_{ij}. At random demand, the total number of deliveries B for the period (year) is estimated. Storage costs and penalties are calculated on the expected reserve (deficit) by the period end. The penalty is determined by the probability of deficit. DSC pays the cost h_{ij} of storage in the warehouse of spare parts of the i-th name for the whole period and delivery g_{ij} size of the spare parts kit. If the reserve does not allow satisfying all requests, it leads to the customer's lost profit, and the DSC pays a penalty for the deficit. The fine includes excess of the cost of unexpected delivery over the usual and the percentage of "frozen" working capital in the DSC due to deficit.

$$d = \sum\nolimits_{i=1}^{N} \sum\nolimits_{j=1}^{M} \left(\frac{\lambda_{ij} B^2}{2 g_{ij}} - \frac{1}{h_{ij}} \right). \tag{1}$$

Time, volume and structure of the spare parts kit are selected in order to minimize the total cost. At the same time, maximum size of delivery should not exceed the capacity of the warehouse, and its cost—the specified total cost:

$$Z = (\tau - t_D) \cdot Z_1 + B \cdot Z_2 + Z_3 \rightarrow \min, \tag{2}$$

where

$$\begin{gathered} Z_1 = \sum\nolimits_{i=1}^{N} \sum\nolimits_{j=1}^{M} \lambda_{ij} h_{ij} k_{ij}; \quad Z_2 = \sum\nolimits_{i=1}^{N} \sum\nolimits_{j=1}^{M} g_{ij} q_{ij}; \\ Z_3 = \sum\nolimits_{i=1}^{N} \sum\nolimits_{j=1}^{M} d_{ij} (S_{ij} + q_{ij}) p_{ij}. \end{gathered} \tag{3}$$

Limitations are:

$$\sum\nolimits_{i=1}^{N} \sum\nolimits_{j=1}^{M} d_{ij} \sum_{i=S}^{\infty} \cdot p_i - h_{ij} \geq 0; \; \sum\nolimits_{i=1}^{N} h_{ij} - d_{ij} \sum\nolimits_{i=S+1}^{\infty} p_i \geq 0. \tag{4}$$

Delivery time is:

$$\tau_j = \sum\nolimits_{i=1}^{N} \left(\left(\frac{q_{ij} + S_{ij}}{\lambda_{ij}} - t_D \right) + g_{ij} \cdot B \right). \tag{5}$$

The efficiency indicator can be the average downtime of spare parts' waiting time:

$$v_j = \sum\nolimits_{i=1}^{M} \frac{\rho_{ij}}{\Lambda_j}, \tag{6}$$

where ρ_{ij} is the average number of query to replace a faulty of the i-th part in BSA of the j-th region; the total flow of serviced requests:

$$\Lambda_j = \sum\nolimits_{i=1}^{N} \lambda_{ij}. \tag{7}$$

Since the cost of the warranty shall not exceed the amount of the reserve, the objective function of forming WSPK is to minimize the total cost of shipping and storage spare parts, penalties amounts due to the lack of spare part required (lost profits of the customer), and take into account its price and the labor input into the repairment of the aggregate. In view of limitation on the integral cost of spare parts, it is hardly possible to create large initial stocks of modules that would guarantee the availability of any spare part should it be needed. On the other hand, the lack of spare part in the WSPK may protract the service time. The balance can be found by the use of specialized software that would effectively organize the logistic processes in the branded service system. The selection algorithm presupposes the availability of several options of cargo delivery along each of the routes. Each of the routes is characterized by a great number of factors that can affect both the quality and time of delivery. Besides, intermodal deliveries may take longer time due to uncoordinated operations of different modes of transport and transfer units. If the database contains ready-made optional solutions, it is possible to make selection from the available variants; otherwise, solutions are searched by the use of simulation. The model limitations are:

$$\sum\nolimits_{i=1}^{N} m_i \cdot n_i \leq U_i; \quad \sum\nolimits_{j=1}^{M} v_j \cdot n_j \leq V_j. \tag{8}$$

The most problematic part of any logistic chain is the road transport due to a large number of restrictions:

- Type of the vehicle's body must suit the type of the cargo.
- When developing a route, it is necessary to take into account, which sections of the road network are allowed to transport goods with such characteristics.
- The cargo must be delivered on time, as breaking the agreed deadlines disrupt obligations to customers DSC.

The objective function is to minimize the total number of vehicles involved in the transportation process and the total distance of transportation:

$$F = \sum\nolimits_{k \in V_M} \sum\nolimits_{(ij) \in U} C_{ij}^{k} \times X_{ij}^{k} \to \min, \tag{9}$$

where C_{ij}^{k}—the cost of transporting goods along the route ij by vehicle k; M—type of vehicle; V_M—the number of vehicles of the same type (in terms of carrying capacity and body type); X_{ij}^{k} can be equal either 1, or 0:

$$X_{ij}^{k} = \begin{cases} 1 & \text{if trip from point } i \text{ to } j \\ 0 & \text{if trip from point } j \text{ to } i \end{cases}, \tag{10}$$

$$C_{ij}^{k} = C_{Mstand} \times (q_{carr} \cdot \gamma_{carr})_M \times n_{iM} \times V_M, \tag{11}$$

where C_{Mstand}—cost (ton-kilometer) of transportation for a vehicle of type M; q_{carr}—load-carrying capacity; γ_{carr}—load factor.

Number of trips per vehicle:

$$n_{iM}^{k} = \left[\frac{T_{iM}}{t_{ij} + t_{ji} + t_{load} + t_{unl}}\right]^{k}, \tag{12}$$

where T_{iM}—daily working time; t_{ij}—direct trip; t_{ji}—return trip; t_{load}—loading time; t_{unload}—unloading time.

4 Conclusion

The practice of doing business shows that the linear model of the economy is inefficient. A positive impact on economic and environmental efficiency can be provided by a paradigm shift in economic development – the transition to a circular economy. In this case, manufacturers must change the logistics system. Reverse logistics will play an important role in such a system. Planning of supply chains should be based on complete information about the possibility of processing the product and restoring it. The article gives an example from the automotive industry. It is shown that it is possible to implement the transition to a circular economy with the help of planning logistic processes in the branded service system of the automotive corporation.

References

1. Pakhomova, N.V., Richter, K.K., Vetrova, M.A.: Transition to circular economy and closed-loop supply chains as driver of sustainable development. St. Petersburg Univ. J. Econ. Stud. **33**(2), 244–268 (2017)
2. Towards the circular economy. https://www.ellenmacarthurfoundation.org/assets/downloads/publications/Ellen-MacArthurFoundation-Towards-the-Circular-Economy-vol.1.pdf
3. Niknejad, A., Petrovic, D.: Optimisation of integrated reverse logistics networks with different product recovery routes. Eur. J. Oper. Res. **238**(1), 143–154 (2014)
4. Lacy, P., Rutqvist, J.: Waste to Wealth. Palgrave Macmillan, London (2015)
5. Cardós, M., García-Sabater, J.P.: Designing a consumer products retail chain inventory replenishment policy with the consideration of transportation costs. Int. J. Prod. Econ. **104** (2), 525–535 (2006)
6. Jha, J.K., Shanker, K.: An integrated inventory problem with transportation in a divergent supply chain under service level constraint. J. Manuf. Syst. **33**(4), 462–475 (2014)
7. Lee, Y.C.E., Chan, C.K., Langevin, A., Lee, H.W.J.: Integrated inventory-transportation model by synchronizing delivery and production cycles. Transp. Res. Part E **91**, 68–89 (2016)
8. Savelyeva, M., Belyakov, S.: Solution of the problem supply chain management in temporal dependence. In: Advances in Intelligent Systems and Computing, vol. 323, pp. 489–496 (2015)
9. Jabali, O., Woensel, T., Kok, A.: Analysis of travel times and CO_2 emissions in time-dependent vehicle routing. Prod. Oper. Manag. **21**(6), 1060–1074 (2012)
10. Bektas, T., Laporte, G.: The pollution-routing problem. Transp. Res. B **45**, 1232–1250 (2011)

11. Makarova, I., Shubenkova, K., Pashkevich, A.: Logistical costs minimization for delivery of shot lots by using logistical information systems. Proc. Eng. **178**, 330–339 (2017)
12. Ćirovića, G., Pamučar, D., Božanić, D.: Green logistic vehicle routing problem: routing light delivery vehicles in urban areas using a neuro-fuzzy model. Expert Syst. Appl. **41**(9), 4245–4258 (2014)
13. Lin, B., Wang, J., Wang, H., Wang, Z., Li, J., Lin, R.: Inventory-transportation integrated optimization for maintenance spare parts of high-speed trains. PloS One **12**(5), e0176961 (2017)
14. Deveci, D.A., Şakar, G.D., Işiktaş, R., Tuna, O.: Intermodal transport-based supply chain developments in Turkish automotive industry: the case of Ford Otosan. In: Piotrowicz, W., Cuthbertson, R. (eds.) Supply Chain Design and Management for Emerging Markets, pp. 225–256 (2015)
15. Reporting Services Reports (SSRS). https://msdn.microsoft.com/en-us/library/bb522712.aspx
16. Makarova, I., Khabibullin, R., Belyaev, E., Mukhametdinov, E., Zhdanov, D.: Planning of the spare parts supply in the system of corporate service automotive company abroad. In: 4th IEEE ICALT 2015, pp. 36–40 (2015)
17. Makarova, I., Khabibullin, R., Belyaev, E., Gabsalikhova, L., Mukhametdinov, E.: Improving the logistical processes in corporate service system. Transp. Probl. **11**(1), 5–18 (2016)

Increasing the Adequacy of Management Decision Making for Choosing Intermediaries in Supply Chains

Valery Lukinskiy(✉), Vladislav Lukinskiy, and Darya Bazhina

National Research University Higher School of Economics (HSE), St. Petersburg, Russia
{lukinskiy, vladas27}@mail.ru, cheefa@inbox.ru

Abstract. To date, a diverse array of expert assessments for quantitative (tangible) and qualitative (intangible) objects based on the analytic hierarchy process (AHP) has been accumulated in various fields of knowledge. The systematization and analysis of the collected data made it possible to put forward the hypothesis that some of the indicators (the eigenvalue of the matrix, the consistency index, and the ration of consistency) can be considered in the form of aggregates of random variables representing an intellectual product and reflecting features of human thinking. The systematization and statistical processing of the results obtained by experts on the basis of the AHP showed that the distribution functions of the values of expert estimates for the consistency indices significantly differ from the similar functions of the generated matrices.

Keywords: Analytic Hierarchy Process (AHP) · Supply chain · Logistic intermediaries

1 Introduction

To date, a diverse array of expert assessments for quantitative (tangible) and qualitative (intangible) objects based on AHP has been accumulated in various fields of knowledge [1–16]. Given the fact that each expert evaluation is obtained and processed according to the unified methodology, it is natural to assume that the results of these studies are depersonalized and presented in the form of combined output indicators: the size of the matrix ($n \times n$), the vector w_i, the eigenvalue of the matrix $\lambda_{\max}$, the consistency index $C.I.$, and the ration of consistency $C.R.$

The systematization and analysis of the collected data made it possible to put forward the hypothesis that some of the above indicators can be considered in the form of aggregates of random variables representing an intellectual product and reflecting features of human thinking. After statistical processing in the form of laws of distribution of a random variable and corresponding parameters (mean values, variances, etc.), 'intellectual' samples can be compared with artificial (generated) samples, taken as a basis for justifying the consistency of expert judgments.

If the difference between the 'intellectual' and the 'artificial' sample is random, they can be combined as one general sample. If the difference is significant, and the

I. Kabashkin et al. (Eds.): RelStat 2018, LNNS 68, pp. 374–388, 2019.
https://doi.org/10.1007/978-3-030-12450-2_36

combination will be in conflict with the statistical criteria, then a new method for evaluating or using the multi-criteria approach will be required.

Obviously, each expert evaluation reflects the individuality of the author, whereas the 'generated' matrix is strictly formalized. At first glance, this is an insurmountable obstacle, but it should be noted that the 'artificial' matrices (sample size: 100 and 500 values) were not subjected to the convergent procedure, whereas each expert matrix undergoes 'self-review' before inclusion in the general sample, i.e. they are inherently closer to the newly formed objects of experts.

2 The Methodology of Forming a Superposition of Distributions of Consistency Indices

Analysis of indicators from various sources allowed the formation of algorithms for their statistical processing:

1. Unification of estimates, e.g. λ_{max}, *C.I.* and *C.R.* for each matrix $n \times n$ (meaning that if only the original matrix is given in the source, then all the indicators are calculated for it, otherwise only the missing ones are computed).
2. Estimation of sharply allocated (extreme) sampling values using various methods, in particular, Arleigh, Irwin, Romanovsky, and others.
3. An estimation of the belonging of a dispersion of two or more samples to one aggregate (the Fisher, Romanovsky, and Cochran test).
4. Estimation of the mean values (by the Student's *t*-test provided that the variance dispersion is random). If the calculated value of the criterion is less than the table value, that the samples can be combined into one set.
5. Determination of the distribution function of the combined sample $F_{\Sigma}(x)$, which includes several distribution functions $F_i(x)$—superposition of distributions. For the calculation of statistical characteristics $F_i(x)$ (mean values $\bar{x}$ and standard deviations σ) the following formulas can be used:

$$\bar{x} = \sum C_i \cdot \bar{x}_i, \tag{1}$$

$$\sigma^2 = \sum C_i \cdot \sigma_i^2 + \sum C_i(\bar{x}_i - \bar{x})^2, \tag{2}$$

   where C_i are the coefficients of superposition; $\bar{x}_i$—mean value of the *i*-th sample; σ_i—the standard deviation of the *i*-th sample.
6. Verification of belonging of expert and generated samples to one set (respectively, variances according to the Fisher criterion and average values by the Student's *t*-test).
7. Probabilistic evaluation of consistency ratio for distributions obtained by expert estimates and their comparison with the results of generated distributions.

Based on this algorithm, calculations were made for 3 × 3 matrices; 4 × 4; 5 × 5; 6 × 6, and 7 × 7. Let us consider the results of these calculations in more detail.

3-by-3 Matrices. The processing of various sources of the literature showed (Table 1) that matrices 3 × 3 are quite often found in the publications of T. Saaty and other specialists [17–29].

Table 1. Results of processing and systematization of 3 × 3 matrices.

Sample	Number of matrices	Experts	Objects	Indicators of *C.I.*	
				Mean	Standard deviation
1	$N_1 = 58(60)$[a]	T. Saaty	Different	0.048	0.038
2	$N_2 = 18(19)$[a]	Specialists	Different	0.053	0.035
3	$N_3 = 100$[b]	Modeling	–	0.358	0.517

[a]Sample size before exclusion of extreme values; [b]Generated sample (Saaty [30])

Analysis of random variables of consistency indices (*C.I.*) of the first sample of $N_1 = 60$ volume showed that the two maximum values ($C.I._{N-1}$ and $C.I._N$) significantly differ from the remaining random values in the range ($0 \div 0.20$), and, therefore, can be attributed to the sharply highlighted observations.

There are two possible options.

The first option is as follows. According to the developed algorithm, a sequential check is made to determine whether the extreme value $C.I._N$ belongs to the main sample, including $C.I._{N-1}$, i.e. volume of $(N - 1)$. If $C.I._N$ does not belong to the sample, then the subsequent verification of the belonging of the value $C.I._{N-1}$ to the sample of the volume $(N - 2)$ is carried out.

The second option provides the possibility of combining two samples. The first, with the volume of $N_{11} = 58$ and the second, which includes the extreme values $C.I._N$ and $C.I._{N-1}$ of $N_{12} = 2$ volume.

The second option is taken into the consideration. As a result of the statistical processing for the sample $N_{11} = 58$, the mean value $x_{11} = 0.048$, the standard deviation $\sigma_{11} = 0.038$ were determined. For the second sample ($N_{12} = 2$), the values equaled to $x_{12} = 0.325$, $\sigma_{12} = 0.21$.

For the estimation of the affiliation of the two samples to one aggregate, Fisher's criterion F can be calculated

$$F = \left(\frac{0.21}{0.038}\right)^2 \approx 30.$$

The maximum table value of the criterion F for the significance level $\alpha = 0.05$ equaled to $F_{0.05} = 9.55$. Since $F = 30 \gg F_{0.05}$, the considered samples cannot be combined into one set.

A similar test was provided for the extreme maximum value of the sample $N_2 = 19$, including expert opinions. As a result, this maximum value was eliminated, which resulted in a decrease in the sample size by a single value ($N_{21} = 18$).

The next step, concerning the test of the possibility of combining two samples $N_{11} = 58$ and $N_{21} = 18$ allowed to conclude:

- The discrepancy of the standard deviations $\sigma_{11} = 0.038$ and $\sigma_{21} = 0.035$ according to the Fisher criterion is random;
- The hypothesis of the possibility of combining the samples into one set of the Student's *t*-test is confirmed.

Further on, the superposition parameters of the distributions were defined by formulas (1) and (2), including the first and second corrected samples with the coefficients $c_1 = 58/76 = 0.764$ and $c_2 = 18/76 = 0.236$: mean value $x_{1\text{–}2} = 0.050$; standard deviation $\sigma_{1\text{–}2} = 0.037$.

Next, we compare the expert ($N_7 = 76$) and generated samples $N_3 = 100$. The performed calculations showed that the given samples do not converge on either the standard deviations ($\sigma_{1\text{–}2} = 0.037$; $\sigma_3 = 0.517$) or the mean values ($x_{1\text{–}2} = 0.050$; $x_3 = 0.358$). Consequently, these samples, representing the consistency indices cannot be combined.

Then, we estimate from the probabilistic point of view the area of critical values of the expert sample, in accordance with the recommendations of T. Saaty. It is woth to remind that the matrix is considered as consistent, provided that *C.R.* $\leq$ 0.10, but also for the matrix 3-by-3 the value *C.R.* $\leq$ 0.05 should be taken into account.

We use the simple rule of determining the distribution law of a random variable based on the coefficient of variation. For an expert sample ($N_7 = 76$), we found that $v = \sigma/x = 0.037/0.050 = 0.74$.

Let us assume that this sample follows the Weibull law with the distribution function

$$F(x) = 1 - \exp\left[-\left(\frac{b_m \cdot x}{\bar{x}}\right)^m\right], \tag{3}$$

where m, b_m are the parameters of the form and scale, respectively, while x is the argument (the consistency index, *C.I.*).

To determine the parameters m, b_m one can use the method of moments or special tables [31] that include the values m, b_m depending on the value of the coefficient of variation. Using the above tables, we found (for $v = 0.74$): $m = 1.35$, $b_m = 0.92$.

For understandability, according to formula (3), we calculate the probability of converging expert matrices provided that *C.R.* $\leq$ 0.05:

$$F(C.I. = 0.05) = 1 - \exp\left[-\left(\frac{0.92 \cdot 0.05}{0.050}\right)^{1.35}\right] = 0.592.$$

In Table 2, the results of calculations *F*(*C.I.*) are presented.

From the analysis of Table 2, it can be concluded that for a maximum value in the sample, the consistency index equals 0.20, the probability *F*(*C.I.*) = 0.996, i.e. almost all expert matrices are consistent.

If we take for the constraint the *C.R.*, recommended by T. Saaty (which is calculated by formula *C.R.* = *C.I.*/*R.I.*, where *R.I.* random (stochastic) consistency index), then we can determine the corresponding probability of consistency ratio *F*(*C.R.*). Taking into account that for the 3 × 3 matrix in different sources [16, 30, 32] three

Table 2. The distribution functions of the consistency index $F(C.I.)$ of 3 × 3 matrices.

Value		Distribution functions		
C.I.	*C.R.*	Sample (1) ($x_{11} = 0.048$; $m = 1.25$; $b_m = 0.93$)	Combined samples (1) and (2) ($x_{1\text{-}2} = 0.050$; $m = 1.35$; $b_m = 0.92$)	Generated sample (3) ($x_3 = 0.358$; $m = 0.73$; $b_m = 1.2$)
0.026[a]	0.050[a]	0.382	0.31	0.15
0.052[a]	0.100[a]	0.592	0.60	0.23
0.100	0.193	0.897	0.90	0.35
0.150	0.289	0.980	0.98	0.44
0.200	0.385	0.996	0.997	0.51

[a]Limit values

values of *R.I.* (0.52, 0.58, 0.66) are given, we accept *R.I.* = 0.52. As for the restriction of consistency, we will carry out the calculations of *C.I.* for two values of *C.R.* (0.05 and 0.10), i.e. $C.I._1 = 0.026$ and $C.I._2 = 0.052$.

The calculated values of $F_1(C.I.)$ are presented in Table 2 and make up $F_1(0.026) = 0.382$; $F_1(0.052) = 0.592$.

4-by-4 Matrices. Based on the generalization of materials from different sources, 5 samples were generated (Table 3). The first sample was compiled on the ground of the materials from two publications [30, 32]. In the second one, the materials had been taken from publications and were analyzed for designing the second sample in consideration with matrices 3-by-3. The third and fourth samples are made based on the results of surveys of master and postgraduate students studying in the specialty 'Logistics and SCM', as well as specialists of various companies. As an object of research, the costs of ordering, transportation, and holding of current and safety stocks for a simple logistical chain were considered.

Table 3. Results of processing and systematization of 4 × 4 matrices.

A number of sample	Number of experts	Experts	Objects	Indicators of *C.I.*	
				Mean	Standard deviation
1	$N_1 = 12(13)$[a]	Specialists	Different	0.100	0.060
2	$N_2 = 21$	Specialists	Different	0.090	0.070
3	$N_3 = 20(22)$[a]	Master students, specialists	Logistics costs	0.250	0.170
4	$N_4 = 7$	Postgraduate students	Logistics costs	0.208	0.051
5	$N_5 = 12$	Modeling	–	1.020	0.650
6	$N_6 = 100$[b]	Modeling	–	0.946	0.658
7	$N_7 = 500$[b]	Modeling	–	0.884	0.612

[a]Sample size before exclusion of extreme values; [b]Generated sample (Saaty [30])

The fifth sample is experimental. Some of the experts, who participated in drawing up the third sample matrices, generated the matrices according to the algorithm described by Saaty [30] on the basis of statistical hypothesis method testing.

Verification of the possibility of combining the first four samples showed that with the allowance for the Fisher and Student's *t*-test it is possible to combine the first and second samples, as well as the third and fourth samples. Thus, for the further processing, two samples were compiled: sample *A* (superposition of samples 1 and 2 with volume $N_A = 12 + 21 = 33$ and parameters $x_A = 0.093$; $\sigma_A = 0.067$; $v = 0.72$) and sample *B* (superposition of samples 3 and 4 with volume $N_B = 20 + 7 = 27$ and parameters: $x_B = 0.245$; $\sigma_B = 0.17$; $v = 0.69$).

Additionally, a combined sample *C* was formed, which is a union of empirical random variables $N_C = N_A + N_B = 60$ with parameters $x_C = 0.160$; $\sigma_C = 0.140$; $v = 0.875$).

Finally, the generated matrices (samples 6 and 7) after the Fisher and Student's *t*-test were also combined into one sample *D* as a superposition of distributions with parameters $x_D = 0.894$; $\sigma_D = 0.620$; $v = 0.694$.

Comparison of samples *A*, *B*, *C* and *D* identified the following conclusions:

- Firstly, the deviation of the variances according to the Fisher criterion for the indicated pairs *A* and *B*, as well as *C* and *D* is significant, i.e. the samples cannot be combined. A similar conclusion can be drawn from the comparison of the mean values by the Student's *t*-test;
- Secondly, the verification of the convergence of the two generated samples (fifth and *D*) showed that they converge by the Fisher and Student's *t*-test;
- Thirdly, all the above samples can be approximated in the first approximation by the Weibull distribution function.

The results of calculating the probabilities of the distribution of *C.I.*, according to the formula

$$C.I. = \frac{\lambda_{\max} - n}{n - 1} \tag{4}$$

are given in Table 4. Considering that the consistency ratio $C.R. = 0.08$ for 4×4 matrices, and the consistency random index, $R.I. = 0.9$ (Saaty [32]), we can find the limit index of the consistency $C.I. = R.I. \cdot C.R. = 0.072$.

The analysis of Table 4 showed that with the probability $F(0.072) = 0.45$, the experts' estimates (sample *A*) are subject to the dependency of $C.R. \leq 0.08$. This value is 3.5 times exceeds the $F(0.072) = 0.13$ on the basis of sample *B* and it almost 20 times exceeds the estimate of $F(0.072) = 0.023$ based on the generated sample *D*.

5-by-5 Matrices. Table 5 shows the results of the systematization and processing of 5×5 matrices in the form of four samples. All extreme values were checked for belonging to the analyzed sample. Subsequent analysis showed that, according to the Fisher and Student's *t*-test, samples 1 and 2 of the *C.I.* can be combined as one set (sample *A* with parameters $x_A = 0.084$; $\sigma_A = 0.070$).

A similar check showed that samples 3 and 4, in which the objects of expert evaluation were the prestige of managerial professions (economist, logistician,

Table 4. The distribution functions of the consistency index $F(C.I.)$ of 4 × 4 matrices.

Value		Sample distribution functions			
C.I.	*C.R.*	*A* ($x_A = 0.093$; $m = 1.4$; $b_m = 0.91$)	*B* ($x_B = 0.245$; $m = 1.5$; $b_m = 0.903$)	*C* ($x_C = 0.16$; $m = 1.15$; $b_m = 0.95$)	*D* ($x_D = 0.894$; $m = 1.45$; $b_m = 0.91$)
0.025	0.028	0.127	0.030	0.110	–
0.050	0.056	0.313	0.080	0.281	0.014
0.072[a]	0.080[a]	0.450	0.130	0.313	0.023
0.100	0.112	0.51	0.200	0.423	0.035
0.200	0.223	0.924	0.473	0.705	0.100
0.300	0.334	0.989	0.690	0.836	0.170
0.400	0.445	–	0.833	0.332	0.242
0.500	0.556	–	–	–	–

[a]Limit values

Table 5. Results of processing and systematization of 5 × 5 matrices.

Sample	Number of experts	Experts	Objects	Indicators of *C.I.*	
				Mean	Standard deviation
1	$N_1 = 14$	Specialists	Different	0.065	0.054
2	$N_2 = 16$	Master students	Geometric figures	0.116	0.083
3	$N_3 = 38$	Master students, specialists	Professions[a]	0.420	0.210
4	$N_4 = 48(51)$[b]	Master and postgraduate students	Professions[a]	0.463	0.248
5	$N_5 = 100$[c]	Modeling	–	1.127	0.528
6	$N_6 = 500$[c]	Modeling	–	1.105	0.525

[a]Economist, logistician, marketing specialist, manager and merchant; [b]Sample size before exclusion of extreme values; [c]Generated sample (Saaty [30])

marketing specialist, manager and merchant) can also be combined as a single sample *B* with parameters $x_B = 0.445$; $\sigma_B = 0.240$.

The third combined sample *C* included the results of processing the generated matrices (Saaty [30]): sampling parameters $x_C = 1.120$; $\sigma_C = 0.526$.

In accordance with the developed algorithm, the parameters of the combined samples *A*, *B* and *C* were calculated using formulas (1) and (2) for the superposition of distributions.

Additional testing showed that each of the combined samples *A*, *B* and *C* should be analyzed independently, i.e. sample *A* and *B*, and also *B* and *C* cannot be combined into a single set.

Table 6 shows the distribution functions of *C.I.* of three samples, for the approximation of which were chosen, respectively, the Weibull distribution (*A*), Rayleigh (*B*), and the truncated normal law (*C*).

Table 6. The distribution functions of the consistency index *F*(*C.I.*) of 5 × 5 matrices.

Values		Sample distribution functions[a]		
C.I.	*C.R.*	*A* ($x_A = 0.084$; $m = 1.2$; $b_m = 0.94$)	*B* ($\sigma_p = 0.363$)	*C* ($x_C = 1.120$; $\sigma = 0.526$)
0.05	0.046	0.400	0.01	–
0.10	0.091	0.680	0.04	0.023
0.11[b]	0.1[b]	0.730	0.07	0.027
0.15	0.135	0.850	0.08	0.032
0.20	0.180	0.930	0.14	0.040
0.30	0.270	0.987	0.29	0.058
0.40	0.360	0.998	0.40	0.085
0.50	0.450	0.999	0.61	0.120

[a]Sample *A*—Weibull distribution, sample *B*—Rayleigh distribution, *C*—the truncated normal law; [b]Limit values

From the analysis of Table 6, it follows that 73% of expert assessments, sample *A*, are consistent with the limitation *C.R.* ≤ 0.1, which ten times more than the same values of sample *B* (7%) and 27 times higher than for the generated sample *C*.

6-by-6 Matrices. Analysis of various sources and special experiments allowed the formation of 4 samples.

The first sample: the number of objects $N_1 = 7$, includes various sources, such as research papers by T. Saaty (mentioned on the introduction to the article) and materials of various studies that have been analyzed in order to design the second sample in consideration with matrices 3-by-3.

Second sample: ten experts $N_2 = 10$; object of research—evaluation of the prestige of the following professions: economist, financier, logistician, marketing specialist, merchant, and manager.

The third sample: twelve experts $N_3 = 12$; for the objects of the second sample, corresponding matrices were generated.

Fourth sample: the group included experts $N_4 = 47$ (the object is the prestige of the profession).

Table 7 shows the results of statistical processing for 4 groups in the form of mean values and standard deviations for *C.I.* After checking for extreme observations, we calculate the criteria for comparing variances and mean values of 6 × 6 matrices. Table 8 shows some results of calculations.

For the first sample, compiled from different sources, there is a significant discrepancy in the variances in all the compared options. We denote it as sample *A*.

Samples 2 and 4 can be combined as a superposition—sample *B*. Also in the form of a superposition of distributions, samples 5 and 6 (sample *C*) are combined.

Table 7. Results of processing and systematization of 6 × 6 matrices.

Sample	Number of experts	Experts	Objects	Indicators of *C.I.*	
				Mean	Standard deviation
1	$N_1 = 7$	Specialists	Different	0.140	0.095
2	$N_2 = 10$	Master students	Professions[a]	0.520	0.356
3	$N_3 = 12$	Modeling[b]	Professions[a]	1.157	0.439
4	$N_4 = 47$	Master and postgraduate students	Professions[a]	0.574	0.410
5	$N_5 = 100$[c]	Modeling		1.032	0.424
6	$N_6 = 500$[c]	Modeling		1.263	0.421

[a]Economist, financier, logistician, marketer, merchant, and manager; [b]Matrices generated by experts; [c]Generated matrix samples (Saaty [30])

Table 8. Results of calculation of the Fisher and Student's *t*-test of 6 × 6 matrices (fragment).

Comparable samples: sample number (size)	Fisher criterion		Student's *t*-test	
	The calculated (F_{ex}) and tabular ($F_{0.05}$) value	Results	The calculated (t_{ex}) and tabular ($t_{0.05}$; $t_{0.01}$) value	Results
2 ($N_2 = 10$) and 3 ($N_3 = 12$)	$F_{ex} = 1.981 < F_{0.05} = 2.90$	Yes	$t_{ex} = 3.49$; $t_{0.05} = 2.09$; $t_{0.01} = 2.845$	No
2 ($N_2 = 10$) and 4 ($N_4 = 47$)	$F_{ex} = 1.330 < F_{0.05} = 2.90$	Yes	$t_{ex} = 0.373$; $t_{0.05} = 1.91$; $t_{0.01} = 2.75$	Yes
1 ($N_1 = 7$) and 2 ($N_2 = 10$)	$F_{ex} = 14 > F_{0.05} = 3.37$	No	$t_{ex} = 2.60$; $t_{0.05} = 2.13$; $t_{0.01} = 2.90$	Yes/No
1 ($N_1 = 7$) and (2 + 4: $N_{2,4} = 57$)	$F_{ex} = 16.86 > F_{0.05} = 3.84$	No	$t_{ex} = 2.91$; $t_{0.05} = 3.0$; $t_{0.01} = 2.6$	Yes/No
5 ($N_5 = 100$) and (2 + 4: $N_{2,4} = 57$)	$F_{ex} = 1.184 < F_{0.05} = 1.74$	Yes	$t_{ex} = 8.7$; $t_{0.05} = 1.96$; $t_{0.01} = 2.57$	No
1 ($N_1 = 7$) and 5 ($N_5 = 100$)	$F_{ex} = 19.93 > F_{0.05} = 3.84$	No	$t_{ex} = 5.6$; $t_{0.05} = 1.96$; $t_{0.01} = 2.57$	No

It should be emphasized that sample 3 generated by experts is consistent with sample 5 generated by Dr. R. Uppaluri, Oak Ridge National Laboratory (Saaty [30]).

Table 9 represents the results of calculations of the distribution functions *C.I.* for samples *A*, *B* and *C*.

Based on the analysis of Table 9, it can be concluded that 52% of expert assessments, sample *A*, are consistent with the limitation *C.R.* ≤ 0.1, which is five times higher than the same values for sample *B* and 100 times greater than for the generated sample *C*.

7-by-7 Matrices. With the increase in the dimension of the judgment matrix, the number of performed calculations using AHP is significantly reduced. Table 10 shows

Table 9. The distribution functions of the consistency index $F(C.I.)$ of 6 × 6 matrices.

Indicator		Sample distribution functions[a]		
C.I.	*C.R.*	A ($x_A = 0.14$; $m = 1.5$; $b_m = 0.903$)	B ($x_B = 0.561$; $m = 1.4$; $b_m = 0.911$)	C ($x_C = 1.225$; $\sigma = 0.43$)
0.05	0.04	0.17	0.029	0.003
0.10	0.088	0.41	0.075	0.0045
0.125[b]	0.10[b]	0.52	0.101	0.0052
0.20	0.16	0.77	0.187	0.008
0.30	0.24	0.93	0.305	0.016
0.40	0.32	0.98	0.420	0.027
0.50	0.40	0.997	0.530	0.045

[a]Sample *A* and *B*—Weibull distribution, sample *C*—normal law; [b]Limit values

Table 10. Results of processing and systematization of 7 × 7 matrices.

Sample	Number of experts	Experts	Objects	Indicators of *C.I.*	
				Mean	Standard deviation
1	$N_1 = 10$	Specialists	Various	0.072	0.062
2	$N_2 = 13(14)$[a]	Master and postgraduate students	Criteria for choosing a supplier	0.315	0.251
3	$N_3 = 100$[b]	Modeling	–	1.470	0.348
4	$N_4 = 500$[b]	Modeling	–	1.330	0.340

[a]Matrices generated by experts; [b]Generated matrix samples (Saaty [30])

data on two samples. The first, which includes materials from various studies, such as research papers by T. Saaty (mentioned on the introduction to the article) and materials of various studies that have been analyzed in order to design the second sample in consideration with matrices 3-by-3.

The second is a survey of a group of graduate and master students, which was carried out for the choice of a supplier by the following criteria: price, reliability, finance, lead time, quality, reputation, and packaging.

Verification of the possibility of combining samples showed:

- The first (*A*) and second (*B*) samples cannot be combined into one set according to the Fisher criterion;
- The third and fourth sample by the Fisher and Student's *t*-test can be combined as a superposition (sample *C* with parameters $x_C = 1.354$; $\sigma_C = 0.343$);
- Samples *B* and *C* cannot be combined (Fisher's criterion $F_{B,C} = 1.87 > F_{0.05} = 1.74$);
- Samples *A* and *B*, in the first approximation, can be approximated by the Weibull distribution, the sample *C* follows the normal law.

The results of calculating the probabilities of the distribution of $F(C.I.)$ are shown in Table 11. Considering that the restriction criterion for the 7×7 matrix is $C.R. = 0.1$, and the random index $R.I. = 1.35$, we can find the limit of the consistency index $C.I. = R.I. \cdot C.R. = 0.135$. From the analysis of Table 11, it can be seen that with the probability $F_A(C.I.) = 0.86$ experts estimations subordinate to dependency $C.I. \leq 0.135$. This value is three times higher than the estimate $F_B(0.135) = 0.27$ and 4 300 times exceeds the estimate $F_C(0.135) = 0.0002$.

Table 11. The distribution functions of the consistency index $F(C.I.)$ of 7×7 matrices.

Indicator		Sample distribution functions[a]		
C.I.	*C.R.*	*A*	*B*	*C* ($x_C = 1.354$; $\sigma = 0.343$)
0.05	0.037	0.46	0.09	–
0.10	0.074	0.75	0.20	–
0.135[b]	0.1[b]	0.86	0.27	0.0002
0.20	0.149	0.95	0.42	0.0004
0.30	0.223	0.99	0.58	0.0011
0.40	0.297	0.999	0.71	0.0027
0.50	0.371	–	0.76	0.0066

[a]Sample *A* ($x_A = 0.072$; $m = 1.17$; $b_m = 0.95$) and *B* ($x_B = 0.315$; $m = 1.25$; $b_m = 0.932$)—Weibull distribution, sample *C*—normal law; [b]limit values

3 Analysis of Probability Estimates of Consistency Indices of Matrices of Different Sizes

From the analysis of Tables 2, 4, 6, 9, and 11, it can be found that for all considered matrices there are three types of samples:

Samples of the first type are individual. They include a family of matrices, each of which is composed for one object by one expert with inherent intelligence, intuition, erudition, and experience;

Samples of the second type are collective. They foresee the consideration of one object by a group of experts, each of which forms a matrix of judgments individually without exchange of information and discussions with other experts;

Samples of the third type are the generated matrices for any impersonal object of a specified dimension $n \times n$ that are processed according to the AHP.

The first conclusion, which can be drawn from examining the statistical parameters of samples is that, according to the Fisher and Student's *t*-test, they cannot be combined into a one set for matrices of different orders.

One of the possible reasons for this phenomenon is the insufficient number of objects for some samples of the first and second types, and the different level of training of experts. In the future, with the accumulation of information, this cause can be eliminated, which, possibly, will result in one combined experimental sample.

Another reason that requires careful verification is the ambiguity of the question of the possibility of combining tangible and intangible objects into one aggregate, which indirectly affects the assessments of experts.

The second conclusion relates to the criteria that are used to assess the consistency of matrices and, consequently, the possibility of making decisions about the ranking of the analyzed objects or the allocation of resources.

In addition to the limiting constraint z_i used in the AHP for the i-th type of matrices for the estimation of the consistency ratio (*C.R.*), i.e. $C.R. \leq z_n$, we use another approach. That is the probability estimation of the affinity of the analyzed matrix to the number of converged with the allowance for the z_n.

Table 12 depicts the probability estimates of the consistency indices *C.I.* for three types of samples and corresponding matrices, taken from Tables 2, 4, 6, 9, and 11. The limiting values of *C.I.* were computed taking into account the random consistency index *R.I.* and the limiting restriction z_n for each type of matrix.

Table 12. Probabilistic evaluation of consistency indices by limiting constraint z_n.

Matrix	Restriction z_n	Limit *C.I.*	Probabilistic estimates from the saples		
			$F_1(C.I.)$	$F_2(C.I.)$	$F_3(C.I.)$
3 × 3	0.05	0.026[a]	0.39	0.31	0.1500
4 × 4	0.08	0.072[b]	0.45	0.31	0.0230
5 × 5	0.10	0.110	0.73	0.07	0.0270
6 × 6	0.10	0.125	0.52	0.10	0.0052
7 × 7	0.10	0.135	0.86	0.27	0.0002

[a]Value of *R.I.* = 0.52; [b]Value of *R.I.* = 0.90

We can try to compare probabilistic estimations by types of samples.

For the first kind of samples, probabilistic estimates range from 0.39 (3 × 3 matrix) to 0.86 (7 × 7 matrix). It means that 40 to 86 percent of expert matrices are consistent. For the second type of samples, respectively, the consistency is from 7% (matrix 5 × 5) to 31% (matrices 3 × 3 and 4 × 4).

An entirely opposite picture is observed for the generated matrices: from the maximum value $F_3(C.I.) = 0.15$ (for the matrix 3 × 3) to $F_3(C.I.) = 0.0002$ (matrix 7 × 7).

The possible first reason for these contradictions is that, for the *R.I.* (random consistency index) the mean *C.I.* (consistency index) value of generated matrices is chosen that are formed without elementary compliance with transitivity requirements between 'judgments' by rows and columns.

The second reason, in our opinion, is the hypothesis accepted by T. Saaty [32] as the absolute truth about the maximum permissible compatibility value that should be 1.10, according to which, the permissible deviation from compatibility should not exceed 10%.

However, taking into account probabilistic estimates, the restriction $C.R. \leq 0.1$ for 6×6 matrices means that $F_1(C.I.)$ exceeds $F_3(C.I.)$ by 100 times, and for 7×7 matrices by 4300 times, respectively.

4 Conclusions

To date, a diverse array of expert assessments for quantitative (tangible) and qualitative (intangible) objects based on AHP has been accumulated in various fields of knowledge.

The systematization and analysis of the collected data made it possible to put forward the hypothesis that some of the indicators (the size of the matrix, the eigenvalue of the matrix, the consistency index, and the ration of consistency) can be considered in the form of aggregates of random variables representing an intellectual product and reflecting features of human thinking.

Obviously, each expert evaluation reflects the individuality of the author, whereas the 'generated' matrix is strictly formalized. At first glance, this is an insurmountable obstacle, but it should be noted that the 'artificial' matrices (sample size: 100 and 500 values) were not subjected to the convergent procedure, whereas each expert matrix undergoes 'self-review' before inclusion in the general sample, i.e. they are inherently closer to the newly formed objects of experts.

Analysis of indicators from various sources allowed the formation of algorithms for their statistical processing.

The systematization and statistical processing of the results obtained by experts on the basis of the AHP showed that the distribution functions of the values of expert estimates for the consistency indices *C.I.* significantly differ from the similar functions of the generated matrices. Therefore, the possibility of choosing the right solution based on the generated matrices requires additional experimental verification.

One of the directions for the future research should be forwarded with extending the generalization of expert assessments representing statistical aggregates of matrices of different dimensions and their comparison with artificial (generated) assessments. Obviously, in this case, in addition to the relation $C.R. \leq 0.10$, it will be possible to obtain probability estimates that will not only rank the criteria and alternatives, but also characterize the reliability (risk) of decisions made regarding the allocation of resources among logistic intermediaries.

References

1. Blair, A., Nachtmann, R., Saaty, T., Whitaker, R.: Forecasting the resurgence of the US economy in 2001: an expert judgment approach. Socio-Econ. Plan. Sci. **36**, 77–91 (2002)
2. Blair, A., Mandelker, G., Saaty, T., Whitaker, R.: Forecasting the resurgence of the U.S. economy in 2010: an expert judgment approach. Socio-Econ. Plan. Sci. **44**, 114–121 (2010)
3. Lukinskiy, V.S., Lukinskiy, V.V.: Analysis of the logistics intermediaries choice methods in the supply chains. Transp. Telecommun. **16**(4), 288–295 (2015). https://doi.org/10.1515/ttj-2015-0026

4. Lukinskiy, V., Lukinskiy, V.: The Analytic Hierarchy Process (AHP): prospects for application in supply chain management. In: Kabashkin, I., Yatskiv, I., Prentkovskis, O. (eds.) Reliability and Statistics in Transportation and Communication, RelStat 2017. Lecture Notes in Networks and Systems, vol 36. Springer, Cham (2018)
5. Niemira, M., Saaty, T.: An analytic network process model for financial-crisis forecasting. Int. J. Forecast. **20**, 573–587 (2004)
6. Ozdemir, M., Saaty, T.: The unknown in decision making. What to do about it. Eur. J. Oper. Res. **174**, 349–359 (2006)
7. Saaty, T., Vargas, L.: Diagnosis with dependent symptoms: Bayes theorem and the Analytic Hierarchy Process. Oper. Res. **46**(4), 491–502 (1998)
8. Saaty, T.: Decision-making with the AHP: why is the principal eigenvector necessary. Eur. J. Oper. Res. **145**, 85–91 (2003)
9. Saaty, T., Ozdemir, M.: Why the magic number seven plus or minus two. Math. Comput. Model. **38**, 233–244 (2003)
10. Saaty, T., Vargas, L., Dellmann, K.: The allocation of intangible resources: the Analytic Hierarchy Process and linear programming. Socio-Econ. Plan. Sci. **37**, 169–184 (2003)
11. Saaty, T.: There is no mathematical validity for using fuzzy number crunching in the Analytic Hierarchy Process. J. Syst. Sci. Syst. Eng. **15**(4), 457–464 (2006)
12. Saaty, T.: Rank from comparisons and from ratings in the analytic hierarchy/network processes. Eur. J. Oper. Res. **168**, 557–570 (2006)
13. Saaty, T., Shang, J.: Group decision-making: head-count versus intensity of preference. Socio-Econ. Plan. Sci. **41**, 22–37 (2007)
14. Saaty, T., Tran, L.: On the invalidity of fuzzifying numerical judgments in the Analytic Hierarchy Process. Math. Comput. Model. **46**, 962–975 (2007)
15. Saaty, T.: Neurons the decision makers, Part I: the firing function of a single neuron. Neural Netw. **86**, 102–114 (2017)
16. Taha, H.: Operations Research: An Introduction, 9th edn. Prentice Hall, Upper Saddle River (2011)
17. Basile, O., Persia, L., Usami, D.: A methodology to assess pedestrian crossing safety. Eur. Transp. Res. Rev. **2**, 129–137 (2010)
18. Beria, P., Maltese, I., Mariotti, I.: Multicriteria versus cost benefit analysis: a comparative perspective in the assessment of sustainable mobility. Eur. Transp. Res. Rev. **4**, 137–152 (2012)
19. Cancela, J., Fico, G., Waldmeyer, M.: Using the Analytic Hierarchy Process (AHP) to understand the most important factors to design and evaluate a telehealth system for Parkinson's disease. BMC Med. Inform. Decis. Mak. **15**(Suppl 3), S7 (2015)
20. Dulange, S.R., Pundir, A.K., Ganapathy, L.: Prioritization of factors impacting on performance of power looms using AHP. J. Ind. Eng. Int. **10**, 217–227 (2014)
21. Gogas, M., Papoutsis, K., Nathanail, E.: Optimization of decision-making in port logistics terminals: using Analytic Hierarchy Process for the case of port of Thessaloniki. Transp. Telecommun. **15**(4), 255–268 (2014). https://doi.org/10.2478/ttj-2014-0022
22. Hamidy, N., Alipur, H., Nasab, S., Yazdani, A., Shojaei, S.: Spatial evaluation of appropriate areas to collect runoff using Analytic Hierarchy Process (AHP) and Geographical Information System (GIS). Model. Earth Syst. Environ. **2**, 172 (2016)
23. Kabashkin, I., Lučina, J.: Development of the model of decision support for alternative choice in the transportation transit system. Transp. Telecommun. **16**(1), 61–72 (2015). https://doi.org/10.1515/ttj-2015-0007
24. Kopytov, E., Abramov, D.: Multiple-criteria choice of transportation alternatives in freight transport system for different types of cargo. In: Proceedings of the 13th International

Conference: Reliability and Statistics in Transportation and Communication (RelStat 2013), 16–19 October, Riga, Latvia, pp. 180–187 (2013)
25. Kopytov, E., Urbach, A., Labendik, V.: Comparative multi-criteria assessment of nano-coating technologies. In: Proceedings of the 13th International Conference: Reliability and Statistics in Transportation and Communication (RelStat 2013), 16–19 October, Riga, Latvia, pp. 207–216 (2013)
26. Maskeliūnaitė, L., Sivilevičius, H., Podvezko, V.: Research on the quality of passenger transportation by railway. Transport **24**(2), 100–112 (2009). https://doi.org/10.3846/1648-4142.2009.24.100-112
27. Rossi, R., Gastaldi, M., Gecchele, G.: Comparison of fuzzy-based and AHP methods in sustainability evaluation: a case of traffic pollution-reducing policies. Eur. Transp. Res. Rev. **5**, 11–26 (2013)
28. Site, P., Filippi, F.: Weighting methods in multi-attribute assessment of transport projects. Eur. Transp. Res. Rev. **1**, 199–206 (2009)
29. Wiethoff, M., Brookhuis, K., Waard, D., Marchau, V., Walta, L., Wenzel, G., Brucker, K., Macharis, K.: A methodology for improving road safety by novel infrastructural and in vehicle technology combinations. Eur. Transp. Res. Rev. **4**, 67–77 (2012)
30. Saaty, T.L.: The Analytical Hierarchy Process. RWS Publications, Pittsburgh (1990)
31. Weibull, W.: Fatigue Testing and Analysis of Results. Pergamon Press, Oxford, London (1961)
32. Saaty, T.L.: Decision Making with Dependence and Feedback. The Analytic Network Process. RWS Publications, Pittsburgh (1996)

The Analysis of Logistics and Supply Chain Management Organizational Structures on Russian Market

Victor Sergeyev(✉)

National Research University Higher School of Economics, Bolshoy Trekhsvytitelsky per. 3, Moscow, Russia
viserg@bk.ru

Abstract. In the article the logistics management and SCM organizational structures research findings that were conducted by the NRU HSE International Centre of Training in Logistics in 2014–2016 are considered. The sample of companies operating on Russian market accounted about 400 organizations. The form for the respondents of online survey was created. Each participant estimated the logistics management organizational structure (logistic and/or SCM department) individually. The assessment was conducted from the perspective of the correspondence between the organizational structure and corporate and logistics strategies, key business processes, organizational structure management efficiency, functional employees allocation and so on in the different aspects of supply chains and logistics controlling. The examples of typical logistics management structures of line and staff, matrix and project oriented structures are shown. Taking into consideration the importance of the SCM best practices and innovative technologies usage, the questions are highlighted in detail and a great variety of examples in SCM departments organizational planning are given.

Keywords: Logistics · Supply chain management · Organizational structure · Survey · Questionnaire · Research methodology

1 Introduction

In the middle of 2017 the practical experience of logistics departments design in Russian companies extends back 15 years. As for the supply chain departments design, it goes back only 5–7 years. In this period the use of international experience was one of the main ways to design logistics and supply chain management organizational structures in Russian companies.

In the development process of the business administration theory and different management aspects several organizational structures segregated. The functional and divisional structures are the main types for operational logistics. Major companies, holding companies and transnational organizations operating on Russian market use the supply chain management (SCM) ideology for providing the management flexibility of complicated economic objects (supply chains are among them) successfully. Meanwhile, the SCM functionality is realized with the help of matrix and project

I. Kabashkin et al. (Eds.): RelStat 2018, LNNS 68, pp. 389–406, 2019.
https://doi.org/10.1007/978-3-030-12450-2_37

organizational structures, network structures and also on the ground of business units, profit (responsibility) centers, etc.

Nowadays only some of Russian companies pay enough attention to the effective logistics and supply chain management organizational structure design, possible risks identification and the implementation of measures that can help to prevent potential logistics problems in the future. In this respect, carrying out the full-scale investigation on the logistics and SCM organizational design condition in Russian companies turns out to be significant from the perspectives of possible problems identification and the best practices usage (including foreign companies which operate on Russian market).

This article is devoted to both general methodological problems of the organizational structures design in companies which operate in different fields and the analysis of such structures design investigation findings in Russia from the perspectives of logistics and SCM. The analysis of the research findings will contribute to understanding the logistics and SCM organizational structures design development tendencies and dynamics in companies which operate on Russian market. Moreover, it will help to bring to light the best practices of such structures design for management and project decision making.

2 General Terms and Literature Review

A company (or a supply chain) is a complicated socio-economic system which consists either of the specific set of structural units (staff) or legally (functionally) independent business-entities of different forms. From the administrative aspect, it is often a hierarchic structure with a variety of formal and informal relations which depend on external factors as well as internal ones. From the managerial aspect, it is a group of people whose actions are coordinated in order to achieve a common goal.

As it is known, company's corporate strategy is implemented through a certain organizational structure. It is a complex of elements (structural units, personnel positions etc.) and established relations between them. For the further description of organizational structure and building of organizational and administrative models (models of organizational behavior) it is necessary to give a definition of *organizational structure* as there are different approaches to terminology.

Organizational structure definitions can be divided into two groups.

The *first group* defines organizational structure as ordered, constant arrangement of related company elements and divisions.

For example, in the paper of Valuev and Ignatieva [1] organizational structure is seen as an ordered set of entities and relations that are developed during the process of production (or any other) management and in the divisions' relations. Milner focuses his attention on the need of division and personnel links formalization during the process of organizational structure design [2].

The *second group* sees organizational structure in the context of the goals set by the company and the tasks that are dealt with in it. For example, Varyas and Kono see the main goal of organizational structure in the efficient allocation of objectives and tasks between the departments and employees at all levels [3, 4]. Mescon et al. see organizational structure as a way of interaction between administration levels and functional parts of the business built to reach the company's goals effectively [5]. Saaty and Kearns focus on the need of coming from the common goal during organizational structure design (by analogy of Balanced Score Card) and then proceed to the objectives decomposition on administration levels and to data flows which connect functions and personnel [6].

Generally, if we use system approach, organizational structure of the company can be defined as a complex of elements linked with management entity by data flows in order to reach corporate goals. Thus, a corporate strategy is a system factor of organizational structure design.

Naturally, organizational structure depends on the key business processes. They determine the effectiveness of the company (supply chain) in a competitive environment to a large extent. Whereas effectiveness of the current organizational structure is determined through the following measures:

- Manageability—ability to respond to top-management and company owner's guidelines and simultaneously supporting accepted corporate and functional strategies;
- adaptivity—on-time company reaction to the internal and external changes;
- specialization—division of powers and responsibilities for the key functions/operations;
- coordination (cooperation)—ability of interaction between the functional divisions of the company in order to reach the common goal;
- administration (transactional) costs.

Let us settle upon some essential points related to the design of logistic and SCM services in organizational structure of the company.

As it is known, before the company starts to design logistics management organizational structure it needs to identify key logistic business-processes in company's SC. Within the concept of SCM, the service (agency, direction, department) of logistics is usually created in the focus company of SC so as to deal with the tasks of strategic planning, controlling, coordination and audit of logistic business-processes. It is important to be aware of the fact that logistics management structure cannot be seen and designed apart from the existing organizational structure which was formed by the company business. Although in a number of cases introduction of logistics (or SCM) concept leads to fundamental changes in existing structure.

The study of organizational structure [7–15] evolution shows that in order to design effective logistics (or SCM department) organizational structure company needs:

- to identify necessary and sufficient logistics functions to manage;
- to make a decision about the outsourcing/insourcing ratio of logistics functions/operations;
- to separate logistics service units according to their management functions that are needed to implement the logistics strategy;
- to establish a system of regulations, document flow and data support.

The first and foremost thing which affects the process of the organizational structure design is logistics strategy (or SC strategy) chosen by the company. Also, there are such factors as identified functional areas and logistics functions, logistic technologies, personnel qualification, staff skills and abilities, information support for logistic services.

A feature of the organizational structure design in the framework of SCM is that each link in SC is a synthesis of the management object and subject. Besides, individual links are the independent business-units which have their own goals and local optimization criteria. Therefore, strategic goal achieving process in SC must be set up by the required level of integration, coordination and directive management in the highest level of SCM of the focus company in the chain. For example, it can be implemented in the form of SCM department.

It should be noted that different levels of logistics development and the introduction of SCM concept caused certain transformations in the companies within logistics and SCM departments' organizational structure. In Russia logistics management works on different principles that reflect industry specific features, company size, staff' level of knowledge, management culture, technology resources, IT-systems, etc. Various types of logistics and SCM departments' organizational structures can exist even within one industry, one competitive environment, meeting the needs of different levels of integration and coordination.

3 The Methodology of Logistics and SC Management Organizational Structure Research

The hypothesis that the evolution of logistics had an effect on organizational design is confirmed by the results of the logistics services controlling study held by the International Center of Training in Logistics in HSE in 2006–2007 and 2014–2015 and by the results of using the grant for the investigation that was conducted by the faculty of business and management in NRU HSE[1] in 2016.

The grant research was conducted in order to clear up the state of logistics in Russian companies and to prove the hypothesis that there are connections between the organizational structure and the effectiveness of company business in big picture. Also, this need was caused by the necessity of Russian companies' top-management to be aware of the key trends and the developments in the area of logistics and SCM structure design and organization of effective supply chains so as to win higher competitive positions.

The aim of the study was to analyze the organizational structures of logistics and supply chain management in the companies that operate on Russian market taking into account the specifics of the country's way of doing business. The study was based on the system analysis of organizational structure of logistics and SCM development state and trends in Russian production, commerce and service companies. Within the scope of the research, there was a survey for top-management about the organizational design which was taken in the companies working in logistics and SCM area. The survey was conducted in written and online questionnaire format. The results were documented

[1] In 2016 Faculty of Business and Management at National Research University Higher School of Economics provided the grant to the Professor Victor Sergeyev under the title "The investigation on the logistics and supply chain management organizational structures of companies operating on Russian market".

and analyzed with the use of a tailored methodology. This survey is surely carried out according to ethical policies.

In the study, it was analyzed if such factors as company size, type of the business, the range of produced or sold products etc. influence the structure type, the staff schedule and personnel's authority in the context of making managerial decisions.

The methodology of the organizational design of logistics and SCM services study is based on the systematic approach (analysis), economic cybernetics methods, operations research, mathematical statistics, optimal decision making, SCM theory.

The objective information for the degree identification of organizational structure development and main problem areas was obtained from the respondents (representatives of logistics services and SCM departments from Russian and foreign companies). The sample was purposive because its elements were selected in a manual way. Logistics and SC top-managers who could give meaningful answers on the specifics of organizational design were the main respondents. Also in this research, a form of purposive sample is intentional because of the easy access to targeted respondents and their motivation to provide detailed and objective answers.

From the perspective of used methods, the research was based on the refutation or the proof of set hypotheses. In other words, the possible relation between research questions which is common to rational company behavior in the context of logistics and SCM was identified firstly. Then on the base of given answers, it was established if it actually was that way. During the study of each hypothesis, a number of statistical data was examined. It was summarized in the tables according to the questions given in hypotheses. The tables showed the number of answers on each question as a percentage of the total number of respondents, the total number of respondents answered on specific question and the total number of each industry representative. It was done for the convenience of further analysis and so it could be possible to compare the number of answers.

The sources of information for the study were: the data from Internet-analysis, literature review on the research problem, written and online questionnaire, interview with respondents—logistics and SC top-managers (directors, heads of logistics and SCM departments and divisions, logistics process coordinators in supply chains, integrated logistics managers, etc.)

4 The Investigation Results

From the perspectives of the organizational design methodology of logistics and SCM functionality, let us consider the results of the research conducted by the International Center of Training in Logistics in NRU HSE in 2014–2016.

The questionnaire was prepared for the online inquiry of the respondents. Each participant could give an individual evaluation for the logistics management (logistics and SCM departments) organizational structure of the company from several perspectives. They are the correspondence of the organizational structure with the corporate and logistics strategies, key business processes, the effectiveness of structure functioning, the allocation of functions between staff etc. in different components of logistics and SC controlling.

The questionnaire heading consisted of general questions concerning the participant (respondent) and the company. All questions (total amount is 50) was grouped into the following parts:

- The perception of logistics and SCM in the company (12 questions).
- The logistics (SC) management organizational structure (7 questions).
- The goal set for logistics controlling concerning the organizational design (9 questions).
- The tools for the organizational structure effectiveness analysis (5 questions).
- The effectiveness measurement—KPI system (4 questions).
- The SC controlling from the perspective of administration (10 questions).
- The barriers and accelerators for the organizational structure development (3 questions).

General facts about the company included the following information: the name, the business area (the economic sector), the annual turnover, the number of employees, the logistic network (SC) configuration and many others.

As seen from the diagram (Fig. 1), the manufacturing companies and the manufacturing and trading companies ranked first with the result of 53%. As for the sales volume (Fig. 2), 38% of companies had the annual turnover ranging from 1 to 10 billion rubles, 27%—from 10 to 100 billion rubles in 2016. It reveals the following fact: rather big companies that operate on Russian market participated in the questionnaire.

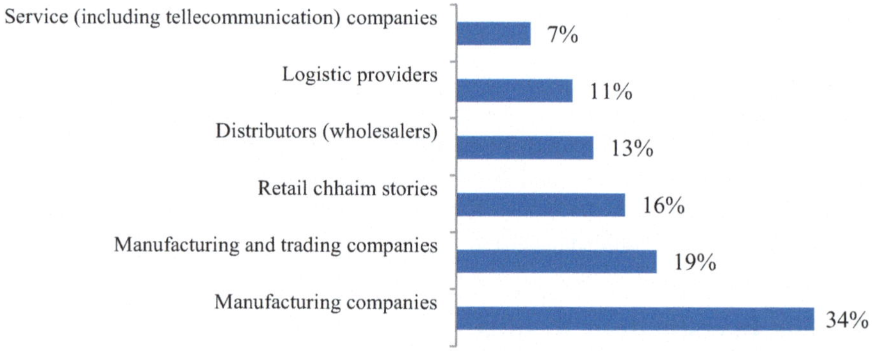

Fig. 1. The allocation between companies by the business area.

Since the fact that the selection numbering 396 companies was formed in a random manner there were both organizations with and without logistics (SCM) department. Moreover, there were some companies where logistics functions are executed piecewise (Fig. 3).

Taking into account the sample representativeness, it is possible to evaluate the number of companies in Russia that has logistics department on the average. It amounts to 81% (Fig. 4). This shows a big progress in the process of understanding the logistics

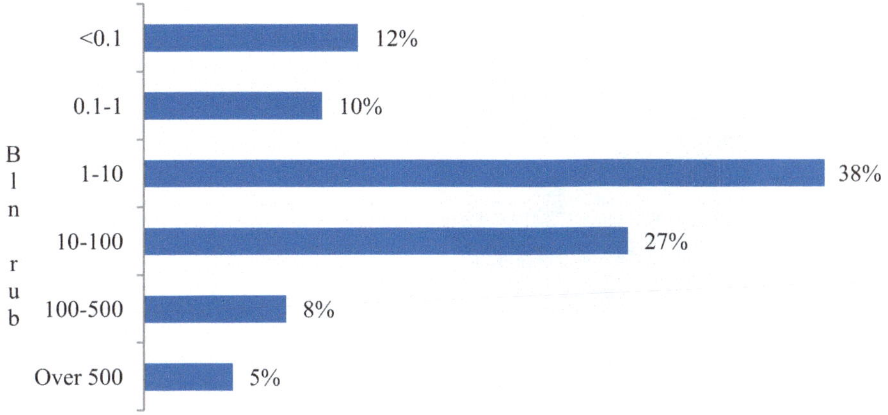

Fig. 2. The allocation between companies by the sales volume (rubles).

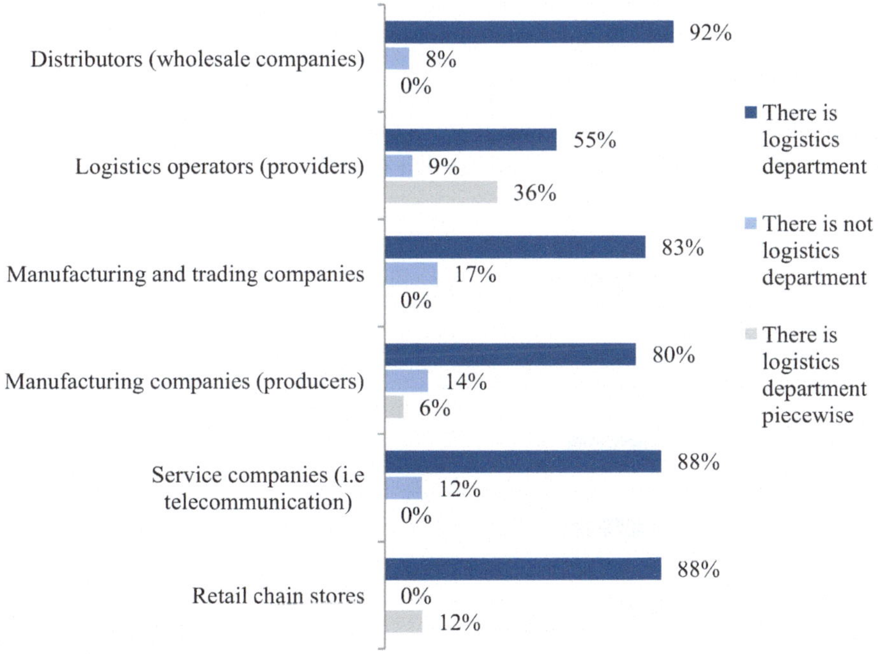

Fig. 3. The logistics department presence in companies operating in different business areas.

role as an essential element for most companies in all business areas. In fact, in 2010 the percentage of companies that had logistics departments in the organizational structure amounted only to 47% [16].

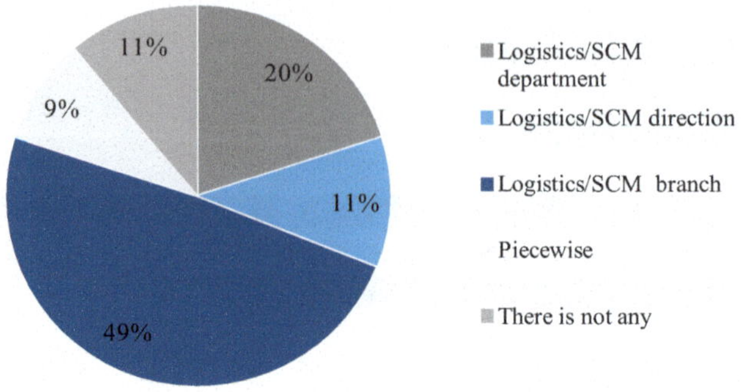

Fig. 4. Logistics (SCM) unit format.

Moreover, it is seen on Figure that there is a logistics department in the majority of companies (except logistic providers which focus on the operational logistics). Meanwhile, most companies that have logistics department in the organizational structure are in the distribution sector. Analyzing Fig. 3, it is important to mention that only in the retail chain stores group there are not companies that do not have logistics unit on the organizational structure.

As the data from the received questionnaire shows (Fig. 5), the logistics unit form depends on the company's size, number of employees and sales volume. In the investigated selection most companies had logistics departments (about 50%). Besides,

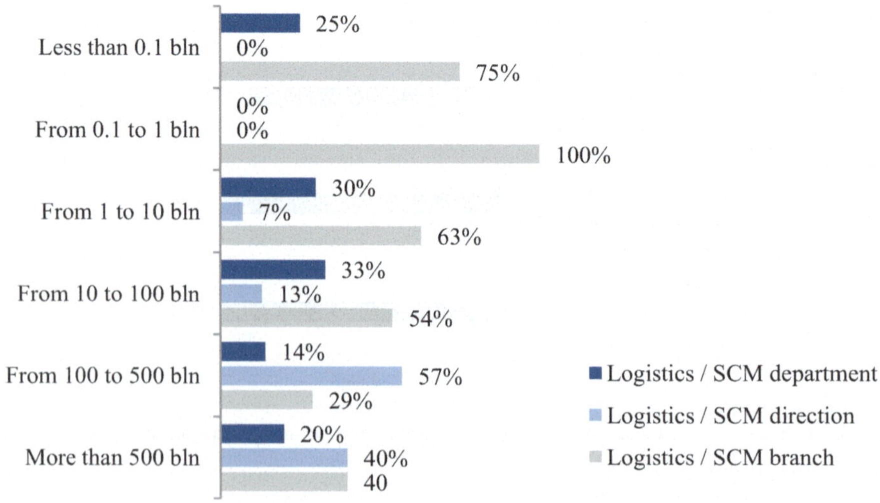

Fig. 5. The allocation of companies from the perspectives of the logistics unit type depending on the annual turnover.

20% of companies had SCM departments: it shows that there are a lot of large companies with big sales volume in the selection. For such companies the department is the most typical form for the units in the organizational structure.

In particular, it is confirmed by the data from the diagram (Fig. 5) where there is the allocation of companies from the selection from the perspectives of annual turnover in 2016. The majority of companies where the logistics (SCM) unit was the branch had annual turnover less than 1 billion rubles, the direction—100–500 billion rubles, the department—1–500 billion rubles. The only exception is the presence of 25% of companies with the logistics (SCM) departments with minimal annual turnover.

Let us consider the most substantial results of the investigation on the questions which were mentioned above (they concerned the logistics (SCM) organizational design). In particular, in the first section respondents were asked to answer the question: "What are the main topics for the company's management in logistics?" The results which can be seen on the Fig. 6 show that only 27% of respondents included the logistics management organizational structure change in the list of prioritized topics. Based on the interview with the top managers of observed companies it was determined that the organizational structure change is a complicated topic which requires more

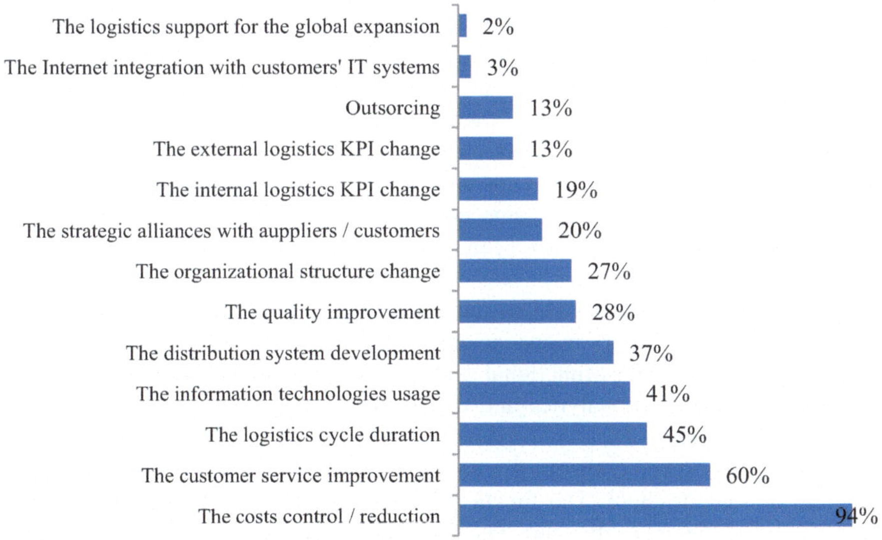

Fig. 6. The main logistics topics (according to respondents).

efforts of top management. Moreover, it does not often lead to any positive results since the company's owners (stakeholders) do not want to make changes in the business administration structure.

To a large extent it is explained by the low level of the integrated approach development for the business administration at whole and logistics in particular. Another reason is the inadequate level of logistics business processes maturity.

As it is known, the company's organizational structure is based on the key business processes and functions. For that matter the term "organizationally functional structure" is frequently used. It consists of several components:

- The organizational unit elements and their subordination (the organization chart).
- The function allocation between the departments.
- The organizational authority allocation between the departments.
- The responsibility allocation between the departments.
- The personnel allocation between the departments.
- The interrelations between the departments.
- The personnel qualification, the enlistment of knowledge/skills.
- The office appliances equipage, etc.

As the majority of surveyed companies have functional organizational structure, let us consider this type's peculiarities from the perspectives of the process of increasing the implemented business processes integration level.

For most companies all over the world logistics has three development stages from the perspectives of the integral paradigm: from operational to coordinating logistics and then to SCM. The peculiarities and duration of coming from one stage to another are extremely variable for different countries. It is necessary to mention that there are mixed types of doing logistics activities in the companies from different business areas.

The questionnaire data processing showed that in 2016 61% of companies had only operational logistics (it means that they did some types of operational functions by themselves (insourcing) or outsource them partially). Transportation/expedition, warehousing, materials handling and associate operations (marking, assortment, bar code marking), export and import customs clearance, etc. are among of them. Meanwhile, 28% of companies delegated some coordination functions such as inventory management, integrated planning, cross-functional logistics coordination, etc. to the logistics departments.

Remained surveyed companies (11%) integrated logistics operations with the help of SCM departments. Inventory management, order cycle management, integrated planning and forecasting, cross-functional and cross-organizational logistics coordination, supply chain system analytics and design, etc. were the main implemented coordination functions (Fig. 7).

The majority of surveyed companies with operational logistics have the functional type of organizational structure. The functional organizational structure is the classic example of the top-down labor division. The functional organizational structure design is based on the company's personnel grouping process according to some business areas (functional areas): production, procurement, marketing, finance, sales, etc. The administration is implemented by the top-down hierarchy from top management to linear (operational) staff of lower echelon. Administrative functions such as planning, controlling, accounting, analysis, etc. are performed by the top management level.

Meanwhile, the logistics organizational structure had rather simple chart. It included such departments which functions were typical for the main operational activities insourcing, i.e. warehousing and transportation departments. If a company had export-import operations than there was the customs clearance department. In case of using outsourcing there was some department that was responsible for planning and

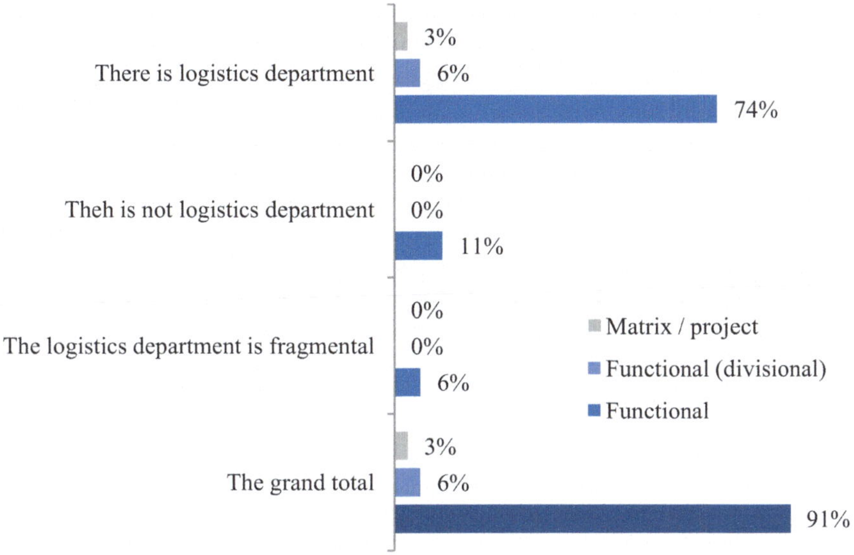

Fig. 7. The companies' allocation concerning the logistics department organizational structure type.

coordination with logistics operators. Niche companies (freight forwarders, expeditors, warehouse operators, insurance companies, customs clearance) and complex operators (3PL providers) are among them.

The number of companies that outsource 100% of functions in operational logistics is relatively slight (less than 3% of the whole selection). It is necessary to mention that most of them are foreign companies. This indicates the insufficient development of logistics outsourcing market in Russia and the unwillingness of many Russian companies to outsource their competencies in operational logistics.

Based on the above, it is possible to design generalized logistics organizational

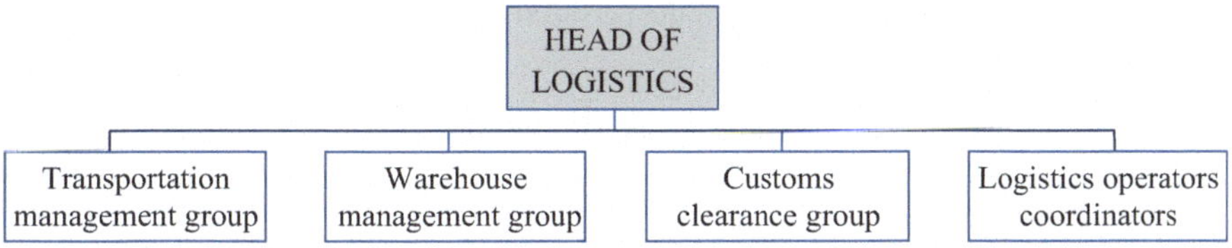

Fig. 8. Generalized logistics organizational chart of manufacturing and trading company.

chart of manufacturing and trading company (Fig. 8). Naturally, transportation and warehouse management groups are responsible for the logistic infrastructure objects: transport subdivisions (car fleet is most commonly) and storage facilities.

During the investigation it was found out that many Russian companies that realize operational logistics subordinate logistic infrastructure objects straight to the head of logistics (department). This practice is not appropriate as there are not logistics specialists in these infrastructure subdivisions. The logistics specialists' goal at the operational level of integration (they should be in transport and warehouse management groups) is the optimization decision-making process. In particular, the decisions can be the following: the minimization of operational logistics costs or finding the balance between costs and service level. For this purpose the analytic and customer service groups (subdivisions) as well as IT-departments for the local support (WMS and TMS systems) are included to the logistics organizational structures.

The subdivisions which are included in the logistics departments in surveyed companies can be seen on Fig. 9. As it is seen from the diagram, the majority of companies have the operational logistics subdivisions. Nevertheless, it is necessary to

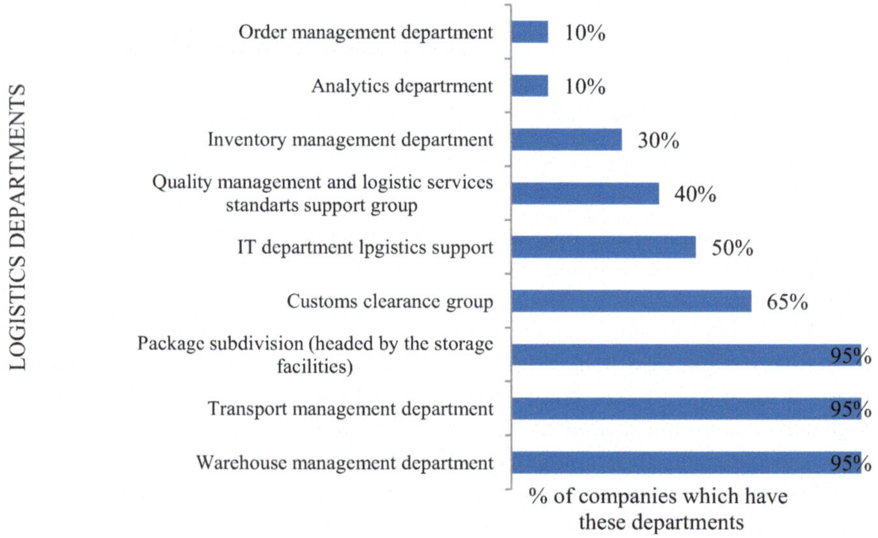

Fig. 9. The allocation of companies from the perspectives of subdivisions presence in logistics department.

mention the increasing number of Russian companies which make the logistics department responsible for the inventory management function (30%) and customer service quality management (40%). This makes it easier for the logistics department to take on the role of the coordinator while managing goods flows.

The analysis of the organizational structure examples for the researched sample showed that there is a number of companies (mostly industrial and manufacturing and trading ones) where external and internal logistics "outlines" are distinguished. Separation of external and internal logistics outlines for industrial companies is often associated with the peculiarities of operational logistics managing specifically on production departments and outside of them, i.e. in supply and distribution.

The divisional logistics structure is typical for the companies that use the corporate strategy of concentrated growth: sales geography development, product range expansion. Simultaneously it can be seen as a combination of individual functional structures which serve a particular market (territory) or implement the policy of product specialization (category management).

As it was noted earlier, the logistics organizational structures development embarked on a course of implementing coordination functions (the most important of which is inventory management) under the authority of logistics departments. The problem of resources optimization during the logistics processes and during the formation of appropriate control actions in supply chain lies in the correct stock replenishment size and periodicity in the relevant links of the chain. Understanding the

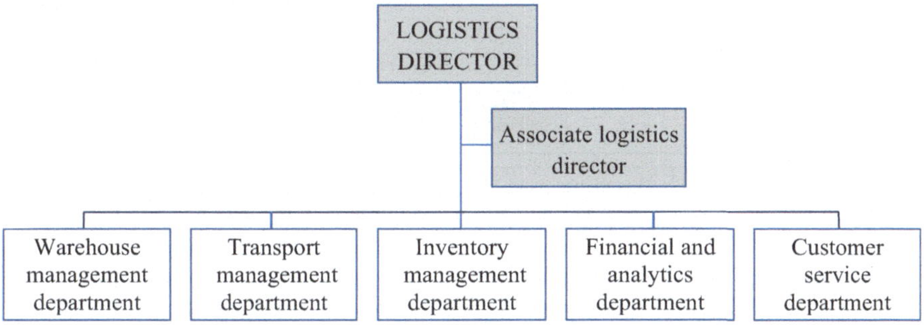

Fig. 10. The organizational structure of "Wimm-Bill-Dann" logistics direction.

importance of giving the inventory management actions to logistics is illustrated in the picture. It shows that 30% of Russian companies have stock management branch in logistics department organizational structure. In the next picture, there is an example of stock management branch inclusion into the directorate structure of "Wimm Bill Dann", one of the biggest companies in the food industry (Fig. 10).

It was mentioned earlier that about one-third of all studied companies gave/included the logistics services functions as well as inventory management and some other coordination functions: integrated planning, services standards and customer's order cycle management, advanced analytics, logistics risks management, etc. However, although inventory management branches growth dynamics was very impressive (from 6% in 2010 to 30% in 2016) other coordination functions were under very small and fragmentary influence of departmentalization.

If we take into account the industry specifics, a coordination role of logistics on Russian market is mostly seen in retail. It the picture there is a summarized divisional structure of logistics in the retail company which sells appliances and electronics. There are all characteristics of functional structures listed above with a look of some coordinated branches.

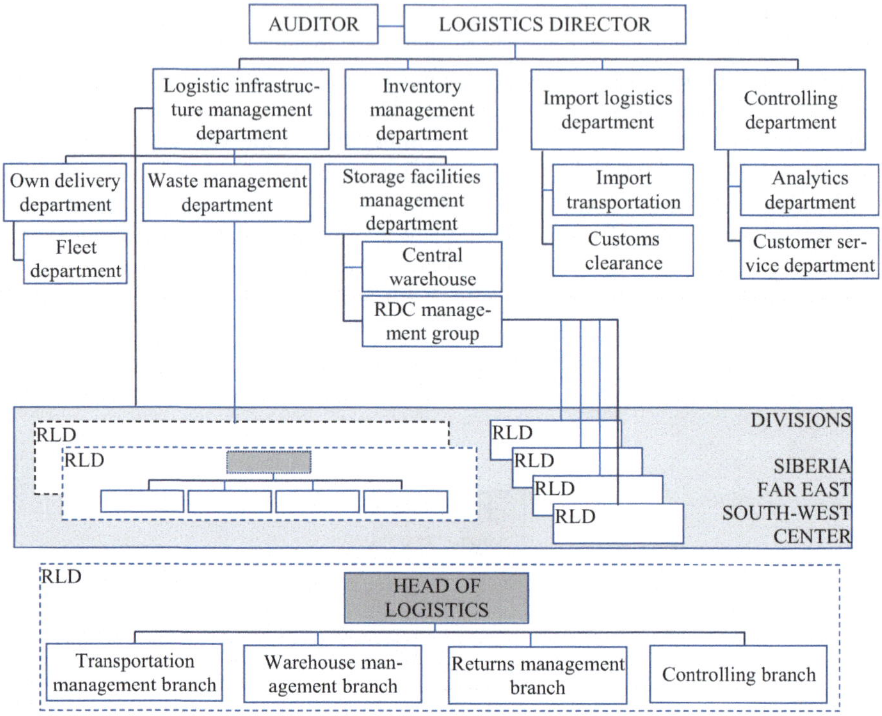

Fig. 11. Generalized divisional structure of retail logistics direction (*Note*: RDC—regional distribution center management group, RLD—regional logistics department).

Interviews with top-managers and logistics directors showed that the addition of the coordination units (planning, analytics, cross-functional coordination, risk management, IT-support, etc.) to the typical logistics service structure was not the result of the measured corporate (or logistic) strategy, but rather of the random factors: shareholders requests, competitors' work imitation, decrease in market position, etc. (Fig. 11).

The logistics organizational structure design analysis also showed that the owners and top-managers (CEO, CFO, CIO) of a number of companies are more willing to delegate coordinating functions not to logistics branches but to SCM departments which are forming not only in large western companies but in Russian ones too. It mostly caused by the fact, that with the increase of logistics branches' status in the context of coordinating managerial functions there are more conflicts with other departments' top-management who traditionally saw logistics as a "cost center". In global practice one of the ways to solve these conflicts was the development of SCM departments which were given coordinating functions as well as the strategic integrated planning and key business-processes controlling functions.

In the end of the organizational structures review let us see the third phase of the organizational design in terms of integrated paradigm—implementation of SCM departments in analyzed samples. As it was mentioned before, the functionality of a majority of the companies which have SCM departments includes most of the coordinating functions listed above as well as operational functions. Departmentalization of the coordinating functions varies widely depending on the business area. Generally, SCM departments (in western companies operating on Russian market as well as Russian companies) consist of such branches as planning (integrated planning, forecasting, demand planning, supply planning, distribution), high-level analytics, controlling, stock management, customer service, risk management, informational integration, etc. Often, the SCM organizational structure consists of procurement and production department and they are mainly subordinate in the outline of planning and controlling.

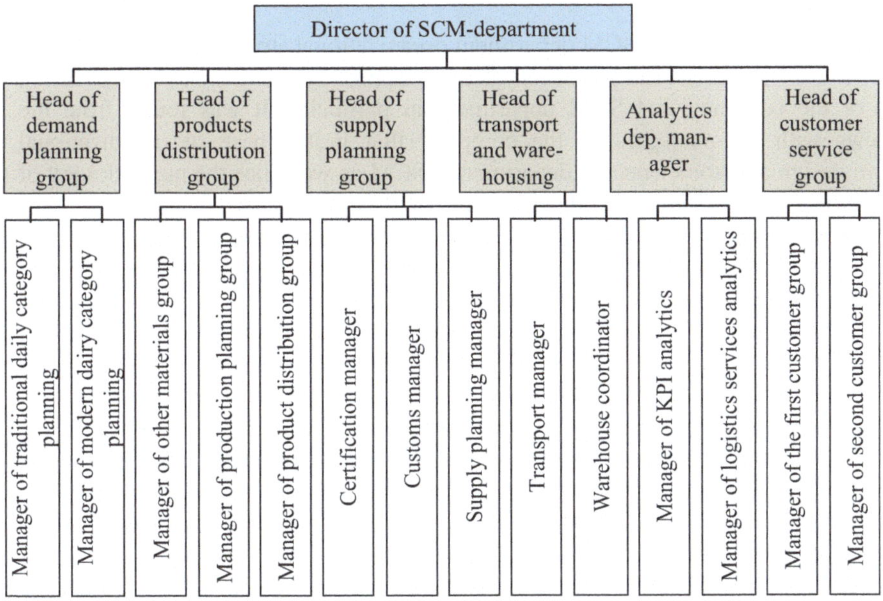

Fig. 12. SCM department organizational structure in “Danone Industry”).

From nearly 400 examined companies 39 of them had SCM department, 3 of them had a branch and 2 of them had a direction. As a representative example, let us see the organizational structure of SCM department in one of the leaders of food industry —“Danone Industry” (Fig. 12).

Let us see the process of SCM department organizational structure design as it is to some extent a showcase example of a systematic approach to organizational structure design of the company and SCM department in particular. It was found from the interviews with top-managers of the company that in the base of organizational structure design and some specific departments (SCM as well) lies the methodology of organizational design which was discussed in details in the paper [17].

In general terms it is built on a consistent hierarchy implementation: the mission, the vision, the corporate strategy (products, needs, markets, competitors, etc.) and policies (including logistics), strategic goals (roadmap), value chain map, systems and processes identification, SC (logistics network) design and analysis, key business-processes and KPI design, the organizational structure development.

SCM department functions were formed based on the principles of the key business-processes integration in procurement, production and distribution, service level control, main material and information flows synchronization in SC, the implementation of the integrated planning and controlling at the top-management level. As the ICTL HSE research showed, the same systematic approach and consistent organizational structure design methodology (in particular, specific logistics and SCM departments) is typical for the majority of large western companies operating on Russian market.

Let us see the process of SCM department organizational structure design as it is to some extent a showcase example of a systematic approach to organizational structure design of the company and SCM department in particular. It was found from the interviews with top-managers of the company that in the base of organizational structure design and some specific departments (SCM as well) lies the methodology of organizational design which was discussed in details in the paper [17].

In general terms it is built on a consistent hierarchy implementation: the mission, the vision, the corporate strategy (products, needs, markets, competitors, etc.) and policies (including logistics), strategic goals (roadmap), value chain map, systems and processes identification, SC (logistics network) design and analysis, key business-processes and KPI design, the organizational structure development.

SCM department functions were formed based on the principles of the key business-processes integration in procurement, production and distribution, service level control, main material and information flows synchronization in SC, the implementation of the integrated planning and controlling at the top-management level. As the ICTL HSE research showed, the same systematic approach and consistent organizational structure design methodology (in particular, specific logistics and SCM departments) is typical for the majority of large western companies operating on Russian market.

As it can be seen from the SCM-department organizational structure, the most typical structural units are presented there. They cover such coordination functions as inventory management, planning, analytics, distribution and production management, customer service and procurement. Logistics support in the company activities is done by the operations department.

Summarizing the logistics and SCM organizational structures research, it should be mentioned that different levels of logistics development and the emergence of SCM concept caused certain transformations in the companies. They appear in the context of organizational structures of logistics and the creation of SCM-departments. As the present study showed, logistics management is built on different principles that show industry specifics, company size, staff' level of knowledge, management culture, technology resources, IT-systems, etc. Given examples are a good demonstration of the fact that even within one industry (in one competitive environment) there are various organizational structures of logistics or SCM departments which correspond to a certain level of integration and coordination.

In the present time, SCM is mostly organized in the focus company as a department, division or a board of directors. Naturally, this SCM organizational design cannot fully reach the goal of SCM in the context of inter-organizational integration and coordination of supply chain contractors. It usually focuses on internal processes of the focus company in the chain. As for the use of managerial supply chain "super-structures" such as 4PL-providers, coordinating councils or committees, including the representatives of key contractors in supply chain, it still has an episodic character.

Organizational arrangements for further integration consist of attempts to implement the key processes' managers into supply chain work, making them SC-managers who are responsible for the whole chain coordination. Some (mostly foreign) companies create supply chain teams and committees with the same goal. Searching for closer cooperation between the customer, the supplier, and the producer (for example, the system supplier, implementation of CRM/SRM macro-processes) can be seen as an example of organizational attempts to increase the level of integration.

At the same time, the objective economic necessity of new cross-functional mechanism implementation which copes with coordinating and integrating decisions internally and in the whole supply chain predetermined the emergence organizational SCM departments as an alternative to coordination function of logistics.

5 Conclusion

As a result of this research following scientifically new findings were collected:

- New classification of logistics and SCM organizational structure types.
- The model of logistics (SCM-department) organizational structure design in the company which is based on systematic approach.
- A multifactorial model of the relationship between business and logistics and SC organizational structure type which allows measuring the influence of the organizational structure type on business and SC effectiveness using KPI.
- Typical organizational structures departmentalization schemes in company's logistics dependent on relationship of personnel's operational and coordination functions.
- New organizational and methodological mechanisms of inter-functional and inter-organizational logistics coordination implementation.
- The models of "insourcing/outsourcing" and "centralization/decentralization" influence in organizational structures of logistics management on business efficiency.
- Perspective forms of organizational design in the context of SCM based on inter-organizational cooperation and outsourcing of 4PL-providers.

Acknowledgements. This article is an output of a research project "Investigation of the logistics and supply chain management organizational structures dynamics of companies in the Russian market", implemented as part of the Research Program of the Faculty of Business and Management at National Research University—Higher School of Economics.

References

1. Valuev, S.A., Ignateva, A.V.: Organisational Management. Neft I Gaz, Moscow (1993)
2. Milner, B.Z.: Theory of Organisation. INFRA-M, Moscow (2007)
3. Varyas, Y.V.: Designing Organisational Management Structure. Znanie, Moscow (1982)
4. Kono, T.: Strategy and Management Structures at the Japanese Companies. Progress, Moscow (1987)
5. Meskon, M., Albert, M., Khedouri, F.: Fundamentals of Management. Delo LTD, Moscow (2001)
6. Saati, T., Kerns, K.: Analytical Planning. Radio I Svyaz, Moscow (1991)
7. Bowersox, D.J., Closs, D.J.: Logistics: The Integrated Supply Chain Management. OLIMP-BIZNES, Moscow (2005)
8. Harris, M., Raviv, A.: Organization Design. Manag. Sci. **48**(7), 852–865 (2002)
9. Kholl, RKh: Organisations: Structures, Processes, Results. Piter, Saint-Petersburg (2001)
10. Lorsch, J.: Organization Design: A Situational Perspective. Organ. Dyn. **6**, 2–14 (1977)
11. Mikltueyt, J., Vuldridzh, A.: Magic of Management. Izdatelstvo AST LLC, Tranzit-kniga LLC, Moscow (2004)
12. Minczberg, G.: Structure at the Fist: Development of Effective Organization. Piter, Saint-Petersburg (2003)
13. O'Shonnesi, J.: Principals of Developing Organization Management. Sirin, Moscow (2001)
14. Stock, J., Greis, N., Kasarda, J.: Logistics, strategy and structure. A conceptual framework. Int. J. Phys. Distrib. Logist. Manag. **29**(4), 2–24 (1999)
15. Stock, J.R., Lambert, D.M.: Strategic Management of Logistics. INFRA-M, Moscow (2005)
16. Sergeyev, V.: The investigation on the logistics controlling in Russian companies. Part 1. Logist. Supply Chain Manag. **4**, 27–36 (2013)
17. Dybskaya, V., Sergeyev, V.: The methodology of logistics and SCM organizational design. Logist. Supply Chain Manag. **6**, 57–68 (2016)

Sustainable Aviation and Maritime Transport

Integrating Air Cargo Road Feeder Services into Green Transport Corridors

Anatoli Beifert and Gunnar Prause(✉)

Wismar University of Applied Sciences, Philipp-Müller-Str. 14, 23966 Wismar, Germany
{anatoli.beifert,gunnar.prause}@hs-wismar.de

Abstract. Although European regional airports contribute considerably to regional accessibility of peripheral and remote regions as well as to socio-economic development, it might be stated that they face severe challenges nowadays. The number of loss making regional airports in Europe keeps increasing. The EU Commission has clearly stated that the public support and financial aids cannot be seen as a sustainable stimulation measure for efficient and durable airports' operations any longer, i.e. the public support to regional and small airports in Europe will be cut. Nevertheless, regional airports can play the role as motors for regional development and growth. On the other hand, the EU introduced and promoted Green Transport Corridor concept as a cornerstone for the development of integrated and sustainable transport solutions based on trans-nationality, multi-modality and a high involvement of public and private stakeholders, including the political level. A limited research has been executed about the integration of regional airports into Green Transport Corridors and which role road feeder services can play in this context. This paper investigates potential role of air cargo and especially road feeder services for the integration of regional airports into Green Transport Corridors as well as favorable conditions for sustainable regional development. Based on the empirical results of these projects, the research focuses on the different concept of Road Feeder Services or "air trucking services" for air cargo and their possible role for the integration of regional airports into Green Transport Corridors. Special attention is paid to the discussion of the tentative role of the air cargo security regulations for third countries in Europe (ACC3) as well as to possible integration air cargo into the concept of Green Transport Corridors. Since the action research of this study was executed in the Baltic Sea region, the presented results includes specifics of this region.

Keywords: Road feeder services · Air trucking · Green Transport Corridor

1 Introduction

Air transport contributes essentially into the European socio-economic development, improving market accessibility to remote regions [1]. Although less than 1% measured in volume, the European aviation industry transports about 40% of Europe's exports and imports value, and transports over 822 million passengers annually [2]. On the other hand, more than 48% of European airports have been identified as loss making,

I. Kabashkin et al. (Eds.): RelStat 2018, LNNS 68, pp. 409–420, 2019.
https://doi.org/10.1007/978-3-030-12450-2_38

regional or national public authorities continue supporting airports in order to provide accessibility and socio-economic obligations [3]. However, according to the new state aid rules for a competitive aviation industry, support for unprofitable European airports shall be cut or considerably reduced. Herewith, the EU Commission made a clear statement that financial subsidies for unprofitable airports cannot be used as sustainable measure any more. The end of the protectionism, strong competition between European airports, liberalization and deregulation in aviation industry, force airports to improve their operational performance [4, 5]. Reduction of public subsidies shall motivate regional airports to consider more flexible planning and other possible alternative development models [1]. Traditional approach in strategic, operational and tactical planning nowadays cannot enhance airports' competitiveness in the constantly changing market conditions [2]. Considering current financial difficulties of the European airports (especially of regional airports) and the relatively high costs in case of airport's infrastructure upgrade, the airports shall increase their ability to have the airports' hard infrastructure as convertible and mutable as possible in order to be able adapt to the current infrastructure without permanent change to current and future needs with minimal financial investments. Flexible airports compared to "non-flexible" may demonstrate a number of competitive advantages to keep their performance results over the operational time period in more sustainable way. This approach shall be of a special importance to the airports and airports' operators, since even the minimal changes in the airport's infrastructure might require considerable financial, personal and time resources. These flexibility options for new business models may offer airports considerable competitive advantage [1]. With the annual air cargo traffic growth up to 4.7% over the next 20 years, the air cargo business can be an important source of revenues and can be seen as a driving force for a airports' profitability [5]. Kasarda and Green stated that airfreight regardless of airports and locations may improve regional and national accessibility through efficient connectivity to distant markets and global supply chains, i.e. efficient air cargo connectivity may be seen as a huge competitive advantage [6]. On the other hand, due to high density of the airports in Europe, the airports' competition for the air cargo is very tough and may be regarded as a "foot-loose" and a tough business [1]. In the framework of the EU funded projects [7], a number of regional airports in the Baltic Sea Region (BSR) have been analyzed; among other things, perspectives of the air cargo business model for the development of regional airports have been evaluated. The findings of the projects showed that the air cargo volume of regional airports in the BSR may be regarded as very insignificant and almost entirely based on the occasional charter flights. However, the growth of the air cargo business for some small and medium airports, that mainly act as feeder-spokes in an airport network system is likely to be based not on pure or dedicated air freighters, but on truck- or road based air services, which is also named as "Road Feeder Services" (RFS). The authors of this study evaluate the potential of the air cargo and especially RFS as a sustainable business model for regional airports in the context of "Green Transport Corridors" (GCT), i.e. network based transnational transportation with "green" or sustainable aspects and high involvement of public and private stakeholders [8] and so called ACC3 regulations (i.e. air cargo and mail carrier operating into the Union from a third country airport), which were introduced by the EU Commission with the aim to ensure that only secured airfreight is transported into the EU, which is valid also for the cargo that is transferred at the last airport before entering the EU [9]. This paper is organized as follows: the theoretical part showcases scientific approaches to regional airports, air cargo and especially the role of RFS for the regional airports' development. The concept of GTC and ACC3 regulations will be presented. The following sections presents methodology and the main results of the investigated cases, including development perspectives through better integration of the air cargo into the Green Transport Corridors by means of RFS, incl. ACC3 regulations as a supplement instrument for airports to generate additional revenue, herewith securing sustainability and profitability.

2 Theory

As it is stated by the EU Commission: "… the regional components of the airport network facilitate access to the core of the network or help to open up peripheral and isolated regions" [10]. European airports may be grouped into: (1) hub airports that provide a full range of services, incl. passenger traffic, domestic, or intercontinental flights as well as airfreight services and (2) regional airports that mainly feed hub-airports with passengers and airfreight, as well as in some cases provide direct flights to other regional airports, herewith fulfilling the connection function to the centers of socio-economic activity [2]. Thus, hub-airports consolidate the air traffic from smaller and regional airports or provide the final-leg air connection to them. This logistics pattern, known also as hub-and-spoke network, has been triggered by the introduction of the Airline Deregulation Act in the USA in 1978 [11–13]. Which considerably promoted hub-and-spoke network development for the airports and airlines operations worldwide, and contributed to costs' reduction both for passenger and air cargo traffic [12, 14]. In the hub-and-spoke network model the flight routes consist of central point that are named as "hub" and a number of end-points that are named as "spokes". Hubs act as consolidation points for passenger and airfreight traffic that move from the various spokes and provide connecting flights to other destination points, which might be other national or international hubs or spokes [11]. The airport-hubs build partnerships with regional airports or airline operator(s) in order to establish a subsidiary-network to remote areas [15]. Regional airports may also benefit from hub-and-spoke pattern through availability of frequent flight connections to hubs with relatively lower fares for passengers, shorter travel times and improved accessibility to remote regions [16]. The airlines of regional airports are also arranged for a hub-and-spoke system, deploying rather smaller aircrafts to feed a hub with the passenger and cargo traffic [17] In spite of decreasing fares and growing demand for the aviation services due to this pattern [5, 18, 19] this situation enabled large airports to become the dominant airport-hubs, herewith increasing competitiveness pressure on regional airports [2, 20]. On the other hand, the ongoing growth of the cargo volume through airport-hubs result may result in congestion and herewith create new opportunities for regional airports and airlines to provide point-to-point transport services, e.g. services of short-haul flights, with no extra quality service at low-fares connecting areas that are poorly served or regional airports that are closed urban areas. This business model known as a "low cost carrier" model (LCC) may provide a sustainable growth opportunity to regional airports and stimulate their development [2, 17]. Acting as "spokes" or feeders for the main airport-hubs, regional airports may fulfil an important role to reduce congestion peaks in the central hub-airports by developing efficient air transport network as well as improving regional accessibility through better air-rail and air-road connections [21]. Air cargo still represents only a single pattern of the aviation value chain and depends on many factors, including location infrastructure, policy regulations, business intensity and competitiveness structure [6, 22]. Airfreight still may improve the connectivity to remote regions and global supply chains in a fast and reliable manner considerably. Representing 1% of the globally transported goods in volume, but 40% in value, the demand for air cargo has grown considerably, air freight has become highly dynamic and flexible, new stakeholders have been entering the market, new concepts and interactions have been developed [2, 5]. Although the air transport demands additional requirements such as security regulations, information, financial intensive infrastructure, in the context of just-in-time or just-in-sequence concepts, potential of air cargo exists especially for time-definite or time sensitive goods to provide efficient solutions in terms of a time-cost ratio; thus, airports' involvement in the air cargo supply may strengthen airport's position considerably [23]. Hübl provided empirical evidences that industrial sites closed to airports are generally more interesting for production companies and their supply chain activities and air cargo may positively contribute to the airports' revenue situation [2, 19]. A number of relevant studies showcased that air cargo part in average may represent 15–30% of airlines' total turnover, whereas the passengers part may be characterized by relatively low margins—up to 7% [2, 5, 24]. The air cargo in the European airport-hubs is mainly carried through the intercontinental network airlines in the passenger planes known as "belly-cargo" [2]. Since it is rather unrealistic that airlines would offer scheduled inter-continental passenger transport services via regional airports, in this context the participation in the air cargo market might be a challenge for regional airports. On the other hand, certain regional airports that prioritized airfreight, may handle large amounts of cargo through so-called dedicated freighters and it might be the niche market for regional airports [18, 25]. In fact, not every flight is a travel option for the air cargo nowadays and the efficient road-network infrastructure in Europe has made the development of alternative services to air transport possible via replacement of the air mode by road through "air trucking" or "road feeder services" (RFS) is a growing trend worldwide [2, 5, 24]. Majority of pickup and delivery locations in Europe are rather at short or middle range distances, deployment of RFS is often the fastest, reliable and the most cost efficient mode of air cargo transport; and provides many advantages such as: higher accessibility e.g. by pick up or delivery services, availability of executive units (i.e. trucks vs. aircrafts), flexibility in terms of official time-schedule applied for the real air crafts in form of the given time slots and routes [2]. The main difference of the RFS and traditional road trucking is that an air-truck service operates between two airports on an official air consignment note or Air Waybill (AWB), which is issued by airlines, thus air-trucking is treated like real aircrafts and have to fulfil same requirements and security regulations set by the relevant authorities as if goods were really flying by air [26, 27]. The RFS have fixed time schedule, e.g. in comparison to traditional trucking services, "air trucks" leave the point of departure to a designated destination at exact time, regardless of its current load capacity [2]. RFS mainly belong to the first or main executive leg in the transport chain and provide for airlines a cost efficient transport alternative from regional airports to an airport-hub. For some European airport-hubs, the air trucking services count for the biggest part of their total airfreight movements [28, 29]. The relevance of the RFS for the air cargo business over the past decade has contributed to a strong decline of traditional scheduled airfreight carried by air in the intra-European scale. The number of airport-to-airport pairs has doubled since 2004 [2]. RFS obtained an important role as a feeder network for the central airport-hubs, RFS are mainly applied to acquire freight from "spokes" to fill up long-haul flights [27]. Further competitive advantage of air trucking especially for regional airports is the fact that deployment of the air trucking "fleet" does not require intensive financial resources compared to the real aircrafts. Thus, the increasing demand for transport of the time-critical goods and a high efficiency of RFS may provide a sustainable business model to regional airports to benefit from the air cargo market [2, 22].

The concept of green transport corridors (GTC) was introduced in 2006 as an initiative of the European Commission in the Freight Transport Logistics Action Plan as the necessity for sustainable and environmental friendly transport sector, providing safer and efficient transportation by reducing accidents, congestions and negative impacts through emissions like noise, light and pollution [10, 30]. The concept of GTC comprises integrated transport concept, whereas see shipping, rail, inland waterways and road transport modes complement each other in efficient and environmentally friendly way [31]. Prause underlined that although there is not a commonly accepted system of criteria for GTC, still a number of characteristics that make a transport corridor "green" have been already recognized by a number of transport initiatives and studies, such as: trans-nationality and co-modality securing environmentally friendly transport along international transport routes, availability of adequate trans-shipment facilities and infrastructure, innovative transport technology and advanced ITS applications are compulsory to achieve environmentally friendly transport solutions. Since CTC concept comprises interdependency between traditional supply chain management and environmental issues, the CTC performance may be analyzed by management control systems for supply chains that includes ecological and the international network environment aspects [32–36]. Furthermore, Prause noted that existing Key Performance Indicators (KPI) for monitoring and management of GTC mainly focus on environmental and operational issues, and do not respect infrastructural, strategic and network-oriented aspects [8]. Thus, in order to secure sustainable, efficient, innovative, safe, environmental friendly and long-term development for green transport corridors a management control system in form of a Green Corridor balanced scorecard may be used by deploying network-oriented KPIs and integrating risk management criteria from supply chain management to the Green Transport Corridor concept [8, 37, 38]. Based on empiric results from relative research projects, Prause proposed a balanced scorecard for GTC, whereby KPIs for monitoring and management of Green Transport Corridors are integrated with cooperative and network-oriented aspects from supply chain management [8].

Prause underlined that in spite that this balanced scorecard focusses on network properties of a corridor and includes the most important perspectives for GTC, the given set of indicators is not complete and the type of measurement of the indicators may be regarded as still open [8]. Since transport multimodality is considered as one of the important criteria of the GCT, the efficient integration of the air cargo and especially air trucking services into the concept of Green Transport Corridors, represents an important issue and is an object of further studies [30]. Although road feeder services may be seen as relatively innovative, secure, and reliable since they are obligatory compliant to all current airfreight regulations, one can hardly classify them as environmentally friendly or "green", since due to fixed time schedule the normal capacity or utilization rate of the "flying trucks" is much lower than regular road-based trucking services [2]. However, sustaining all basic air cargo characteristics as time efficiency and security, the RFS may be still perceived as more environmentally friendly compared to the airlines or real aircraft operations [2, 27]. Another important criterion that is valid both for GTC and the air cargo supply chain is the security. The EU Commission has introduced in 2015 new security regulations that are known also as ACC3 regulation that refer to every air cargo transported to the EU from the third countries, which is defined as the prerequisite for all flights that carry cargo or mail for transfer, transit or unloading and is applied for all airports of the European Union member states as well as for Iceland, Norway, Switzerland and the UK. Focusing here rather on the objectivity and functionality of the ACC3 regulations than on geographical scope of so called "green", "white" or "red" third countries, the main aim of ACC3 regulations is to ensure that only secured cargo is transported into the EU, meaning that each air cargo carrier needs a special ACC3 designation certificate for cargo flights into the EU for each airport that is situated outside the EU zone. Thus, air cargo in particular and the whole air cargo supply chains in general must fulfil a set of security requirements in order to be allowed into the EU zone [9]. In this context, the main impact of the ACC3 framework is an increased security of the air cargo and the whole air cargo supply chain for Europe. Furthermore, since road feeder services officially replace or take over the role of air cargo service operators, the air-trucking along the GTC shall be compliant to the ACC3 regulations. On the other hand, in order to achieve ACC3 standards, the involved stakeholders expect an increase in costs, time and bureaucracy [2, 20, 39].

3 Methodology and Research Design

Although a number of qualitative research studies and definitions are available that relate to airports and airline business [22], connectivity in air transport networks [40], air cargo and its role in the aviation business [17] supply and demand of the air cargo services [1, 2]. However, too less attention has been paid to special operational requirements of the regional airports and possible role of road feeder services as a possible business model to enter air cargo market, especially in the context of integration opportunities into the GTC. Based on the theoretical approach for hub-and-spoke airport networks of Sugiyanto et al. [17]; strategies' evaluation approach of relevance of air cargo business of Reis [5]; on the qualitative analysis of competitive rivalry in the air cargo industry suggested by Zondag [41]; as well as on Green Corridor Balanced Score Card of Prause [8], the authors of this study propose the following assessment matrix on the matter of the involvement potential of regional airports into the air cargo market by means of air-trucking services and its possible compliance to the concept of Green Transport Corridor (Table 1).

Table 1. RFS and GTC compliance model assessment.

Main criteria element	Assessment indicator
GTC balanced scorecard	Sustainability Growth perspective Cooperation intensity (data exchange, coordination needs) Cooperation quality
Air cargo demand side	Regional business concentration indicator Customers' awareness about air cargo services Air cargo volume potential in airport's catchment area
Air cargo supply side	Quality of regional road infrastructure and landside access Quality and availability of infrastructure facilities Availability of logistics service providers; Multimodality competences and competition environment Airport's infrastructure expansion potential Level of airports' air cargo marketing Compliance to ACC3 security regulations

The applied weighting scale of the assessment indicators (incl. the weight distribution between evaluation criteria) has been based on the overall compilation of the experts' evaluations and the results of the experts' interviews fulfilled in the framework of the BACN, EWTC-II, BSR Transport Cluster, BB projects. Based on the assumptions that regional airports may be positioned as "spokes" or feeders in the air transport network and that the RFS basically fulfil the "feeding" function, it might be assumed that the regional airports may also benefit from air cargo sector. In the framework of this study the following research questions are investigated: can air trucking services of regional airport be integrated or compliant to the concept of Green Transport Corridor and be utilized as a sustainable business model? With the regard to the specifics of the air cargo market in the BSR, the empirical data includes qualitative expert interviews and surveys produced in the framework of the EU funded research projects: BACN, BB, EWTC-II. The empirical data was collected form of experts' statements involved into the projects' activities, external experts' evaluations, project documentation and observations gathered though interviews and surveys. Following target groups and relevant stakeholders participated in the surveys and expert interviews (a) representatives from Transport Ministries and Airport Management; (b) representatives from Transport and Logistics companies from participating regions; (c) representatives from the academic side, (d) expert from aviation sector, air cargo security and air cargo freight sector. In terms of the presented investigated case studies, 67 qualitative interviews were conducted and evaluated.

4 Findings

In the framework of the EU research projects "Baltic.AirCargo.Net", "Baltic Bird" and "East West Transport Corridor II" the applicability of the concept of Green Transport Corridors and the role of regional airports in the corridors has been investigated. The presented case study has been chosen here by the authors since it has been identified in the research projects that the air cargo business model may be a potential driver for sustainable development, road feeder services have been named as the first realistic step in the air cargo sector especially for this airport. Due to the fact, that Grodno Airport is situated outside the EU area, it has been selected for the case study analysis, since the relevance of ACC3 regulations may be applied and investigated here.

Case Study—Grodno Airport (Belarus). Grodno Airport belongs to five regional airports in the Republic of Belarus that is situated near Grodno city that is located close to the Polish and Lithuanian boarder. Road is the most used transport mode for the passengers and the cargo transport in the region. Only one regular flight to Kaliningrad two times per week is offered at the moment. Grodno Airport is a 100% state-owned airport operated by national Transport Ministry—BELAERONAVOGATSIA. Navigation services are the main revenue source of Grodno Airport. Collected empirical data demonstrated considerable deviations between internal and external evaluations, e.g. airport's internal stakeholders tend to overestimate airport's own tangible and intangible criteria; parallel to this a certain underestimation of risks or inadequate quality of cooperation criteria has been noted. Although, the readiness to take appropriate risks is considered as indispensable prerequisite for the sustainable and well-established business attainment nowadays [9]. An obvious underestimation of performance indicators may distort decision-making process for the business development plans of the given regional airport and imply severe and irreversible impact for the business sustainability with long-lasting negative consequences [42].

Following assessment scale of the given criteria was applied (very good: 5; adequate: 4; sufficient: 3; insufficient: 2; very poor or absent: 1) (Table 2).

Table 2. RFS and GTC compliance model assessment of Grodno airport.

Criteria	Assessment indicator	Sore
GTC compliance indicators (GTC Balanced Scorecard)	Sustainability	2
	Growth perspective	1
	Cooperation intensity	2
	Cooperation quality	1
Air cargo demand side	Regional business concentration	2
	Customers' awareness about air cargo services	2
	Air cargo volume potential	3
Air cargo supply side	Quality of regional road infrastructure	4
	Quality and availability of infrastructure facilities	3
	Availability of logistics service providers	3
	Multimodality	3
	Airport's infrastructure	2
	Level of airports' air cargo marketing	2
	Compliance to ACC3	1

Preliminary analysis of Grodno Airport has shown a number of gaps in the evaluation between internal stakeholders and external experts. Grodno Airport's

Management and the airport's stakeholders have identified that IT-based organizational process and the quality of the offered intermediate services can be considered as distinctive resources. External experts have identified rather low cooperation intensity and cooperation quality. Moreover, the cross-referencing of the gained results has demonstrated a tendency of overestimation of the assessment criteria by the internal stakeholders. The external experts identified the following diversification criteria in Grodno Airport as poor:

- Poor growth perspectives and low level of innovation activities;
- Poor level of cooperation and competition sophistication, mainly due to national regulations imposed by National Air Line, i.e. Belavia;
- Low level of value added services, including deficit of specialized services and support;
- Low level of logistics services;
- Absence of adequate cargo facilities and cargo terminal(s).

The study results showed that a close collaboration with relevant national authorities, home logistics companies, business lobby and foreign airlines is a necessary precondition to increase compliance indicators for Green Transport Corridor. Furthermore, the development and implementation of the RFS by deploying ACC3 compliant air cargo destinations outside EU via Grodno airport with the EU airports, might be the first realistic step for Grodno Airport to enter the air cargo market. For those air cargo destinations outside the EU without ACC3 certification Grodno Airport can be developed to a long haul air cargo base due to its proximity to the European green transport corridors. The business model for Grodno Airport can be to operate as an air cargo link to non—ACC3 destinations where incoming and outgoing cargo is forwarded by normal truck or rail, e.g. between "Rail Baltica" and "EWTC" and Grodno via the Belarus border. This solution may offer a cost efficient and quick air cargo connection link between the EU and long haul destinations that do not comply with the ACC3 regulations. However, it may require establishment and implementation of policy regulations for the air-trucking services, enabling border crossing procedures between the EU and Belarus as it is foreseen by the RFS, i.e. with no ordinary customs control on the road cross-border pints, but in the respective customs area of the connected airports.

5 Conclusions

Regional airports are facing a strong competition in the Baltic Sea Region and the opportunities that are being offered by the air cargo market, the concept of Road Feeder Service and Green Transport Corridor may strengthen the economic viability of the airports and the regions. Results of the study demonstrated that air cargo business may offer to the regional airports an additional option to the passenger traffic and secure airport's sustainability. Furthermore, the analysis showed that although regional airports may mainly act as feeders for airport-hubs, integration in the air cargo supply chain and considering the perspectives of GTC may bring many advantages not only to airports' stakeholders, but also to the regional community in terms of economic and social benefits, as well as improved accessibility to a region, with a precondition that relevant national authorities, logistics companies and airlines cooperate in order to increase compliance indicators for GTC. The study also showed that belly cargo concept is rather inappropriate for regional airports, since it is based on the intercontinental passenger flights; on the other hand—RFS might be at this time the only one realistic concept for regional airports when they are trying to access to the air cargo market, especially utilizing advantages of GTC. The above-presented results showcase that regional airports shall accept the usability of the RFS or air-trucking concept in the context of GTC much stronger than they do it today. Thus, it is of a vital importance to raise public and political awareness of the integration potential of RFS into GTC concept and its role in the international logistics network; it shall be more efficiently adapted to the EU and BSR transport policies as well as to regional development strategies.

References

1. Beifert, A.: Regional airports' potential as a driving force for economic and entrepreneurship development. Entrepreneurship Sustain. **3**, 228–243 (2016). https://doi.org/10.9770/jesi.2016.3.3(1)
2. Beifert, A.: Role of air cargo and road feeder services for regional airports—case studies from the baltic sea region. Transp. Telecommun. **17**(2) (2016). https://doi.org/10.1515/ttj-2016-0008
3. COM: Keep Europe moving—sustainable mobility for our continent. MID-term Review of the European Commission's 2001 Transport White Paper. Commission of European Communities, Brussels (2006)
4. Fewings, R.: Provision of European airport infrastructure. Avmark Aviation Economist, 18–20 July 1999 (1999)
5. Reis, V., Macário, R.: Spectrum of competitiveness of passenger airlines on air cargo market. In: Air Transport Research Society Conference, 27–30 June, Abu Dhabi (2009)
6. Kasarda, J.D., Green, J.D.: Air cargo as an economic development engine: a note on opportunities and constraints. J. Air Transp. Manag. **11**, 459–462 (2005)
7. EWTC: Green Corridor Manual—Task 3B of the EWTC II Project (2012)
8. Prause, G.: A green corridor balanced scorecard. Transp. Telecommun. **15**(4), 299–307 (2014). Transport and Telecommunication Institute, Latvia. https://doi.org/10.2478/ttj-2014-0026
9. Hallside, M., et al.: Implementing Regulation (EU) No 1082/2012 amending Commission Regulation (EU) No 185/2010 in Respect of EU Aviation Security Validations—Evaluation report (2012)
10. EU Commission: Community Guidelines on Financing of Airports and Start-up Aid to Airlines departing from Regional Airports. European Commission, Brussels (2005/C 312/01) (2005)
11. Dennis, N.: Airline hub operations in Europe. J. Transp. Geogr. **2**(4), 219–233 (1994)
12. White, L.J.: Economies of scale and the question of "National Monopoly" in the airline industry. J. Air Law Commer. **44**(3), 545–573 (1979)
13. FTLAP: Communication from the Commission: Freight Transport Logistics Action Plan. Commission of European Communities, Brussels, 18 October 2007 (2007)

14. Wei, S., Yanji, M.: Hub-and-spoke system in air transportation and its implications to regional economic development-a case study of United States. Chin. Geogr. Sci. **16**(3), 211–216 (2006)
15. Rietveld, P., Brons, M.: Quality of hub-and-spoke networks; the effects of timetable co-ordination on waiting time and rescheduling time. J. Air Transp. Manag. **7**, 241–249 (2001)
16. Morrison, S., Winston, C.: The Economic Effect of Airline Deregulation. Brookings Institution, Washington, D.C. (1986)
17. Sugiyantoa, G., Santosaa, P.B., Wibowoa, A., Santib, M.Y.: Analysis of hub-and-spoke airport networks in Java Island, based on cargo volume and freight ratio. In: The 5th International Conference of Euro Asia Civil Engineering Forum (EACEF-5), Procedia Engineering, vol. 125, pp. 556–563 (2015)
18. Barrett, S.D.: Airport competition in the deregulated European aviation market. J. Air Transp. Manag. **6**(1), 13–27 (2000)
19. Hübl, L., Hohls-Hübl, U., Wegener, B., Schaffner, J.: Hannover airport—Ein Impulsgeber für die Region. Beiträge zur regionalen Entwicklung, Heft Nr. 84. Kommunalverband Großraum, Hannover (2001)
20. Bootsma, P.D.: Airline flight schedule development; analysis and design tools for European hinterland hubs. University of Twente, Utrecht (1997)
21. Urbatzka, E., Wilken, D.: Future airport capacity utilisation in Germany: peaked congestion and/or idle capacity. J. Air Transp. **9**(2), 111–132 (2004)
22. Derigs, U., Kurowsky, R., Vogel, U.: Solving a real-world vehicle routing problem with multiple use of tractors and trailers and EU-regulations for drivers arising in air cargo road feeder services. Eur. J. Oper. Res. **213**(2011), 309–319 (2013)
23. Ihde, G.B.: Transport, Verkehr, Logistik. Vahlen, Munich (2001)
24. Boeing: World Air Cargo Forecast 2014–2015 (2015). http://www.boeing.com/resources/boeingdotcom/commercial/about-our-market/cargo-market-detail-wacf/download-report/assets/pdfs/wacf.pdf. Accessed Aug 2017
25. Boonekamp, T.: Network quality and catchment area analysis in the air cargo industry. SEO Economic Research, Amsterdam (2014)
26. Grandjot, H-H., Roessler, I.: Air Cargo Guideline, Munich, vol. 271, pp. 86–88 (2007)
27. Heinitz, F., Hirschberger, M., Werstat, C.: The role of road transport in scheduled air cargo networks. In: 2nd Conference of Transportation Research Group of India (2013)
28. Vahrenkamp, R.: Der Gütertransport im internationalen Luftverkehr. Internationales Verkehrswesen. March(55), pp. 71–75 (2003)
29. Copenhagen Airport: Air cargo statistics of Copenhagen Airport (CPH, 2016) (2017)
30. BSR Transport Cluster: Transport research project in the framework of the INTERREG IVB Programme (2012). http://transportcluster.eu
31. Prause, G., Hunke, K.: Sustainable entrepreneurship along green corridors. J. Entrepreneurship Sustain. **1**(3), 124–133 (2014). https://doi.org/10.9770/jesi.2014.1.3(1)
32. Lee, P.: Measuring supply chain integration: a social network approach. Supply Chain Forum. Int. J. **6**(2), 58–67 (2005)
33. Gladen, W.: Performance Measurement—Controlling mit Kennzahlen, 3rd edn. Gabler-Verlag, Wiesbaden (2005)
34. Sarkis, J.: Introduction. Greener Manag. Int. **35**(3), 21–25 (2001)
35. Hunke, K., Prause, G.: Management of green corridor performance. Transp. Telecommun. **14**(4), 292–299 (2013). https://doi.org/10.2478/ttj-2013-0025
36. Schröder, M., Prause, G.: Risk management for green transport corridors. J. Secur. Sustain. **4** (4) (2015)
37. Prause, G., Schröder, M.: KPI building blocks for successful green transport corridor implementation. Transp. Telecommun. **16**(4), 277–287 (2015)

38. Sydow, J., Möllering, G.: Production in Networks, 2nd edn. Franz Vahlen, Munich (2009)
39. https://www.cph.dk/en/about-cph/b2b/airline-sales/cargo1/trafikstatistik/. Accessed October 2017
40. Veldhuis, J.: The competitive position of airline networks. J. Air Transp. Manag. **3**(4), 181–188 (1997)
41. Zondag, W.J.: Competing for air cargo—a qualitative analysis of competitive rivalry in the air cargo industry. Master thesis, Free University of Amsterdam, Amsterdam (2006)
42. Machina, M., Viscusi, W.: Economics of Risk and Uncertainty (v. 1), pp. 67–75. Elsevier's Science, Oxford (2014)

Condition Monitoring of Helicopter Main Gearbox Planetary Stage

Aleksey Mironov[1(✉)] and Deniss Mironovs[2(✉)]

[1] D un D centrs, 7 Margrietas Str., 1046 Riga, Latvia
info@ddcentrs.lv

[2] Institute of Materials and Structures, Riga Technical University, 6a Kipsalas Str., 1048 Riga, Latvia

Abstract. Planetary stages of gearboxes remain one of the main issue of helicopter health monitoring because of its complicated vibration signature containing multiple modulated signals. Authors analysed different vibration techniques allowing to step ahead in planetary gearboxes diagnostics and found some limitations for its industrial application. Based on the overview new requirements are considered to advance technique for planetary gears diagnostics. Aiming to develop diagnostic abilities of planetary gears, the paper discusses basic principles of its time-spatial model and main stages of data development, including: data separation by narrowed time window, vibration frequency range choice and transfer functions, spatial mesh modelling and Spatial Time Domain Distribution (STDD). Samples of STDD application on a helicopter gearbox are considered. There are conclusions about practical application availability of STDD for helicopter gearbox diagnosis.

Keywords: Vibration diagnostics · Planetary gears · Fault identification

1 Introduction

Main gearbox of a helicopter is a critical part of power transmission. Any damage happened in flight, especially in main drive may cause fatal consequences. To increase flight safety and to reduce operation costs the Health and Usage Monitoring Systems (HUMS) are used since early 1990. Vibration analysis techniques are applied in HUMS as the main tool for mechanical units' condition monitoring. In regards to main gearbox the principal task of vibration diagnostics is to detect in time fatigue cracks in gears and other rotating parts. Planetary stages of gearboxes became the most problem because of its complicated vibration signature that contains multiply modulated signals [1].

There are different approaches are applied for vibration analysis and damage detection in planetary stages. Many works consider diagnostic indicators of gears problems in frequency domain. Starting from simplest spectral analysis [2] different sources consider the spectral characteristics of vibration signal [3], energy ratio based on difference spectra [4], non-iterative deconvolution approach called Multipoint Optimal Minimum Entropy Deconvolution Adjusted [5]. The Harmonic Index feature in [6] is defined as the amplitude sum of all apparent sidebands of a specific gear meshing harmonic of the raw data, and the Intra-Revolution Energy Variance feature is

I. Kabashkin et al. (Eds.): RelStat 2018, LNNS 68, pp. 421–430, 2019.
https://doi.org/10.1007/978-3-030-12450-2_39

defined as the variance of the vibration energy around a specific gear meshing harmonic within an individual revolution [6]. An analysis of experimental data in frequency domain is also used for diagnostic indicators search, like [7] or [8] combining the techniques of enveloping, Welch's spectral averaging and data mining-based fault classifiers. Using the Short Time Fourier Transform [9] an amplitude-frequency modulation is observed on the run up of a gear. In [10] authors apply cyclo-stationary analysis and application of spectral correlation techniques on mining vibration signals. Also [11] uses cyclic load variations extracted from the instantaneous speed estimated via a time-frequency spectrogram of vibration signals. To separate modulations in planetary gearbox, a modulation signal bi-spectrum based sideband estimator was developed and used to achieve a sparse representation for the complicated signal contents, which allows effective enhancement of various sidebands for accurate diagnostic information [1, 12]. Assuming that the gearbox and the sensor revolve inversely at the speed of planet carrier, the problem can be transformed into two easier parts: research on fixed-shaft gearbox signal model and research on influence of sensor spinning [11, 13]. An integrated time-domain averaging (TDA) is proposed under the background, which introduces a pre-processing step—demodulation in traditional TDA and the revolution period estimation if no phase reference available [14]. Though frequency domain based techniques are widely used, its application depend strongly from structural specialties of gearboxes.

The technique using time windows for initial vibration separation [15–17] found the widest application for helicopter planetary gearbox. Time Signal Averaging (TSA) acknowledged earlier for simple gear vibration is applied for planetary ones after preliminary separation by time window. The width of the vibration window is equal to single tooth mesh period. This window opens each time when the planet gear passes to capture the vibration signal section. This captured signal length conforms to the time window width. The phase position of the window is opening at the moment the selected tooth of specific planet passes the transducer attached to the annulus (ring) gear. As the window width is small in relation to the carrier rotation period, the transfer function between the zone of tooth contact and the transducer can be considered constant [1]. Averaging vibration signal in time sections corresponding the same tooth of the same planet passes the transducer zone, someone may accumulate data for further averaging. Averaged by such way data presents the waveform of transducer response to dynamic excitation of the selected planet tooth. Varying phase position of time window someone may estimate waveforms of other teeth for all planet gears. By the same approach sun gear teeth may be surveyed.

Later TSA followers [18, 19] offer to apply multiple time windows according to fixed ring gear teeth number. By this way transfer features of vibration paths between each tooth and a transducer to be taken into account. Another development of TSA technique was offered [20] where windowed vibration signals are recombined specifically for the targeted gear component by utilizing the geometric properties.

Cumulative effect from different vibration analysis techniques allowed breakthrough in planetary gearboxes diagnostics. However, its wide application discover some limitations there:

- TSA and familiar algorithms of vibration data development require high resolution phase signal. Unfortunately, most helicopters have no phase indicators and only tach generators instead there are for rotation speed measurement.
- Practically all above techniques consider to use only part of measured signals that are separated by windows. Therefore, for evaluation of all planetary gears above techniques require significant time for vibration data collection t.
- Very few techniques consider single measurement direction of vibration that limits technique's resolution.
- Planetary gears with fixed annulus (ring) gear are only one type of different planetary and differential gearboxes. There are many gearbox types where all gears rotate.

Taking above into account the main goals of advanced planetary gears diagnostics are considered as:

- coverage of all gearbox types, including differential and multistage ones;
- processing vibration data of gearboxes using tachometers only without phase indicators,
- diagnostics of helicopter gearboxes in-flight.

Aiming to achieve above targets this paper considers the model of planetary gear interactions able to provide:

- capability to monitor helicopter gearbox in real time,
- optimal utilization of vibration signal by:
- using time-continuous measured vibration signal without exemptions,
- 3-axial measurement of vibration signal on a gearbox housing,
- detection of any potential failures of planetary/differential gears, including planet driver cracking, assembling errors, etc.

2 Spatial-Time Model of Gear Interactions

2.1 Data Separation

From diagnostic point of view the most valuable time segment of planetary gear is one revolution period of slowest gear of a gearbox. Vibration data within this period could be accumulated and averaged if to find the way how to extract the data caused by each specific tooth interaction from raw signal. During this longest period of gear rotation each tooth multiply meshes, so accumulation of these interactions and then averaging would be sufficient for the gear characterization. Accuracy of such characterization depends of specific tooth mesh number within the period. For instance, one planet interacts with all ring teeth during its one revolution thus data enhancing for each planet could be sufficient enough. Opposite, ring gear teeth have limited meshes during a period (as planet number in this planetary gear) so, less enhancement is available and less resolution in one period. Anyway, even from data of single period some characterization of planetary gears could be done in case of proper data mining. To reduce errors the data of consecutive periods could be averaged.

Further consideration will base on the sample of planetary gearbox consisting of N fixed planet gears (z_g teeth each) and both rotating ring (z_r) and sun (z_s) gears. Typical TSA approach for vibration signal separation uses time window that width is equal to interval of single tooth mesh with planet and ring gears. Using planet teeth number z_r and rotation speed ω_r of ring this interval is determined as

$$\Delta T_r = 2\pi/z_r\omega_r. \tag{1}$$

Such approach works well when the transducer is fixed to immovable ring next to mesh zone. In this case the closest mesh energy reaches a transducer almost undamped but opposite mesh energy is damped multiply. Unfortunately, for common cases, where vibration paths from different teeth vary slightly, effectiveness of such windowing is limited. The problem is that actually within time interval ΔT_r all N planets interacts simultaneously with both ring and sun. It means 2 N meshes occurs during typical TSA window ΔT_r. So, collecting data within such time interval someone summarizes responses of all 2 N meshes.

To separate responses of above 2 N mesh the width of window must be 2 N times less than in (1)

$$\Delta T_w = \pi/Nz_r\omega_r = {}^{T_r}/_{Nz_r}. \tag{2}$$

So, separation of vibration data by single mesh window ΔT_w provides data for each elementary interaction of gears [1].

2.2 Gear Mesh Modelling

Impulse meshing. For second step to vibration data separation the impulse nature of tooth mesh to be recognized. Tooth stress analysis made in [21, 22] for spur gears illustrated by diagram on Fig. 1a. Stress cycles measured on loaded side of the sun tooth typically have short positive impulse part and then smooth descent to negative zone. Short impulse part is evident shock, the smooth one is “rolling” of teeth under the descending load. A shock amplitude is related to tangential velocity difference between teeth before its contact. Backlashes in gears cause this difference and in depends on its sign (positive or negative) strain gauge may sense it as bending stress growth or down. Considered interactions could be schematically modelled as on Fig. 1b. Impact is shown as fast growing force $F(t)$ on a tooth that starts since moment t0 (before contact) till a moment t_1 of maximal force $F_{\max}$ (both teeth velocities are equal). Then force is going down until teeth disconnection. Impact duration $\Delta t_{im} = t_1 - t_0$ depends of loaded tooth stiffness but maximal force value $F_{\max}$—of velocity difference before mesh contact $\Delta v_i^{r-g} = v_i^r - v_i^g$.

In a planetary gear with N satellites there are more than 2 N mesh points as a planet simultaneously is in mesh with sun and ring by more than one tooth each. There is a problem to separate summarized response of a transducer to smooth phases from 2 N mesh points if its vibration paths differ a few. However, impulse phases of above meshes follow each other with average time shift ΔT_w (Eq. 2) that allows response separation in time.

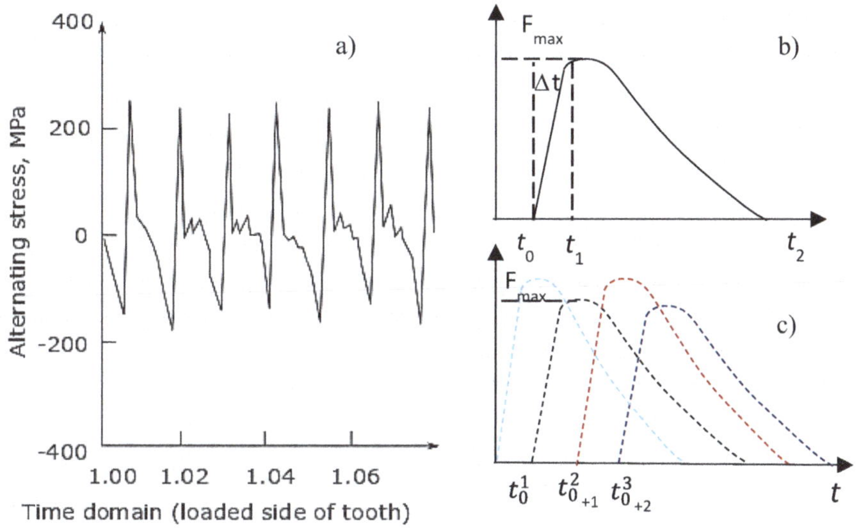

Fig. 1. Impulse nature illustration for spur gears meshing: **a** Fig. 6 from [22]; **b** impact and smooth parts of single teeth mesh and **c** multiple teeth mesh.

Figure 1c illustrates modeling of force $F(t)$ impulses from sequense of meshes. So, a transducer fixed on a gearbox housing may seize signals from each teeth mesh.

Another important case is that meshes follow according to kinematics of the gear. Therefore, from one side phase shift between meshes of planetary gears allows separation in time the signals of gear elements and from other side the knowledge of the gear kinematics provides opportunity for identification of above signal sources.

Vibration paths and frequency range. Transfer properties of vibration paths between mesh points and the transducer determine the response of last one. Transfer properties for each of K vibration paths from source S to transducer T (Fig. 2a) are described by transfer function H_s^k.

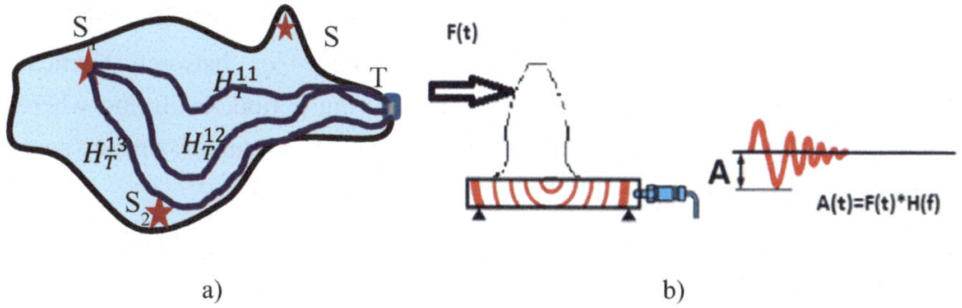

Fig. 2. Vibration paths and signal propagation.

Vibration paths differ each from other by length, materials, joints, etc. Integral transfer function H_{Σ} may describe combination of all paths between the source and the transducer.

$$H_{\Sigma}^{1}(f) = \sum\nolimits_{k=1}^{K} H_{T}^{k}(f). \tag{3}$$

Each $H_{1}^{k}(f)$ describes specific path through which wideband frequency signal (impulse) propagates from mesh point S_1 to transducer. Impulse signal propagates in material as stress waves and achieving a transducer actuate its electric signal generation (Fig. 2b). Waves shape and velocity varies along the path, so $H_{T}^{k}(f)$ is essentially nonlinear and demonstrate many resonances.

Captured amplitude A of the response (Fig. 2b) may be used for excitation estimation, for instance measuring peak-peak and total energy of the response in time window ΔT_w. To allow that the sampling frequency F_d must exceed maximal frequency of vibration. That is why frequency structure of measured vibration to be considered. Taking for instance, planetary gear with N fixed satellites and z_r teeth of the ring gear the number of excitation impulses (meshes) in a period is $2Nz_r$. So, meshing frequency F^{imp} could be estimated as $F^{imp} = 2Nz_r\omega_r$ that for typical helicopter gearbox is within 3–10 kHz. Stress waves carrying energy of impulse excitation have combined structure containing both longitudinal and transverse waves. By analogy with radio engineering these waves may be considered as carrier modulated by impulse excitations of mesh. Different materials and joints of typical vibration path nonlinearly transform this wave carrier so measured response has wide frequency range. According to experimental research conducted by D un D centrs main vibration energy of vibration paths of typical gearboxes was observed in 3–7 kHz band.

Thus, in depend of gearbox type both excitation and its carrier frequency could be the maximal frequency to be sampled. Assuming, for instance, 10 digital values of vibration as minimal for wave response parameters estimation the sampling frequency for above case to chosen about 100 kHz.

Identification of gear tooth mesh. To use measured data entirely the mesh window ΔT_w of Eq. (2) to be applied Nz_r times for one period T_r of the ring revolution. Time shift between multiple windows is the same as window width $\Delta t = \Delta T_w = {}^{T_r}/_{Nz_r}$. So, applying mesh windows to vibration someone will have for whole period the data of Nz_r mesh windows with a step ${}^{T_r}/_{Nz_r}$ between. Vertical line on Fig. 3a schematically reflects time vector of one ring period revolution where Nz_r mesh windows follow each other with $\Delta t = {}^{T_r}/_{Nz_r}$ step. Based on known step someone may calculate position $p_j = j\frac{T_r}{Nz_r}$ of jth mesh within the period.

By this way p_j mesh located in specific position (n_{ij}, r_j, s_j) of the planetary gear has also geometrical coordinates as the point of solid body having transfer function $H_j = H(n_{ij}, r_j, s_j)$.

Therefore, based on kinematics rules and known data of the planetary gear each single mesh window can be definitely related to specific teeth and the mesh point can have transfer function to a transducer [1].

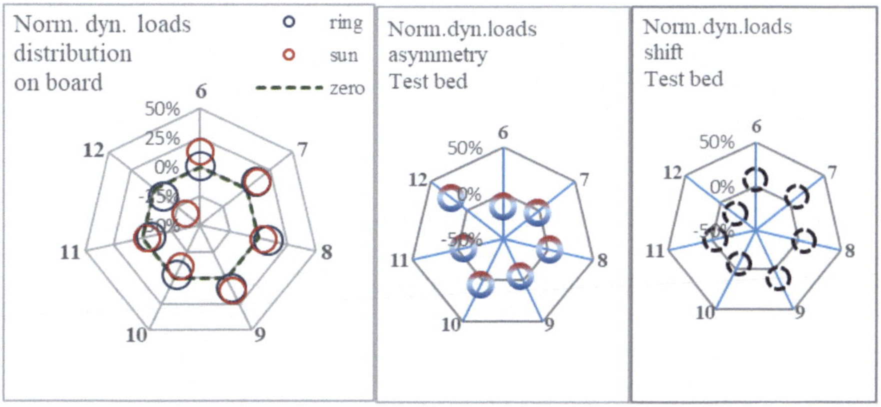

a) b) c)

Fig. 3. Normalized dynamic loads distribution diagram for closing stage of differential gear: **a**—planet loads from ring and sun; **b**—loads asymmetry; **c**—loads shift.

3 STDD Application on a Helicopter Main Gearbox

The sample of STDD technique trial application on helicopter main gearbox is described below. This differential 2-stage epicyclical gearbox of medium size helicopter could have a problem with abnormal distribution of dynamic loads between planets that may lead both to planet carrier cracking and to teeth and bearings failure.

Unevenness of dynamic loads distribution between sun, ring and satellites may appear during assembling of new or overhauled gearbox caused by misalignment, technical requirements violation, etc. Also external factors may cause difference in loads. If so, dynamic loads reassign between satellites, sun and ring that means loads reduction for some teeth and higher forces on others. Excessive loads on any teeth, gears and bearings means its abnormal operation and accelerated wearing.

The STDD—technique applied to operating gearbox decides two specific tasks:

- separation of dynamic loads unevenness caused by external factor,
- estimation of unevenness loads distribution between planets of the gear as quantitative appraisal.

Separation of external factor influence is important to avoid mistake in gearbox quality estimation. Axis misalignment between ring (sun) shaft and a torque load simulator when a gearbox mounted on a test bed may cause an asymmetry of planet loads.

For further consideration it is assumed that misalignment of concentric gears causes asymmetric loads but any irregularity of planet axes leads to arbitrary loads distribution.

Based on STDD two values of normalized dynamic loads for each planet are calculated: for planet–ring mesh ($\bar{a}_n^{p-r}$) and planet–sun ($\bar{a}_n^{p-s}$). Each value is peak-peak amplitude $a_n^{\max} - a_n^{\min}$ selected from k values of STDD transformed vibration signal referred to average mean of these k values

$$\bar{a}_n^p = \left(a_n^{\max} - a_n^{\min}\right)/_{a_n^{av}} = \left(\max\left(a_n^i\right) - \min\left(a_n^i\right)\right)\Big/\sum\nolimits_{n=1}^{N} a_n^i, \tag{4}$$

where $i = 1, 2, \ldots, k$.

"Radar" diagram having N numbered rays (Fig. 3a) is suit well for imaging normalized estimates of dynamic loads distribution in the gear with N planets. Small circles on rays of radar diagram denote values of STDD calculated load parameters, where each ray corresponds to according planet. Normalized dynamic load value for planet–ring mesh ($\bar{a}_n^{p-r}$) are outlined by blue and planet–sun ($\bar{a}_n^{p-s}$) by red circles. As a sample for consideration on Fig. 3 there is the dynamic loads distribution diagram for closing stage of two stages differential gear tested on the bed.

Presented diagram demonstrates each planet has different loads in mesh with ring and sun. This difference is the biggest for planets 6 and 12 while planets 7–9 and 11 have minimum scatter of loads. Another thing is a "shift" of loads values on diagram to right. This is because average means of inner and outer mesh normalized loads of planets 6–9 are positive but others—negative. To characterize above noted features of loads distribution there are two parameters considering two factors of loads distribution between planets interacting both ring and sun gears:

asymmetry of planet loads from ring and sun—normalized difference

$$\delta\bar{a}_n^{av} = \left(\bar{a}_n^{p-r} - \bar{a}_n^{p-s}\right)/\left[\frac{1}{2N}\sum\nolimits_{n=1}^{N}\left(\bar{a}_n^{p-r}; \bar{a}_n^{p-s}\right)\right]; \tag{5a}$$

unevenness distribution between planets - normalized average characterizes

$$\delta a_n^{av} = \left(a_n^{p-r} + a_n^{p-s}\right)/\left[\frac{1}{N}\sum\nolimits_{n=1}^{N}\left(a_n^{p-r}; a_n^{p-s}\right)\right]. \tag{5b}$$

Radar diagrams on Fig. 3 illustrates above two factors of loads distribution calculated using Eq. 3b and c. Planet 12 on Fig. 3b shows "outstanding" behaviour meaning 24% higher dynamic loads in ring mesh while planet 6 contrary overcharged by sun for 13% only. This diagram shows the difference between planets loads from ring and sun sides. Figure 3c illustrates some shift to right side charging more planets 6–9. Misalignment between gearbox output shaft (related to ring) and the shaft of breaking system may cause such shift in parameter distribution. Above diagrams have utilitarian functions during testing. If overloading of some planets (Fig. 3a) exceeds a threshold the diagram indicates the wrong planet to be treated. Second diagram may work as a tool for shafts aligning to avoid overcharging of gear teeth during the run-in process on a test bed.

By this way STDD based diagnostic technique allows accurate estimation of loads distribution abnormality in a gearbox separating ambient and structural factors influence.

4 Conclusions

To get round limitations existing for condition monitoring of planetary gears the spatial-time model of gear interactions is developed. The model identifies gear meshes of a planetary gear as spatially distributed impulses of wide frequency range. Each impulse (mesh) relates to interaction between the tooth of specific planet and sun or ring gear. Taking into consideration the construction of helicopter main gearbox the model may account vibration paths between transducer and gear meshes. Using above model the Spatial Time Domain Distribution technique was developed and applied to actual helicopter main gearboxes. Practical application cases on helicopter gearbox demonstrate practical capabilities of spatial-time domain approach and STDD for helicopter gearbox diagnosis.

Acknowledgement. The paper uses materials related to research No. 18 “The study of planetary gearbox vibration diagnostic technology and creation of its demonstrator” of the project “Establishment of Engineering Systems, Transport and Energy Competence Centre” in cooperation with Central Finance and Contracting Agency of Latvia (1.2.1.1/16/A/008). This work contains the results of the research study, experimental testing and development vibration diagnostic techniques demonstrator.

References

1. Mironov, A., Mironovs, D.: Vibration diagnostics of planetary gearbox of engineering and transport machines based on spatial model application. In: 17th International Scientific Conference Engineering for Rural Development, 23–25 (2018).
2. Lahdelma, S., Juuso, E., Immonen, J.: Advanced condition monitoring of epicyclic gearboxes. In: 10th International Conference on Condition Monitoring and Machinery Failure Prevention Technologies, CM 2013/MFPT Poland (2013)
3. Feng, Z., Zuo, M.: Vibration signal models for fault diagnosis of planetary gearboxes. J. Sound Vib. **331**(22), 4919–4939 (2012)
4. Le, Y., et al.: Two new features for condition monitoring and fault diagnosis of planetary gearboxes. J. Vib. Control **21**(4), 755–764 (2015)
5. McDonald, G.L., et al.: Multipoint optimal minimum entropy deconvolution and convolution fix: application to vibration fault detection. Mech. Syst. Signal Process. **82**, 461–477 (2016)
6. Biqing, W., Abhinav, S., Romano, P., Vachtsevanos, G.: Vibration monitoring for fault diagnosis of helicopter planetary gears. In: Proceedings of 16th IFAC World Congress, Prague (2005)
7. Zhuang, L.: Experimental analyses of vibration and noise of faulted planetary gearbox. In: Proceeding of Inter-noise 2014, Melbourne, Australia (2014)
8. Yoon, J., He, D., Van Hecke, B., Nostrand, T., Zhu, J., Bechhoefer, E.: Vibration-based wind turbine planetary gearbox fault diagnosis using spectral averaging. Wind Energy **19**(9), 1733–1747 (2015)
9. Hammami, A., Alfonso, F., Viadero, F., Chaar, F., Haddar, M.: Dynamic behaviour of two stages planetary gearbox in non-stationary operations. In: Conference Surveillance 7, France (2013)

10. Bartelmus, W., Zimroz, R.: Vibration diagnostic method for planetary gearboxes under varying external load with regard to cyclostationary analysis. In: 527/2009 Wrocław (2009)
11. Miao, Q., Zhou, Q.: Planetary gearbox vibration signal characteristics analysis and fault diagnosis. Shock Vib. **2015**, Article ID 126489, 8 p. (2015)
12. Tian, X. et al: Diagnosis of combination faults in a planetary gearbox using a modulation signal bispectrum based sideband estimator. In: Proc. of 21st International Conference on Automation & Computing, University of Strathclyde, Glasgow, UK (2015)
13. Deshpande, L., Randall, R., Smith, W.: Diagnostics of planet gear faults from the effects of meshing with the ring and sun gears. Int. J. Cond. Monit. **3**, 36–41 (2014)
14. Jing, L.: An integrated time domain averaging scheme for gearbox diagnosis. In: Proceedings of 8th International Conference on Reliability, Maintainability and Safety, ICRMS (2009)
15. McFadden, P.D.: A technique for calculating the time domain averages of the vibration of the individual planet gears and the sun gear in an epicyclic gearbox. J. Sound Vib. **144**(1), 163–172 (1991)
16. Howard, I.M.: Epicyclic transmission fault detection by vibration analysis. In: Australian Vibration and Noise Conference 1990: Vibration and Noise-measurement Prediction and Control; Preprints of Papers, pp. 171–178. ACT: Institution of Engineers, Barton (1990)
17. McFadden, P.D., Howard, I.M.: The detection of seeded faults in an epicyclic gearbox by signal averaging of the vibration. In: Propulsion report 183, Melbourne: DSTO (1990)
18. Liang, X., Zuo, M.J., Chen, W.: Dynamics-based vibration signal modeling for tooth fault diagnosis of planetary gearboxes. In: Fault Diagnosis and Detection, IntechOpen (2017)
19. Liang, X., et al.: Vibration signal modeling for a planetary gear set. Eng. Fail. Anal. **48**, 185–200 (2015)
20. Yoon, J., He, D., Van Hecke, B., Nostrand, T., Zhu, J., Bechhoefer, E.: Planetary gearbox fault diagnosis using a single piezoelectric strain sensor. In: Proceedings of the 2014 Annual Conference of the Prognostics and Health Management Society, United States, pp. 118–127 (2014)
21. Cornelli, R.W., Westerveltt, W.W.: Dynamic tooth loads and stressing for high contact ratio spur gears. J. Mech. Des. **100**(1), 69–76 (1978)
22. Krantz, T.L.: Gear tooth stress measurements of two helicopter planetary stages. In: 6th International Power Transmission and Gearing Conference sponsored by the American Society of Mechanical Engineers. NASA Technical Memorandum 105651 (1992)

Research and Practical Application of Vibration Transfer Functions in Diagnostics of Jet Engine

Aleksey Mironov[1(✉)], Pavel Doronkin[1(✉)], and Deniss Mironovs[2(✉)]

[1] D un D centrs, 7 Margrietas Str., 1046 Riga, Latvia
info@ddcentrs.lv
[2] Institute of Materials and Structures, Riga Technical University, 6a Kipsalas Str., 1048 Riga, Latvia

Abstract. The paper considers vibration paths properties of gas turbine engine non-rotating parts aimed for application for vibration diagnostics, specifically in Vibropassport™ as the methodical and algorithmic platform for practical application. Authors discuss reasons for transfer features research and application in vibration diagnostics. Some specialties of transfer features are considered for different types of excitation sources, like a whole rotor or separate gear. Low frequency range and wide frequency band vibrations generated by such sources need different approach to experimental testing techniques. The general view is given on experimental techniques for testing of integral aggregates and smaller excitation sources of an engine. There is analysis of transfer features of concentrated and distributed multiple sources as well as some samples of experimental testing and transfer coefficients determination. Main areas of transfer properties in application of engine vibration diagnostics are considered. The paper includes conclusions regarding discussed matters and tasks for further works.

Keywords: Transfer function · Aviation engine · Vibrational diagnostics · Modal properties

1 Introduction

There are problems in vibration diagnostics of aviation engines due to its multiple interactions between rotating and static parts as well as structural and operation specialties. Above difficulties still limit state of the art diagnostic systems in their ability to detect faults of all engine aggregates. Advanced vibration diagnostic system, like presented in [1] have to be capable for detection any failures, including bearings, compressor and turbine stages, combustion chamber, gears and many more.

To provide above mentioned tasks advanced diagnostic system of jet engine applies diagnostic parameters based on wide frequency band vibration [1, 2]. Vibration models substantiating such system consider vibration $a(t)$ as a function of two independent arguments as non-linear functions: dynamic loads inside engine $P(t)$ and stator's transfer function H:

I. Kabashkin et al. (Eds.): RelStat 2018, LNNS 68, pp. 431–439, 2019.
https://doi.org/10.1007/978-3-030-12450-2_40

$$a(t) = f[P(t), H(M, E, \zeta)]. \quad (1)$$

There are plenty of sources that actuate engine stator vibrations, including mechanical (engine parts with different relative velocity vectors) and gas dynamic (blades-flow-vanes) interactions. All dynamic interactions in jet engine are non-linear and non-stationary because of high complexity of processes. Technical state of engine aggregates influences dynamic interactions *P*(*t*) that causes modification of vibration structure. Typically, accelerometers measuring vibrations are mounted on a housing, therefore structural response on excitation is to be defined for technical state determination [3]. That is why vibration diagnostics application needs the transfer features of the engine $H(M, E, \zeta)$ to be defined both for identification of dynamic interactions in the engine and for structural properties of the housing. Elasticity (E), damping (ζ) and mass distribution (M), being functions of construction specialties and applied materials, define modal properties (modal frequencies, shapes and damping) of the vibration path.

The goal of this work is to analyze and to study experimentally the transfer functions of typical vibration paths of a jet engine in depend of excitation source types. There are some basic assumptions taken in this study.

Housing response on interior actuation is considered as time independent for most diagnostic tasks. Such consideration is based on assumption about steadiness of mechanical properties of structural elements in operational mode [4]. Modal properties of engine structure non-linearly depends of excitation frequency and energy. Frequency Response Functions (FRF) are typically used for modal properties characterization of a structure.

2 Transfer Function Dependence on Excitation Source Type

Multiple vibration sources of gas-turbine engine have different vibration signatures in time and frequency domains. Practically all interactions in jet engine have complex nature so, to simplify modeling of engine vibration we consider three basic kinds of interaction. *Quasiperiodic* interactions related to rotors operation are first type, deterministic *impulse* (short time) interactions are the second and interactions appeared as *random* are considered as the third type. Energy of vibration related to the first type interactions appears mainly in low frequency band, while impulse and random interactions generate vibration in wide frequency range.

For instance, energy transformation from gas flow to kinetic energy of rotor in jet engine is followed by impulse interactions of blade-vanes, gears and bearings. All together these interactions generate wide-band excitation applied to most engine parts.

Massive rotors on elastic supports respond to above excitation in low frequency range mainly. Mechanical modification of rotors and its supports influences its modal properties. Therefore, using modal properties representations (like FRF) for vibration diagnostics it is possible to detect rotor failures. Identification of modal properties of engine structure includes modeling and experimental stage that will be considered briefly below.

Separate load impulse generated by single interaction, like blade-vane interaction or teeth impacts has too small energy to oscillate engine rotor. However, its excitation propagates as waves through engine structure disseminating important information about operation of interacting elements. To collect this information the wide frequency range (up to ultrasound) to be considered as opposed to quasi periodical excitation appeared in low frequency band. To characterize mechanical properties in wide frequency band the transfer functions are used. Transfer function of each vibration path from a source to an accelerometer depends of geometry, material, number and types of joints.

There are different approaches for estimation of transfer functions of engine structure but experimental modal analysis techniques are common for any. Although presentation of transfer functions can be different the most typical is matrix method [5, 6]. This study considers the research of test engine transfer functions aimed for Vibropassport™ application. Frequency range and number of vibration paths of specific engine part define the order of applied vibration model and corresponding transfer function. Frequency range depends on interactions type described by the model. For instance, oscillation model of engine rotors typically considers 10–1000 Hz range but bearing and blade-vane interaction models may include range up to 100 kHz or even more. Vibration paths number is related to structural properties of engine, like vanes number or stator carrier. The simplest transfer function of 1st order relates transfer coefficient to frequency, the second order function describes dependence to frequency and path number, 3rd order transfer function includes also spatial dimensions.

Thus, for vibration diagnostics of all jet engine aggregates variety of transfer functions are required. For instance, to diagnose engine integral parts, like a fan or a turbine, algorithms of Vibropassport™ uses 1st order transfer matrix (vector) in low frequency band. For diagnostics of separate stage of compressor, where wide frequency range interactions, at least 2nd order transfer function to be used.

3 Experimental Techniques of Transfer Function Identification for Vibration Diagnostics

Taking above mentioned into account two stages of transfer function definition are needed for the test engine. First one is modal properties estimation and identification for integral parts of test engine. The second is transfer function measurement of multiple vibration paths. The techniques applied in the described experiment study require minimal engine disassembling for application of testing accessories.

3.1 Modal Properties of Integral Aggregates

Typical engine condition monitoring system uses 1st harmonic of rotor vibration as principal diagnostic parameter. Such parameter allows monitoring of integral engine units as fan, turbine and gas generator. Considering most of airliners with middle and large engines the frequency band for above vibration measurement does not exceed 500 Hz, which is not enough for diagnostics of impulse and random interactions.

For effective monitoring the set of modal properties of integral units and engine-frame system has to be identified. The technique for modal properties identification in low frequency band is well known and considers both testing excitation of rotor and response measurement on the housing. Controllable generator allows rotor excitation by shaker with sequentially variated frequency while load cell measures the force level. Figure 1a shows partially disassembled engine of typical construction, where core turbine is excited by the shaker.

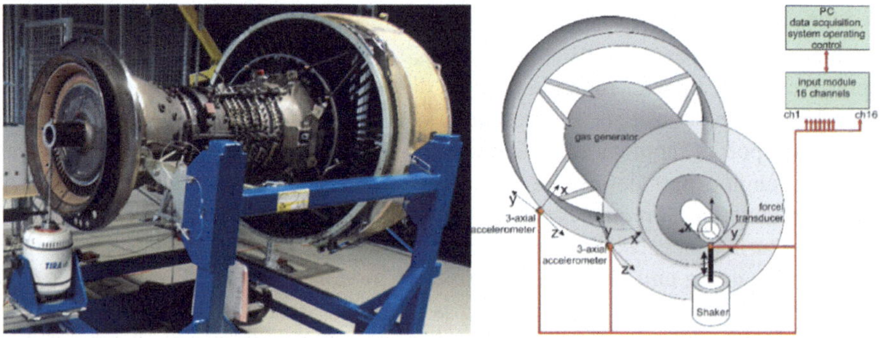

a) b)

Fig. 1. (a) Experimental study of gas generator modal properties; (b) Measurement setup.

Figure 1b shows measurement set up for this case. The measurement system includes multichannel data input module, controlled by computer with specialized software installed, two 3-axial accelerometers, mounted on the turbine and compressor housing. The load cell fixed on the rod connected to the shaker.

Response of the engine housing on test excitation is measured using 3-axial accelerometers providing spatial vector of vibration. Based on test results FRF is computed describing response to forced rotor oscillation in operation mode range. Actual demonstrator of vibration diagnostic technologies based on two shafts engine allows application of above experimental technique of modal properties estimation. As the sample of above technique application Fig. 2 shows the transfer functions of the engine gas core tested in frequency range 80–400 Hz by harmonic excitation force 10 N. FRF are computed using vibration data from two 3-axial accelerometers: at the turbine and at the compressor stators.

The diagrams illustrate how essentially vibration on the housing depends on transfer properties determined by excitation frequency. For instance, turbine accelerometer (Fig. 2a) at frequency 80 Hz responds 0.5 ms^{-2}/N only, while at turbine resonance at about 230 Hz (supposedly 1st rotor critical speed) FRF reach 3.0 ms^{-2}/N that is 6 times higher. For compressor (Fig. 2b) the ratio between maximal and minimal FRF values appears even more and exceeds factor of 10.

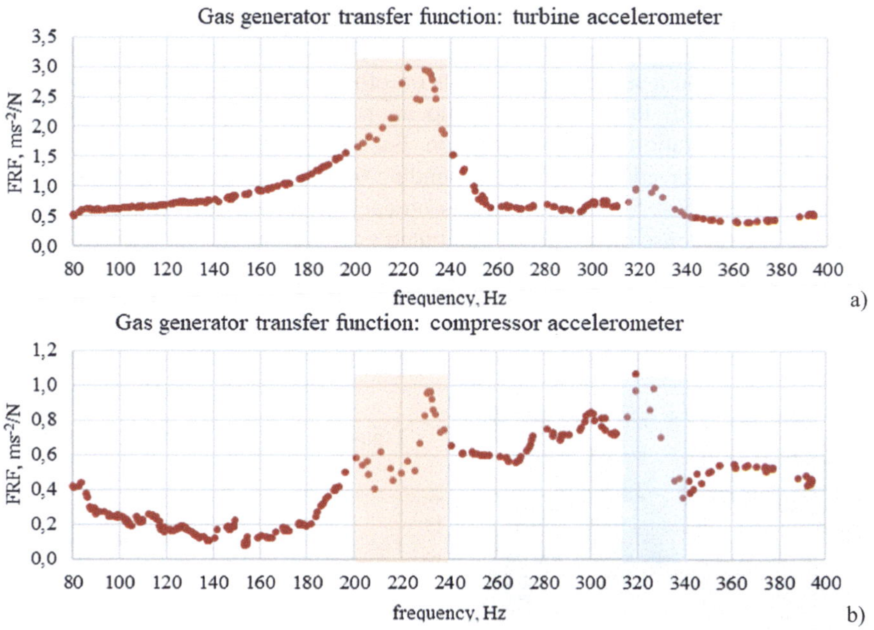

Fig. 2. FRF of gas generator excited from turbine side and vibration measured on casing of: turbine (a); compressor (b).

3.2 Transfer Function of Vibration Paths

Concentrated sources of wide band vibration. Vibration models of Vibropassport™ consider two typical types of impulse excitation in jet engine. Gears are typical *concentrated* sources which impulses propagate through bearings and then pass some stator parts. Bearings are also considered as concentrated sources. Guide vanes in compressor and turbine are typical *spatially distributed* sources. Such sources generate impulses of dual nature: from one side sources are correlated, because excited by wakes of same blades, from another side each vane generates its lift pulses independently from others. Though both types generate wide-band frequency excitation, there is difference in definition of its transfer function.

Experimental research of transfer function of wide-band sources differs from conventional approach to modal parameters estimation considered in above section. To study wide-band properties of specific vibration path the impulse excitation must have energy and structure compatible with parameters of original source. That is why the study begins from research of original excitation sources using both modeling and experimental works. Such works requires specific test rig of jet engine for impulse interactions simulation and appropriate equipment for measurement and analysis, like in [7]. When parameters of original sources of excitation has clarified then actuation type to be chosen. Any type of actuation considers measurement of actuating force as well as controllable parameters of actuation.

"Roving" hammer testing method is widely applied technique for experimental FRF definition. In discussed application this technique considers one accelerometer mounted on one side and a set of impact points on other side of vibration path, where force impulse (measured) to be attached. Allocation of these points on the engine housing depends of vibration path configuration. Each point of impact may have one, two or three Degree Of Freedom (DOF) in depending on vibration path model. The test procedure includes at least 5 repeated impacts on each DOF and registration of both actuation force and response on it. FRF is computed as power spectra ratio of response to impact. Figure 3a illustrates the single impact case using hammer with embedded load cell applied for experimental estimation of Frequency Response Function (FRF) of satellite gear of aviation planetary gearbox.

a)

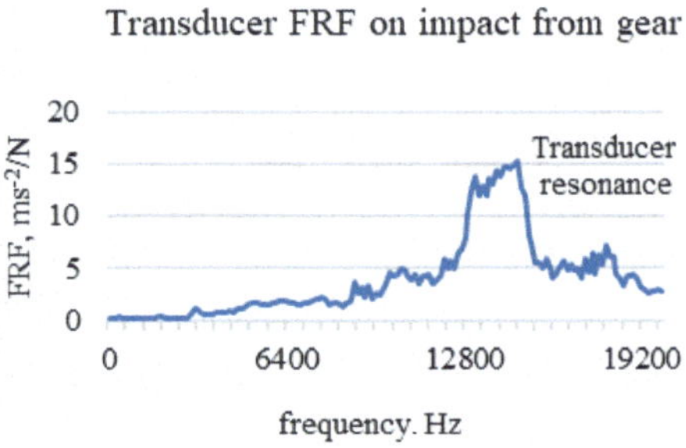

b)

Fig. 3. Testing the vibration path from satellite gear to transducer: impact hammer with load cell (a); FRF of vibration path (b).

Experimentally estimated FRF of vibration path from the gear to the accelerometer considers both transfer function of the structure (gear-bearing-sleeve-planet carrier) and amplitude-frequency response of the measurement system. A transducer with its mounting piece typically is the part of such system. Common transfer function in wide frequency band could have essential resonance related to transducer body (Fig. 3b). For instance, FRF value at gear meshing frequency is 0.28 ms^{-2}/N however, in transducer resonance zone FRF reaches 15 ms^{-2}/N.

Depending on vibration model the diagnostic parameters of Vibropassport™ need various versions of transfer function for vibration data processing. For instance, diagnostic parameter of gas generator rotor utilizes narrow band transfer function $H(f)$ as function to frequency, like on Fig. 3b. Gear diagnostic parameters typically need single value transfer coefficient k_Σ considering all frequency range of signal measurement. Considering above testing case, where the single load impulse of 0.2 ms length was applied, the summarized response computed in 0–1000 Hz range was 0.83 ms^{-2}/N but in wider range 0–20,000 Hz—69 ms^{-2}/N. So higher transfer properties in wide frequency range increase fault resolution in one-two orders higher than the parameter considering narrow frequency zone next to the teeth meshing frequency.

That is why vibration diagnostic parameters for impulse sources technical state estimation use wide frequency range. The simplest version of such parameter is integral transfer coefficient that considers sum response in wide frequency range.

Modal survey of each engine aggregate includes modeling and experimental identification of separate elements of any vibration path. The model considers transfer coefficient of vibration path as a product of separate element functions, like

$$k_{\Sigma} = \sum k_{joint} * \sum k_{dist}, \tag{2}$$

where k_{joint}, k_{dist}—decay coefficients for joints and distances accordingly.

For example, the transfer coefficient of planetary gear considered above (from gear to the transducer) was modeled using data of modal tests

$$k_{\Sigma}^{G} = 0.115e^{-0.61x}, \tag{3}$$

where coefficient 0.115 describes decay between gear tooth and the carrier and 0.61 exponent index reflects decay depending on distance (x) along the carrier body.

Spatially distributed vibration sources. Most of engine vibration sources are distributed around the engine and mutually related in contradiction to concentrated sources. Turbine guide vanes that generate impulse lift forces from nonstationary flow are typical sample. Each wake of blade upstream actuates single vane and amplitudes of wakes may reach half of nominal flow velocity so, impulse forces acting on vanes are essential. Vanes of one compressor/turbine stage are identical in theory however, vibration paths from each vane to transducer may differ a lot. Taking that into account diagnostic parameters for guide vanes have to estimate its individual coefficients for every stage.

To illustrate above, Fig. 4 shows the outline of compressor housing cross-section and result of its FRF experimental study. The study includes 26 tests (as vanes

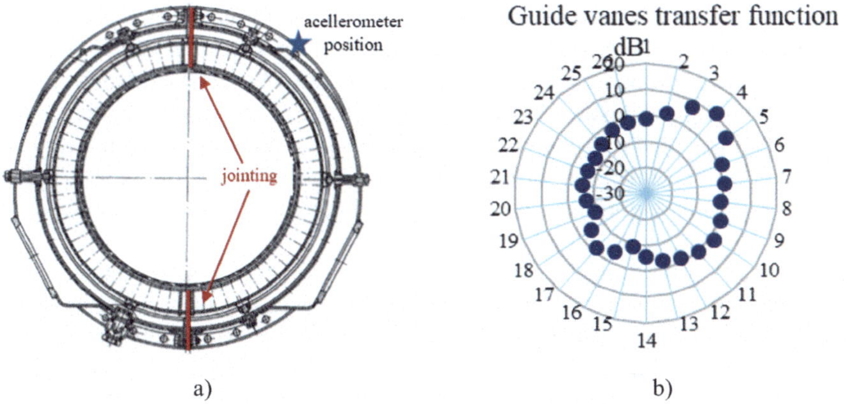

Fig. 4. Guide vanes transfer coefficients: cross section of compressor housing of two halves vertically jointed (a); radar diagram of transfer coefficient values for each guide vane (b).

quantity) to measure experimentally FRF (as the set of coefficient values) of each vane to accelerometer on the housing flange (Fig. 4a). The structure of compressor housing has specialties, like two halves jointed in vertical plane (Fig. 4a) as well as peripheral mounting nodes of vanes grid.

The transfer function on diagram Fig. 4b is presented by k_{Σ} coefficient values for each of 26 vanes according to its position in the construction. It is evident, transfer coefficient is non-linear function of distance from vane to accelerometer. 4th vane lying next to the accelerometer has maximal transfer value (>10 dB) but opposed vanes have 20 dB less values basically because of decay s. Some specialties of transfer values distribution are caused by compressor halves joints (next to vanes 1st and 13th) and fixtures of guide vanes grid as well as other factors of compressor housing anisotropy. So, although remoteness from transducer is the main factor for transfer function of distributed concentrated sources, the combination of joints plays important role also.

4 Transfer Functions in Vibration Diagnostics

There are three main areas in vibration diagnostics of jet engines where appropriate transfer function is required. These are:

- modeling of vibration for engine units and aggregates,
- development of diagnostic algorithms and its validation,
- optimal allocation of transducers around an engine.

Adequate vibration models consider relations between parameters of both: dynamic excitation of engine aggregates and vibration measured on a housing. In depend of specific engine aggregate above relations may be expressed in terms of modal analysis for quasi-periodical loads or as transfer functions for any concentrated or multiple distributed sources. The last one may be presented as multi-dimensional matrices describing relations in time, frequency and spatial domains.

Data development algorithms requires knowledges of vibration transfer functions of specific engine unit. As it is considered in previous sections, for gears and bearings the better algorithm uses as much wider frequency range as possible including even resonance zone. Transfer properties for above algorithms must be obtained from experimental tests of engine aggregates and its final verification. Algorithms facilitated by preliminary identified modal properties of engine integral units allow monitoring of operating engine in real time. For instance, wear of engine mounts on aircraft pylon may be diagnosed using algorithm considering modal parameters modification (caused by wearing) in comparison to initial condition.

Frequency range of a transducer and its amplification factor in resonance zone appear as important factor for measurement and computation of some diagnostic parameters. So, depending on diagnostic tasks the transducer with proper resonance frequency and amplification factor could be chosen. The knowledge of transfer functions from critical engine aggregates may assist to optimize number and allocation of applied transducers without losing diagnostic system resolution.

5 Conclusions

High effectiveness of advance vibration diagnostic techniques, like applied in Vibropassport™, requires knowledge of transfer properties of vibration paths from typical sources of dynamic forces to the transducer on engine housing. Experimental research study of transfer functions demonstrates its dependence from excitation sources nature, frequency band of signal measurement, modal properties of structural parts and other factors. Application of experimental modal analysis technique for integral engine units shows effective approach for modal properties identification in laboratory and industrial conditions. Based on experimental results, the set of basic techniques for estimation of vibration transfer properties for aviation jet engines was described. For instance, known technique of harmonic excitation remains the basic tool for engine units, such as compressor, turbine, etc. Another approach like impulse techniques was admitted as more effective tool for transfer properties identification aimed for diagnostics of blades, vanes, gears and bearings as wide-band vibration sources. The type and force of impulse actuation must be chosen considering the scale of natural excitations that must be preliminary studied on operating engine.

Acknowledgments. The paper uses materials related to research 1.2.1.1/16/A/008 "The Study of Dynamic Properties of Aviation Engine Aggregates for Vibration Diagnostic Technology Research and Demonstrator Development" of the project "Establishment of Engineering Systems, Transport and Energy Competence Center" in cooperation with Central Finance and Contracting Agency of Latvia. This work contains the results of the research study, experimental testing and development vibration diagnostic techniques demonstrator.

References

1. Mironov, A., Doronkin, P., Priklonsky, A., Badekha, V.: Advanced vibration diagnostic techniques for overhaul costs saving of helicopter engines. In: 42nd European Rotorcraft Forum 2016, Lille (2016)
2. Wade, D., et al.: Measurement of vibration transfer functions to inform machine learning based HUMS diagnostics. In: AHS 72nd Annual Forum, 17–19 May, West Palm Beach, Florida, USA (2016)
3. Fabry, S., Ceskovic, M.: Aircraft gas turbine engine vibration diagnostics. Mag. Aviat. Dev. **5** (4), 24–28 (2017)
4. Masahiro, A.: Vibration transfer path analysis for combine harvester using multibody dynamics for engine hybrid method combining CAE and experiment. In: YANMAR Technical Review, 10 April 2015
5. Bilosova, A., Bilos, J.: Vibration diagnostics. Investment in education development. In: Report CZ.1.07/2.2.00/15.0132, Ostrava (2012)
6. Tcherniak, D., Schuhmacher, A.P.: Application of transmissibility matrix method to NVH Source contribution analysis. In: Conference and Exposition on Structural Dynamics XXVII, Orlando, Florida (2009)
7. Mironov, A.: Modeling of variable aerodynamic forces in turbine based on experimental data. In: Proceedings of the 12th International Symposium on Unsteady Aerodynamics, Aeroacoustics and Aeroelasticity of Turbomachines, ISUAAAT12, Imperial College London, UK (2009)

Measurements of the Parameters of a Broadband Satellite Data Channel in the SEVSAT Ship System

Aleksandr Krivchenkov(✉) and Aleksej Skrunds

Transportation and Telecommunication Institute, Lomonosova str. 1, Riga 1019, Latvia
krivchenkovs.a@tsi.lv, aleksej.skrunds@rambler.ru

Abstract. At present time SEVSAT is in use on board of ships for the providing of broadband communication channel between the ship and the coast via satellites in geostationary orbits. In the measurements the parameters of satellite channel, depending on various external affecting factors, were obtained. The experiments were made on real equipment with various combinations of ship position, sea state, weather conditions. Based on the results of the experiments the statistical analysis of the influence of various factors was performed.

Keywords: Satellite channel · SEVSAT · Broadband access · QoS measurement

1 Introduction

The goal of the measurements was to analyse the parameters of satellite channel, depending on various external affecting factors. This channel is provided in Ship Equip Very Small Aperture Terminal (SEVSAT) Satellite System. At present time SEVSAT is in use on board of ships for the organization of broadband communication channel between the ship and the coast via satellites in geostationary orbits (GEO).

Many important aspects of such systems are described in the fundamental [1] but till this time it is a lack of relevant experimental data about parameters of SEVSAT channels. The system providers give figures only for satellite channel bandwidth with a rather unreliable accuracy. Also, the impact of various external factors that have undoubtedly a significant influence on traffic of different types is not sufficiently studied.

To address this shortcoming, it was necessary to conduct series of experiments on real equipment with various combinations of ship position, sea state, weather conditions. For weather forecasts many parameters: sea waves high, wind force, pressure, temperature, humidity, precipitation, cloud cover, dew points, etc. can be taken into account.

In experiments, the parameters of data traffic were also changed. Data packets size and generated bitrate on transport level were variable. In the experiments the most important network characteristics for determining the quality of network services

I. Kabashkin et al. (Eds.): RelStat 2018, LNNS 68, pp. 440–449, 2019.
https://doi.org/10.1007/978-3-030-12450-2_41

(QoS) were measured: channel bandwidth or maximal data rate, delay in packet delivery, variation in packet delivery delays (jitter), percentage of errors in packets.

Based on the results of the experiments the statistical analysis of the influence of various factors was performed. The quantitative importance of the influence of various factors on the parameters of the satellite channel was shown.

2 Inmarsat Services for Ships

Inmarsat Company offers global mobile services. It provides telephone and data services to users worldwide, via portable or mobile terminals which communicate with ground stations through GEO satellites. For use on ships are offered services for broadband data communications [2].

More modern service Fleet Express based on Global Express platform (I-5 satellites launched in 2013–2017 years) "guarantees" global bandwidth, business intelligence, enhance efficiency, and crew welfare. It is obvious that for such application in Fleet Express different types of data traffics need to be supported. Due to Fleet Express presentation such characteristic as Guaranteed Performance means "committed data rates backed-up by service level agreements guarantee that the customer always gets what they pay for" [3]. This long-standing "blue dream" in communication, which can only be achieved if QoS is implemented on the global network.

More mundane and popular service FleetBroadband launched on I-4 group satellites in 2005–2013 years is described as [4]:

- provides cost-effective voice and data through a compact antenna;
- provides thousands of vessels of all sizes and in all maritime sectors for communications such as email, phone and applications that help them do everything from plotting the most fuel-efficient route to diagnosing mechanical faults by video link;
- boosts morale on board by enabling crew to stay in touch with family and friends, use social media, and watch news and entertainment.

In this case the service also deals with the applications traffics of different types and with platform on which global QoS is not implemented. The performance parameters are defined only in the most general form:

- FB150 delivers global voice, IP data up to **150 kbps** and SMS texting with less need for high bandwidth applications;
- FB250 offers faster data speeds up to **250 kbps** and live streaming for applications like video conferencing;
- FB500 is fastest FleetBroadband service, delivering speeds of up **to 432 kbps** and streaming rates up to 256 kbps for multiple user operational and crew communications, anywhere at sea, in any conditions.

Please pay attention that mentioned data rates are given for some data stream consisting from packets of infinite size and that condition in the environment around the ship is also important. Practical experience shows that there are a wide class of applications poorly operating on such a ship's broadband satellite channel (for an example, cloud service speedtest.net or Remote Desktop Connection aps based on RDP protocol). So, the main goal of our research was to clarify the real characteristics of satellite channel and influence on them an environment conditions.

3 Ship Equip VSAT System

The general topology of SEVSAT system is presented in the next Fig. 1. And configuration of typical ship equipment is displayed in Fig. 2.

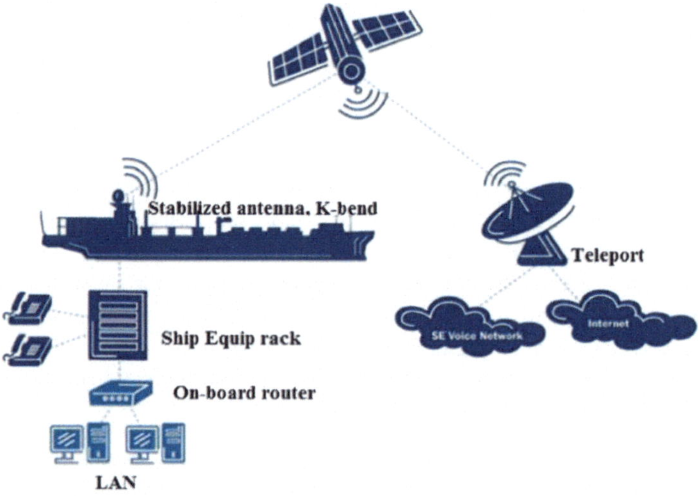

Fig. 1. SEVSAT system topology as FleetBroadband platform.

GEO satellites VSAT move at an altitude of 36,000 kilometres. The speed of the satellite's motion coincides with the speed of rotation of the planet itself. In this connection, such a satellite is at the same point of the sky all the time.

The antenna is mounted on the deck. The antennas of the marine VSAT are equipped with a moving part—a gyro stabilization platform, by means of which the position of the antenna is controlled relative to the horizon and the North. All this allows the antenna at any time to be configured with a satellite, to find and not lose a signal when the vessel manoeuvres and pitching.

The Inmarsat-4 satellite constellation provides 4 coverage areas. Wide zonal beams and narrow zonal beams are also defined. In Fig. 3 areas and zones of coverage are illustrated on the user interface of Control Unit.

If consider the background of coverage providing in SEVSAT system and antennas control one can come to conclusion of possible sufficient dependence of channel quality from zone to what ship is connected or ship coordinates.

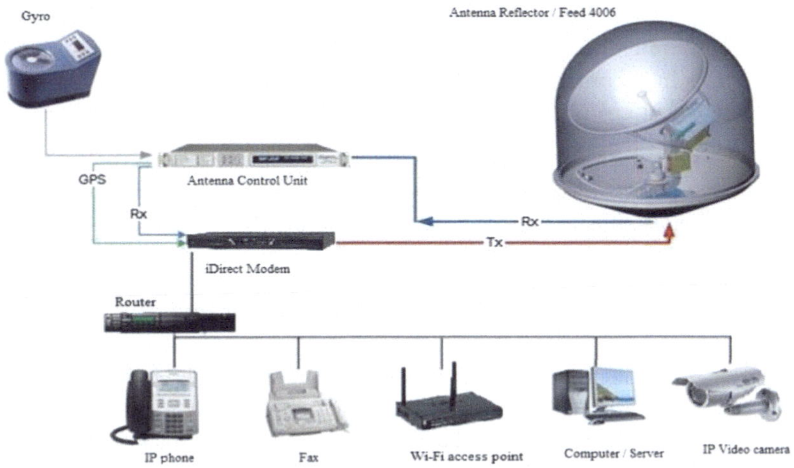

Fig. 2. Configuration of SEVSAT ship equipment.

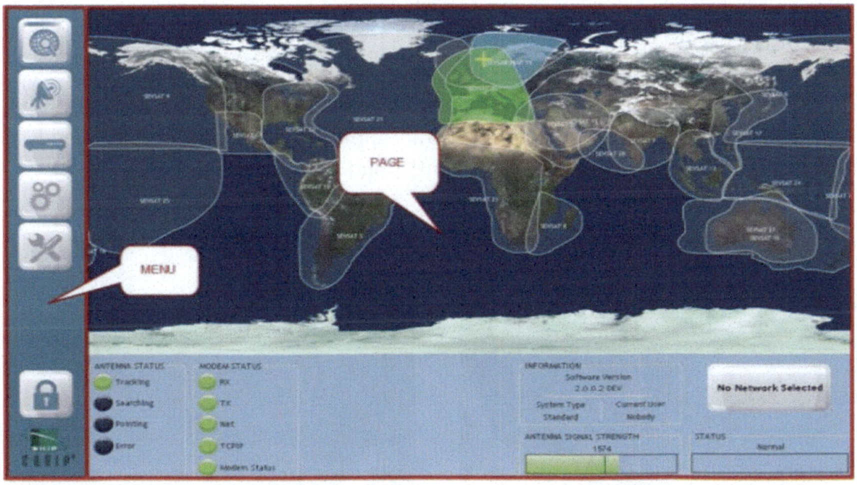

Fig. 3. SEVSAT control unit user interface, zones of coverage are defined.

4 Measured Characteristics of the Channel

Different applications may make different demands to satellite channel characteristics. The next characteristics of the channel are important for high level of QoS:

- channel bandwidth or maximal possible data rate, measured in bit per second;
- delay in packet delivery is defined as the time of delivery of the packet between the source and the receiver for all packets successfully transmitted, as well as affected by errors;

- variation in packet delivery delays or jitter; as jitter has different definitions, in our investigation "interarrival jitter" was calculated; the method of measurement and calculation is in accordance with the recommendations [5];
- the percentage of errors in packets.

Obtaining the values of the characteristics listed above allows us to answer the question of what classes of applications the capabilities of our channel will correspond to. Compliance is established in the recommendations [6].

5 Factors Influencing on the Channel Characteristics

Some environmental factors, or a combination of these factors, indirectly or directly affect the quality of a broadband satellite Internet connection in ship operation conditions. In our research we will not be able to consider such factors as Sun activity mentioned in [1] but try to take into account weather conditions as: wind strength, waves heights, cloud cover, precipitation.

Along the experiments the next external factors are fixed:

- position of the ship: with the help of GPS;
- zone of coverage: with the help of SEVSAT control unit;
- network path of packets: with the help of Ping Plotter utility [7];
- sea state: with the help of Beaufort Scale; this twelve-level scale is adopted by the World Meteorological Organization for an approximate assessment of wind speed from its impact on objects in high seas [8] and Fig. 4;

a b

Fig. 4. Sea view for Beaufort scale 2—(a), 5—(b).

- meteorological data: with the help of zyGrib [9] service for visualization of meteorological data, which generates of weather forecasts and sea waves with the application of many parameters: wind, pressure, temperature, humidity, precipitation, cloudiness, dew points, etc.;

- precipitation intensity: is classified by the relative amount of precipitation per unit time.

6 Field Experiments and Results

In the period 06.2017–12.2017 when the ship was in several sees (East-China Sea, Yellow Sea, Seto Inland Sea, South-China Sea, Tokyo Bay) the 16 sets of field experiments on the broadband satellite channel were performed. In every experiment the real connection between two nodes are tested. In one node a console client program was used for traffic generation with given characteristics. This client was connected to the ship LAN with route to the VSAT terminal. The second node in server mode was used for receiving the traffic from Internet. The server was configured on computer with a real static IP address (DNS stdomino-01.tsi.lv with IP 87.110.232.73).

In both cases (client and server) the utility iPerf designed to test the bandwidth of a communication channel was used [10]. It is a generator of TCP or UDP traffic and is used to determine the traffic characteristics between two or more nodes.

The experiments parameters: transport protocol UDP; the length of packets coming on transport layer 150, 200 and 250 bytes; generated bitrate on application level 50, 60, 70, 80, 90, 100, 150, 200, 250 Kbps. This set of bitrates is taken because for the real channel bandwidth with the generated bitrate less than 50 kbps, the percentage of packets received with errors is almost zero, and with the bitrate greater than 250 kbps, the percentage of error packets is almost always greater than 50%.

The next Fig. 5 presents the conditions of experiments conducted on 19.06.2017 from board of the ship in East-China Sea. Corresponding data received from experiments are presented in Fig. 6.

Date:	19.06.2017	
Local Time:	23:00	
Protocol:	UDP	
Distance AP -	GEO	
Atm. pressure:	1004.3 kPa	
Precipitation:	0.68 mm/h	
Relavive Humidity:	90%	
Cloud Cover:	95%	
Wind:	5.4 m/s	
Ship position LAT:	29'14,05' N	
Ship position LON:	126'32,06' E	East-China Sea
Sea State:	3 Bf	-slight
Sattellite in use:	No: 50 Gateway: 202.5.182.1 speedcast.net	Australia
Round Trip Time:	1114 ms	
UDP buffer size:	41 kByte	

Fig. 5. Factors of experiments on satellite channel.

						Calculated relative parametrs			
Generated Client Bit Rate	App. Packet size	Received Server Bit Rate	Jitter	Packet received with error	Percentage of errors	Relative bitrate	Relative jitter	Mean errors	Max Received Server Bit Rate
Kbps	byte	Kbps	ms	pkts / total		dimensionless	dimensionless	dimensionless	Kbps
50	150	50,5	32,048	0 / 418	0,00%				
50	200	51,3	41,341	0 / 314	0,00%				
50	250	50,3	48,206	0 / 250	0,00%	1,01	0,035	0,000	50,70
60	150	62,1	26,662	0 / 501	0,00%				
60	200	60,6	39,735	0 / 376	0,00%				
60	250	59,9	55,284	0 / 301	0,00%	1,01	0,031	0,000	60,87
70	150	63,7	77,002	9 / 584	1,50%				
70	200	70,1	39,452	8 / 439	1,80%				
70	250	66,6	42,969	12 / 352	3,40%	0,95	0,042	0,022	66,80
80	150	53,4	32,426	176 / 668	26,00%				
80	200	76,7	29,829	17 / 501	3,40%				
80	250	62	36,198	96 / 402	24,00%	0,80	0,030	0,178	64,03
90	150	65,7	65,026	155 / 751	21,00%				
90	200	83,4	22,866	37 / 564	6,60%				
90	250	82,4	30,501	35 / 451	7,80%	0,86	0,035	0,118	77,17
100	150	97,4	65,874	12 / 834	1,40%				
100	200	97	21,473	20 / 626	3,20%				
100	250	100	27,831	0 / 501	0,00%	0,98	0,028	0,015	98,13
150	150	132	61,101	53 / 1251	4,20%				
150	200	159	13,321	59 / 939	6,30%				
150	250	133	17,292	84 / 752	11,00%	0,94	0,023	0,072	141,33
200	150	128	49,629	491 / 1669	29,00%				
200	200	187	10,444	104 / 1251	8,30%				
200	250	147	17,192	273 / 1001	27,00%	0,77	0,021	0,214	154,00
250	150	106	53,897	125 / 2084	54,00%				
250	200	115	21,217	831 / 1563	53,00%				
250	250	115	22,189	666 / 1252	53,00%	0,45	0,024	0,533	112,00

Fig. 6. Data of experiments on 19.06.2017.

Similar experiments were conducted from June to Dcember (Fig. 7).

Later statistical processing of the results of experiments has been carried out. The analysis has followed to [11] and it is Multivariate analysis of variances.

The main purpose of the analysis is to calculate the effect of each individual factor affecting the parameters of the channel. The assessment of the impact of such factors will help to draw conclusions about the degree of their influence on the quality of the satellite channel. We describe an example of some parameters and factors analysis.

Two main parameters were selected: the maximum bitrate received in the server (it may be also defined as "bandwidth") and the average "relative error rate". The following factors were analysed: "zone number" of satellite zone depending on the position of the ship; "sea state" according to the Beaufort scale at the time of the experiment; and presence of "precipitation" during the experiment.

The measure of the impact or condition that a factor can take is the so-called different "levels" of exposure. For example, the state of the "precipitation" factor can take the values: "No Rain" or "Light Rain" (experiments were carried out at

Levels of factors for different experiments or Channel parametrs figures						
	19.06.2017	24.06.2017	26.06.2017	01.07.2017	07.07.2017	12.07.2017
Factor or Channel parametrs						
Sattellite in use:	No: 50	No: 50	No: 50	No: 37	No: 50	No: 50
Gate	202.5.182.1	202.5.182.1	202.5.182.1	216.139.177.97	202.5.182.1	202.5.182.1
Round Trip Time:	2,8	2,7	2,5	2	2,5	2,5
Sea State:	3 Bf	2 Bf	1 Bf	1 Bf	4 Bf	5 Bf
Precipitation:	0.68 mm/h	0.00 mm/h	1.63 mm/h	1.63 mm/h	0.02 mm/h	0.00 mm/h
Cloud Cover:	95%	60%	95%	95%	60%	40%
Wind:	5.4 m/s	2.7 m/s	1.4 m/s	1.1 m/s	6.4 m/s	8.2 m/s
Calculated Bandwidth	112	117	122	188	160	78
Calculated Bandwidth from Regression	114	101	85	117	91	39
Calculated Ralative Jitter from Regression	0,033	0,028	0,035	0,023	0,029	0,041
Calculated Errors from Regression	0,056	0,002	0,035	0,004	0,000	0,212
Max. Received server bitrate	154	165	139	188	160	78
Max Jitter	77	64	73	65	73	62

Fig. 7. Experimental results overview for 6 dates.

Table 1. Parameters of SEVSAT channel for various factors.

Zone number	Sea state Beaufort scale	Precipitation			
		No rain		Light rain	
		Bandwidth Kbps	Relative error rate	Bandwidth Kbps	Relative error rate
50	1	175.0	0.005	138.7	0.035
	2	165.0	0.002	133.3	0.040
	3	134.7	0.010	160.9	0.031
	4	92.8	0.023	159.7	0.000
	5	78.3	0.212	139.7	0.013
37	1	139.6	0.018	187.7	0.004
	2	209.3	0.037	159.0	0.011
	3	153.0	0.018	174.0	0.000
	4	234.0	0.001	148.0	0.018
	5	141.7	0.064	104.2	0.000

precipitation intensity <2.5 mm/h). Thus, the results of the experiments are presented for analysis in the following Table 1.

The results of the main stage of the variance analysis of the presented data are collected in Table 2.

Further analysis of statistical data, examples of which are given, for example, in the work [1], allows to draw the conclusions formulated in the next section.

Table 2. Variance analysis of data.

Component	Sum of Squares		Percentage of Variation	
	Bandwidth Kbps	Relative error rate	Bandwidth Kbps	Relative error rate
Individual value	483719.97	0.056		
Overall mean	458595.7	0.015		
Total variations	22124.3	0.041	100%	100%
Main effects:				
Sea state beaufort scale	6549.8	0.01	26.07%	25.18%
Precipitation	16.6	0.003	0.07%	6.59%
Zone number	3714.2	0.002	14.78%	5.01%
First-order interactions:				
Sea state—precipitation	2491.9	0.015	9.92%	35.38%
Sea state—zone number	2194.3	0.005	8.73%	11.39%
Precipitation—zone number	1828.8	0.000	7.28%	0.07%
Standard error	45.6	0.041		

7 Conclusions

The main results of this work are:

- As a result of testing the satellite Internet connection from the ship it was concluded that the main factor influencing the broadband access is determined by the Sea state in the corresponding region. The disturbance of the sea in such experiments can be estimated on the Beaufort scale basis. Using proposed technique and combinations of factors affecting the quality of a satellite channel its QoS can be studied further.
- The measure of Sea state factor affecting and variation analysis show for Bandwidth 26.07% and for Relative error rate 25.18% of total variations. The main reason—lacks in antenna proper stabilization.
- The satellite Zone used affects with a intensity of 14.78%, and it is the next main factor on which channel characteristics depend. The main reason seems to be different gateways and routes in ground part of the network.
- The presence of precipitation, and in experimental conditions, they were insignificant, little effect on the characteristics of the satellite channel.
- Data from the 16 sets of field experiments of the SEVSAT system in the period 06.2017–12.2017, when the ship was in various Seas, allow to analyse many other parameters and factors for satellite channels.

References

1. Kota, S., Pahlavan, K., Leppanen, P.: Broadband Satellite Communications for Internet Access. Springer, New York (2004)
2. Maritime: Inmarsat homepage. https://www.inmarsat.com/maritime. Accessed July 2018
3. Fleet Express: Inmarsat homepage. https://www.inmarsat.com/service/fleet-xpress/. Accessed July 2018
4. FleetBroadband: Inmarsat homepage. https://www.inmarsat.com/service/fleetbroadband/. Accessed July 2018
5. RFC 1889, from IETF. https://tools.ietf.org/html/rfc1889. Accessed July 2018
6. Y.1541: Network performance objectives for IP-based services, December 2011. https://www.itu.int/rec/T-REC-Y.1541/en. Accessed July 2018
7. PingPlotter software: Retrieved from PingPlotter homepage. https://www.pingplotter.com/. Accessed July 2018
8. Beaufort scale: Retrieved, from ENCYCLOPEDIA BRITANNICA. https://www.britannica.com/science/Beaufort-scale. Accessed July 2018
9. zyGrib—GRIB File Viewer Weather data visualization, from zyGrib homepage. http://www.zygrib.org/. Accessed July 2018
10. iPerf—The ultimate speed test tool for TCP, UDP and SCTP, Retrieved from iPerf homepage. https://iperf.fr/. Accessed July 2018
11. Jain, R.: Art of Computer Systems Performance Analysis Techniques for Experimental Design Measurements Simulation and Modeling. Wiley, Hoboken (1991)

Implementation of a Multiple Remote Tower

Darius Bazaras(✉), Vaidotas Kondroška, Giedrė Rakauskienė, and Jonas Vytautas Jusionis

Vilnius Gediminas Technical University, Plytinės 27, LT-10105 Vilnius, Lithuania
darius.bazaras@vgtu.lt

Abstract. The number of passengers carried by air transport has been growing every year in the world during the last decade. The average annual growth rate of air passengers has been 8.4% since 2008 [1]. Such an increasing demand for air traffic services has posed plenty of challenges, such as larger airports or increased network congestion, to the management of air traffic.

"Oro navigacija" manages air traffic in all four Lithuanian international airports.

The Lithuanian State Enterprise "Oro navigacija" needs to take acute decisions to substantially upgrade its competitive advantage in today's extremely challenging environment for the management of air traffic. A multiple remote tower could be a right way forward. Its main merit is cost-efficiency: it has a potential to bring about sizeable cost savings. It would also allow for the more even distribution of the workload of air traffic controllers among all the airports in Lithuania. In addition, a multiple remote tower would turn "Oro navigacija" to a modern R&D knowledge centre.

The preliminary cost benefit analysis indicates that it would be beneficial financially to establish a multiple remote tower for providing air traffic services to three national airports in Lithuania.

Keywords: Logistics · Aviation · Transport · Air control · Remote tower

1 Introduction

The number of passengers carried by air transport has been growing every year in the world during the last decade. The average annual growth rate of air passengers has been 8.4% since 2008 [1]. Such an increasing demand for air traffic services has posed plenty of challenges, such as larger airports or increased network congestion, to the management of air traffic [2]. In Lithuania, the State Enterprises "Oro navigacija" and "Lithuanian Airports" are the companies facing those challenges directly, as they both operate in the industry of air traffic. "Oro navigacija" manages air traffic in all four Lithuanian international airports, while "Lithuanian Airports" runs three of those airports. Despite the fact that air traffic is increasing each year the main challenge for both State Enterprises "Oro navigacija" and "Lithuanian Airports" is to ensure service in low traffic airports (e.g. Palanga (PLX), Kaunas (KUN), Šiauliai (SQQ)).

"Lithuanian Airports" can be regarded as an example of a very successful Lithuanian company, which is currently beating market benchmarks due to an

I. Kabashkin et al. (Eds.): RelStat 2018, LNNS 68, pp. 450–454, 2019.
https://doi.org/10.1007/978-3-030-12450-2_42

insightful strategic decision taken earlier. Last year, the growth rate of the number of passengers travelling via airports in Lithuania was 10%, in spite of the major reconstruction of the runway at the largest national airport in Vilnius [3]. According to the CEO of "Lithuanian Airports", one reason for the recent success of airports in Lithuania has been courage to start actively engaging with low cost airlines at the time when many European airports were still rather ignorant about them. He claims the decision of turning to low cost airlines was very timely, since today low cost airlines are already dominating in Europe, while traditional airlines have shifted towards longer-distance flights [4].

As a result, a question arises if "Oro navigacija" could follow success achieved by "Lithuanian Airports". The successful example of "Lithuanian Airports" hints that insightful decisions can bring success to a company even in an extremely challenging business environment. So, is there any insightful strategic decision that could potentially result in the notable upgrade of the competitive advantage of air traffic management in Lithuania? Turns out that a multiple remote tower could be an answer.

2 The Development of Remote Tower Services

There are several essential concepts that define remote tower services. Remote tower services are the air management services that are performed from a distant location rather than from a tower placed in the territory of an airport. Remote tower services can be performed either from a single or a multiple remote tower [5]. Air traffic controllers in a single remote tower conduct air traffic management only for one airport at a time. Even though if several modules are placed in a tower, the air traffic controllers of a single remote tower can perform air traffic management for different airports strictly sequentially, i.e. only after the management for a previous airport has been ended. Conversely, air traffic controllers in a multiple remote tower conduct air traffic management for more than one airport simultaneously [6].

The development of remote tower services has already accelerated in Europe. The world's first remotely-operated airport was an Oernskoeldsvik airport in Sweden. Air traffic has been managed from a remote tower there since April 2015 [7]. Later, in early 2017 the Irish Aviation Authority completed the world's first trials of managing the air traffic of two low-volume airports (Cork and Shannon) by a single air traffic controller from a multiple remote tower. The director of the air traffic management of the Irish Aviation Authority, P. Kearney, was extremely optimistic after those trials: "It will almost certainly be possible to allow one controller to simultaneously provide air traffic management services for more than one low-volume aerodrome" [8]. In late 2017 it was the first time when the air traffic of three airports—Budapest, Papa and Debrecen—was managed simultaneously from a multiple remote tower (http://www.remote-tower.eu/wp/?p=184). Following those trials, "Hungaro Control" has declared its ambitions to turn the Budapest airport to the first world's medium-sized airport operated remotely full-time by 2019 [9]. All those examples mentioned in the paragraph were very important cases for the development of remote tower services in Europe, but the trials of remote tower services are not limited to only those examples [10, 11]. It is also important to mention that there are several countries in Europe (Norway, Finland, Germany, Estonia) already considering implementation solution of remote tower.

Moreover, "SESAR JU" ("Single European Sky Air Traffic Management Research Joint Undertaking")—a public-private partnership founded by the EU and Eurocontrol in 2007—has also been investigating the conduct of remote tower services. Since 2016, SESAR JU has been carrying out the project "PJ05 Remote Tower" with the aim to adequately reflect operational needs and technical solutions for the operation of remote tower services. The project is intended to be finished by the end of 2019 (http://www.remote-tower.eu/wp/?page_id=9) [12, 13].

3 The Advantages and the Potential Challenges of a Multiple Remote Tower

The most apparent advantage of exploiting a multiple remote tower is cost efficiency. According to some calculations, the introduction of a multiple remote tower could reduce costs by 30–40% [14]. The savings would come mainly from markedly lower fixed costs of air traffic management. Firstly, only one physical tower building would be needed instead of tower buildings in every single airport. Secondly, the number of air traffic controllers could potentially be reduced, because one air traffic controller could replace several others if those others were not given a full workload in their previous low-volume airports. As a result, the maintenance costs of physical towers as well as the training fees for air traffic controllers would also go down [9].

Besides cost efficiency there are several other advantages a multiple remote tower could provide. At first, a multiple remote tower would allow for the smoother distribution of workload among air traffic controllers. Under the remote management of air traffic, workload could be smoothed not just within a single airport, but across several airports. This would surely lead to the better working conditions of air traffic controllers, as many of them would not have to work excessive hours anymore. In addition, a multiple remote tower would undoubtedly be an R&D knowledge centre, promoting innovations and also contributing to the economic progress of a whole country. Finally, the safety of air traffic management would increase with the implementation of a multiple remote tower, since the introduction of such a tower requires the installation of the most sophisticated and reliable IT infrastructure, which can provide air traffic controllers with smart visibility via a computer screen which is much better compared to the plain eye-visibility through a window [9].

On the other hand, reforms always entail some challenges, and the introduction of a multiple remote tower is not an exception. The most prominent challenge is the need to further enhance cybersecurity, since the matter of life or death in several airports simultaneously would depend on the reliable operation of IT infrastructure in just one tower building. In addition, the thorough assessment is needed on how the work performance of air traffic controllers would change in a multiple remote tower, since theoretically there is a possibility that human performance could be affected due to the management of air traffic at several airports with potentially different air conditions at the same time. In the end, legal issues should be sorted out, because the spread of multiple remote towers could spur cross-border shopping for looser regulatory standards. Therefore, the usage of multiple remote towers needs to be uniformly regulated at least at a European level [5].

4 The Case of Lithuania

A multiple remote tower would cope with the current challenges of "Oro navigacija" perfectly well. At first, expenses for the management of air traffic controllers comprise a huge share—even 40%—of all the costs of "Oro navigacija". Therefore, if it was possible to manage air traffic controllers more efficiently, sizeable savings could be realized. In addition, a multiple remote tower would let distribute the costs of "Oro navigacija" related to different Lithuanian airports more evenly. Currently, direct costs for one unit of service at the airports of Kaunas, Palanga and Šiauliai are 4 to 7 times higher than at Vilnius airport. Furthermore, the better distribution of not only costs, but also workload among the air traffic controllers managing air traffic in all Lithuanian airports would be feasible with the exploitation of a multiple remote tower. At the moment, the average number of take-offs and landings managed by one air traffic controller at Vilnius airport is 2 to 10 times larger compared to the controllers at the other three airports in Lithuania [15]. It is important to mention that Vilnius airport is not listed as high intensity airport in Europe. Therefore, the visibility from the tower right now is very limited, so in any case qualitative service cannot be provided without high quality video cameras system.

The process of the introduction of a multiple remote tower in Lithuania could be gradual. In particular, it does not have to be a hasty all-or-nothing reform. In the beginning, a geographic location for a multiple remote tower has to be determined. However, having been built the tower could start operating not as a multiple, but as a single remote tower managing air traffic of one low–volume Lithuanian airport. After some time, if the exploitation of the single remote tower was successful, the tower could be transformed to a multiple remote tower by expending its responsibilities to manage the air traffic of a second small airport. Eventually, the coverage of the multiple remote tower could be broadened to include the management of the air traffic of three or maybe even all four airports in Lithuania [15].

5 Conclusion

The Lithuanian State Enterprise "Oro navigacija" needs to take acute decisions to substantially upgrade its competitive advantage in today's extremely challenging environment for the management of air traffic. A multiple remote tower could be a right way forward. Its main merit is cost-efficiency: it has a potential to bring about sizeable cost savings. It would also allow for the more even distribution of the workload of air traffic controllers among all the airports in Lithuania. In addition, a multiple remote tower would turn "Oro navigacija" to a modern R&D knowledge centre. Furthermore, several European countries have already started their acquaintance with remote tower services. So, Lithuania can't afford itself to carelessly miss the opportunity to advance with the most promising contemporary innovations in the very high value-added sector of aviation.

Remote tower solution should be considered as the most appropriate option to ensure qualitative aerodrome air traffic service for low intensity airports 24/7.

The preliminary cost benefit analysis indicates that it would be beneficial financially to establish a multiple remote tower for providing air traffic services to three national airports in Lithuania. This does not mean that Lithuania needs to immediately start with a multiple remote tower covering the management of air traffic over the whole country. However, the introduction of a single remote tower as early as possible would be a very insightful initial step.

References

1. World Bank: Air transport, passengers carried (2018). https://data.worldbank.org/indicator/IS.AIR.PSGR
2. European Commission: Periodic Reporting for period 3—COMPAIR (COMPetition for AIR traffic management) (2018). https://cordis.europa.eu/result/rcn/194963_en.html
3. Vilnius airport: In 2017 Lithuanian airports had a record high of arrived passengers and number of flights. Statistical information. (In Lithuanian: "2017–ieji Lietuvos oro uostams atskraidino rekordinę keleivių ir skrydžių gausą") (2018). http://www.vilnius-airport.lt/lt/naujienos/?id=2361868
4. Radžiūnas, V.: CEO of Lithuanian airports: we will not have a national carrier. Interview. (In Lithuanian: "Lietuvos oro uostų vadovas: mes neturėsime nacionalinių avialinijų") (2016). https://www.delfi.lt/verslas/transportas/lietuvos-oro-uostu-vadovas-mes-neturesime-nacionaliniu-avialiniju.d?id=73298408
5. ECA: Remote Tower Services. ECA Position Paper (2014)
6. Expert Group on the Human Dimension of the Single European Sky: The Human Dimension in Remote Tower Operations. Position Paper (2017)
7. SAAB: Sweden First in the World with Remotely Operated Air Traffic Management. Press Release (2015). https://saab.com/news-and-media/2015-04/sweden-first-in-the-world-with-remo-tely-operated-air-traffic-management
8. Croft, J.: Irish Remote Tower: Two Airports with One Controller. AviationWeek.com (2017) http://aviationweek.com/commercial-aviation/irish-remote-tower-two-airports-one-controller
9. HungaroControl: Remote Tower for Multiple Airports. Presentation (2017)
10. CANSO: Remote Virtual Tower from FREQUENTIS in Jersey to be the first of its kind in the British Isles (2017). https://www.canso.org/remote-virtual-tower-frequentis-jersey-be-first-its-kind-british-isles
11. NLR: Remote Tower Live Trials in the Netherlands (2017). http://www.nlr.org/news/remote-tower-live-trials-in-the-netherlands
12. Remote Tower: First Multiple Remote Tower validation for three Airports at DLR Braunschweig (2017). http://www.remote-tower.eu/wp/?p=184
13. Remote Tower: About PJ05 (2018). http://www.remote-tower.eu/wp/?page_id=9
14. Avinor: Remote Towers (2018). https://avinor.no/en/avinor-air-navigations-services/services/remote-towers
15. Oro Navigacija: Remote Aerodrome (Tower) Service Provision. Company's Information. Presentation (2017)

Reliability, Safety and Risk Management

Safety Status on Road Transport System in the European Union

Gabriel Nowacki[1], Krzysztof Olejnik[2(✉)], and Bartosz Zakrzewski[3]

[1] Cybernetics Faculty, Military University of Technology, Gen. Witolda Urbanowicza 2, 00-908 Warsaw, Poland
gabriel.nowacki@wat.edu.pl

[2] Faculty of Production Engineering and Logistics, Opole University of Technology, WIPIL PO, Gen. Kazimierza Sosnkowskiego 31, 45-272 Opole, Poland
k.olejnk@po.opole.pl

[3] Scientific Information and Publications Section, Motor Transport Institute, Jagiellońska 80, 03-301 Warsaw, Poland
bartosz.zakrzewski@its.waw.pl

Abstract. The main paper problem is to discuss safety status on road transport system in the European Union and how to improve it, especially by reduction number of victims. In theoretical part refers Sect. 2, the following research methods were used: analysis, analogy, synthesis, induction and deduction. Characteristics of road transport infrastructure, counties of the most and smallest total length of motorways, and the statistics of road accidents, dead and wounded people in the European Union were presented. Furthermore the paper discussed the characterization of dangerous goods in the EU and the advantages and disadvantages of road transport. In practical part refers Sect. 3, apart from mentioned theoretical, the empirical methods were applied: modelling and tests. Based the initiatives of the EU for road transport safety, some solutions of research projects in Poland were described, especially program of analyzing data from digital tachograph and smart cards, functional structure of National Automatic Toll Collection System with proposition of OBU and also methodic and prototype of eCall system. The paper resolved the main problem, the number of victims in the EU was significantly reduced from 54,900 in 2001 to 25,300 in 2017, by initiatives, legislative procedures and implemented Intelligent Transport Systems.

Keywords: Road transport · ITS · Threats assessment · Safety management

1 Introduction

Road accidents cause a huge number of fatalities, but they are not recognized in many countries at a sufficient level. According to the World Health Organization (WHO), approximately 1.2 to 1.3 million people die because of the road accidents yearly, whereas 30–50 million are injured [1].

I. Kabashkin et al. (Eds.): RelStat 2018, LNNS 68, pp. 457–467, 2019.
https://doi.org/10.1007/978-3-030-12450-2_43

The major of the victims come from developing countries. In 79 countries, including Poland, the number of accident victims has been reduced in the last three years, while in 68 countries it has increased [2].

Currently, road accidents rank 9th in the hierarchy of the most common cause of deaths, it is expected that in 2030 it may be approximately 1.9 million, which will rank road accidents at 5th place.

Every year a very large number of children are killed in road accidents (about 186,000). Moreover, injuries caused by road accidents have become the main cause of death of people aged 15–29. Almost half (46%) of people who die on the roads worldwide are "defenseless road users": pedestrians (21%), motorcyclists (14%), cyclists (8%) and drivers of scooters (3%).

The WHO estimates that only 15% of countries have a full range of laws and regulations regarding the five major sources of threat: speeding, driving under the influence of alcohol, lack of use of helmets, seat belts and child seats.

The data provided by the European Commission shows that European roads remain the least dangerous in the world. In 2016, 2500 people lost their lives, which is 600 less than in 2015, and 5.5 thousand less than in 2010.

The Commission also estimates that on EU roads 135,000 people have been seriously injured. Social costs, such as rehabilitation, health care, material damage, etc., of people injured or killed in road accidents are estimated at approximately 100 billion Euro [2].

2 Characteristics of Road Transport in the EU

2.1 Transport Infrastructure

Transport—the movement of people, cargo, energy and information in space using appropriate resources and all activities aimed at this purpose. It is immanently related with the world functioning around us—especially the autonomous transport.

The infrastructure of the determined transport consists of: means of transport, transport routes, media being transported, as well as people organizing, supervising and carrying out transport [3, 4].

The TEN-T trans-European transport network includes road transport, railways, airways, maritime transport and river transport constituting the most important connections from the point of view of the EU development, as well as point elements of infrastructure in the form of sea, air, inland and road-rail terminals.

In accordance with the Council Regulation No. 1108/70 [5] and **Commission Regulation No. 851/2006/EC** [6], transport infrastructure, including the trans-European transport network, consists of the following transport infrastructure: rail, road, maritime, air, multimodal, inland waterways, equipment and intelligent transport systems.

2.2 Road Net in the EU

The system of the marking of European routes was introduced on the basis of the European Agreement on Main International Traffic Routes (AGR), concluded in Geneva on November 15th, 1975. The European route means an international car road in Europe, marked with a single, double or triple-digit number, preceded by the letter E (white letters on a green background).

The numeration of roads is preserved when the road is crossing national borders. The UN Economic Commission for Europe (UNECE) is responsible for overseeing the European route network and proposing changes in its course.

The total length of the road network in the EU in 2017 is approximately $5.5 \cdot 10^6$ km, including 77,730 km of motorways. The total number of motorways in Spain is 16,214 km, Germany—12,993, France 11,559, Italy—6844.

Speed limits on motorways for passenger cars in the EU countries vary (e.g. Poland—140 km/h, Austria, Italy, Slovakia, Slovenia—130 km/h).

On the 1st of May 2017, the obligation to mark all cars with identifiers for Euro 3, Euro 4, Euro 5 and Euro 6 emission categories has come into force.

For example, cars with Euro 1 and Euro 2 emission class can no longer drive through Tyrol. Special ecological stickers will only be available at authorized points for around € 25–50. Fine for the lack of the ecological sticker is 2180 euros.

2.3 Advantages and Disadvantages of Road Transport

Road transport is one of the main types of transport, where passengers and goods travel through the means of transport (cars) on land roads. Like all transport types, it has both disadvantages and advantages.

The disadvantages include, above all, dependence to a certain extent on atmospheric conditions and the fact that it greatly contributes to pollution of the natural environment, causing, among other things, the greenhouse effect. In addition, road transport has a high accident rate, especially when compared to air, sea and rail transport—Table 1.

Table 1. Death hazard in 2010–2017 in respect of transport means [7, 8].

Mean of transport	Victims on million passenger-kilometer
Air transport	0.101
Maritime transport	0.120
Railway transport	0.156
Bus	0.443
Car	4.450
Two-axle vehicle	52.593

The advantages of this type of transport include:

- guaranteed access to every town, any place where there is no access for rail, ship, plane, etc., can be reached,
- it is beneficial in transporting people over short distances,
- makes people lives easier, cheaper car prices,
- specialist rolling stock adapted to transporting cargo of various transport susceptibility,
- adaptation of the means of transport to the transport of almost all types of cargo,
- the best spatial availability resulting from the highest network density of all transport means,
- the most favorable adaptation of the road network to the distribution of supply and markets,
- the best possibility of transport to the carriers of different means of transport,
- reaches the destination, easily accessible to everyone.

The disadvantages of this type of transport include:

- heavy traffic, which often causes traffic jams in large agglomerations—dependent on the weather,
- contributes significantly to the environmental pollution (including the greenhouse effect),
- longer transport time compared to the air transport,
- small capacity of lorries,
- sometimes high maintenance costs, expensive fuel,
- a large number of road accidents and fatalities, terrible condition of roads.

The statistics show that transport is the most frequently attacked area by terrorists, which is related to the fact that the objects and means of transport focus large number of travelers and are easily accessible to terrorists.

In the last period of time, trucks were used several times to carry out terrorist attacks—Table 2.

Table 2. Terrorist fatalities involving vehicle attacks in the EU between July 2016 and June 2017 [9].

Data	Place	Fatalities
July 14, 2016	Nice, France	86
August 16, 2017	Barcelona, Spain	13
December 19, 2016	Berlin, Germany	12
June 3, 2017	London, United Kingdom	8
April 7, 2017	Stockholm, Sweden	5
March 22, 2017	London, United Kingdom	4
June 19, 2017	London, United Kingdom	1

Due to the development of new technologies, autonomous cars have been constructed. The FBI warns that these vehicles can be used as a carrier of deadly loads [10].

Terrorists can use them for terrorist attacks. After loading explosive charges (bombs), vehicles of this type will drive themselves to their destination and cause an explosion at the scheduled time.

If the terrorists decide to change the safety function in this type of vehicle, as a result the autonomous vehicles might ignore traffic lights or speed limits.

Autonomous cars, equipped with radar equipment, lasers, cameras and GPS receivers, based on the collected data will create 3D terrain maps (buildings, roads, road users—cars, pedestrians, etc.).

The onboard computer of the autonomous vehicle will be programmed in such a way, that the vehicle will comply with the rules and regulation of the road system, without collision or accident, reaching its final destination.

2.4 Victim Statistics on Road Accidents in the EU

In 2017 in the European Union, the main number of victims of road accidents have occurred in rural area—55, 37% in urban area and 8% in motorways.

After two years of stagnation (2014 and 2015), the number of road fatalities was reduced by 2% in 2016, and by another 2% in 2017. What is more, when comparing the data from 2001 and 2010, it should be stated that there has been a 43% drop in the number of victims (Fig. 1).

In 2017, the EU Member States counted 49 road fatalities per one million inhab-

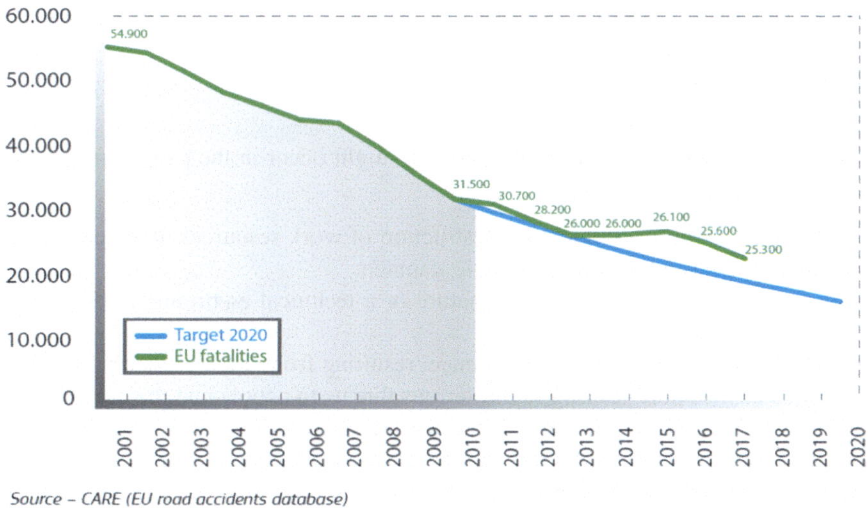

Fig. 1. Victims of road accident in 2001–2016 in EU [11].

itants, and eight of them stood below 40 [11]. Member States with the best road safety scores were Sweden (25) and the UK (27), followed by the Netherlands (31), Denmark (32), Ireland (33) and Estonia (36). On the other hand, Member States with the highest fatality rate were Romania (98) and Bulgaria (96).

The statistics of road traffic fatalities depends on area of the world, the most in Africa—26.6 victims, the best in Europe—9.3 victims per 100,000 population (Fig. 2).

Fig. 2. Road traffic fatalities all over the world [12].

2.5 Dangerous Goods Transport

In the EU Member States, dangerous goods are transported through:

- road transport (ADR)—65.8%.
- rail transport (RID)—27.4%,
- inland transport (AND)—6.8%.

The following events (divided into categories) might occur in the case of transport of hazardous materials:

- loss: damage to health or loss of life, destruction of work resources, damage to the natural environment, as well as economic damage,
- ordinary threat: a physical situation in nature or a technical environment that may cause losses,
- catastrophe: this is an emergency occurrence, resulting from uncontrolled processes in the course of transport, leading to the serious threats to people and the occurrence of losses, which cannot be controlled by local means,
- accident: a danger of a threat, which can be controlled by local means,
- incident: an accident that does not cause losses,
- failure: uncontrolled changes leading to losses in industrial activity (in the process of production, storage, transport).

Within the European Union Member States, approximately 72% of all freight is transported by roads. It turns out that 33% of robberies that target to steal the freight, takes place in the parking lots during the stopover.

This is a very worrying phenomenon, due to the fact that the applicable legal provisions impose on the driver the obligation to use the break, daily or weekly rest in a timely manner.

Due to the frequent occurrence of the incidents involving theft of freight, and the possibility of terrorist attacks in the future, the EU has taken legislative steps to safeguard car parks and exchange data in this area [13, 14].

The EC recommended that there should be 5000 car parks on the network of the highest category roads that meet the monitoring requirements. This allows, in most cases, the delivery of goods from the place of shipment directly to the place of collection without the need for time-consuming transshipment operations.

3 The EU and Polish Safety Initiatives of Road Transport

3.1 Digital Tachograph System

In 2006, the EU imposed Directive 561 [15] to refer Digital Tachograph. It stressed that transport firms has to download and store data in archives: mass memory of tachograph and driver card data. This allows a complete record of activities to be obtained for drivers and vehicles for evaluation to support improved fleet management.

PC Net Service with the Motor Transport Institute developed the TachoScan solution [16]. The program of analyzing data from digital tachograph and smart cards can generate a driver's general report, which shows distance, working hours, waiting hours and rest hours, start and stop work hours per day for each individual driver. Software supports the data export to Excel, which allows the use of this information in various financial and accounting applications.

3.2 European Electronic Toll Service—EETS

In 2004, the EU has imposed directive 2004/52/EC refers to EETS [17]. It allows tolling of vehicles using a single on-board unit, one user account and one contract with the chosen EETS Provider in all EETS domains. The on-board unit has to meet certain technical requirements—it should be interoperable and capable of communicating between electronic toll collection systems, deployed in different toll domains. It must be compatible with all of the technologies specified in Directive 2004/52/EC (i.e. DSRC, satellite system GNSS, GSM).

Motor Transport Institute has developed research project of functional structure of National Automatic Toll Collection System—R10 0001 04, from 2008 to 2010. It consists of the following elements: four intelligent on-board units (OBU's), two control gates and laboratory model of Centre [18]. OBU automatically calculates the amount of charge due, taking into account the vehicle category (admissible weight, number of axles), the emissions class and road distance. OBU is equipped with GPS, GSM and DSRC module, so it is interoperable with other electronic toll systems in the EU member states and meets requirements of directive 2004/52/EC and EC Decision of 6 October 2009.

Figure 3 shows that the tested OBU with the number of measurements were respectively 90% of perfect values, and 8% excellent values. PDOP (Position Dilution of Precision): 1 (perfect), 2–3 (excellent), 4–6 (good), 7–8 (moderate), 9–20 (poor), above 20 (bad). The horizontal axis (X) depicts values for PDOP. The vertical axis (Y) depicts the number of measurements (in percentages) during which a given value of PDOP was obtained. The statistics were calculated based on 4627 measurements of position.

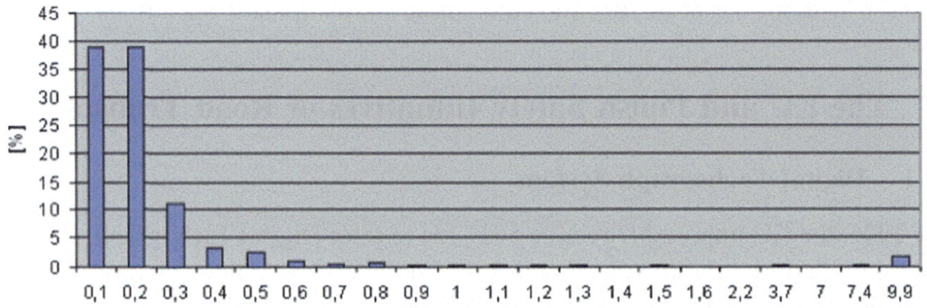

Fig. 3. Distribution of PDOP for all OBU [19].

The number of satellites used for measurements of all OBU devices is presented in Fig. 4. For the purposes of NATCS it was assumed that GPS receiver in OBU should track at least 5 satellites, for more accurate calculations and in the event of the loss of

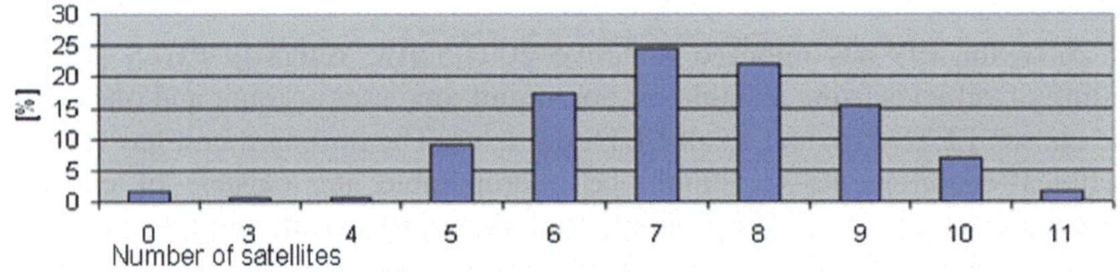

Fig. 4. Number of GPS satellites used for location measurement [19].

signal from one of them.

The presented data shows that the maximum number of satellites used for the purpose of location was 11, and in the case of 99% of measurements at least 5 satellites were The NATCS has recognized not only four OBU of mentioned system, but also OBU from Toll Collect (Germany) and Passango from France which proved interoperability in the EU. The efficacy of automatic detection of number plates was 99.9%.

3.3 Pan-European eCall System

Pan-European eCall is linked to the e-safety' initiative, which is part of a comprehensive strategy of the EC to stay safe on the roads and improve transport efficiency in the EU.

Costs and benefits estimates of the eCall system carried out within the e-merge project and other research indicate the possibility of avoiding, on the European scale, deaths of up to 2500 people a year and reducing the impact of accidents by 15%.

The onboard eCall system devices will be installed in all newly registered vehicles in the EU, started from April 2018 [20].

After launch, the eCall on-board device will connect with the emergency services using the cellular network, enabling the transfer of electronic data and making a voice call. Notifications will be picked up by the emergency notification centers (ENC) operating within the emergency call network 112.

The device will send a minimum data set (MSD) regarding the accident in electronic form. The minimum set of data will contain information such as: time of the accident, exact geographical location of the vehicle at the time of the accident, vehicle data (including VIN identification number), eCall status (at least information about whether the call was started manually or automatically) and information regarding the operator providing the services.

Apart from the EU Member States, some countries have decided to implement the eCall system: Iceland, Norway, Switzerland, Turkey.

Motor Transport Institute has developed research project on methodic evaluation of eCall system—R10 0016 06/2009, from December 1, 2009 to December 31, 2011. During the project realized tasks:

- Laboratory apparatus place to research eCall board unit, according to formal, legislation and technical requirements,
- Prototype of eCall unit simulator.

Test results approved eCall unit simulator to come all requirements, the manual call was activated by the driver of the vehicle, while the automatic call was activated under certain conditions after the impact, thanks to the sensors mounted in the vehicle (air bag launch or in the case of overloads).

4 Conclusions

Programs implemented by the EU institutions and EU Member States to reduce the number of accidents, victims and injuries, bring tangible results, but there is still much to do in this regard.

According to the EC decision, from April 2018 all new vehicles must be equipped with the eCall on-board devices.

The threat of the land transport sector with terrorist attacks in the world is quite significant—so far they have focused over 23% of all attacks that have been carried out.

A significant increase in the carriage of goods by road in the EU, including dangerous goods, increases the threats related to the theft of cargo, as well as related to the possibility of a terrorist attack.

The EU has introduced minimum requirements for TEN-T network security management (including security audits) and requirements for the transport of dangerous goods.

New regulations on offenses committed by drivers abroad and vehicle technical tests were also introduced to reduce the number of technical failures, and thus the number of road accidents.

Furthermore, modern transport system has to be as little as possible susceptible to threats and permanent in economic, social and environment protection reasons.

The research projects of Motor Transport Institute in cooperation with other institutions, have improved significantly road transport safety in Poland, especially to reduce number of victims from 5534 in 2001 to 2831 in 2017.

References

1. WHO: New Global Status Report on Road Safety. http://www.who.int/en/news-room/detail/19-10-2015-despite-progress-road-traffic-deaths-remain-too-high, http://www.who_new_-global_status_report_on_road_safety. Accessed 19 Oct 2015
2. European Commission: Mobility and Transport. Road safety. Statistics accidents data. http://ec.europa.eu/transport/road_safety/specialist/statistics_en. Accessed 02 Apr 2017
3. Global Status Report on Road Safety 2015: World Health Organization Library Data. http://www.who.int/violence_injury_prevention/road_safety_status/2015/en/
4. Intermediate report on the development of railway safety in the European Union. European Railway Agency (2013)
5. Council Regulation No. 1108/70 of 4 June 1970 introducing an accounting system for expenditure on infrastructure in respect of transport by rail, road and inland waterway. OJ: JOL_1970_130_R_0004_005
6. Commission Regulation No. 851/2006 of 9 June 2006 specifying the items to be included under the various headings in the forms of accounts shown in Annex I to Council Regulation (EEC) No. 1108/70. OJ L 158, pp. 3–8, 10 June 2006
7. Transport statistics. https://ec.europa.eu/eurostat/statistics-explained/index.php/Transport. Accessed 9 Julay 2018
8. Sitarz, M., Chruzik, K., Grabon, M., Gamon, W.: Intermediate report on the development of railway safety in the European Union. European Railway Agency, 45 9/2013, p. 45 Statistics, Eurostat (2013)
9. Vehicular attacks. https://www.telegraph.co.uk/news/0/westminster-nice-vehicle-rampage-attacks-uk-europe/. Accessed 18 Aug 2018
10. The Guardian/International: FBI warns driverless cars could be used as 'lethal weapons', Wednesday, 16 July 2014. http://www.theguardian.com/technology/2014/jul/16/google-fbi-driverless-cars-leathal-weapons-autonomous
11. Road Safety Statistics: What is behind the figures? Brussels. http://europa.eu/rapid/press-release_MEMO-18-2762_en.htm. Accessed 10 Apr 2018
12. WHO Statistics. https://www.sciencedirect.com/science/article/pii/S0386111216300103, http://www.who.int/gho/road_safety/mortality/en/19. Accessed Oct 2015

13. Olejnik, K., et al.: Optimization of Conditions of the Flow of People, Goods and Media in logistics. Technical University of Opole Publishing House, Opole (2015)
14. Olejnik, K.: Safety in Motor Transport—Chosen Problems. ITS Publishing House, Warsaw (2009)
15. Regulation (EC) No. 561/2006 of the European Parliament and of the Council of 15 March 2006 on the harmonization of certain social legislation relating to road transport and amending Council Regulations (EEC) No. 3821/85 and (EC) No. 2135/98 and repealing Council Regulation (EEC) No. 3820/85. OJ L 102, pp. 1–14, 11 April 2006
16. Nowacki, G., Mitraszewska, I., Kamiński, T.: TachoScan PC NET service digital tachograph equipment. WIT Trans. Built Environ. **2007**(96), 587–592 (2007). https://doi.org/10.2495/UT070551
17. Directive 2004/52/EC of the European Parliament and of the Council of 29 April 2004 on the interoperability of electronic road toll systems in the Community. OJ L 166, pp. 124–143, 30 April 2004
18. Nowacki, G.: The Polish scientific researches on electronic toll collection area. In: Road and Rail Infrastructure, pp. 149–154 (2012)
19. Nowacki, G.: Research project of functional structure of national automatic toll collection system—R10 0001 04, from 2008 to 2010. MTI, Warsaw (2011)
20. The European Parliament Recommendation: April 2017. https://ec.europa.eu/digital-single-market/en/news/ecall-all-new-cars-april-2018

Information Support Systems for Crisis Management

Bogusław Jagusiak[1(✉)] and Anna Chabasińska[2]

[1] Cybernetics Faculty, Military University of Technology, Gen. Witolda Urbanowicza 2, 00-908 Warsaw, Poland
boguslaw.jagusiak@wat.edu.pl

[2] Administration and National Security Faculty, The Jacob of Paradyz University, Teatralna 25, 66-400 Gorzów Wielkopolski, Poland
annaadamk@wp.pl

Abstract. The main goal of the research is to determine the level of importance of the IT decision support systems for the efficiency of crisis management processes, and system's operational continuity. Such detailed research problem determines the specific goals involving: (1) identification of the key processes for the continuity of operation within the system; (2) establishing the relationship between the decision time and system continuity; (3) exposing the methods and means proper for optimizing the decision-making time in crisis management situations. The final development of the research problem required the use of inquiry procedures, prior to analytical involvement, centered on the theory of strategy, as well as on the SWOT techniques, and desk research techniques. Certain primary, as well as derivative propositions of the theory of autonomous systems (TSAs) have been accepted as already defined. The terminological conventions contained within the TSA were also used, along with the derivative concepts significant to the specific rules of defining the theory. The innovative goal of such scrutiny, extending the existing state of scientific knowledge in security research, which is at the same time a praxeological directive, is to show that information processes are the foundation of the system's autonomy, i.e. maintaining the business continuity. Nonetheless, decision-making time is a key parameter in crisis management. Therefore, streamlining information flow (communication), and shortening the decision-making time, at the level of management and contractor duties, requires the implementation of the IT systems that are able to receive and handle the data in real time.

Keywords: Decision support systems · Crisis response · Social cybernetics

1 Introduction and Formulation of the Research Problem

Contemporary approach to study of the control systems presents the ample opportunities for growth of the versatile scientific syntheses, with application potential to evaluate large social systems, in order to optimize information flow solutions. This approach goes beyond guessing, intuition or conjecture, and can be respective of decision making processes, since based on knowledge stemming from duly justified

I. Kabashkin et al. (Eds.): RelStat 2018, LNNS 68, pp. 468–478, 2019.
https://doi.org/10.1007/978-3-030-12450-2_45

claims of science. Such knowledge seems to be the key factor in times of scientific and informational revolutions, and during dominance of intensive data expansion [1].

In view of the above, the research objective was to focus on the probabilistic systems, i.e. systems where the new evaluative state can be predicted with probability significantly different from 1. In consideration were those accountable material objects, that besides the location in time and space would have assigned mass m and energy E, as in:

$$e_n(x_1, x_2, x_3, t, m, E), \tag{1}$$

with inclusion of the tangible-active relations (with feedback) between occurring objects with mass and energy flow (interacting)—what is otherwise understood as the dynamic relations. In this research the probabilistic systems composed of people, the elementary subjects, and their interrelations, are conventionally considered as the social systems. Such an approach is justified by the fact that human body performs functions of an autonomous system, with an impact on the environment, including information and energy retrieval from the environment, with processing and storage of information and energy, and the persistence of functional balance (homeostasis). Hence, one can consider society as an autonomous super-system, constituted of humans as the autonomous subsystems.

On those grounds, the main goal of study was to formulate the value of basic conditions (sufficient and necessary) ensuring the continuity of social systems, under threats or ongoing crisis's, and conflicts.

The main research problem is expressed here by the question (Q0): what are the essential features of the functioning social systems models, and systems supporting the management, in the situation of crisis's and conflicts?

This question brings the main hypothesis formulated as follows (H0): In the crisis and conflict conditions the management support systems and models strongly determine the communication processes, and the continuity of social systems.

In order to verify the above hypothesis three specific research problems have been formulated, and are included as the following intrinsic questions:

Q1: What system's fundamental processes define its continuous activity?

Q2: In what way decision making time affects the continuous activity of a system.

Q3: What means (methods) should be implemented for optimizing the decision making time and system's operational continuity, in the crisis and conflict conditions. As follows, and analogically to the number of research problems, the authors formulate the following three specific hypotheses:

H1: Fundamental for the continuity of system's activity are the information flows, that is the adequate communicative process occurring between the management and contractors.

H2: The decision-making time affects both the performance of contractors and the permanence of organizational structure of a system (otherwise the continuity of operation).

H3: Optimization of decision-making time and system's operational continuity under the conditions of crises and conflicts is possible, thanks to the use of models and management support systems.

According to the authors, the theoretical methods should be used to extract the research problem, as formulated above. In this scientific research the method of inference or deductive reasoning has been used, since for certain type of statements it is the consequential value that counts the most (theoretical strategy prior to the research). Hence, some primary as well as derivative propositions from the theory of autonomous systems (TAS) and information theory (quantitative and qualitative, in general—IT) were accepted as the certain premises. Also included according to the rules of definition, was the conventional terminology of theory derivatives.

2 The Elements of the Cybernetic Theory of Security

Both, the living organisms and the societies, are not only the probalistic systems, but also the autonomous systems, i.e. they have the ability to control themselves in their own interest, and are able to counteract the loss of their ability, hence their existence can be seen as a control process, the aim of which is to counteract disruption of the self-steering ability [2]. That is, the ability of a system to maintain its state at the given time t, where the key processes (functions) are implemented. This is considered to be the process continuity assurance. However, the external or internal impact (potential or real) that causes, or may cause disruption to the implementation of the key functions of the system, are considered by authors as a "threat". Occurence of a threat due to particular system susceptibility may cause a sudden increase in the entropy of the system (chaos, disorder). Such a sudden increase in entropy is considered here as the "crisis situation", and its climax (with the highest value of entropy)—the "crisis". On the other hand, the "conflict" is understood as a relation where the control program goals of either system, or subsystems, are incompatible. In this understanding a "conflict" is a special case of a "crisis situation".

In order to counteract disturbances of the continuity of operation, a number of feedback occurs in the autonomous system, which manifests itself in the two-phase process of protection. The first phase concerns prevention (prophylaxis), i.e. it focuses on removing susceptibility to interference of possible certain sources of danger (controllable), while the second is manifested in the response (therapy), i.e. the removal of disturbances and their effects caused by threats.

The protection optimization process should lead towards the prevalence of prophylaxis over therapy. Being able to prevent crisis is the general goal, since with it there is definitely no threat of losses and disruptions to the essential operational activities. This underlines the accepted basic axiom of the theory of security, upon which the theory of security is built, as a derivative of TAS. And this axiom can be accentuated as the first law of the system security theory (SST) and has its counterpart in the folk proverb that says "prevention is better than cure". Other axioms of the security theory can also be formulated, for example: the operating systems can secure a specific system to varying degrees depending on the interest of the system organizer (second law of SST); only the autonomous systems have the ability to self-secure and are able to defend this ability (third law of SST); cutting off the system from its protective systems prevents its further functioning (third law of TBS) [3]. The autonomous system security was presented in Fig. 1.

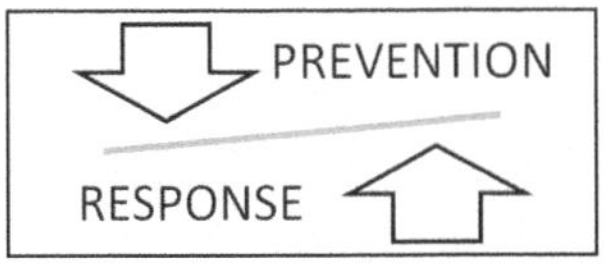

Fig. 1. Autonomous system security

The self-steering ability is the essence of the autonomous system, but in order to be controllable it must maintain the structure enabling control [4]. The interconnection of self-balancing processes through homeostat, which, like every organ strives to maintain its own functional balance, enables the autonomous system to maintain its structure despite the various, even contradictory, self-balancing processes and environmental impacts taking place within and without. For this purpose, the homeostat must be able to influence the energy track and the information track. Home information processes exert influence on material processes and energy processes through homeostat. They influence the information processes. In fact, it is a feedback between the correlator and the homeostat, and between the battery and the homeostat.

Let us assume that the State, as a special case of organized structure, can be simplified and considered as the two-stage structure, consisting of a central steering center with the intermediate executive centers subjected to it, and those subjected to the contractor [5]. The most important factor in the stability of the structure is the close correspondence between decisions of the central management center (C.M.C) and the decisions of the intermediate management centers (I.M.C), and finally, between the decisions of the intermediate management centers and the contractors' activities [6].

Moreover, we assumed that: through the decision-making efficiency of the management center we can understand the ratio of the number of decisions made by this center, in reference to the time it takes to make a decisions.

Hence, the time to make a decision was labeled the *decision-time*.

If the number of decisions taken is D and the decision-time t_d, then the decision-making efficiency of the steering center—which we designate as W_d—will be expressed in the formula [7]:

$$W_d = \frac{D}{t_d}.$$

By executive performance, we mean the ratio of the number of decisions made by the management center, which can be converted into appropriate activities (responses) to the given executive center, until the time the processing takes place, i.e. until the time of their execution.

Hence, the time of decision processing actions was labeled the *executive time*.

If by D we mark the number of decisions of the management center processed by the executive center, and the executive time is marked by t_w, then the executive performance—which we denote W_w, will be expressed by the formula [7]:

$$W_w = \frac{D'}{t_w}.$$

It is worth taking a look at the three relationships that may exist between decision-making performance and executive performance.

The first case concerns a situation where the decision-making capacity of the management center is greater than the performance of the centers under its control [7]:

$$W_d > W_w.$$

In such a situation, the executive centers not being able to process all decisions of the management take appropriate actions. Then it is necessary to select decisions, and implement those at the discretion of the contractors. The occurrence of the indicated dependence causes the instability of the structure, i.e. the lack of close dependence between the decisions of the management, and the activities of the contractors. After some time, the unfulfilled management decisions build up, and the executive bottlenecks occurs.

The second case concerns a situation where the decision-making capacity of the management center is smaller than the performance of the centers under its jurisdiction [7]:

$$W_d < W_w.$$

In this situation, the executive centers are able to process all management decisions but have some excess decision-making time, which in the worst case can be used for actions that are not in line with management directives. Hence, the system's operational continuity within the organizational structure is lowered due to decision diffusion.

The third case concerns a situation where the decision-making capacity of the management center is equal to the executive performance (subordinate centers) [7]:

$$W_d = W_w.$$

In this situation the executive centers are able to process all management decisions and do not have any excess decision-making time. In this case we observe close correspondence between management's decisions and contractor's activities. Therefore, the above equation describes the system's operational continuity within the organizational structure [2].

When considering a two-tier organizational structure, it is possible to introduce a condition of higher operational continuity, with:—the decision-making capacity of the central management center,—the decision-making capacity of intermediate decision centers,—the performance of contractors. Therefore, the condition of higher level operational continuity is given by the formula:

$$W_d^c > W_d^p.$$

On the other hand, the condition of lower-level operational continuity will take the form:

$$W_d^p > W_w.$$

The decision-making capacity of a two-tier organizational structure is "the ratio of the number of decisions made by the central management center, to the time when these decisions are processed into the decisions of the contractors and their actions".

3 The IT Systems and Technologies Used for Crisis Management

Under the modern conditions of scientific and informational revolution, decision-making centers are forced to process large amounts of various types of information [8]. This is because every decision made in one area of life has different effects in many other areas. Therefore, comprehensive solving of decision-making problems requires taking into account the possible effects of one decision on a number of other decisions in other areas of social life.

The use of IT and ICT systems in data processing streamlines and accelerates the process of analysis and evaluation of information necessary to determine a decision [9]. However, the preparation of data that can be processed using computer technology requires increasingly more time. The same applies to the total decision-making time, which is the sum of time needed to prepare the data, and to finalize a decision. In addition, IT decision support systems can not just alone replace management centers in decision-making.

Because of this, the total decision time needed to take majority of decisions increases. It also often becomes longer than the time of implementation of the actions to which the decision relates. On this basis, it can be anticipated that decision-making time is now becoming the basic factor that determines the continuity of social activities and the efficiency of the entire system. It is assumed that the contractors have the appropriate energy-related resources necessary to perform specific activities, and their performance depends on the frequency of appropriate control signals sent to them by the management centers. Thus, the performance of performers is inversely proportional to the length of the decision-making time of the management. Shortening the decision-making time of the management causes a corresponding increase in the performance of contractors.

Decision-making time is particularly important in the crisis management response phase. "Speed of action, making certain decisions in response to a threat, is a kind of struggle against the time of decision makers in the crisis management center" [10]. Extending the decision-making time also causes the system's exposure time to have a destructive impact. Thus, greater losses and delays in resource recovery (reconstruction phase) are generated. Improving information flow (communication) and shortening the decision-making time in the system enables modern systems obtaining and handling data in real time [11].

Data sources in assistance systems can be divided into personal and non-personal (technical). In the case of most countries, the tasks of personal sources are performed, for example, by an interview, counterintelligence (generally state services), mass media, and diplomatic missions. Non-personal (technical) sources are used for quantitative and qualitative analysis of parameters and characteristics of devices emitting energy (MASINT—Measurement and Signature Intelligence). The measurement and signature recognition includes, among others, computer, acoustic, radiolocation, radiation, infrared, chemical and biological, nuclear, beam, laser and electromagnetic intelligence [12] (Fig. 2).

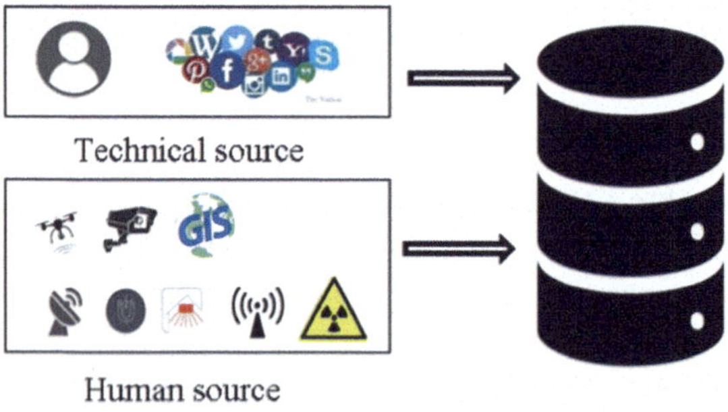

Fig. 2. Data sources in monitoring systems

For the needs of crisis management institutions, dedicated OLTP systems (On-line Transactional Processing) are used [13]. These are transactional data processing systems, i.e. operations performed in the current (production) enterprise database used for daily work (e.g. data from monitoring systems). OLTP class systems are optimized for maximum transaction efficiency, high parallelism and availability. Requirements for such systems are following:

- Required access to current data (without delay).
- Performing a large number of simple queries from many users (even hundreds of queries per second).
- The database system should be optimized for quick data retrieval.
- Frequent operations of adding, removing modifications to individual records.

In order to support large, variable and diverse data sets, Big Data technologies and a real-time data system are used (particularly critical requirements in emergency response) [14]. The main components are (Fig. 3):

- Data analysis techniques such as A/B tests, machine learning and natural language processing, artificial intelligence,
- Big data technologies such as business intelligence, data processing in the cloud, as well as databases, data warehouses,
- Visualizations, such as charts, graphs and other forms of data presentation.

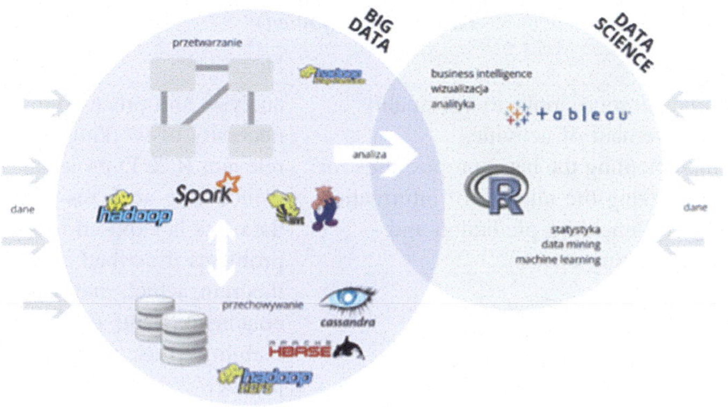

Fig. 3. Big data and data science technology [15]

The OLAP system (On-line Analytical Processing) is also important in crisis management. It is mainly used to create reports, including tabular statements and charts, (ability, means, resources), as well as forecasts and simulations. This kind of processing is usually intended for various types of users: management, analysts, administrators. Requirements for such systems are, among the others:

- A small number of queries, but for large amounts of data.
- These systems basically read information only from the database.
- Not a full data calculation, which may be available with a delay where the calculation may take from a few seconds to many hours.

An important element related to communication during the crisis and conflict is the combination of real-time visualization of data from geoanalysis. This solution helps in seeing and understanding data, with decisions made faster [16].

Table 1 sets out a synthetic combination of strengths and weaknesses (SWOT) for the application of Big Data technology in support of crisis management processes [17].

Table 1. SWOT analysis of big data applications in crisis management [18]

	Positive	Negative
Internal	Strengths: The potential to increase involvement in prevention, not just the response to the emergence of the crisis Virtually infinite data and information resource The possibility of using such elements as crowd sourcing, or customer—	Weaknesses: The continuous increase in the amount of data causes a huge demand for equipment that makes possible cataloguing and analyzing larger sets for data In the case of Unstructured Data, there is a problem with the preparation of appropriate assumptions for their

(*continued*)

Table 1. (*continued*)

	Positive	Negative
	administration applications, quality improvement of activities Implementing the basic propositions for maximizing the number of information sources, enabling prediction and counteracting crises	analysis and processing, this adds necessity of working out primary relevant R & D projects Conducting analyzes of this type of data sets is exposed to a number of problems described in the HACE theorem, which may lead to wrong conclusions and, consequently, the inability to use the effects of work in practice
External	Chances: Large development potential and the ability to search and run new analytics concepts Relatively new trend, which means no rigid and existing tools and ways to work in this area Complementarity with data warehouses and Data Mining cause evolutionary rather than revolutionary implementation The lack of clearly defined Big Data components gives practically unlimited development opportunities in the direction of including further types of data into the set	Threats: Due to the many data sources Big Data not collected in the central database for the needs of crisis management are useless in the case of discontinuities of infrastructure elements such as cellular networks and the Internet The use of extensive (growing continuously) and distributed databases may prevent the use of Data Mining elements and procedures, for instance due to the costs of data transfer and storage The collection and processing of sensitive data that forms a large part of Big Data causes many doubts of a fundamental nature and affects the negative perception of such data sets as a source of public surveillance In relation to the above, many countries try to regulate the access and storage of such collections legally, which may lead to the overregulation and uselessness of such residual information

4 Conclusions

1. In view of the conducted research, the main research hypothesis and detailed hypotheses can be confirmed. In reference to the fundamentals of the cybernetic theory of autonomous systems (axiomatic theory in the abstract stage) authors confirm that system's operational continuity constitutes the foundation of the system's autonomy by maintaining proper level operational continuity, although with specific variations. This was symbolically shown by various formulas of system's operational continuity.

2. However, one common factor would determine property of internal processes of lower or higher continuity of operations. This would be the *decision time* as a key parameter in crisis activities (and not only in those). Its shortening enables modeling and management of the support systems. And also, allows preparation for potential threats by proper channeling of informational flow.
3. Efficient analysis and exploration of data, with its visualization, is the basis for maintaining continuity of operations in the times of the information revolution. The implementation of management support systems allows the efficiency of the entire organization to be increased, by improving communication between management and contractors. This allows, by controlling system's operational continuity—through ability to properly adjust informational flow, to enhance steering capability of the entire given system.

References

1. Chaładyniak, D., Czarnecki, A.: Analysis of selected tools for scanning IT systems. Zeszyty Naukowe WWSI **17**(11), 89–109 (2017)
2. Kossecki, J.: Elements of Modern Knowledge About Controlling People. WZiA, Kielce (2001)
3. Konieczny, J.: The Engineering of Systems. WNT, Warsaw (1983)
4. Mazur, M.: Cybernetic Theory of Autonomous Systems. PWN, Warsaw (1966)
5. Zaskórski, P., Szwarc, K.: Modeling of security and continuity of government processes. Nat. Secur. Stud. **12**(2017), 335–364 (2017)
6. Zaskórski, P.: Information and business continuity of a firm. Zeszyty Naukowe WWSI **5** (2011), 217–232 (2011)
7. Kossecki, J.: Social Cybernetics. PWN, Warsaw (1981)
8. Serwer of address points. https://www.swdst.pl/swd-st-spa-serwer-punktow-adresowych/. Accessed 14 May 2018
9. Janowski, R.: Analysis of selected tools for information collection about the compromised IT system. Zeszyty Naukowe WWSI **17**(11), 69–87 (2017)
10. Najgebauer, A.: Models of Threats to the Urban Agglomeration Together with the Crisis Management System Based on the Example of the Capital City of Warsaw. WAT, Warsaw (2009)
11. Smadi, S., Aslam, N., Zhang, L.: Detection of online phishing email using dynamic evolving neural network based on reinforcement learning. Decis. Support Syst. **107**, 88–102 (2018)
12. Nowacki, G.: Satellite Recognition of the USA and the Russian Federation (Rozpoznanie satelitarne USA i Federacji Rosyjskiej). AON, Warsaw (2002)
13. Najgebauer, A.: IT systems for decision support in conflict situations. WAT, Warszawa (1999)
14. Marz, N., Warren, J.: Big data. Best practices for building scalable data systems in real time. Helion, Gliwice (2016)
15. Wazka project. http://wazka.wat.edu.pl/pl/wyniki. Accessed 13 May 2018
16. Zaskórski, P., Pokorski, G.: Geospatial data in threat monitoring and analyzing. Nat. Secur. Stud. **12**(2017), 365–378 (2017)

17. Distance, big data. http://datascience.ii.pw.edu.pl/bigdata.html. Accessed 14 May 2018
18. Drosio, S., Stanek, S.: The big data concept as a contributor of added value to crisis decision support systems. In: Sammon, D. (eds.) Proceedings of the 2016 Open Conference of the IFIP WG 8.3, "Big Data, Better Decisions, Brighter Future", University College Cork, Cork, pp. 228–239 (2016)

Dimensions of Security System Complexity and Risk Detection Praxis

Arnold Warchał(✉)

Military University of Technology, ul. Gen. Witolda Urbanowicza 2, 00-908 Warsaw, Poland
arnold.warchal@wat.edu.pl

Abstract. The multidimensional spectrum of security predicaments forces a question about compatibility and complexity of various solutions to threats and dangers. The number of problem responsive solutions grows, yet, the ability of countering is not always a guarantee. The State institutions responsible for security in various fields, must deal with those problems strategically, but this is also an issue in any field of operation, including business and management, or others. The author probes those problems stemming from following question: How complexity of a system, or systems using different approaches, technologies and techniques retains unity of the whole, and its capability in practical response to threats. This is considering the fact, that innovative techniques and new technologies give us ability for fast response, but also for multiplications of false results, and new threats within the security system as a whole. In view of this author, through comparative, analytical and phenomenological methods, examines factors linked to theoretical propositions, where base terminology used rests on security studies and political science, but is generative of scientific fields. Author suggests that the multiplication of complex system solutions brings about new variables, and therefore, also possible failures. The scope of research is bound by outlook based on processes and scientific solutions, aiming at synthetic view of possible threats, and solution propositions worked out by the researchers of the Military University of Technology in Warsaw.

Keywords: Security system · Threats assessment · Complex systems · Innovations

1 Introduction

The contemporary quandary of various security definitions, categories, outlooks, and ranges of possible explanations for issues associated with the external threats, and the internal risks, as accounted in preparation for, and creation of a complete prevention system, is nowadays submerged in multidimensional evaluative descriptions. Those may, or may not, provide answers to proper functioning, as well as to understanding the hindrances in development of security systems, and individual security factors. From the scientific or academic perspectives we can grasp those because of proper intellectual "measurement", becoming one of the comparative elements in the general field of security, if it is done with concise approach of theoretical, that is logical, precision [11], or leads to theory and praxis building based on proper identification of security

I. Kabashkin et al. (Eds.): RelStat 2018, LNNS 68, pp. 479–488, 2019.
https://doi.org/10.1007/978-3-030-12450-2_46

phenomena [12]. Presented here philosophical generalization points out to essential elements of secure system, and system's theory, as based on the idea of the most common denominator, that author names the "molecular" component in the process of security building. This deliberation is in an essay form, but rests on scientific, and security studies concepts, that are comparative with intrinsic variables, and can be synthesised towards the analogical fractal solution.

While studying many arrays of concepts describing security phenomena and different security systems as an academician, and from even more personal perspective—presenting studies on security, as the co-editor of the *National Security Studies* (published in Polish as: *Studia Bezpieczeństwa Narodowego*), printed by Military University of Technology's, Cybernetics' Faculty, being responsible for its content, the author understands security as a field of proper value activity, of various nomenclatures, bounded by intellectual normative constructs, in leading to theory—praxis symmetry or model building, often leading to the system creation, and at the same time bounded by physical phenomena. However, it is understood that the occurrence of a phenomena and description of it must be relative to the area of subjective interests, with objective outcome. And by "objective" it is not only meant a rational common understanding, but a methodical use of subjective knowledge to override the differences, what allows grasping or arriving at the most common denominator at the "molecular" level of any discussion about a system or its components, and anomalies.

Problem being, when we accept such an approach, of an author describing the very complexity of a nature of security, it will become the essence of compatibility of phenomena depiction. It is after all both, of a tangible and intangible character, and pertains to holistic end, not just the accidental occurrences—understood more in terms of anomaly than continuum or system matrix value, however, changing "measurements" of the whole. And it comes in a way that word symmetry comes to mind. Whatever the nature of occurrence in description its informative nature will represent the interest of a particular observation, as well as, at least, it will represent its measure in accordance to whole system (that may, or may not, be influenced by the intellectual character of description, with overt or covert residues). This type of endeavour, whatever its practical outcome, has to be explained by some kind of a theory, in general, or methods leading to applicable model building stemming from norms permeating the whole system, with the finale in praxis.

Therefore, this study has a multidimensional spectrum and concerns question of compatibility and complexity of a system, and its adeptness to ever changing natural and human made environments. In direct hypothesis, the end number of security dimensions, if not rationalized to practical point of unison, may create both risks for system functioning, and may become insensitive to threats, that it is supposed to recognize and react against, if necessary. It is also understood, that the field of security dimensions will create problems of complexities, hindering the flow of information needed to guarantee the risk-free system, able to react to threats. Hence, the compatible units will have to be applied, in order to sustain the connections between units responsible for levelling the differences.

As it is implied in the title, a secure system is defined by its operational activity with aim of proper functioning, what resembles the cybernetic view on system unity and reactive control mechanisms [6], but also as is understood by philosophical definition of an unified systems [13], or even in public administration and management [16], not to mention considerations about political systems. In proper perspective, this will also refer to general definition of ability to react of any autonomous system, where the sum of elements and its activities determines reaction of the whole. In any sense, we don't know however, if it always happens and weather common continuous reactive modes, with practical outcome of infrastructure built and institutions specialized, will not accidentally: (a) become outdated, because of cumulative values of repetitious information, (b) narrow the field of observation to obvious focus, leaving out the rest of field or dimensions uncontrolled, (c) create risks for threat knowledge sensing and reactivity. Those after all determine the process of survival at any level of natural and artificial, including human and created, life.

The survival process is really what encompasses the various definitions of security, threats and risks, dangers to autonomy, unity or effectiveness, in relation to a system. If this is understood then common mistake in considering what it means to have a secure system can be avoided. Security, if we are to observe its proper nature, is not, according to the author, a constant for any type of natural or artificial "mechanism" or "structure" we can geometrically define or prescribe for it ontologically unchanging form. Distinctiveness of, for example, a geometrical or ontological means of description is considered here as not the essence, but the end product of conscious reactivity to repetitious values of created models of life itself, where the end game is for general life, not just its components (i.e. geometric or ontological outlook about it). This is an intersubjective knowledge, and its value can be simplified without normative explanation to symbolic 0–1 status of conscious being. Problem being is that 0–1 status will become complex because of energy rhythm flow, which is considered here as the interchange between proper energy—mass count, that can be on molecular level, where the particular resembles the whole at its closest, and is important because of compatibility between different levels of proper values, transposed as knowledge—information duality. What in any form structures the identity of a system, and it is rational for any type of a system.

Given the multidimensional scope of a problem, both, the various specific for the subject analysis modes of description, and the phenomenological method of research—looking for the separate values of an analysed element, and for possible underlying common denominator (putting in parentheses the connection between the given separate unit and the whole, or other possible components, what shields from abstraction by extra component/s). This allows for independent analysis and latter comparison of separately analysed components, and synthesis presenting holistic process, where the security phenomena occurs. The view of it is multidimensional in a sense, that the secure system will have within it components necessary for survival, similar in various dimension in operative purpose, and those less important, often abstracted from the environment and focused on individual functions. Their interaction will create internal risks because of the difference in "rhythm"—number count of its processes, and compatibility with internal other processes or external dimensions of similar activity, can be in danger of dissonance. So both, on the level of processual operations, and in between the dimensions of system's operational goals, lack of compatibility between cohesive units will stress the system with abstract values that do not belong to a set. It is important to understand the weight of different, new, and abstract variables, since those may affect already complex system not only in particular points of operation, but possibly changing it, or bringing system failure to the whole.

The inter-subjective literature discussing the problems of security is used here only in the exemplificatory sense. It is not a base of research, however inclusively the literature is being used, as an reference to the discussed topic. This discussion is based on few hundred articles covering various security problems, that for obvious reasons can not be presented here, published in the National Security Studies—numbers 1–12. And can be considered: a generalization on security studies problems, from academic perspective.

2 Complexity and Compatibility of a Security System

The complexity denotes the many variables of changing or stable nature, and refers in mechanical outlook to functional operation, on many levels of possible interaction on different levels of a system, where each level has two goals, that seem to be in conflict: provision for a system and profusion for individual functions. The conflict can be resolved quickly by adding symbolic value to chosen important factor, from the level of system control, specifying potential modes of decision process, where the two functions—provision for a system, and for individual activity, are becoming the same. Already on this basic levels of the system control, system function, and system components, there is a solution for internal risks associated with potential divergent modes of system's operations. All that has to be done is the transposing of the energy flow of changing variable into properly understood information. Such model of information is operative at the "molecular level". The inherent value of this is not stemming from any normative observation of a phenomena, and application of a creative solution, but from nature itself.

This is the complexity of nature, its biological dimension, that applies simplification to its functions, in respect to its environment, or becomes more complex, because of changing environment, that is multiplication of information necessary for system's survival/continuum. Whatever the case, dimension of biological environment presents a system proper tasks for staying within a spectrum of life, where the process itself becomes obvious but, given the proper input and output of energy, it can regulate itself relatively well. Once it provides for internal simplification of its functions, the complexity problem disappears, while the factors leading to simplification of the external factors do not have to be considered for the self regulating or autonomous system, that we observe in natural setting. In natural sciences such explanation is visible within general evolution theory, where "evolution" means adaptation through multiplication or simplification. Observations made from such perspective are proven by comparison between, for example, plants. Looking at features of the remains of tree from times of dinosaurs, one can notice its more complex nature, than present tree's structure. All because of surrounding environment, that was more complex by itself. And at this level, where the process of interaction between the external and internal factors becomes visible, and functions can be adapted symmetrically to changing ratio of outside factors, the risk of "internal organs" or elements' malfunctions can be eliminated. This is the molecular level since on this level the part of a system resembles the whole system, at least when it comes to functions of system's structural or other type of stability.

The complexity of a system creating risk for its general task accomplishment occurs when adding simplification elements to autonomous functions puts in risk its compatibility mode with the other elements of a system. Problem being is that this is visible at the internal level of system's operation, and externally it becomes obvious only after the entire system becomes stressed or malfunctions. Since our creativity rests on observation of natural factors, and synthesizing into a symbolic value energy measurement, that makes life "material", it is still understood that the process of energy exchange is as important, and not less, even if the structural organic activity provides with it, its internal rhythm for the ratio of energy exchange. Transposing this into general mathematical symbolic value provides the measuring entity with the ability to build models of structural energetic interchange. Those models in turn provide reference examples for secure system creation. Those models based on biological dimension seem to be rather internally focused, and the outside threat is "measured" because of internal incompatibility of a particular element overdosed with unknown information, or when the system itself is overloaded with the processing of multiplying information, and its energy thread becomes exhausted.

However, the biological systems are not only internally active, with constant diagnostics of elements, and compatibility equations, but also externally active. This comes from the molecular symmetry. The organs at the molecular level stop interacting when the compatibility comparison is not equal to the proper ratio of known and unknown elements. That is, when according to the operating model visible only symbolically, based on exchange of biological information, the sum of information does not resemble in proper ratio the individual count of understood and misunderstood information. At this moment, the molecular stability resting on direct ratio resemblance between part and the whole becomes blurred. And the system slowly starts to change its goals, due to differences in compatibility of the elements. This in turn hinders its other goals, to the moment of total failure. Before the failure, however, the internal informational conflict becomes at given moment apparent, and this is the moment when any system biological or other, has to be taken over with control resting on proper information flow. This is when the other "dimension" of operational activity because of control center, becomes apparent. Understanding the process of informational exchange, when a model of security system is based on biological dimension, makes it possible to create artificial, yet reflective of natural ways of solving the possible problems of elements compatibility and systems integrity. This "outgrowth" of natural ability to self-diagnose and internally redirect the flow of energetic information becomes a model for other activities. Understanding the informational process transforms the human activity in different fields, and can become helpful in informatics, decision making, or conflict situation solutions [8].

We have to remember of course, that the aim of a security system in its autonomous activity is not just the diagnostics and observation of system's internal activity, so as to discover threats, created by the system itself, or by external elements changing the internal information structure. The process itself is "ecological", that is, is based on the interaction of parts within the whole in an exchange mode with a goal, if any, of preservation of system's stability in measure with the external factors. Meaning that the "life" of a system does not just mean simple structural integrity and constancy. The "ecology" of life proves that the life system is processual, where process implies both activity and reactivity. And this is where, at the molecular level, activity and reactivity have to become compatible, if the integrity and the proper functioning of the entire system can be kept stable, where instability of a system implies threat to the whole or its parts. Aim of this prolonged paragraph discussion on natural system was to shown that molecular level of system functioning is where the symmetry is possible and this is the level where any changes can be done to a system, without destabilizing its nature of activity. What this "symmetry" implies is, of course, a different level of discussion, perhaps engraved in symbolic value of the "philosopher's stone".

3 External Threat Factors and Dimensions of Security

In the context of differences in categories referring to various dimensions of security, where term dimension is used loosely to focus on group of elements that create, by establishing common categories, a whole "ecological" unit, considered as a sum of particular activity and reactivity of the secure system. The secure system is not understood as a closed system with boundary walls eliminating the outside sources of change, but system that is able to incorporate the outside information for the purpose of the particular goal, that a system develops while in operative mode. The secure system is open, while the elements of the system are able, by natural or directive cause, to make proper alternation allowing the informational flow of energy to be utilized by new means worked out by adaptation, or directed in a manner deemed necessary by the controller [15]. This way, when we the fear of complexity as the enemy of security [4], it can be subdued, if not directly endangering the system as a whole. The main components here are the compatibility and adaptation of a system, in a similar way the evolution does on the universal level.

Let's keep in mind, nonetheless, that the universal dimension is a system resembling, from our perspective, a closed system with its own "ecology"—parts exchange process within a universal unit. We deal with such systems only on the theoretical level and in most cases, the theoretical propositions can not always be adequately measured without particular theory and methodology combination, as in scientific empirical endeavors. Within universal dimension of elements' interchangeability we will find those, of the molecular type—repeating the general energy information in a same proportional notion for a different level of activity, where at the molecular level it is only truly visible, whether the system will continue its energetic information processing, or whether it will be hindered by inadequate amount of necessary elements, or overloaded, with potentiality to malfunction. Based on observable differences in system's operational rhythm, what can be measured mechanically by means of energy and mass distribution account, or by compatibility factors of interchangeable elements with particular activity assigned to proper units.

This will lead to changes in characteristics of the whole unit, however if the adaptability and compatibility of the particular elements will be in unison, the internal risk will be minimal. Problem being left, nonetheless, is the external nature of information received and adequate reaction to new variables that do not fit into the system itself. In this case, depending on the mechanism recognizing properly labeled differences, the outside threat is known at the moment of surpassing the amount of data expected to be exchanged or adapted. Understanding the proper values will allow for model creation, and in turn will lead to model testing, leading to proper theory—praxis evaluation, and to practical solutions needed to operate safe system or to properly react to the unknowns, or unaccountable. However, this has to be worked out on the trial bases. One of the main focus has to be made on simplification and compatibility.

Systems of those quality, reflecting molecular level of empirical energy measured can be seen as innovative for closed end systems. This innovation is not referring to new mechanisms, however, but to new ways of security process adaptability within a changing environment. The only innovation required is the adjustment of one matrix, i.e. molecular level of compatibility that serves as the middle point in processual exchange. It can serve its goals anywhere, where the complexity of a system is observed, with negative effects on the field of actions. It is after all the interchangeability that can be labeled as "innovation", not a new process in itself. Let's remember though, that the universal systems may be too large to observe and adapt the observation to appropriate unison, of the molecular level. Aforementioned biological system is one of the few where it can be verified with instruments, as well as with an open eye. This process of security entanglement within the realms of typical and necessary threat protection can be visible at the other dimension, the social sphere of human activities, where the praxis of various models are worked out on daily bases, due to different application of the same paradigm of security approach, regardless of its theoretical value of scientific knowledge, serving more as a background, than a practical tool in general.

It seems that the best observation, where many variables can be used under one, in this case, legal background, and where the tangible and intangible means of securing the political system can be measured and used interchangeably. And we know, that the same problem of many variables determining the system's input, output, and overall quality, can be much better observed looking at the national and international interaction, where already creating a system of interaction makes possible the security of the whole [10]. There are of course many threats to national and international system of security [5], of many and various dimensions [9], both overt and covert, as in the examples of deception tactics used against the integrity of a political and military systems [14], yet, when the "innovation" of molecular level reflective mode can be used, the most secure system will be at the same time, the most natural. And in nature, biological or other, we can look for solutions to outside threats or internal risks to the system as a whole. The same applies to the regional [3] or global level [1], where the security is bound by the stable, legal processes, and unstable number of variables extending the amount necessary to upkeep the molecular level affirmation.

The important factor one has to consider, when looking at the sensitivity of a system towards the outside perimeters, is its ability to predict on the outset the amount of information adaptable, and to calculate the unknowns with predictability, as to probable outcome of verification and adaptability of the unknowns, in comparison to known variables. This makes it easier to adapt such approach to different analysis necessary for predictive purposes, such as creating regional security strategy [2], or national security strategy planning [7]. Whatever the sphere of social life and human activities, finding the most common denominator based on the most universal molecular level similarity between the various dimensions, leads to risk compatibility resolution and sensitivity towards the external factors, anticipating threats.

4 Conclusions

The deliberations on the subject of complexity and risk detection praxis are not bound by any particular field of observation. Phenomenologically they are applicable to many fields. We must look for its molecular reflective (repetitious on micro and macro levels) measures, unifying tangible and intangible (or unknown) elements, creative of a system and system creating, in response to external factors—accepted or rejected based on similarity or lack thereof. The best example here would be the fractal analogy, it is a form creating but will not allow any changes at the molecular level—its known value. Let's remember that the Security studies use knowledge of many fields, and this "knowledge", also theories from other subjects, including fractal theories (given here only as an example of molecular level processual activity of tangible and intangible unifying metrics) become the "praxis" for security systems, in a longer, or sometimes shorter run. This presents one general problem, if applicability is to be useful, it has to become the common/molecular/(fractal) denominator, not just the functional mechanism of security structure, of the entire system and its generative values. Based on those, security system organizers create system models. The useful application of this, in view of empirical data and its variables (keeping in mind also the "unknowns", as in "x" equations) must resemble the molecular repetition of interaction between the entire system and its external factors, to be predictable. The "models" accept the values of various dimensions, if the constant transformation "molecular value flow" is accepted in accordance to the same symbolic value of the established organizational interactivity.

Because of its general scope, upon finding the molecular level resemblance between different objects or values interconnecting dimensions of interaction, the compatible elements can be used through the entire matrix of operational unity. It is this compatibility that upon detection of internal differences in already complex system, through any kind of diagnostics—ability to discern the constancy and the rhythm of operational information flow, the risk to internal functions will be reduced by a factor of possible changes. Those changes however may be risk creating, if the proper common denominator, the molecular repetitions of a quality in all the particular parts, will not be reflected. The internal focus has to be harmonized, if the risk detection will be an autonomous for the system's internal activity. In a similar ways, as it is with the biological organisms and their energy flow and molecular repetition (or its mathematical equivalent).

The same molecular compatibility factor has to be accounted for the multi dimensions of system's practical activity, referring to differences in general internal or external phenomena, and to various fields of measurement, however, in unison acting by the same directive, even if sometimes from different "control centers". Such unity can be created by understanding of the most common denominator that permeates the various dimensions of security phenomena. It is the binary symbolic understanding of temporal activity that allows to differentiate between the known and unknown elements of the information flow, in reference to observed phenomena. However, it is the similarity of harmonious activities that allows to observe the differences, that is the "rhythm" of system's general and particular activities. This "molecular base" provides fundament for practical modeling and evolutionary praxeology not just by rejecting or destroying "threats", but by absorbing, using or neutralizing threat energy.

References

1. Alberts, D.S., Czerwinski T.J.: Complexity, Global Politics, and National Security. National Defense University Washington, D.C. (1997). http://www.dodccrp.org/files/Alberts_Complexity_Global.pdf. Accessed 20 Feb 2018
2. Bendiek, A., Kaim, M.: New European Security Strategy—The Transatlantic Factor, "SWP Comments 34". German Institute for International and Security Affairs, June 2015
3. Chmutina, K., et al.: The complexity of security dimensions: a comparison of the North-West and South-East European regions. In: Beyerer, J., Meissner, A., Geisler, J. (eds.) Proceedings of Future Security, Berlin, 15–17 September, pp. 169–176 (2015). https://dspace.lboro.ac.uk/dspace-jspui/bitstream/2134/19142/1/Chmutina%20etal_Future%20Sec.pdf. Accessed 5 Apr 2018
4. Goldfarb, J.: A complexity is the enemy of security, "SecurityWeek" (2015). https://www.securityweek.com/complexity-enemy-security. Accessed 16 Apr 2018
5. Hardenberg, M.: National Security Challenges, National-Security-Challenges-White-Volume-July-2012-FINAL.pdf (2018). http://pcnlab.asc.upenn.edu/wp-content/uploads/2017/07/National-Security-Challenges-White-Volume-July-2012-FINAL.pdf. Accessed 10 Mar
6. Kawalerski, P.: Bezpieczeństwo w perspektywie cybernetycznych sprzężeń zwrotnych (The National Security Through Perspective of the Cybernetic Feedback). "Studia Bezpieczeństwa Narodowego", no. 12/2017, Wojskowa Akademia Techniczna, pp. 83–108 (2017)
7. Koziej, S.: EU global strategy: opportunities and challenges of further implementation, "Pulaski Policy Papers", Warsaw (2017)
8. Najgebauer, A.: Informatyczne systemy wspomagania decyzji w sytuacjach konfliktowych. WAT, Warszawa (1999)
9. Nowacki, G.: Transformacja struktur operacji psychologicznych Stanów Zjednoczonych spowodowana nowymi zagrożeniami bezpieczeństwa międzynarodowego (Transformation of Psychological Operations Structures in the United States as a Result of New Threat to International Security). Studia Bezpieczeństwa Narodowego (National Security Studies), no. 9/2016, Wojskowa Akademia Techniczna, pp. 181–200 (2016)
10. Scheffran, J.: The complexity of security (2008). https://onlinelibrary.wiley.com/doi/pdf/, https://doi.org/10.1002/cplx.20242. Accessed 16 Apr 2018
11. Stańczyk, J.: Formułowanie Kategorii Pojęciowej Bezpieczeństwa (Formulating the Concept Categories of Security). FNCE, Poznań, pp. 85–197 (2017)
12. Stańczyk, J.: Poszukiwanie reguł definiowania bezpieczeństwa (In Search of Imperatives Defining Security). Studia Bezpieczeństwa Narodowego (National Security Studies), no. 12/2017, Wojskowa Akademia Techniczna, pp. 15–28 (2016)
13. Świniarski, J.: Filozofia bezpieczeństwa (The Philosophy of Security). AON, Warszawa (2004)
14. Warchał, A., Żukowski, P.: The Undisclosed Front—Aspects of Structural Pattern and Action Networks of Subversive-Reconaissance Activities in Contemporary Conflict Areas.

Open Sources Analysis. Studia Bezpieczeństwa Narodowego (National Security Studies), no. 9/2016, Wojskowa Akademia Techniczna, pp. 221–242 (2016)
15. Zaskórski, P., Pokorski, G.: Dane geoprzestrzenne w monitorowaniu i analizie zagrożeń (Geospatial Data in Threat Monitoring and Analyzing). Studia Bezpieczeństwa Narodowego 12/2017 (National Security Studies), Wojskowa Akademia Techniczna, pp. 365–378 (2017)
16. Zaskórski, P., Szwarc, K.: Modelowanie procesów zapewniania bezpieczeństwa i ciągłości działania organizacji administracji publicznej (Modeling of Security and Continuity of Government Processes). Studia Bezpieczeństwa Narodowego 12/2017 (National Security Studies), Wojskowa Akademia Techniczna, pp. 335–364 (2017)

Testing of the Drivers Ability to Assess the Distance on the Road While Driving

Krzysztof Olejnik[1(✉)], Marcin Łopuszyński[1], Gabriel Nowacki[2], and Bartosz Zakrzewski[3]

[1] Faculty of Production Engineering and Logistics (WIPiL), Opole University of Technology (PO), Gen. Kazimierza Sosnkowskiego 31, 45-272 Opole, Poland
k.olejnik@po.opol.pl, m.lopuszynski@po.opole.pl
[2] Cybernetics Faculty, Military University of Technology, Gen. Witolda Urbanowicza 2, 00-908 Warsaw, Poland
gabriel.nowacki@wat.edu.pl
[3] Motor Transport Institute, Jagiellońska 80, 03-301 Warsaw, Poland
bartosz.zakrzewski@its.waw.pl

Abstract. The article refers the study of the behaviors of drivers while approaching the intersections with traffic lights. It is an element of the active safety system of traffic participants. This study presents the results of testing the ability (accuracy) of the distance assessment on the road by drivers, when approaching the stop line of the intersection. The existence of a so called "dilemma zone", its extent and location relative to the stop line is not physically determined prior to the intersection. When the traffic light signal is changing, the driver must assess his distance to the intersection. What is more, the driver must also accurately assess and determine the location of the 'dilemma zone' and decide, whether to continue driving or start the stopping of the vehicle. When approaching an intersection with a changing signal, the driver finds himself in a risky situation. Sudden braking might surprise the following driver and lead to the rear end collision. If the drive is continued, the driver might enter the intersection after the red light signal has already turned on. Making the right decision depends, among other things, on the accurate assessment of the distance of the vehicle from the intersection with a minimum error. The driver must do it in a very short time. Extending the decision-making time is detrimental to the driver. Pilot surveys were conducted on a group of over 200 people. For this purpose, a questionnaire was prepared and four vehicle positions with known distances to the stop line were proposed. The vehicle positions symbolized the set of traffic bollards next to the road. The respondents assessed the distance between individual bollards, as vehicles in front of the stop line. The results confirmed the initial hypothesis that there is a large error of distance assessment made by drivers. This situation is encouraging to committing offenses and the occurrence of collisions and road accidents, so the need for additional training of drivers was indicated. In addition, it is necessary to consider marking the critical points on the road, e.g.: the "dilemma zone", enabling and facilitating the correct assessment of distance and choosing the right behavior.

Keywords: Dilemma zone · Distance assessment · Active safety

I. Kabashkin et al. (Eds.): RelStat 2018, LNNS 68, pp. 489–499, 2019.
https://doi.org/10.1007/978-3-030-12450-2_47

1 Introduction

To eliminate accidents and collisions in road transport, among other things, the knowledge and skills of drivers should be verified. They include: knowledge of applicable regulations and the ability to apply them in practice. A special situation is the vehicle passing through the intersection with traffic lights. Changes in the light signals in the tri-color signaling are sudden and surprising to road traffic participants. The knowledge of the drivers (about how to behave after such change) is unsatisfactory.

The authors have previously conducted research on the drivers [2] in the discussed area. In addition, proper conduct depends not only on the knowledge of correct behaviors. As the time-space coincidence is of a great importance in avoiding accidents, it is also important to be able to assess the distance on the road. It is extremely important to correctly localize by the driver e.g. the dilemma zone before the intersection and the distance of the car from the stop line. Proper assessment of the distance together with the knowledge of how to behave allows the driver to avoid undesirable events. Entering the intersection after the red signal turns on as a result of a wrong assessment of the distance and duration of the yellow signal is at least an offense. Rapid braking (in the dilemma zone) makes it possible for the following vehicle to crash into the stopping car leading to a collision or accident. The ability to locate the dilemma zone before the intersection is one of the elements determining the driver's correct behavior (Figs 1 and 2).

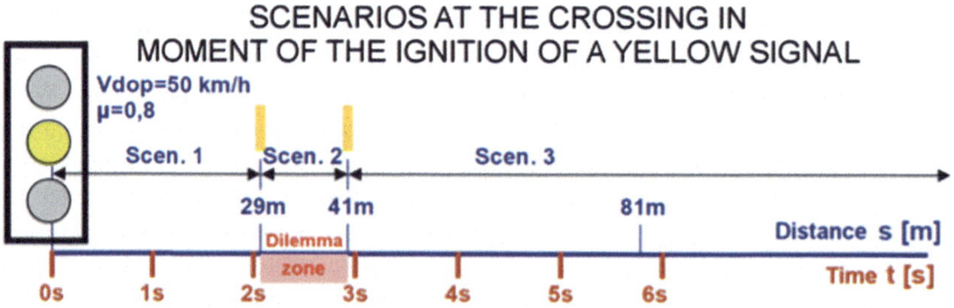

Fig. 1. Possible scenarios before the intersection in a time-space domain [2].

Fig. 2. Distance scheme before the intersection to assess how to behave [2].

The authors conducted studies on the state of knowledge of the drivers [2] on how to behave in various areas on the approach before the intersection. The diagram and collective results are presented below [2].

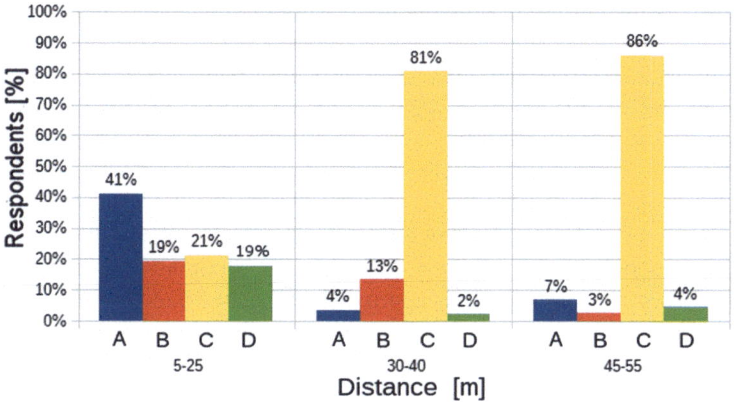

Fig. 3. Distribution of respondents decisions when driving the car, divided into three zones (I, II, III) and A, B, C, D—according to Table 1 [2].

The results of the questionnaire were divided into three zones (Fig. 3), where:

- zone I is an area within 30 m from the intersection (*answers from the range 5–25*),
- zone II is an area within 30–40 m from the intersection (*answers from 30–40*),
- zone III is an area within 40–55 m from the intersection (*answers from 45–55*).

This analysis showed that about 59% of the respondents would behave incorrectly when the traffic light onset was changing from green to yellow in zone I. 41% would proceed through the intersection, i.e. behave correctly. Approximately 19% would accelerate. Unfortunately, about 40% of respondents would try to stop the car which is unacceptable in this zone. The results obtained in zone II are very concerning. 81% of respondents believe that the driver should break non-abruptly. This might result in entering onto the intersection with a speed lower than permissible while the perpendicular traffic has already started its movement.

The answers obtained for the zone III are the most optimistic. 86% of respondents believe that they should break non-abruptly, which is the correct behavior. The other 14% of respondents will behave incorrectly.

2 Test Characteristics

2.1 Test Method

This paper presents the results of pilot studies on the scale and value of errors in distance assessment by vehicle drivers approaching the intersection when the traffic lights signal changes. The pilot study was to answer the question of whether the research problem posed was significant. The test results show a large inaccuracy of locating (the vehicle they drive) in different areas before the intersection, in relation to the stop line.

The error of distance assessment affects the decisions made regarding the behavior of the traffic participant. Thus, it affects the transport safety system. This is particularly important when bad weather conditions occur (large road slippery) and, as a result, a significantly longer stopping distance. Similar problems occur on dry road and good weather conditions for public transport drivers when driving public transport. In trams, trolleybuses, buses, some passengers travel in a standing position. For these vehicles, the braking delays should not exceed $a = 1.5\ \text{m/s}^2$. The stopping distance of such a vehicle in built-up areas ($v = 50$ km/h) is approximately $sZ = 81$ m. Proper assessment of the distance while driving the car allows you to take effective measures to avoid an accident.

2.2 Description of the Pilot Studies

To test the ability of assessing the distance by the driver, pilot studies were conducted on a group of over 200 people. They were students, university employees, participants of seminars of the Opole University of Technology. The study was attended by 68% of students and 32% of people with long experience in driving a car. The authors are convinced that large-scale studies will confirm poor state of knowledge in the field of distance assessment on the road. In the population of drivers of cars for the age group of students (not students think worse in abstract). In the study each participant completed the questionnaire after observing the marked area before the intersection. At the approach to the intersection, four cones (symbolizing cars in this place) were placed in given distances (known to the investigators) to the stopping line. Figure 4 shows the actual deployment of bollards on the intersection approach. Participants of the study assessed the distance to the stop line of consecutive cones (simulating the location of the car).

These distances were arbitrarily chosen (they took into account the location of the "dilemma" zone), they were not known to the participants of the study. Each participant determined the distance (to the stop line) of individual bollards based on their knowledge and experience. They depicted particular positions of cars approaching the intersection. In addition, the participants of the study determined the margin of error in the questionnaire that they felt they were making when determining the distance of each bollard to the stop line. The collected test results are presented in Tables 2 and 3 and in diagrams (Figs. 5, 6 and 7). The questionnaire is presented below.

2.3 The Results of Pilot Studies

Obtained results of the assessment of the distance made by the study participants were grouped into quanta with a value of $k = 5.0$ m. The numerical values of estimates and percentages in the scale of all responses of the survey participants were determined. The results of percentages were rounded to one. The research was aimed at making a preliminary assessment of the phenomenon and verification of the need for in-depth testing. The margin of error estimated by the respondents of individual distances was grouped according to the following values: $\Delta = 1$ m; $\Delta = 2$ m; $\Delta = 3$ m; $\Delta = 4$ m;

QUESTIONNAIRE

Scheme of deployment of bollards before the intersection.

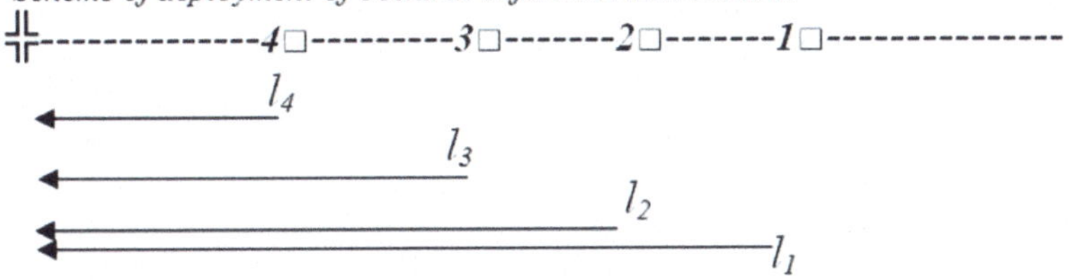

Specify the location of individual bollards, giving the values of the distance l_i of the bollard to the stop line. Specify the margin of error that was made when determining the distance values of each bollard.

N.	Distance l_i	Value [m]	Margin of estimated error Δ [m]
1.	l_1		
2.	l_2		
3.	l_3		
4.	l_4		

Fig. 4. A view of the rated section to the intersection with ballots.

Δ = 5 m; Δ = 10 m. Table 2 presents the distances estimated by the respondents. Table 3 presents the estimated error margins estimated by the respondents for estimating the distance values.

2.4 Discussion of the Obtained Results

The distance l_1 of bollard (the farthest from the stop line, the actual distance was 65.0 m)—70% of the participants in their assessment depressed the distance to the stop line, of which 30% gave the distance below the real distance. In contrast, 30% of respondents heavily overstated the value of the distance to the intersection (from 1.5 times more to 2.4 times more—in relation to the actual value).

Distance l_2 of bollard (the second in the order furthest away from the stopping line, the actual distance was 42.0 m)—70% of respondents in their assessment depressed the distance to the stop line, of which 40% reported the distance below the real distance.

Table 1. Summary of answers—percentages [2].

No	Distance [m]	Proceed with driving **A** [%]	Brake abruptly **B** [%]	Do not brake abruptly **C** [%]	Other, e.g. accelerate **D** [%]
1	5	80	1	1	18
2	10	65	6	4	25
3	15	34	26	14	26
4	20	17	35	36	12
5	25	11	28	53	8
6	30	4	20	72	4
7	35	2	13	83	2
8	40	4	7	88	1
9	45	6	2	88	4
10	50	6	3	86	5
11	55	8	3	83	4

However, 10% of respondents heavily overstated the value of the distance to the intersection (2.0 times more—in relation to the actual value).

The distance l_3 of bollard (the third in the order furthest from the stop line, the actual distance was 29.0 m)—90% of the surveyed in their assessment depressed the distance to the stop line, of which 50% gave the distance below the real distance. However, 10% of respondents heavily overstated the value of the distance to the intersection (over 2.0 times more—in relation to the actual value).

Distance l_4 of bollard (nearest to the stop line, the value of the actual distance was 15.0 m)—90% of the respondents in their assessment depressed the distance to the stop line. 60% reported the distance value being one-third of the actual distance and 30% reported the distance value being two-thirds of the actual distance. On the other hand, 10% of respondents heavily overstated the value of the distance to the intersection (they gave nearly 2.0 times the value—in relation to the actual value).

The dilemma zone (for drivers of passenger cars and lorries, under these atmospheric conditions) was in the area from 29.0 m to 41.0 m to the stop line before the intersection [2]. In the study, the vehicle drivers definitely misjudged their distance to the intersection. They did not know about the existence of the "dilemma" zone and thus were unable to accurately assess and determine its location. In real traffic, the consequence of this is to make the wrong decision to continue driving or start stopping the vehicle. In this situation, when approaching the intersection with a changing signal of the traffic lights, they act in the area of increased risk. Suddenly braking, they may surprise the following driver and cause rear end collision. If driving is continued, the driver may enter the intersection after the red signal has already been lit and at least an offence is committed. The results of the conducted research confirmed the initial thesis about a large scale of incorrect assessment of the distance made by the drivers. This causes wrong decisions before the intersection. In particular cases, this may cause adverse events. Up to date there are no markings on the roads (that would be helpful for

Table 2. The values of estimation of the length of the distance to the intersection declared by the participants of the study (own work).

Distance l_j	Percentage [%]											
	Value [m]											
	Up to 5 m	From 5 m to 10 m	From 10 m to 15 m	From 15 m to 20 m	From 20 m to 25 m	From 25 m to 30 m	From 30 m to 35 m	From 35 m to 40 m	From 40 m to 45 m	From 45 m to 50 m	60 m/[1] and 90 m/[2]	100 m/[3] 120 m/[4] 160 m/[5]
l_1	0	0	10	0	10	10	0	20	0	20	0	/[3]10 /[4]10 /[5]10
l_2	0	10	20	10	10	10	10	0	20	0	/[1]10	0
l_3	0	50	0	30	10	0	0	0	0	0	/[2]10	0
l_4	60	30	0	0	0	10	0	0	0	0	0	0

Table 3. Margin of distance estimation error declared by study participants (own work).

No	Distance l_j	Number of answers / Percentage [%]					
		Error value [m]					
		$\Delta = 1$ m	$\Delta = 2$ m	$\Delta = 3$ m	$\Delta = 4$ m	$\Delta = 5$ m	$\Delta = 10$ m
1.	l_1	41 / 20	19 / 10	39 / 20	20 / 10	81 / 40	0 / 0
2.	l_2	59 / 30	61 / 30	21 / 10	39 / 20	0 / 0	20 / 10
3.	l_3	81 / 40	59 / 30	39 / 20	21 / 10	0 / 0	0 / 0
4.	l_4	99 / 50	61 / 30	19 / 10	0 / 0	21 / 10	0 / 0

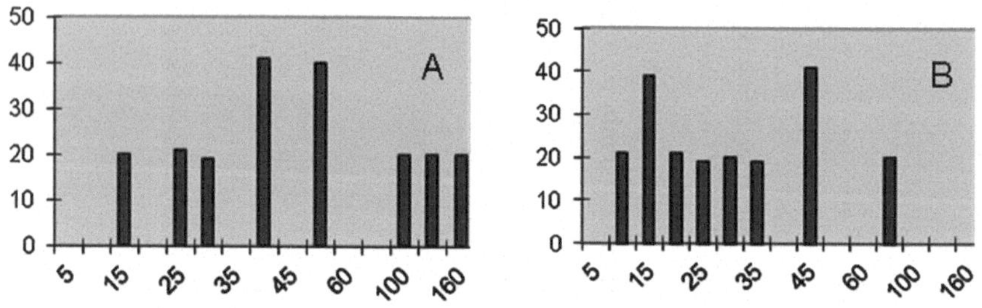

Fig. 5. A—numerical distribution of the value of the estimation of the length l_1 of distance to the intersection declared by the participants of the study; B—Numerical distribution of the value of the estimation of the length l_2 of distance to the intersection declared by the participants of the study.

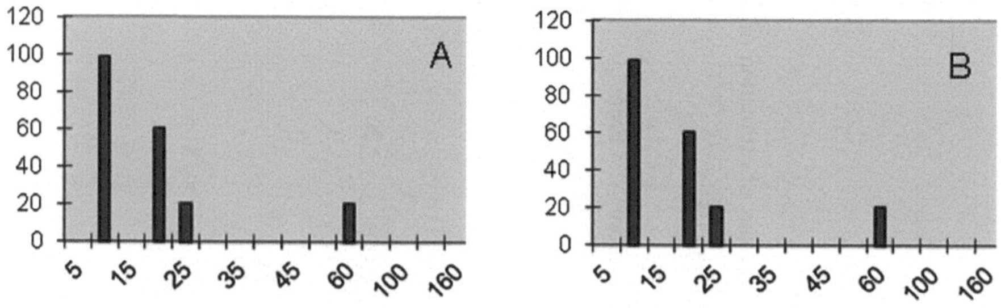

Fig. 6. A—numerical distribution of the value of the estimation of the length l_3 of distance to the intersection declared by the participants of the study; B—numerical distribution of the value of the estimation of the length l_4 of distance to the intersection declared by the participants of the study.

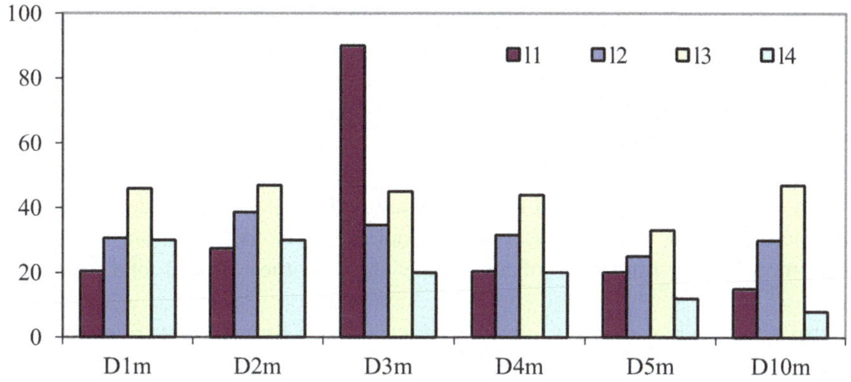

Fig. 7. Numbers of the error margin estimate declared by the study participants.

the driver) that sets critical distances, making it easier to take correct actions. This situation is conducive to committing offenses and the occurrence of collisions and road accidents. It is necessary and desirable to train the drivers. In addition, it is necessary to consider marking critical points on the road, e.g.: "dilemma zone", enabling and facilitating the correct assessment of distance and choosing the right behavior. "Dilemma zone" in given atmospheric conditions is an area on the road in which the manager makes decisions with high risk of losses, for good atmospheric conditions, it is between 29 m ÷ 41 m before the intersection [2]. The margin of error when estimating the distance given by the participants of the study was definitely small in relation to the actual error made when assessing the distance.

3 Conclusions

The research is conducted as a pilot study. The pilot sample includes better educated traffic participants. It can be expected that in the sample representative for the population there will be even worse results.

All drivers should be helped by conducting additional training and by using additional devices on the roads.

The respondents knew the conditions of the test, the state of the atmosphere described as ideal. For bad weather conditions, the test results will deteriorate. A pilot study is usually carried out to estimate: the needs of testing, feasibility, time, cost, etc., and this was the goal of the pilot studies. The main objective of this pilot study was to demonstrate the necessity for extensive research to justify the need for change.

The acceleration maneuver while approaching the intersection (frequently observed drivers behavior) will cause the increase of the speed above the permissible speed limit. It will be an offense against the applicable law. In some (selected) situations, such behavior, non-compliance with binding law, may be the choice of a lesser evil, e.g. avoiding a collision or accident [1].

Transition states and ambiguous areas in transient states are required to be specified. Users should have a limited and minimized possibility to interpret them freely [5, 6]. The requirements of binding regulations should contain a clear and precise way of dealing with situations that raise doubts about the assignment: is it still the current set or should it go into the other one [3, 4].

Currently existing or created requirements in the regulations do not take into account such situations. They should be anticipated and included in the description of the occurrence. Unclear boundaries, blunted, out of focus delimitation, imprecise definitions, moment of transition of one state into another should and must be eliminated.

If the driver of the rear car (e.g. when front car intends to turn right) is warned by the front car (e.g. by turning on the indicator in advance), rear car driver increases vigilance, increases the distance between vehicles, decreases the speed etc. Such use of a flashing yellow light signal should be used in controlling the traffic at the intersection. Then the flashing yellow signal (lighting up in the final phase of the green signal) would warn the drivers about the danger related to the red signal lighting up and prohibiting the entry onto the intersection. The forthcoming clear and comprehensible information (change of the current state to another) given accordingly ahead of time, allows to prepare for proper conduct (for a given situation and accompanying conditions). There is a clear need to educate and inform traffic participants about the characteristics of proper behavior. Both those who have long been granted a driving license as well as a part on courses of driving license. In terms of the correct assessment of distance in the road traffic, drivers so far have not been trained and tested.

Investigated and analyzed abilities (validity of the distance assessment) of the drivers should rise anxiety of traffic engineers responsible for the road safety system. It is unacceptable, that the drivers should be expected the proper conduct, if a significant part of this group does not understand the theoretical assumptions of the course of the vehicle stopping phenomenon. In addition, they are not able to locate the distances of the areas on the road necessary to make correct decisions in the road traffic.

Undoubtedly, this situation may contribute to the occurrence of adverse events in specific situations. It is necessary to supplement the knowledge of traffic participants through appropriate training—also during the driving license course. It is also necessary to consider the necessity of marking the areas before the intersection that are important in the stopping process. It is necessary to provide the driver with advance information about the change of green to yellow signal. The enormous technical progress and equipment development observed create opportunities to repair this system flaw associated with a change of traffic lights surprising to the drivers at the intersection. The authors plan to make an extended research using different research methodologies.

References

1. Gugerty, L., McIntyre, S.E., Link, D., Zimmerman, K., Tolani, D., Huang, P., Pokorny, R.: Effects of intelligent advanced warnings on drivers negotiating the dilemma zone. Hum.

Factors: J. Hum. Factors Ergon. Soc. **56**(6), 1021–1035 (2012). https://doi.org/10.1177/0018720814525438
2. Olejnik, K., Łopuszyński, M., Koszałka, G., Grenčík, J.: Behavior of vehicle drivers while approaching an intersection after the traffic light signal changes from green to yellow. In: Conference Automotive Safety 2018, Častá—Papiernička, Slovakia (2018)
3. Wua, M., Maa, W., Lia, L.: Characterize dilemma zone and minimize its effect at coordinated signalized intersections. Procedia—Soc. Behav. Sci. **96**(2013), 766–778 (2013)
4. Poliak, M., Poliakova, A., Mrnikova, M., Jaskiewicz, M., et al.: The competitiveness of public transport. J. Competitiveness **9**(3), 81–97 (2017). ISSN 1804-171X
5. Machiani, S.G., Abbas, M.: Safety surrogate histograms (SSH): a novel real-time safety assessment of dilemma zone related conflicts at signalized intersections. Accid. Anal. Prev. **96**, 361–370 (2016)
6. Lina, X., Chenga, L.: Engineering countermeasures to reducing unintentional yellow-light running owing to dilemma zone. Procedia—Soc. Behav. Sci. **96**, 900–904 (2013)

Fatigue Life Dependence on Non-uniform Hardening Effect After Surface Rolling

Aleksei Kurushin, Vitalijs Zaharevskis(✉), and Olegs Kovzels

SIA Centre Composite, Darzciema Str. 60, Riga, Latvia
{aleksej.kurushin, vitalijs.zaharevskis, olegs.kovzels}@composite.lv

Abstract. Fatigue failure of high strength steel shafts in different mechanisms leads to expensive replacement of shaft or the entire mechanism. Fatigue failure usually takes place in stress concentration areas. One of the effective methods to improve shafts fatigue characteristics is surface hardening by rolling [1]. In order to reach maximum fatigue life of the shaft the process of calibration, selection of rolling parameters (roller section radius, curvature, rolling force, etc.) [2] and non-uniform rolling influence are described in this article. The task was to design a bolted joint that has to fail by shear under specified load conditions. To ensure these requirements two grooves with certain geometry are made on bolt's surface. Grooves decreased fatigue life of the bolted joint several times. In order to repair fatigue characteristics, groove areas were hardened by surface rolling. Several computer models were made in order to select optimal hardening parameters and understand how non-uniform hardening effects fatigue life. Numerical results were verified by experimental data. As a result necessary rolling parameters were selected and fatigue life of shafts with stress concentrations (grooves) increased no less than 3 times.

Keywords: Rolling · Material hardening · Fatigue

1 Introduction

Surface hardening by plastic deformation is commonly used to improve fatigue strength of the material. Surface hardening methods are divided into two groups—static and impact methods. During static hardening, hardening equipment interact's with material's surface under constant load, and interaction (contact) point is moving smoothly, covering all the hardening surface. Inertial forces have no effect on that type of hardening. In impact methods, hardening equipment or hardening environment repeatedly interacts with hardening surface or it's part, wherein hardening load is changing from zero to maximum value. In case of impact hardening of local zones interaction point could consistently move (like in static methods) through all hardening surface. Rollers or balls can be used in static hardening. In impact hardening metallic fraction, balls from steel, glass or plastic are used. Liquid with abrasive particles also can be used as working environment during surface hardening. Surface rolling with a ball or roller and shot peening are the most commonly used surface hardening technics. The main case of this study is surface rolling [3] with a roller.

I. Kabashkin et al. (Eds.): RelStat 2018, LNNS 68, pp. 500–506, 2019.
https://doi.org/10.1007/978-3-030-12450-2_48

The research object of this article is steel bolt in double shear joint, with grooves in shear-critical sections. Radius of the groove is 3 mm, depth—2.3 mm. Picture of the bolt is shown in Fig. 1, joint loading conditions in Fig. 2a. Joint was loaded cyclically with the symmetric cycle. Testing equipment is shown on Fig. 2b. Aim of this research was to understand the effect of different rolling parameters on hardening quality and how does it affects the fatigue life of the shaft. Rolling parameters were chosen in accordance with recommendations from [4, 5]. After performing computer simulation of rolling process and after first rolling tests were made, rolling parameters were specified. Proper hardening by rolling could return the endurance limit of notched specimen to the level of endurance limit of unnotched specimen [6].

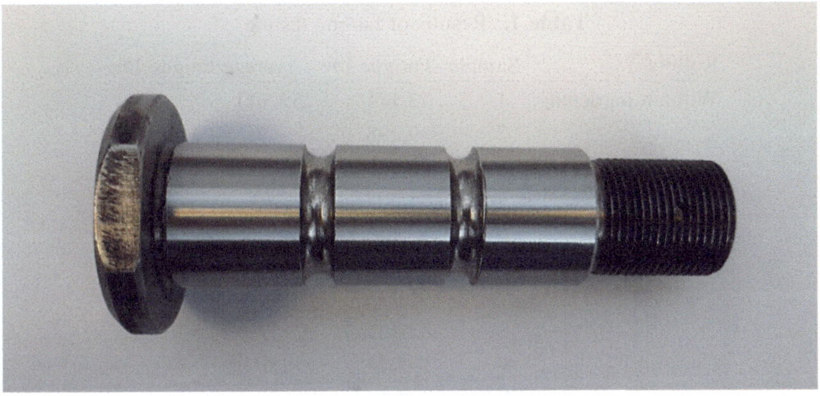

Fig. 1. Research object.

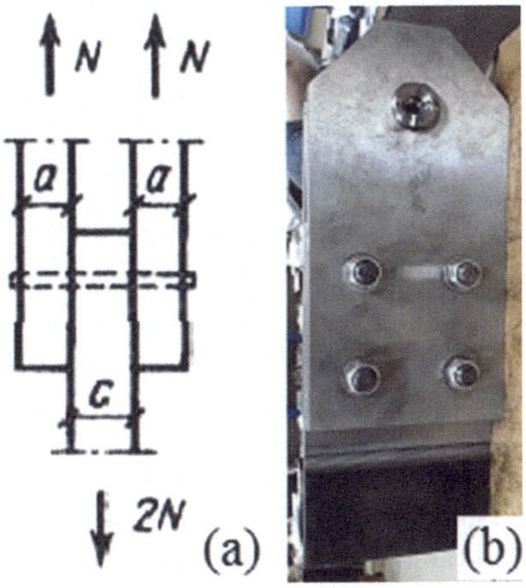

Fig. 2. Loading conditions (a), testing equipment (b).

During this study, endurance tests of 12 bolts were performed, bolts were tested till fracture and the grooves were periodically inspected on the existing cracks. Results are presented in Table 1. Bolts without hardening showed the average fatigue life of 52,500 cycles. To improve fatigue life of the bolt, groves were hardened by rolling. At first attempt, rolling was performed with roller which had the same section as the groove has. Radius of roller's section was 2.95 mm. Rolling with those parameters increased the average fatigue life up to 69,137 cycles or by 31.7%, however the dispersion of the results stayed at high level. Whereas in bolts without hardening, initial detectable crack was always inspected at 30,000 cycles, but in bolts after hardening, crack was inspected between 30,000 and 50,000 cycles.

Table 1. Results of fatigue testing.

Rolling	Sample	Fatigue life	Average fatigue life
Without hardening	1	43,171	52,500
	2	39,298	
	3	72,228	
	4	55,303	
Non uniform	5	60,053	69,137
	6	42,299	
	7	91,775	
	8	82,421	
Uniform	9	132,445	>180,000
	10	234,192	
	11	>180,000	
	12	>180,000	

During the rolling process, initial non coincidence gap (Fig. 3) between roller and the bolt radiuses, was discovered. The gap appears because of the imperfection of groove's and roller's geometry, manufactory tolerance. Average size of the gap was 0.02 mm.

The gap had a negative effect on the hardening quality. As the result, the critical section of the bolt (groove's center) wasn't hardened.

2 Methodology

To get the distribution of residual stress and strain after rolling with the initial gap, this experiment was simulated using finite element method (FEM) [7] in MSC Marc solver. Model, loading scheme and boundary conditions are shown on Fig. 4. The distribution of residual plastic strain in the groove's section is shown on Fig. 5.

One can see, that roller with section radius 3 mm doesn't harden sides and center of the groove, only half of the groove's arc is hardened after rolling.

Fig. 3. Initial gap between bolt's groove and roller during rolling process.

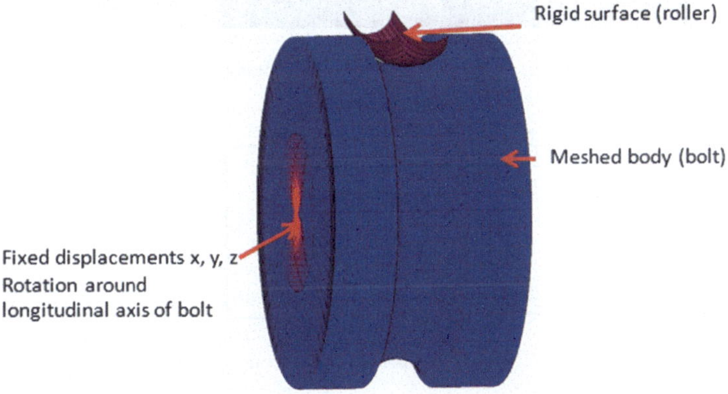

Fig. 4. FE model.

Properties of plastically deformed area in this case depends on the following factors: size of the initial gap and contact area between roller and shaft. To improve the rolling process, roller's section radius was lowered from 3 to 2 mm which allows to roll the whole groove in few steps—starting from the sides and going to the centre.

To perform groove's uniform hardening, the step by step rolling scheme was chosen. On the first step groove's sides are being hardened. After that, the position of rolling equipment is changed and intermediate parts of the groove are being hardened. After that, the position of rolling equipment is changed one more time and groove's centre is being hardened. Altogether, rolling consists of 5 steps which are shown on Fig. 6, rolling equipment is shown on Fig. 7.

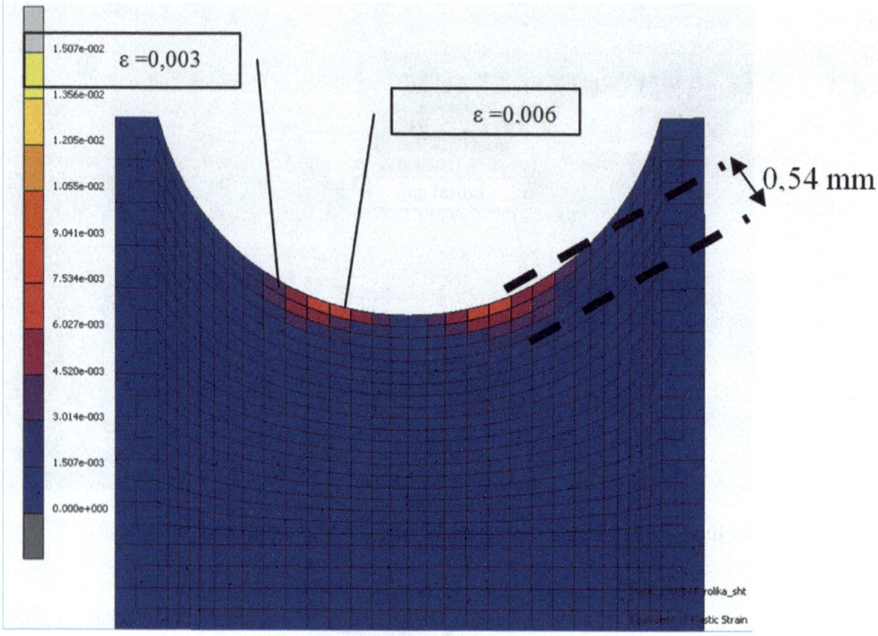

Fig. 5. Equivalent of residual plastic strain.

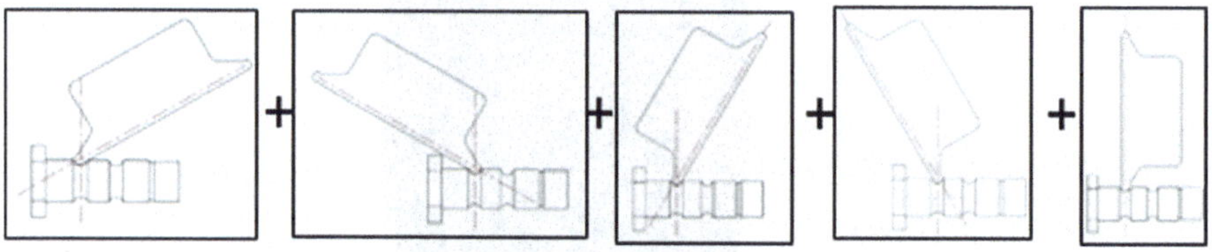

Fig. 6. Step by step rolling procedure scheme.

After step by step hardening, bolts showed much better results in the experiment than previous bolts. Cracks in grooves were detected much later—after 90,000 cycles. New hardening scheme increased fatigue life to average value of 180,000 cycles or on 242.9% with minimum fatigue life of 132,445 cycles.

To get the distribution of residual stress and strain after rolling, rolling process was simulated using FEM. The distribution of residual plastic strain in the groove's section, and an average depth of residual strain are shown on Fig. 8.

On the Fig. 8 you may see, that plastic deformation is uniformly distributed along groove's arc and residual plastic strain level and depth is higher, compared to in previous case.

Fig. 7. Intermediate position of rolling equipment.

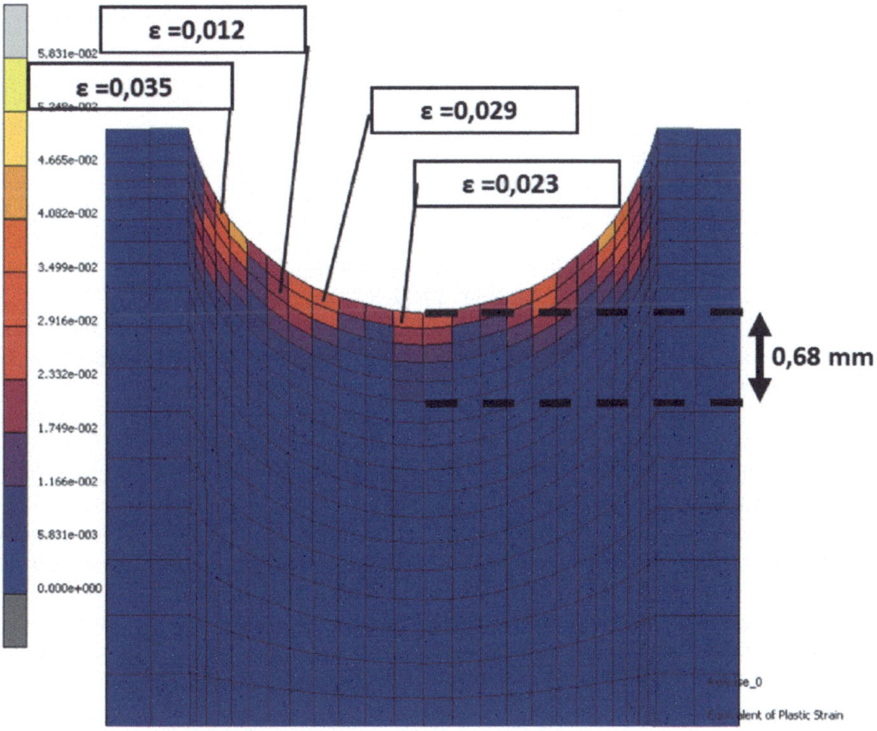

Fig. 8. Equivalent of residual plastic strain.

3 Results

Experiment results of fatigue life are shown in Table 1.

4 Conclusions

Experiment data and calculation results show, that rolling quality has a critical effect on the fatigue strength of detail. Rolling with optimal parameters leads to increase of fatigue life for about 3 times, and get more stable results, compared to parts without hardening.

Step by step rolling is 6 times more time-consuming process than rolling in one step, although, bolts hardened by that technology show 3 times higher fatigue life and more stable results and this is the critical benefit.

Acknowledgements. The research leading to these results has been supported by the European Regional Development Fund project "Development of Competence Centre of Mechanical Engineering", contract No. 1.2.1.1/16/A/003 signed between Competence Centre of Mechanical Engineering and Central Finance and Contracting Agency, Research No. 1.12. "Fatigue life improvement of high-loaded parts such as shaft by surface hardening with rolling".

References

1. Nalla, R.K., Altenberger, I., Noster, U., Liu, G.Y., Scholter, B., Ritchie, R.O.: On the influence of mechanical surface treatments—deep rolling and laser shock peening on the fatigue behavior of Ti-6Al-4V at ambient and elevated temperatures. Mater. Sci. Eng. A **355** (1–2), 216–230 (2003)
2. Juijermi, P., Altenberger, I.: Fatigue performance enhancement of steels using mechanical surface treatments. J. Met. Mater. Minerals **17**(1), 59–65 (2007)
3. Altenberger, I.: Deep rolling—the past, the present and the future, In: International Conference on Shot Peening 2005. In: Proceedings of 9th International Conference on Shot Peening, pp. 6–9. IITT-International, Noisy-le-Grand (2005)
4. Drozd, M.C.: Engineering Calculations of Elastoplastic Contact Deformation. Mashinostroeenie, Moscow (2004)
5. Boiko, N.I.: Resource-Saving Technologies for Surface Layers of machine Details Quality Improving. Marshrut, Moscow (2006)
6. Blason, S., Rodríguez, C., Belzunce, J., Suarez, C.: Fatigue behaviour improvement on notched specimens of two different steels through deep rolling, a surface cold treatment. Theor. Appl. Fract. Mech. **92**, 223–228 (2017)
7. Hassani-Gangaraj, S.M., Carboni, M., Guagliano, M.: Finite element approach towards an advanced understanding of deep rolling induced residual stresses and an application to railway axes. Mater. Des. **83**, 689–703 (2015)

Identification of Reliability Models for Non-repairable Railway Component

Jaroslaw Selech(✉) and Karol Andrzejczak

Poznan University of Technology, Piotrowo 3, 60965 Poznan, Poland
{jaroslaw.selech,karol.andrzejczak}@put.poznan.pl

Abstract. The reliability analysis of technical object, particularly determine the lifetime distributions over time, requires knowledge of the course of functional characteristics of elements included in a given system. Having a database of damage times for individual components, it is easy to determine the average working time to damage the element and the standard deviation for this time, the problem arises when selecting the appropriate form of distribution. Incorrect determination of the type of damage distribution of technical object elements may cause a large error in the results of the reliability and durability assessment of the system. Not having enough data to verify the hypothesis of the form of distribution, commonly, depending on the case considered, the form of the density function of damage or cumulative distribution is assumed. Knowing the course of the probability of damage over time, preventive actions can be taken before the risk of damage reaches the limit value, thus avoiding financial losses and ensuring an adequate level of security. The paper presents methods for assessing the reliability of tested elements for the first failure using the most commonly used distributions in the reliability theory. The results of calculations were compared for several cases. The calculations were carried out for selected components of the rail vehicle.

Keywords: Failure distribution · Censored observation · Non repairable component

1 Introduction

In the reliability analysis of technical objects, different probability distributions can be used to model damage data [1–3]. Nevertheless, the most commonly used probability distributions include normal distribution, exponential distribution and Weibull distribution [4–8]. In the presented research results, apart from the mentioned distributions, other less frequently used distributions were used for the reliability analysis, such as logarithmic-normal distribution, gamma distribution, generalized gamma distribution, logistic, log-logistic and Gumbel distribution. The quality of matching the probability density function to empirical data for the analysed cases is better than the in case of the usual distributions. Estimation of parameters of these distributions can be carried out by analytical methods as well as numerically using specialized IT tools [9, 10]. The estimation of the distribution parameters is the data modelling that requires determining the distribution that best fits the data and estimating the parameters of this distribution

I. Kabashkin et al. (Eds.): RelStat 2018, LNNS 68, pp. 507–518, 2019.
https://doi.org/10.1007/978-3-030-12450-2_49

(shape, scale, position). Different methods of estimating parameters are used, which include both numerical and graphical methods. The most commonly used methods are moment method, the method of maximum likelihood, the least-squares method, fitting distribution to grids, probability plot correlation coefficient (PPCC) and others [11].

In this work, the Likelihood Value (LKV) test was used to determine the distribution of the studied data, and the maximum likelihood estimation (MLE) was used to estimate the parameters of the selected distribution. The reliability analysis was performed on the components of rail vehicle fleet that damage most often and, et the same time, are the most-influencing in terms of costs of corrective maintenance [12, 13]. Its purpose was to identify the best-fit probability distributions, the knowledge of which may affect not only the level of transport safety, but also may provide the theoretical basis for planning rational maintenance schedules and reducing the costs of maintaining the vehicle fleet [14]. Maintenance activity can be divided into two groups: scheduled maintenance and unscheduled maintenance. Scheduled maintenance consists of preventive maintenance inspection, being itself a planned servicing of vehicles. Unscheduled maintenance is work resulting from component failures, which may necessitate road calls, being more costly [15]. The main objective of this paper is unscheduled maintenance based on the reliability study presented herein.

2 Subject of Research

The tests were carried out on a fleet of 45 urban rail transport vehicles. The scope of data included five initial years of fleet use, including two years of a guarantee period and three subsequent years with a maintenance contract [16]. All vehicles were used in similar conditions, i.e. the same track infrastructure and similar daily runs as well as the same schedule and scope of preventive maintenance. Four components from the operational database that includes information on damages were selected from various construction groups. These components were generating the largest costs as part of corrective maintenance in the analysed 5-year fleet usage period. This group includes:

1. Hydraulic aggregate K140 mounted on a trailer bogie, a part of the brake system responsible for emergency braking.
2. Main monitoring module, an electronic component responsible for controlling all cameras in the vehicle.
3. Driver's seat.
4. Drive controller used for starting and braking the tram.

Hydraulic aggregate K140 is a device used to create pressure in the hydraulic system of brakes and is responsible for emergency braking, it is located on the trailer bogie. In each unit there is, among others, an electric hydraulic pump, a tank with hydraulic oil (brake fluid), overflow and proportional valves, and sometimes also braking system controllers [17]. The hydraulic aggregates are powered with the on-board voltage from the accumulators, which are charged from the on-board converter. Thanks to the use of accumulator power supply in the event of a possible failure of the converter, the power supply of the devices is ensured. One of the most common causes of damage to the pressure aggregate in use was a leak, which caused oil to enter the engine compartment and to damage it. The monitoring module is responsible for controlling all cameras, both internal and external. The main cause of the damage was the incorrectly displayed image from the mirror cameras, usually it was rippling or there was a complete disappearance of the displayed image. The drive controller is a multi-steady switch designed for switching windings of electric motors or auxiliary elements and is responsible for starting and braking the tram. In the drive controller, contactors were wiped and caused no reaction at the given position. In the case of a driver's seat, the main cause of damage, causing replacement for a new one, was the inability to adjust the height of the seat, which was caused by the failure of the pneumatic cylinder.

All damaged components in the analysed period were replaced with new ones,

Table 1. The most cost-intensive corrective maintenance during the 5 years of operation of the fleet of 45 vehicles

No.	Component name	Number of corrective maintenance	Maintenance cost [PLN]	Total cost (all maintenance) [PLN]	Percentage of costs for all corrective maintenance
1	Hydraulic power unit	27	22,600	610,200	7.32%
2	Main monitoring module	54	5,328	287,712	3.45%
4	Drive controller	23	9,059	208,357	2.50%
5	Driver seat	15	8,412	126,180	1.51%

which generated large costs of corrective maintenance. The presented research considered total costs of damage removal, i.e. costs of parts and costs of man-hours related to the replacement of these parts. In some cases, damage to the tested components in a given vehicle occurred two or even three times, the number of recorded damages in the tested time interval for all tested components is given in Table 1.

3 Example of Right Censored Failure Time of Data

As example of life data for 5 years of operation for driver seat c is shown in the Table 2, includes exact failure time (in kilometers) and suspension time (in kilometers). Suspension time is right censored data that did not faile by the end of the test and in all the studied cases, the suspected failure mechanism is fatigue.

Table 2. Example of failure time statistical driver seat

Life/(km)	F/S	Life/(km)	F/S	Life/(km)	F/S	Life/(km)	F/S
1	2	3	4	5	6	7	8
327,707	S	109,490	F	328,878	S	94,611	F
272,671	S	182,548	S	314,589	S	113,068	S
360,404	S	214,243	F	268,743	S	314,146	S
243,532	S	10,603	F	164,253	F	271,384	S
226,971	F	114,646	S	155,480	S	316,331	S
57,863	S	281,302	F	328,480	S	286,225	S
185,466	F	10,964	S	328,168	S	182,861	F
183,970	S	319,088	S	279,118	F	106,664	S
314,522	S	352,654	S	8,004	S	254,974	S
339,907	S	145,296	F	142,851	F	290,198	S
255,320	F	162,447	F	154,501	S	299,174	S
56,359	S	2,770	S	271,868	S	264,787	S
342,241	S	305,087	S	244,005	S	265,863	S
328,494	S	335,015	S	100,210	F	294,895	S

4 Method of Selection the Best Fitting Distribution Model

ReliaSoft's WEIBULL++ Distribution Wizard can provide guidance in selecting a distribution based on statistical tests [18]. The Distribution Wizard uses three factors in order to rank distributions: the Kolmogorov-Smirnov (K-S) test, a normalized correlation coefficient and the likelihood value. This article will show how these rankings are calculated [4, 11]. The Distribution Wizard in Weibull++ ranks the selected distributions in terms of the fit to the data entered. In order to determine the ranking, the three tests are used in conjunction with weights assigned to each test. In the Table 3, the second column, AVGOF, contains values obtained using the Kolmogorov-Smirnov (K-S) test. The third column, AVPLOT, provides the results of the second test, which is a normalized correlation coefficient (rho). The fourth column, LKV, contains the likelihood values. The fifth column contains models distribution ranking which are ranked according to how well they fit the data, with rank 1 being the best fit, in some cases more than 1 choice is proposed (Table 3).

On the Intermediate sheet of the Analysis Details page, these values are then weighted and combined into one overall value DESV which is given by [19]:

$$\begin{aligned} DESV = & (AVGOF\ Rank \times AVGOF\ Weight) \\ & + (AVPLOT\ Rank \times AVPLOT\ Weight) \\ & + (LKV\ Rank \times LKV\ Weight). \end{aligned} \tag{1}$$

Table 3. Rank of distributions for driver seat

Distribution	AVGOF	AVPLOT	LKV	Ranking
1P-Exponential	5.44354E-06	2.75	−220.11	8
2P-Exponential	3.89795E-06	1.98	−222.31	7
Normal	1.68843E-07	2.41	−220.38	6
Lognormal	9.99978E-11	2.21	−219.63	4
2P-Weibull	**9.1999E-08**	**1.29**	**−220.34**	**1**
3P-Weibull	9.99978E-11	1.07	−219.89	3
Gamma	**2.56571E-07**	**0.89**	**−235.74**	**1**
G-Gamma	0.263986715	2.78	−225.20	5
Logistic	9.99978E-11	2.25	−219.68	10
Loglogistic	4.324466506	3.34	−229.53	9
Gumbel	5.44354E-06	2.75	−220.11	11

Software allows user specify different weights depending on whether the parameter estimation method is rank regression or MLE [19].

5 Results from Analysis

Analysis Method uses the maximum likelihood estimation (MLE) for estimating the parameters of the chosen distribution. Rank Method uses the Median Ranks (MED), Confidence bounds Method uses the Fisher Matrix (FM) [4]. In this case of hydraulic power unit, the 3P- Weibull distribution is the suggested model, for monitoring module and drive controller G-Gama distribution, and for driver seat 2-parameter Weibull distribution and Gamma distribution are the highest ranked.

The following figures present the cumulative distribution function (cdf) and the probability density function (pdf) for all tested components and give the calculated parameters for the selected distributions. The probability plot shows the trend in the probability of failure over time, the unreliability vs. time plot shows the probability of failure of the product over time and the probability density function plot shows the probability density function of the data over time, all these plots allow visualize and better analyse the distribution of the data set. Using the Distribution Wizard's recommendations, for hydraulic power unit Weibull distribution with 3 parameters (scale shape, location) is the best fitted and was used for calculations. Their values are: $\eta = 392{,}848$ km, $\beta = 2.0833$, $\gamma = -55{,}320$ and failure rate $h = 0.0000047288$/km. Computed graphs are further presented (Figs. 1 and 2).

For the monitoring module, the best fit distribution was the two-parameter exponential distribution, where the scale parameter and the location parameter obtained the following values: mean time to failure = 238,341 km, $\gamma = 2{,}749$, and failure rate $h = 0.0000041957$/km. To illustrate the fit of the selected distribution, the probability function and the probability density function are shown in Figs. 3 and 4.

In the case of the drive controller, the generalised Gamma distribution was found to be the most suitable, for which three parameters were estimated $\mu = 13.13$, $\sigma = 0.4346$, $\lambda = 1.95$ and the failure rate $h = 0.0000033657$/km. Details on the distribution fitting and probability density are presented in Figs. 5 and 6.

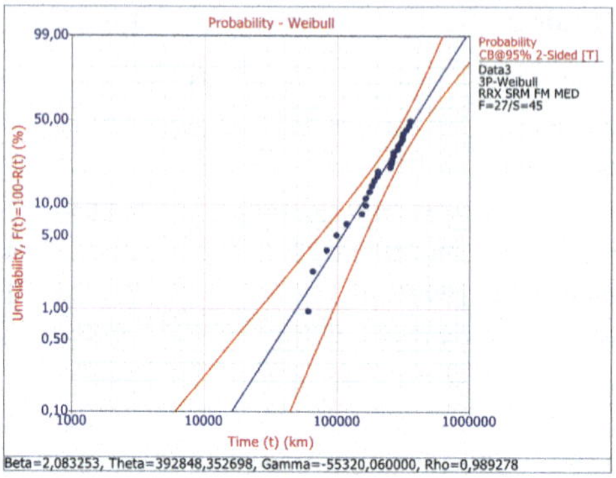

Fig. 1. 3P-Weibull cdf graphic representation for hydraulic power unit

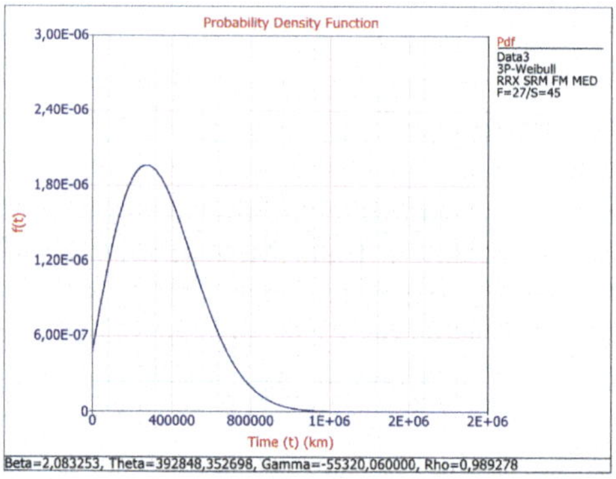

Fig. 2. 3P-pdf plot for hydraulic power unit

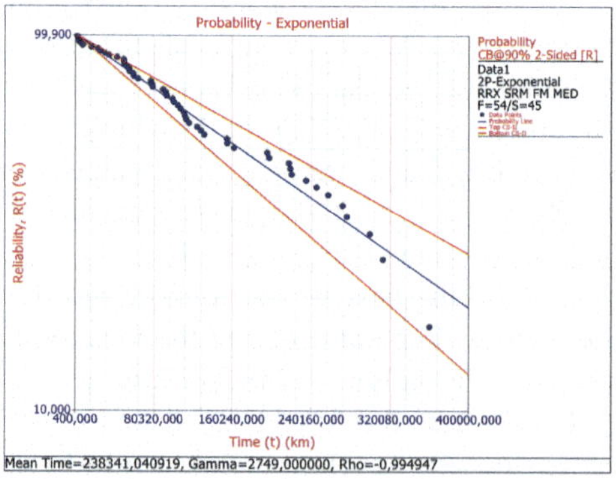

Fig. 3. 2P-Exponetial cdf graphic representation for monitoring module

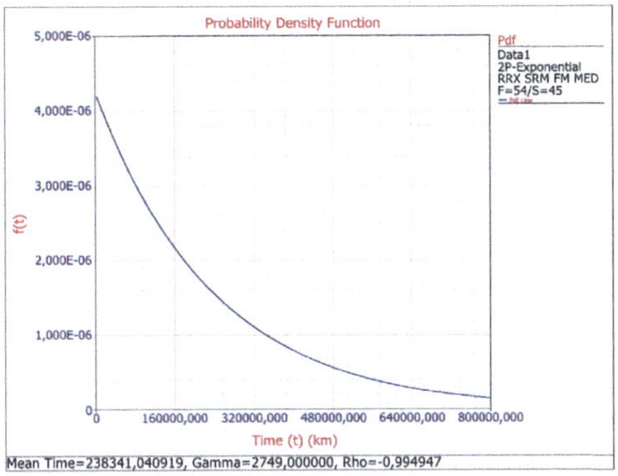

Fig. 4. 2P-Exponetial pdf plot for monitoring module

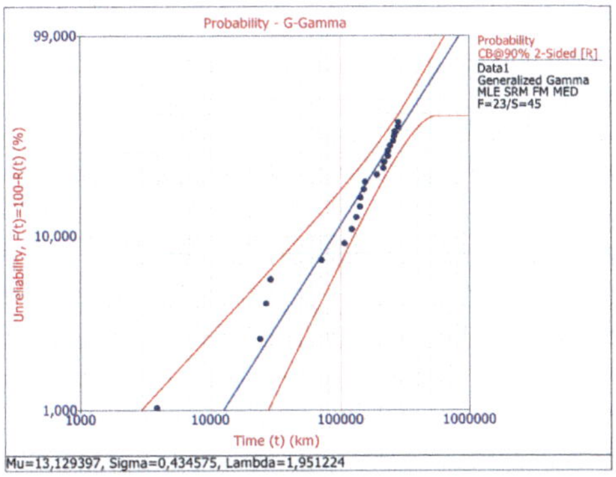

Fig. 5. Generalized Gamma cdf graphic representation for drive controller

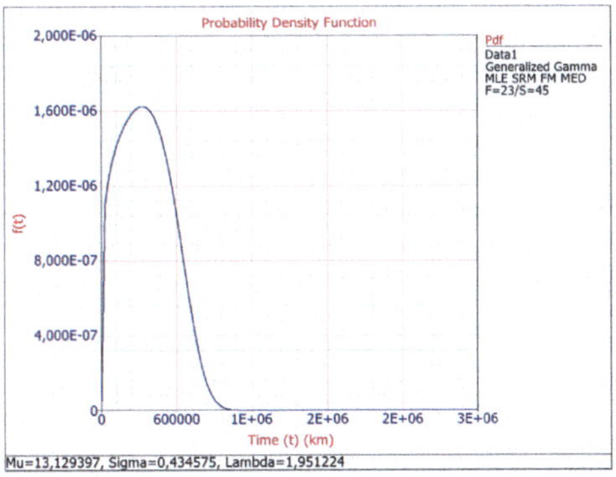

Fig. 6. Generalized Gamma pdf plot for drive controller

The family of three-parameter generalized gamma distributions is given:

$$f(t|\kappa,\beta,\theta) = \frac{\beta}{\theta\Gamma(\kappa)}\left(\frac{t}{\theta}\right)^{\kappa\beta-1} e^{-\left(\frac{t}{\theta}\right)^{\beta}}, \quad t > 0. \tag{2}$$

In the case of G-Gamma distribution Weibull++software uses a reparameterization with parameters θ, κ and β, where $\theta > 0$ is a scale parameter, $\beta > 0$ and $\kappa >$ are shape parameters, as shown in [20], where:

$$\mu = \ln(\theta) + \frac{1}{\beta} \cdot \ln\left(\frac{1}{\lambda^2}\right), \tag{3}$$

$$\sigma = \frac{1}{\beta\sqrt{\kappa}}, \tag{4}$$

$$\lambda = \frac{1}{\sqrt{\kappa}}, \tag{5}$$

where $-\infty < \mu < \infty$, $\sigma > 0$, $\lambda > 0$.

The last tested component of the vehicle was the driver's seat. Data included in the analysis are included Tables 2, 3, 4, and 5. In this case, Weibull++ software indicated two possible distributions as the most relevant to the data. Nevertheless, by carefully analysing the diagrams of the probability function for both of the indicated distributions, it can be pointed out that the two-parameter Weibull distribution is a distribution that better reflects the recorded damage times. For the two-parameter Weibull distribution, the scale parameter was determined as η = 699,005 km, the shape parameter was β = 1.32048 and the failure rate was h = 0.0000015278/km. Details on the probability function and the probability density function are shown in Figs. 7 and 8.

Table 4. Data analysis details

Distribution	AVGOF	AVPLOT	LKV	DESV	Ranking
1P-Exponential	7	10	7	760	8
2P-Exponential	6	8	4	580	7
Normal	5	4	8	570	6
Lognormal	3	7	6	470	4
2P-Weibull	1	5	1	180	**1**
3P-Weibull	2	3	5	310	3
Gamma	1	2	3	180	**1**
G-Gamma	4	1	11	550	5
Logistic	8	9	9	850	10
Loglogistic	1	6	2	230	9
Gumbel	9	11	10	970	11

Table 5. Rank of distributions for all analyzed components

Distributions	Hydraulic power unit	Monitoring module	Drive controller	Driver seat
1P-Exponential	8	11	9	8
2P-Exponential	5	①	8	7
Normal	4	7	5	6
Lognormal	10	5	10	4
2P-Weibull	7	9	3	①
3P-Weibull	①	4	2	3
Gamma	3	6	4	①
G-Gamma	2	2	①	5
Logistic	6	8	7	10
Loglogistic	9	3	6	9
Gumbel	11	10	11	11

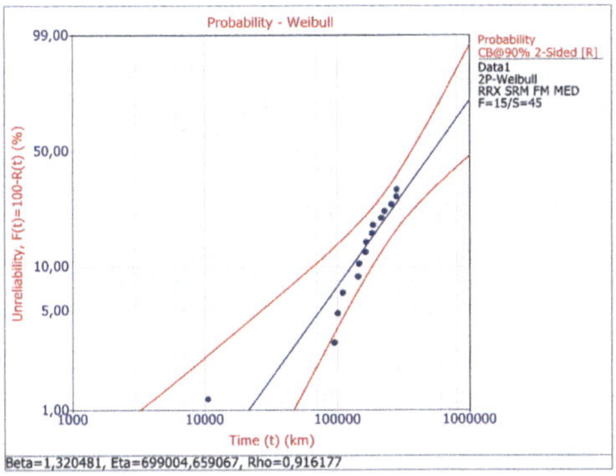

Fig. 7. 2P-Weibull cdf graphic representation for driver seat

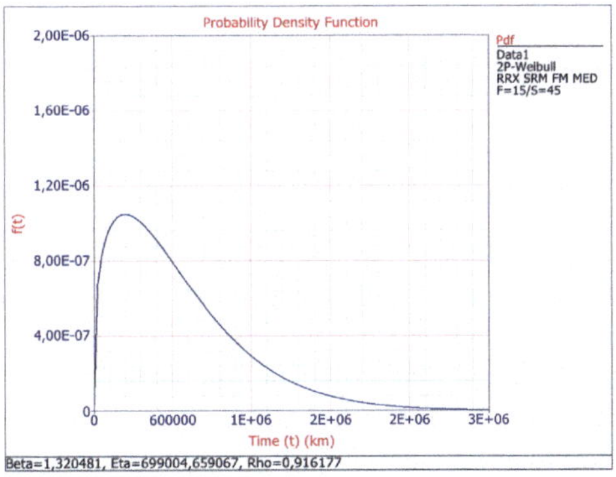

Fig. 8. 2P-Weibull pdf density plot for driver seat

For the Gamma distribution, which in the case of the driver's seat analysis was classified along with the two-parameter Weibull distribution as best suited, the following parameters were estimated $\mu = 12.34$, hence $e^{\mu} = 228,662$ is a scale parameter, $\kappa = 2.15$ as a shape parameter and failure rate $h = 0.0000024471$/km. Details on the cdf and the pdf are shown in Figs. 9 and 10.

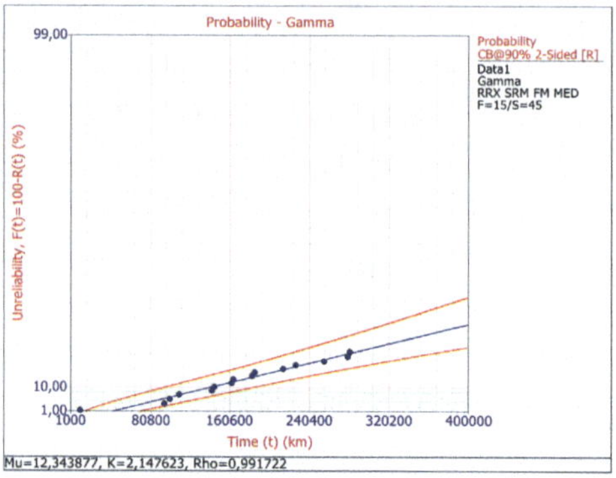

Fig. 9. Gamma cdf graphic representation for driver seat

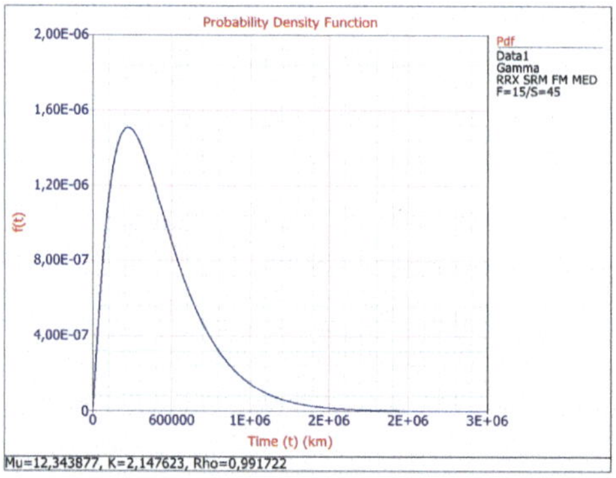

Fig. 10. Gamma pdf plot for hydraulic driver seat

6 Conclusions

The analysis of the determination of the best-fitting distribution for selected four rail vehicle components presented in the paper was carried out using the Likelihood Value (LKV) test to determine the rank of distributions, and the maximum likelihood estimation (MLE) was used to estimate distribution parameters. The obtained results constitute an important argument for the possibility of using the proposed method of predicting the time of damage of vehicles in railway systems. Predictions made with classic methods using a match of determined models of random variable distributions can give a prediction error. An important factor that supports their use is the possibility to draw conclusions about the causes of damage or the way how the vehicle was operated and maintained. The research confirmed the assumptions that there are methods and technical possibilities with help of which the reliability of the components used in the vehicles can be predicted accurately enough. Correct determination of the prediction allows to avoid some damages and related accidents. The data on which the prediction is built come directly from the actual operating system, therefore the obtained prediction results take into account the conditions and actual operation of a specific rail vehicle system. Benefits resulting from the correct determination of the random variable distribution that describes the functionality period of a renewable object such as a rail vehicle are significant due to the costs generated by not using the full durability of the actual component as well as losses due to damage and downtime resulting from the repair. Obtained results of the determination are aimed at indicating procedures useful for predicting the reliability of rail vehicles. The results will be used to assess the accuracy of prediction and the applicability of the tested procedures in real vehicle operation systems. Prediction of the ongoing reliability processes during operation will allow for a better understanding of the causes of vehicle reliability deterioration, and at a later stage of the research it will enable the estimation of costs related to corrective maintenance of vehicles. In addition, identification and analysis of factors affecting reliability will enable the development of more accurate and effective methods of predicting reliability indicators. To improve time efficiency, fast computation is required. In this paper a fast calculation program was presented, developed by Reliasoft, along with a brief description of the methodology. In parallel tests carried out, the identified probability distributions of damage to individual components are used to predict their failures in the further operation of rail vehicles. In these studies, conditional probability distributions of damage to individual components are used. Damage prediction results of selected drive controllers are prepared for publication entitled 'Generalised gamma distribution in the prediction of corrective maintenance of fleet vehicles'.

Acknowledgements. The research has been financed by National Centre for Research and Development—Applied Research Program PBS III/B6/30/2015 and PUT 05/51/DSPB/3551 and PUT 04/43/DS PB/0096.

References

1. Gertsbakh, L.: Reliability Theory with Applications to Preventive Maintenance. Springer, Heidelberg (2000)
2. Lewis, E.E.: Introduction to Reliability Engineering, 2nd edn. McGraw-Hill, New York City (1994)
3. Cox, S., Tait, R.: Safety, Reliability and Risk Management: an Integrated Approach, 2nd edn. Butterworth Heinemann, Great Britain (1998)
4. Andrzejczak, K., Selech, J.: Quantile analysis of the operating costs of the public transport fleet. Transp. Probl. **12**(3), 103–111 (2017). https://doi.org/10.20858/tp.2017.12.3.10
5. Ascher, Harold, Feingold, Harry: Repairable Systems Reliability—Modeling, Inference, Misconceptions, and Their Causes. Marcel. Dekker, New York (1984)

6. Hajkowski, J., Popielarski, P., Sika, R.: Prediction of HPDC casting properties made of AlSi9Cu3 alloy, In: Advances in Manufacturing, pp. 621–631. Springer, Manufacturing (2017). https://doi.org/10.1007/978-3-319-68619-6_59
7. Gill, A.: Optimisation of the technical object maintenance system taking account of risk analysis results. Eksploatacja i Niezawodnosc—Maint. Reliab. **19**(3), 420–431 (2017). http://dx.doi.org/10.17531/ein.2017.3.13
8. Fuc, P., Rymaniak, L., Ziolkowski, A.: The correlation of distribution of PM number emitted under actual conditions of operation by PC and HDV vehicles. WIT Trans. Ecol. Environ. **174**, 207–219 (2013). ISBN: 978-1-84564-718-6. WIT Press
9. https://www.reliasoft.com/Weibull. Accessed 2018
10. Andrzejczak, K., Młyńczak, M., Selech, J.: Computerization of operation process in municipal transport. In: Zamojski W., Mazurkiewicz J., Sugier J., Walkowiak T., Kacprzyk J. (eds.) Contemporary Complex Systems and Their Dependability. DepCoS-RELCOMEX 2018. Advances in Intelligent Systems and Computing, vol. 761, pp. 13–22. Springer, Cham (2019). https://doi.org/10.1007/978-3-319-91446-6_2
11. Młyńczak, M.: Analiza danych eksploatacyjnych w badaniach niezawodności obiektów technicznych, Zeszyty Naukowe Wyższej Szkoły Oficerskiej Wojsk Lądowych im. gen. T. Kościuszki, nr. 1 (2011)
12. Patra, A.P., Söderholm, P., Kumar, U.: Uncertainty estimation in railway track lifecycle cost: a case study from Swedish National Rail Administration. In: Proceedings of IMechE, vol. 222 Part F: J. Rail and Rapid Transit (2008)
13. Wang, N., Chang, Y-C., El-Sheikh, A.A.: Monte Carlo simulation approach to life cycle cost management. Struct. Infrastruct. Eng.: Maint. Manag. Life-Cycle Des. Perform., 739–746 (2012). https://doi.org/10.1080/15732479.2010.481304j. Dolce, E.: Analytical Fleet Maintenance Management, SAE International, SUA (1994)
14. Trojanowska, J., Dostatni, E.: Application of the theory of constraints for project management. Manag. Prod. Eng. Rev. **8**(3), 87–95 (2017). https://doi.org/10.1515/mper-2017-0031
15. Research Project "Increase in the efficiency of functioning of public means of transport as a result of implementation of LCC and RAMS concepts in accordance with the IRIS standards based on integrated information technology system" financed by Polish National Center for Research and Development. No. PBS3/B6/30/2015
16. Walus, K.J.: Driver's strategy and braking distance in winter, transport means 2017. In: Proceedings of the 21st International Scientific Conference, Juodkrante, Lithuania, 20–22 September 2017, Part 2, pp. 505–509 (2017). ISSN 1822-296 X, e-ISSN 2351-7034
17. http://www.weibull.com/hotwire/issue91/relbasics91.htm. Accessed 2018
18. https://www.reliasoft.com/Weibull. [dostęp: 2018]
19. Lawless, J.F.: Statistical Models and Methods for Lifetime Data. Wiley, New York (1982)

Generalised Gamma Distribution in the Corrective Maintenance Prediction of Homogeneous Vehicles

Karol Andrzejczak(✉) and Jarosław Selech

Poznan University of Technology, 5 M.Skłodowska-Curie Square, 60-965 Poznan, Poland
{karol.andrzejczak, jaroslaw.selech}@put.poznan.pl

Abstract. Vehicles of municipal transportation are complex, multi-component technical objects that require high reliability and safety level and, on the other hand, ensure high level of quality and low cost of service. These requirements and limitations need reliable information for making reasonable decisions. Operators tend to use advanced information concerning daily usage, maintenance, reliability, costs, and many other aspects. Paper shows possibility of application of the generalized gamma distribution family in prediction of the corrective maintenance of fleet vehicles. In this paper, the mileage expressed in kilometres is used as the lifetime of individual tram components. To illustrate the method of adaptation of operational data concerning the fleet of 45 trams, assessments of unknown parameters of the distribution of mileage to damage of a very important tram component, i.e. the drive controller, were determined. The estimation of parameters includes censored data prepared according to the developed method of processing the operational database. As the next step, on the basis of this distribution and the knowledge about the mileage of the currently used drive controllers, the conditional their unreliability were determined in the subsequent periods of using the tram fleet. The results obtained can be used by fleet operators to plan vehicle corrective maintenance.

Keywords: Corrective maintenance · General gamma distribution · Censored data

1 Introduction

This paper is dedicated to prediction of corrective maintenance for a complex technical object such as a fleet vehicle. Corrective maintenance restores the functionality of the vehicle by replacing damaged components. A component damage can be produced by different causes that can be classified either as internal or external [1–3]. Internal damages occur due to the inner structure of components (e.g. ageing and quality of materials). External damages often occur due to the environmental conditions in which vehicles operate (e.g. difficulties of routes, driver's driving quality, accidents, vibrations, humidity, and pollution). Generally, component damages can be divided into two groups. The first group are component damages that may be predicted by one or several condition monitoring indicators. This damage is referred to as a 'gradual damage'. It

I. Kabashkin et al. (Eds.): RelStat 2018, LNNS 68, pp. 519–529, 2019.
https://doi.org/10.1007/978-3-030-12450-2_50

also called ‘soft’ (or ‘degradation’) damage. The current research and developments in degradation models are reviewed and summarised in [4]. The second group are component damages that cannot be predicted by either condition monitoring indicators or by measuring the age of the asset. This kind of component damage is referred to as a ‘sudden damage’. However, in the case of long-life systems, sudden component damage can depend on the time of its use. This paper presents the results of research on the prediction of sudden damage to vehicle components based on the operational data of a fleet of homogeneous vehicles.

In the investigation of vehicle failure processes resulting from the damage of its individual components, a family of the Generalised Gamma distribution (GG) becomes particularly important due to its high flexibility [5]. This family includes a number of known sub-families of distributions (see [6–8]). Estimation of life time parameters of selected vehicle components is based on the operational database of a fleet of homogeneous urban transport vehicles from the first five years of their operation. The operational database is composed of tens of thousands of records for detailed information on corrective maintenance of several hundred different components exchanged at least once in five years of using a fleet of 45 vehicles.

During operated time, the costs of all the maintenance performed with the vehicles will exceed the purchase costs of the fleet [9]. In addition to the cost of scheduled maintenance, a significant part of the costs will be incurred for unscheduled maintenance caused by sudden damage to vehicle components. These unscheduled maintenance operations are the subject of reliability tests carried out by the authors of this paper. The identification of the mileage distribution was made for most expensive several dozen vehicle components. It should be noted that due to the ongoing process of operating the vehicle fleet, the so-called right-censored times [10] are used in estimation of parameters of selected distribution families.

The purpose of the estimation of GG parameters for the selected component is determining its distribution of the remaining mileage and the probability of damage in the subsequent useful periods counted in weeks, months or even years of its further operation. In this point and interval prediction a conditional GG distribution is used. Among the dozen components tested, the mileage distribution of the drive controller was characterized by the most similar to the GG distribution [11]. Hence the drive controller was chosen to predict its probability of damage during selected mileage of the further use the fleet of vehicles.

The paper consists of six sections. Introduction is the first section. Assumptions and symbols are given in Sect. 2. Section 3 presents a family of GG distributions as a model for the mileage of vehicle components. The main novelty of the conducted research consisting in the development of the method of adaptation of the operational database for the determining remaining useful mileage for chosen component is given in Sect. 4. Section 5 presents the results of the application of the developed adaptation method and the GG distribution family to the damage prediction of the tram drive controller during selected mileage of its further use. Such information is necessary in planning inspections and assessing the costs of exchanging parts of vehicles in the further process of fleet use. At the end of the paper are given conclusions.

2 Assumptions and Symbols

Vehicle breakdowns occur at random times of their use and at random mileages. The consequences of a vehicle breakdown depend on the type of damage. We assume that the cause of a vehicle breakdown during its use is the damage of a component which is replaced as a whole. The vehicles consist of several thousand components and over a long period of use, i.e. 25–30 years, even several hundred of them may be damaged, despite scheduled maintenance. In this paper, we assume that the lifetimes of individual components are calculated by mileage of the vehicles they are used in. These mileages express the operational use of vehicle components and are referred to as their 'useful mileages'. After each vehicle breakdown, damaged components are replaced for new ones as part of corrective maintenance. During the process of operating a fleet of vehicles, the Operational DataBase (ODB) is constantly kept up-to-date. In the ODB are registered: vehicle identifier, date and time of vehicle breakdown, name and identifier of the damaged component, mileage of the vehicle at which it was damaged, damage circumstances, maintenance cost including materials and labour, consequences of vehicle breakdown and other information [12]. The ODB on vehicle breakdowns is maintained by the operator from the moment of purchase of a new fleet of homogeneous vehicles, and more precisely from the date of putting the vehicles into service. Let the day t_0 means the day, when the service of the fleet vehicles starts. The time unit is assumed to be a day of vehicle being in operation and for simplicity we assume that $t_0 = 0$. We denote the set of fleet vehicles with the symbol J. We assume that this set consists of at least several dozen homogeneous vehicles used in the same operating regime. All fleet vehicles before being put into service for the first time have to pass the first maintenance check, so for the day t_0 they are functional. This means that all components of vehicles allowed for operation are fully functional. When the vehicle starts to be used, the useful mileage of its components begins. Vehicles are complex repairable objects, while their components are assumed to be not repairable and each damaged component is replaced by a new one as part of corrective maintenance. Scheduled maintenance and preventive maintenance carried out together with scheduled maintenance are not covered by the conducted research. The research covers only the useful mileages of those components that are damaged during the use of vehicles. Damage of this type is usually sudden and generates unexpected additional operating costs. With the fleet's lifetime, the ODB develops with new data on mileage, breakdowns and corrective maintenance of vehicles. After some time of fleet operation, on a specific day t_1 the operational database contains detailed information on vehicles' maintenance in chronological order. We denote the ODB from this particular day as $B(t_0, t_1)$. On the basis of this database, the history of the vehicle fleet operation and maintenance is reconstructed and the reliability characteristics of the most frequently damaged components from those that generate high costs of corrective maintenance are estimated. Let K denotes the set of indices of those vehicle components that are the subject of the research. We assume that vehicle failures caused by damage to their components from the set K are independent. A probabilistic model of the useful mileage of the k-th component $(k \in K)$ is a non-negative random variable L_k of a continuous type for which we assume that the following function exist: probability density function (pdf) $f_k(l)$, cumulative distribution function (cdf) $F_k(l)$ and survival function $S_k(l)$.

For a sufficiently long period of use of the vehicle fleet, there may be components that have been replaced by the new ones at least once. So in the ODB at the moment t_1 there are current vehicle mileage and vehicle mileages at which replacements of damaged components were made. The method of converting vehicle mileage into the mileage of individual vehicle components is developed in the fourth part.

3 The Family of GG Distributions as a Model of the Useful Mileage for Vehicle Components

To search for a probabilistic model of the useful mileage of individual components from the set K, a family of the GG distributions was used. This family was introduced by Stacy [5]. The GG family is flexible in that it includes several well-known subfamilies of distributions used in studies of the duration of various events or phenomena [13–15]. The advantage of the GG family is its ability to behave like other families of lifetime distribution and thus fit a narrower family of distributions to a specific set of operating data. Adequate selection of parameters leads to such subfamilies of distributions as: exponential distribution, two-parameters gamma distribution, Weibull distribution, half normal distribution, generalized normal distribution, Rayleigh distribution, Maxwell-Boltzman distribution and chi distribution [16]. Maximum-Likelihood Estimation (MLE) of the parameters and quasi maximum likelihood estimators for its subfamily can be found in [17–20].

Despite the mathematical complexity and the requirement to have large samples of data, the GG family is increasingly used to model lifetime of technical objects. This was due to the development of computer support for reliability analyses, in particular software for estimating the parameters of the family of GG distributions and its subfamilies [21, 22].

In order to probabilistically describe random events occurring during the operation of the vehicle fleet, we introduce random variables $L_{j,\,k,\,i}$, $j \in J$, $k \in K$, $i = 1, 2, \ldots$, that characterize the useful mileage of the i-th instance of the k-th component in the j-th vehicle. In addition, we assume that for a given component $k \in K$ there is a distribution of its useful mileage L_k which is the same regardless of the vehicle in which it is installed, from the novelty of the vehicle or after its corrective maintenance. So to estimate the parameters of the random distribution of the mileage of the k-th component, we use useful mileages of all instances of this component that were or are still on day t_1 installed in fleet vehicles. Due to the assumption of the independence of damages of individual vehicle components, the estimation of the parameters of their useful mileages is performed separately for each of them. Identification of the unconditional useful mileage distribution is necessary to determine the conditional remaining useful mileage of the already use components.

The GG family is a four-parameter distribution family that we refer to as $\mathrm{GG}(\alpha, \beta, \theta, \gamma)$. Let us assume the parametrisation in which the pdf of this family is defined by the formula:

$$f(l|\alpha,\beta,\theta,\gamma)=\frac{\beta(l-\gamma)^{\alpha\beta-1}}{\theta^{\alpha\beta}\Gamma(\alpha)}\exp\left(-\left(\frac{l-\gamma}{\theta}\right)^{\beta}\right),\ l>\gamma, \tag{1}$$

where $\theta > 0$ is a scale parameter, $\alpha > 0$ and $\beta > 0$ are shape parameters, γ is continuous location parameter, and $\Gamma(\alpha)$ is gamma function of α. For $\gamma = 0$ we get family of the three-parameter generalized gamma distributions with pdf:

$$f(l|\alpha,\beta,\theta)=\frac{\beta}{\theta\Gamma(\alpha)}\left(\frac{l}{\theta}\right)^{\alpha\beta-1}e^{-\left(\frac{l}{\theta}\right)^{\beta}},\ l>0. \tag{2}$$

Further discussion will focus on these family of distributions and these family is denoted by $\mathrm{GG}(\alpha,\beta,\theta)$. The important role in statistical inference and reliability theory are played the moments of the family GG distributions. If for $k \in K$, $L_k \sim \mathrm{GG}(\alpha_k,\beta_k,\theta_k)$, then the expected useful mileage $\mathbb{E}(L_k)$ and variance $\mathbb{D}^2(L_k)$ are given by the following formulas:

$$\mathbb{E}(L_k)=\frac{\theta_k\Gamma\left(\alpha_k+\frac{1}{\beta_k}\right)}{\Gamma(\alpha_k)},\ \mathbb{D}^2(L_k)=\frac{\theta_k^2\Gamma\left(\alpha_k+\frac{2}{\beta_k}\right)}{\Gamma(\alpha_k)}-\left(\frac{\theta_k\Gamma\left(\alpha_k+\frac{1}{\beta_k}\right)}{\Gamma(\alpha_k)}\right)^2. \tag{3}$$

In order to improve the identification of a distribution subfamily, the following re-parametrisation of parameter α,β,θ is applied, as given by Lawless in [8]:

$$\mu=\ln(\theta)+\frac{1}{\beta}\ln\left(\frac{1}{\lambda^2}\right),\ \sigma=\frac{1}{\beta\sqrt{\alpha}},\ \lambda=\frac{1}{\sqrt{\alpha}}, \tag{4}$$

where $-\infty<\mu<\infty, \sigma>0, \lambda>0$. Such parametrised family of GG distributions for k-th component of the vehicle we denote $\mathrm{GG}(\mu_k,\sigma_k,\lambda_k)$. With this parametrisation, the reliability function of the useful mileage $S_k(l)$ in the system of new parameters takes the form:

$$S_k(l)=1-\Gamma_I\left(\frac{\exp\left(\lambda\left(\frac{\ln(l)-\mu}{\sigma}\right)\right)}{\lambda^2};\frac{1}{\lambda^2}\right), \tag{5}$$

where $\Gamma_I(\alpha;x)$ is incomplete gamma function of α and x, which is given by:

$$\Gamma_I(\alpha;x)\underline{\underline{\mathrm{def}}}\frac{1}{\Gamma(\alpha)}\int_0^x s^{\alpha-1}e^{-s}ds,\ \alpha>0, x>0. \tag{6}$$

For GG distribution the useful mileage $l_{s,k}$ for the k-th vehicle component for ensuring the reliability at the level $s \in (0,1)$ assuming that this component has just been replaced, is given by:

$$l_{s,k} = \exp\left(\mu_k + \frac{\sigma_k}{\lambda_k}\ln\left(\lambda_k^2\Gamma_I^{-1}\left(1 - s; \frac{1}{\lambda_k^2}\right)\right)\right), \tag{7}$$

where Γ_I^{-1} is inverse function to incomplete gamma function.

4 Adaptation of the Operational Database

The reliability tests of the useful mileage of vehicle components are carried out on the basis of the ODB $B(t_0, t_1)$ containing information on corrective maintenance of all vehicles of the tested fleet up to the day t_1. Let $n_{jk}(t_1)$ indicate the number of corrective maintenances j-th vehicle due to replacement of the k-th component, whereas $l_{j,k,i}(j \in J, k \in K, i = 1, 2, \ldots, n_{jk}(t_1) + 1)$ denotes the useful mileage of the i-th instance of the k-th component in the j-th vehicle. We will consider various situations of component replacement. If $n_{j,k}(t_1) = 0$, then up to the day t_1 k-th component of the j-th vehicle has never been replaced yet and its useful mileage $l_{j,k,1}(t_1)$ is equal to the mileage of the j-th vehicle for the day t_1,, i.e. $l_{j,k,1}(t_1) = l_j(t_1)$. Mileage $l_{j,k,1}(t_1)$ registered on a day t_1 is then censored, because this component of the vehicle is still functional.

If $n_{j,k}(t_1) = 1$ then up to the day t_1 k-th component of the j-th vehicle was once replaced with a new one. The day of loss of functionality $\tau_{j,k,1}$ and the mileage of the vehicle $l(\tau_{j,k,1})$ to this event are recorded in the database. Useful mileage $l_{j,k,1}(t_1)$ of the original k-th component in the j-th vehicle is equal to the mileage of the vehicle $l(\tau_{j,k,1})$. At this stage, the component was replaced with a new one, which remains useful until the day t_1. Useful mileage $l_{j,k,2}(t_1)$ of the replaced component for the day t_1 is censored and equal to the difference

$$l_{j,k,2}(t_1) = l_j(t_1) - l(\tau_{j,k,1}). \tag{8}$$

In case of $n_{jk}(t_1) \geq 2$, i.e. when the number $n_{jk}(t_1)$ of corrective maintenances of the j-th vehicle due to replacements of the k-th component to the day t_1 of vehicle usage is at least 2, then the days of loss of functionality and mileages $l_j(\tau_{j,k,1})$, $l_j(\tau_{j,k,2}), \ldots, l_j\left(\tau_{j,k,n_{jk}(t_1)}\right)$ [in km] have been registered in the database $\tau_{j,k,1}$, $\tau_{j,k,2}, \ldots, \tau_{j,k,n_{j,k}(t_1)}$ of a maintained vehicle due to replacements of this component. Let $l_{j,k,1}, l_{j,k,2}, \ldots, l_{j,k,n_{j,k}}(t_1)$ mean useful mileages of the k-th component for its subsequent replacements in the j-th vehicle. For the day t_1 of usage of the j-th vehicle, in this vehicle is installed the $(n_{jk}(t_1) + 1)$-th instance of the k-th component. Mileage $l_{j,k,i}(t_1)$ of the k-th component in the j-th vehicle to the i-th replacement with a new one is equal to the difference in mileage of the vehicle

$$l_{j,k,i}(t_1) = l_j(\tau_{j,k,i}) - l_j(\tau_{j,k,i-1}) \text{ for } i = 1, 2, \ldots, n_{j,k}(t_1) \text{ and } l(\tau_{j,k,0}) = 0 \tag{9}$$

The last of the mentioned instances of the k-th component, i.e. the $\left(n_{jk}(t_1)+1\right)$-th instance is for the day t_1 still functional and its mileage $l_{j,k,n_{j,k}+1}(t_1)$ is censored. This mileage is equal to the difference in vehicle mileage:

$$l_{j,k,n_{j,k}+1}(t_1) = l_j(t_1) - l\left(\tau_{j,k,n_{jk}(t_1)}\right). \tag{10}$$

The presented adaptation of the ODB has been used to prepare statistical data and identification the distribution of the useful mileage for the selected component.

5 Application of the Family GG Distributions to the Damage Prediction

The authors of this paper have ODB of a fleet of $|J| = 45$ vehicles used for 5 years and during that time dozens of expensive corrective maintenances of vehicles were registered due to the replacement of their components that determine the functionality and safety of their use. Based on the ODB $B(t_0, t_1)$ evaluation $\hat{\mu}_k, \hat{\sigma}_k, \hat{\lambda}_k$ of unknown distribution parameters $\mathrm{GG}(\mu_k, \sigma_k, \lambda_k)$ for the k-th component is performed. Then based on the designated distribution $\mathrm{GG}\left(\hat{\mu}_k, \hat{\sigma}_k, \hat{\lambda}_k\right)$ and knowledge about $l_{j,k,n_{j,k}+1}(t_1)$, i.e. about the useful mileage $\left(n_{jk}(t_1)+1\right)$ of the k-th instance of the component in the j-th vehicle of the fleet, the prediction about its damage in the further use of the j-th vehicle is made. The WEIBULL++ computer software was used for estimation of the parameters of the GG distribution and for prediction of the probability of damage to the selected vehicle component.

One of the key components of the tram is a hand-held drive controller similar to a joystick. It is a basic control device for the driver. By moving the controller handle forward the wagon will move. The further the driver moves the handle, the more the wagon will accelerate. Returning the controller to the middle (neutral) position will cause coasting (without driving force). When the driver pulls the handle of the controller towards him, the braking will start, and the braking force depends on how much is the handle tilted from the neutral position. When the handle is pulled beyond the marked section, the emergency braking will be activated.

During the time covered by the research, $n_k(t_1) = 23$ corrective maintenance were registered due to the change of the drive controller. The total costs of these services amounted to approximately 208 thousand PLN (approximately 50 thousand EUR). Using adaptation of the ODB for the needs of the reliability tests conducted, data on the drive controller's mileage for all fleet vehicles up to the day t_1 was compiled in the form of pairs $(l_i; \delta_i)$, $i = 1, 2, \ldots, n_k(t_1) + |J|$, where l_i is expressed in kilometres of useful mileage of the i-the instance of the drive controller, $n_k(t_1) + |J| = 68$ is the number of data records about the controller and

$$\delta_i = \begin{cases} 1, & \text{if } l_i \text{ is the observed operational mileage,} \\ 0, & \text{if } l_i \text{ is the censored operational mileage.} \end{cases} \tag{11}$$

The method developed in part four for the entire vehicle fleet, 68 pairs $(l_i; \delta_i)$ were obtained, and in 45 cases the drive controller was functional on the day t_1 of the interruption of observation:

(327707; 0), (156048; 1), (116623; 0), (360404; 0), (243532; 0), (216144; 1), (68690; 0), (235268; 1), (134168; 0), (314522; 0), (339907; 0), (311679; 0), (262359; 1), (79882; 0), (328494; 0), (354100; 0), (244106; 1), (24143; 1) (23789; 0), (339492; 0), (292266; 0), (258866; 1), (60222; 0), (352654; 0), (132841; 1), (177672; 0), (305087; 0), (335015; 0), (308114; 0), (328878; 0), (143058; 1), (142315; 1), (29216; 0), (268743; 0), (108027; 1), (29092; 1), (71490; 1), (111124; 0), (235029; 1), (93451; 0), (328168; 0), (219815; 1), (67307; 0), (283162; 1), (14190; 0), (271868; 0), (122934; 1), (121071; 0), (284525; 0), (207679; 0), (152554; 1), (161592; 0), (265598; 1), (5786; 0), (316331; 0), (286225; 0), (282281; 1), (3889; 1), (3355; 0), (254974; 0), (290198; 0), (26807; 1), (272367; 0), (264787; 0), 265863; 0), (192863; 1), (102032; 0), (315371; 0).

When the GG distribution is fitted to this data using MLE, the following values for parameters are obtained: $\hat{\mu} = 13.129$, $\hat{\sigma} = 0.435$, $\hat{\lambda} = 1.951$. The information obtained was used to determine the function of unconditional failure $\hat{F}_k(l) \underline{\underline{\mathrm{def}}}\, 1 - \hat{S}_k(l)$ for the drive controller. This function takes the form:

$$\hat{F}_k(l) = \Gamma_I\left(\frac{\exp\left(1.951\left(\frac{\ln(l)-13.129}{0.435}\right)\right)}{3.807275}; \frac{1}{3.807275}\right). \tag{12}$$

The results of the estimation of unconditional failure probabilities along with the 90% confidence intervals determined using the Fisher information matrix for the useful mileages of the drive controller are summarised in Table 1. The values of the selected useful mileages are resulting from the predicted mileage of trams on average per quarter at $15,000 \pm 3,000$ [km].

Table 1. Estimation of the damage probability for the new tram drive controller.

Mileage l [km]	Lower bound	$\hat{F}_k(l)$	Upper bound
15,000	0.0040	0.0124	0.0380
30,000	0.0114	0.0280	0.0669
60,000	0.0329	0.0634	0.1187
120,000	0.0928	0.1435	0.2153
240,000	0.2434	0.3244	0.4174

At the end of this paper, the results of estimation $\Pr(L_k \le l | L_k > l_1)$ are presented for $l > l_1$, i.e. the conditional probability of damage (conditional unreliability) of the drive controller which has already the useful mileage of l_1 on the day t_1 and is still in use. Day t_1 results from a five-year period of collection of fleet operational data. The predicted probabilities of future damages of tram drive controllers, which already have useful mileages of 5,000; 30,000; 60,000; 120,000 [km] for further mileages of vehicles equipped with these drive controllers that amount to: 30,000 and 60,000 [km] are presented in Tables 2 and 3.

Table 2. Estimation of conditional probabilities of tram drive controllers' damage during the next 30,000 [km] of their mileage.

Conditional mileage l_1 [km] on day t_1	Lower bound	$\hat{F}_k(l \vert L_k > l_1)$	Upper bound
5,000	0.0141	0.0303	0.0641
30,000	0.0222	0.0364	0.0591
60,000	0.0287	0.0414	0.0594
120,000	0.0371	0.0504	0.0681

Table 3. Estimation of conditional probabilities of tram drive controllers' damage during the next 60,000 [km] of their mileage.

Conditional mileage l_1 [km] on day t_1	Lower bound	$\hat{F}_k(l \vert L_k > l_1)$	Upper bound
5,000	0.0369	0.0665	0.1170
30,000	0.0504	0.0764	0.1141
60,000	0.0615	0.0855	0.1177
120,000	0.0754	0.1026	0.1381

Tables 2 and 3 summarise conditional probabilities of damage only for four drive controllers with different mileages. For the fleet operator, the presented research method allows to obtain important information about the probability of damage for currently used driving controllers.

6 Conclusions

The method of preparing statistical data presented in this report using ODB and identification the useful mileage distribution can be used for various randomly damaging vehicle parts. Any of the available computer programs for statistical analysis based on censored data may be used to identify the distribution. Knowing the distribution of the component unreliability over mileage, preventive actions can be taken before the risk of damage reaches the limit value, thus avoiding financial losses and ensuring an adequate level of safety. The choice of a generalized gamma distribution resulted from its high elasticity. The calculations show that not only for relatively new components, predictions of failure probabilities are not subject to a large error. A larger error appears for parts used for many years. In the case of multiannual projections, the presented method is characterized by an unacceptable error. In this case, it is recommended to diagnose the elements in the meantime. Fortunately, in practice the predictions of components damage are more interesting for weekly or monthly periods of use of vehicles. In such cases, the confidence intervals as a measure of the quality of prediction of conditional probabilities of damage to the vehicle component are at an acceptable level.

Acknowledgements. The research work financed with the means of statutory activities PUT 04/43/DS/PB/0104 and 05/51/DS/PB/3551.

References

1. Montoro-Cazorla, D., Perez-Ocon, R.: Reliability of a system under two types of failures using a Markovian arrival process. Oper. Res. Lett. **34**(5), 525–530 (2006)
2. Rausand, M.: Reliability centered maintenance. Reliab. Eng. Syst. Saf. **60**(2), 121–132 (1998)
3. Yang, G.: Environmental-stress-screening using degradation measurements. IEEE Trans. Reliab. **51**(3), 288–293 (2002)
4. Gorjian, N., Ma, L., Mittinty, M., Yarlagadda, P., Sun, Y.: A review on degradation models in reliability analysis. In: Proceedings of the 4th World Congress on Engineering Asset Management, Athens, Greece, pp. 369–384 (2009)
5. Stacy, E.W.: A generalization of the gamma distribution. Ann. Math. Stat. **33**, 1187–1192 (1962)
6. Andrzejczak, K.: Selected methods of analysis of censored product lifetime (in Polish). Acta Universitatis Lodziensis. Folia Oeconomica **156**, 115–133 (2002)
7. Jaggia, S.: Specification tests based on the heterogeneous generalized gamma model of duration: with an application to Kennan's strike data. J. Appl. Econ. **6**, 169–180 (1991)
8. Lawless, J.F.: Statistical Models and Methods for Lifetime Data, 2nd edn. Wiley, Hoboken, New Jersey (2003)
9. Andrzejczak, K., Selech, J.: Quantile analysis of the operating costs of the public transport fleet. Trans. Prob. **12**(3), 103–111 (2017)
10. Cox, D.R., Oakes, D.: Analysis of Survival Data. Chapman & Hall, New York (1984)
11. Selech, J., Andrzejczak, K.: Identification of reliability models for non-repairable railway component. Submitted to RelStat-2018 (2018)
12. Andrzejczak, K., Młyńczak, M., Selech, J.: Computerization of operation process in municipal transport. In: Zamojski, W., Mazurkiewicz, J., Sugier, J., Walkowiak, T., Kacprzyk, J. (eds.) Contemporary Complex Systems and Their Dependability. DepCoS-RELCOMEX 2018. Advances in Intelligent Systems and Computing, vol. 761, pp. 13–22. Springer, Cham (2019). https://doi.org/10.1007/978-3-319-91446-6_2
13. Allenby, G.M., Leone, R.P., Jen, L.: A Dynamic model of purchase timing with application to direct marketing. J. Am. Stat. Assoc. **94**, 365–374 (1999)
14. Heinrich, M., Pilch, R., Smolnik, M., Szybka, J., Wiązania, G.: Reliability estimation using a three-parameter gamma distribution (in Polish). Logistyka **6**(2015), 1047–1054 (2015)
15. Yamaguchi, K.: Accelerated failure-time regression models with a regression model of surviving fraction: an application to the analysis of "permanent employment" in Japan. J. Am. Stat. Assoc. **87**, 284–292 (1992)
16. Khodabin, M., Ahmadabadi, A.R.: Some properties of generalized gamma distribution. Math. Sci. **4**(1), 9–28 (2010)

17. Harter, H.L.: Maximum-likelihood estimation of the parameters of a four parameter generalized gamma population from complete and censored samples. Technometrics **9**, 159–165 (1967)
18. Hirose, H.: Maximum likelihood parameters estimation by model augmentation with application to the extended four-parameters GG distribution, Japan (1999). http://hdl.handle.net/10228/717
19. Stacy, E.W., Mihram, G.A.: Parameter estimation for a generalized gamma distribution. Technometrics **7**, 349–358 (1965)
20. Stacy, E.W.: Quasimaximum likelihood estimators for two-parameter gamma distribution. IBM J. Res. Dev. **17**, 115–124 (1973)
21. Corporation, R.: Life Data Analysis Reference, Tucson. ReliaSoft Publishing, AZ (2015)
22. Roy, R., Stark, R., Tracht, K., Takata, S., Mori, M.: Continuous maintenance and the future – Foundations and technological challenges. In: CIRP Annals – Manufacturing Technology, vol. 65, pp. 667–688 (2016)

Impact of Sulfate Reducing Bacteria on Biocorrosion of Pipeline Steels

Miroslava Polutrenko[1], Pavlo Maruschak[2(✉)], Olegas Prentkovskis[3], Anatoliy Tymoshenko[4], and Olena Maruschak[1]

[1] Ivano-Frankivsk National Technical University of Oil and Gas, Karpatska 15 str., 76019 Ivano-Frankivsk, Ukraine
[2] Ternopil National Ivan Pul'uj Technical University, Ruska 56 str., 46001 Ternopil, Ukraine
Maruschak@tu.edu.te.ua
[3] Vilnius Gediminas Technical University, Plytinės g. 27, 10105 Vilnius, Lithuania
olegas.prentkovskis@vgtu.lt
[4] Open International University of Human Development, Lvivska 23 str., 03115 Kyiv, Ukraine

Abstract. Samples obtained from natural gas pipelines were used directly for determine the regularities in corrosion steels 17G1S-U and 20 under the influence of sulfate-reducing bacteria (SRB) of type Desulfovibrio sp. strain Kyiv-10 and their relation to the microstructure of steel in condition that simulated a natural gas pipeline environment. Microbiological corrosion induced by SRB of pipeline steel 17G1S-U and steel 20 were studied. The impact of the inhibitors nature on the specimen protection degree under the development of anaerobic corrosion under the influence of SRB is investigated.

Keywords: Main gas pipeline · Damage · Steels 17G1S-U and 20 · Biocorrosion

1 Introduction

The gas transportation system of Ukraine consists of gas mains, distribution networks, gas storages, compression and gas measuring stations, is the second largest in Europe and one of the largest in the world. The length of the gas mains of the gas transportation system of Ukraine is 37,600 km.

A long-term exploitation of the gas transportation system led to the degradation of its materials (steel grades 17GS, 17G1S-U, 09G2S, 20). In addition to the corrosive-mechanical component, the damage to the gas main is also caused by the biological factor, which leads to the destruction of the protective insulating coating under the influence of the corrosive and dangerous microorganisms found in the soil. It is known that biocorrosion is one of the most widespread degradation processes that occur in oil and gas mains [1–3]. Biocorrosion accounts for about 50% of all corrosion damage to the gas transportation system. The methods of its inhibition are not always effective, since most of the dangerous microflora shows enhanced adhesive ability to the walls of

I. Kabashkin et al. (Eds.): RelStat 2018, LNNS 68, pp. 530–538, 2019.
https://doi.org/10.1007/978-3-030-12450-2_51

the pipelines and turns to a biofilm in the form of sticky mucus under a layer of corrosion [4–6]. The greatest biodamage is caused by sulfate reducing (sulfate renewing) bacteria (SRB). SRB are actively involved in anaerobic biocorrosion [7, 8]. They are heterotrophic microorganisms that form corrosively active metabolites (H_2S, NH_3, CO_2, carboxylic acids) [9, 10]. They also cause the destruction of oil and gas mains. Of particular danger is that corrosive lesions are localized and can develop into transient ulcers. In addition, being located in the lower part of the pipe under the layer of corrosion products is typical of them [2, 3].

These bacteria change the corrosion activity of the soil due to the accumulation of products of their life, in particular, hydrogen sulfide and iron sulfide [1, 11]. In addition to the localized corrosion, these compounds increase the risk of sulfide corrosion cracking, or hydrogenation of the pipe wall [12].

The purpose of this research is to determine the regularities in corrosion steels 17G1S-U and 20 under the influence of SRB of type Desulfovibrio sp. strain Kyiv-10 and their relation to the microstructure of steel.

2 Methods of Mechanical and Physical Research

Specimens of pipe steels 17G1S-U and 20 with the size of 10 × 30 mm were investigated. The chemical composition of the investigated steels is given in Table 1.

Table 1. Chemical composition of the investigated steels*.

Steel grade	Chemical composition, %										
	C	Si	Mn	Ni	S	P	Cr	V	N	Cu	As
17G1S-U	0.15–0.2	0.4–0.6	1.15–1.55	>0.3	>0.035	>0.035	>0.3	>0.12	>0.012	>0.3	>0.08
20	0.17–0.24	0.17–0.37	0.35–0.65	>0.25	>0.035	>0.035	>0.25	–	–	–	–

*in accordance with the manufacturer's certificate.

SRB cells of type Desulfovibrio sp. strain Kiev-10 were grown on a liquid Postgate's B medium in a thermostat at a temperature of 28 °C for 14 days. Pure colonies of sulfate reducers were obtained on a semi-liquid Postgate's B medium by seeding a tenfold dilution. The variants with additional injections into the system of inhibitors were also studied. Inhibitors were organic nitrogen-containing compounds:

- inhibitor 1—1, 8-dioxo-3, 3, 6, 6, 9-pentamethyl-10-phenyl-1, 2, 3, 4, 5, 6, 7, 8, 9, 10-decahydroacridine;
- inhibitor 2—hexamethyldodecylammonium chloride.

The control medium was Postgate's B sterile nutrient medium for the cultivation of SRB bacteria. The exposure period of steel specimens was 62 days. Upon completion of the research, the metallic specimens were subjected to the mechanical and chemical treatment to remove corrosion products from their surface.

The corrosion rate of metallic specimens was determined by the gravimetric index of the corrosion rate ($K_{gr.}$):

$$K_{gr.} = m_0 - m/S\,\tau, \left(\text{mg/dm}^2 \cdot \text{day}\right), \quad (1)$$

where m is the final mass of the specimen, mg; m_0 is the specimen mass before corrosion, mg; S is the surface area of the specimen, dm^2; τ is the exposure period, day.

The bactericidal properties of the inhibitors studied were determined according to the DSTU 3999-2000 [13]. The concentration of inhibitors was 0.5 and 1.0%. The effectiveness of the inhibitors studied was characterized by the degree of protective action of inhibitors (Z), calculated by the formula:

$$Z = K_{gr} - K_{gr1}/K_{gr} \cdot 100\,\%, \quad (2)$$

where: $K_{gr.}$ is the corrosion rate in the uninhibited medium, mg/dm^2•day; K_{gr1} is the corrosion rate in the presence of inhibitors, mg/dm^2•day.

For microstructural and fractographic studies, the electron microscope REM-106I and raster electron microscopy methods were used.

3 Results and Discussion of Experimental Findings

It is known that the structure affects the kinetics of biocorrosion, therefore, we performed metallographic analysis of cross-sections of the investigated steels. Both steels had a ferrite-perlite structure (Fig. 1). Steel 17G1S-U has more perlite, which looks like stripes, Fig. 1a. Steel 20 is characterized by a structure in the form of ferrite and perlite grain conglomerates, the content of which is approximately the same, Fig. 1b. Layer by layer grinding of the specimens showed that their structure is uniform across the inner volume, without a significant gradient. Structural defects in the form of cracks, stratifications or pores were not found.

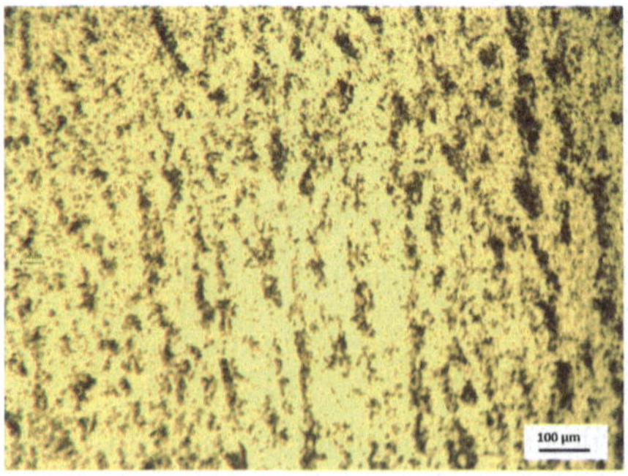

a

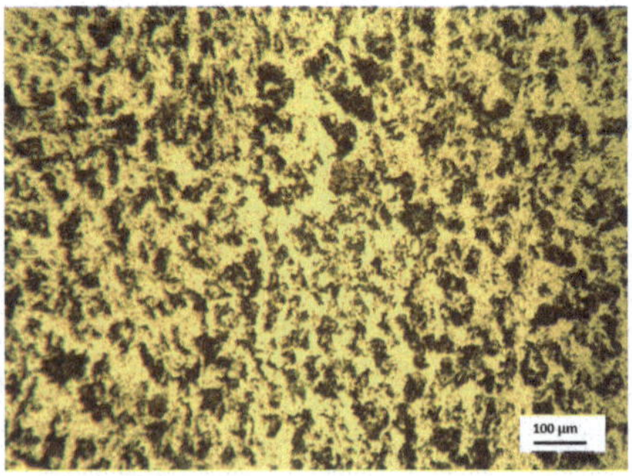

b

Fig. 1. Microstructure of steel 17G1S-U (a) and 20 (b).

At the same time, it should be noted that in comparative test conditions, the growth activity of SRB depends on the grade of steel. Thus, for the 17G1S-U steel, the rate of biocorrosion caused by SRB is almost 40% higher than for steel 20. The analysis of the experimental studies on the growth suppression activity of SRB of type *Desulfovibrio sp.* strain Kyiv-10 during the model laboratory studies has shown that in the sterile Postgate's B medium, in the case of steel 17G1S-U, no traces of corrosion were detected, and the medium was unchanged, whereas in the case of steel 20, visible traces of the external corrosion were noticed on the metal surface. In Postgate's B medium inoculated by SRB, there is an active growth of bacteria, as evidenced by greater losses of metal. The surface of steel specimens was covered with a black precipitate, and the cloudiness of the corrosive medium was observed, indicating the development of biocorrosion with the participation of SRB, which results in the formation of biogenic sulfide of iron. When the specimens were dowsed into the solution of hydrochloric acid, the sulfide film was destroyed causing the release of hydrogen sulfide. As a result, the surfaces of specimens from steel 17G1S-U and steel 20 were covered with corrosion damage under the influence of SRB. The influence of organic corrosion inhibitors on the growth of SRB was estimated by the corrosion rate inhibition activity of the steel specimens [14], (Fig. 2).

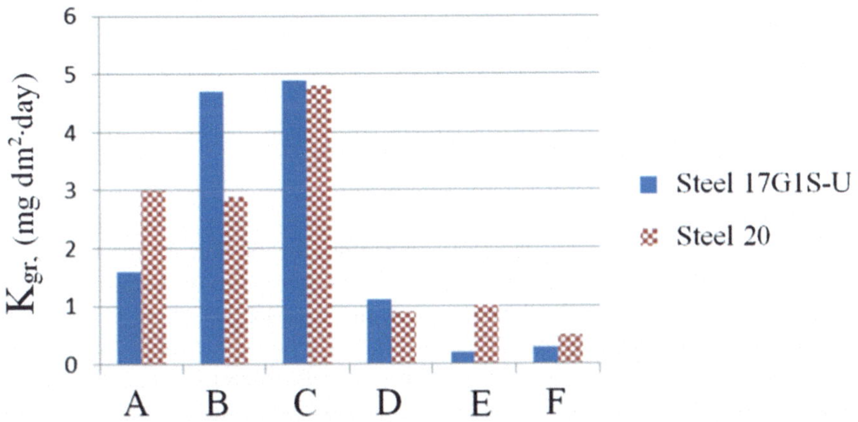

Fig. 2. Corrosion rate of specimens from steel 17G1S-U and steel 20 in Postgate's B medium in the presence of *Desulfovibrio sp.* strain Kyiv-10: A—Control; B—Postgate's B medium + SRB; C—Postgate's B medium + SRB + inh. No.—0.5%; D—Postgate's B medium + SRB + inh. No. 1—1.0%; E—Postgate's B medium + SRB + inh. No. 2—0.5%; F—Postgate's B medium + SRB + inh. No. 2—1.0%.

Table 2 presents the experimental results on the effect of inhibitors on the protection from biocorrosion caused by SRB for steel 17G1S-U and steel 20.

Table 2. Impact of the inhibitor type on the degree of protection of specimens from steel 17G1S-U and steel 20.

Steel grade	Test scheme	Inhibitor	Inhibitor concentration, %	Degree of protection, Z, %	Specimen surface damage morphology
17G1S-U	B	–	–	4.7	The surface has a metallic luster
	C	1	0.5	Absent	Local "corrosive islands" are formed
	D	1	1.0	76.6	Uniform corrosion damage and shallow local corrosion damage
	E	2	0.5	95.74	The corrosive effect is insignificant
	F	2	1.0	93.62	
20	B	–	–	2.9	Darkening of the surface, corrosion film is formed
	C	1	0.5	Absent	Localized corrosion of the surface, separate areas, merging of corrosive areas
	D	1	1.0	68.97	Uniform corrosion damage and shallow local corrosion damage
	E	2	0.5	65.52	The corrosive effect is insignificant
	F	2	1.0	82.76	

4 Use of Inhibitors to Protect Pipe Steels

The morphology of corrosion damage caused to the surface of specimens from steel 17G1S-U and steel 20 was analyzed and compared with the data on the mass loss in order to evaluate the kinetics and mechanisms of biocorrosion. In the initial state, the specimen surface had a metallic luster, and at the micro level, there were clear-cut stripes resulting from milling. For tests according to scheme A (Table 1), no damage was found on the specimen. Scheme B indicates a uniform corrosion of the surface of specimens from steel 17G1S-U, and a significant damage to the surface of specimens from steel 20, Fig. 3a, d.

Effect of inhibitor 1. When inhibitor 1 at a concentration of 0.5% was added to the medium (Scheme C), the mass loss was the greatest. At a concentration of 0.5%, this inhibitor showed no inhibitory effect, but, on the contrary, acted as a stimulant of biocorrosive processes caused by SRB. At the end of the experiment, an increase in the corrosion rate of steel 20 by 65–66% was observed. However, for steel 17G1S-U, the growth was about 4–5%. "Corrosion islands" of 40–100 μm in length and 50 μm in width were formed on the specimen surface from steel 17G1S-U, and "spots" of the same size were formed on the surface of steel 20, Fig. 3b, e. These damaged areas have a random orientation and are combined into a network that crosses the initial deformed relief at different angles. Depressions between islands (and spots) are the areas, in which the corrosion process is localized (macrolocalization). The difference in microcorrosion processes is that for the 17G1S-U steel, despite the "islands", corrosion

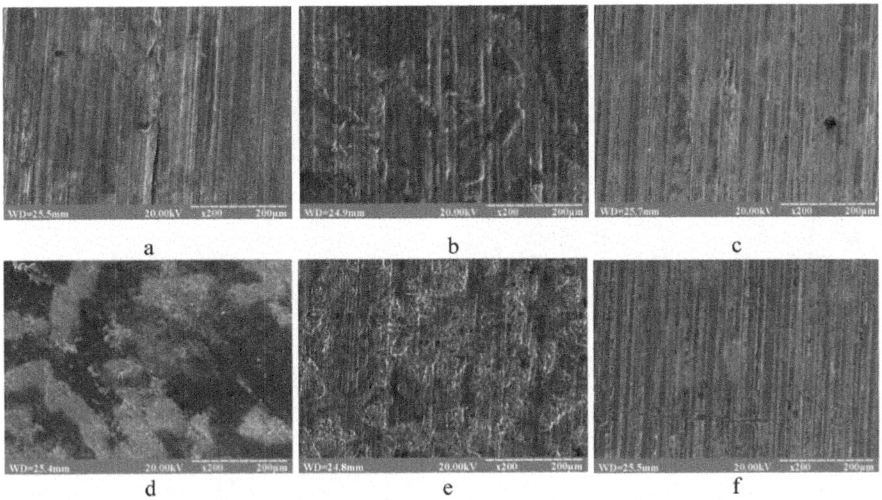

Fig. 3. Morphology of specimen surface from steel 17G1S-U (a–c) and steel 20 (d–f) under test scheme B (a, d), C (b, e) and D (c, f).

developed fairly evenly in the depressions. For steel 20, corrosion pits have different depths, indicating a significant localization of damage [15]. At the same time, an increase in the number of SRB cells was observed, which may be due to the adaptation of the SRB to the action of the toxicant (in particular, due to the attachment of microorganisms to the steel surface with the formation of a protective biofilm), which caused an intensification of the corrosive processes. A gradual increase in the number of SRB is probably due to the formation of less toxic metabolites compared with the initial inhibitor. Increasing the concentration of inhibitor 1 to 1.0% (Scheme D) resulted in the suppression of the sulfate-reducing activity, as evidenced by a change in the mechanism of corrosion. Corrosive manifestations on the specimen surface indicate the uniformity of the process, (Fig. 3c, f), and minor corrosion indicates the effectiveness of the corrosion protection. The degree of metal protection against biocorrosion was 76.60% for steel 17G1S-U, and 68.97% for steel 20.

Effect of inhibitor 2. At a concentration of 0.5%, this inhibitor almost completely suppressed the growth of SRB on the surface of steel 17G1S-U, the degree of metal protection against corrosion was 95.74%. An increase in the concentration of inhibitor 2 in the nutrient medium up to 1.0% slightly reduced the effectiveness of inhibition of the SRB growth activity, which is confirmed by a decrease in the protective effect of the inhibitor to 93.62%. The corrosive effect was insignificant. The surface of specimens from steel 17G1S-U was almost in its original state, with some corrosive microdamage, Fig. 4a, b.

For steel 20, using inhibitor 2 at a concentration of 0.5% significantly reduced the effectiveness of inhibition of the SRB growth activity. This is evidenced by the degree of metal protection against corrosion—65.52%. In the visual inspection, traces of corrosion were found on the specimen surface, Fig. 4b, however, the corrosion was evenly distributed over the surface, without any localization.

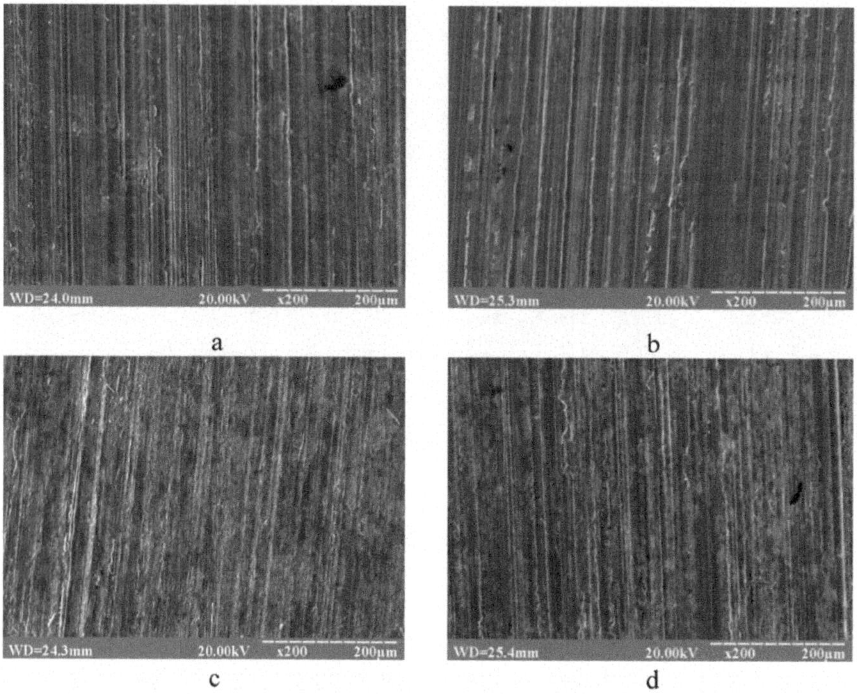

Fig. 4. Morphology of specimen surface from steel 17G1S-U (a, b) and steel 20 (c, d) under test scheme E (a, c) and F (b, d).

Increasing the concentration of inhibitor 2 to 1.0% (Scheme F) resulted in a decrease in SRB growth, and the degree of metal protection against biocorrosion increased to 82.76%. In this case, a high degree of metal protection against biocorrosion caused by the SRB is confirmed by the condition of the outer surface of the specimen, which has almost no corrosion damage, Fig. 4d.

The results obtained have shown that the maximum inhibitory effect to suppress the growth of SRB was provided by inhibitor 2 (Schemes E, F) for both steel grades studied. A visual analysis of the specimen surface after the experiments has shown that in the presence of inhibitor 2 at a concentration of 0.5 and 1.0%, no corrosion damage was detected on the steel surface. In the presence of the investigated inhibitor, there was no clouding of the Postgate's B medium, no sulfide film was formed on the steel specimens, and a gray-black precipitate dropped to the bottom of the specimens, indicating the death of a significant number of SRB cells and the slowdown of the microbial corrosion of the steel. A high degree of metal protection against biocorrosion caused by SRB indicated the bactericidal properties of the studied inhibitor. The results obtained are consistent with the results presented in [11]. In our opinion, a significant protective effect of inhibitor 2, which belongs to quaternary ammonium salts, is associated with the fact that under conditions of microbial corrosion of steels, a thin sulfide film is formed, which is coated with a monomolecular layer of the inhibitor. It stops its further growth and inhibits the anode dissolution of the metal. Inhibitor 2 can be considered as a promising component in the development of new inhibitory compositions for the protection of steels against biocorrosion caused by SRB, in particular, oil- and gas pipelines.

Taking into account the results obtained, it should be noted that the most effective method for reducing the probability of biocorrosion development in the gas transportation systems is not diagnostics and repair of pipes, but taking preventive measures that restrict the formation and growth of defects, one of the most effective of which is the use of optimal inhibitors.

5 Conclusions

The rate of biocorrosion processes is largely determined by the specific rate of growth of bacteria, from which the biofilm is formed on the surface of the gas main. Corrosion rates of specimens from steel 17G1S-U and steel 20 were detected in the presence of organic inhibitors in a sterile Postgate's B medium, which was inoculated by SRB bacteria. The impact of the inhibitor nature on the specimen protection degree under the development of anaerobic corrosion under the influence of SRB is investigated. High bactericidal properties of inhibitor 2 and a significant protective effect against microbial corrosion caused by SRB have been proved.

References

1. Andreyuk, E.I., Kozlova, I.A.: Litotrophic Bacteria and Microbiological Corrosion. Naukova Dumka, Kiev (1977). (in Russian)
2. Liu, H., Fu, C., Gu, T., Zhang, G., Lv, Y., Wang, H., Liu, H.: Corrosion behavior of carbon steel in the presence of sulfate reducing bacteria and iron oxidizing bacteria cultured in oilfield produced water. Corros. Sci. **100**, 484–495 (2015)
3. Liu, H., Cheng, F.: Mechanistic aspects of microbially influenced corrosion of X52 pipeline steel in a thin layer of soil solution containing sulphate-reducing bacteria under various gassing conditions. Corros. Sci. **133**, 178–189 (2018)
4. Usher, K.M., Kaksonen, A.H., Cole, I., Marney, D.: Critical review: microbially influenced corrosion of buried carbon steel pipes. Int. Biodeterior. Biodegradation **93**, 84–106 (2014)
5. Xu, D., Li, Y., Gu, T.: Mechanistic modeling of biocorrosion caused by biofilms of sulfate reducing bacteria and acid producing bacteria. Bioelectrochemistry **110**, 52–58 (2016)
6. Herrera, L.K., Videla, H.A.: Role of iron-reducing bacteria in corrosion and protection of carbon steel. Int. Biodeterior. Biodegradation **63**(7), 891–895 (2009)
7. Hamilton, W.A.: Sulfate-reducing bacteria and anaerobic corrosion. Ann. Rev. Microbiol. **39**, 195–217 (1985)
8. Javed, M.A., Neil, W.C., McAdam, G., Wade, S.A.: Effect of sulphate-reducing bacteria on the microbiologically influenced corrosion of ten different metals using constant test conditions. Int. Biodeterior. Biodegradation **125**, 73–85 (2017)
9. Pogrebova, I.S., Kozlova, I.A., Purish, L.M., Gerasika, S.E., Tuovinen, O.H.: Mechanism of inhibition of corrosion of steel in the presence of sulfate-reducing bacteria. Mater. Sci. **37**(5), 754–761 (2001)

10. Noor, N.M., Yahaya, N., Abdullah, A., Tahir, M.M., Sing, L.K.: Microbiologically influenced corrosion of X-70 carbon steel by Desulfovibrio vulgaris. Adv. Sci. Lett. **13**, 312–316 (2012)
11. Davey, M.E.: Microbial biofilms: from ecology to molecular genetics. Microbiol. Mol. Biol. Rev. **64**(9), 847–867 (2000)
12. Polutrenko, M.S., Maruschak, P.O., Prentkovskis, O.: The role of the biological factor in the corrosion damage of the metal of underground oil and gas pipelines. In: Proceedings of the 20-th International Conference on "Transport Means", Juodkrantė, Lithuania, 5–7 October, pp. 424–427 (2016)
13. DSTU 3999-2000. Protective coatings are polymeric, oilbituminous and coal-based. Methods of laboratory testing for bioresistance, Kyiv, "Derzhstandart Ukrainy", 16 p. (2000). (in Ukrainian)
14. Kashkovskiy, R.V., Igoshin, R.V., Sayapin, A.O.: Assessment of inhibiting features of some commercial bactericides. Int. J. Corros. Scale Inhib. **7**(2), 112–125 (2018)
15. Polutrenko, M., Maruschak, P., Tymoshenko, A., Sorochak, A.: Influence of soil microorganisms on metal corrosion of underground pipelines. Koroze a Ochrana Materiálu **62**(2), 54–59 (2018)

Sustainable Transport Interchange

Conceptual Models for Better Interoperability Between Road and Rail Transport in Lithuania

Aidas Vasilis Vasiliauskas[1(✉)], Virgilija Vasilienė-Vasiliauskienė[2], and Jolanta Sabaityte[2]

[1] The General Jonas Žemaitis Military Academy of Lithuania, Šilo Str. 5A, Vilnius LT 10322, Lithuania
aidas.vasiliauskas@mil.lt

[2] Vilnius Gediminas Technical University, Saulėtekio Str. 11, Vilnius LT 10221, Lithuania

Abstract. One of the main strategic goals declared in the common European transport policy—development of efficient, environment-friendly, sustainable transport system. The key prerequisites for the realization of this ambitious goal is development of national and international co-modal transport systems, that are based on the interoperability of different transport modes. Interoperability itself is also hard to implement without proper arrangement of legal framework, organizational measures and, finally, interconnected transport infrastructures of different transport modes (common multimodal transport network). The goal of this article is to present the results of the study aimed at identification of key shortages of transport network in Lithuania. The tasks were set in order to reach the goal: to identify the factors that led to such situation, and to discuss conceptual models that might be useful in changing current situation.

Keywords: Interoperability · Multimodality · Transport network · Transport infrastructure · Transport management

1 Introduction

The last decades of 20th and beginning of the 21st century is associated with the process of globalization that had touched different aspects of society. Globalization is associated with the emergence of world-wide market. In other words, globalization gave a chance to every country to compete honestly and secure their competitive position in these world-wide markets. The only real measure that assure competitiveness is provision of high quality services. Countries, in which industry of services takes a huge percentage share of their national economy, are far well developed, more competitive and create more work places for their inhabitants. The conclusion can be made that at current stage of world economy development, industry of services becomes one of the most important driving forces that assure economic stability and higher competitiveness of the country. One of the most important branches of service industry is transport sector. Currently transport services receive more and more attention, since scientists and business representatives comes to a common conclusion that transport is important factor determining current as well as future conditions of

I. Kabashkin et al. (Eds.): RelStat 2018, LNNS 68, pp. 541–550, 2019.
https://doi.org/10.1007/978-3-030-12450-2_52

countries economy and quality of inhabitant's life. Many studies were conducted and scientific papers were written about the quality and efficiency of transportation services. Majority of them form background for political initiatives and strategic development plans of transport systems of different countries. Lithuania is not an exemption in this respect. National transport policy strategy (which is in line with major EU Transport policy documents) anticipate that by the year 2025 Lithuania will develop modern, well-balanced, multimodal transport system. Its technical parameters, quality of offered services and level of safety will equal to those that are present in old-timer EU member countries. It is also anticipated that full realization of foreseen measures will allow Lithuanian transport sector to become important element of Southern Baltic region transport system, which serves Lithuanian and common EU needs. To reach that, key transport system development directions are nominated. The most important are: development of transport network (infrastructure), development of multimodal services, development of IT and ITS, implementation of environmental protection measures, implementation of safety measures, etc. At the same time there is a clear identification that mentioned above directions can be treated as the fields which need most attention. It should be no surprise that despite huge financial investments made and construction works carried out during the last few decades, there is still a lack of proper infrastructure links and nodes. This cause serious problems to interoperability of different transport modes, and, therefore, impede interaction between different transport modes, and creation of multimodal services (multimodal transport system) as well. That's why the main object discussed in the given article are obstacles preventing efficient interoperability between the road and rail transport modes in Lithuania. The main goal of this article is to present the results of the study aimed at identification of key shortages of transport network in Lithuania. In order to reach the goal, tasks were set: to identify factors that led to such situation and to discuss the conceptual models that might be useful in changing current situation. Article starts with short theoretical introduction discussing concept of interoperability and multimodal transport networks which supports interoperability. This is followed by the presentation of results of the study depicting status of Lithuanian transport network problems and reasons behind them. The third part of the article is dedicated to presentation of conceptual models that could be applied to eliminate identified problems. Article ends with summarizing conclusions and recommendations.

2 Multimodality and Main Elements of Multimodal Transport System

In the process of globalization rates of international trade between different countries (especially with further dislocated countries) are growing. At the same time requirements for faster delivery of products also increase. This, in turn, leads not only to greater transportation distances, but also calls for fast and reliable services. Services provided by road freight transport in most cases would be the only real option that meets mentioned above requirements. However, scarcity of energetic resources (inefficiency in terms of energy consumption) and emergence of geographical barriers (inefficiency in terms of delivery time) calls for alternative modes of transportation in global supply chains. The concept of multimodality in such circumstances naturally emerged as a potential solution [1]. This concept has a lot of different definitions. Sometimes multimodality is defined as integration of different transport modes [2]. Also, multimodality can be defined as transportation of goods by at least two different modes of transport while goods stay in the same (multimodal) loading unit during the entire process of transportation [3]. Nowadays such transportation alternative becomes more and more popular [4, 5]. Some scientists argue that multimodal services are of great importance to our society. It has a huge impact on the economy of given country or region, as well as on its sustainable development [6]. At the same time scientists agree that the main prerequisite for successful implementation of "intermodality in the reality" is adequate level of interoperability between different transport modes.

Interoperability is defined as the ability of two, or more, transport systems to operate effectively and efficiently together [7].

Three main dimensions of interoperability can be distinguished as follows: technical, organizational and juridical.

Technical interoperability is achieved when different transport systems are linked in ways which effectively and efficiently extend the network of services. Technical interoperability requires the various systems of physical infrastructure in different transport systems to interface effectively and efficiently.

Organizational interoperability occurs when different organizations are willing and able to co-operate to provide transport services.

Legislation, in particular the differences between European and national legislation, may cause impediments to juridical interoperability.

Using these dimensions of interoperability, full interoperability is achieved when transport systems are deemed to have technical, organizational and juridical interoperability [8].

Knowing the key idea of interoperability, assumption, that fully integrated and well-performing multimodal network is a main prerequisite for efficient multimodal services, can be made. Therefore, from this point onwards this article will concentrate on the technical interoperability, i.e., infrastructure and transport networks of different transport modes.

Transport infrastructure, understood as mainly including roads, railroads, airports and seaports, has often been claimed as an important determinant of productivity and economic growth. Lack of adequate transport infrastructure is a significant inhibitor to local economic performance.

The widely accepted wisdom of 'road leads to prosperity' has encouraged policy-makers to make huge investments in expanding the nation's transport network. Transport infrastructure investment is believed to have a capability of increasing productivity and fostering economic growth.

Where transport infrastructure is poor, the development of multimodal transport may not be easy. In order to be able to gain maximum benefit from multimodal transport, infrastructure that is capable assuring interoperability between different transport modes must be in place [9].

Progressive thinking governments invest into the development of multimodality through various programmes and various funds [4]. Majority of these investments are dedicated to development of infrastructure (links and nodes of different transport modes), since it is the background of transport system assuring interoperability of different transport modes [10].

United Nations Agencies, such as the Economic Commission for Europe (ECE) has also provided a framework for inter-governmental co-operation and agreement aimed at trade facilitation and integrated transport. These agreements are at the core of a simplified, normalized and harmonized European transport system. The following are a few examples:

1. European Agreement on Main International Traffic Arteries (AGR) of 1975 provides all member countries with the international legal framework for the construction and development of a harmonized international road network.
2. European Agreement on Main International Railway Lines (AGC) of 1985 provides an international legal framework for the development of a coherent international rail network with a view to facilitate and develop international rail traffic.
3. European Agreement on Important International Combined Transport Lines and Related Installations (AGTC) of 1991, provides the legal framework for the development of international combined transport infrastructure and services, particularly combined road/rail transport infrastructure and services, and for the improvement of their efficiency. The idea behind the AGTC is that door-to-door transport is only as strong as its weakest link therefore the AGTC sets minimum standards for rail lines and terminals as well as for inland waterways and terminals.
4. European Agreement on Main Inland Waterways of International Importance (AGN) of 1996, provides a legal framework for the establishment of an internationally agreed European network of inland waterways and ports, as well as the uniform infrastructure and operational parameters to which they should conform [11].

Development of national multimodal transport network starts from the integration of main links of different transport modes on the certain territory. It usually happens because of integration of rail and road transport links through the multimodal transport terminals. Therefore, density of these roads as well as number of such terminals should be sufficient [12]. Later, these national multimodal networks must be integrated with similar networks on the territories of neighboring countries, in order to cover the territory as wide as possible.

Mentioned above requirements lead to an assumption that there are several infrastructure features that determine capabilities of multimodal transport network and efficiency of multimodal transport services:

Density of road transport network;

- Density of rail lines;
- Density of multimodal terminals on a given territory;
- Capabilities of multimodal terminals.

These characteristics might serve as core measurements that define level of development of multimodal network.

3 Identification of Transport Infrastructure Problems in Lithuania

Infrastructure sets the degree of mobility. No major change in the transport sector will not be possible without an adequate network support and a more intelligent use of its. Overall, investment in transport infrastructure boosts economic growth, create wealth and jobs and promote geographical accessibility, trade and mobility. It must be planned to maximize the positive impact on economic growth, minimizing negative environmental impacts.

As it was mentioned in the introduction of this article, by the year 2025 Lithuania is expected to develop modern, well-balanced, multimodal transport system. To reach that, key transport system development directions are nominated. At the same time there is a clear identification that mentioned above directions can be treated as the fields which need most attention.

To understand present situation of all transport infrastructure elements that are necessary to assure smooth development of multimodal transport network and multimodal transport services in Lithuania, research covering this topic was carried out. The main aim of this research was to get a clear understanding of condition of infrastructure which determines capability of network to assure multimodal services. To simplify the study, only infrastructure assuring interoperability of land transport modes (i.e. road and rail) was examined. The proper questionnaire consisting of 16 questions based on the conducted literature analysis was created and distributed among the experts (total number 10) working in the position of directors and vice-directors of state enterprises responsible for the development of rail-road terminals and multimodal services. This questionnaire is surely carried out according to ethical policies.

After processing of obtained data, the following summing up results can be presented:

- All respondents agreed on the fact that Lithuania needs development of multimodal transport services (and, therefore, supporting infrastructure as well).
- Though all respondents are united and state that to develop multimodal services Lithuania needs adequate network of roads, rail lines and multimodal terminals, they also recognize that even despite the sufficient investments being made into the mentioned above elements, present condition of these elements is not acceptable to assure necessary level of interoperability.
- As the main problems that prevent development of rail-road multimodal services, experts identified the lack of terminal infrastructure and lack of professional management of transport infrastructure (Fig. 1).
- When asked about most probable reasons that led to such situation (i.e. what stands behind the two most problem aspects), experts gave several possible options, which are presented in the Table 1.

Conclusion can be made, that currently Lithuanian transport infrastructure supposed to assure multimodal rail—road transport services is not sufficient. This is mainly because of lack of multimodal transport terminals and lack of professional management of infrastructure. Also, it is worth mentioning that there are identified several important reasons that led to emergence of such problems. Therefore, elimination of these causal

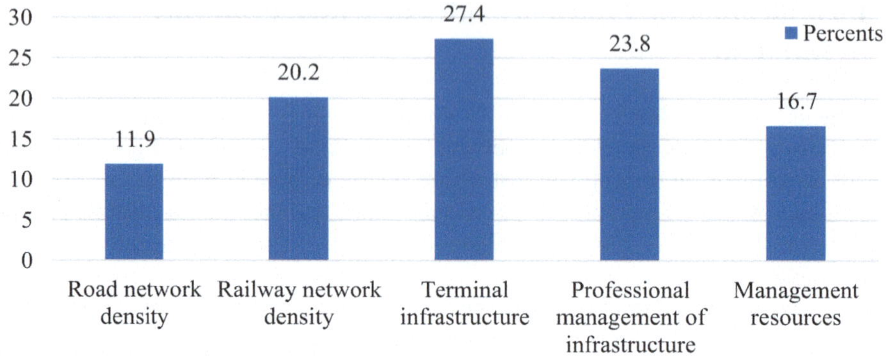

Fig. 1. Main problem aspects of multimodal transport infrastructure in Lithuania (%).

Table 1. Reasons that conditioned main identified problems.

Reasons	Problems	
	Lack of terminal infrastructure	Lack of professional management of infrastructure
1	Ignoring needs of users	Lack of strategy
2	Complicated access	Wrong allocation of funds
3	Lack of space	Lack of responsibility
4	Too complicated collaboration condition	Lack of competence
5	Wrong selection of geographical territories	Wrong selection of partnering institutions

sources could serve as starting point for the elimination of problems impeding development of multimodal services on the territory of Lithuania.

Next chapter presents conceptual model intended to eliminate identified problems (lack of terminal infrastructure and lack of professional management of transport infrastructure) and assure better interoperability between road and rail transport modes in Lithuania.

4 Identification of Transport Infrastructure Problems in Lithuania

Development of the model is based on the idea of elimination of causes that led to emergence of the main problems (lack of terminal infrastructure and lack of professional management of transport infrastructure). The main assumption here is the fact that every problem emerged due to some decisions taken in the past, i.e., the fact that there are some reasons that led to current status of the problems. These reasons were presented in the Table 1.

During the conducted study experts were asked to name possible solutions for elimination of mentioned above reasons. The summarizing answers are presented in the Table 2. These solutions are supposed to be key measures to be applied in order to change current problem situation with lack of terminal infrastructure and lack of professional management of transport infrastructure.

Table 2. Suggested solutions for elimination of the main problems.

Problems	Suggested solutions	Problems	Suggested solutions
Lack of terminal infrastructure		Lack of professional management of infrastructure	
Ignoring needs of users	Match the activities of different parties	Lack of strategy	Regulation of relationships through the improvement of services and quality
Complicated access	Application of adequate technical solutions	Wrong allocation of funds	Increase of investments
Lack of space	Development of infrastructure in accordance to market requirements	Lack of responsibility	Increase of competences of responsible persons
Too complicated collaboration condition	Simplification of collaboration agreements	Lack of competence	More careful selection of staff, evaluation of staff's competence
Wrong selection of geographical territories	Development of lacking infrastructure links	Wrong selection of partnering institutions	Application of stricter criterions for partner selection

Putting all these items into one common chart can form certain "fish-bone" diagram which could serve as a background for conceptual model. However, prior to that, certain approbation of the conceptual model was conducted in order to eliminate possible shortages and missing solutions. Method of expert evaluation was applied and 5 top level experts from the field of transport sector were selected. Processing of expert's answers provided following results:

- All 5 experts agree that suggested conceptual model could eliminate two main problems, and, also agree that application of such model could lead to solution of road-rail transport interoperability problems in Lithuania;
- 4 of 5 experts agree that solutions suggested to eliminate separate causes of identified problems are adequate and correct;
- Experts suggested several supplementary solutions that could be worth taking into account and could increase the quality of conceptual model. These additional suggestions are presented in the Table 3.

Table 3. Additional solutions suggested by the experts.

Solutions	Problems	
	Lack of terminal infrastructure	Lack of professional management of infrastructure
1	Creation and analysis of future visions	Quick adaptation to changing market conditions
2	Efficient publicity of possibilities	Motivation of employees
3	Development of e-services	Improvement of education of specialists
4	Application of best practice	Increase of competence of collaboration with partners

When these solutions are incorporated into initial models, the final two conceptual models (the first one for solving problem of lack of terminal infrastructure and the second one for solving problem of lack of professional management of transport infrastructure) may be presented (Figs. 2 and 3).

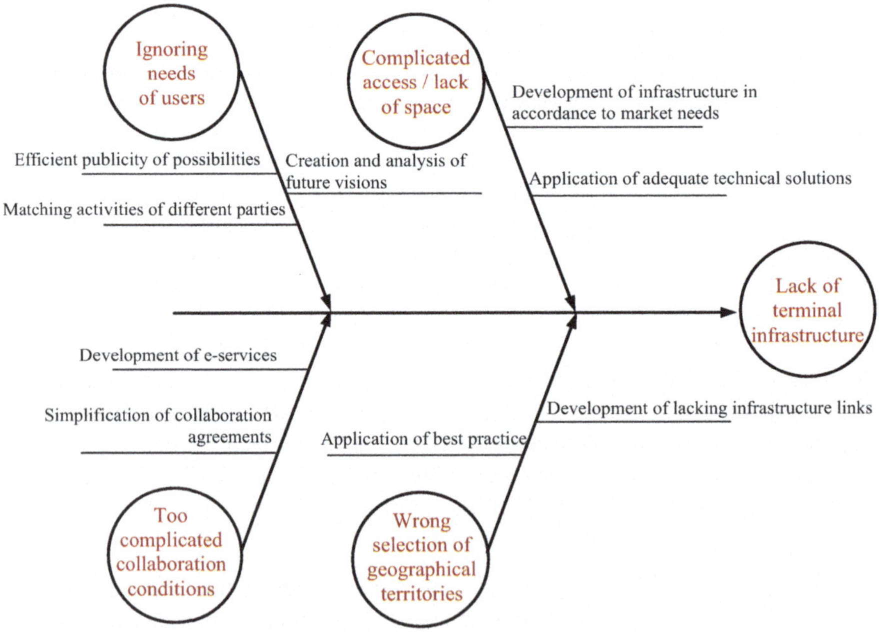

Fig. 2. Conceptual model for solving problem of lack of terminal infrastructure.

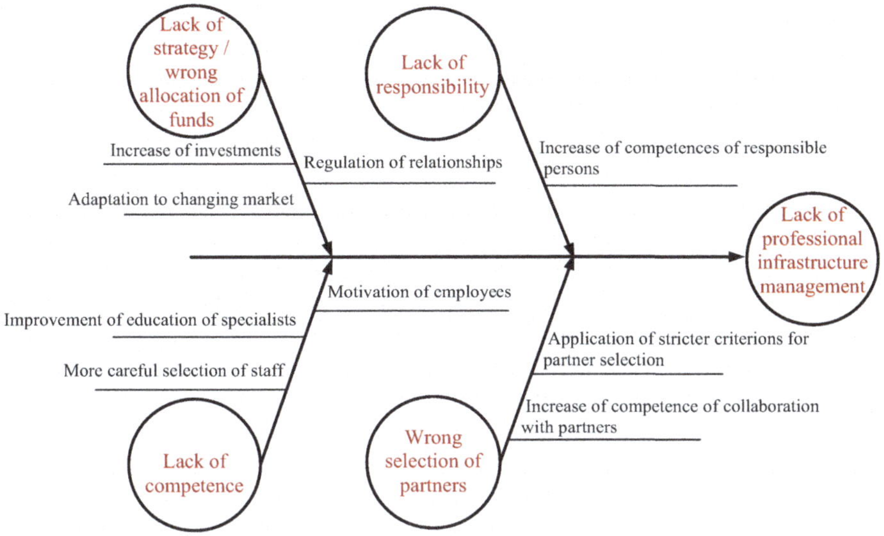

Fig. 3. Conceptual model for solving problem of lack of professional management of infrastructure.

5 Conclusions

The carried out research allowed to understand that currently transport services receive more and more attention. Based on the carried out research, the following conclusions are made:

1. It is relevant to analyze the concept of the multimodality in the context of interoperability, distinguishing main dimensions: technical, organizational and juridical. The research was concentrated on the technical interoperability, i.e. infrastructure and transport networks. The analysis of the multimodality transport support programs and various funds was made. The analysis revealed that development of national multimodal transport network starts from the integration of main links of different transport modes on the certain territory. Later, these national multimodal networks must be integrated with similar networks on the territories of neighboring countries, in order to cover the territory as wide as possible. The factors for description of multimodal transport capabilities were set.
2. The analysis of Lithuanian national transport policy strategy revealed that by the year 2025 Lithuania will develop modern, well-balanced, multimodal transport system. Key transport system development directions are nominated. The most important are: development of transport network (infrastructure), development of multimodal services, development of IT and ITS, implementation of environmental protection measures, implementation of safety measures, etc. However, despite huge financial investments made during the last few decades, there is still a lack of proper infrastructure. These cause serious problems to interoperability of different transport modes.

3. Empirical research was carried out in order to understand present condition of Lithuanian transport infrastructure. The results revealed that Lithuania needs development of multimodal transport services. Current condition of supporting transport infrastructure is not acceptable to assure necessary level of interoperability of different transport modes and efficient intermodal services. As the main problems that prevent development of interoperability between rail and road transport modes experts identified the lack of terminal infrastructure and lack of professional management of transport infrastructure.
4. Besides the identification of these two core problems, reasons that led to emergence of such situation were identified as well. Last but not least, possible solutions that may be helpful in elimination of identified reasons were discovered as well. Putting all these items into one common chart can form certain "fish-bone" diagram which could serve as a background for conceptual models dedicated to solving problem of lack of terminal infrastructure and solving problem of lack of professional management of transport infrastructure, which in turn may lead to development of necessary transport infrastructure, which, in turn, can assure better interoperability between road and rail transport, and more efficient multimodal transport services.

References

1. UNECE: Terminology of combined transport. New York and Geneva (2001)
2. Reis, V., Meier, J.F., Pace, G., Palacin, R.: Rail and multi-modal transport. J. Res. Transp. Econ. **41**, 17–30 (2013)
3. Wagener, N.: Intermodal transport in Europe—opportunities through innovation. Sci. J. Logist. **10**(4), 371–382 (2014)
4. Lowe, D.: Intermodal Freight Transport, pp. 304–306. Butterworth-Heinemann Publisher, Oxford (2005)
5. Crainic, T.G., Kim, K.H.: Intermodal Transportation. Handbook, pp. 30–140. Elsevier B.V. (2007)
6. Balint, A.O.: Connect to compete the link between intermodal transport and logistics. J. VASE Buchares **6**(1), 3–4 (2012)
7. Rodrigue, J. P., Comtois, C., Slack, B.: The geography of transport systems. Routledge Publisher, 4th ed, p. 454 (2016)
8. Tsamboulas, D.: Development strategies for intermodal transport in Europe, design and policy, pp. 2–4. Edward Elgar Publishing, Cheltenham (2008)
9. Banister, D.: Transport and economic development: reviewing the evidence. Transp. Rev. **32** (1), 1–2 (2012)
10. Beškovnik, B.: Managing and organizational changes of intermodal network in transition regions: the case of south-east Europe. J. Intereuropa, Glob. Logist. Serv. **5**(2), 38–46 (2010)
11. Muller, G.: Intermodal Freight Transportation. Eno Transportation Foundation, Inc., Washington (1999)
12. Lakshmanan, T.R.: The broader economic consequences of transport infrastructure investments. J. Transp. Geogr. **19**, 1–12 (2011)

Techniques for Smart Urban Logistics Solutions' Simulation: A Systematic Review

Ioannis Karakikes[1(✉)], Eftihia Nathanail[1], and Mihails Savrasovs[2]

[1] Department of Civil Engineering, University of Thessaly, Pedion Areos, 38334 Volos, Greece
iokaraki@uth.gr

[2] Transport and Telecommunication Institute, Lomonosova Street 1, LV-1019 Riga, Latvia

Abstract. Today, cities devise their own Sustainable Urban Logistics Plan (SULP) to improve the sustainability of their distribution system. Modern SULPs, following the development of technology, consider smart solutions e.g. pick-ups and deliveries by electric vehicles, bicycles or drones, city lockers, ITS systems for planning/routing, crowdsourcing services and other, which aim at mitigating the negative effects of the freight transport in the urban area. The effectiveness of these solutions, however, is not for sure, since their performance relies on particularities of cities' urban freight transport system as well as the level of infrastructure, cooperation and policy adoption. To better understand and assess the impacts of a proposed solution in a city context, ex-ante evaluation through modeling is advised.

This study synthesizes the types of simulation techniques that are used to model the impacts of innovative smart urban freight solutions. A systematic literature review was performed in Web of Science Core Collection, SCOPUS and JSTOR databases to identify records that tackle with modeling smart urban freight solutions and present real case study results. Having gathered all relevant records through a query-based identification process, a screening process was adopted to keep only those that have an essential contribution to the topic. Eighty-two full papers met the criteria and were included in the qualitative analysis. Analysis' key findings were that (1) the majority of studies use custom-made techniques for the evaluation of urban freight solutions, (2) there is growing tendency from 2015 onwards for such studies and, (3) "ITS for freight monitoring and planning/routing" is the most prominent solution in such studies.

Keywords: City logistics · Last mile distribution · Evaluation · Modeling · PRISMA

1 Background

Urban freight transport is a main component of sustainable transport networks and one of the main contributors on traffic congestion and environmental pollution in the cities. Trends as urbanization, consumerism and the ever-increasing e-commerce cause a vast demand of products and services and make the distribution of goods within urban areas an essential priority for public authorities as policy makers, while it affects greatly all

I. Kabashkin et al. (Eds.): RelStat 2018, LNNS 68, pp. 551–561, 2019.
https://doi.org/10.1007/978-3-030-12450-2_53

involved stakeholders (shippers, carriers and consumers). This increased demand of goods renders the upkeep of a high traffic and living quality in cities a challenging process.

Over the last decades, many smart logistics solutions have been developed to allay cities' problems related to distribution of goods. These solutions in some cases have been modeled to enable ex-ante evaluation before their actual implementation.

This specific study is a continuation of the work by Karakikes and Nathanail [1], with a focus on the identification of available simulation models in the evaluation of urban freight solutions. Specifically, the objective of this study is to provide a systematic literature review to answer the following research question:

RQ: What are the most prevalent simulation techniques regarding the evaluation of innovative urban freight solutions?

2 Methods

2.1 Search Strategy

To answer the research question, the Web of Science Core Collection, SCOPUS and JSTOR were searched to retrieve articles focusing on simulation in urban logistics after year 2003. This period of almost 16 years was selected to guarantee the innovation of the modeled solutions. The selection of these three databases was made as all contain core journals and conference proceedings related to urban freight transportation. The papers were screened following the Preferred Reporting Items for Systematic Reviews and Meta-Analysis (PRISMA) guidelines.

The keywords were formulated by the authors to identify papers on relevant simulation techniques on urban logistics solutions. Keywords such as "city logistics simulation", "urban freight simulation" and "last mile simulation" specify the issues addressed.

2.2 Paper Inclusion Criteria

The criteria for inclusion in this review were that studies addressed the research question, meaning that the paper should employ at least one simulation technique for the evaluation of minimum one city logistics solution. Additional filters imposed were the English language and the type of the document. The included document types were the (1) Journal, (2) Conference Proceedings and (3) Book chapter, see Table 1.

3 Results

Initially, 623 records were spotted (Fig. 1). 19 of these records were overlapped or were registered two or three times in the same database (in a separated conference and journal publication). Out of the 604 remaining records, 444 were excluded as the length of the full text was short (less than 4 pages) or irrelevant to the research question set. The screening process at this stage was performed following the steps below:

Table 1. Queries used for the different databases.

Database	Search queries
Scopus	TITLE-ABS-KEY (urban AND freight AND simulation) OR TITLE-ABS-KEY (city AND logistics AND simulation) OR TITLE-ABS-KEY (last AND mile AND simulation) AND (LIMIT-TO (PUBYEAR, 2019) OR LIMIT-TO (PUBYEAR, 2018) OR LIMIT-TO (PUBYEAR, 2017) OR LIMIT-TO (PUBYEAR, 2016) OR LIMIT-TO (PUBYEAR, 2015) OR LIMIT-TO (PUBYEAR, 2014) OR LIMIT-TO (PUBYEAR, 2013) OR LIMIT-TO (PUBYEAR, 2012) OR LIMIT-TO (PUBYEAR, 2011) OR LIMIT-TO (PUBYEAR, 2010) OR LIMIT-TO (PUBYEAR, 2009) OR LIMIT-TO (PUBYEAR, 2008) OR LIMIT-TO (PUBYEAR, 2007) OR LIMIT-TO (PUBYEAR, 2006) OR LIMIT-TO (PUBYEAR, 2005) OR LIMIT-TO (PUBYEAR, 2004) OR LIMIT-TO (PUBYEAR, 2003)) AND (LIMIT-TO (LANGUAGE, "English")) AND (EXCLUDE (SUBJAREA, "PHYS") OR EXCLUDE (SUBJAREA, "MATE") OR EXCLUDE (SUBJAREA, "MEDI") OR EXCLUDE (SUBJAREA, "ART") OR EXCLUDE (SUBJAREA, "AGRI") OR EXCLUDE (SUBJAREA, "BIOC") OR EXCLUDE (SUBJAREA, "ARTS") OR EXCLUDE (SUBJAREA, "ECON") OR EXCLUDE (SUBJAREA, "PSYC") OR EXCLUDE (SUBJAREA, "CHEM") OR EXCLUDE (SUBJAREA, "CENG") OR EXCLUDE (SUBJAREA, "NEUR") OR EXCLUDE (SUBJAREA, "PHAR") OR EXCLUDE (SUBJAREA, "NURS") OR EXCLUDE (SUBJAREA, "IMMU")) AND (EXCLUDE (DOCTYPE, "cr") OR EXCLUDE (DOCTYPE, "re")) AND (EXCLUDE (SUBJAREA, "MATH")) AND (LIMIT-TO (LANGUAGE, "English")) AND (EXCLUDE (LANGUAGE, "Spanish") OR EXCLUDE (LANGUAGE, "Polish"))
Web of science	TITLE: (urban freight simulation) OR TITLE: (city logistics simulation) OR TITLE: (last mile simulation) AND YEAR PUBLISHED: (2003–2018) AND LANGUAGE: (English) AND DOCUMENT TYPES: (Article OR Book Chapter OR Proceedings Paper) Timespan: 2003–2018. Indexes: SCI-EXPANDED, SSCI, A&HCI, ESCI
JSTOR	ti: ("urban freight simulation" OR "city logistics simulation" OR "last mile simulation") AND y: (2003–2018)

1. Search records' titles that contain the words: "bandwidth", "mesh networks", "WiMAX", "optical augmentation" and "Gateway" and exclude them, as they were captured wrongly under the keyword "last mile simulation".
2. Search line by line for records in irrelevant domains.
3. Read remaining records' abstracts and exclude those referring to:

- Optimization models regarding the development or testing of new approaches (i.e. algorithms) without showcasing a real case study.
- Extreme logistics solutions in special events (natural disaster, earthquakes, tsunami etc.).

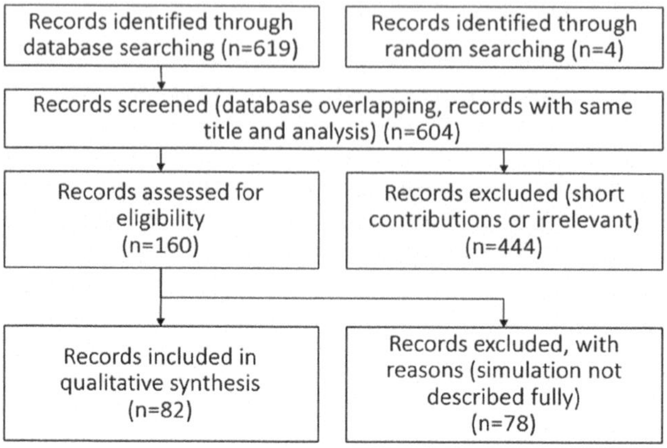

Fig. 1. PRISMA flow diagram of assessment procedure and results: number of records included and excluded and reasons.

From the 160 eligible records 82 were the essential and included in the qualitative analysis, mainly considering the completeness of the simulation process description. The structure of the review results is shaped in three columns which describe the simulation technique, the logistics solution and the reference. The considered solutions can be found in Table 2 and the catalogue of records included in the qualitative analysis in Table 3. It has to be noted that four more emerging solutions have been added to the number of considered solutions stemming from NOVELOG project (http://novelog.eu/): (23), (24), (25) and (26) (see Table 3).

Table 2. Smart logistics solutions according to NOVELOG project [2].

Urban freight solutions	
(1) Multimodality for urban freight	(13) Access by load factor
(2) Urban consolidation centres	(14) Multi-users lanes
(3) Trans-shipment facilities	(15) Enforcement and ITS adoption for control and traffic management
(4) ITS for freight monitoring and planning/routing	(16) Businesses recognition scheme
(5) Home deliveries system	(17) Public transport indirect promotion for shopping
(6) E-commerce system for small shops	(18) Urban planning measures
(7) Cargo bikes for B2B and B2C	(19) Harmonization and simplification of city logistics rules
(8) Electric vehicles diffusion in businesses (zero-emission transport)	(20) Off peak deliveries
(9) Reverse logistics integration into supply chain	(21) Public transport for freight
(10) City lockers	(22) Freight travel plans
(11) Loading/Unloading areas and parking	(23) Crowdsourcing
(12) Access: time windows, emission zones	(24) Autonomous vehicles
	(25) Drone deliveries
	(26) 3D-printing

Table 3. Catalogue of records.

Technique	Solution (solution no.): times that this solution is met in such studies	Reference
Systems Dynamics (SD)	(6): 1, (12): 1, (13): 1, (15): 1, (16): 1	[3, 4]
Agent-based Simulation (ABS)	(1): 1, (2): 1, (4): 3, (7): 1, (8): 1, (11): 3, (12): 2, (16): 1, (23): 1	[5–16]
Traffic simulation (time discrete or continuous)	(2): 5, (3): 1, (4): 1, (7): 2, (8): 3, (10): 1, (11): 4, (12): 2, (15): 1, (19): 1, (24): 1	[17–34]
Discrete-Event Simulation (DES)	(1): 1, (2): 1, (4): 5, (8): 1	[35–39]
Hybrid modeling	(4): 4, (8): 1, (9): 1, (11): 1, (12): 1, (13): 1, (20): 1	[40–45]
Misc. (custom-made)	(1): 7, (2): 4, (3): 1, (4): 10, (5): 2, (6): 1, (7): 1, (8): 2, (10): 1, (11): 5, (12): 2, (14): 1, (15): 2, (18): 1, (20): 1, (23): 1, (24): 1	[46–84]

Note 1 Some logistics solutions have been adjusted to achieve a common nomenclature.
Note 2 All identified records ($n = 623$) can be found here: http://alliance-project.eu/wp-content/uploads/2018/09/Simulation-Techniques-Repository.xlsx.

Custom made simulation models have been the leading simulation technique over the last 15 years (Fig. 2). Traffic simulation, agent based simulation and discrete event simulation are also techniques that have been used frequently over the last year, whereas few hybrid models have been detected mainly during the last two years. It is worth noting here that only after 2010 we see a clear trend for using simulation for the evaluation of real case studies in urban freight transport. This can be attributed on the one hand to the data provision and acquisition easiness level which is growing year by year and the development of simulation domain (user friendly interfaces, computers' growing processing capacity, etc.) and on the other hand to the better structure of literature databases.

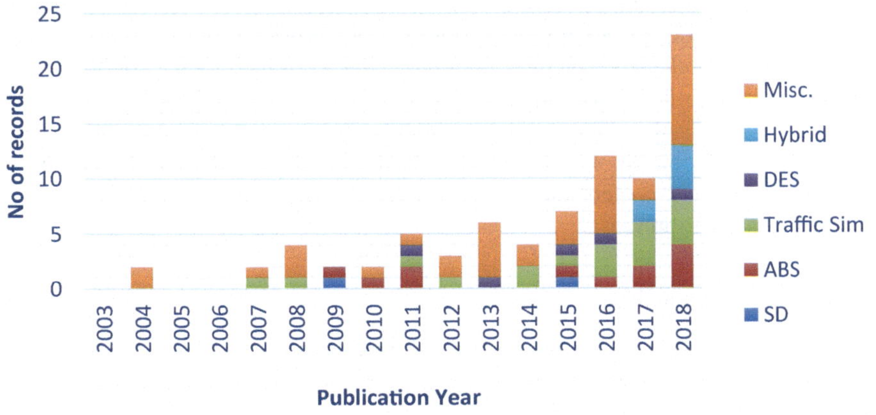

Fig. 2. Number of records per simulation technique and publication year.

"ITS for freight monitoring and planning/routing" was the most prominent urban freight solution of all (Fig. 3). This can be attributed to the high number of full papers testing customized simulation-optimization algorithms for dynamic routing of freight vehicles, stating that this could significantly impact the negative effects induced by urban freight transport. "Loading/Unloading areas and parking", "Urban consolidation centers", "Multimodality for urban freight" and "Access: time windows and emission zones" complement the "top-5" urban freight solutions modeled mainly by miscellaneous methods.

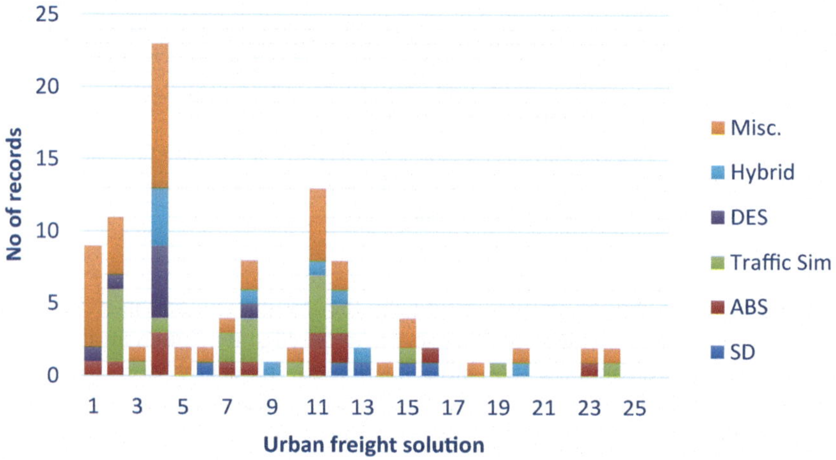

Fig. 3. Number of urban freight solutions per simulation technique and publication year.

4 Discussion

4.1 Limitations

This systematic review has specific limitations. Firstly, the keywords were formulated by the authors in order to capture all relevant studies, however the spotted records may not depict a comprehensive picture. Secondly, the examined years 2003–2018 may leave out some significant contributions. However, as mentioned earlier, the scope of this review was to gather innovative solutions in the sense that innovation is inextricably connected with the latest years of technological development. Thirdly, the exclusion of abstracts, posters and other document types (long reviews) may also leave out significant contributions.

4.2 General Conclusions

The overview of this paper indicates that the evaluation of urban freight solutions through modeling is gaining year by year more attention, which is evident based on the actions and initiatives taken by European Commission in the field. Especially, the last five years, researchers and practitioners tend to publish more their studies and custom made models. This trend can assist stakeholders to learn from others' experiences and understand how new and innovative urban freight solutions can contribute to their urban transport system.

Another general conclusion is that ITS and cooperative oriented solutions are clearly in the top preferences of city logistics studies. Spatial restrictions and economic sustainability encourage the optimization of the current network rather than small infrastructural interventions.

Acknowledgements. This work has been supported by the ALLIANCE project (http://alliance-project.eu/) and has been funded within the European Commission's H2020 Programme under contract number 692426. This paper expresses the opinions of the authors and not necessarily those of the European Commission. The European Commission is not liable for any use that may be made of the information contained in this paper.

References

1. Karakikes, I., Nathanail, E.: Simulation techniques for evaluating smart logistics solutions for sustainable urban distribution. Proc. Eng. **178**, 569–578 (2017). Elsevier. 16th International Conference Reliability and Statistics in Transportation and Communication, RelStat 2016, Riga, Latvia (2017). http://dx.doi.org/10.1016/j.proeng.2017.01.110
2. NOVELOG: Deliverable D3.2. Multi stakeholder multi criteria decision making tool (2016)
3. Tan, C.K., Blanco, E.E.: System dynamics modeling of the SmartWay transport partnership. In: Second International Symposium on Engineering Systems. MIT, Cambridge, 15–17 June 2009
4. Qiu, Y., Shi, X., Shi, C.: A system dynamics model for simulating the logistics demand dynamics of metropolitans: a case study of Beijing, China. J. Ind. Eng. Manag. JIEM **8**(3), 783–803 (2015). https://doi.org/10.3926/jiem.1325
5. Schroder, S., Dabidian, P., Liedtke, G.: A conceptual proposal for an expert system to analyze smart policy options for urban CEP transports. In: 2015 Smart Cities Symposium Prague, SCSP 2015 (2015)
6. Fikar, C.: A decision support system to investigate food losses in e-grocery deliveries. Comput. Ind. Eng. **117**, 282–290 (2018)
7. Alho, A., Bhavathrathan, B.K., Stinson, M., Gopalakrishnan, R., Le, D., Ben-Akiva, M.: A multi-scale agent-based modelling framework for urban freight distribution. Transp. Res. Proc. **27**, 188 (2017)
8. Bean, W.L., Joubert, J.W.: A systematic evaluation of freight carrier response to receiver reordering behaviour. Comput. Ind. Eng. **124**, 207–219 (2018)
9. Baindur, D., Viegas, J.M.: An agent based model concept for assessing modal share in inter-regional freight transport markets. J. Transp. Geogr. **19**(6), 1093–1105 (2011)
10. Chen, P., Chankov, S.M.: Crowdsourced delivery for last-mile distribution: an agent-based modelling and simulation approach. In: IEEE International Conference on Industrial Engineering and Engineering Management, p. 1271 (2018)
11. Yang, H., Sun, L., Lan, S., Yang, C.: Freight group behavior under freight traffic restriction policy: the case of Beijing city distribution. Ind. Manag. Data Syst. **117**(10), 2287–2304 (2017)
12. Boussier, J.-M., Cucu, T., Ion, L., Estraillier, P., Breuil, D.: Goods distribution with electric vans in cities: towards an agent-based simulation. In: 24th International Battery, Hybrid and Fuel Cell Electric Vehicle Symposium and Exhibition 2009, EVS 24, p. 1851 (2009)

13. Sopha, B.M., Siagian, A., Asih, A.M.S.: Simulating dynamic vehicle routing problem using agent-based modeling and simulation. In: IEEE International Conference on Industrial Engineering and Engineering Management, p. 1335 (2016)
14. Boussier, J.-M., Cucu, T., Ion, L., Breuil, D.: Simulation of goods delivery process. Int. J. Phys. Distrib. Logist. Manag. **41**(9), 913–930 (2011)
15. Kin, B., Ambra, T., Verlinde, S., Macharis, C.: Tackling fragmented last mile deliveries to nanostores by utilizing spare transportation capacity—a simulation study. Sustainability (Switzerland) **10**(3), 653 (2018)
16. Wang, H., Winter, S.: Utilizing taxi empty cruise time to solve the short distance trip problem. In: 17th ITS World Congress (2010)
17. Haas, I., Friedrich, B.: An autonomous connected platoon-based system for city-logistics: development and examination of travel time aspects. Transportmetrica A: Transp. Sci. (2018). https://doi.org/10.1080/23249935.2018.1494221
18. Singhania, V.R., Marinov, M.: An event-based simulation model for analysing the utilization levels of a railway line in urban area. Promet—Traffic—Traffico **29**(5), 521–528 (2017)
19. Lee, G., You, S.I., Ritchie, S.G., Saphores, J.-D., Jayakrishnan, R., Ogunseitan, O.: Assessing air quality and health benefits of the Clean Truck Program in the Alameda corridor, CA. Transp. Res. Part A: Policy Pract. **46**(8), 1177–1193 (2012)
20. Marquez, L., Salim, V.: Assessing impacts of urban freight measures on air toxic emissions in Inner Sydney. Environ. Model Softw. **22**(4), 515–525 (2007)
21. Lopez, C., Gonzalez-Feliu, J., Chiabaut, N., Leclercq, L.: Assessing the impacts of goods deliveries' double line parking on the overall traffic under realistic conditions. In: 6th International Conference on Information Systems, Logistics and Supply Chain, ILS 2016 (2016)
22. Sárdi, D.L., Bóna, K.: Developing a mesoscopic simulation model for examination of freight traffic of shopping malls in Budapest. In: 2017 Smart Cities Symposium Prague, SCSP 2017—IEEE Proceedings (2017)
23. Aditjandra, P.T., Galatioto, F., Bell, M.C., Zunder, T.H.: Evaluating the impacts of urban freight traffic: application of micro-simulation at a large establishment. Eur. J. Transp. Infrastruct. Res. **16**(1), 4–22 (2016)
24. Melo, S., Baptista, P.: Evaluating the impacts of using cargo cycles on urban logistics: integrating traffic, environmental and operational boundaries. Eur. Transp. Res. Rev. **9**(2), 30 (2017)
25. Holguín-Veras, J., Sánchez-Díaz, I.: Freight demand management and the potential of receiver-led consolidation programs. Transp. Res. Part A: Policy Pract. **84**, 109–130 (2016)
26. Alho, A.R., de Abreu e Silva, J., de Sousa, J.P., Blanco, E.: Improving mobility by optimizing the number, location and usage of loading/unloading bays for urban freight vehicles. Transp. Res. Part D: Transp. Environ. **61**, 3–18 (2018)
27. Bhuiyan, M.F.H., Awasthi, A., Wang, C.: Investigating the impact of access-timing-sizing regulations on urban logistics. Int. J. Logist. Syst. Manag. **20**(2), 216–238 (2015)
28. Zhang, L., Matteis, T., Thaller, C., Liedtke, G.: Simulation-based assessment of cargo bicycle and pick-up point in urban parcel delivery. Proc. Comput. Sci. **130**, 18 (2018)
29. Vonolfen, S., Affenzeller, M., Beham, A., Wagner, S., Lengauer, E.: Simulation-based evolution of municipal glass-waste collection strategies utilizing electric trucks. In: Proceedings of the 3rd IEEE International Symposium on Logistics and Industrial Informatics, LINDI 2011, p. 177 (2011)
30. Marcucci, E., Danielis, R.: The potential demand for a urban freight consolidation centre. Transportation **35**(2), 269–284 (2008)
31. Magniol, S., Lopez, C., Gonzalez-Feliu, J., Chiabaut, N., Leclercq, L.: The searching time to measure the freight loading zone accessibility using microscopic traffic simulation. In:

Proceedings of the Information Systems, Logistics and Supply Chain, ILS 2018, p. 406 (2018)
32. Aschauer, G.J., Starkl, F.: Time, cost and carbon dioxide benefits-rescheduling urban freight operations. Proc. Inst. Civil Eng.: Transp. **167**(6), 393–399 (2014)
33. Nourinejad, M., Wenneman, A., Habib, K.N., Roorda, M.J.: Truck parking in urban areas: application of choice modelling within traffic microsimulation. Transp. Res. Part A: Policy Pract. **64**, 54–64 (2014)
34. Karakikes, I., Mitropoulos, L., Savrasovs, M.: Evaluating smart urban freight solutions using microsimulation. In: Kabashkin, I., Yatskiv, I., Prentkovskis, O. (eds.) Reliability and Statistics in Transportation and Communication, RelStat 2017. Lecture Notes in Networks and Systems, vol. 36. Springer, Cham (2018). https://doi.org/10.1007/978-3-319-74454-4_53
35. Gattuso, D., Cassone, G.C.: A statistical analysis for micro-simulation of UDC operativity. Proc. Eng. **21**, 114 (2011)
36. Fatnassi, E., Chaouachi, J.: Discrete event simulation of loading unloading operations in a specific intermodal transportation context (2016)
37. Lebeau, P., Macharis, C., van Mierlo, J., Maes, G.: Implementing electric vehicles in urban distribution: a discrete event simulation. World Electr. Veh. J. **6**(1), 38–47 (2013)
38. Makhloufi, R., Cattaruzza, D., Meunier, F., Absi, N., Feillet, D.: Simulation of mutualized urban logistics systems with real-time management. Transp. Res. Proc. **6**, 365 (2015)
39. Behiri, W., Belmokhtar-Berraf, S., Chu, C.: Urban freight transport using passenger rail network: scientific issues and quantitative analysis. Transp. Res. Part E: Logist. Transp. Rev. **115**, 227–245 (2018)
40. Costa, Y., Duarte, A., Sarache, W.: A decisional simulation-optimization framework for sustainable facility location of a biodiesel plant in Colombia. J. Clean. Prod. **167**, 174–191 (2018)
41. Simoni, M.D., Claudel, C.G.: A fast simulation algorithm for multiple moving bottlenecks and applications in urban freight traffic management. Transp. Res. Part B: Methodol. **104**, 238–255 (2017)
42. Simoni, M., Claudel, C.: A simulation framework for modeling urban freight operations impacts on traffic networks. Simul. Model. Pract. Theory **86**, 1–204 (2018)
43. Elia, V., Gnoni, M.G., Tornese, F.: Improving logistic efficiency of WEEE collection through dynamic scheduling using simulation modeling. Waste Manag. **72**, 78–86 (2018)
44. Marcucci, E., Le Pira, M., Gatta, V., Inturri, G., Ignaccolo, M., Pluchino, A.: Simulating participatory urban freight transport policy-making: accounting for heterogeneous stakeholders' preferences and interaction effects. Transp. Res. Part E: Logist. Transp. Rev. **103**, 69–86 (2017)
45. Karakikes, I., Hofmann, W., Mitropoulos, L., Savrasovs, M.: Evaluation of freight measures by integrating simulation tools: the case of Volos Port, Greece. Transp. Telecommun. J. **19** (3), 224–232 (2018). https://doi.org/10.2478/ttj-2018-0019
46. Lin, X., Chen, Y.-H., Zhen, L., Jin, Z.-H., Bian, Z.: A crowdsourcing matching and pricing strategy in urban distribution system (2018)
47. Pan, S., Chen, C., Zhong, R.Y.: A crowdsourcing solution to collect e-commerce reverse flows in metropolitan areas. IFAC-PapersOnLine **28**(3), 1984–1989 (2015)
48. Zhao, Y., Ioannou, P.A., Dessouky, M.M.: A hierarchical co-simulation optimization control system for multimodal freight routing. In: Proceedings of the IEEE Conference on Intelligent Transportation Systems, ITSC, p. 1 (2018)
49. Barceló, J., Grzybowska, H., Orozco, J.A.: A simulation based decision support system for city logistics applications. In: 15th World Congress on Intelligent Transport Systems and ITS America Annual Meeting 2008, p. 377 (2008)

50. Gonçalves, M., Jiménez-Guerrero, P., Baldasano, J.M.: Air quality management strategies in large cities: Effects of changing the vehicle fleet composition in Barcelona and Madrid greater areas (Spain) by introducing natural gas vehicles (2008)
51. Iwan, S., Kijewska, K., Johansen, B.G., Eidhammer, O., Małecki, K., Konicki, W., Thompson, R.G.: Analysis of the environmental impacts of unloading bays based on cellular automata simulation. Transp. Res. Part D: Transp. Environ. **61**, 104–117 (2018)
52. Motraghi, A., Marinov, M.: Analysis of urban freight by rail using event based simulation. Simul. Model. Pract. Theory **25**, 73–89 (2012)
53. Oliveira, R.L., Lima, R.S., Lima, J.P.: Arc routing using a geographic information system: application in recyclable materials selective collection (2014)
54. Hassall, K.: Bi-modal terminals—shrinking urban freight exposure through a quantum leap in freight productivity. In: 33rd Australasian Transport Research Forum, ATRF 2010 (2010)
55. van Duin, J.H.R., Kortmann, R., van den Boogaard, S.L.: City logistics through the canals? A simulation study on freight waterborne transport in the inner-city of Amsterdam. Int. J. Urban Sci. **18**(2), 186–200 (2014). https://doi.org/10.1080/12265934.2014.929021
56. Comi, A., Rosati, L.: CLASS: A DSS for the analysis and the simulation of urban freight systems. Transp. Res. Proc. **5**, 132 (2015)
57. Durand, B., Mahjoub, S., Senkel, M.-P.: Delivering to urban online shoppers: the gains from "last-mile" pooling. Supply Chain Forum **14**(4), 22–31 (2013)
58. Gonzalez-Feliu, J., Morana, J., Grau, J.-M.S., Ma, T.-Y.: Design and scenario assessment for collaborative logistics and freight transport systems. Int. J. Transp. Econ. **40**(2), 207–240 (2013)
59. Zeimpekis, V., Giaglis, G.M., Minis, I.: Development and evaluation of an intelligent fleet management system for city logistics. In: Proceedings of the Annual Hawaii International Conference on System Sciences (2008)
60. Letnik, T., Farina, A., Mencinger, M., Lupi, M., Božičnik, S.: Dynamic management of loading bays for energy efficient urban freight deliveries. Energy **159**, 916–928 (2018)
61. Schau, V., Apel, S., Gebhardt, K., Kretzschmar, J., Stolcis, C., Mauch, M., Buchholz, J.: ICT for urban area logistics with electric vehicles compared within simulated and real environments (2017)
62. Chen, D., Ignatius, J., Sun, D., Goh, M., Zhan, S.: Impact of congestion pricing schemes on emissions and temporal shift of freight transport. Transp. Res. Part E: Logist. Transp. Rev. **118**, 77–105 (2018)
63. Greasley, A., Assi, A.: Improving last mile delivery performance to retailers in hub and spoke distribution systems. J. Manuf. Technol. Manag. **23**(6), 794–805 (2012)
64. Muñuzuri, J., Cuberos, M., Abaurrea, F., Escudero, A.: Improving the design of urban loading zone systems. J. Transp. Geogr. **59**, 1–13 (2017)
65. Schau, V., Apel, S., Gebhardt, K., Kretzschmar, J., Stolcis, C., Mauch, M., Buchholz, J.: Intelligent infrastructure for last-mile and short-distance freight transportation with electric vehicles in the domain of smart city logistic. In: Proceedings of the 2nd International Conference on Vehicle Technology and Intelligent Transport Systems, VEHITS 2016, p. 149 (2016)
66. Taniguchi, E., Shimamoto, H.: Intelligent transportation system based dynamic vehicle routing and scheduling with variable travel times. Transp. Res. Part C: Emerg. Technol. **12** (3–4), 235–250 (2004). Special Issue
67. Pinto, R., Golini, R., Lagorio, A.: Loading/unloading lay-by areas location and sizing: a mixed analytic-Monte Carlo simulation approach. IFAC-PapersOnLine **49**(12), 961–966 (2016)

68. Sárdi, D.L., Bóna, K.: Macroscopic simulation model of a multi-stage, dynamic cargo bike-based logistics system in the supply of shopping malls in Budapest. In: 2018 Smart Cities Symposium Prague, SCSP 2018, p. 1 (2018)
69. Ramsay, E.D., Bunker, J.M.: Management of competing demands on urban freight corridors. Road Transp. Res. **16**(3), 3–15 (2007)
70. Crainic, T.G., Errico, F., Rei, W., Ricciardi, N.: Modeling demand uncertainty in two-tier city logistics tactical planning. Transp. Sci. **50**(2), 559–578 (2016)
71. Nuzzolo, A., Comi, A.: Modelling the demand for rail in an urban context: some methodological aspects. In: European Transport—Trasporti Europei, no. 57 (2015)
72. Abadi, A., Ioannou, P., Dessouky, M.M.: Multimodal dynamic freight load balancing. IEEE Trans. Intell. Transp. Syst. **17**(2), 356–366 (2016)
73. Kay, M.G., Jain, A.: Pricing a public logistics network. In: IIE Annual Conference and Exhibition 2004, p. 439 (2004)
74. Muñuzuri, J., Cortés, P., Onieva, L., Guadix, J.: Simulating the effects of pedestrianisation on urban freight deliveries. In: European Transport—Trasporti Europei, no. 54 (2013)
75. Arnold, F., Cardenas, I., Sörensen, K., Dewulf, W.: Simulation of B2C e-commerce distribution in Antwerp using cargo bikes and delivery points. Eur. Transp. Res. Rev. **10**(1), 2 (2018)
76. Perboli, G., Rosano, M., Saint-Guillain, M., Rizzo, P.: Simulation-optimisation framework for City Logistics: an application on multimodal last-mile delivery. IET Intell. Transp. Syst. **12**(4), 262–269 (2018)
77. Fonseca, A.G., Oliveira, R.L., Lima, R.S.: Structuring reverse logistics for waste cooking oil with geographic information systems. In: Proceedings of CUPUM 2013: 13th International Conference on Computers in Urban Planning and Urban Management—Planning Support Systems for Sustainable Urban Development, p. 1 (2013)
78. Rizet, C., Cruz, C., Vromant, M.: The constraints of vehicle range and congestion for the use of electric vehicles for urban freight in France. Transp. Res. Proc. **12**, 500 (2016)
79. Moen, O.: The five-step model—procurement to increase transport efficiency for an urban distribution of goods. Transp. Res. Proc. **12**, 861 (2016)
80. Holguín-Veras, J., Hodge, S., Wojtowicz, J., Singh, C., Wang, C., Jaller, M., Aros-Vera, F., Ozbay, K., Weeks, A., Replogle, M., Ukegbu, C., Ban, J., Brom, M., Campbell, S., Sanchez-Díaz, I., González-Calderón, C., Kornhauser, A., Simon, M., McSherry, S., Rahman, A., Encarnación, T., Yang, X., Ramírez-Ríos, D., Kalahashti, L., Amaya, J., Silas, M., Allen, B., Cruz, B.: The New York city off-hour delivery program: a business and community-friendly sustainability program. Interfaces **48**(1), 70–86 (2018)
81. Agatz, N., Campbell, A., Fleischmann, M., Savelsbergh, M.: Time slot management in attended home delivery. Transp. Sci. **45**(3), 435–449 (2011)
82. Yu, J.J.Q.: Two-stage request scheduling for autonomous vehicle logistic system. In: IEEE Transactions on Intelligent Transportation Systems (2018). https://doi.org/10.1109/TITS.2018.2849091
83. Wang, X., Shang, Y.: Urban freight service capacity dynamic coordination system. Inf. Technol. J. **12**(20), 5589–5594 (2013)
84. Russo, F., Comi, A.: Urban freight transport planning towards green goals: synthetic environmental evidence from tested results. Sustainability (Switzerland) **8**(4), 381 (2016)

Possible Consequences of the Implementation of Transport Integration in the Riga Planning Region

Julius Uhlmann(✉)

Bauhaus-Universität Weimar, Marienstraße 13D, 99423 Weimar, Germany
julius.uhlmann@uni-weimar.de

Abstract. This paper deals with the question of how measures of transport integration can benefit the public transport system in the Latvian capital Riga and its surrounding area. The public transport system is currently highly fragmented with a variety of operators which lack tariff and timetable coordination. Two concepts with different levels of transport integration are developed and evaluated for their consequences on various fields. The consequences are expected to unfold in slightly higher transport volumes and higher customer satisfaction. While the costs of organization are likely to decrease due to centralization, new integrated tariff models might reduce the revenues. Measurable environmental effects will not unfold in a short period.

Keywords: Public transport · Transport integration · Public Transport Authority

1 Introduction

1.1 Background and Motivation

The Latvian capital Riga and surrounding area together form the economic and political centre of the Republic of Latvia and the biggest urban agglomeration in the Baltic states. It faces some distinct challenges in terms of public transport (PT).

The years after Latvia regained its independence in 1991 were characterized by a modernization of the country and an integration into the European Union. While the PT fleet has gradually modernized and modern means of fare collections have been introduced with some operators in recent years, the PT services still lag behind most Western European cities in terms of transport integration (intermodal journeys with a unified ticket, coordinated timetables, etc.). The road infrastructure is congested during peak hours, especially the bridges crossing the river Daugava, while PT passenger numbers are decreasing [1].

The City of Riga and the municipalities in the surrounding area also lack common planning and coordination in the transportation field, which reflects the current trends of suburbanization in the area [2].

An approach to improve PT in Riga is transport integration, aiming to make PT more convenient to use and therefore more attractive to the city's inhabitants while, ideally, saving resources due to efficient collaborations.

I. Kabashkin et al. (Eds.): RelStat 2018, LNNS 68, pp. 562–571, 2019.
https://doi.org/10.1007/978-3-030-12450-2_54

This paper develops two approaches to transport integration, representing different levels of integration, and evaluates the possible consequences for passengers, transport volumes, cost and revenues, and the environment.

1.2 Characterization of Research Area

The investigation area for this thesis is composed of the City of Riga and the statistical region *Pierīga* (Fig. 1), which is also the territory of the Riga Planning Region (*Rīgas plānošanas reģions* or RPR).

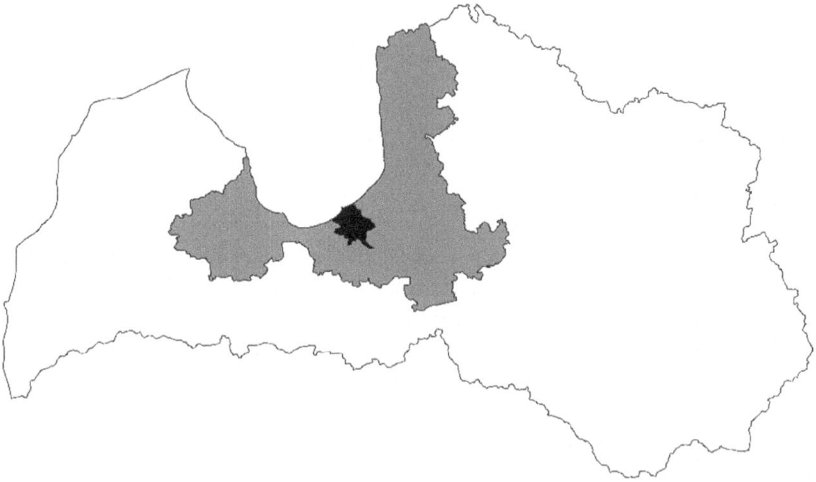

Fig. 1. Location of the City of Riga (black) and the statistical region *Pierīga* (grey) in Latvia.

In the investigation area, six modes serve the PT: tramway, trolleybus, local bus, minibus, regional bus and regional train. Operations are carried out by several independent operators, of which a majority is fully or partly owned by the municipality or the state.

Currently, the PT operators in the RPR use different ticketing and pricing systems. Ticketing systems include different smart cards, cash ticketing, and mobile ticketing. The pricing is either based on a flat fare for single rides (most local transport) or on a zonal (regional trains) or distance-based tariff (regional buses). Prices for the same relation can differ between the modes. For tramways, trolleybuses, local buses and minibuses in Riga a unified ticketing system called *e-talons* is used.

Demand elasticities in relation to price changes observed in Riga have been found to be minimal: Despite the high relative price change (+91.7%) for single ride tickets in Riga in 2015, passenger numbers only decreased by 4.7%, equalling an elasticity of −0.052 [3], much lower than demand elasticities for PT found in other settings [4, 5]. This could be attributed to the structure of fare payers: Only 18% are paying the full fare while 48% are eligible for free travel and the remaining share uses discounted or monthly tickets [6].

1.3 Current Planning State

Measures promoting transport integration have been discussed by planning authorities and stakeholders before and were embraced in regional development plans and strategy papers of PT companies [7–9]. However, no precise planning providing conceptual designs and focusing on the effects of transport integration has been done.

The furthest planning for PT integration can be found in the 2010 "Riga and Pieriga Mobility Plan" (RPMP) where the authors identify "enforcing the coordination of regional public transport in Pieriga, including heavy rail, to avoid parallel lines or competition between transport modes" as one of the major measures to improve the management of PT in Riga and Pieriga. The estimated financial benefits of the proposed introduction of a PTA are a reduction of current organization cost by 20%, a reduction of the current PT cost by 1% and an increase in revenue by 0.5% [10].

2 Literature Review

2.1 Fields of Transport Integration

Transport Integration is a multi-step process including several fields that might or might not be related or presupposed to each other. Terminology varies strongly between different authors. Therefore, a framework is developed in which the different approaches of different authors are combined. Five fields of integration are considered: information integration, ticket integration, tariff integration, network integration and wider integration.

Information Integration. Information integration refers to all measures aiming to unify the outward appearance of a transport system. This can include a corporate design for publications, common advertisement campaigns, a common website [4, 11] or the issuing of common timetables. Regardless of the common appearance, the individual PT operators might still plan and carry out their operations and tariffs individually, but the information is shared with other operators and presented to the passengers as one.

Ticket Integration. Ticket integration refers either to the usage of a unified ticket or the common acceptance of each other's tickets on certain routes [4, 12, 13]. The aim is to make travel easier for the passenger. Mutual acceptance can be based on a concept of remuneration or on goodwill. The underlying tariff system does not have to be integrated as well, which might lead to different prices for the same service, as ticket integration refers more to the physical ability to use one (physical) ticket.

Tariff Integration. Tariff integration refers to a common tariff for a transportation service offered by several operators. This way a journey has the same price regardless of the operator that sold the ticket [4, 11, 13]. The tariff might be integral (meaning that only the common tariff is offered) or partial (meaning that the common tariff is applied only to journeys that require a change of operator, while for journeys with only one operator a different tariff is applied) [4]. Tariff integration generally entails questions about how the common revenue from ticket sales should be collected and distributed.

Network Integration. Network integration, also referred to as service integration, comprises all measures of coordination and cooperation in the actual operation, infrastructure usage, and development of future infrastructure [4, 11, 12]. The goals are seamless travel through different PT operators and modes and a good utilization and effective planning of infrastructure. The seamless travel is achieved through coordination and cooperation between the different PT operators regarding routing, service frequency, scheduling, and the creation of interchange points. Infrastructure-related planning also involves the local spatial planning authorities.

Wider Integration. The term wider integration comprises a multitude of possible approaches for transport integration. They range from the participation of interest groups and concerned people in the planning process [12] over the integration of individual modes of transportation [4] to fully holistic approaches of a system of, throughout all fields cross-linked, planning authorities [11].

2.2 Effects of Transport Integration

In the past, the implementation of measures of PT integration has resulted in a multitude of effects including increases in PT ridership and the modal share of PT. Transportation systems and the environments they operate in differ markedly between the case studies. The results of a case study in one city may therefore not be universally valid. Also, the effects occurring after an implementation can only in a few cases be traced back precisely to the implementation itself, as the integration is often accompanied by several other measures.

Transport Volumes. A major indicator of the effects of transport integration is the development of transport volumes, as an increase in the number of passengers is often one of the goals of the implementation. Various case studies refer to the changes in transport volumes after the implementation [14–16]. The introduction of transport integration measures, in many cases, led to an increase in transport volumes. A study conducted in 69 PT systems between 1991 and 2000 found an increase in transport volumes of 2% in the short run and 12% in the long run as a result of the introduction of tariff integration. The increase was even bigger (5% in the short run and 25% in the long run) when an urban integrated tariff system was extended to a bigger area including intercity systems [17].

While not directly being an implementation of transport integration, the results of the implementation of the fare-free public transport (FFPT) scheme in Tallinn in 2013 can give valuable insights, as both cities are somewhat comparable: The short-term (four months) result of the implementation was an increase of 3% in PT ridership [18]. Over the three years after the implementation, the number of PT rides increased by 6.7% [19]. Which part of the long-term effects can be attributed solely to the FFPT and which to the accompanying measures remains unclear. The resulting demand/fare elasticity of −0.067 is comparable to the one found in Riga.

Modal Choice. Besides passenger numbers, the overall behavior of modal choice can also be influenced by transport integration measures. In a study conducted by Evangelios and Schütze [20] in the area of the Public Transport Authority (PTA) in the Leipzig-Halle Metropolitan Area (*Mitteldeutscher Verkehrsverbund*), a 15–20% increased likelihood of modal choice for PT was the result of the existence of a PTA. The authors attribute this increase mostly to the benefits of implementation (unified ticket and unified timetable). The need to buy a second ticket during a journey was found to strongly reduce the likelihood of a modal choice for PT [20].

Passenger Satisfaction. An OECD study comparing the level of customer satisfaction in 35 cities in Europe found a 13 percentage points higher share of citizens satisfied with the PT system in cities with a PTA (76% of citizen satisfied) than in cities without (63%) [21]. According to Pawlasova [22], service continuity, referring to the possibility of easy interchanges, is one of the main factors influencing the satisfaction of customers. Timetable coordination as a measure of network integration increases the service continuity.

Environment. Drawing a direct connection between transport integration and changes in the environmental situation is complicated. Transport integration might lead to an increase in the attractiveness of the PT system, which could be followed by a modal shift resulting in lower volumes of motorized traffic, which then is reflected in lower levels of air and noise pollution. Those changes can only hardly be traced back to the transport integration measures but are more likely the result of a package of measures. A holistic planning approach, as part of wider integrated planning that also includes coordination with land-use planning, could lead to a higher efficiency of planning with positive results in terms of land consumption, while integrated spatial and transportation planning could reduce travel times and distances.

Studies focusing on the effects of transport integration on the environment are rare. One study was conducted in 263 metropolitan areas within the OECD, focusing on cities with comparable population levels and density and per capita gross domestic product. It found that cities with a Public Transport Authority (PTA) have approximately 9% lower levels of particulate matter (PM 2.5) than cities without a PTA [21].

3 Methodology

3.1 Scenario Design

Two scenarios for an implementation of measures of transport integration for the Riga Planning Region are developed based on best practice examples from other PT systems in Europe:

Scenario A: Ticket Integration. The implementation of a system of ticket integration for the RPR using a unified smart card, while ticket prices, discounts, and planning authority stay with the current authorities responsible for them. Small discounts are granted for intermodal tickets. The Estonian nationwide PT ticketing system *Ühiskaart* [23] and the existing integration of tickets of other bus companies on the electronic ticket of *Rīgas Satiksme* [24] showed the general possibility of the implementation of this scenario.

Scenario B: Information, Ticket, Tariff and Network Integration. The implementation of a Public Transport Authority that acts as governing body regulating and unifying the ticketing and pricing system for all modes of PT in the RPR as well as being responsible for marketing, timetable coordination, and financial compensations. Operations still are carried out by individual PT companies. The base for the fare calculation will be a newly introduced zonal tariff similar to the PTA in Hamburg, Germany (*Hamburger Verkehrsverbund*) [25].

3.2 Estimation of Effects

For the estimation of the quantifiable effects and consequences of the implementation, different methodologies were used:

- When applicable, effects or costs of directly comparable projects were taken for the estimation. This was the case for the estimation of costs related to the new ticketing system for which records of costs from a similar project exist.
- When planning documents already mentioned a quantification of effects these effects were taken directly. This was the case for the estimation of changes in organizational costs, for which calculations in the RPMP exist.
- For the estimation of the increase in passenger numbers findings from the literature review were taken as a base and taken into context with findings from Tallinn, while also considering the elasticities and the structure of fare payers found in Riga.
- When no data from comparable projects exist or the data found cannot be put into context with data from Riga, only a range or tendency for the expected consequences is mentioned.
- Exemplary pricing systems were developed for each scenario to monitor the price change on a defined set of routes.

4 Findings

4.1 General Consequences

In both scenarios, Passengers will profit from a system that is easier and more comfortable to use. For all modes of PT within the RPR, the same smart card will be valid. Using a “balance”-based system for the smartcards, the passengers only need to top up one card.

The assumed ticketing solution in both scenarios is the use of the e-talons smartcard currently used in Riga, as a large part of the fleet is already equipped with the necessary technology. The implementation of the system in Riga in 2008 to 2009 cost 14 million Latvian Lats (€19.92 million) [26]. Equipping all other transport vehicles would lead to costs in the same range.

Assessing the short-term effects on the environment for both scenarios is difficult. A modal shift could reduce the number of cars on the main streets in Riga and reduce noise and air pollution, for which road traffic is a major source. However, this cannot be achieved by simple means of transport integration; it must be the result of a holistic approach on mobility planning, which needs to include not only the Public Transport Authority but also other departments of the City of Riga and the municipalities in the RPR.

4.2 Consequences of Scenario A

The sole implementation of ticket integration will most likely not have a significant effect on the passenger numbers, as the changes for the passenger are only small. In combination with accompanying measures, changes could be possible.

The expected losses in revenue due to the new ticket system are small, as only small discounts for users of intermodal monthly tickets are granted. Nevertheless, the losses will exist as a risk and need be taken over by the state or the municipality through increased funding.

4.3 Consequences of Scenario B

Additional to the easier ticketing in Scenario A, the implementation of Scenario B will lead to an easier understandable tariff compared to the currently existing distance, zonal, and flat fares in operation. In particular, the attractiveness of multimodal journeys will increase, as an interchange will no longer require the purchase of a new ticket. This has been shown to be valued by the passengers.

The change towards a further integrated system, as proposed in Scenario B, can have a higher effect on passenger numbers, as the improvement for the customer is more considerable (easier ticketing and tariff systems, free interchanges, coordinated schedules). Still, the short-term increase in passenger numbers will not be substantial. Abrate et al. in [17] found the short-run effect of tariff integration to be around 2 to 5%, which appears reasonable taking the results from Tallinn into consideration, where an increase in passenger numbers of only 6.7% was found while PT was made totally fare-free there [19]. The structure of passengers in Riga with its high share of passengers traveling for free or with discounts and the observed fare elasticities in Riga support this argumentation, as a substantial share of the PT passengers, at least in Riga, is not sensitive to PT price. Chen et al. in [27] found an asymmetric reaction on a price decrease giving evidence for small reactions to the new improved pricing.

Increased cost of operation would accrue, due to less efficient rotations depending on the operational changes due to the network integration. The centralization of several fields of business in a PTA could lead to reduced costs as parallel structures would be reduced. In the RPMP a cost reduction of 20% equalling €200,000 was estimated [10].

The new "end-to-end" tariff will have lower prices for multimodal journeys, leading to revenue losses while increasing transport volumes in the PT can also increase the revenue. The increase will be in the same range, although the additional rides might be attributed to new ticketing models to some extent, so the increase in revenue will be smaller than the one in transport volumes.

4.4 Comparison

Table 1 presents a comparison of the expected consequences of the implementation of the described scenarios:

The consequences of Scenario B are more extensive as more fields are integrated and the resulting changes, for the passengers as well as for the PT operators, are more substantial than in Scenario A.

Table 1. Consequences of scenarios.

Consequences for	Scenario A	Scenario B
Transport volumes	Only minor changes	2–5% increase
Costs	Cost for new ticketing system (~ €20 million) Same cost of operation Same cost of organization	Cost for new ticketing system (~ €20 million) Higher cost of operation Lower cost of organization
Revenues	Only minor changes	Losses due to new tariffing system Increases due to increased transport volumes
Environment	No short-term effects	No short-term effects

5 Conclusions

Transportation integration would improve the PT system in Riga and the surrounding area, as it would lead to a PT system that would be easier and more convenient for the passengers. The extent of the improvement would depend on the measures taken. As the demand elasticity towards the price for the PT ticket is quite low in Riga, the increases in transport volumes would be small while costs for the implementation and losses in revenue due to new tariffing could occur. Those costs must be taken over by the municipalities and the state. Therefore, the motivation for the implementation of transport integration should rather be the aim to create an attractive PT system to improve the traffic situation in the RPR than commercial interests.

The short-term environmental effects are expected to minimal, but a long-term modal shift resulting from the transport integration could have positive effects.

Transport integration should not be treated as a singular measure but be included in a holistic framework of coordinated planning in several fields of the Riga Planning Region.

Further research will be needed as soon as the local government laid out more precise scenarios than the two presented in this paper.

Acknowledgments. The author would like to thank the Riga City Development Department, especially Evelīna Budiloviča and Kristaps Niedols, for the cooperation in the research process and the LEONARDO office Dresden, Germany for funding of the project.

The project was submitted as a diploma thesis supervised by Prof. Udo J. Becker and Stefan Huber of the Chair of Transportation Ecology at the Dresden University of Technology.

References

1. RP SIA: "Rīgas satiksme": Infografika: Tramvaju, trolejbusu un autobusu pasažieru pārvadājumi (2016). gadā. https://www.rigassatiksme.lv/lv/aktualitates/infografika-tramvaju-trolejbusu-un-autobusu-pasazieru-parvadajumi-2016-gada/. Accessed 11 Oct 2017

2. OECD: OECD Economic Surveys: Latvia 2017. OECD Publishing, Paris (2017)
3. Uhlmann, J.: Implementation of concepts of transport integration in Riga and Pieriga and its consequences on passengers, transport volume, costs and revenue, and environment. Diploma thesis, Technische Universität Dresden (2018)
4. Hilferink, P., Roest Crollius, A., van Elburg, J.-C.: Integration and regulatory structures in public transport. Final report, Rijskwijk (2003)
5. Holmgren, J.: Meta-analysis of public transport demand. Transp. Res. Part A: Policy Pract. **41**(10), 1021–1035 (2007). https://doi.org/10.1016/j.tra.2007.06.003
6. RP SIA: "Rīgas satiksme": 2016. gada pārskats (2017)
7. RP SIA: "Rīgas satiksme": Rīgas pašvaldības SIA "Rīgas satiksme" ilgtermiņa stratēģija no 2012. gada līdz 2033. Gadam
8. Rīgas domes pilsētas attīstības departments: Transporta attīstības tematiskais plānojums, Riga (2017)
9. Ķīlis, R.: Sustainable Development Strategy of Latvia Until 2030 (2010)
10. Tensen, D.K., Sluis, C.M.: Riga and Pieriga mobility plan. Final report, Riga (2010)
11. Preston, J.: What's so funny about peace, love and transport integration? Res. Transp. Econ. **29**(1), 329–338 (2010). https://doi.org/10.1016/j.retrec.2010.07.042
12. Potter, S., Skinner, M.J.: On transport integration: a contribution to better understanding. Futures **32**, 275–287 (2000)
13. Kummer, S.: Einführung in die Verkehrswirtschaft, 2nd edn. fakultas.wuv, Wien (2010)
14. Passenger Transport Executive Group: The Benefits of Simplified and Integrated Ticketing in Public Transport, United Kingdom (2009)
15. Sharaby, N., Shiftan, Y.: The impact of fare integration on travel behavior and transit ridership. Transp. Policy **21**, 63–70 (2012). https://doi.org/10.1016/j.tranpol.2012.01.015
16. Auerbach, L.: Network, Timetable and Tariff Integration in Verkehrsverbund Oberelbe. Public Transport Case Study, Dresden (2009)
17. Abrate, G., Piacenza, M., Vannoni, D.: The impact of integrated tariff on public transport demand: evidence from Italy. Reg. Sci. Urban Econ. **39**(2), 120–127 (2009)
18. Cats, O., Reimal, T., Yusak, S.: Public Transport Pricing Policy—Empirical Evidence from a Fare-Free Scheme in Tallinn, Estonia, Stockholm (2014)
19. Tallinna Linnakantselei: Statistika aastaraamat Tallinn arvudes 2017: Statistical Yearbook of Tallinn 2017, Tallinn (2017)
20. Evangelios, C., Schütze, M.: Zur Frage des Nutzens von Verkehrsverbünden: Eine empirische Analyse im Gebiet des MDV. Zeitschrift für Verkehrswissenschaft **84**(1), 50–68 (2013)
21. Ahrend, R., Gamper, C., Schumann, A.: The OECD metropolitan governance survey: a quantitative description of governance structures in lager urban agglomerations. In: OECD Regional Development Working Papers, 2014/04. OECD Publishing, Paris (2014)
22. Pawlasova, P.: The factors influencing satisfaction with public city transport: a structural equation modelling approach. JOC **7**(4), 18–32 (2015). https://doi.org/10.7441/joc.2015.04.02
23. Rindango, A.S.: Ühiskaart. https://www.pilet.ee/viipe/uhiskaart/persod/perso/?lang=en. Accessed 3 Mar 2018
24. SIA: "Ekspress Ādaži": SIA "Ekspress Ādaži" autobusos, kas kursē Ādažu novada teritorijā, no 1. februāra braucienam var reģistrēties ar e-talonu (2017)
25. Hamburger Verkehrsverbund: Gemeinschaftstarif des Hamburger Verkehrsverbundes (HVV) für die Verbundverkehrsunternehmen, Hamburg (2018)
26. Dienas bizness: Daudz iebildumu par viedkartes nenolasīšanu sabiedriskajā transportā. http://www.db.lv/citas-zinas/daudz-iebildumu-par-viedkartes-nenolasisanu-sabiedriskaja-transporta-19

3946?utm_campaign=article_title&utm_source=categorypage_kolonnaa&utm_medium=article name. Accessed 11 Dec 2017

27. Chen, C., Varley, D., Chen, J.: What affects transit ridership?: a dynamic analysis involving multiple factors, lags and asymmetric behaviour. Urban Stud. **48**(9), 1893–1908. https://doi.org/10.1177/0042098010379280 (2010)

Environmental Friendly Transport Interchanges: Active Travel Accessibility and Policy

Vissarion Magginas[1], Eftihia Nathanail[1], Giannis Adamos[1](✉),
Maria Tsami[1], Irina Yatskiv (Jackiva)[2],
and Evelina Budilovich (Budiloviča)[2]

[1] Traffic, Transportation and Logistics Laboratory, University of Thessaly,
Pedion Areos, 38334 Volos, Greece
giadamos@civ.uth.gr
[2] Transport and Telecommunication Institute, Lomonosova Street 1,
Riga 1019, Latvia
Jackiva.I@tsi.lv

Abstract. This paper focuses on active travel accessibility and the required actions to achieve a satisfying level of non-motorized access to urban interchanges. To this end, a systematic literature review was conducted, related to active travel policy measures and actions, as well as to measures aiming at the promotion of walking and cycling and the improvement of access to public transport terminals. This process was necessary in order to identify good practices and successful interventions implemented in Europe, but also to indicate potential legal, operational and infrastructure gaps and bottlenecks. Based on the above, the public transport system of Riga was investigated in terms of legislation, infrastructure, safety and space availability, addressing active travel accessibility to and from the main urban interchanges: Riga International Coach Terminal, Riga Central Railway Station and Riga Passenger Port Terminal. The critical assessment of the literature review findings and the analysis of the Riga transport system facilitated the drafting of recommendations for stakeholders and decision makers, who wish to put together action plans geared towards tackling the issue of active travel accessibility at urban interchanges.

Keywords: Soft transport modes · Interchange accessibility · Strategic plans · Good practices

1 Introduction

Environmental pollution is one of the greatest problems of contemporary society. Its effects can be experienced in each aspect of every-day life, and its causes can be traced in a multitude of human activities, including transport. The need for transport has grown as technology has advanced, as is the ease with which it can be facilitated. This has led to an increase in environmental impacts, like air and noise pollution. The effects can be mediated by reducing the need for transport by private vehicles [1]. This is more

I. Kabashkin et al. (Eds.): RelStat 2018, LNNS 68, pp. 572–581, 2019.
https://doi.org/10.1007/978-3-030-12450-2_55

easily achieved in large-scale transport through facilities, like transport interchanges. However, it is a lot more difficult to achieve in respect to the access to these facilities. This requires a targeted approach and specialised infrastructure and measures to encourage active travel (i.e. walking and cycling) accessibility to the interchanges [2].

This type of travel, in addition to being environmentally friendly, is also promoting a healthy way of life that has exercise at its core. However, active travel has certain limitations. A person cannot easily use this type of travel to reach destinations that are further than a few kilometres away. A solution to this problem would be to motivate the person to conduct at least part of its journey by cycling or walking and continue further by using a public transport mode. For this reason, motivating people to cycle or walk to a public transport station can have a significant impact on reducing car use and alleviating other unsustainable transport behaviours. To achieve this, one could use soft or hard policy. Hard policy consists of more hands-on measures, like infrastructure construction or improvement, financial support to soft modes, or penalties for using unsustainable transport modes. Soft policy is more focused on influencing the individual steering it towards more sustainable transport habits.

The goal of this paper is to find a balance between hard and soft policy measures aiming to promote access to transport interchanges using active travel and provide indicative guidelines that effectively promote its use for this purpose in everyday life.

2 Method

The methodological approach consisted of three steps. The first step included a thorough state-of-the-art review regarding the relevant hard and soft policy measures that can be used to promote active travel, as well as the European Union communications and policies that relate to or support this type of travel. In addition, certain case studies across Europe were reviewed to extract useful insight in regards to the results that such measures can provide. The second step was related to the analysis of the active travel situation in Riga, Latvia and the city's public transport system active travel accessibility. Lastly, a series of recommendations were formulated, in order for the Riga public transport system to improve its active travel accessibility.

3 State-of-the-Art Review

3.1 Active Travel

Active travel is the act of travelling through non-motorized means and concerns those trips made out of necessity (for example daily work commuting) in contrast with doing the activities above for recreational purposes [3].

This type of travel helps significantly in improving public health and mitigating the effects of environmental pollution through the reduction of motorised vehicles' use [4–6]. Its adoption, however, is dependent of different factors, such as land use planning, distance to be covered, available infrastructure and facilities, as well as traveller behaviour and population and cultural characteristics [4, 6, 7].

Active travel accessibility to public transport systems is also a very important factor in promoting active travel. Djurhuus et al. in [3] studied the relationship between active commuting and the accessibility of the public transport system of Copenhagen. Results showed that the farthest commuters live from a public transport stop, the less likely they are to be active travellers.

In 2010, the NICHES+ coordination action published a handbook containing guidelines for constructing innovative cycling facilities at transport interchanges [8]. The handbook pointed out the characteristics of these facilities and the benefits they could offer to travellers and included case studies from the Dutch town of Zutphen, the multi-modal transport interchange of the Swedish city of Lund and the cycle parking of the Budapest West Railway Station.

ADONIS project [9] was a European Union (EU) project that ran from 1996 to 1998 with the goal of substituting short car trips with cycling and walking. A sub-project of this program was the development of best practices to promote cycling and walking. A catalogue of measures was formulated, containing both hard and soft measures based on 102 examples from Spain, Denmark, Belgium and the Netherlands. These measures are categorised based on their focus and their impact area in order to be easy for the interested party to select and implement them.

In 2013, Tsami et al. [2] conducted a study regarding the sustainable design of transport interchanges. The study focused on the New Railway station of Thessaloniki, Greece and consisted of a questionnaire survey and the assessment of 51 indicators by the station's users. Furthermore, some intervention scenarios were developed and were assessed by travellers. Part of the questionnaire and four intervention scenarios were dedicated to active travel. The results indicated the lack of essential infrastructure in the station facilities, making accessing the station by active travel very difficult or impossible. The results concerning the effect of the intervention scenarios showed that their implementation would result in an increase of travellers using active travel and thus reverting towards a more sustainable way of transport.

3.2 European Union Policy

EU policy and legislation are more focused on hard measures, as they are easier to be implemented and they can have an immediate impact. An example of this approach can be found in COM (1998) 431 [10], where there is a clear focus on hard measures, like spatial and land use planning, infrastructure construction and modification and the enforcement of additional transport charges to promote sustainable mobility. In recent years, however, EU is becoming more open to the prospect of soft policy by issuing communications and directions encouraging measures of a soft nature. This is evident in COM (2007) 551 [11], where an alternative approach to urban mobility education is proposed, based on traveller awareness campaigns and initiatives, guided by entities like the CIVITAS network. In COM (2009) 279 [12] it can be seen that although the focus of the EU policies remains on hard measures, the soft policy has gained some traction since there are also parts referring to traveller awareness and behaviour, as well as public participation in developing new transport policy. COM (2009) 490 [13] emphasises again on hard measures, like infrastructure and technology, but it also includes actions on improving information provision and access, as well as more campaigns on sustainable mobility behaviour. It also includes an action focusing on energy efficient driving training, as well as an action promoting the formulation of a common platform for public transport officials to exchange opinions and ideas. COM (2013) 527 [14] is related to the adoption of Intelligent Transport Systems (ITS) for the promotion of sustainable transport. In regards to soft modes, the document supports the use of ITS for the promotion of bike-sharing schemes, thus supporting cycling, as well as for the protection of vulnerable road users, such as pedestrians. Moreover, the communication suggests measures for the stricter implementation of restricted and charged access plans, thus reducing motorised vehicle usage.

In addition to the above communications, the European Union also supports active travel through the White Papers of 2001 and 2011 [15, 16] and COM (2013) 913 final [17]. These policy directions are among the most important the European Union has published about sustainable urban mobility, and they are indirectly related to active travel. The 2013 Communication also details the formulation and adoption of Sustainable Urban Mobility Plans (SUMPs), which are schemes supporting and promoting the adoption of sustainable urban mobility, by using policy measures and soft and hard actions (awareness campaigns, infrastructure development). Since cycling and walking form a large part of sustainable mobility, active travel is also addressed in this policy document, and it holds a significant role in SUMPs.

3.3 Soft Policy

Furst (2014) used a segmentation approach to formulate, through individualised marketing, policy measures in order to promote environmentally friendly access in some Germany and Austria-based academic institutions. The author divided passengers into six heterogeneous clusters which he named "eco-friendly travelers", "see-saw-travelers" (indifferent to the environmental situation, valuing level of service above all), "pragmatic, but environmentally open-minded, travelers", "pragmatic travelers who dislike public transportation", "prestige-oriented travelers" and "car-oriented travelers" based on their responses in a pre-tested online questionnaire. The study concluded that it is possible, through the integration of market segmentation and identified target groups into comprehensively sustainable mobility schemes, to retain and raise the percentage of travellers choosing to use sustainable means of transport [18].

Price and Leather (2011) suggested soft measures like cycle training, special events promoting soft modes, special working plans conforming with a potential increase in travel time the use of soft modes could result in and social marketing and information provision schemes [19].

Richter et al. published a research report in 2009, concerning soft transport policy and the results of its implementation [20]. The study uses the classification above of soft policy measures and proceeds to present two well-known initiatives of soft policy for transport, Travel Blending and IndiMark, as well as some programs that have been used in Japan, Australia, the United Kingdom (UK) and other countries. The measures identified, fell mainly into categories 1–5 of the above classification. The results of the meta-analysis study showed an effect size of 0.24, 0.08 and 0.11 for categories 1, 2 and

3–5 respectively. Finally, the study comments on the different approaches used by the countries.

In 2008, Bamberg et al. published a review paper of soft transport policy measures based on behavioural theory [21]. In this paper, the researchers attempted to pinpoint the weaknesses of previous studies and explain the rationale behind soft policy measures, by analysing four behavioural models (the Norm-Activation model, the Theory of Planned Behavior, a combination of these models and the Self-Regulation model of behavioural change) and the studies based on them. The authors proposed concentrating on developing and testing new intervention ideas, as well as evaluation tools for soft policy measures under realistic field conditions.

3.4 Hard Policy

City-Hub project (2013) identified some potential (hard) measures promoting active travel access at interchanges [22]. These included better signposting and wayfinding, direct pedestrian and bike routes to and from the stations, safe cycle parking, better information provision, better traffic control, security measures and information provision for pedestrians and cyclists.

Price and Leather (2011) suggested measures like a dedicated cycle and pedestrian lanes, improved footways, substitution of car parking spaces with bicycle ones and actions related to land use (like pedestrianisation of certain streets) [19].

4 Implementation of Policy Measures

In the following section, some of the most relevant and distinct cases are analysed, and methods and results of this simultaneous implementation are discussed.

In 2007, the European Commission [23] published a report, which summarised the majority of the programs and projects promoting the use of soft modes. Along with some regulations implemented in Belgium, France, the Netherlands and the UK, there were also good practices from Copenhagen, Camden, Lyon, Geneva and various Norwegian regions. Indicatively, in the case of Copenhagen, it was estimated that 82.5 million Euros were invested between 2002 and 2012 in the new and improved cycling and walking infrastructure.

Galloway et al. (2009) [24], in the framework of CIVITAS Initiative, published a deliverable regarding the implementation and results of a combined program consisting of both soft and hard measures in the cities of Preston, Ploiesti and La Rochelle, targeting cycling. More specifically, in La Rochelle, a new 14 km long cycle path was created and supported by cycling organisations and local stakeholders, while in Preston the new cycle paths exceeded the targets set for their length (14.5 km instead of 11.5 km). In Preston, an increase of 75% in cycling was also noted, supported by the implementation of a TravelSmart program. Finally, in Ploiesti, 8 km of new cycle tracks were constructed.

A deliverable published by the MIDAS program in 2009 summarised soft and hard measures implemented in various cities [25]. In this report, good practices such as the bicycle training program that took place in Liverpool were referenced, as was the pedestrian and cycling infrastructure that was improved or constructed in Bologna, Clermont-Ferrand, Cork and Suceava. The results of the implementation of the measures are also summarised, and they show an increase of 40% in the usage of bike hire service in Bologna, 9% increase of walking in Cork, 32% increase in cycling in Liverpool, but no changes in the case of Suceava. The implementation of the MIDAS project, also encouraged a change in the attitudes of the citizens towards alternative transport modes. More specifically, in Cork 9% of the participants were positively influenced by the measures, while in Liverpool 13% of the respondents that did not cycle at all, started cycling after the program. Finally, in Suceava, the program resulted in a 50% increase in awareness.

In 2010, Santos et al. [26] published a paper where cases and results of the successful implementation of hard and soft policies were summarised. In regards to cycling and walking best practices from the Netherlands, Denmark, Germany, Colombia, England, France and Spain were presented. Specifically, in the Netherlands, as a result of about 227 million Euros of investment during the period 1994–2004, the average person cycles roughly 2.5 km per day, while Denmark and Germany achieved similar results through similar programs. In Bogota, 300 km of cycle tracks were constructed, while London, Paris and Barcelona made significant progress by implementing innovative bike hire schemes. In the case of Paris, in 2008 there was one bike every 135 citizens.

Based on an IndiMark project implemented in Goteborg during 1998 and 1999, Almgren in [27], presented a series of figures and results that showed a 6% decrease in car usage, a relative increase of 45% for cycling, and a reduction of 2150 tons of CO_2 per year. The cost of the project was equal to 270 million Swedish Crowns.

In 2014, Boschetti et al. [28], as part of the CIVITAS Vanguard, published a report on innovative urban transport solutions, which included good practices and examples of measures implemented in various European cities. In Donostia-San Sebastian, four extra kilometres were pedestrianised, while the cycling network was expanded by 22 km. The overall program resulted in 22 km of newly built pedestrian roads and 80 km of new bike lanes, and a large number of new bicycle parking spaces.

Kumar et al. (2010) presented some interchange accessibility practices related to active travel that have been implemented in Delhi. Specifically, these included bike-sharing schemes and pedestrian ways design [29].

Gill (2018) discussed the bicycle policy of Singapore and described the design and implementation of dedicated bicycle paths and lanes, while also bringing up the case of Copenhagen, where the bicycle is fully integrated into the city's transportation system [30].

A summary of these measure implementations can be found in Table 1. In most cases, soft and hard measures are implemented together, while infrastructure construction and improvement take up a large part of these implementations. The scope of each case was in the majority local, while most of the cases focus on a combination of cycling and walking. Regarding stakeholders, in most cases, these included mainly the travellers and the local authorities, while the time horizon of each project was usually between one and three years.

Table 1. Summary of measures' implementation.

Measure	Approach	Modes	Stakeholders	Area	Ref.
Friendly spaces, speed limit reduction, barriers' removal, information campaigns	Soft/hard	Cycling, walking	Travellers, academia, local authorities	EU cities	[23]
Cycle parking, park and ride service, bicycle bridge	Hard	Cycling	Travellers, local authorities, market analysts	Sweden, Hungary, Spain, the Netherlands	[8]
Direct pedestrian and cycling routes, information provision, traffic control	Soft/hard	Cycling, walking	Travellers, local authorities	Preston, La Rochelle, Ploiesti	[24]
Pedestrianisation, personalised travel plans, active travel promotion	Soft/hard	Cycling, walking	Travellers, local authorities	London, Abu Dhabi	[19]
Travel plans, school travel awareness plans, facilities	Soft/hard	Cycling, walking	Travellers, local authorities	EU cities	[25]
Traffic calming, bike rental services, the connection of cycle paths with bus rapid transit system	Soft/hard	Cycling, walking	Travellers, local authorities	Various cities	[26]
Individualised marketing	Soft	Cycling, walking	Traffic and public transport authorities	Goteborg	[27]
Bike sharing, parking spaces, extended lanes	Soft/hard	Cycling	Travellers, academia, local authorities	EU cities	[28]

5 Active Travel Accessibility in Riga Public Transport System: Situation Analysis and Recommendations

In the case of Riga, although as much as 52.1% of its population cycles periodically, only 9.9% of citizens use the bicycle for their daily commuting. The main reason for this is the bad connectivity between the bicycle and the Riga public transport system, as well as the bad connectivity between different transport modes. Along with the increased number of traffic accidents involving cyclists and certain land use planning issues that pose restrictions in the interventions that can be implemented (mainly those of a hard nature, such as infrastructure development or expansion) and make cycling as a commuting option harder.

This comes at odds with the fact that the majority of the city's population live within cycling distance of the city centre (89%), which is the effective distance of cycling. Specifically, 42% of citizens live within 5 km (20 min) of the city centre, which has a particularly competitive ride distance indicating great cycling potential.

Accessibility should enable citizens to walk, cycle or use public transport, which is the friendlier transportation mode, in terms of environment and health [31]. However, the "Rules of the road" of the Republic of Latvia [32] foresee that cyclists need to use the road network or the infrastructure for bicycles, and only in the case that the conditions are dangerous for their life, they can also use the sidewalk.

The public transport provider of Riga city municipality (RP SIA "Rīgas satiksme") allows carrying bicycles free of charge in public transport (trams, trolleybuses, buses) by the rules of procedure. Regarding trains, the conditions of the train provider for passenger (JSC "Pasažieru vilciens") for the carriage of bicycles stipulate that a luggage ticket must be purchased, if there are no bicycle holders in the wagon. Railway wagons that are fitted with bike holders are marked with a label at the doors [33].

Regional buses are not fitted with bike holders to facilitate bicycle carriage and improve mobility while providing easy transportation of bicycles. Since public transport is subsidised from the state budget, the State Tax Administration "ATD" concludes contracts with carriers and set conditions for the provision of buses with bicycle holders, assessing the routes where this would be needed and where there is the greatest demand for bicycle transportation [34].

Riga provides hard and soft policies for cyclists. Representative hard policies include government laws and regulations, such as the Rules of the road [32], the Road traffic law [35], the Latvian administrative violations code [36] and other.

In terms of soft policies, the Road Traffic Safety Authority organises informative events for children in schools. Also, students of the fifth class can obtain a cyclist license, after passing the respective safety exam. Inhabitants are regularly informed for such activities and initiatives through mass media and booklets.

The main Riga interchanges provide bicycle parking free of charge: Riga International Coach Terminal at the entrance and the car parking, Riga International Airport at the main airport area and the car parking, and Riga Central Railway Station at the entrance and the car parking of two nearby malls. On the other hand, such facilities are missing from the Riga Port Terminal.

Lastly, as part of this attempt, from 2016, the Riga Municipality Development Department includes cyclists in its traffic volume counting as a separate category.

The review of active travel cases and the Riga public transport system situation showed that for active travel to gain significant traction, it mainly needs a steady legal framework and a combination of soft and hard measures. That can promote physically involving ways of travel both through behavioural change and through the direct support of infrastructure provision, expansion and improvement.

First of all, it needs support in the form of a proper legal framework, which can be constructed based on the European Commission's directives and communications that were mentioned and analysed previously. Moreover, certain soft and hard measures have to be implemented. Individualised marketing and travel awareness campaigns can be utilised to affect travellers' travelling decisions and steer them towards active travel, while hard measures, especially infrastructure improvement and construction (for example bicycle parking, bike holders, dedicated cycling and walking lanes) can promote active travel in a more direct way.

6 Conclusions

Active travel is known to help improve active travellers' health and protect them from conditions like obesity and coronary heart disease. It also significantly contributes to the mitigation of climate change and pollution.

As shown earlier, several factors contribute to the adoption of active travel, including land use planning, distance to be covered, available infrastructure and facilities, as well as traveller behaviour and population and cultural characteristics. In order for these factors to be exploited and the barriers overcome, a series of interventions need to take place in order to make the public aware of the benefits and opportunities and at the same time support active travellers through the development of adequate infrastructure. Finally, another important element that can help in supporting active travel is national and European legislation.

Acknowledgements. This paper is based on the research and work that has been conducted in the framework of the ALLIANCE project (http://alliance-project.eu/), which has received funding from the European Union's Horizon 2020 Research and Innovation Programme. The authors would like to thank both the consortium of the project and the European Commission.

References

1. Garling, T., Bamberg, S., Friman, M., Fujii, S., Richter, J.: Implementation of soft transport policy measures to reduce private car use in urban areas. In: Panels of the Energy Efficiency and Behaviour Conference (2009)
2. Tsami, M., Adamos, G., Nathanail, E.: Sustainable development design for the transformation of the Thessaloniki's railway station into a city hub. In: European Transport Conference 2013, Frankfurt, Germany (2013)
3. Lavin, T., Metcalfe, O., Higgins, C.: Active travel-healthy lives. The Institute of Public Health in Ireland (2011)
4. Barton, H., Horswell, M., Millar, P.: Neighbourhood accessibility and active travel. Plann. Pract. Res. **27**(2), 177–201 (2012)
5. Djurhuus, S., Hansen, H.S., Aadahl, M., Glumer, C.: Individual public transportation accessibility is positively associated with self-reported active commuting. Front. Public Health **2**, 240 (2014)
6. Vale, D.S., Saraiva, M., Pereira, M.: Active accessibility: a review of operational measures of walking and cycling accessibility. J. Transp. Land Use **9**(1), 209–235 (2016)
7. Song, Y., Preston, J., Ogilvie, D.: New walking and cycling infrastructure and modal shift in the UK: a quasi-experimental panel study. Transp. Res. Part A **95**, 320–333 (2017)
8. Monigl, J., Berki, Z., Szekely, A.: Guidelines for implementers of innovative cycling facilities for interchanges. NICHES+ (2010)
9. Nilson, P.K., la Cour Lund, B.: Promotion of cycling. In: ADONIS (1998)

10. European Commission: COM 431 (1998)
11. European Commission: COM 551 (2007)
12. European Commission: COM 279 (2009)
13. European Commission: COM 490 (2009)
14. European Commission: COM 527 (2013)
15. European Commission: White Paper-European Transport Policy for 2010—Time to Decide/COM/2001/370 final/, Brussels (2001)
16. European Commission: White Paper-Roadmap to Single European Transport Area—Towards a Competitive and Resource Efficient Transport System/COM/2011/0144final/ (2011)
17. European Commission: Together Towards Competitive and Resource-Efficient Urban Mobility/COM/2013/913 final/, Brussels (2013)
18. Furst, E.: Making the way to the university environmentally sustainable: a segmentation approach. Transp. Res. Part D **31**, 1–12 (2014)
19. Price, D., Leather, A.: Transport Mobility Management: Small Changes—Big Impacts: Understanding TMM in the Urban Context. JOURNEYS (2011)
20. Richter, J., Friman, M., Garling, T.: Soft Transport Policy Measures 1: Results of Implementation. Karlstadt University Studies 31 (2009)
21. Bamberg, S., Fujii, S., Friman, M., Garling, T.: Evaluation of soft transport policy measures based on behavioral theory. Transp. Policy **18**(1), 228–235 (2011)
22. City-Hub: Innovative Design and Operation of New or Upgraded Efficient Urban Transport Interchanges. City-Hub Project (2013)
23. European Commission: Sustainable Urban Transport Plans: Preparatory Document in Relation to the Follow-Up of the Thematic Strategy on Urban Environment (2007)
24. Galloway, M., Graindorge, M., Monteanu, S.: Innovative Soft Measures: Deliverable 11 of the Success Project. CIVITAS (2009)
25. MIDAS: Soft Measures for Sustainable Mobility: Lessons from Case Studies in Aalborg, Bologna, Clermont-Ferrand, Cork, Liverpool and Suceava (2009)
26. Santos, G., Behrendt, H., Teytelboym, A.: Policy instruments for sustainable road transport. Res. Transp. Econ. **28**(1), 46–91 (2010)
27. Almgren, C.: Effects on Mode Choice with Individualised Marketing (IndiMark) in Goteborg. In: ECOMM 2003, Karlstad, Sweden (2003)
28. Boschetti, F., Maurizi, I., Cre, I.: Innovative urban transport solutions: CIVITAS makes the difference. CIVITAS (2014)
29. Kumar, P., Jain, S.S., Kulkarni, S.Y., Parida, M.: Accessibility to transit station in multi modal transport framework for Delhi. In: ICSBE-2010, Kandy, 13–14 December 2010
30. Gill, A.: Changing Cities and Minds for Active Mobility. Lee Kuan Yew School of Public Policy (2018)
31. Veloconcepcija 2015: Rīgas pilsētas velosatiksmes attīstības koncepcija 2015–2030 gadam. Riga city Trafic department (2015)
32. Rules of the road N.571. "Ceļu satiksmes noteikumi", MK noteikumi Nr. 571. https://likumi.lv/doc.php?id=90585
33. Pasažieru vilciens. https://www.pv.lv/lv/informacija-pasazieriem/vilciena-ar-velosipedu/. Accessed 4 Oct 2018
34. Cyclist infrastructure development plan: Velosatiksmes attīstības plans 2018–2020 gadam, Riga city (2018)
35. Road traffic law: Ceļu satiksmes likums. https://likumi.lv/doc.php?id=45467
36. Latvian administrative violations code: Latvijas Administratīvo pārkāpumu kodekss. https://likumi.lv/doc.php?id=89648

A Cross-case Analysis of Riga Interchanges' Information Services and Technologies

Irina Yatskiv (Jackiva)[1(✉)], Evelina Budilovich (Budiloviča)[1], Iveta Blodniece[1], Eftihia Nathanail[2], and Giannis Adamos[2]

[1] Transport and Telecommunication Institute, Lomonosova 1, LV, 1019 Riga, Latvia
Jackiva.I@tsi.lv
[2] University of Thessaly, Pedion Areos, 38334 Volos, Greece

Abstract. The design of an optimal interchange depends on the local environment and individual preferences of users on modal choices. Various issues in quality are important for good interchange management and operation, and priorities may significantly differ among different interchanges. This paper aims to investigate and assess the performance level of four interchanges in Riga, Latvia: Riga International Coach Terminal, Riga Central Railway Station, Riga International Airport and Riga Passenger Terminal that serves cruise ships.

To achieve the aim of this study, a structured form was used for face-to-face interviews, and key stakeholders of each interchange were invited to give their feedback in selected items.

Keywords: Public transport · Terminals · Information services · Comparison analysis

1 Introduction

The European Commission (EC) in 2001 built a path towards upgrading urban interchanges to increase public transport usage [1]. Various issues in quality are important for efficient interchange management and operation, and priorities may significantly differ among different interchanges. The design of an optimal interchange depends on the urban form and users' preferences, needs and priorities.

The nature of public transport (PT) makes it necessary for people to change transport modes and good interchanges minimise the burden of changing and therefore, the operation phase should focus on passenger needs. The integration of information systems and ticketing, as well as the application of other Intelligent Transport Systems (ITS) and services for users, are crucial for the improvement of the travel experience and the appeal of the public transport terminals. Technological innovations at transport terminals significantly increase the availability of information and useful data. Accurate, valid and timely information enhances the level of users' convenience and improves the efficient operation of the transport system [2].

This paper aims to investigate and assess the performance level of four interchanges in Riga Latvia: Riga International Coach Terminal (RICT), Riga Central Railway Station (RS), and Riga International Airport (RIX), Riga Passenger Terminal (PORT)

I. Kabashkin et al. (Eds.): RelStat 2018, LNNS 68, pp. 582–592, 2019.
https://doi.org/10.1007/978-3-030-12450-2_56

that serves cruise ships, in terms of provided information systems, ticketing and other ITS services.

A structured form was used to conduct face-to-face interviews, and key stakeholders of each interchange were invited to give their feedback in specific items to achieve the aim of this study.

The structure of the paper is as follows. The next section reviews the state of the art. Section 3 introduces the proposed methodology for the data collection and the cross-case analysis. Section 4 gives short descriptions of the studied interchanges; Sect. 5 presents the analysis of the survey results and in Sect. 6 the main findings are discussed. Lastly, Sect. 7 presents the main conclusions of this paper.

2 State of the Art

The Action Plan for the Deployment of Intelligent Transport Systems [3] and the most recent White Paper on Transport Policy [4] consider that multimodal transport should be based on three pillars: people, integration and technology. More recently, the EC [5] highlighted that the focus should be placed on integrated transportation services, through information provision and intermodal coordination, where different transportation modes are interconnected [6].

Developed standards EN13816 [7] and EN15140 [8] have defined services, which should be investigated when analysing an interchange. According to these standards, the main components formulating the quality of services are: availability, accessibility, information, time, customer care, comfort, security and environment.

Many EU projects have investigated information services successfully functioning in the systems of public transport in European countries in the last decade: for instance, POLITE [9] and TTRANS [10] etc.

Yatskiv et al. [11] analysed the information systems for public transport in Latvia. Findings revealed that clear information systems are needed for the provision of easy, efficient and seamless information. This information should be integrated with different operators and modes and especially if it is provided to multimodal passenger terminal users [12].

Adamos et al. [13] after the analysis of the stakeholders and travellers of RICT point of view suggested that the main quality services for users are information systems, ticketing and other ITS services.

Travel information and intermodal services are spread across all interchange zones, such as access-egress zone, arrival-departure-transfer and the facilities-retailing [6]. It is highlighted in CIHT [14] that it's necessary to set clear directions to where connecting services and information about the possibility to transfer or change the transport mode can be found. Door-to-door transport including interchanges enables easier access to these facilities and integrated ticketing can reduce costs. Platform numbers, services and facilities are required to enable users to find their way in the less possible time, reducing the interchange penalty.

Information provides the following benefits to multi-modal travel: assisting the planning of journeys; supporting the promotion of public transport by raising awareness of alternatives; helping traveller to reduce the risks of travelling and manage the consequences of delays that occur during the journey. Examples of information systems include:

- telephone call-centre services;
- online rail journey planners, which is essentially single modal travel, but they can also provide information about transport services available at stations;
- multi-modal planners; real-time information [15];
- live mobile updates, e.g. text messages, enabling travellers to know, for example, when the next bus is expected before starting their journey or even enroute information via smartphones.

Preston et al. [16] and Crockett et al. [17] suggested that it will be necessary to ensure that information and journey planning services are designed and targeted to meet the needs of different users and journey purposes. Information provision also plays a significant role in 'smarter travel' behavioural change programmes, such as workplace travel plans and personalised travel planning [6].

Integrated smart ticketing facilitates the easiness to transfer and promotes the use of public transport. Ticketing includes ticket offices, vending machines or automated machines for ticket selling, ticket via the internet (online services) and "app" stores.

The implementation of the concept "Open Data" is becoming more common in transportation: integration of different timetables is available in some cases; coverage of all operators and modes is needed; a centralised web service or mobile application for all options can make transferring and use of interchanges much easier [12].

3 Methodology

Methodology for this research was developed in the following way: (1) Stakeholder definition; (2) Questionnaire preparation; (3) Face to face survey; (4) Cross-case analysis. Firstly, the stakeholders that are appropriate for answering the questions were defined and contacted by the authors. RICT was represented by the Chairwoman of the Board of Directors, the Airport by a member of the Board and the Riga Central Railway Station by a member of the Organisation. In the case of the PORT, two persons, one from the terminal board and the second from the ferry operator side, gave their feedback.

The questionnaire consists of 11 questions about the information provision, information integration and uniformity, ticketing and ITS systems at the interchange (ITS) [6]. The question about ITS systems is divided into 23 short questions, and the stakeholder was asked to answer if he/she considers whether a number of services are "in use", "needed" or "not needed."

The face-to-face survey was carried out in August and September 2018. The purpose of the survey was to:

- Investigate the level and quality of the information provided to travellers throughout the total journey, in the trip planning phase and during the trip especially at the interchange points, through previous research and case study analysis.

- Assess whether integrated online information is given at the interchange points including incident information.
- Check if there is a cooperation among the information systems of various operators.
- Understand integrated ticketing issues for the use of public transport and acceptance of intermodality, i.e. changes during the trip.
- Define the possibility of the ticket purchasing systems, especially mobile solutions.
- Investigate if it is possible to separate the emergency information given at the interchange points from the daily incident and breakdown information.

A cross-case analysis was applied amongst the four interchanges. Information about ticketing, disabilities, weakest and strongest parts of each interchange was analysed. Also, the indoor and outdoor services were considered in the analysis.

4 Description of the Interchanges

Riga, as the capital of Latvia, is a central node of the country transport star-designed network with a total territory of 303,996 km^2 and a population of over 641,423 registered residents in 2017.

Riga Urban Public Transport System (RUPTS) consists of 54 bus lines, 9 tram lines, 19 trolleybus lines, 21 minibus lines, passenger railway with 22 stations, 9-night buses. Railway covers all parts of the city and consists of many passengers and cargo stations. Some suburban train lines connect Riga region with city of Riga. Although, the railway is fully electrified and connects all parts of the Riga city, still the railway passenger services are not included in the RUPTS.

The public transport network that it is very dense in the city centre has a thinning network towards the borders of the city. In general, the coverage of the PT network (walking distances to stop) can be evaluated as good/very good. Only 5–7% of inhabitants of Riga and 3–5% of employees need more than 5 min to reach the nearest PT stop.

There are four main transport hubs in Riga. RICT, the Railway Station (RS) and the PORT are situated in the old town, which is listed in UNESCO protected areas. On the other hand, the Airport is geographically located in another municipality but still it can be considered as a transportation hub of the city of Riga. RICT and RS provide regional, long-distance and international trips, while the Port and Airport only international. A brief description of each interchange is provided below:

1. Riga International Airport (RIX) is the largest international aviation company in the Baltic area and the main air traffic centre in this region offering regular passenger, cargo and postal delivery to the cities of Europe and the world. RIX renders both aviation and non-aviation services. It serves both national and international airlines becoming one of the few European airports that facilitate both full service and low cost airlines.
2. Riga International Coach Terminal (RICT) provides services for passengers and road passenger transportation companies that operate regional passenger inter-city, regional, local and international routes. Coach terminal location provides easy interfaces to other transport modes, located at the heart of the capital.

3. Riga Passenger Terminal Ltd. (PORT) was founded in August 2002 to manage and develop the passenger port of Riga and its territories. Riga Passenger terminal serves cruise ships, ferries, super yachts, sailing yachts, navy vessels and other non-cargo ships.
4. Riga Central Station (RS) is the main railway station in Riga that operates all passenger trains within Latvia. It is known as the main point of Riga due to its central location, and many public transport stops are located in this area.

The description of the four main interchanges regarding passengers, area, working hours, governance and transport modes for the last mile is represented in Table 1.

Table 1. Interchanges' description.

PT interchanges	Passengers (2017)	Total area (m^2)	Working hours	Governance	Transport modes for the last mile
RIX	6,097,434	17,340	00:00–24:00	State	Car, taxi, bus, minibus, bike*
RICT	1,661,529	13,000	00:00–24:00	Private	Car, taxi, tram, bus, minibus, bike*
PORT	87,384	~ 1,000	09:00–18:00	Private	Car, taxi
RS	1,733,380	38,000	05:00–24:00	State	Car, taxi, bus, trolleybus, minibus, bike*

bike*—exist parking and absent cycling roads.

Two of interchanges RIX and RS are owned by state, while, PORT and RICT can be characterized as private. Regarding passengers' flow, RICT and RS serve approximately two million passengers per year. RS is the biggest one, and PORT is the smallest regarding of territory coverage. RIX and RICT operate daily, while PORT works from 9 till 18 o'clock and RS from 5 till 24 o'clock. Last mile services provided from/to interchange are served by all transport modes in RIX, RICT and RS. However, the accessibility of the PORT is classified as relatively poor, and the possibility to reach from/to this interchange is exclusively by individual transport (car) or taxi.

5 Survey Result Analysis

The representatives of the four main interchanges were interviewed, and their answers were generalised and analysed by cross-case methods.

The analysis of the ticketing services of all interchanges is given in Table 2. "Mobile applications", "internet", "ticket machine", "ticket offices" and "No Near Field Communication (NFC) Payment" are the most commonly services used in RICT, RIX and RS. RIX offers additional ticketing services such as "Airport operator" and "Tourism companies". In PORT the purchase of tickets is possible via "internet", "ticket offices", "ticketing machines" and "No NFC Payment".

Regarding the availability of services for people with disabilities (Table 3) RICT meets five out of six standards analyzed: "request help 36 h before the trip" (other

Table 2. Ticketing services in the interchanges.

Ticketing services	RICT	PORT	RIX	RS
Mobile applications	Bezrindas.lv, Mobility.lv		AirBaltic air tickets, sky-scanner	Pasazieru vilciens, Mobility.lv
Internet	×	×	×	×
Homepage	×		×	×
Ticket offices	×	×	×	×
Ticket machines	×	×	×	×
No NFC payment	×	×	×	×
By phone	×			
Airport operator			×	
Tourism companies			×	

interchanges have 48 h), "apply for service by phone or email", "informative video about assistance available", "tactical guidelines for visual impaired" and "Three specialised summon boards". In the case of PORT there are only two possibilities of this kind of services, such as "request help 48 h before the trip" and "apply for service by phone or email". It has to be mentioned that only RS provides the service "Boarding the departing train and disembarking from the arriving train using mobile platforms".

Table 3. Provided services for people with disabilities in the interchanges.

Services for people with disabilities	RICT	PORT	RIX	RS
Request help 48 h before the trip	× (36 h before)	×	×	×
Apply for service by phone or email	×	×	×	×
Informative video about assistance available	×		×	×
Has tactical guidelines for visual impaired	×			
Three specialised summon boards	×			
Boarding the departing train and disembarking from the arriving train using mobile platforms				×

When assessing the indoor-outdoor services provided in the interchanges (Table 4), it was observed that RICT is more convenient in the indoor services, while RIX provides exclusively the option of "Smart ticketing". Outdoor service "Facilities and layout are available on the internet" is available in all interchanges. The option

Table 4. Indoor-outdoor provided services in the interchanges.

Indoor-outdoor services	RICT			PORT			RIX			RS		
	In	Out	In/out	In	Out	In/out	In	Out	In/out	In	Out	In/out
Electronic departure time displays and disturbance information	×			×			×			×		
Multi-language information	×			×			×			×		
Smart ticketing							×					
Information centre with personal service	×						×			×		
Guidance and warning surfaces for the visually impaired (VI)	×									×		
Tactile maps of the interchange for the VI	×											
Intelligent automated passenger or people counting	×						×			×		
Intelligent security systems (e.g. CCTV)	×			×			×			×		
Journey planner for long-distance PT for pre-trip planning		×									×	
Facilities and layout available on the internet		×			×			×			×	
Matrix barcodes (e.g. QR-codes)		×										
Guidance and warning surfaces for the VI								×				
Area or terminal fleet management with the aid of cameras			×			×			×			×
Tactile maps of the interchange for the VI			×						×			×

"Journey planner for long-distance PT for pre-trip planning" is provided by RICT and RS, and "Matrix barcodes" only in RICT.

Special services that can be defined as indoor/outdoor like "Area or terminal fleet management with the aid of cameras" are met in all interchanges. "Tactile maps of the interchange for the visually impaired' are used in RICT, RIX and RS, but not in PORT.

Based on the analysis of the ITS systems (Table 5), it can concluded that 12 services are in use in all interchanges, one service is needed at all interchanges (Intelligent Indoor-Navigation System) and one service is not needed in any interchange (Information with hearing aids (e.g. "T-coil")).

Table 5. Intelligent system/services in the interchanges: (1) in use; (2) needed; (3) not needed.

Services	RIX			RICT			PORT			RS		
	1	2	3	1	2	3	1	2	3	1	2	3
1	2	3	4	5	6	7	8	9	10	11	12	13
Journey planner for local PT for pre-trip planning		×				×			×	×		
Journey planner for long-distance PT for pre-trip planning		×		×					×		×	
Information for interchange facilities and layout available on the internet (or via call centre) for pre-trip planning (important especially for the disabled)	×			×			×			×		
Smart ticketing [speeds up transfer]	×				×				×			×
Electronic departure time displays based on *timetables* (for multiple stops)	×			×			×			×		
Electronic departure time displays based on *timetables* (at stops)	×			×			×			×		
Electronic departure time displays based on *real-time information* (for multiple stops, incl. fleet monitoring systems)	×			×			×			×		
Electronic departure time displays based on *real-time information* (at stops)	×			×			×			×		
Departure times via audio calls	×			×			×			×		
Real-time disturbance information provided via *displays*	×			×			×			×		
Real-time disturbance information provided via *audio calls*	×			×			×			×		
Multi-language information	×			×			×			×		
Public access information kiosk/internet kiosk restricted for PT information (not for open internet surfing)				×					×	×		
Information centre with personal service	×			×			×			×		
Audio services for the visually impaired (e.g. a special dedicated information area with a push button)			×	×			×			×		
Guidance and warning surfaces for the VI	×			×				×			×	
Tactile maps of the interchange for the VI	×			×				×		×		
Information with hearing aids ("T-coil")			×			×			×			×
Matrix barcodes (QR-codes) for additional information with mobile phones (for departure times for a specific stop or platform)	×				×				×			×
Intelligent Indoor-Navigation System		×			×			×			×	
Intelligent security systems (e.g. CCTV)	×			×			×			×		

(*continued*)

Table 5. (*continued*)

Services	RIX			RICT			PORT			RS		
	1	2	3	1	2	3	1	2	3	1	2	3
1	2	3	4	5	6	7	8	9	10	11	12	13
Area or terminal fleet management with the aid of cameras, in-vehicle systems, variable message signs etc. for guiding buses, taxis, park & ride etc.	×			×			×			×		
Intelligently automated passenger or people counting (infrared, video, thermal etc.)	×			×				×		×		

Table 6. Strengths and weaknesses of the studied interchanges.

	Strengths	Weaknesses
	1	2
RIX	1. Fast way to travel for those who use smart ticket 2. Technological advancement and investment in information technology 3. Infrastructural development 4. Service for passengers with reduced mobility 5. Mobile solutions for ticket purchasing 6. Location and access to PT	1. No journey planner services; 2. Tickets are not valid for multiple modes and operators; 3. No warning surfaces for visually impaired inside the airport
RICT	1. Integrated information (15 + operators) 2. Mobile solutions for ticket purchasing 3. Journey planner for long-distance PT 4. Central location to downtown, access to PT 5. Terminal adjusted for passengers with disabilities	1. Infrastructural development 2. Tickets are not valid for multiple modes and operators 3. No smart ticketing
RS	1. Infrastructure contains several systems that allow avoiding unpleasant incidents 2. The most punctual trains in Europe 3. Mobile solutions for ticket purchasing 4. Central location to downtown, access to PT 5. Call centre available (24/7) 6. Mobile platforms for passengers with disabilities	1. No journey planner services 2. Tickets are not valid for multiple modes and operators 3. One operator
Port	1. Infrastructural development 2. Central location to downtown.	1. Terminal need upgrade 2. One main operator 3. No journey planner services 4. No mobile solutions for ticket purchasing 5. Tickets are not valid for multiple modes and operators 6. Less accessible by PT 7. Working hours

6 Discussion

The cross-case analysis facilitates the identification of strengths and weaknesses of the Latvian interchanges (Table 6).

7 Conclusion

The comparison of the Latvian interchanges studied in this paper revealed clear differences in the level of information systems currently implemented. As the interchanges are different regarding transport modes and operators involved, age, size and their role in the wider urban environment, it is clear that the needs and benefits of various intelligent transport systems and services are also different. Considering that all terminals are characterised by multiple stakeholder involvement, a clear governance model, addressing the provision of reliable information to travellers, is very important for good centralised management that is needed to ensure that all crucial issues are included in daily interchange operations.

Concluding, only by integrating and linking data of all stakeholders, it is possible to combine a traveler's individual mobility demand with mobility services and provide the basis for more efficient, accessible and affordable mobility.

Acknowledgements. This paper is based on the research and work that has been conducted in the framework of the ALLIANCE project (http://alliance-project.eu/), which has received funding from the European Union's Horizon 2020 Research and Innovation Programme. This study is a part of an EU project that is surely carried out according to ethical policies. The authors would like to thank both the consortium of the project and the European Commission.

The authors would like to thank the representatives of Riga International Coach Terminal and, especially Dr. V. Gromule; M. Ozols (Riga Railway Station); A. Saveljevs (Riga International Airport); J. Dzenitis (Riga Passenger Terminal).

References

1. European Commission: White Paper—European Transport Policy for 2010—Time to Decide, Brussels: /COM/2001/370 final/ (2001)
2. Grotenhuis, J.W., Bart, W.W., Rietveld, P.: The desired quality of integrated multimodal travel information in public transport: customer needs for time and effort saving. Transp. Policy **14**, 27–38 (2007)
3. European Commission: The Framework for the Deployment of Intelligent Transport Systems in the Field of Road Transport and for Interfaces with Other Modes of Transport, Brussels: 2008/0263/COD (2010)
4. European Commission: White paper—Roadmap to a Single European Transport Area—Towards a Competitive and Resource Efficient Transport System, Brussels: / COM/2011/0144 final/ (2011)
5. European Commission: Together Towards Competitive and Resource-Efficient Urban Mobility, Brussels: /COM/2013/913 final/ (2013)

6. Monzon, A., Ciommo, F.: City-HUBs Sustainable and Efficient Urban Transport Interchanges (2016)
7. EN13816: Transportation—Logistics and Services—Public Passenger Transport—Service Quality Definition, Targeting and Measurement, Brussels (2002)
8. EN15140: Public passenger Transport—Basic Requirement and Recommendations for Systems that Measure Delivered Service Quality, Brussels (2006)
9. Polite-project.eu (2018). http://www.polite-project.eu/. Accessed 8 Dec 2018
10. T-TRANS Upbringing ITS Products and Services to Market (2018). http://www.ttransnetwork.eu/ttrans/. Accessed 8 Dec 2018
11. Yatskiv, I., Gromule, V., Pticina, I.: Analysis of different aspects of infomobility for public transport in Latvia. In: Zamojski, W., Mazurkiewicz, J., Sugier, J., Walkowiak, T., Kacprzyk, J. (eds.) Proceedings of the Tenth International Conference on Dependability and Complex Systems DepCoS-RELCOMEX (2015)
12. Adamos, G.: Operation and Management of Intermodal Transport Systems: Passenger Interchanges. 2nd Summer School, Project Alliance, Riga (2018). http://alliance-project.eu/
13. Adamos, G., Builovich, E., Yatskiv, I., Nathanail, E., Tsami, M., Magginas, V.: Quality of service in urban transport interchanges: understanding the gaps between users and operators expectations. In: European Transport Conference, Dublin (2018)
14. CIHT: Institute of Highways & Transportation, Guidelines for Providing for Journeys on Foot, 0902933302, 9780902933309 (2000)
15. Palmer, D., James, C., Jones, M.: The door to door journeys. A Report Produced by Transport Research Laboratory for the Campaign for Better Transport, 55 p. (2011)
16. Preston, J., Wall, G., Whiteing, T.: Delivery of user needs: final report. Rail Research UK report: RRUK/C2/06 (2006)
17. Crockett, J., Beecroft, M., McDonald, M., Whitein,g T., Whelan, G., Nash, C., Fowkes, T., Wardman, M.: Rail user needs and delivery mechanisms. Rail Research UK report RRUK/C2/02 (2004)

Optimization of Interaction of Automobile and Railway Transport at Container Terminals

Vladimir Shepelev[1(✉)], Leonid Zverev[1], Zlata Almetova[1], Ksenia Shubenkova[2], and Elena Shepeleva[1]

[1] South Ural State University, 76 Lenin Avenue, Chelyabinsk 454080, Russia
shepelevvd@susu.ru
[2] Kazan Federal University, 10a prospekt Syuyumbike, Naberezhnye Chelny, Russia

Abstract. The article deals with the main processes occurring at the automobile-railway container terminal. The main economic entities of the automobile-railway communication are distinguished, including the owner of the railway infrastructure, the operator of the railway rolling stock, the container terminal and a road carrier. A technique is proposed for determining optimal parameters for the operation of terminal complexes, which is based on minimizing the amount of costs of all the participants in the transportation process. There is a need to revise the criteria for optimizing the calculation of the optimal amount of loading and unloading mechanisms, taking into account the costs for over-normal idle hours of motor vehicles under loading (unloading), over the planned time of work performance and costs due to idle time of the railway rolling stock on public roads in anticipation of loading (unloading). This technique assumes that the arrival of containers is a random stream of events, and the entire complex of terminal hardware is a service channel. The incoming flow of containers is classified based on the overall and weight characteristics. The results of the study show that when determining the optimal amount of loading and unloading mechanisms, it is necessary to take into account the costs associated with interdependent idle time of loading and unloading mechanisms, rolling and automobile stock. The additional time spent on the cargo deconsolidation, as well as the possibility of delivering lightweight containers in pairs, significantly affect the processing of the incoming container flow at the terminal. As a result of the study, it has been established that the interdependent idle time of transport and loading and unloading mechanisms is significantly reduced, taking into account the volumetric and weight characteristics of containers, the probabilistic arrival nature and the assessment of the transport capacity of vehicles used to deliver containers to the final consumers. It has been established that the savings of the total costs of the container terminal of Chelyabinsk Cargo Station and the losses of carriers when organizing loading and unloading operations based on the presented methodology will amount to 9%.

Keywords: Container transportation · Loading and unloading mechanisms · Railway transport operators container terminal parameters

I. Kabashkin et al. (Eds.): RelStat 2018, LNNS 68, pp. 593–602, 2019.
https://doi.org/10.1007/978-3-030-12450-2_57

1 Introduction

A significant proportion in the value chain is the cost of logistics. On average, the cost of transport services at enterprises is in the range of 5–35%, depending on such factors as the volume of output, geographic location, the resource used, type of business, etc. Therefore, enterprises performing transport services face the task of the development of advanced logistics technologies, the introduction of modern technical means and methods of transportation in order to reduce transportation costs. The trend towards a geographical shift of production from the point of origin of industry to the East [1] led to the development of intermodal transport systems.

The movement of goods by enlarged cargo units, in containers, facilitates the integration processes of transport systems and facilitates the interaction of automobile and railway transport, thereby speeding up and reducing the cost of transportation. The development of a network of container terminals on the territory of Russia fully meets the goals and objectives of the Transport Strategy of the Russian Federation for the period until 2030 [2].

The creation of an intermodal transport chain leads to a number of issues of strategic, tactical and operational level of planning and operations management [3–5]. Moreover, in comparison with other areas of research, the area of interaction of modes of transport remains less studied and in many cases there are no generally accepted methods and settings [6–8].

One of the reserves to reduce costs in the field of container transportation is the optimization of the interaction of automobile and railway transport at transport hubs [9, 10]. A lot of studies have been devoted to the issues concerning tactical and operative planning of operations at land container terminals. Most studies on the maintenance of terminal capacities consider minimizing the costs of container processing and an efficient use of storage facilities [3]. A number of early studies contributed to the understanding of the problems associated with the distribution of space. In works [11, 12], the main properties associated with the processing of containers at warehouses were analyzed. The problems of cost reduction and optimization of technological parameters by determining the storage location of containers were solved in [13]. The tasks of distribution and dispatching of loading and unloading machines were dealt with in [14, 15]. The solutions are, as a rule, reduced to a minimum of the total mileage, total waiting time or general equipment delays. Let us also note the work of Sadovskaya [16], in which sound recommendations were proposed to improve the operational planning and management of loading and unloading facilities for the processing of large-capacity containers at the container site, which allow reducing idle hours of automobile transport during the processing of containers.

From the perspective of the logistical approach, the task of organizing loading and unloading operations in transit terminals is to determine the amount and productivity of loading and unloading complexes ensuring the effect in the form of minimum economic losses for all the participants in logistical processes: carriers of related modes of transport, owners of the transport infrastructure, warehouse complexes, etc. The problematic character of this situation is that, on the one hand, excessive unloading facilities lead to their downtime, a decrease in the efficiency of container terminals and a low return on investment in their creation. On the other hand, the shortage of loading mechanisms leads to the increase of rolling stock downtime.

The existing contradictions determine the need for additional studies to establish an optimization criterion in determining the operating parameters of terminal complexes.

2 Research Methods

In the present research, the container terminal of Chelyabinsk Cargo Station was chosen as the object of study. As a result of statistical studies carried out, it is noted that the average number of certain states during a working day is subject to significant fluctuations. At some point in time, a large accumulation of rolling stock may occur at the container site, waiting for the execution of technological operations. The low-predicted approach of both automobile and railway transport affects the mode of operation of the entire container terminal, which causes significant costs associated with idle waiting hours. Representing the container terminal area as a system that receives the flow of containers for processing (loading into a carriage or a car, as well as unloading from a carriage or a car), it is possible to identify the main production resources that limit the processing capacity: the capacity of the container site, the capacity of loading and unloading mechanisms, the capacity of loading and unloading ways, the amount of automobile transport.

Thus, it was decided to use the interdependent approach widely used by domestic and foreign scientists to determine the optimal amount of transport and loading and unloading means. An attempt was made to develop this approach by including previously unrecorded factors.

Previous studies have not provided the opportunity to scientifically justify the optimal amount of loading and unloading mechanisms, taking into account the processing capacity of different types of containers. This was reflected in the fact that the basis of the technical complex of the container terminal during the creation and development of the container transport system in the Russia consisted of gantry cranes of 20–24 tons load capacity at the seizure, while the number of containers 40 ft. long with a maximum gross weight of 30–48 tons increases on the network [17]. Under these conditions, the utilization coefficient of loading and unloading mechanisms load capacity is approximately equal to one, which led to their low operation reliability. At the same time, when calculating the time for container processing at the terminal, it should be borne in mind that the technical characteristics of automobile transport at most do not allow a gross mass of more than 24 tons for the transportation of a container due to restrictions on load capacity and maximum axle load [18]. In this way, the loaded containers arriving at the station require the deconsolidation of the cargo for delivery by automobile transport to the consignee's warehouse. At the same time, 20-foot containers, either empty or laden with lightweight cargoes, may be delivered to the address of one consignor or consignee in pairs by one transport vehicle. Thus, the productivity of the fleet of machines directly depends on the weight characteristics of containers.

Studying the terminal operation statistical data allows us to distinguish the main trend of container receipt taking into account their weight characteristics (Fig. 1). In addition, the time of arrival or delivery of containers to the terminal is random.

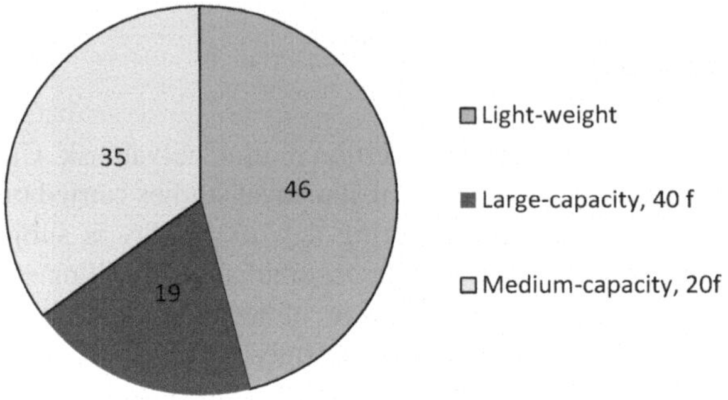

Fig. 1. Distribution of arrived containers.

The process of the arrival of laden containers from the shipper to the station is random due to the fact that it depends on a number of factors, including the date of conclusion of trade contracts, the readiness of the goods for shipment, the availability of goods in the buyer's warehouse, the seasonality of transportations; and at sea terminals it also depends on the weather at the ports of departure and destination, which provides direct influence on the approach of the vessel and the work of the port. The delivery of loaded containers by automobile transport to the terminal is not subject to accurate calculation because it is impossible to calculate the date of container loading by each particular sender. As the cargo is ready for shipment, the sender orders the necessary container from the owner of the latter or hires a freight forwarder who organizes the transportation. But the date of shipment in any case remains random (Fig. 2).

To take into account the unevenness of the arrival of container traffic, the methods of queuing theory were used. The QT deals with random quantities, which are largely present in the operation of any container site. Thus, for example, the mass of a cargo in a container lifted by a loading and unloading mechanism is random. The type of container that the consignor will need is also indeterminate in advance: it will be either a 20-ft. or 40-ft. container, which remains unknown until the moment of submitting an application for container loading. Each container site contains all the informal details of the QT models: the source of applications, the incoming demand stream, the service channel, the waiting queue, idle hours of the channel in anticipation of receipt of the application [19].

Cost optimization criteria are generally used in determining the parameters of loading and unloading complexes, but the analysis of scientific works shows that the peculiarities of the formation of costs for performing loading and unloading operations and losses of vehicle owners due to idle hours under these operations or the expectation of their implementation are not fully taken into account, namely, the losses of operators of the railway rolling stock due to idle transport vehicles in anticipation of loading and

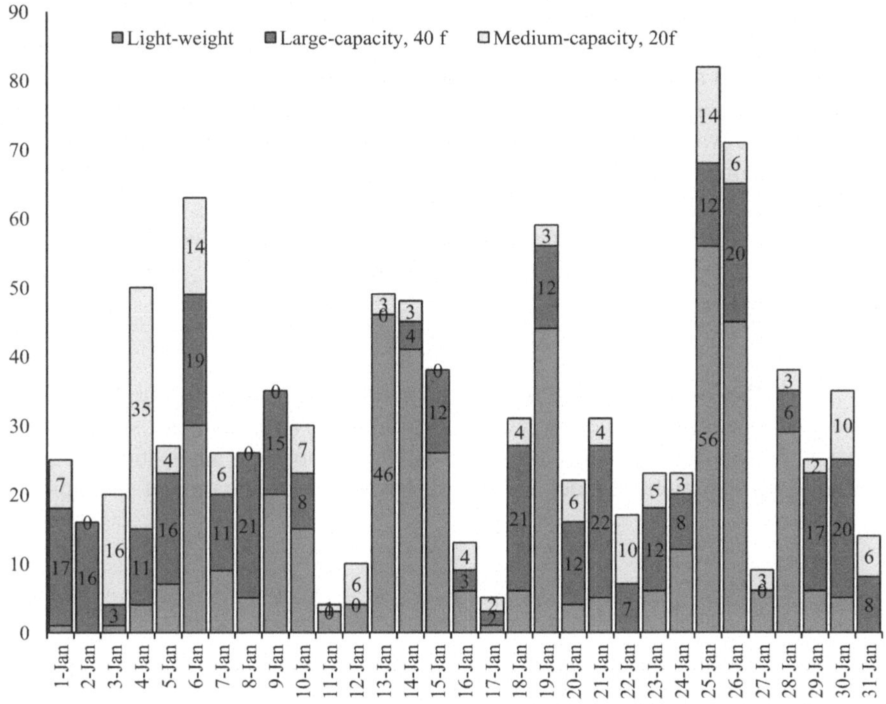

Fig. 2. Monthly irregularity of container arrival.

unloading operations. In this case, the infrastructure owner will be charged for each container for every hour of idle time on public roads [19].

Calculation of the optimal amount of loading and unloading facilities (*n*), under which it is possible to provide the minimum aggregate operating costs and unproductive idle hours of loading and unloading complexes, as well as road carriers' losses due to idle rolling stock under the appropriate operations (C_{nm}) is:

$$C_{nm} = C^{A}_{nm} + C^{RW}_{nm} + C^{CD}_{nm} + C^{CW}_{nm} \rightarrow \min, \tag{1}$$

where C^{A}_{nm} is costs for over-normative idle hours of automobile vehicles for loading (unloading) over the planned time of work, roubles; C^{RW}_{nm} is costs due to idle time of the railway rolling stock on public roads in anticipation of loading (unloading) operations, roubles; C^{CW}_{nm} is the cost of operating one loading and unloading mechanism for the entire period of work, rubles; C^{CD}_{nm} is costs due to the forced idle hours of each loading and unloading mechanism in the absence of transport vehicles for the whole period of idle time, roubles.

As it was said above, the investigation of the functioning of the systems under consideration was carried out by the methods of queuing theory, which is characterized by two parameters: the parameter of the incoming flow of service applications and the service parameter. These and other parameters are random depending on the distribution law and the structure of the research system itself and the parameters of its functioning. Since under the service channel is meant the whole complex of technical means presented at a particular container site, then let us take the number of containers brought to the container platform as the incoming flow of requirements, as well as those loaded on the railway rolling stock or unloaded from it to the platform by loading and unloading machines. The random size of the incoming container flow is distributed according to Poisson's law.

Thus, the costs for over-normal idle hours of automobile vehicles under loading (unloading) over the planned time of work performance (C_{nm}^{A}) is:

$$C_{nm}^{A} = t_{lt}(n) \cdot Q_c \cdot C_h, \tag{2}$$

where $t_l(n)$ is the average waiting time of the automobile vehicle for the beginning of technological operations, which is calculated from the *QT* formulas for single-channel systems with expectation (M/M/1/m); Qc is the number of applications requiring maintenance by automobile vehicles; C_h is the cost of 1 h of idle time of an automobile vehicle.

$$Qc = 2k_1 + k_2 + \frac{1}{2}k_3 + Q, \tag{3}$$

where k_1 is the number of large-capacity containers (FEU); k_2—number of medium-capacity containers (TEU); k_3—number of lightweight containers; Q is the number of applications remaining from the previous day.

$$C_{nm}^{RW} = N_w \cdot C_{rw}, \tag{4}$$

where N_w is the number of railway rolling stock that the system is unable to service:

$$C_{rw} = \frac{e_r \cdot (1 + r)k_c \cdot L_r}{365 \cdot 24 \cdot 1000}, \tag{5}$$

where e_r is the expense rate for the maintenance and amortization of 1 km of station tracks for the year, roubles; L_r is the length of station tracks occupied by one unit of railway rolling stock, m; k_c is the coefficient of occupancy of station tracks for the parking of wagons; r is the coefficient of profitability.

$$C_{nm}^{CW} = P_w C_w nT, \tag{6}$$

$$C_{nm}^{CD} = C_d P_0 \sum_{n=0}^{n-1} \frac{(k-n)\left(\frac{\alpha}{\mu}\right)^n}{n!}, \tag{7}$$

where P_w is the probability that loading and unloading mechanisms are busy; P_0 is the probability that loading and unloading mechanisms are vacant; C_w, C_d is the cost of an hour of loading and unloading mechanisms work and the cost of an hour of idle time, respectively.

As a result of fluctuations in individual components of the car's turnover, the number of trips that one car can perform during the shift will also change. The average time of a car's turnover, taking into account loading and unloading operations and traffic with a load, is described by the normal distribution law with the specified parameters:

$$Z(n) = \frac{k_a T}{t_{lt}(n) + t_l + t_m^0 + t_m^c + t_{un} + t_{unt}}, \tag{8}$$

where t_l is the idle time of the car under loading; t_m^c is the time spent by the car for traffic with cargo (taking into account the average technical speed of traffic in the city); t_m^0 is the time spent by the car on driving in empty condition (taking into account the average technical speed of traffic in the city); t_{un} is the time to unload the car; t_{unt} is the waiting time for unloading; k_a is the coefficient of vehicle use by time (0.94); T is the duration of the work shift.

$$t_{unt} = \frac{\rho t_{un}(\gamma'^2 + \gamma_0^2)}{2(1-\rho)} \tag{9}$$

where ρ is the level of loading mechanisms; γ' is the coefficient of variation of vehicle arrival intervals; γ_0 is the coefficient of variation of the service time of the transport unit.

3 Findings

It is worth noting that the optimal estimated number of loading and unloading mechanisms cannot always be realized in practice due to the presence of restrictions on the simultaneous placement of the maximum allowable amount of unloading devices at one post. It is assumed in the calculations that the location of the loading-unloading fronts and their length allow the simultaneous operation of several loading and unloading mechanisms, and the cleaning of the railway rolling stock is carried out at night, without affecting the intensity of the work of loading and unloading mechanisms [20].

Practical studies have shown that queuing systems have good functioning parameters with a ratio λ/μ significantly less than one, which is clearly seen from the graph (Fig. 3).

In this case, it is necessary to take into account two situations that arise when the number of the flow of incoming applications changes:

- the number of applications is too large—in this case, there is long waiting time or an insufficient amount of servicing equipment;
- an insufficient number of applications enter the container site, which leads to an increase in equipment idle time or its excess.

The average daily amount of containers m_0 arriving at the terminal was determined within the framework of the study. When the options were compared (Fig. 4), it was established that the lowest total costs for a given average amount of arriving containers m_0 are reached at n = 2 units.

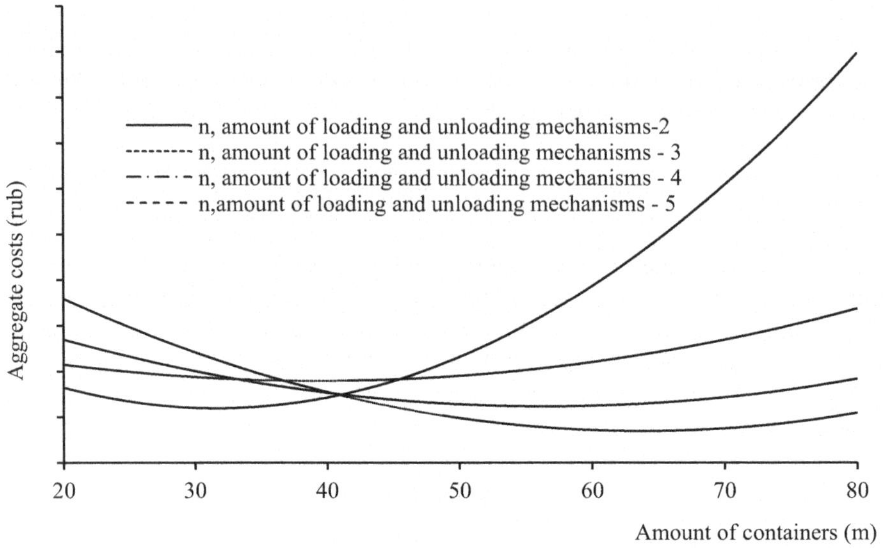

Fig. 3. Complex costs associated with loading and unloading operations and idle time of rolling stock depending on the number of containers being handled.

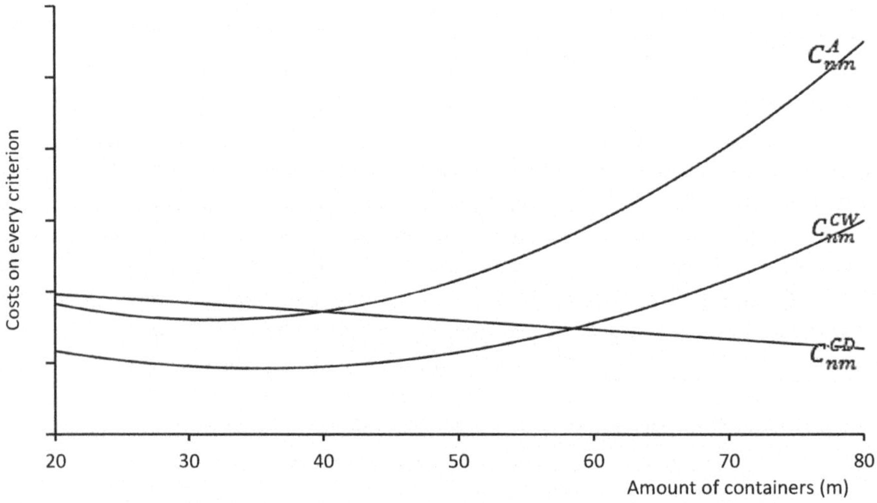

Fig. 4. Graph of changes in costs at *n*, the amount of loading and unloading mechanisms—2.

An increase in the amount *n* of loading and unloading means allows reducing losses due to the idle time of transport vehicles C^{A}_{nm}, C^{RW}_{nm}, but there will be losses due to the forced idle time of loading and unloading mechanisms C^{CD}_{nm}, which leads to an increase in total costs C_{nm} (Fig. 4).

4 Conclusion

As a result of analyzing the research on the problem of the efficiency of using transport and loading and unloading facilities at container terminals, it has been established that the proposed methods do not fully take into account the volumetric and weight characteristics of containers, the time spent on cargo deconsolidation and the cost of the involved infrastructure. The proposed model allows us to reduce the losses associated with the limit-exceeding idle hours of railway and motor transport and loading and unloading facilities, and shorten the delivery time, taking into account the probabilistic nature of the interaction of the participants in the transportation process at the container terminal and the reassessment of the transport capacity of the automobile transport. The savings of the total costs of the container terminal of Chelyabinsk Cargo Station and the losses of carriers when organizing loading and unloading operations based on the presented methodology will amount to 9%.

From the viewpoint of the owner of the infrastructure and the carrier, the application of the methodology will make it possible to effectively use the available capacities of the railway network and to plan their development taking into account the minimum investment.

References

1. Islam, D.M.Z.: Barriers to and enablers for European rail freight transport for integrated door-to-door logistics service. Part 1: barriers to multimodal rail freight transport. Transp. Probl. **9**(3), 43–56 (2014)
2. Order of the Government of the Russian Federation of November 22, 2008 No. 1734-r: On the Transport Strategy of the Russian Federation. http://legalacts.ru/doc/rasporjazhenie-pravitelstva-rf-ot-22112008-n-1734-r/. Accessed 17 May 2018
3. Caris, A., Macharis, C., Janssens, G.K.: Planning problems in intermodal freight transport: accomplishments and prospects. Transp. Plan. Technol. **31**, 277–302 (2008)
4. Pekin, E., Caris, A., Crépin, T., Macharis, C., Jourquin, B., Janssens, G.: Integrated decision support tool for intermodal freight transport. In: Nectar Cluster Meeting on Freight Transport and Intermodality (2008). http://hdl.handle.net/1942/8352. Accessed 17 May 2018
5. Tsamboulas, D., Vrenken, H., Lekka, A.: Assessment of a transport policy potential for intermodal mode shift on a European scale. Transp. Res. Part A: Policy Pract. **41**(8), 715–733 (2007)
6. Crainic, T.D., Kim, K.H.: Intermodal transportation. Transportation **14**, 467–537 (2005)
7. Hassall, K.: Freight exposure and how to measure it. In: Proceedings of the TRANSLU 2008, Bucharest, Romania (2008)
8. Hopp, W., Spearman, M.: Factory Physics: Foundations of Manufacturing Management, 2nd edn. McGraw-Hill, Boston (2001)
9. Marinov, M.: Rail and multimodal freight: a problem-oriented survey (part II-1). Transp. Probl. **4**(2), 73–83 (2009)
10. Bontekoning, Y.M., Macharis, C., Trip, J.J.: Is a new applied transportation research field emerging?—A review of intermodal rail-truck freight transport literature. Transp. Res. Part A: Policy Pract. **38**(1), 1–34 (2004)
11. Taleb-Ibrahimi, M., Castilho, B., Daganzo, C.F.: Storage space vs handling work in container terminals. Transp. Res. Part B: Methodol. **27**(1), 13–32 (1993)

12. Kim, K.H.: Evaluation of the number of rehandles in container yards. Comput. Ind. Eng. **32** (4), 701–711 (1997)
13. Kim, K.H., Park, K.T.: Dynamic space allocation for temporary storage. Int. J. Syst. Sci. **34** (1), 11–20 (2003)
14. Bish, E.K.: A multiple-crane-constrained scheduling problem in a container terminal. Eur. J. Oper. Res. **144**(1), 83–107 (2003)
15. Kim, K.H., Bae, J.W.: A dispatching method for automated guided vehicles to minimize delay of containership operations. Int. J. Manag. Sci. **5**(1), 1–25 (1999)
16. Sadovskaya, O.L.: Improvement of operational planning and management of cargo road transport and loading and unloading operations at container points. Cand. Sci. Diss. Moscow, 153 p. (1984)
17. Mamontov, I.Yu.: Organization of the functioning of the technical complex of the container terminal on the basis of optimization of the fleet of transshipment facilities. Moscow (2013)
18. Decree of the Government of the Russian Federation of April 15, 2011 No. 272: On approval of the rules for the carriage of goods by road. http://legalacts.ru/doc/postanovlenie-pravitelstva-rf-ot-15042011-n-272/. Accessed 23 June 2018
19. Resolution of the Federal Energy Commission of Russia of June 19, 2002 No. 35/12: On approval of the rules for the application of rates for the use of wagons and containers of the federal railway transport (Tariff Guide No. 2). http://legalacts.ru/doc/postanovlenie-fek-rf-ot-19062002-n-3512. Accessed 20 June 2018
20. Ostrovskiy, A.M., Druzhinina, M.G., Kuz'mina, A.A.: Interaction of the operator companies with industrial enterprises and the railway. Russ. Railways **2**, 61–63 (2011)

Unsupervised Learning-Based Stock Keeping Units Segmentation

Ilya Jackson[1(✉)], Aleksandrs Avdeikins[1,2], and Jurijs Tolujevs[1]

[1] Transport and Telecommunication Institute, Lomonosova Iela 1, Riga, Latvia
{jackson.i, tolujevs.j}@tsi.lv
[2] Trialto Latvia LTD, "Dominante", Kekavas p., Kekavas n. 2123, Latvia

Abstract. This paper reports on the unsupervised learning approach for solving stock keeping units segmentation problem. The dataset under consideration contains 2279 observations with 9 features. Since the "ground truth" is not known, the research aims to compare such clustering algorithms as *K*-means, mean-shift and DBSCAN based only on the internal evaluation, thus, this research may be considered as descriptive cluster analysis. Besides that, several preprocessing techniques are utilized in order to improve the result.

Keywords: Clustering · Inventory segmentation · Inventory clustering · Data mining · Unsupervised machine learning · Principal component analysis

1 Introduction

An average inventory system contains immense number of stock keeping units (SKUs). In general case, it is computationally impossible to consider each item individually and manage it under individual inventory policy. As far back as late 80s, an essentially important question has arisen: "how to aggregate stock units into groups so that the resulting inventory policies are sufficiently close to those policies that would have been generated if every unit was treated individually?" [1].

Nowadays the development of the efficient methodology for defining SKU's groups is still relevant. In previous research we managed to develop a metaheuristic technique for finding a nearly-optimal reorder policy in stochastic multiproduct inventory systems [2]. However, dimensionality of real-world problems requires a segmentation of an assortment in such a way that each segment is relatively homogeneous and may be treated under a common inventory policy. Thus, it becomes an extremely tempting opportunity to take advantage on the state-of-the-art unsupervised machine learning approaches in order to tackle this long-standing problem.

Early attempts to group SKUs by cluster analysis may be tracked back to the study of Srinivasan and Moon [3]. The researchers introduced a hierarchical clustering-based methodology for supporting inventory systems in supply chains. Several heuristics were applied to identify the relationships between items and take into account product features with a significant impact on supply chain. The Calinski-Harabasz index was used to validate the result of the average linkage clustering.

I. Kabashkin et al. (Eds.): RelStat 2018, LNNS 68, pp. 603–612, 2019.
https://doi.org/10.1007/978-3-030-12450-2_58

In 2007 the K-means-based SKU segmentation methodology was proposed [4]. The research aimed to reduce the time required to compute the inventory-control parameters in large-scale multi-echelon inventory system. The segmentation methodology was tested on different multi-echelon inventory systems in order to understand an effect of resulting penalty costs. Three years later the similar k-means-based approach by Egas and Masel was applied to determine storage assignments [5]. The paper concludes with the statement that the method managed to reduce the number of aisles to retrieve orders by 20–30% compared to a demand-based assignment strategy.

It is also important to emphasize the recent paper [6]. The authors conducted a study on the application of a constrained clustering method reinforced with principal component analysis. According to the research, the proposed method is able to provide significant compactness among item clusters.

This study discusses an application of various clustering algorithms to solve the SKU-aggregation problem. Since the "ground truth" is not known, such algorithms as K-means, mean-shift and DBSCAN are compared based only on the internal evaluation. In this regard, this study may be considered as a descriptive clustering analysis. The research utilizes dataset provided by the "Trialto Latvia LTD", the third-party logistics operator. Since SKU's groups should take into account all attributes with a sufficient impact on the certain inventory operation, considered features include information beyond the inventory cost and volume that are used in classical ABC analysis. Besides, the work pays special attention to feature-scaling, anomaly-detection and validation approaches. In order to perform all necessary calculations and data transformations we wrote a script in Python 3.6 [7]. The script uses a free machine learning library "scikit-learn" [8] as a core complementing it with several algorithms.

2 Dataset Description

The initial dataset consists of 2279 SKUs (observations) with 9 features. Selected features include only numerical data and comprise a lot of information beyond that utilized by a classical ABC analysis (Table 1).

Table 1. Selected features.

ID	Unit price (EUR)	Expire date (days)	Total outbound (units)	Number of outbound orders	Pallet weight (kg)	Pallet height (cm)	Units per pallet
1	0.058	547	2441	9	105.6	1.56	1920
2	0.954	547	0	0	207.68	1.00	384
3	2.385	547	23	12	165.78	1.02	108
...	...	...	...	...	...	...	...
2278	2.02	730	710	354	322.56	1.19	288
2279	1.99	730	765	363	322.56	1.19	288

All the features have an undeniable impact on the inventory management and constitute two core groups: handling-related and turnover-related. Such features as expire date, pallet weight, pallet height and number of units per pallet determine the speed and subtlety of handling. On the other hand, total outbound and number of outbound orders indicate how tradable a particular SKU is. The total outbound and the number of outbound orders is represented as different attributes despite the fact of sharing some mutual information. It is done on purpose, since both the demand size and the demand frequency are important for the research. It is also worth to note that the feature "number of outbound orders" is calculated based on arisen demand from 2017-02-06 to 2018-02-13 (537,791 orders in total).

3 Methodology

3.1 Data Preprocessing

The first relevant problem we noticed was the excessive presence of missing data. Such features as "unit price", "pallet gross weight", "pallet height" and "units per pallet" contain 710(31.1%), 371(16.3%), 787(34.5%) and 295(12.9%) missing values respectively. Moreover, 1070(47%) observations include at list one missing value and all four features are missing in 208(9.1%) observations. Since each observation stands for a real tradable product, we cannot afford to drop such precious information. Since such naive methods for treating missing data as replacement by the feature mean or may insert a bias [9], we have decided to impute missing values using the nearest neighbor imputation (NNI). NNI, justifying its name, imputes missing values using values calculated from the k nearest neighbors. The nearest neighbors are found by minimizing a distance function, in our case the Euclidean distance [10].

According to Chen and Shao, NNI has two pivotal benefits [11]. Firstly, the method provides asymptotically unbiased and consistent estimators for population means. Secondly, the NNI method is expected to be more robust against model violations than methods based on parametric models, such as ratio imputation and regression imputation. An important parameter for the NNI method is the value of k. In this study we used k of 10 which is suggested by Batista and Monard [12].

Since anomaly detection and clustering algorithms incorporated to this research utilize distances between points as a metric, attribute scaling becomes a necessary prerequisite. This research applies Z-score attribute standardization relying on the research by Mohamad and Usman, which concludes with the statement that the application of Z-score standardization prior to K-means clustering leads to more accurate result compared to decimal scaling and min-max normalization [13].

In light of the fact that clustering algorithms, which use distances between points as a metric, such as k-means, are extremely sensitive to presence of anomalies in dataset, it is crucial for quality of cluster analysis to identify and remove such anomalies. For this purpose, the local outlier factor (LOF) was applied. Despite the fact that LOF is a relatively old algorithm, Campos, et al. conclude with the statement that even after 16 years of active research LOF remain the state of the art, especially in datasets with possibly larger amounts of outliers [14].

The distinguishing feature of LOF is utilization of reachability distance as an additional measure. Let *k*-distance (*A*) be the distance of the *A* to the *k*-th nearest neighbor. Based on that the local reachability density (*lrd*) is defined as follows:

$$lrd(A) = \frac{|N_k(A)|}{\sum_{B \in N_k(A)} \max\{k - distance(B), d(A,B)\}}, \tag{1}$$

where $N_k(A)$ is the set of *k* nearest neighbors. Eventually the local reachability densities are compared with those of the neighbors in order to calculate LOF:

$$LOF(A) = \frac{\sum_{B \in N_k(A)} \frac{lrd(B)}{lrd(A)}}{|N_k(A)|}. \tag{2}$$

The LOF should be interpreted the following way, a value around 1 or less indicates an inliner, while values significantly larger than 1 clearly indicate an outlier.

3.2 Cluster Analysis

In this research we take advantage on the pivotal mechanics of mean-shift clustering to discover the number of "clots" and potential clusters respectively. In fact, mean-shift is a centroid-based algorithm, which works by iteratively sorting out candidates for centroids to be the mean of the points in the considered region. It starts with an initial estimate *x*. Then the kernel function $K(x_i{-}x)$ calculates the weight of nearby observations for further reestimation of the mean. After that such candidate-centroids are filtered in order to eliminate duplicates forming the final set of centroids. The weighted mean of the density determined by *K* is Eq. (3) [15].

$$m(x) = \frac{\sum_{x_i \in N(x)} K(x_i - x)x_i}{\sum_{x_i \in N(x)} K(x_i - x)}, \tag{3}$$

where $N(x)$ is the neighborhood of *x*. The algorithm iteratively assigns $m(x)$ to *x* repeating the estimation until $m(x)$ converges. That is important to note that the flat kernel was used in the research Eq. (4).

$$K(x) = \begin{cases} 1, & if\, x \leq 1 \\ 0, & if\, x > 1 \end{cases}. \tag{4}$$

The key advantage of mean-shift is application-independence and clustering without assumptions on predefined shape of clusters.

The number of discovered "clots" is further used as an initial value of parameter k in *k*-Means clustering algorithm; one of the oldest and, nevertheless, commonly used methods for partitioning the observations. The algorithm attempts to clusters data separating a set of data observations into k clusters minimizing the Euclidean distance-based objective function. It repeatedly proceeds two pivotal steps, namely assignment each data point to the cluster with closest centroid and recalculation of the centroids as the mean of all the observations in that cluster until the algorithm converges forming the Voronoi diagram.

Since k-means cannot find non-convex clusters, DBSCAN (density-based spatial clustering of applications with noise) is also incorporated to the research for an extra insurance. DBSCAN is a density-based data clustering algorithm that finds a number of clusters by estimation of density distribution of corresponding observations. Unlike k-means, DBSCAN determines clusters based on the concepts of "ε-neighborhood" and "density reachability" viewing clusters as areas of high density separated by areas of low density [16].

3.3 Clustering Validation

Since the ground truth is not known, abovementioned algorithms are compared using internal cluster validation tests. Namely, silhouette index, Calinski-Harabasz index and Dunn index were calculated in order to compare clustering results based on such properties of clusters as density, shape and separability.

Silhouette validity index refers to a method of internal clustering consistency validation. The index reflects cohesion and separation of clustered data and may be defined as follows Eq. (5).

$$s(i) = \frac{b(i) - a(i)}{\max\{a(i), b(i)\}}, \tag{5}$$

where $a(i)$ is the average distance between an observation i and all the other observations, $b(i)$ is the lowest average distance between i and observations in clusters of which i does not belong to [17]. The values of index are bounded in range $-1 \leq s(i) \leq 1$. The score is generally higher for dense and well separated clusters, which corresponds to a concept of a cluster.

Calinski-Harabasz (CH-index), also known as pseudo F-statistics, is a traditional and computationally fast method for internal clustering validation. For k clusters the CH-index $s(k)$ is defined as the ratio of the between-clusters dispersion mean and the within-cluster dispersion:

$$s(k) = \frac{tr(B_k)}{tr(W_k)} \times \frac{N - k}{k - 1}, \tag{6}$$

where $tr(B_k)$ is the trace of the between-clusters dispersion matrix and $tr(W_k)$ is the trace of within-cluster dispersion matrix [18]. As a side note, the value of CH-index is generally higher for convex clusters which make it biased towards DBSCAN and other density-based clustering techniques.

Dunn index (DI) measures a compactness and well-separation of clusters. Namely, the score is higher if clustering produces a small variance between observations within one cluster keeping mean values of different clusters sufficiently far apart:

$$DI = \min_{i=1,\ldots,n_c} \left\{ \min_{j=i+1,\ldots,n_c} \left(\frac{d(c_i, c_j)}{\max_{l=1,\ldots,n_c} diam(c_k)} \right) \right\}, \tag{7}$$

where $d(c_i, c_j) = \min_{x \in c_i, y \in c_j} d(x, y)$ stands for the intercluster distance between clusters c_i and c_j, $d(x, y)$ is the Euclidean distance between observations x and y, $diam(c) = \max_{x,y \in c} d(x, y)$ is the diameter of a cluster [19]. The DI values lie in the interval from 0 to infinity such that the higher values correspond to better clustering.

3.4 Dimensionality Reduction

In this research PCA gave hope to improve the current result for two main reasons. Firstly, some attributes are interrelated and correlate with each other sharing some mutual information (Table 2). Secondly, it is quite natural for a real-world data to contain some portion of noise, for instance, due to the measurement error. Additionally, some noise may be inserted during NNI procedure. Based on that PCA could be useful as a noise-reduction tool.

Table 2. Pearson correlation coefficient of attributes.

	Expire date	Total outbound	Number of outbound orders	Pallet weight	Pallet height	Units per pallet
Unit price	−0.08	−0.07	−0.09	−0.09	−0.09	−0.04
Expiredate	1.00	0.08	0.07	−0.35[a]	−0.36[a]	0.04
Total outbound	–	1.00	0.86[b]	0.04	−0.04	−0.03
Number of orders	–	–	1.00	−0.04	0.01	0.00
Pallet weight	–	–	–	1.00	0.28[a]	0.06
Pallet height	–	–	–	–	1.00	−0.04

[a] indicates moderate linear relationship
[b] indicates strong linear relationship

PCA may be defined as a common statistical procedure that maps the data to a new coordinate system such that the first coordinate contains the greatest variance, the second coordinate contains the second greatest variance and so on. This study incorporates PCA for two major purposes. Firstly, PCA is applied after all preprocessing and clustering for data-visualization. For this purpose, exactly two principal components are derived, such that each principal component corresponds to an axis on 2-d plot. Secondly, we tried to improve the result obtained from k-means relying on noise-reduction and feature-extraction properties of PCA.

4 Experiment

4.1 Data Preprocessing

Applying the NNI we varied the value of *k* in range from 3 to 10 and did not observe a significant difference, thus, it was concluded to use *k* of 10 as suggested by Batista and Monard [12]. As the result, all the missing values were successfully imputed. Right after that the data was transformed using *Z*-score standardization.

Since clustering algorithms that use distances between points as a metric, such as *k*-means, are extremely sensitive to presence of anomalies in dataset, the LOF algorithm was applied iteratively varying k in reasonable range from 5 to 30. What was interesting, in each run exactly 114 outliers (5.1%) have been detected. Roughly speaking, an outlier is an observation that deviates too much from other observations as to arouse suspicion that it is generated by a completely different mechanism. However, for inventory management outliers are the most precious pieces of data. In this particular case identified outliers stand for small SKUs with very high unit price, SKUs with high total demand, but low demand frequency, SKUs with extremely high or extremely low pallet weight and so on. These findings were reported to the representative of the "Trialto Latvia LTD", however, they were also dropped from the dataset prior to cluster analysis for the purpose of clustering validity. As the result 2165 observation remained.

Right after that mean-shift with a flat kernel is applied to estimate the number of blobs and potential clusters respectively. The algorithm detected 26 potential clusters, however, the vast majority of these clusters (18) consists of 3 or less observations. These observations are, in fact, outliers that were able to survive the anomaly detection with LOF. We trimmed 35 new-found outliers and repeated the procedure iteratively. In the second iteration 28 more outliers were removed and 10 potential clusters were discovered with no single cluster containing less than 6 observations. Eventually, 63 additional anomalies missed by LOF (2.9%) were detected and trimmed. As the result 2102 observation remain in the dataset.

4.2 Cluster Analysis

10 clusters defined by mean-shift (Fig. 1) were a starting point. Mean-shift served the research well discovering existing blobs and potential outliers, however, the cluster analysis by itself is quite poor, according to all the indexes (Table 2).

DBSCAN showed even worse results due to the fact that the dataset contained quite convex clusters with large differences in densities. This also implies the problem that DBSCAN considers a significant portion of the observations as noise. For instance, the algorithm with ε (local radius for expanding clusters) of 0.747 defined 10 clusters classifying 173 observations (8.2%) as noise (Fig. 1). Since each observation stands for a product and we were already forced to trim 177 outliers, we cannot afford to lose any more precious data. Besides that, the clustering result is generally very poor based on all the indexers.

Since *k*-means clustering requires some prior guessing about the number of clusters [20], we take key advantage on mean-shift assuming that the number of found "clots" corresponds to a nearly-optimal value of *k* parameter in *k*-means clustering. In our case this guess turned out to be correct, namely *k*-means with k of 10 results the best segmentation based on the silhouette index and DI, holding one of the largest value of the CH-index at the same time. It may also be observed that k of 9 demonstrates a quite promising result (Fig. 2).

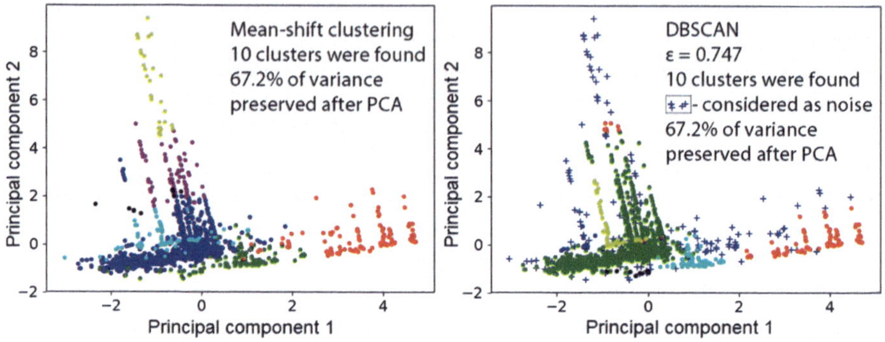

Fig. 1. Cluster analysis by mean-shift and DBSCAN visualized via PCA.

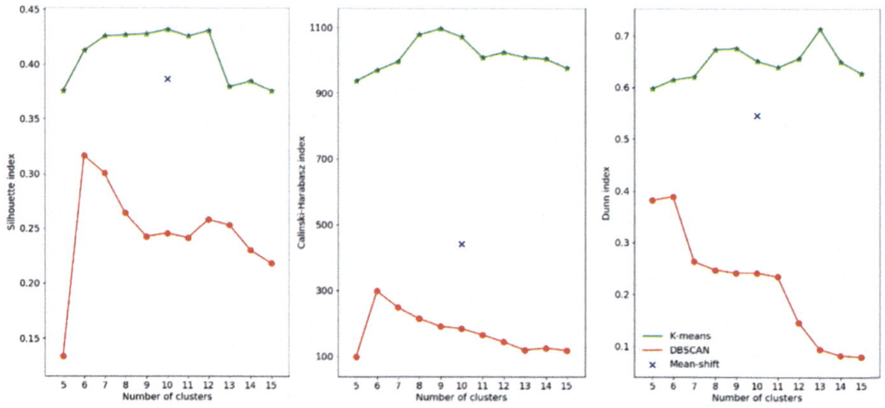

Fig. 2. Values of validity indexes depending on the number of clusters.

4.3 Further Improvements via PRINCIPAL Component Analysis

Eventually, the incorporation of PCA was indeed a right decision, since gradually lowering the number of principal components to 4 we managed to improve the initial result on 11% according to silhouette validity index preserving more than 95% of variance (Table 3, Fig. 3). It is also important to emphasize that such indexes as *CH* and *DI* rose even more. However, these figures are less representative due to excessive sensitivity of such indexes to dimensionality.

Table 3. 10-means after PCA.

Number of components	Silhouette	*CH*	*DI*	Variance preserved (%)
Prior to PCA	0.432	1072.0	0.639	100
6	0.441	1120.1	0.662	99.10
5	0.452	1163.5	0.667	97.82
4	**0.475**	**1368.3**	**0.750**	**95.11**
3	0.456	2173.1	0.746	82.09
2	0.454	3641.7	0.581	67.21

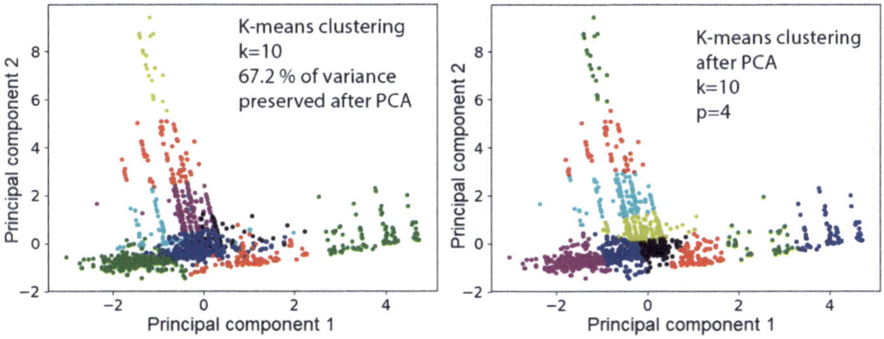

Fig. 3. *K*-means clustering prior to and after PCA with 4 principal components.

5 Conclusion

In conclusion, it is worth to mention that with the combination of PCA and *k*-means we managed to achieve decent SKUs' segmentation, according to internal validity tests. On the other hand, due to the fact that the dataset comprises quite convex clusters with large differences in densities, DBSCAN was quite inefficient.

Taking into account the fact that the cluster analysis incorporated features with undeniable impact on the inventory management beyond that utilized by a classical ABC approach, each cluster is homogeneous enough to be treated under a common inventory policy. Thus, the proposed methodology is expected to be efficient for real-world inventory control problems of high dimensionality.

References

1. Ernst, R., Cohen, M.A.: Operations related groups (ORGs): a clustering procedure for production/inventory systems. J. Oper. Manage. **9**(4), 574–598 (1990)
2. Jackson, I., Tolujevs, J., Reggelin, T.: The combination of discrete-event simulation and genetic algorithm for solving the stochastic multi-product inventory optimization problem. Transp. Telecommun. J. **19**(3), 233–243 (2018)

3. Srinivasan, M., Moon, Y.B.: A comprehensive clustering algorithm for strategic analysis of supply chain networks. Comput. Ind. Eng. **36**(3), 615–633 (1999)
4. Egas, C., Masel, D.: Determining warehouse storage location assignments using clustering analysis. In: 11th IMHRC Proceedings on Progress in Material Handling Research, Milwaukee, Wisconsin, USA, pp. 22–33 (2010)
5. Desai, V.L.: Evaluating Clustering methods for multi-echelon (r, Q) policy setting. In: Proceedings of the IIE Annual Conference, p. 352. Institute of Industrial and Systems Engineers (2007)
6. Yang, C.L., Nguyen, T.P.Q.: Constrained clustering method for class-based storage location assignment in warehouse. Ind. Manage. Data Syst. **116**(4), 667–689 (2016)
7. GitHub repository "pumpy". https://github.com/Jackil1993/pumpy. Accessed 10 July 2018
8. Pedregosa, F., Varoquaux, G., Gramfort, A., Michel, V., Thirion, B., Grisel, O., Blondel, M., Prettenhofer, P., Weiss, R., Dubourg, V., Vanderplas, J.: Scikit-learn: machine learning in Python. J. Mach. Learn. Res. **12**(1), 2825–2830 (2011)
9. Zloba, E., Yatskiv, I.: Statistical methods of reproducing of missing data. Comput. Model. New Technol. **6**(1), 51–61 (2002)
10. Jonsson, P., Wohlin, C.: An evaluation of k-nearest neighbour imputation using likert data. In: 10th International Symposium on Software Metrics, pp. 108–118 (2004)
11. Chen, J., Shao, J.: Jackknife variance estimation for nearest-neighbor imputation. J. Am. Stat. Assoc. **96**(453), 260–269 (2001)
12. Batista, G.E., Monard, M.C.: A study of K-nearest neighbour as an imputation method. HIS **87**, 251–260 (2002)
13. Mohamad, I.B., Usman, D.: Standardization and its effects on K-means clustering algorithm. Res. J. Appl. Sci. Eng. Technol. **6**(17), 3299–3303 (2013)
14. Campos, G.O., Zimek, A., Sander, J., Campello, R.J., Micenková, B., Schubert, E., Assent, I., Houle, M.E.: On the evaluation of unsupervised outlier detection: measures, datasets, and an empirical study. Data Min. Knowl. Disc. **30**(4), 891–927 (2016)
15. Comaniciu, D., Meer, P.: Mean shift: a robust approach toward feature space analysis. IEEE Trans. Pattern Anal. Mach. Intell. **24**(5), 603–619 (2002)
16. Tran, T.N., Drab, K., Daszykowski, M.: Revised DBSCAN algorithm to cluster data with dense adjacent clusters. Chemometr. Intell. Lab. Syst. **120**, 92–96 (2013)
17. Rousseeuw, P.J.: Silhouettes: a graphical aid to the interpretation and validation of cluster analysis. J. Comput. Appl. Math. **20**, 53–65 (1987)
18. Caliński, T., Harabasz, J.: A dendrite method for cluster analysis. Commun. Stat. -Theory Methods **3**(1), 1–27 (1974)
19. Yatskiv, I., Gusarova, L.: The methods of cluster analysis results validation. Transp. Telecommun. **6**(1), 19–26 (2005)
20. Cheng, Y.: Mean shift, mode seeking, and clustering. IEEE Trans. Pattern Anal. Mach. Intell. **17**(8), 790–799 (1995)

Transport Economics and Policy

The Socio-Economic Impact of Green Shipping: A Holistic View from the Baltic Sea Region

Gunnar Prause[1,2(✉)], Karin Reinhold[1], Marina Järvis[1], Eunice Olaniyi[1], and Piia Tint[1]

[1] School of Business and Governance, Tallinn University of Technology, Akadeemia tee 3, 12618 Tallinn, Estonia
gunnar.prause@ttu.ee
[2] Wismar Business School, Wismar University, Philipp-Müller-Str. 14, 23966 Wismar, Germany

Abstract. Green shipping enjoys high attention on the environmental agenda of the Baltic Sea Region (BSR). After the implementation of the SECA regulation in BSR in 2015, the next step towards cleaner shipping will be the NECA regulation from 2021. Both emission control measures represent a partial but important position in the socio-economic impact of green shipping comprising occupational health, safety and environmental issues.

It is still not clear if green jobs are safe for workers and this poses new challenges for occupational health and safety—such as new emerging risks related to new technologies, dangerous chemicals, processes, workforce and demographic changes. Traditionally, shipping is perceived to be a relatively dangerous industry since maritime sector workers are often exposed to long hours of demanding working conditions, isolation, rigid organisational structures and high levels of psychological stress and physical fatigue that could lead to occupational diseases, accidents and harm.

The presented research is empirically validated by survey results, primary data analysis and expert interviews and explores the possible occupational health risks of the Baltic Sea shipping workers to determine their work-health related challenges. The results are further discussed vis-a-vis the impact of SECA regulations on maritime workers.

Keywords: SECA regulations · Clean shipping · Maritime occupational health · Socio-Economic risks

1 Introduction

Scientific studies show that air pollution from international shipping accounts for approximately 50,000 premature deaths every year in Europe [1], an annual social cost of more than €58 billion [2] implying that ship emissions are dangerous to health. When in the air, SO_2 and NO_x from ships emissions convert to fine particles, sulphates and nitrate aerosols and together with black carbon increase the health impacts of shipping pollution. These particles are linked to premature deaths because when these

I. Kabashkin et al. (Eds.): RelStat 2018, LNNS 68, pp. 615–624, 2019.
https://doi.org/10.1007/978-3-030-12450-2_59

particles get into the lungs, they diffuse into the bloodstream and trigger inflammations which eventually cause heart and lung failures and could lead to cancer [3].

Consequently, the International Maritime Organization (IMO) and the European Parliament (EP) decided in 2005 and 2012 to establish the Sulphur Emission Control Areas (SECA) in Northern Europe where operating ships from 2015 must use fuel with a low sulphur content not exceeding 0.1% [4]. At the end of 2015, European Union approved in the frame of the BSR Interreg Programme the "EnviSuM (Environmental Impact of Low Emission Shipping Measurements and Modelling Strategies) project" to assess the technical efficiency and the socio-economic impacts of clean shipping solutions in the Baltic Sea Region (BSR). After two years of project activities, some of the published results of the socio-economic impacts of the regulation focused majorly on emission measurements, the efficiency of abatement technologies and economic impacts of SECA regulations on the maritime stakeholders [5, 6].

However, the transition towards a green economy and the constant pressure on the economy and environmental issues has opened up new challenges for Occupational Health and Safety (OSH) which includes potential exposure to traditional occupational risks, emerging risks from new technologies, working processes and workforce demographic changes (diverse workforce) [7]. This means that there is need to ensure that green jobs contribute to the protection of ecosystems and biodiversity, reduce energy, minimise materials and water consumption through high-efficiency strategies, decarbonise the economy and minimise generation of all forms of waste and pollution [6]. In addition, green jobs must symbolise the quality and stable jobs, sound safety and healthy working conditions, adequate wages and good career perspective as well as ensuring workers' right, social dialogue and protection [7].

Occupational safety issues in maritime sector have been discussed from the point of maritime accidents [8], analysis and prevention [9] and safety culture assessment [7, 10]. A number of previous studies [i.e. 11, 12, and 13] demonstrated that maritime safety depends on humans, organisations and technical components of the ship system, the marine environment factors as well as in the implementation of the proactive safety management system. In spite of this, there are still limited studies on occupational health (i.e. health risk assessment in shipping) when compared to other sectors. This is surprising since traditionally, shipping is perceived to be a relatively dangerous industry that requires proper management of occupational health and safety, security and environmental issues that are focused on risk assessment and analysing control measures [12, 14, 15 and 11].

In order to prevent risks or to be able to handle them, an efficient risk management process has to be implemented which is based on the generic management process. This encompasses the following steps: risk identification, analysis, handling and control.

The risk identification step is often considered as the most important step since only identified risks can be managed [15] and this is the major focus of this study.

During risk analysis, especially in transportation management, Schröder and Prause explained that the gathered risks are first assessed by indicating the likelihood of occurrence and the possible damage [16]. The risks are further prioritised in the order of risk handling management. Methods supporting risk identification and analysis are commonly brainstorming and the failure mode and effects analysis (FMEA). Risk handling is the third step of the risk management process and entails drawing up strategies to handle risks for example; risk avoidance, reduction, transfer, and sharing or full risk embrace.

In this vein, the study aims to identify the health risks associated with working on the ship in the Baltic Sea by exploring what characterizes as hazards or risks introduced by being a maritime worker in the Baltic Sea Region. It further examines if and how the SECA regulations have influenced the health status conditions of these seafarers. The result has the potential to enhance the understanding of enabling and inhibiting factors that can foster ensuring the safety and wellbeing of maritime workers.

The work is organised as follows: after the introduction, next section is the theoretical background of SECA regulations and sulphur related rules as well as the importance of occupational health and safety (OSH) and environmental factors in ports area and in shipping vessels. Then, the system of methods used for this research is described which is followed by the analysis of the primary and secondary data complemented by ex-pert interviews. The results and their implications are discussed in section four and in the last section, the article was concluded.

2 Literature Review

Maritime working environment faces physical, ergonomic, chemical, biological, psychological and social factors that may be responsible for occupational diseases, accidents and other harm [11]. The shipping industry is a complex system—a closed 'social milieu' with all necessary competence on the board [17, 18]. This is an indication for unusually long period aboard exposing the workers to heavy workloads, long shifts, demanding working conditions, isolation, rigid organizational structures and high levels of psychological stress and physical fatigue [19, 20].

SO_X and NO_X emissions not only cause various harmful effects on human health, they affect agricultural crops, ecosystems and material corrosion, and in turn, reduce the actual supply of labour, increase the user cost of capital and decrease productivity. The social costs of these emissions are the costs estimates of the monetized damages associated with their harmful effects [21].

In the shipping industry, the enforcement of the legislative framework is complex and weak due to the organisational international structure that includes more than 50 International conventions of Maritime Organisation (IMO) and International Labour Organisation (ILO). As a result, the European Framework Directive on Safety and Health at Work (1989/391/EEC) establishes the general principles for managing safety and health [22]. Yet, there are still limited data about applications of the relevant guidelines for practice [4].

The benefits of managing health and safety at workplaces are widely recognised and the SIRC (2018) in several studies concluded that it is still not clear if green jobs are safe for workers, which poses new challenges for occupational health and safety—such as new emerging risks related to new technologies, dangerous chemicals, processes, workforce and demographic changes [15].

Occupational health and safety issues are usually subsumed into environmental and social risks. The health risks of the population impacted by shipping operations, safety issues of maritime workers on-board or ashore including also the exposure to shipping emissions are all inclusive in the environmental risks. On another hand, an important subcategory of social risks represents "safety deficiencies" due to unknown safety regulations or/and lack of awareness about the relevant regulations or different legal systems (called "lack of knowledge") [23].

Taking a bottom-up approach, the authors, clustered the risks in main categories describing the sources of risks. The main categories were chosen in accordance with the three pillars of sustainability, i.e. economical risks, environmental risks and social risks as well as the risks sources used by Schröder and Prause [16, 24].

Occupational Environment: Operations, facilities and different activities may create adverse effects on the environment [12]. Effective maritime environmental management needs to focus also on the potential impact of the work environment on occupational risks. In Baltic Sea region, cancer has been said to be very common among ship workers compared to land-based occupations, especially lung cancer and mesothelioma (various general diseases such as cardiovascular diseases, work-related stress and depression, hearing impairments have also been diagnosed among seafarers) [23]. Studies have also highlighted the negative safety effects of stress, fatigue, staffing level, alienation and heavy load as the essential causes of operation errors common in maritime transport [i.e. 1, 25]. Collisions, fires/explosions, personal occupational accidents, homicide, suicide and diseases are also major occupational hazards for seafarers [17].

Occupational Safety: Occupational accidents in maritime sector includes human factors, prevention measures and interventions [18, 22]. Diverse studies i.e. Hetherington et al. and Nævestad have addressed the different issues within maritime safety and security and have demonstrated the importance of the effective combinations of different related factors. These factors include; human factors, organisational factors (social organisation of the personnel on board, operating level), economic pressure, teamwork, social skills, knowledge and training, languages skills, situation awareness, decision-making, communication, automation and technological innovations and safety culture. These factors will help to build the framework related to the identified risk [17, 26].

Occupational Health: Healthy workers are a vital prerequisite for social, environmental and economic development and sustainability [7]. The benefits of having healthy workers and safe working conditions are related to the high possibility to recruit skilled workers, an increase of labour productivity and workers' motivation, increased operational performance, the reduction of social and economic costs of occupational accidents and diseases [12, 16]. Taking into account these issues, green jobs should be decent; in particular, they should be a healthy and safe environment most especially for the workers' health and their well-being [24].

A conspicuous gap in the maritime sector is the lack of reliable and comparable data on the exposure to occupational hazards of the workers, how the possible occupational health risks-safety regulations integrate the everyday operation management into the complex legislative framework, and how this framework is implemented to ensure safe and healthy working environment [23]. According to WHO, a surveillance system in the field of occupational health must collect, analyse and integrate data which enables monitoring workers' health and ensure early detection of significant change caused by working conditions, environment and pollution, an organization of work [2]. In other words, occupational health and safety (OSH) systems in different countries must be able to integrate different activities such as workers' health surveillance (analysis of mortality, occupational disability, occupational accidents and diseases, absenteeism, lifestyle and unhealthy behaviour), and monitoring of working conditions (risk assessment and management) and the first step to this is the identification of the associated risks.

3 Methodology

First, a literature review identified the existing research gap for this study and it targets the identification of risks and the impact on health and works safety of sea workers in the context of the SECA regulations. For this purpose, the research design constitutes as follows:

The primary empirical evidence in this paper is based on the qualitative research style according to Talja [14]. Here, the complexity of the research requires personal interviews because risk management addresses a sensitive issue of a workplace and it was important to build trust with the different respondents for data gathering.

Four (4) experts in the field of maritime OSH: occupational health physician, labour inspector specialized in the maritime sector, a statistician in occupational accidents and diseases from Labour Inspectorate, and head of the Estonian Seamen's Independent Union were interviewed. The interviews were face-to-face, structured and open–ended due to the exploratory nature of the research. The purpose of these interviews was to gain knowledge on working environment, health and safety of maritime workers, identify potential risk factors, gather data on possible health problems and identify the steps to-wards progressive improvement in the field of OSH.

To identify applicable accounts for data interpretation, content analysis of the interview data was made. Each response was put in a grid to classify the account of each respondent. Accumulated reflective impression of the synopses and evaluations of the data were prepared to understand the relationship between the multiple sources of evidence. This was followed by the interpretation and narration of the data [14].

Lähteenmäki-Uutela et al. drafted a methodology for the assessment of the social-economic impact of the SECA regulations on the BSR for the EnviSuM project, which was used to for the empiric measures of this studies and comprising a BSR-wide survey, secondary data analysis, measurements, expert's structured interviews and case studies [27].

4 Findings

4.1 Occupational Risks from Working on a Ship

The interviews with specialists highlighted the main occupational hazards in the maritime sector. From the bottom-up clustering approach (i.e. occupational environment, safety and health), the identified risks are categorised as follows:

Occupational Environment: This category revealed noise, especially in cold periods (icy conditions), vibration, particularly during stormy weather and while arriving at the port, 3) physically strenuous work, psychologically strenuous work. This effect forms the occupational environment also include vibrations from the hull, exhausts and skin contact with oils and physical workload. According to a statistician and occupational health physician, the main hazards to diagnose environmental occupational diseases among seafarers to be a physical strain, repetitive movements and compulsory postures.

Occupational Safety: Occupational accidents mainly occur while slipping, tripping and falling (sprains, dislocations, strains), while doing manual handling of loads (mainly back injuries) or working with chemicals (mainly dermal corrosions or eye damages). Interview with labour inspector revealed that there is a lack of resources to conduct comprehensive inspection over the compliance with requirements of OSH in the maritime sector. The head of the Estonian Seamen's Independent Union confirmed the statement and added that working conditions are observed only occasionally while the most complicated conditions occur during arriving at the port.

In larger vessels, the safety management is usually efficient and regular working environment monitoring is evident. In smaller vessels, the conditions may be considerably harder and workers may be exposed to different health risks. Additionally, the experts report the formal status of workers' representatives in the maritime sector—often, they are elected but no practical activities are followed and performed by them. As a result, workers have no possibility to express their concerns about the working conditions to trustworthy persons.

Occupational Health: The interview with occupational health physician revealed that the ordinary health check-ups (to receive health certificate for working in the maritime sector) do not examine the workers' health problems in-depth. The examination takes place after every 2 years and largely based on worker's health declaration. It usually lasts about 30–45 min depending on patient and problems. During the visit, visual check, audiometry, blood and urine tests are conducted. When serious problems are detected, the occupational health physician involves specialists.

The toxic damages from chemicals are tracked with haemogram with leukogram (hB-CBC-5Diff). When changes are detected, medically, the reason behind may be complex making it difficult to conclude if the changes have occurred because of the working environment or a person's life-style. Toxic damages are detected from urine sample as well as changes in liver and/or kidney functions are indications of health problems but also remains unclear whether the changes have occurred due to alcohol consumption or from exposure to toxic chemicals at work or in the environment. In the vessels, there can be various occupational exposure to toxins and chemicals, which may contribute to cancer development. There are a few cases per year when the seamen's health indicators do not meet the requirements but anomalies are only detected in the occurrence of other diseases like diabetes, cancer, liver and kidney failure. Some other health issues are either hearing or visual impairment.

4.2 SECA Regulations and Air Quality Measurement Results

Regarding the impact of the SECA regulations on the health and safety of the maritime workers, the empirical results of the EnviSuM project revealed new results. Barregard and Molnar made investigations in the port area of Gothenburg using the measurement results during the project at different places and were able to estimate the number of reduced extra deaths due to SOx emissions in BSR to be about 1000 deaths [28]. Germany and Russia are the countries with the highest numbers of reduced deaths due to their high population.

Generally, the investigations of the EnviSuM project brought to light the air quality in BSR improved by at least 70% after then enforcement of SECA regulations. Repka and Jonson gave a detailed analysis of the results that show a decrease in air pollution from shipping to be 71% in Norway and around 6% in Russia [19, 29]. In total, the sulphur deposition in the BSR has decreased by 50% from 2014 to 2016. Borkowski analysed the air quality in the urban area of the Polish Tri-city (Gdynia–Sopot–Gdansk) and calculated the SOx pollution caused by port operations from ships and identified significant SOx concentrations in close ports areas which are up to 20 times higher than in distance-port areas affirming that close proximity to shipping increases exposure to pollutants [30].

5 Discussion

The result from the interviews regarding maritime occupational hazards and risks confirms a previous study on a Swedish merchant ship by Forsell et al. Their studies show that the most commonly reported problems among seafarers on this ship are noise (70% among deck respondents, 83% for engine and 71% for service), the risk of an accident (deck 67%, engine 77% and service 64%) [10].

From the result of this study, there appears to be a weak connection between the health inspection and workplace risk assessment. It is evident that workplace risk assessment may not always be adequate. For example, if health inspection is conducted according to Maritime Safety Act which requires a health certificate to confirm that the state of health of person wishing to enter a seafarer's contract of employment is good, then a follow up should be made to monitor the state of health of these employees. This is contrary to what was discovered that most of the time, there are no connections to the particular vessel, working conditions and the health of the workers.

In like manner, it appears as if maritime accidents is being underreported due to weak a OSH system where self-employed entrepreneurs are not motivated to report accidents to the Labour Inspectorate as well as to analyse the possible causes. Corroborating other studies [i.e. 9, 31] underreporting of maritime accidents could be a problem for relevant authorities trying to enhance maritime safety through legislation as well as for other stakeholders using maritime casualty statistics for their analysis. Using Estonia as an example, even though the occupational disability is integrated into the Estonian Health Insurance Fund responsible for the general health care and pension scheme, there appears to be a weak motivation from OSH legislation for the employers to deal with OSH issues and ensure good working conditions for workers at sea. OSH is still not a priority for a company's management and seen separately from their strategic management. There is no cost-benefit analysis conducted by the employers who do not systematically analyse the costs of occupational accidents or diseases as it relates to their employees.

The hazards and risks identified do not have enough evidence of occupational origin, making it difficult to explore in-depth and connections with occupational hazards determined, the result from the air quality measurement from EnviSuM has made a strong connection with the impact of close proximity with shipping activities. In this regard, if the concentration of SOx pollution in the areas close to the ports were about 20 times higher than in far areas from the ports, it would not be farfetched to suggest that indeed close proximity to shipping increases exposure to pollutants especially for workers who work long and tedious hours abroad. Additionally, a reduction in deaths associated with SOx pollution in the BSR could also imply a significant positive impact on the health of the maritime workers from the SECA regulations. This summation also correlates with the significant drop in the SOx pollution caused by port operations.

6 Conclusion

One of the objectives of this paper was an attempt to gain knowledge about occupational hazards and diseases for seafarers in the BSR. In addition, because effective maritime environmental management needs to have potential impact on work environment and on occupational risks the work also aim to find out the effect of the SECA regulation on health and wellbeing of seafarers.

Study revealed different risks that the authors classified using a bottom-up clustering approach of potential sources such as occupational environment, occupational safety and occupational health. The study also revealed that the areas in close proximity to the ports are 20 times high in SO_2 concentration than far distance area, thus seafarers' that work days and nights around and on the ships are definitely in danger of high exposure from shipping emissions and are most likely susceptible to emissions induced diseases. However, if the air quality in BSR has improved by at least 70% after the enforcement of SECA regulations, then, it is no gainsaying to infer that the SECA regulations may have positively affected the health and safety of maritime workers.

The results also revealed the importance of having a holistic approach to occupational health and safety assessment and what need to be considered in the maritime sector. These issues are essential not only to comply with relevant legislation and requirements but also to diminish possible costs and to safeguard the environmental protection and ensure healthy and safe working environment.

The causes and mechanisms behind the most serious cases would certainly need further clarification and research as well as maritime risks prioritisation and handling management.

Acknowledgements. This work is a part of the ongoing EnviSuM (Environmental Impact of Low Emission Shipping: Measurements and Modelling Strategies) project sponsored by the European Regional Development Fund. The referred results of the EnviSuM project are elaborated by the EnviSuM consortium and belong to all project members.

References

1. CEEH: Assessment of health—cost externalities of air pollution at the national level using the EVA model system, CEEH Scientific Report No. 3. Roskilde (2011). ISSN 1904-7495
2. WHO: Health in green economy: measuring health gains from sustainable development. Geneva: World Health Organisation—Public Health and Environment Department, Health Security and Environment Cluster (2012)
3. Boffetta, P., Dosemeci, M., Gridley, G., Bath, H., Moradi, T., Silverman, D.: Occupational exposure to diesel engine emissions and risk of cancer in Swedish men and women. Cancer Causes Control **12**(4), 365–374 (2001)
4. Knapp, S., Franses, P.H.: Does ratification matter and do major conventions improve safety and decrease pollution in shipping? Mar. Policy **33**, 826–846 (2009)
5. Pukkala, E., et al.: Occupation and cancer-follow-up of 15 million people in five Nordic countries. Acta Oncol. **48**(5), 646–790 (2009)
6. Reinhold, K., Kalle, S., Palju, J.: Exposure to high or low frequency noise at workplaces: differences between assessment, health complaints and implementation of adequate personal protective equipment. Agron. Res. **12**(3), 895–906 (2014)
7. Olaniyi, E.O., Atari, S., Prause, G.: Maritime energy contracting for clean shipping. Trans. Telecommun. **19**(1), 31–44 (2018). https://doi.org/10.2478/ttj-2018-0004
8. Akyuz, E., Celik, M.: Utilisation of cognitive map in modelling human error in marine accident analysis and prevention. Saf. Sci. **70**, 19–28 (2014)
9. Håvold, J.I.: Safety-culture in a Norwegian shipping company. J. Saf. Res. **36**, 441–458 (2005)
10. Forsell, K., Eriksson, H., Järvholm, B., Lundh, M., Andersson, E., Nilsson, R.: Work environment and safety climate in the Swedish merchant fleet. Int. Arch. Occup. Environ. Health **90**, 161–168 (2016)
11. Olaniyi, E., Prause, G., Boyesen, J.: The impact of SECA regulations on clean shipping in the BSR. In: Ölçer, A, Kitada, O., Dalaklis, D., Ballini, F. (eds.) Trends and Challenges in Maritime Energy Management Book Series of WMU Studies in Maritime Affairs, vol. 6, no. 1. Springer International Publishing (2018a). https://doi.org/10.1007/978-3-319-74576-3
12. Antão, P., Calderón, M., Puig, M., Michail, A., Wooldridge, C., Inov, R.M.: Darbra: identification of occupational health, safety, security (OHSS) and environmental performance indicators in port areas. Saf. Sci. **85**, 266–275 (2016)
13. Devanney, J.: Uses and Abuses of Ship Casualty Data. Centre for Tanker Excellence (2009)
14. Talja, S.: Analyzing qualitative interview data: the discourse analytic method. Lib. Inf. Sci. Res. **21**(4), 459–477 (1999)
15. SIRC: Studies on the various issues related to the transnational composition of ship crews are available on the Seafarers' International Research Centre Website (2018). http://www.sirc.cf.ac.uk/SIRC_free_online_reports.aspx. Accessed 01 May 2018
16. Schröder, M., Prause, G.: Risk management for green transport corridors. J. Secur. Sustain. Issues **5**(2), 229–239 (2015). https://doi.org/10.9770/jssi.2015.5.2(8)
17. Hetherington, C., Flin, R., Mearns, K.: Safety in shipping: the human element. J. Saf. Res. **37**, 401–411 (2006)

18. ILO: Appendix IV: Draft Guidelines concerning a statistical definition of employment in the environmental sector. In: The Nineteenth International Conference of Labour Statistics, ICLS 2013. ILO, Geneva (2013)
19. Repka, S.: Benefits of SECA compliance. In: XVII International Environmental Forum "Baltic Sea Day", 22–23 March 2016, St. Petersburg, Russia (2018)
20. Rosendahl, K.E.: Social costs of air pollution and fossil fuel use—a macroeconomic approach, Statistics Norway (1998). ISBN 82-537-4542-7
21. Sanchez, A.B., Poschen, P.: The social and decent work dimensions of a new agreement on climate change: a technical brief. International Labour Office (2009)
22. Moreira, S., Vasconcelos, L., Santos, C.S.: Sustainability of green jobs in Portugal: a methodological approach using occupational health indicators. J. Occup. Health **59**, 374–384 (2017)
23. Transport & Environment: Air pollution from ships (2017). https://www.transportenvironment.org/what-we-do/shipping/air-pollution-ships. Accessed 28 June 2018
24. Schröder, M., Prause, G.: Transportation of dangerous goods in green transport corridors - conclusions from Baltic Sea Region. Trans. Telecommun. **17**(4), 322–334 (2016). https://doi.org/10.1515/ttj-2016-0029
25. Allen, P., Wadsworth, E., Smith, A.: Seafarer's fatigue: a review of the recent literature. Int. Marit. Health **59**, 81–89 (2008)
26. Nævestad, T.-O.: Safety culture, working conditions and personal injuries in Norwegian maritime transport. Mar. Policy **84**, 251–262 (2017)
27. Lähteenmäki-Uutela, A., Repka, S., Haukioja, T., Pohjola, T.: How to recognize and measure the economic impacts of an environmental regulation: case SECA. J. Clean. Prod. **154**, 553–565 (2017)
28. Barregard, L., Molnar, P.: Premature deaths due to emissions from shipping in the Baltic—some preliminary results. In: XVII International Environmental Forum "Baltic Sea Day", 22–23 March 2016, St. Petersburg, Russia (2018)
29. Jonson, J.E.: Regional air quality modelling. In: XVII International Environmental Forum "Baltic Sea Day", 22–23 March 2016, St. Petersburg, Russia (2018)
30. Borkowski, T.: Changes in air quality in the Tri-City (Gdynia, Sopot and Gdansk) related to the operation of seaports and changes in SECA requirements. In: Maritime University of Szczecin, EnviSum Project meeting in Tallinn, 11–12 October 2017 (2017)
31. Ek, Å., Runefors, M., Borell, J.: Relationships between safety culture aspects—a work process to enable interpretation. Mar. Policy **44**, 179–186 (2014)

Impact of Joining the European Union on the Development of Transport Policy in the Republic of Latvia

Juris Kanels(✉)

Transport and Telecommunication Institute,
1 Lomonosova Str., 1019 Riga, Latvia
kanels.j@tsi.lv

Abstract. Over the past 25 years Latvia has rapidly evolved into a modern European state. Dramatic changes in the political and economic system of Latvia also accompanied the reforms in the field of public administration. Public administration reforms started in 1990 and the current planning period (2014–2020) is already the fourth most significant stage in the development of public administration policy. However, in the author's opinion the greatest influence on the development of the state administration system was left to the process of preparation for Latvia's accession to the European Union and becoming a member of the EU.

Transport is one of the central elements of the process of European integration, which helps to create an internal market conducive to employment and economic growth. Transport policy is one of the common policies of the European Union that has existed since the beginnings of the EU, as it was considered essential to guarantee three of the four freedoms of the common market set out in the Treaty of Rome in 1957—the freedom of movement of people, services and goods.

There is no doubt that also Latvia's transport sector has evolved as a member of the European Union. The implementation of the EU legislation, technical, social and environmental standards has taken place, and very significant investments have been made for infrastructure development.

At the same time, it is difficult to find any comprehensive report on the particular impact of accession to the EU on the Latvian transport sector and the consequences of this accession. Therefore the author made a research trying to answer following questions: What changes in the improvement of the normative base and the development of infrastructure have taken place in the transport sector of the Republic of Latvia from 1999 to 2016? How has the transport policy changed in the time after Latvia's accession to the European Union? What effects of economic, social and environmental impact in the transport sector have caused Latvia's accession to the European Union?

Current paper is a short summary of this research.

Keywords: Public administration · Integration in the EU · Impact assessment · Transport policy

I. Kabashkin et al. (Eds.): RelStat 2018, LNNS 68, pp. 625–636, 2019.
https://doi.org/10.1007/978-3-030-12450-2_60

1 Introduction

In assessing Latvia's accession to the European Union and its body of legislation *acquis communautaire* one can fully agree with the following expert conclusions: "Analysing the overall course of the reform of Latvian public administration, it follows that since 1997–1998 the progress in this regard has mainly occurred due to external pressure as a reaction to the criticism expressed by the European Union and other international organisations or donor countries or to the established conditions for benefiting from aid programmes, which cannot be evaluated unambiguously. Also, there is a mixed attitude towards the tendency that Latvia's politicians have gone the easiest way in the EU integration process—not timely preparing Latvia's position on a certain issue, but waiting for expert recommendations, and then implementing them …, i.e. Latvia has often taken a strongly reactive position [1].

In turn, the journalist and former Saeima deputy Edvīns Inkēns has described the time when Latvia joined the EU in the following way: … *even if Latvia did not enter the European Union, everything that has been done in the process of modernisation of the state is already an achievement in itself* [2].

The membership in the European Union undoubtedly also affects the functioning of the public administration system. The EU encourages the public administration to become more competitive in order to maximise the rational use of human resources, ensure the efficient functioning of institutions, provide modern, high-quality and convenient services to the population, while reducing the administrative burden and simplifying the regulatory framework. So, even if the integration process has occurred as a reaction to "external pressure", this process can still be considered useful for the Latvian state.

At the same time, the assessment of the impact of integration is rather hypothetical, since the assessment of the actual impact of joining the European Union has been largely fragmented. At this paper author focuses on analysing the development of the policy for the transport sector in Latvia as a result of accession to the EU. The analysis of this transport policy of action is carried out by using the method of ex-post impact assessment.

1.1 Use of the Impact Assessment in Public Administration

In the context of the political system, the term *"public administration"* refers to a part of the executive power divided into political government and public administration. Peters has concluded that "public administration by its nature is one of the main political institutes, because it directly ensures the functioning of the decisions made by the political elite" [3]. It fulfils the functions of the state at various levels of manifestation of power and acts in accordance with the adopted laws and the existing standards.

During the last quarter of the twentieth century, in several developed countries of the world, various administration system reforms were launched under different names, but the objectives were often similar—to improve the efficiency of services provided to citizens. However, the most popular and most widely used model is the decentralised administration system model—*"New Public Management"* (hereinafter also *"NPM"*) created in the United Kingdom in the 1980s and 1990s and actively supported in Australia, New Zealand and Canada. It envisaged replacing the classic hierarchical model of public administration characterised by unified centralised management with the wider delegation of functions to individual institutions subject to the government and the use of management methods that are usually applied in the private sector [4].

Putting reforms of the public administration system into practice, in particular, implementing the principles of the New Public Management, the evaluation and assessment becomes an essential elements of the administrative activity, which through methods and procedures designed in a certain way enables the obtaining of data that can help to understand whether specific policies and activities implemented as a part of them are executed well or not and whether these activities drive the process towards the intended results. It can be said that performance evaluation is a part of a broad, specific culture layer, also consisting of functions such as audit, inspection and quality control.

The *impact assessment* is characterised as an important strategic element in the policy-making and administrative management process in order to analyse a priori the possible positive or negative, direct and indirect, planned and accidental impact on individuals, institutions or the environment, which will be generated or have been generated by the decisions to be made, actions to be taken or projects to be implemented [5]. It relates to the main areas of public sector activities—programmes, legislation and the budget through which the government institutions implement their policies and thus affect the life of society.

The impact assessment method has also found its way of being applied to the process of European integration. The document *"Improving policy instruments through impact assessment"* prepared through the Joint Initiative SIGMA of the OECD and European Union states that, in the context of European integration, the impact assessment aims to provide the government and the private sector with information on the political, economic, social and environmental impact of implementing the Community policy or certain regulations and directives. Ideally, such an assessment should result in quantitative indicators showing the costs of implementing the EU legislation and their impact on national, local government and private sector budgets. However, even if the quantitative information cannot be obtained, qualitative information can still play an important role. For example, it is important to assess the impact of the decisions to be made on the functioning of certain institutions without specifying exactly what the costs of these changes will be. An important thing in itself is the understanding of the problems to be solved, the need for staff training and business continuity [6].

1.2 Description of Research Methods

The paper focuses on the type of impact assessment based on a combination of facts and values aimed at assessing the impact of the actions performed (interventions). In order to specify the objectives of the policy assessment and to get a clearer idea of the possible results, it is important to determine the most precise criteria for the fulfilment of the task. The assessment criteria determine the values that will be used in the assessment.

The assessment criteria should be perceived as "concepts" to be taken into account in the assessment of the social and economic impact of policy measures. In turn, by carrying out a specific assessment for each of the generally defined criteria, specific criteria should be identified, coordinating them with the key stakeholders, for example, criteria and/or standards that are consistent with the type and context of the programme being implemented [7].

In order to answer the research questions which are mentioned earlier, an impact assessment has been carried out by using the benchmarking method and the assessment criteria against which to measure the outcomes. Author has chosen to apply the following range of *qualitative criteria* to assess the economic and social impact of joining the European Union on the development of the Latvian transport sector:

- Fulfilment of the objective—have significant policy outcomes been achieved?
- Suitability (usefulness): what is the extent to which the objectives of the planned activity have been fulfilled, taking into account their relative importance?
- Relevance: what is the extent to which the objectives of the activity performed are relevant to the state priorities and policies?
- Sustainability: what is the benefit from continuing the activity performed after the completion of the project?

The paper deals with case research using the "comparative research on public management policy" proposed by Barzelay [8]. The main task of using this method is to benchmark the analysed similar cases and to explain their immediate or long-term effects (*outcomes*) in order to try to understand why the results and consequences of the action are exactly, as they can be effectively identified, and why under similar conditions the consequences are different or vice versa—what these different cases have in common. In particular, with regard to the subject of this paper, the research goal of the analysis is to identify and understand the changes that have taken place in Latvia's transport sector after our country's accession to the European Union, and further to make generalisations and evaluate the consequences of the actions by analysing the country's activities, individual cases and their outcomes.

Barzelay suggests using one of the models—the Kingdon's model, the Baumgartner and Jones's model, or Levitt and March's model allowing one to analyse the public policy making process. The first two of these models are basically focused on analysing public policy making, while the third is focused on the organisation's ability to learn and change in the policy making process.

Separate elements of each of them have been used during the research to a greater or lesser extent. For example, the author has used Kingdon's model for the analysis in terms of the aspects of determining the political agenda in Latvia in relation to joining the EU, but it is not about assessing the possible alternatives, because it often did not take place in practice. Baumgartner and Jones's model contributes to the analysis by providing an opportunity to use the element of the conflict expansion process (for the purposes of this paper it is the period when Latvia began to prepare for membership of the EU), during which the previous forms and means of policy implementation were discredited (denied), as well as to use the element of creating faith in the necessary future structural changes (while being a part of the European Union). In this way, by using the external pressure factor, the ground for significant policy reforms was also provided in Latvia. In turn, the elements of Levitt and March's model help to capture the changes that have taken place in the transport sector in Latvia after joining the European Union.

2 Main Outcomes of the Study

2.1 Importance of the Transport Sector in the EU

Transport is one of the central elements of the European integration process, which helps to create an internal market conducive to employment and economic growth. In the context of the European Union's internal market development project, it is essential for all of the 28 EU Member States to have proper transport connections. This means the need to build the missing connections and eliminate the many technical and administrative obstacles hampering smooth traffic and trade flows and creating unnecessary congestion in the European transport system. In many cases, it is also necessary to harmonise the different transport policies of the Member States which may distort competition and to prevent barriers to market access [9].

Over the last 60 years, the transport sector has grown considerably in the EU, and it significantly contributes to prosperity and employment in Europe. The transport sector currently employs around 10 million people, i.e. 4.5% of all workers in the EU, and it accounts for about the same percentage of gross domestic product. Good transport connections are also very important to the European Union's economy in terms of exports—transport operations provide for 90% of the EU's external trade. Today, households in the European Union spend 13.5% of their income on transport goods and services, such as seasonal train tickets and recreational or business flights, therefore transport is the second largest household budget line after housing expenses [10].

2.2 Impact of the Development of the Transport Sector on the Overall Macroeconomic Indicators of the Republic of Latvia

In Latvia, as with elsewhere in Europe and in the world, transport plays an important role in the economy and in providing accessibility. During recent years, the share of transport in Latvia's GDP has been around 10%, with around 9% of the population employed in the sector. The changes that have taken place in the transport sector over the last 20 years are very significant, and they can be characterised by a variety of indicators in both qualitative and quantitative terms. Unfortunately, it cannot be denied that the contribution of the transportation and storage industry to the overall structure of the Latvian national economy has decreased in recent years. According to the Central Statistical Bureau, in 2016 the transportation and storage sector only generated 8.7% of the total value added of Latvia's GDP. For comparison, in 2000, the sector's share was 11.9%, but in 2001 it was 13.2% (Fig. 1) [11].

Latvia's integration into the European Union is definitely not the only or even the main factor affecting these sectoral development indicators. Accession to the EU creates stability, a favourable legal framework and a positive institutional environment, as well as promotes investment. These factors have positive impact factors, but apart of them there are also other, negative, influential factors, mainly, outside the European Union. In the coming years, the prospects for the transport sector of Latvia, especially for the transit in East–West direction, are likely to be negative, as Russia gradually reduces its export share crossing ports of other countries. Only with the creation of the railway line *Rail Baltica* will the North-South transport direction gain more significance. However, this is still an issue for a distant future.

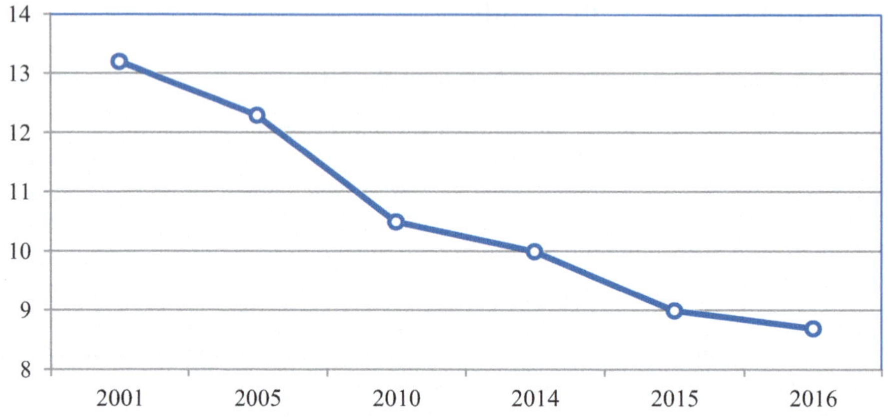

Fig. 1. Share of the transport and storage sector in the total value added of the GDP (% at current prices).

The issue of energy consumption in the transport sector as a whole and the environmental impact of this consumption are also significant. According to the statistical data, over the whole period after the restoration of the national independence, the total energy consumption in the transport sector of Latvia has remained unchanged (with a certain decline in the middle of the period). Despite the relative stability of the total energy consumption, the positive thing is that to some extent, the structure of the energy resources has changed and the use of renewable energy resources has increased in our national transport sector. Although, in 2015, the share of these energy resources in the total transport sector consumption of Latvia was only 3.9%, it is still 1.6 times more than ten years ago [12].

As an example of the positive contribution of the industry to the national economy development, the author can mention the aviation sector. According to a detailed analysis of macroeconomic data, in 2016, the total contribution of the aviation sector to the GDP of Latvia comprised 2.5% or EUR 644 million, including a direct share of the sector amounting to 0.6% of GDP or EUR 141 million, but 0.7% or EUR 172 million was ensured by the indirect or induced impact (related to goods and service supply chains), while 1.3% or EUR 331 million was due to the catalyst effect on other sectors (related to the availability of high-quality aviation services—for tourism, mail, etc.).

After Latvia's accession to the European Union, EU funds have become an important driving force behind economic development, and total EU funds used in the national economy during the period of 2007–2013 accounted for 3% of the national GDP. It is precisely this factor that is of particular importance for the transport sector, as also evidenced by data from the Ministry of Finance on the use of EU funds by sectors. As of May 2017, transport is in first place among all other sectors of Latvia, and investments made in this sector account for 26.7% of all European Union funds absorbed by the country. When comparing these figures with public funding, it can be concluded that the use of EU funds makes it possible to invest 3 times more in the transport sector than would have been possible with the state budget funds only [13].

2.3 Assessment of the Outcomes and Consequences of the Republic of Latvia's Accession to the European Union in the Transport Sector

After joining the European Union, the Latvian transport sector as a whole has obviously developed in regard to several key indicators (Table 1). The number of employees in the sector, the number of companies and their turnover are gradually increasing. But indeed, this development is not linear, since the activities of the Latvian transport sector are largely subject to the influence of various external conditions, therefore the direct impact of integration into the European Union is difficult to assess. It should not be forgotten that the performance of the transport sector, especially in internationally orientated areas of activity, such as international road transport, ports and transit, maritime transport, is affected by many interacting positive and negative factors (political or economically motivated decisions, the economic situation in the world and in individual countries, participation in the single EU market, decisions by cargo owners, the international situation, etc.).

Assessing the past activities towards the internationalisation of the national transport sector, it should be noted that attention has been focused on two aspects of integration—harmonisation of Latvian legislation in accordance with international standards, which is one way of eliminating institutional barriers between EU member states, as well as a set of activities for the implementation of certain projects supported by the European Union, in order to facilitate the integration of the national transport sector into the international transport system.

Table 1. Indicators characterising the Latvian transport sector [14].

Year Indicators	2006	2010	2012	2014	2014 versus 2006, %
Number of employees (thousands of people)	68.2	68.2	72.8	77.9	114%
Number of companies	4,701	5,570	6,303	6,951	148%
Annual turnover (EUR million)	n/a	4,008	5,402	5,383	134% (vs. 2010)

Creation of a legal framework for the activity of the sector. The main difficulties encountered in the accession negotiations in the transport sector were due to the need to agree on an extension of the time limits for the introduction of certain EU requirements in the field of road transportation. In Latvia, within the transport section, for 3 cases an agreement was reached over transition periods for the implementation of EU requirements, all of which were related to road transportation:

- Latvian road carriers, after two years of transition, had the right to freely transport domestic cargo within other EU Member States, the same as for foreign carriers in the territory of Latvia;
- for companies engaged in domestic transportation, a transitional period was established until 31 December 2006 to meet the financial security requirements of the operation;
- a transitional period was set for equipping the vehicles engaged in domestic transportation with a tachograph until 1 January 2005.

The most important problems that arose when implementing EU transport conditions were the following:

- a part of the legislative acts could not be introduced or amended in line with EU requirements until the Republic of Latvia had joined the EU (for example, on competition, cabotage transportation in the field of road transport), because by opening up its market Latvia could not receive the same rights in other EU countries;
- there were some obstacles to the drafting and introduction of regulatory enactments, for the implementation of which funding was needed;
- the issue of the implementation of European Union standards, especially in the railway sector, became topical;
- in Latvia, the transport sector had a high share of outdated machinery that did not meet certain international standards, for example, in the field of aviation and railways.

The fulfilment of the criteria set by the European Union, as well as the process of accession negotiations itself was quite successful for the country, and already on 21 December 2001, negotiations on Latvia's accession to the European Union resulted in the conclusion of the Transport Policy section of the negotiations. On 9 October 2002, the European Commission positively assessed Latvia's overall progress on the implementation of reforms in its progress report and recommended concluding the accession negotiations with Latvia in December of that year. Thus, the previous phase of the relationship ended, with Latvia becoming a fully-fledged Member State of the European Union, recognising Latvia's ability to assume the responsibilities of a Member State, including in the transport sector [15].

Impact indicators. Latvia's membership in the European Union has various positive benefits. These can be mentioned in terms of national security and involvement in global processes, investments in the country's sustainable development, as well as in terms of increasing the welfare of people. The impact of the European Union can be measured more directly by using several indicators, which, among other things, also apply to the transport sector. For each euro paid into the EU budget, Latvia recovers on average four euros (for example, in 2014, the Latvian contribution amounted to EUR 270 million while the received funding for Latvia was EUR 1,062.2 billion). The availability of the EU's structural and Cohesion funds has contributed to an annual increase in Latvia's GDP by 1.1%. Fund investments during the financial and economic crisis accounted for 70% of the total public investments in Latvia, substantially facilitating crisis management [16].

The positive effect of accession to the EU was that twofold opening up of the market occurred and Latvian companies had the right to operate freely in other EU Member States. With this opportunity, the development of transport business was promoted. There were improvements in the work quality of transport companies, road and rail safety, as well as working conditions for employees. Significant improvements were made in the social field, as joining the European Union provided workers, especially vehicle drivers, with social rights to a certain guaranteed rest period, and set a limited time for driving. In the field of aviation, significant benefits were enjoyed by airline passengers—the quality and safety of transportation improved, as Latvian airlines, which until then used aircraft not complying with EU standards, were no longer allowed to do so. Expansion of competition in the aviation market has caused a significant drop in ticket prices. Due to the access to the free labour market, transport specialists from all sectors have acquired the opportunity to choose the country of employment (especially in the maritime field), although this factor has both positive and negative effects because people leave Latvia for better job opportunities.

In almost all sub-sectors of transport, as the main negative factors of the accession to the European Union should be mentioned additional costs for businesses in order to meet the requirements of EU regulatory enactments and technical regulations, as well as increasing competition operating under the conditions of the single market. Significant financial resources were required for purchasing and equipping adequate vehicles and training, examinations and salaries for drivers, controllers and security consultants. The introduction of social guarantees for employees in the public transport sector and the imposition of EU tax rates on fuel increased the cost of public transport services with the prices of public transport services growing accordingly [17].

3 Conclusions

In response to the three research questions outlined earlier the author concludes that a direct positive correlation between Latvia's accession to the European Union and the development of the national transport sector can primarily be determined by looking at the introduction of the EU legislative and technical requirements in the field transportation and attracted EU funds to meet the country's priority needs and investments in the development of the transport infrastructure. From 2004 to 2013, total investments in the transport sector from the European Regional Development Fund and the Cohesion Fund of the EU amounted to almost EUR 1.5 billion. Without integration in the EU, none of these funds would be available, and in this case, it is very difficult to imagine, first of all, what the condition of roads in this country would be if not for the EU funds that have allowed to maintain and develop the roads at least at a 75% level, after which their condition is still not good enough. Likewise, the infrastructure of railways, ports and air traffic in Latvia would certainly be in a worse condition than it is today.

Fulfilment of performance indicators of the actions taken during the EU accession process and their impact are undoubtedly significant (Table 2). In a short period of time Latvia adopted the European Union legislation, which has facilitated the consolidation of the rule of law in Latvia and given the state the opportunity to work with the regulatory framework and technical standards that are in line with international standards. This, in turn, leads to the possibility of fully participating in the largest single European single market. Both the investments made and the legislative norms established certainly have a positive impact on the ecological situation in Latvia.

Despite some of the aforementioned negative factors, the author nevertheless believes that accession of Latvia to the European Union have mainly positive impact on the development of the whole country and particularly on its transport sector. This is evidenced both by the statistics referred to in the paper and by the process of adopting the EU legislation in the country, which ended with formal recognition by the other EU Member States—the decision to allow the Republic of Latvia to join the EU because it is able to fulfil the obligations and responsibilities of a Member State.

Table 2. Economic and social impact of accession to the EU on the development of the Latvian transport sector.

No	Criterion	Positive impact	Negative or neutral impact
1	2	3	4
1.	Fulfilment of the objective (outcome)	– Latvia has demonstrated its ability to assume and fulfil the obligations of an EU Member State and apply the common legislation – The transport sector has a significant share in the structure of GDP and employment in the country; a number of economic indicators characterising the sector's work (total number of employees and number of companies) are increasing – Considerable EU funds have been absorbed	– Basically, under the influence of external factors, the sector's share in the total added value of Latvia's GDP is decreasing – Despite the participation in the EU single market and available investments from EU funds, not all transport sub-sectors have successfully developed, as evidenced by the deteriorating condition of roads, worsening of rail and port performance, maritime transport becoming an economically insignificant field of activity
2.	Suitability (usefulness): the extent to which the objectives of the planned activity have been fulfilled	– There have been successful developments in certain sub-sectors such as aviation, road transport, demonstrating positive results and exceeding the intended objectives	– The international competition has grown, and the work of transport companies has been negatively influenced by the technical requirements and limitations imposed at the EU level
3.	Compliance with the needs of consumers and state priorities (relevance)	– The activities carried out in the sector as a whole comply with the state priorities and needs of society—Latvia's integration into the single EU transport system has taken place, traffic safety and the protection of passenger rights have improved	– Decisions on the use of EU funds have often been politically motivated, favouring certain sub-sectors (railways, ports) in comparison to the rest

(continued)

Table 2. (*continued*)

No	Criterion	Positive impact	Negative or neutral impact
4.	Sustainability	– No negative impact on the country's sustainable development caused by accession to the EU has been detected in general – Investments in the transport sector from EU funds have allowed significant improvement of the state of the infrastructure – There has been an increase in the share of renewable energy sources in the energy consumption of the sector	– The state budget resources have been used less than before (for example, the Road Fund was liquidated) for development of the sector, and a "dependence" on EU funds has formed. Consequently, it is currently not known how the construction (maintenance) of roads will be financed after 2020 -With increasing volumes of road transportation, the amount of harmful CO_2 emissions is increasing, thus contradicting the EU and Latvian policy priorities

In the author's opinion, the research significance of this paper is that for the first time in Latvia there has been an attempt to summarize and analyze the results and consequences of the state's accession to the European Union, although only in one—in the transport sector. This analysis is done using impact assessment method which is widely used in different countries. Thus, it is also possible to share internationally recognizable information about the most significant positive and negative economic and social impact of the accession to the European Union on the development of the transport sector of Latvia and to compare these results with other countries' data.

References

1. Jansone, D., Reinholde, I., Ulnicāne, I.: ES ietekme uz Latvijas publiskās pārvaldes reformu. Jaunā pārvalde. Publication by the State Civil Service. Riga 6(48), 9 (2003)
2. Inkēns, E.: Parlamenta iesaiste valsts sagatavošanā dalībai Eiropas Savienībā. Atgriešanās Eiropā Latvijas prezidentu, premjerministru, ministru un diplomātu esejās, pp. 247–264. Zinātne, Riga (2016)
3. Peters, B.G.: The politics of bureaucracy, 4th (edn.), pp. 1–360. Longman Publishers, New York (1995)
4. Cassese, S.: The age of administrative reforms. In: Hayward, J., Menon, A. (eds.) Governing Europe, pp. 128–138. Oxford University Press, Oxford (2003)
5. World Bank. https://ieg.worldbankgroup.org/evaluations
6. Project SIGMA. Improving policy instruments through impact assessment, pp. 1–52. Paper No. 31, Paris (2001). http://www.oecd.org/officialdocuments/publicdisplaydocumentpdf/?doclanguage=en&cote=ccnm/sigma/puma (2001) 1

7. Rogers, P.: Overview of Impact Evaluation. Methodological Briefs: Impact Evaluation 1. Florence: UNICEF Office of Research, pp. 1–16 (2014)
8. Barzelay, M.: The New Public Management. Improving Research and Policy Dialogue, pp. 1–172. University of California Press, Berkeley and Los Angeles, California (2001)
9. European Union Publications Office. Ieskats Eiropas Savienības politikā. Transports. Information leaflet, Luxembourg, pp. 1–20 (2014)
10. Ibid
11. Central Statistical Bureau. Value added structure and changes by kind of economic activity; transportation and storage. Database—Economy and Finances, 26 March 2018. http://data.csb.gov.lv/pxweb/lv/ekfin/ekfin_ikgad_ikp/IK10_0060.px/table/tableViewLayout2/?rxid=073f5e9f-0c35-49b3-8288-f030e2e1580a
12. Eurostat. Energy, transport and environment indicators. Statistical books (2017). http://ec.europa.eu/eurostat/documents/3217494/8435375/KS-DK-17-001-EN-N.pdf/18d1ecfd-acd8-4390-ade6-e1f858d746da
13. Ministry of Finance of the Republic of Latvia. ES fondu ieguldījums pa reģioniem un nozarēm. Riga (2017). http://www.esfondi.lv/es-fondu-investiciju-progress-latvijas-regionos-un-pasvaldibas
14. Eurostat. Mobility and Transport. Statistical pocketbooks 2009–2017. https://ec.europa.eu/transport/facts-fundings/statistics/pocketbook-2017_en
15. Cabinet of Ministers of the Republic of Latvia. Par Eiropas Komisijas 2002. gada progresa ziņojumu par Latviju. Approved by Order No. 601, 29 October 2002
16. Ministry of Foreign Affairs of the Republic of Latvia. Ieguvumi no Latvijas dalības Eiropas Savienībā. http://www.mfa.gov.lv/arpolitika/eiropas-savieniba-arpolitika/sabiedribas-informesana-es-jautajumos/ieguvumi-no-latvijas-dalibas-eiropas-savieniba. Accessed 05 Mar 2018
17. Ministry of Transport of the Republic of Latvia. Par iestāšanās līguma anotāciju. Letter to the Ministry of Affairs, Riga, 03.03.2003, No. 03-5/17 (2003)

Evaluation of a Necessity for Subsidies for Electric Vehicle Purchases in Latvia: 2013–2017

Aivars Rubenis[1(✉)], Aigars Laizāns[2], and Andra Zvirbule[2]

[1] Ivory Group, SIA, Skolas 15-7, Ikskile, Latvia
aivars.rubenis@ivorygroup.eu
[2] Latvia University of Life Sciences and Technologies, Liela iela 2, Jelgava, Latvia

Abstract. This article is evaluating if as of the 2017 there is an objectively justifiable necessity for governmental subsidies for purchasing electric vehicles (hereinafter—EV) in Latvia. The analysis was carried out using Total Cost of Ownership (*TCO*) approach and comparing all costs for electric and internal combustion engine vehicles using pair analysis. The time period was chosen to assess how the technological and market developments had affected the requirement for government subsidies for EV purchase since the development of Latvia National Electromobility plan in 2013. The study demonstrated that there was a significant difference in *TCO* and subsequently in the justification for purchase subsidies between various classes of vehicles as well as even within a single class of cars, depending on the vehicle manufacturer, hence, several possible approaches for subsidies were examined to recommend the most effective form for subsidizing EV purchases.

Keywords: Electric vehicles · TCO · EV policy · EV purchase subsidies

1 Introduction

Since the signing of Paris Agreement in 2015 [1] the need for decreasing greenhouse gas emissions for most countries has stopped being just a goodwill gesture and more of a binding target—including Latvia this reduction is 40% by 2030 compared to 1990 [2]. Transportation sector is still lagging behind other sectors, responsible for 27% of all CO_2 emissions in EU in 2017 [3] and 22.17% in Latvia [4]. GHG emissions from transportation have increase for three years in a row [5], and electric cars (EVs) are seen as one of the most feasible ways, to achieve the goals. As such their purchase and use is subsidized by many countries [6].

As part of Latvia National Electromobility plan developed in 2013 [7], for a brief period Latvia also subsidized EV purchases to the amount ranging from 35 to 55% of EV purchase value, but not more than 18,500 EUR [8]. As a result 174 EVs were purchased within a couple of month, while the program lasted [9]. However after the end of subsidies the sales of EVs have been disappointing, reaching the 25 electric vehicles sold in 2017, which is 0.1% of all new vehicles sold in Latvia [10]. In many

I. Kabashkin et al. (Eds.): RelStat 2018, LNNS 68, pp. 637–646, 2019.
https://doi.org/10.1007/978-3-030-12450-2_61

countries electric cars have turned from exotic vehicles to everyday means of transport, battery electric cars reaching 20.8% of sales in 2017 in Norway [11]. The research in Norway has shown, that vehicle buyers feel that subsidies is an effective way to increase EV sales [12]. This has also been supported by survey in Latvia, done by Road traffic safety department in 2016, where 60% of the respondents named EV price as the main obstacle for EV purchase [13], which can be offset by EV purchase subsidies.

There have been numerous studies done on the methodology and the comparative costs of EVs, however predominantly those have been done in US or western EU countries. This research will specifically address the situation in Latvia, considering the local market conditions. So far, there has only been one public study in Latvia in this are done in 2011 by Berjoza et al. [14], when many factors of EV use were unknown yet and had to be assumed. Furthermore, that study was mostly focusing on possible economic effects of EV introduction for the country in general.

Therefore, a research was commissioned in 2017 by the government of Latvia to evaluate the means to increase EV sales and, in particular, to assess necessity for subsidies for electric vehicle purchases.

Thus, the aim of this research was to determine whether since the end of previous subsidy program in 2013 there was still substantial cost differences between electric vehicles and internal combustion engine (ICE) vehicles in Latvia, which would justify granting purchase subsidies from government to encourage EV sales in Latvia. In addition, the research had to assess various vehicle market segments to recommend the appropriate amount and allocation method for the EV purchase subsidy.

2 Materials and Methods

2.1 General Considerations

The practical aim of the research determines the approach for the calculations. Determining the necessary amount of the subsidies would depend on whether one aims to determine objective differences in costs between EV and ICE vehicles, or to determine the overall costs for a country between using unpolluting modes of transportations; or to assess the individual purchaser's perception when selecting one's new vehicle.

In order for the subsidies to be effective, final value for a customer from EV purchase should be bigger than from that of ICE vehicle. Alternately costs $C_{\mathrm{Subj,EV}}$ should be lower that $C_{\mathrm{Subj,ICE}}$ and the subsidies should be big enough to cover the difference. On the other hand, subsidies the government has to provide S_{EV} should not be larger than the additional macroeconomic effects gained from people using EVs $C_{\mathrm{Macro},EV}$: this includes lower costs for health services by improving the environment and ecology; the benefits of improving the foreign trade balance by replacing imported fossil fuels with locally produced electricity; benefits of the emergence of a new local industry in the field of electromobility; loss from excise duty on fuel, among others.

Therefore, the optimum subsidy should satisfy both customer and the government.

$$\begin{cases} C_{\mathrm{Subj,EV}} - S_{EV} < C_{\mathrm{Subj,ICE}} \\ S_{EV} < C_{\mathrm{Makro},EV} \end{cases}. \tag{1}$$

While the subjective value for the customer consists of both objective component and subjective impressions $C_{\mathrm{Subj,EV}} = C_{EV} + C_{Subj}$ (e.g. likes, dislikes, costs of inconvenience of moving to different technology, charging, driving range limitations etc.) This research focuses only on the first part: determining the objective necessity for subsidies.

2.2 Total Cost of Ownership Approach

The costs of vehicles cannot be compared only on purchase price basis—some cars can cost initially more, but then be cheaper to run because of better fuel economy, or they could cost less, but then break down more and cost more in the maintenance. Therefore, to compare two assets objectively Total Cost of Ownership (*TCO*) method is used extensively in the research [15], and which is used in this research to evaluate the true costs of both EV and ICE vehicles:

$$TCO = PC + \sum\nolimits_{t=0}^{n} OC_t - RV_n. \tag{2}$$

Thus, all the costs are accounted during vehicle lifetime, starting from purchase costs (*PC*), including all operation costs (*OC*) during vehicle operation each year (*t*), up to income from the sale of the vehicle at year *n*, which been used as the retaining value (*RV*). The calculations are modified to include net present value (*NPV*) of costs over the time of vehicle use, by discounting *TCO* at year *t* using discount rate *r*.

$$NPV(t, r) = \sum\nolimits_{t=0}^{n} \frac{TCO_t}{(1+r)^t}. \tag{3}$$

Studies have used country-specific social discount rate [16] for discount factor (*r*) —Hagman et al. [17] uses monthly bank interest rate to calculate the total interest cost. In this research the WACC approach is used, adapted for individual use which is more suitable approach to evaluate necessity for subsidies from purchaser's point of view.

$$r = \frac{E}{A} * r_e + \frac{D}{A} * r_d, \tag{4}$$

where E/A is equity proportion of the vehicle purchase (down payment for the lease), and D/A is the debt (lease) proportion, subsequently r_e—discount rate for own capital and r_d—lease interest rate. Using the average of typical offers from lease companies, the total discount rate for EV buyer in Latvia was calculated to be 5%.

2.3 Time Frame for *TCO* Calculations

The time horizon in used vehicle *TCO* calculations is of utmost importance, as it determines the remaining value of the vehicles, and usually is not linear, and as such it has been pointed by most researchers as most influential factor for *TCO* calculation. Furthermore, it also affects many of the cost calculations: e.g., distance driven during the vehicle lifetime and hence the energy (fuel) costs, battery replacement costs, servicing costs, etc.

Arguably the way to compare the *TCO*s for EVs and ICE vehicles would be using full vehicle lifetime, which would include all investment and operational costs of each type of vehicle. Twelve years have been used for full lifetime EV *TCO* calculations [16, 18]—this approach however involves a lot of assumptions, as the data for electric vehicles currently is available only for up to 6 years.[1] Another approach, used in this research, is to use shorter investment horizon and include retaining value (i.e. vehicle resell value), assuming, that market forces are correctly compensating for the rest of the vehicle lifetime. Unlike most research done in US and Europe which use lifetime horizons as short as three years for a new car [17], five years have been chosen for the *TCO* calculations in this research for Latvia market. The reasons for this is that the real vehicle usage time differs significantly in Latvia from that in the EU. The average vehicle age in EU is 10.7 years [19], while in Latvia that is 16.3 years,[2] with average first registration vehicle age being 9.4 years [20]. While even five-year period does not reflect typical vehicle use in Latvia, it has been chosen because of the limitations of the available data, as the first EV in Latvia became available for sale in June 2013 [21] and the data is even less available for the rest of the vehicles used in analysis.

The average annual mileage for all vehicles in Latvia in 2016 was 16,456 km, however for EVs the annual mileage was 9299 km, with only 18% of EVs driving more than 16,000 km a year[2]. In this research we have used total average driving distance in Latvia (16,456)[2], assuming, that one of the reasons for EV annual mileage being less was lack of charging infrastructure in Latvia, which is being rolled out in 2018–2019.

2.4 Pairwise Comparison

For the analysis we're using pairwise comparison of EVs with ICE vehicles with similar parameters as described in [22]. In some cases, it is easy, as the same model has both ICE and EV versions, in other cases we have used similar models of the same manufacturer (Tesla being an exception), with the entry-level configuration (Table 1).

2.5 Cost Structure

Several of the costs in *TCO* calculations laid out in Table 2 may vary among countries, or even between different user categories within one country, thus presenting different results. E.g. VAT can be reclaimed by companies in EU, and thus should be exempt from *TCO* but not for private persons. On the other hand, in Latvia companies pay

---

[1] Here we are talking about series production EVs from major manufacturers, which can be used for pairwise comparison; using Nissan Leaf as a milestone, released in the U.S. and Japan in December 2010.

[2] Data: Latvia Road Safety department as of 01.10.2017, previously unpublished.

company car tax, which depends on the vehicle weight and power, and unlike annual vehicle operation tax, EVs are not exempt from it [23].

Table 1. Vehicles used for pairwise comparison of EVs with ICE vehicles in Latvia.

EV	Available in Latvia	ICE
VW e-UP!	Autumn, 2013	VW Up! 1.0 BMT
BMW i3	11.2013	BMW 118i
VW e-Golf	03.2014	VW Golf 1.4 TSI
Nissan Leaf	2011	Nissan Juke 117 CVT
KIA Soul EV	06.2014	KIA Soul 1.6 GDi
Tesla Model S 75D	02.2012	Audi A7 3.0 TDI Quatro

Table 2. Cost structure for comparison of EVs with ICE vehicles in Latvia.

Costs	Type	Vehicle		Notes
		ICE	EV	
Capital costs (Purchase Costs)				
Vehicle rice	*One time*	×	×	
Registration costs	*One time*	×		
(–) Purchase subsidies	*One time*		×	
Operational costs				
Use of vehicle				
Energy costs	Distance Dependent	×	×	
Vehicle insurance	Annual	×	×	Depends on various factors
Third party liability insurance	Annual	×	×	Depends on the owner
Repair and service costs	Distance dependent	×	×	
Parking costs	Special	×		Depends on city policies
Battery costs	Special		×	
Taxes				
VAT	*One time*	×	×	Private persons only
Vehicle registration:	*One time*	×	×	
Vehicle operation tax	Annual	×		
Company car tax:	Annual	×	×	Companies only
Other				
Technical inspection	Annual	×	×	
Retaining value				
(–) Vehicle resale value	*One time*	×	×	

Detailed analysis of assumptions and data for each cost account has been carried out based on the market information and the data from electric vehicle dealers in Latvia, however it is beyond the size limitations of this article. The summary of all the costs are presented in the results section.

2.6 Driving Distance and Energy Costs

Driving distance is one of the most influential factors apart from the price. As EVs are cheaper to drive than ICE vehicles, then the higher the distance, the better results for EVs.

Average annual distance for ICE vehicles is 16,456 km.[3] The current driving distances for EVs are much lower—only 18% of all EVs in Latvia currently drive more than 16,000 km annually.[4] This can be explained by lack of charging infrastructure, which is being rolled out in 2018–2019. Therefore, in this research an average driving distance has been assumed for both EVs and ICE vehicles, to evaluate the costs of potential equal use, not the costs for actual current use.

The energy costs have been calculated using actual corrected real life energy consumption (30% above NEDC data [24]) and average fuel costs 1.1 EUR/l as per time of the research without expected inflation.[5]

3 Results

The results presented in Table 3 demonstrate that in 2017 most of the EVs available in Latvia are objectively more expensive to use than their ICE counterparts. The *TCO* differences between vehicles varies; from EV being 81% more expensive than ICE vehicles (VW-eUP!) to EVs being 13% less expensive (Tesla Model S). These results also confirm the conclusions from previous research in other countries, that this difference is more pronounced in small vehicle segment than in premium vehicle segment.

Table 3. Pairwise comparison of EVs with ICE vehicles.

Vehicle used		Purchase price, EUR		Resale value (5 years) EUR		Total cost of ownership, EUR			
EV	ICE	EV	ICE	EV	ICE	EV	ICE	dTCO	dTCO, %
VW e-UP!	VW Up!	27,154	12,839	14,373	6796	23,808	13,374	10,434	81
BMW i3	BMW 1 series	40,390	26,450	15,629	7590	33,660	30,184	3476	13
VW e-Golf	VW Golf	40,415	20,473	18,914	7444	31,671	22,205	9466	46
Nissan Leaf	Nissan Juke	35,890	24,550	11,990	8200	32,278	26,045	6233	25
KIA Soul EV	KIA Soul	29,490	17,240	15,720	8374	22,563	19,219	3344	19
Tesla Model S	Audi 7 series	86,563	76,600	48,000	29,000	54,637	64,609	−9972	−13

[3] Data: Latvia Road Safety department as of 01.10.2017, previously unpublished.

[4] Data: Latvia Road Safety department, 2017, previously unpublished.

[5] Data: Neste Latvia Diesel Neste Futura D. 01.10.2017, This might skew the results in favour of ICE vehicles, as on 16.07.2018 respective diesel costs in Latvia are 1.197 EUR/l.

However, in Latvia the differences between manufacturers are more distinct than between segments—so both Volkswagen EVs demonstrate the highest cost difference: 81% for e-UP! and 46% for e-Golf.

This difference can be partially explained that the VW ICE vehicles are the first and the third cheapest, which makes the EV cost increase relatively much larger than for more expensive ICE vehicles. KIA however is an example that despite being the second cheapest ICE vehicle, it needs the smallest subsidy (only 3.3 thousand EUR) for KIA Soul EV to be economically equal (Fig. 1).

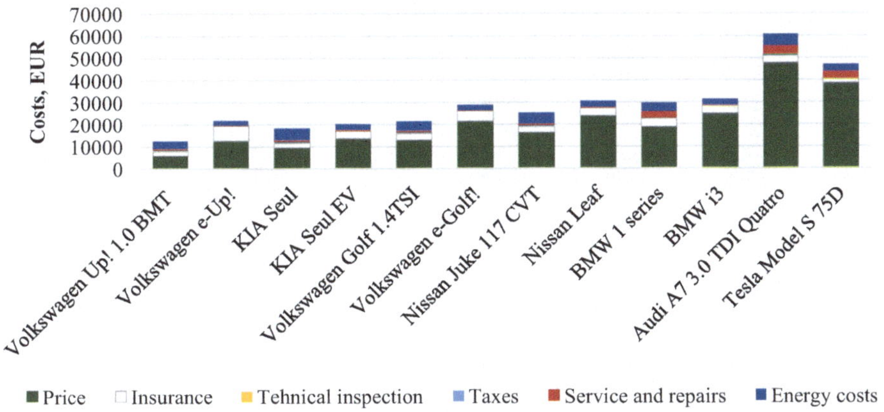

Fig. 1. Cost breakdown for selected EVs and their ICE counterparts.

Several observations were hard to explain, and they could be attributed to the very limited sample size. E.g. insurance for e-UP! is proportionally much too high, reaching

Table 4. Main cost breakdown of EVs with ICE vehicles by vehicle segments.

Cost comparison	Small		Medium		Premium	
	ICE (%)	EV (%)	ICE (%)	EV (%)	ICE (%)	EV (%)
Price	47	58	60	75	78	75
Insurance	16	30	12	14	5	12
Technical inspection	1	0.5	0.5	0.4	0.2	0.4
Taxes	5	0.0	2.3	0.0	1.2	0.0
Service and repairs	4	1	5	1	6	3
Energy costs	27	10	21	10	9	9
Total	100	100	100	100	100	100

30% of vehicle costs—this might be because in Latvia most of e-UPs are owned by municipalities or universities (purchased within state grant programme), and as such they might carry higher insurance risks, but this is an unverified assumption (Table 4).

4 Discussion

The results show that the premium EV segment is objectively already equal to the ICE vehicles, and does not need additional subsidies. Current Latvian legislature sets maximum amount of vehicle price that can be depreciated for tax purposes to be 50,000 EUR [25]. The same upper EV price limit could be also applied for subsidies for EV purchases (Fig. 2).

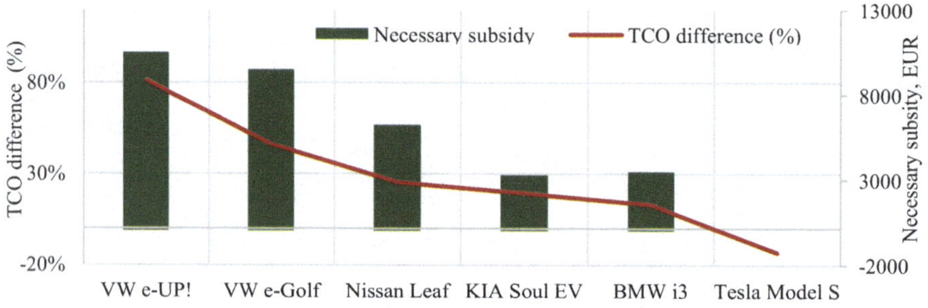

Fig. 2. *TCO* difference and calculated objective necessary subsidies for selected EVs.

As the relatively cheapest EVs need larger subsidies, the subsidies cannot be directly tied to the cost of the vehicles but rather should be set as a lump sum. The objective minimum subsidy amount for EVs is about 3.5 thousand EUR, which would set KIA Soul EV and BMW i3 on equal monetary terms with their ICE counterparts and make Nissan Leaf just slightly more expensive to use than Nissan Juke. Currently VW electric vehicles are noticeably more expensive than their ICE counterparts and the difference this big cannot be covered without over subsidising other electric vehicles.

Further research will be carried out in 2018–2019, after introduction of fast charging infrastructure in Latvia would let evaluate the impact on the real world EV driving patterns in Latvia compared to ICE vehicles, and to see if that would lead to changes in *TCO* evaluation.

5 Conclusions

The research indicated, that based on 2017 market data in Latvia, *TCO* of EVs are higher that their ICE counterpart for most of the vehicle classes and the governmental subsidies still are needed to offset the cost difference.

The analysis also revealed, that:

1. The purchase price and *TCO* difference is most distinct in the small car segment, where EVs are up to 81% more expensive than equivalent ICE vehicles; while the difference is insignificant, or indeed negative, in premium car segment and requires no subsidy. There is also marked difference in *TCO* among manufacturers, even

within one vehicle segment, some requiring large subsidies, while others being almost on par;

2. Small class vehicles are relatively more exposed to price increase; therefore, purchase subsidies should not be proportionally linked to the price of the electric car but should be considered as lump sum, thus benefiting small vehicle class relatively more;
3. In 2013 Latvian EV grant programme the subsidy for purchase of electric vehicles was set to 18,500 EUR. As of mid-2017, the suggested subsidy amount to match EV lifetime costs to ICE vehicles in Latvia is approximately 3500 EUR.

Acknowledgements. This paper has been published as a part of the research contract with Latvia Road Safety department (contract No. 11-EVKD/933).

References

1. UNFCCC Conference of the Parties: Paris Agreement, pp. 1–16. United Nations, Geneva (2015)
2. European Union: Intended Nationally Determined Contribution of the EU and Its Member States No. October 2014, pp. 1–7 (2015)
3. European Environment Agency: National emissions reported to the UNFCCC and to the EU Greenhouse Gas Monitoring Mechanism—European Environment Agency. National greenhouse gas inventories (IPCC Common Reporting Format sector classification) (2018). https://www.eea.europa.eu/data-and-maps/data/national-emissions-reported-to-the-unfccc-and-to-the-eu-greenhouse-gas-monitoring-mechanism-14. Accessed 02 July 2018
4. Latvia's National Inventory Report 2018 for Submission under UNFCCC and the Kyoto Protocol, Riga (2018)
5. European Evironment Agency: Small cut in EU's total greenhouse gas emissions in 2016 but transport emissions keep increasing—European Environment Agency (2018). https://www.eea.europa.eu/highlights/small-cut-in-eus-total#tab-related-publications. Accessed 02 July 2018
6. Hardman, S., Chandan, A., Tal, G., Turrentine, T.: The effectiveness of financial purchase incentives for battery electric vehicles—a review of the evidence. Renew. Sustain. Energy Rev. **80**(March), 1100–1111 (2017)
7. Ministry of Transport of Republic of Latvia: Electric Mobility Development plan 2014–2016, Riga (2014)
8. Vides aizsardzības un reģionālas attīstības ministrija: Klimata pārmaiņu finanšu instrumenta finansēto projektu atklāta konkursa Siltumnīcefekta gāzu emisijas samazināšana transporta sektorā—atbalsts elektromobiļu un to uzlādes infrastruktūras ieviešanai nolikums. LR Ministru Kabinets, Riga (2014)
9. Vides aizsardzības un reģionālas attīstības ministrija: Klimata pārmaiņu finanšu instrumenta finansēto projektu atklāts konkurss 'Siltumnīcefekta gāzu emisijas samazināšana transporta sektorā – atbalsts elektromobiļu un to uzlādes infrastruktūras ieviešanai.' http://www.varam.gov.lv/lat/darbibas_veidi/KPFI/projekti/?doc=17817. Accessed 20 May 2016
10. Road Safety Direcotorate of Latvia: Vehicle Statistics in Latvia. Vehicle Statistics in Latvia (2018). https://www.csdd.lv/transportlidzekli/transportlidzekli-vizualizacija. Accessed 02 July 2018

11. Norwegian Electric Vehicle Association: Norwegian EV market. Norwegian EV Market (2018). https://elbil.no/english/norwegian-ev-market/. Accessed 02 July 2018
12. Bjerkan, K.Y., Nørbech, T.E., Nordtømme, M.E.: Incentives for promoting battery electric vehicle (BEV) adoption in Norway. Transp. Res. Part D Transp. Environ. **43**, 169–180 (2016)
13. Rubenis, A., Berjoza, D., Grīslis, A., Francis, I.: Analysis of the development of the Latvian national electric vehicle charging network (in Latvian), Riga (2016)
14. Berjoza, D., Jurgena, I.: Economic effect of electric vehicles. In Engineering for Rural Development, pp. 179–184 (2011)
15. Ellram, L.M.: Total cost of ownership. Int. J. Phys. Distrib. Logist. Manag. **25**(8), 4–23 (1995)
16. Bubeck, S., Tomaschek, J., Fahl, U.: Perspectives of electric mobility: total cost of ownership of electric vehicles in Germany. Transp. Policy **50**, 63–77 (2016)
17. Hagman, J., Ritzén, S., Stier, J.J., Susilo, Y.: Total cost of ownership and its potential implications for battery electric vehicle diffusion. Res. Transp. Bus. Manag. **18**, 11–17 (2016)
18. Macharis, C., Lebeau, P., Van Mierlo, J., Lebeau, K.: Electric versus conventional vehicles for logistics : a total cost of ownership, pp. 1–10 (2013)
19. European Automobile Manufacturers Association: Average Vehicle Age|ACEA—European Automobile Manufacturers' Association. Average Vehicle Age (2016). http://www.acea.be/statistics/tag/category/average-vehicle-age. Accessed 17 Oct 2017
20. European Commission: Consumer market study on the functioning of the market for second-hand cars from a consumer perspective, No. October (2014)
21. Whatcar.lv: Opel Ampera var nopirkt Latvijā. Auto Blitz (2013). http://www.whatcar.lv/auto-zinas-raksts/5235/opel-ampera-var-nopirkt-latvija-zinama-ari-cena. Accessed 15 July 2018
22. Zsuza, C., Drossinos, P., Thiel, Y.: The effect of fiscal incentives on market penetration of electric vehicles: a pairwise comparison of total cost of ownership. Energy Policy **105**, 524–533 (2017)
23. Republic of Latvia: Law on the Vehicle Operation Tax and Company Car Tax. Saeima of Republic of Latvia, Riga (2017)
24. Cauwer, D.C., Mierlo, V.J.: Electric vehicle use and energy consumption based on real-world electric vehicle fleet trip and charge data and its impact on existing EV research models
25. Republic of Latvia: Enterprise Income Tax Law. Saeima of Republic of Latvia, Riga (2017)

Assessment of the Influence of Social-Cultural Environment in the Context of Global Logistics

Darius Bazaras(✉), Ramūnas Palšaitis, Kristina Čižiūnienė, Artūras Petraška, and Karolis Kaminskas

Vilnius Gediminas Technical University, Plytinės 27, 10105 Vilnius, Lithuania
darius.bazaras@vgtu.lt

Abstract. The modern aspects of the supply and logistics chains cannot be restricted by national borders. The strategists of the national transport policy must address new challenges of globalization to attain common pursuits by harmonizing actions when advanced international transport chains and corridors are established. Constantly changing and improving business environment demands re-assessing and re-analysing the importance of logistics and marketing as well as its functions and interaction. The links between logistics and marketing are significant as merging these functions and adapting it into company's activities may increase profits and improve the quality of services. On the other hand, the important factor is the environment in which all these matters emerge. In this case, the awareness of socio-cultural environment becomes vital.

The aim of this research—to conduct an assessment on the role of the socio-cultural environment in the context of global logistics.

The analysis of company's activities from the perspective of socio-cultural environment involves essential factors influencing these activities: traditions, language awareness, negotiation skills, attitudes towards overtime work, etc. However, the awareness of socio-cultural environment is not of the least importance. The article provides assessment results on the role of the socio-cultural environment in the context of global logistics. As suggested by results, the socio-cultural environment has a major impact on global businesses, especially logistics. Diverse regions are distinguished for its characteristic socio-economic environment which often causes certain difficulties in international negotiations that are common in logistics activities.

Keywords: Social-cultural environment · Global logistics · Employees · Regions

1 Introduction

It is very important to take into account that modern supply and logistics chain requirements cannot be restricted by the state borders. Policymakers developing the national transport policy must evaluate and meet the recent globalisation challenges, and reach, in the international space, a common position in harmonising their actions aimed to establish progressive international transport chains and corridors.

The constantly changing and increasingly developing modern business environment requires to re-estimate the importance of logistics as a marketing instrument, and

I. Kabashkin et al. (Eds.): RelStat 2018, LNNS 68, pp. 647–657, 2019.
https://doi.org/10.1007/978-3-030-12450-2_62

to carry out a detailed analysis of functions and their interaction. Relationship between the logistics and marketing is very important since pooling of these functions and their application in the undertaking activity gives a possibility to increase profitability via the improvement of quality of services provided by an undertaking. On the other hand, another major factor is the environment in which operations take place. In this respect the awareness of a social-cultural environment becomes a very important aspect.

The aim of the Article is to evaluate the impact of the social-cultural environment in the context of global logistics.

The objectives set to achieve the aim:

- To carry out the analysis of scientific literature sources by evaluating the importance of social-cultural environment for business;
- To present the findings of the survey on social-cultural environment in the logistics sectors of Europe, Scandinavia, Russia and North America.

Methodology:

- Analysis and structuring of the scientific literature;
- Quantitative survey.

2 The Importance of Social-Cultural Environment for Business

People are affected by multiple internal factors, as well as by the external factors. The social environment has huge impact on human judgements. According to McEwen [1], social environment has powerful effect on the body and the brain, as well as on the judgements of an individual and physical functions. This could cause even more serious changes in the human body or in the entire evolutionary chain. According to Toyokawa et al. [2], the social environment has such a huge effect on an individual that it can even result in the gene mutation and affect the ability of the next generation to evaluate social environment and the related problems, for instance the inappropriate nutrition or inability to concentrate.

These disorders can, and are often reflected in a working environment, for instance, in the logistics sector, where concentration and focus are of utmost importance. Inability to focus attention might cause huge loss, especially in the air transport where even a minor mistake might cost lots of money; therefore, in recruiting employees it is very important to evaluate their resistance to stress and distancing from the surrounding environment. In the majority of cases men are more resistant to these factors than women, but this is not determined genetically and is more related to the environmental impact affecting both genders.

According to Sharp and Ganong [3], women are more affected by the inadequate social environment compared to men; women‘s resistance to stress and environmental impact is also lower than that of men. Women are concerned that, at a certain age (according to the prevailing public norms) they must marry, give birth to a child and achieve professional career success. Men cope with the environmental impact more easily, because for them family problems (according to the same public norms) play a minor role—it is the role of a woman to take care of home and family.

The survey carried out by Hatzenbuehler [4] in 34 countries revealed that social environment has such a huge impact on the mental health of people that they start behaving inadequately; this phenomenon is especially obvious among young people when they, affected by the social environment, can be assailed by suicidal thoughts. Young people, after graduating from higher institutions or still studying and experiencing stress both in the working and neutral environment, become more vulnerable and less active at work compared to their peers (aged 15–22) whose social environment is considered as positive. According to Hipp [5], the society considers that social environment is not adequate when even a short stay in such an environment forces an individual who was not formerly involved in the crime, to act against and violate the rules.

Therefore, it is necessary to head young minds in the right direction, since their first job might have huge impact not only on their career but also on their family life. Inadequate micro-climate in the job place (especially the first job) might give negative impression about the entire performance of a relevant sector. If this is the case, a young person is inclined to leave such a company and later, when choosing the future activity, avoid the employers of the entire sector. This issue is especially relevant for Lithuania's region and for the countries where the sector of transport accumulates major economic benefits directed toward transit or distribution of goods or service flows. Reluctance of employees to work in the above area is of major concern and results in the competition for qualified workers. Small and medium-size undertakings face big problems in developing their business, because their work remuneration will be less competitive than that of larger companies involved in the analogous activity.

The authors Whitworth and Friedman [6] presented the model on the impact of social environment on individuals (Fig. 1).

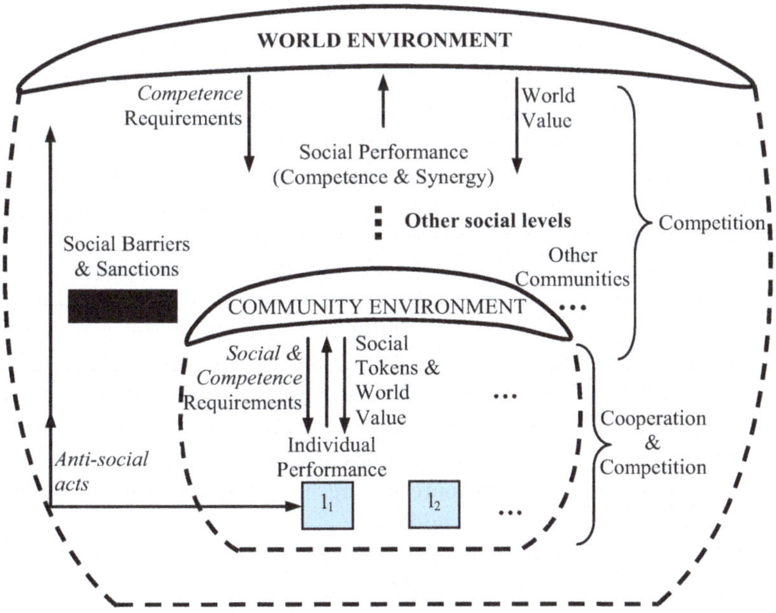

Fig. 1. Model of social environment [6].

They choose two levels on how an individual is affected by the world environment and by the community environment. According to the researchers, the communication is always a two-sided phenomenon. The first level is related to the competition between individuals, later between the communities which can also join and create a more robust structure. In their turn, individuals might disseminate anti-social, i.e. negative actions and, consequently, influence the entire surrounding environment, thus, affecting other communities; whereas the latter affect the first and vice versa. In other words, community as a derivative affected by positive or negative factors, is filtering and processing the information and making decisions. This is the basis of any logical/intellectual system (data accumulation/conclusions/actions). Depending on what kind—good or bad—information is received by the system/community, the community performance model will send relevant response to the primary information source and to the new community. Individuals and communities also determine changes in the global environment. Individuals compete for the norms established in the community (e.g. the role of the leader, achievements of family members). Following the same principle, communities led by the individuals with higher achievements, compete with each other for a social status in the remaining, wider environment which is identified by the authors as the world environment. If not stopped by a stronger individual, even one negative person might affect a bigger group of people; the latter affects a wider group of people and, finally, the entire local community. This causes a stir among the community members and the community changes in response to the received signals. In the worst case, the community conveys negative information to another community; thus logically, the chain expands and reaches a macro-level, thus changing the life of the entire communities and individuals, as well as the understanding of good and evil. The synergy effect is not always positive, but the above worst scenario is not necessarily the only scenario. In any chain, starting from an individual and finishing with small macro environment levels, when two equal groups collide, the stronger group will win; this means that negative information transferred to an individual's community can be defeated by the positive information, but only when the positive one is substantial and supported by the environment. According to the author, the entire surrounding environment is inter-related. It does not matter how we treat this environment; we must see a wider picture, i.e. observe it from distance, yet not omit the fact that general picture consists of minor details.

According to the author Sharot et al. [7], the cultural environment is created by individuals themselves, i.e. by individuals who have been living in a certain area for a long time. This environment might be positive or negative depending on the level of education of individuals and on their theoretical and practical knowledge acquired in childhood. According to Roshita et al. [8], the knowledge gained in childhood and determined by the cultural environment, later determines an individual's decision to work or not to work, what road of life to choose, what is right and what is wrong etc. Later, when an individual becomes more mature, it is very difficult to change his/her thinking and life norms acquired in childhood.

When an employer recruits an employee from an unknown country or region and has no knowledge about the cultural traditions of that country, it is quite difficult to evaluate all possible risks related to that employee. Only after a certain period of time, depending on the specific nature of an undertaking, it is possible to say whether a job is a good or a bad fit for an employee. Taking that into account, initially the employees are subject to an employment probationary period. Yet, this might cause other problems: during the probationary period an employee might experience huge pressure and simulate the qualities not characteristic of him/her in order to demonstrate that an appropriate image about him/her complies with the employer's expectations.

Thereby the efforts are made to avoid a cultural shock. According to Chen et al. [9], the cultural shock is inevitable if an individual is not well aware of the environment where he/she is planning to work or perform other activity requiring certain mental skills.

3 Findings of the Logistics Sector Survey Carried Out in Europe, Scandinavia, Russia and North America

The survey includes the interviews with 45 respondents from four different regions (Europe, Scandinavia, Russia and Northern America), from them 15 women and 30 men. The survey revealed that women and men differently evaluate certain aspects. The aim of the survey was to define the reasons determining the respondents' desire to work in the transport/logistics sector. According to the findings, the main factors are money and desire of self-realisation; less important are carrier pursuit and public perception. This leads to the conclusion that transport/logistics sector is a place with the continuous/ongoing activity, competition and movement of major financial flows making the ambitious people choose work in this sphere.

Competition is an important prerequisite for the development of the European single market and globalisation of the economy, whereas the competition policy affects the structure of markets and behaviour of its participants with a view to retaining and developing effective competition in the single market [10].

Until the end of 19th century most economists viewed competition as a behaviour and as a dynamic process of the competitors' contest during which everybody tries to gain advantage [10, 11]. This competition theory is built in the works of Adam Smith (1776), David Ricardo (1817), John Stuart Mill (1848) and other representatives of the British political economy school; they try to identify factors determining prices and, correspondingly, distribution of a social product to different public groups [10, 12]. According to Ginevičius et al. [13], the national economic needs can be investigated in a broad and narrow sense. According to the latter, it is considered that this is a specialist in a certain knowledge and skills environment, corresponding with the modern and future undertakings' requirements. In a broader sense, the economic needs are related to the increase of needs of various economic sectors, and to their development. Taking this into account, it could be stated that one way or another a human factor is inevitable; it determines the competitiveness of single enterprises, regions or even of the entire country.

Therefore, the survey aimed to specify the skills and qualities of employees to be recruited in the logistics sector.

The survey established the following major factors of human resources involved in the logistics sector: interpersonal competences, the knowledge of foreign languages, buying/selling skills; less important factors included ambitiousness and dynamism.

However, taking into account common undertaking's performance indicators and in order to retain competitiveness of an undertaking both in local and international markets, trust, sound micro-climate and strict leaders are of utmost importance. That is probably why the survey results indicated that most important personal qualities in the logistics activity are: non-standard thinking, the search of environmental friendliness and altruism.

The global business involves more and more companies and clients from different religions, with different mentality and culture, and, thus, increases the complexity of business transactions. Following the above analysis, Čižiūnienė et al. [14] is of the opinion that social and cultural differences between inter-continental employees in the sector of transport/logistics have huge impact on this business. Therefore, the knowledge of the subtleties of the social-cultural environment can considerably facilitate performance. For instance, in Eastern Europe the overtime is a normal practice, as well as work during weekends or even holidays, whereas in Southern Europe the employees are not inclined to work overtime, during weekends or holidays [14]. In Scandinavia employees are hard working and committed, but, as is the case with the Western European region, they avoid hard physical work. Therefore overtime is not a frequent phenomenon. In this region the employees are respected as family members [14]. In Russia the employees do not work hard; if they have a possibility not to work, they will take this chance. Earlier, to leave a working place or not to show up was a norm for quite a long time. But times are changing, and changes are unavoidable. Russia is a country which cannot be treated unambiguously. Its citizen in certain regions, as well as their work ethics, differ completely, but in general this is a country with huge potential. In North America self-realisation is not the purpose of employees; their goal is to earn more money so as to be able to spend it as they want [14]. In view of this, the survey aimed to clarify whether the overtime is a positive phenomenon in the logistics activity. According to the findings, 83.33% of men said "Yes", and 66.67% of women provided a negative answer. Taking this into account and assuming that the sector of transport/logistics promotes competition among the employees (motivated by the bonus pay to the basic salary), an assumption can be made that men, considering a possibility to work and earn more, justify the overtime hours. Whereas women are negative toward overtime by highlighting that they have to take care not only of their career but also of their family well-being and children. The survey findings confirmed the above statements. To the question: "Is family a more important factor than your career?", 100% of women answered "Yes" (compared to 60% of men). When answering the question "What could determine your decision to work overtime?", the respondents unanimously noted the necessity to finish the started activity, and the ability to earn money. Whereas the issues like self-realisation and the importance of being acknowledged by others were absolutely irrelevant and not motivating.

Despite the differences of geographical regions, the employees of the transport/logistics sector are most of all concerned about: work results (44.45%), work arrangements/order (22.22%) and support of the colleagues (22.22%). Less important are: colleagues' work results in the analogous activity (6.67%) and the opinion of the authorities (4.44%). The breakdown of priorities was a bit different in analysing the important elements both, in the family life and in business (see Table 1).

Table 1. The differences of geographical regions.

Europe	Scandinavia	Russia	North America
Attention	Transparency	Compliance with the rules and norms	Fast decision-making
Fast decision-making	Attention	Established/prevailing work order	Established/prevailing work order
Compliance with the rules and norms	Compliance with the rules and norms	Attention	Compliance with the rules and norms
Transparency	Established/prevailing work order	Fast decision-making	Transparency
Established/prevailing work order	Fast decision-making	Transparency	Attention

According to the survey results, the Compliance with the Rules and Norms is in the third place as the general main criteria. This suggests that compliance with the laws and common business dogmas is equally important for all the regions.

On the other hand, the respondents were also unanimous in answering the question what criteria are most important in pursuing activity in the logistics sector (Fig. 2).

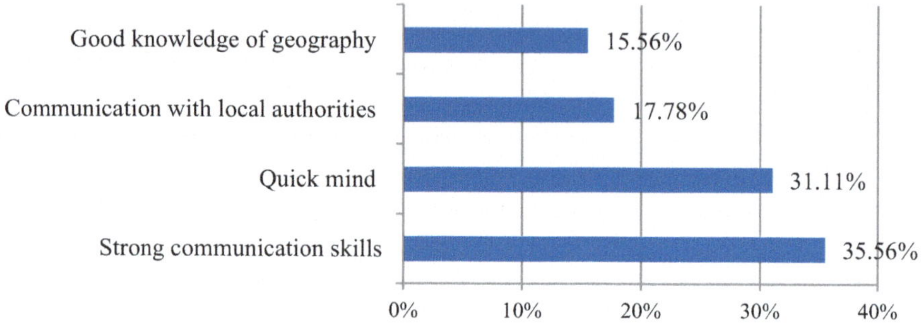

Fig. 2. Criteria relevant for work in the logistics sector.

Pursuant to the results, the most important criteria are strong communication skills and quick mind.

Trying to clarify what qualities are necessary in order to avoid misunderstandings and miscommunication, the opinions of the respondents differed, therefore compatibility of the opinions was evaluated (Table 2).

Table 2. Table of the obtained evaluation ranks (*Source* developed by the authors).

No. of respondent	Factor encryption symbol (m = 4)*			
	a	b	c	d
$\sum_{i=1}^{n} R_{ij}$	90	100	133	127
$\bar{R}_j = \frac{\sum_{i=1}^{n} R_{ij}}{n}$	2	2.22	2.96	2.82
$\sum_{i=1}^{n} R_{ij} - \frac{1}{2}n(m+1)$	−22.5	−12.5	20.5	14.5
$\left[\sum_{i=1}^{n} R_{ij} - \frac{1}{2}n(m+1)\right]^2$	506.25	156.25	420.25	210.25

Criteria coding: Knowledge of the traditions of the region where an individual is employed (a); Knowledge of the religion holidays and usual days-off (b); Awareness and avoidance of cultural traditions determined by family ties and other relations (c); Respect of local habits and traditions (d).

The concordance coefficient is calculated according to the formula, when there are no related ranks.

$$W = \frac{12S}{n^2(m^3 - m)} = \frac{12 \times 1293}{45^2(4^3 - 4)} = 0.1277.$$

The weight of the concordance coefficient is calculated according to the formula; a random mean is received.

$$\chi^2 = n(m - 1)W = \frac{12S}{nm(m + 1)} = \frac{12 \times 1293}{45 \times 4(4 + 1)} = 17.240.$$

χ^2 random mean 17.240 was higher than critical χ^2_{kr} (equal to 7.81473) mean; therefore, the opinion of respondents is considered as coordinated, whereas the average ranks indicate a common opinion of experts.

According to the formula, the least mean of the concordance coefficient $W_{\min}$, is calculated; it states that the opinion of all 45 respondents about the major 4 factors which are important in order to avoid misunderstandings and miscommunication, is still considered as coordinated.

$$W_{\min} = \frac{\chi^2_{v,\alpha}}{n(m - 1)} = \frac{7.81473}{45(4 - 1)} = 0.0579 < 0.1277.$$

The estimations demonstrated that the opinion of 45 respondents about 4 major factors which are important in order to avoid misunderstandings and miscommunication, is the same, and that the opinion of experts is coordinated.

The importance indicators Q_j are calculated in relation to factors which are important in order to avoid misunderstandings and miscommunication for work in the logistics sector. The obtained data is presented in Table 3.

Table 3. Rank evaluation table (*Source* developed by authors).

Sign of indicator	Factor encryption symbol				Sum
	a	b	c	d	
$\bar{q}_j$	0.2	0.222	0.296	0.282	1
d_j	0.8	0.778	0.704	0.718	3
Q_j	0.267	0.259	0.235	0.239	1
Q'_j	0.3	0.278	0.204	0.218	1
Breakdown of factors	1	2	4	3	

Table 3 presents all the factors and their breakdown (from most to least important).

According to the respondents' evaluations and performed calculations, the factors which are important in order to avoid misunderstandings and miscommunication for work in the transport/logistics sector, were distributed as follows:

1. Knowledge of the traditions of the region you are employed in;
2. Knowledge of the religious holidays and usual days-off;
3. Respect of local habits and traditions;
4. Understanding and avoidance of cultural traditions determined by family ties and other relations.

Evaluating the impact of the social-cultural environment on the global logistics, the conclusion could be made that this impact does exist and that in order to avoid misunderstandings, it is very important to be aware of the traditions and culture of the region where one is employed or with which labour relations are maintained.

To sum up the results of the research and the analysis of literature sources, the conclusion could be made that the majority of authors analyse a social environment as certain qualities, habits, religions etc. of relevant regions. This suggests that focus is given to the impact of the global environment on an individual and to the communal activities. Yet, in the scientific works presented in rather provocative titles, the authors like Toyokawa et al. [2], Sharp and Ganong [3], Hatzenbuehler [4], Hipp [5] investigate not only social, but also relevant physiological and psychological phenomena having afterwards huge effect on the labor market. The studies and insights of these authors suggested that social environment has major influence on an individual, and, according to Toyokawa et al. [2], might affect even gene mutation and influence evaluation of the social environment of future generations which, pursuant to the authors of the article, might affect activities in the logistics sector requiring major attention and concentration.

Pursuant to Sharp and Ganong [3], men cope easier with the environmental impacts, because family problems for them are secondary (this statement has also been confirmed in the surveys), whereas women do not tolerate overtime hours and have to take care of not only working activity, but also the welfare of family and children.

The survey executed by the authors of this Article, confirmed the main ideas of all the authors under analysis, and demonstrated that not only the global environment, communities or stereotypes have major impact, but also psychological, physiological and similar factors.

This study, interview and survey are surely carried out according to ethical policies.

4 Conclusion

The research results have shown that the role of the socio-cultural environment in global logistics is perceived equally important in all regions, however there are differences in respondents' attitudes when it comes to gender, such as: male respondents tend to be more in favour of the overtime work in comparison to their female counterparts; according to male respondents, religion and diverse celebrations do not have a substantial impact on logistics business while female respondents think the opposite. Despite geographical differences, the most valued aspects by the employees of transport/logistics sector are as follows: work results, order of performing operational functions, and support from their co-workers. Slightly different priorities can be seen whilst examining important elements in family life and business: the third most important criterion in Europe, Scandinavia and North America is obeying to the rules and norms, while the second most important criterion in Russia and North America is the established order. Taking into account the most important criteria in logistics sector, all respondents agree that qualitative communication skills and rapid response are fundamental in logistics business. This once again proves that in different regions the impact of socio-cultural environment on global logistics is equally valued.

References

1. McEwen, B.S.: Brain on stress: How the social environment gets under the skin. Proc. Natl. Acad. Sci. **109**(Suppl. 2), 17180–17185 (2012). https://doi.org/10.1073/pnas.1121254109
2. Toyokawa, S., Uddin, M., Koenen, K.C., Galea, S.: How does the social environment "get into the mind"? Epigenetics at the intersection of social and psychiatric epidemiology. Soc. Sci. Med. **74**(1), 67–74 (2011). https://doi.org/10.1016/j.socscimed.2011.09.036
3. Sharp, E.A., Ganong, L.: I'm a Loser, I'm Not Married, Let's Just All Look at Me: ever-single women's perceptions of their social environment. J. Fam. Issues **32**(7), 956–980 (2011). https://doi.org/10.1177/0192513X10392537
4. Hatzenbuehler, M.L.: The social environment and suicide attempts in lesbian, gay, and bisexual youth. Pediatrics **127**(5), 896–903 (2011). https://doi.org/10.1542/peds.2010-3020
5. Hipp, J.R.: Resident perceptions of crime and disorder: how much is "Bias", and how much is social environment differences? Criminology **48**(2), 475–508 (2010). https://doi.org/10.1111/j.1745-9125.2010.00193.x
6. Whitworth, B., Friedman, R.: Reinventing academic publishing online. Part I: rigor, relevance and practice. First Monday **14**(8) (2010). http://firstmonday.org/htbin/cgiwrap/bin/ojs/index.php/fm/article/view/2609/2248
7. Sharot, T., Korn, C.W., Dolan, R.J.: How unrealistic optimism is maintained in the face of reality. Nat. Neurosci. **14**(11), 1475–1479 (2011). https://doi.org/10.1038/nn.2949
8. Roshita, A., Schubert, E., Whittaker, M.: Child-care and feeding practices of urban middle class working and non-working Indonesian mothers: a qualitative study of the socio-economic and cultural environment: child-care and feeding practices of Indonesian mothers. Matern. Child Nutr. **8**(3), 299–314 (2012). https://doi.org/10.1111/j.1740-8709.2011.00298.x
9. Chen, A.S., Lin, Y., Sawangpattanakul, A.: The relationship between cultural intelligence and performance with the mediating effect of culture shock: a case from Philippine laborers in Taiwan. Int. J. Intercult. Relat. **35**(2), 246–258 (2011). https://doi.org/10.1016/j.ijintrel.2010.09.005
10. Klimašauskienė, D.: Konkurencijos samprata ekonomikos teorijoje. Ekonomika **79**, 109–123 (2007). Vilniaus universitetas
11. Wagner, R.: Competition as a rivalrous process: Attilio da Empoli and the years of high theory that might have been. J. Econ. Stud. **28**(4–5), 337–345(9) (2001)
12. Cook, P.: Competition and its regulation: key issues. In: Centre on Regulation and Competition Working Paper Series. Paper No. 2 (2001). http://www.competition-regulation.org.uk/publications/working_papers/
13. Ginevičius, R., Ginevičienė, V.B.: The compliance of master's degree studies with the economic needs of the country. Technol. Econ. Dev. Econ. **15**(1), 136–153 (2009)
14. Čižiūnienė, K., Kaminskas, K., Petraška, A.: Socialinės-kultūrinės aplinkos vaidmuo globalioje logistikoje: ekspertinis vertinimas. Aukštųjų mokyklų vaidmuo visuomenėje: iššūkiai, tendencijos ir perspektyvos = Role of Higher Education Institutions in Society: Challenges, Tendencies and Perspectives **1**(5), 53–58 (2016). Alytus: Alytaus kolegija. ISSN 2029-9311

Innovative Economics

Who are Latvian Women Entrepreneurs of Small Businesses? A Preliminary Study

Kristine Uzule, Irina Kuzmina-Merlino(✉), and Ishgaley Ishmuhametov

Transport and Telecommunication Institute, 1 Lomonosova Street, LV-1019 Riga, Latvia
{Uzule.K,Kuzmina.I,Ishmuhametovs.I}@tsi.lv

Abstract. Statistical data suggest that Latvian women entrepreneurs are amongst the most active business founders in Europe. According to the Stockholm School of Economics in Riga, 10% of women aged between 18 and 64 in the country have started their own businesses of which the majority were found to lead small or micro-companies, which ultimately results in about 40% of all small and micro- size Latvian companies being run by women. What makes these women different from other professionals? Given the fact that many women reside in similar contexts but not all of them become entrepreneurs, **the aim** of this research was to identify personality characteristics of women entrepreneurs in Latvia. In order to translate the global aim into research questions, the Big Five model of personality, which is a dispositional model, was deployed in a questionnaire format. As for the specifics of the **research questions**, they were created based on a review of a few Big Five Personality Tests and targeted verification of previous research findings on women entrepreneurs in other countries. Earlier women were found to score higher on Conscientiousness and Openness to Experience and lower on Agreeableness and Neuroticism. Based on earlier research findings on Extraversion, this trait was suggested to play a moderate role. Since personalities reside in a social context, **another aim** was to identify one most powerful factor beyond the personality dimensions that has encouraged them to engage in entrepreneurial activities. To attain this aim, participants were asked a question of the most influential factor that affected their decision to start and sustain their enterprises. The obtained data were discussed in a broader context.

Keywords: Women entrepreneurs · Big five personality model · Latvia · Interviews

1 Introduction

Recent times have been marked by women gaining more rights to participate in economic activities that are not only associated with traditional female roles, such as teachers and nurses, but also with activities initially dominated by men—politicians, artists, scientists, neurosurgeons, lawyers, financial analysts, pilots, priests etc. More and more women are also establishing their own enterprises, thus, becoming entrepreneurs. According to Global Entrepreneurship Monitor Report 2016/2017 conducted

I. Kabashkin et al. (Eds.): RelStat 2018, LNNS 68, pp. 661–669, 2019.
https://doi.org/10.1007/978-3-030-12450-2_63

in 74 countries across the globe, 274 million women worldwide were running their own entrepreneurial activities [1]. As for the European context, the European Commission declares that of all entrepreneurial start-ups 30% are owned by women [2].

Consistently with international studies, also Latvian entrepreneurs have been becoming a more powerful force on the employment market. According to Global Entrepreneurship Monitor Report 2016/2017, female entrepreneurs in Latvia significantly contribute to economic activities, reaching almost 10% of all business activities in the country [1]. The Stockholm School of Economics in Riga claims that 40% of all entrepreneurs in the country are females [3], of whom almost 60% are in the age range of 25–49, whereas slightly more than 32% are between 50 and 64 years old [4]. Female entrepreneurs mostly start their businesses because of available opportunities rather than the necessity [1]. Female entrepreneurs most likely operate under the same conditions as their male counterparts due to the equal ratio of 0.5 that was obtained for female/male participation in entrepreneurial activities [1]. Such data suggest that Latvian women entrepreneurs are an important group to study because they noticeably contribute to the Latvian economy.

There are different approaches to how a particular group could be studied. The authors of this study were interested in how women entrepreneurs were different from other groups given the fact that many aspects of social, economic and educational environment are shared with other professional groups. One discipline that addresses differences between people is social psychology, and in particular, personality psychology within which the Big Five Personality Model, as a dispositional model, has widely been used, which is why this Big Five model has been chosen for this study. Through the application of the Big Five approach, the identified personality traits were hypothesized to point to specific features of women entrepreneurs which would set them apart from others and which would allow them to successfully conduct and manage their businesses in the long term perspective.

2 Overview of the Big Five Personality Model

One of the earliest tests of the Big Five Personality Model was developed by Goldberg and tested personality traits along continua of opposites traits (also known as dimensions, facets or factors)—extraversion, neuroticism, agreeableness, conscientiousness and openness to experience [5]. Lexical descriptions or components of these five dimensions can be found in various sources [6–8] (Table 1).

This approach to the study of interaction of personality traits and business factors has been widely used, for example, when researching professional burnout [9, 10], job satisfaction [11, 12], workaholism [13] and job effort [14]. Entrepreneurs and their professional behavior has also been studied by applying the Big Five Personality model [15–18]. Some findings are summarized in Table 2.

Scores are typically obtained by participants' completing questionnaires that are structured along a Likert scale, consistently with which participants have to indicate the degree to which they agree with a statement. In order to reduce the time that participants are expected to spend completing a questionnaire, shorter versions of the Big Five test, containing 30–50 items, have been developed [6, 8, 19] and successfully tested even on aboriginal farmers of Amazonia [20]. In the following section, research methodology is discussed.

Table 1. Lexical descriptions of personality factors [6–8].

Big Five dimensions	Lexical descriptions
Extraversion	Gregarious, talkative, outgoing, energetic, enthusiastic, boisterous, assertive, eager, friendly, sociable, lively, social, etc.
Neuroticism	Neurotic, depressed, blue, agitated, stressed, tense, worried, emotional, unstable, upset, obsessed, indecisive, maladjusted, uneasy, irritable, angry, not self-confident, etc.
Agreeableness	Agreeable, helpful, unselfish, altruistic, forgiving, trusting, warm, friendly, considerate, kind, polite, cooperative, easygoing, accommodating, complying, not stubborn etc.
Conscientiousness	Conscientious, thorough, accurate, reliable, organized, diligent, persevering, efficient, planning, persistent, focused, careful, selfless, caring empathetic, duty-oriented etc.
Openness to experience	Original, seeking novelty, curious, different, ingenious, active, imaginative, inventive, artistic, aesthetic, reflective, sophisticated, musical, open, creative, adventurous etc.

Table 2. Outcomes of the Big Five personality model in relation to the entrepreneur profile.

Dimensions	[15]	[16]	[18]	Summary of scores
Extraversion	Higher scores	No effect	No effect	Lower/moderate
Neuroticism	Lower scores	No effect	Low scores	Lower
Agreeableness	Lower scores	No effect	Low scores	Lower
Conscientiousness	Very high scores	Higher scores	Higher scores	High
Openness to experience	Very high scores	Lower scores	Higher scores	Moderately high

3 Research Methodology

3.1 Research Aim and Hypothesis

Within the social psychology framework and in particular within the research area of individual differences, the Big Five model of personality, as a dispositional model, has been used for this study. The *research aim* is to identify some key factors affecting women's decision to open and sustain their own companies.

Considering the possibility that these factors might be common to many more women than there is a number of businesses opened and led by women, another key aspect to consider was the personality of these women. This research used a 30-item Big Five questionnaire, which was combined with 25 face-to-face interview questions designed by these authors in an effort to consider the Big Five personality traits in relation to business and general questions. Interviews were chosen as research method, which is why the ranking or score research was discarded from the consideration. The interview methodology allows to identify the presence of a component and reasons for its existence.

Considering earlier findings on entrepreneurship summarized in Table 2, the main research question was: Do Latvian women entrepreneurs have features of conscientiousness, openness to experience, extraversion, neuroticism and agreeableness?

The same findings have formed the foundation for the research hypothesis: Based on the research summary of earlier authors, Latvian women entrepreneurs are expected to produce information that might indicate high levels of features of conscientiousness, openness to experience, moderate levels of features of extraversion and low levels of features of neuroticism and agreeableness.

3.2 Research Participants and Procedure

Research participants were 16 female entrepreneurs, owners of micro-size businesses in the areas of consultancy, information technologies, restaurant, construction, retail, and tourism. The age of participants varied from 38 to 64, with most participants having been in the age range of 40–55. All participants are the owners of their own small businesses in Latvia. All participants had higher education. The participants were approached through personal connections or via recommendations. The participants were not compensated for their participation in the study, which ultimately limited the duration of interviews and the number of questions that could be asked.

Participants were interviewed at their workplaces during February 2018—May 2018. In the beginning of an interview which was conducted by a member of this author team, each participant received a printed list of questions, which were asked by the interviewer. Each participant was asked for the consent to participate in the interview and to be recorded. Participants were answering questions over the duration of 45–78 min, which depended on their talkativeness. Interviews were recorded. This survey was conducted in accordance with ethical policies.

3.3 Research Design

The limited number of participants imposed restrictions on the research design, which resulted in the need to deploy the qualitative method as a quantitative method requires a considerably higher number of participants, the number that was inaccessible to the authors of this paper. The chosen type of a qualitative research design was an interview.

Given the fact that participants were not compensated for their participation, the number of questions to be included into the interview had to be relatively low to reduce the time of the interview. Earlier attempts to adopt questionnaires for interview purposes were undertaken by Lang and colleagues with the resulted 15 questions to answer [21]. However, Ryser reported that shorter versions of the Big Five Personality test, consisting of 10–15 questions, were found problematic for Italian, German and French-speaking residents of Switzerland [22]. This is why the foundation of the interview questions was the analysis of the 30-item questionnaire of Soto and John [8].

Thus, the resulted number of questions for this study was 25, which meant that 5 questions per personality dimension were designed. Approximately half of the questions focused on participants' personal experience, whereas another half focused on participants' values regarding business operations and employees. Such answers were considered to help tap into more comprehensive representations of dimensions. Another division of questions focused on direct tapping into a particular dimension and indirect questions, when questions seemed to focus on another topic, not directly related to personality. The answers that were produced to these indirect questions were assumed to produce secondary information that could be interpreted as support for a specific or a number of specific personality dimensions. The fact that some questions were directly targeting personality dimensions was assumed sufficient to activate concepts related to personality. These direct questions functioned as some sort of prime for indirect questions, which is why, if indirect questions were able to elicit answers related to personality dimensions, these answers were considered to be tapping into real personality representations at the conceptual level. Thus, the combination of direct and indirect questions was thought to boost the reliability of answers.

All questions in relation to personality dimensions were randomized to ensure that participants were not pre-programmed to produce similar answers on the same dimension. Each participant received the same order to questions.

4 Research Results

Research data were analyzed using discourse analysis principles that are based on lexical analyses of utterances, the frequency of repetition of related concepts as well as hesitations before and after the mentioned concepts. The collected data are summarized in Table 3.

Table 3. Research data summary.

Personality dimensions	Degree of presence (number of participants)			
	High	Relatively high	Relatively low	Low
Conscientiousness	15	–	1	–
Openness to experience	–	12	4	–
Extraversion	–	11	5	–
Neuroticism	–	3	3	10
Agreeableness	–	1	10	5

The degree of presence (Table 3) was assessed based on (1) lexical frequency of mentioned concepts in relation to other concepts mentioned within the same interview, (2) adjectives and adverbs describing the concepts or actions, (3) a range of descriptors, including mutually exclusive, and (4) typical vocabulary and concepts used with reference to personality dimensions in other papers, such as [6, 8] and others.

5 Discussion and Conclusion of Results

The obtained results allow to answer the research question asked in Sect. 3.1, specifically, Latvian women entrepreneurs contain traits that are consistent with conscientiousness, openness to experience, extraversion, neuroticism and agreeableness because all answers contained answers and comments on these dimensions.

As for research hypothesis, stated in Sect. 3.1, it was confirmed, specifically, answers suggested that Latvian women entrepreneurs contain features indicating that *conscientiousness* displayed the highest levels of features as all remarks apart from one indicated the importance of the feature through the produced vocabulary, such as, *important, significant, absolutely necessary* when talking about the importance of having a workplace well-organized, completing work by the deadline, not procrastinating, thorough planning, completing work promptly even when work requires effort because it is not liked, etc.

Another dimension, *openness to experience,* was concluded to display relatively high features because about 14% of remarks produced answers that questioned or contradicted features that might be viewed as consistent with openness to experience. The first part of the conclusion derived from remarks indicating that participants valued diversity of experience, multidimensional and multidisciplinary education, interests in arts, music, foreign languages and self-development as important and very important. The second part of the conclusion was produced based on remarks that did not appreciate diversity of experience, importance of multidisciplinary education, arts, music, foreign languages etc. The latter remarks were produced with reference to two questions only, one focused on transformation of educational system in Latvia and another one on weaknesses of college graduates.

The answers to questions pertinent to *extraversion* produced conflicting results, indicating the presence of *extraversion* and *introversion* traits, which is why it was concluded that this dimension displayed the moderate-level features. Some examples of conflicting results are the following. “In a new event I approach new people, particularly at a business event or other important event, however, I do not feel comfortable about it. But I do it because I have to, I need to. I meet several new people, not just one or two.” “Yes, I like new events, be it an exhibition or a seminar, workshop. I meet one, two, three new people and I like it. You can always learn something new from these people! I enjoy the experience! What is the greatest number of new people that I met at an event? Well, I have not counted them, but… I don’t know, maybe, four… let’s say, four… Out how many people? It was at a seminar… Probably there were 30 participants.”

Remarks that focused on the dimension of *neuroticism* overwhelmingly indicated low levels of traits associated with this facet, which is why it can be concluded that participants produced answers consistent with *high emotional stability*. For example, “I continued jumping with the parachute even after one of such jumps resulted in broken bones and I had to learn to walk again and my backbone hurt for months. Was I not afraid? Um… no, not really, I enjoyed it, it’s the rush of adrenaline.” “After an unpleasant meeting, negotiations are completed, I do not worry about them anymore. It is not productive. I do take care of myself. I want to live longer.” “After a conflict situation, I worry about it maximum till next morning. In the morning I have a new day, new responsibilities, tasks, objectives, they have to be completed. That conflict or difficulty is history now." Only one remark indicated some features of neuroticism.

The last dimension was *agreeableness*. The obtained results might be considered conflicting. The view of participants of themselves was quite positive—they thought they were quite agreeable—"I do not like conflicts. I always try to solve them, which is why I talk to my employees if there are conflicts amongst them. Conflicts are not productive." "I am open to talking to my workers. They can always come to me and we can discuss things." "I agree with Steve Jobs that once you have hired people, you should consider their opinion, you should trust them, their judgement." "I am never in conflict with my business partners. We negotiate things." However, some remarks pointed to lower levels of agreeableness, rather some lack of agreeableness. For example, "I don't like conflicts, particularly at work, so, I talk to my employees if there are conflicts among them. But I solve it once, maybe, maybe, twice, and if it reoccurs, I fire them. Who exactly? That depends." "When I hire people, I tell them that I do not tolerate conflicts at work. I understand that they have families, other issues, but they must be left at home. Work is work. If they are unhappy, they can go." "If there is a conflict at work with one person, that is, most team members complain about that person's character, they say that person is arrogant, ignores others' work, I am not going to fire that person if that person is a good specialist. And if conflicts continue, as long as work is done properly, it's not my problem. I am not going to fire that specialist. I am going to talk to this person just once."

The fact that Latvian women entrepreneurs have been found to display features consistent with higher levels of conscientiousness and openness to experience and lower levels of neuroticism might point to some possibility that these features at some level might be universal to women entrepreneurs because of the need to secure the existence of their businesses, which rely on good levels of organization, excellent planning and ability to develop different types of scenarios for business operations depending on different confounding variable that might quickly change on volatile markets particularly those with lower levels of population solvency and small customer groups, such as the case in Latvia. The necessity to secure the existence of business and maybe even its development requires the ability to make difficult decision, solve conflicts and function in the environment of possible uncertainty—these features are associated with lower levels of neuroticism. Often staring a business requires a vision, and this vision is linked to creativity, which is part of openness to experience.

However, the obtained results on agreeableness are somewhat different from some of the previous studies. The conflicting nature of the results might be attributed to the context of entrepreneurial activities in which Latvian women entrepreneurs have to exist, for example, such as a limited employee market, limited partner market etc., which encourage them to be open to negotiations and talks with employees, at least to a certain degree. Obviously, strong presence of low features of agreeableness in some remarks might be associated with deadlines, organizational planning, in other words, higher features of conscientiousness that secure business existence on the market.

The results of extraversion, taken in the context of previous studies, which indicated either no effect or significant effect, might be interpreted as features that might not necessarily be of the universal character. They might depend on the local context, which might set particular constraints on the mechanisms and steps that ensure the existence of a business. For example, there is a limited number of potential partners which need to be found.

6 Conclusion

Within the social psychology framework and in particular within the research area of individual differences, the Big Five model of personality, as a dispositional model, has been used for this study. As for the specifics of the research question, it was hypothesized that successful Latvian women entrepreneurs would score higher on Conscientiousness and Openness to Experience and lower on Agreeableness and Neuroticism.

Finally this study finds that Latvian women entrepreneurs similar to women entrepreneurs in other countries in that they have produced answers pointing to high presence of Conscientiousness, relatively high degree of presence of Openness to experience, lower degrees of Neuroticism and Agreeableness. All of participants have produced answers consistent with a higher degree of Extraversion because all questions on Extraversion received positive answers on this qualitative characteristic.

For the women surveyed, it was not the first employment. Social factors facilitated their start-ups. The key feature of their character is the ability to take risk and responsibility for the people. Everyone has very strong motivation "I wish."

Consistently with the hypothesis, women entrepreneurs in Latvia scored: (1) High on Conscientiousness and Openness to Experience; (2) Low on Agreeableness and Neuroticism. However, the answers suggest that participants scored high also on Extraversion. So, the hypothesis of this research is partly confirmed.

Finally, a few remarks should be made on limitations of this study. To create a reliable profile of women entrepreneurs in Latvia, women entrepreneurs of various sizes of businesses would have to be considered. Women entrepreneurs should also complete an in-depth Likert scale questionnaire, which ultimately requires not only more resources but also a considerably larger pool of participants, which in the context of Latvia, a small economy, might become a real challenge. However, even of participants completed such a traditional questionnaire, the obtained results might point to the subjective rather than objective reality, the reality based on participants' perceptions.

References

1. Kelley, D.J., Baumer, B.S., Brush, C., Greene, P.G., Mahdavi, M., Majbouri, M., Cole, M., Dean, M., Heavlow, R.: Global entrepreneurship monitor. Women's entrepreneurship 2016/2017 report (2017). https://www.babson.edu/Academics/centers/blank-center/global-research/gem/Documents/GEM%202016-2017%20Womens%20Report.pdf. Accessed 22 Aug 2018
2. Female entrepreneurs. Internal market, industry, entrepreneurship and SMEs. European Commission. https://ec.europa.eu/growth/smes/promoting-entrepreneurship/we-work-for/women_en. Accessed 20 Aug 2018

3. Latvian women most entrepreneurial in Europe—new study. The Baltic Times (2014). https://www.baltictimes.com/latvian_women_most_entrepreneurial_in_europe_-_new_study/. Accessed 10 June 2018
4. European commission report: Statistical data on women entrepreneurs in Europe (2014). http://eige.europa.eu/resources/statistical%20data%20on%20women%20entrepreneurship%20in%20europe.pdf. Accessed 20 Aug 2018
5. Goldberg, L.R.: An alternative "description of personality": the Big-Five factor structure. J. Pers. Soc. Psychol. **59**(6), 1216–1229 (1990)
6. John, O.P., Srivastava, S.: The Big Five trait taxonomy: history, measurement, and theoretical perspectives. In: Pervin, L.A., John, O.P. (eds.) Handbook of Personality: Theory and Research, pp. 102–138. Guilford Press, New York (1999)
7. Kwantes, P.J., Derbentseva, N., Lam, Q., Vartanian, O., Marmurek, H.H.C.: Assessing the Big Five personality traits with latent semantic analysis. Pers. Individ. Differ. **102**, 229–233 (2016)
8. Soto, C., John, O.P.: Short and extra-short forms of the Big Five Inventory-2: the BFI-2-S and BFI-2-XS. J. Res. Pers. **68**, 69–81 (2017)
9. Canadas-De la Fuente, G.A., Vargas, C., Luis, C.S., Garcia, I., Canadas, G.R., De la Fuente, E.I.: Risk factors and prevalence of burnout syndrome in the nursing profession. Int. J. Nurs. Stud. **52**, 240–249 (2015)
10. Kokkinos, C.M.: Job stressors, personality and burnout in primary school teachers. Br. J. Educ. Psychol. **77**, 229–243 (2007)
11. Judge, T.A., Heller, D., Mount, M.K.: Five-factor model of personality and job satisfaction: a meta-analysis. J. Appl. Psychol. **87**(3), 530–541 (2002)
12. Sutin, A.R., Costa Jr., P.T., Miech, R., Eaton, W.W.: Personality and career success: concurrent and longitudinal relations. Eur. J. Pers. **23**, 71–84 (2009)
13. Burke, R.J., Matthiesen, S.B., Pallesen, S.: Personality correlates of workaholism. Person. Individ. Differ. **40**, 1223–1233 (2006)
14. Merino-Tejedor, E., Hontangas-Beltran, P.M., Boada-Grau, J., Lucas-Mangas, S.: Hardiness as a moderator variable between the Big Five model and work effort. Pers. Individ. Differ. **85**, 105–110 (2015)
15. Brandstatter, H.: Personality aspects of entrepreneurship: a look at five meta-analyses. Pers. Individ. Differ. **51**, 222–230 (2011)
16. Ciavarella, M.A., Buchholtz, A.K., Riordan, C.M., Gatewood, R.D., Stokes, G.S.: The Big Five and venture survival: is there a linkage? J. Bus. Ventur. **19**, 465–483 (2004)
17. Leutner, F., Ahmetoglu, G., Akhtar, R., Chamorro-Premuzic, T.: The relationship between the entrepreneurial personality and the Big Five personality traits. Pers. Individ. Differ. **63**, 58–63 (2014)
18. Zhao, H., Seibert, S.E.: The Big Five personality dimensions and entrepreneurial status: a meta-analytical review. J. Appl. Psychol. **91**(2), 259–271 (2006)
19. Costa Jr., P.T., McCrae, R.R.: Domains and facets: Hierarchical personality assessment using the revised NEO personality inventory. J. Pers. Assess. **64**(1), 21–50 (1995)
20. Gurven, M., Von Rueden, C., Massenkoff, M., Kaplan, H., Vie, M.L.: How universal is the Big Five? Testing the Five-Factor model of personality variation among Forager farmers in the Bolivian. J. Pers. Soc. Psychol. **104**(2), 354–370 (2013)
21. Lang, F.R., John, D., Ludtke, O., Schupp, J., Wagner, G.G.: Short assessment of the Big Five: robust across survey methods except telephone interviewing. Behav. Res. Methods **43** (2), 548–567 (2011)
22. Ryser, V.-A.: Psychometric properties of extra-short Big Five personality measures in multi-topic surveys. In: Documenting personality traits in the SHP and MOSAiCH. FORS Working Paper Series, vol. 5, pp. 1–35 (2015)

Considering and Rationalizing the Establishment of a New Distribution Channel for Export Travel-Related Services in Latvian Market

Kristina Mahareva(✉)

Doctoral Programme "Regional Economics and Economic Policy",
Baltic International Academy, Riga LV-1003, Latvia
kmah@tutanota.com

Abstract. The tourism trends analysis explores the main trends for Latvian inbound and domestic tourism providing statistical data collected over a ten-year period. The research uses the official statistical absolute data and relative data stemmed from the author's assessment. The paper outlines low consumption of Latvian hotel offers, non-resident accommodation demand with the number of nights spent annually, showing a sluggish acceleration of inbound tourism revenue, with a general decrease in non-resident spending per day and popularity of rural tourism services. The conducted statistical data analysis results in determining a new way to encourage the export of Latvian travel-related services by establishing a new distribution channel grounded on service aggregation in the framework of the New Distribution Capability model introduced by the IATA.

Keywords: Inbound tourism · Statistics · Regional services · New distribution method

1 Introduction

Inbound and domestic tourism are priority spheres of Latvian national economic development providing a significant share of the Gross Domestic Product (GDP). The export sector of inbound tourism is harmoniously connected with domestic tourism. The Latvian legislative documents [1, 2] in this sphere of tourism establish the following purposes:

- to focus on the export of tourist services and formation of export-oriented travel-related products;
- to increase the number of foreign tourists staying in places of accommodation for 4 days or more, up to 1.5 million per year;
- to qualitatively increase the average occupation of places (beds) in tourist accommodation institutions beyond the bounds of the summer season.

Problems, tendencies, scenarios, the history of regional, border and rural tourism development in economic, social, ecological and human value aspects have been

I. Kabashkin et al. (Eds.): RelStat 2018, LNNS 68, pp. 670–683, 2019.
https://doi.org/10.1007/978-3-030-12450-2_64

widely studied in various works of Latvian authors. However, firstly, the indicators and trends of inbound tourism, the implementation of the above-mentioned tourism sector's directive indicators in dynamics remained unconsidered. Secondly, the use of new information technologies for regional travel-related service aggregation and increasing of such service export have not been investigated yet.

The purpose of this study is to provide rationale for the necessity to establish a new channel for travel-related service export regarding the local context. It is expected to be based on new information technologies of the air transport for selling of ancillary non-aeronautical services (AS). In order to achieve this purpose, the following tasks are set:

- to analyze and identify trends in inbound and domestic tourism indicators on the ground of the official statistical information;
- to demonstrate that the state administrative and marketing resource for Latvian inbound tourism development has been almost exhausted by now;
- to propose a method arranging export for regional tourism services based on adjustment and implementation of the New Distribution Capabilities (NDC) project of the International Air Transport Association (IATA) in Latvia;
- to develop an organizational structure for distribution of the Latvian regional tourism services as ancillary services in the global market of passenger air transport, the central element of which is regional aggregator.

2 Analysis of Inbound and Domestic Tourism Parameters

2.1 Analysis Procedure

The analysis procedure is based on the following approach.

1. To carry out the analysis, a ten-year period (2007–2016) of statistical observation was selected, including absolute statistical indicators of the pre-crisis, crisis (2008–2009) and post-crisis years [3, 4]. The relative indicators are calculated by the author. The procedure of obtaining the indicators and specifying their terminology correspond to those described on the site [4]. To identify the trend, the regression equations obtained with the help of Trendline Excel tools are applied. The type of equations from among the presented in Trendline Excel is selected according to the highest value of the determination coefficients (R^2). The regression equations allow to make a short-term forecast of the indicators with a reasonable error. The available published actual data of the results dated by 2017 [5] were used to estimate the forecast error. It should be noted, that forecasting for each subsequent year requires the regression equations should be adjusted, regarding the published statistics of the previous year.
2. Accommodation establishments are divided into different groups. The group "Hotels and other accommodation establishments" (Hotels & others) includes those with the quantity of beds more than 10; the group "Hotels and similar places of accommodation" (Hotels), comprises only establishments with regular cleaning on a daily basis (hotels, motels, rehabilitation centers, resort hotels, apartment hotels and guesthouses); the group "Guesthouses and other short-stay accommodation

establishments" (Guesthouses) includes tourist and recreation centers, holiday homes and cottages, visitor flats, youth hostels; the group "Campings", consists of camping grounds, recreation vehicle parks, trailer parks. The establishments of the group "Rural tourism" provide accommodation services in small-capacity places of accommodation outside Riga (public and private houses), that do not refer to the "Hotels & others accommodation establishments", and their indicators are examined separately.

2.2 Indicators of Number of Visitors and Non-residents in "Hotels and Others Total"

The official statistics on the total number of visitors (domestic and non-residents), including the number of non-residents located in "Hotels & others" (section TUG01 [3, 5] and 1.5 [4]), are summarized in a diagram in Fig. 1. Over the observed period non-residents and residents did not practically change in number, on average, account for 67% and 33%, respectively, for the period under review.

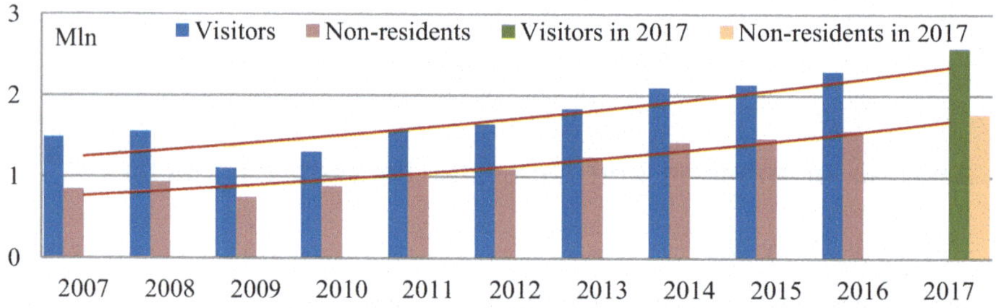

Fig. 1. Dynamics of the number of visitors and non-residents staying in "Hotels & others" establishments.

Exponential trends in the number of visitors, including non-residents

$$\text{Visitors: } y = 1.1811e^{0.0629x},$$

$$\text{Non-residents: } y = 0.7117\,e^{0.0789x}$$

with very close relevance to the original data (R^2 = 0.6897 and R^2 = 0.872), calculated for the period 2007–2016, give a forecast error for 2017 of about 5% and 1%.

2.3 Indicators for the Number of "Hotels and Others" Accommodation Establishments

The dynamics in the number of "Hotels", "Hotels & others" indicators and the number of accommodation establishments for different category groups for the period 2007–2016 is shown in Fig. 2 (the initial data of sections TUG01 [3, 5] and 1.1 [4]).

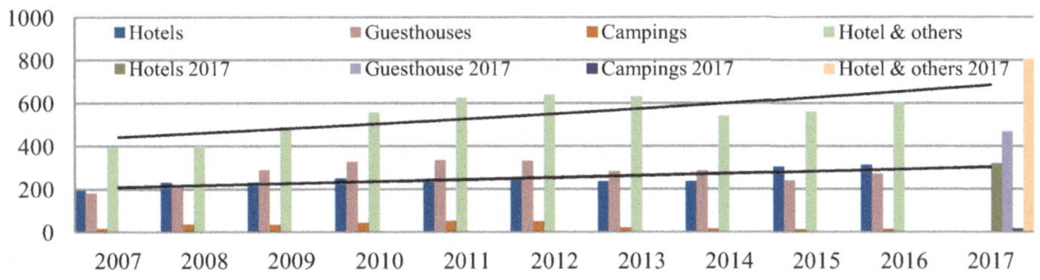

Fig. 2. Dynamics of the number of accommodation establishments for different category groups.

Regression equations with acceptable $R^2 = 0.5359$ and $R^2 = 0.6729$ respectively, show positive dynamics in the growth of the total number of establishments:

$$\text{Hotels: } y = 422.95\, e^{0.0438x},$$

$$\text{Hotels \& others: } y = 203.23\, e^{0.037x}.$$

Possible forecast errors for these indicators, in comparison with the actual data for 2017, account for about 13% and 6%, respectively.

The diagram in Fig. 2 also shows:

- unstability of the number of "Guesthouses";
- evidently decreasing dynamics of the "Campings" number (17 in 2017) and their extremely modest share (below 3%) in the total number of "Hotels & others"

2.4 Indicators of Establishments' Admission in "Hotels and Others" Group

The analyzed data of this set of indicators are the statistical data of sections TUG01, TUG03 [3, 5] and 1.3 and 1.10, 1.11 [4]. The dynamics of the total number of beds in the establishments of all accommodation categories and, in general, for "Hotels & others" is presented in Fig. 3. The trends represented by the regression equations with the values of the determination coefficients $R^2 > 0.5$ for all indicators, except for the "Campings", show a growing trend in the quantity of beds in general:

$$\text{Hotels: } y = 0.6376x + 18.053, R^2 = 0.8239;$$

$$\text{Guesthouse: } y = 7.0266x^{0.2077}, R^2 = 0.7054,$$

$$\text{Hotels \& others: } y = 28.561e^{0.0268x}, R^2 = 0.5179.$$

The forecast errors make less than 3% for all the establishments, except for the group "Guesthouse", where it is more than 15% due to the large number of beds introduced in 2017. However, in the future this outlier will certainly be attenuated. The quantity of beds in "Campings" has significantly reduced in recent years and has become insignificant as compared to the total number of beds in "Hotels & others".

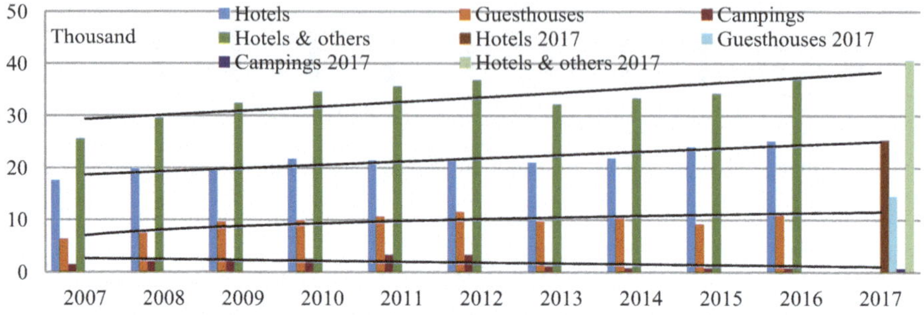

Fig. 3. The dynamics of the number of beds in different "Hotels & others" establishments and in general.

2.5 Indicators of Bed Occupancy Rate in "Hotels and Others" Accommodation Establishments

The indicators of bed occupancy rate, in our opinion, are the most important, as they determine the engagement degree of "Hotels & others" establishments' resources, and marketing of these resources directly forms these establishments' income. Figure 4 shows the dynamics of changes in this indicator for establishments of all categories "Hotels & others" countrywide in Latvia and, particularly, in Riga for the period 2007–2016 (the data from Sect. 1.12 [4]).

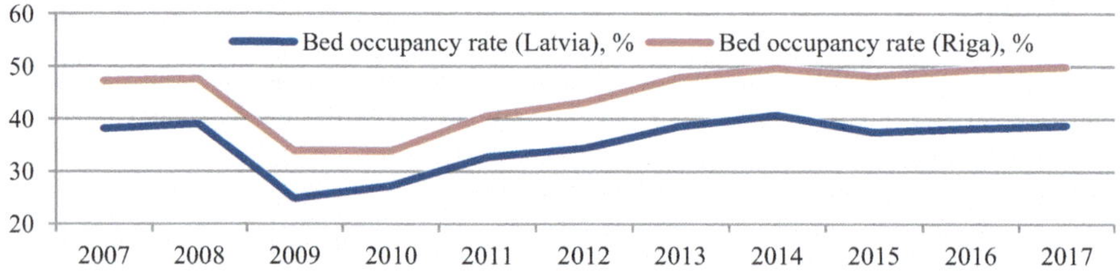

Fig. 4. "Hotels & others" accommodation establishments. Bed occupancy rate, %.

The rate dynamics is almost identical both for Latvian and Riga establishments. The indicator of the establishments' average occupation in the "Hotels & others" group throughout the period of 2007–2017 is less than 37% and 47% for Latvia and Riga, respectively, i.e. the engagement of the "Hotels & others" establishments' resources in Riga is, on average, 10% higher than that of Latvia. For the period of 2013–2017, this indicator for both Latvia and Riga settled at an average annual level of 39% and 49% respectively, which roughly corresponds to the 2007 and 2008 levels. Overall, these figures are not encouraging. Hotel resources account for less than 50%; this indicates that, on the one hand, there is a significant potential for increasing the number of visitors, and, on the other hand, the hotel business is witnessing significant losses and unused opportunities.

Figure 5 presents self-drawn charts of the indicators of hotel room occupation for different categories in the “Hotels” group for 2007–2016, calculated by the author on the basis of the initial statistical indicators from Table TUG01 [3, 5] and Sects. 1.3–1.5, 1.12 [4].

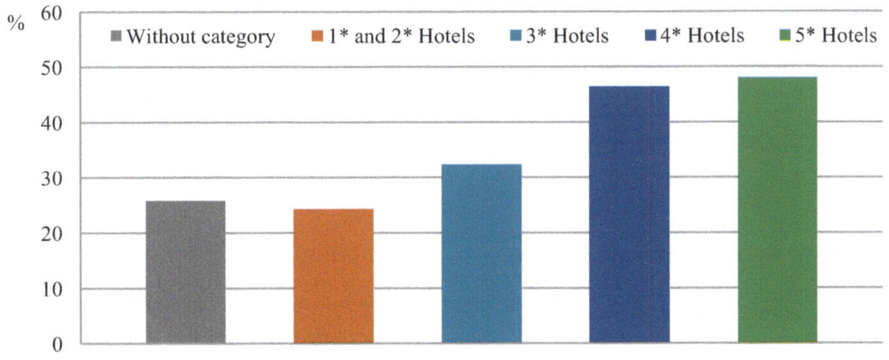

Fig. 5. The averaged indicators of bed occupation in accommodation establishments of the group “Hotels” in Latvia over the period 2007–2016.

The diagrams in Figs. 4 and 5 show as a whole:

- the existence of a significant potential of accommodation resource in establishments of all categories for the hotel business development without any special need to erect new establishments;
- unsatisfactory average engagement indicators of establishments’ accommodation resource, which is averagely below 50% for establishments of all categories and significantly affects the performance of the hotel business;
- the greatest engagement of the 5* hotels among all hotel categories and extremely modest engagement of the accommodation resource at the one-star and two-star hotels, as well as non-categorized hotels.

2.6 Indicators of the Average Number of Nights Spent by Non-residents in Accommodation Establishments

The graphs showing the average number of nights spent by non-residents in “Hotels & others” establishments and in all places of accommodation in Latvia, including staying with friends or relatives at their private flats and houses, for 2007–2016 are represented in Fig. 6 (the data of sections TUG03 [3, 5] and 1.10, 3.2 [4]). This indicator is very significant, as the main directive documents for the development of Latvia, in general, and its export of tourist services, in particular, provide for the need to increase it up to 4, approximately, for all the establishments of the “Hotels” group. However, the value of this indicator for non-residents in "Hotels" over the period 2007–2016 does not vary practically from year to year and amounts, on average, only to 2.03 nights.

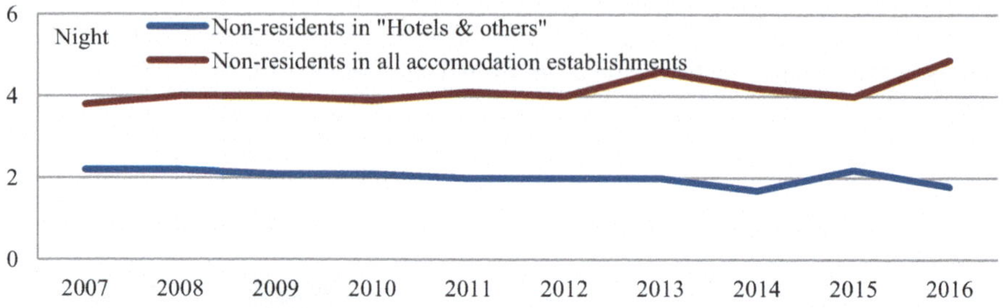

Fig. 6. The average number of nights spent by non-residents in accommodation establishments "Hotels & others" and in all allocation places of Latvia.

2.7 Indicators of "Rural Tourism" Accommodation Establishments

Figures 7 and 8 respectively represent charts of the non-resident number in the "Rural & others" accommodation establishments and the number of rural tourism accommodation establishments (sections TUG06 [3] and 2.1 [4]). The official data for 2017 had not been published by the time of the article's publication.

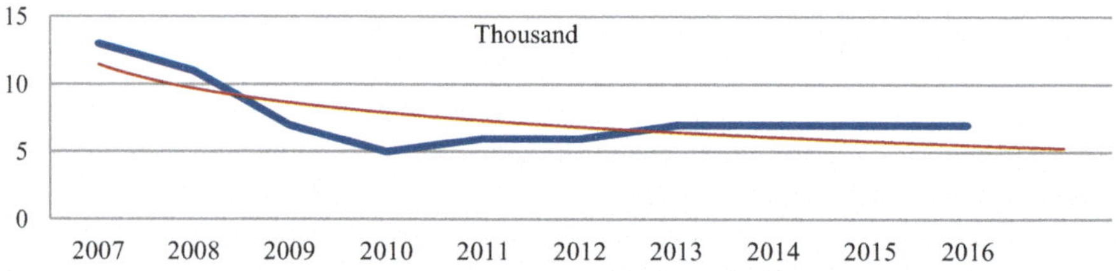

Fig. 7. The number of non-residents in "Rural tourism" accommodation establishments.

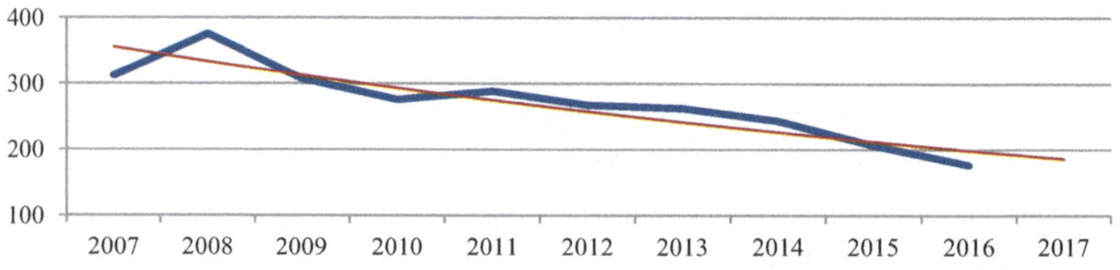

Fig. 8. The number of "Rural tourism" accommodation establishments.

The regression equations for the number of non-residents in "Rural tourism' and the number of "Rural tourism" accommodation establishments respectively

$$y = -2.577\ln(x) + 11.493,$$

$$y = 380.45\,e^{-0.065x}$$

with R^2 = 0.5905 and R^2 = 0.8378 showing a decreasing trend of the indicators under consideration.

In general, the diagrams in Figs. 7 and 8 provide the basis for the following conclusions:

- an extremely low and non-changing value of non-resident number in "Rural Tourism", which is about 7000 people in recent years, and practically does not affect the overall flow of inbound tourists;
- a decreasing trend in the number of rural tourism accommodation establishments;
- the insufficient efficiency of the taken measures and activities in rural tourism.

Indicators of Expenditures during Non-Resident Overnight Travellers' Trips.

The total foreign travellers' expenditures are made by the cost of goods and services, which are bought in advance during the preparation for the trip and during the trip itself. The expenses of foreign travellers visiting Latvia do not include: capital investments or transactions that relate to real estate and works of art; roundtrip transportation costs to Latvia and domestic trips. The official data on the indicators relating to 2017 had not been published by the time of the article preparation. Figure 9 shows the dynamics and changing trend in total expenditures of non-residents staying in Latvia longer than overnight for the post-crisis period from 2010 to 2016 (Sect. 3.2 [4]).

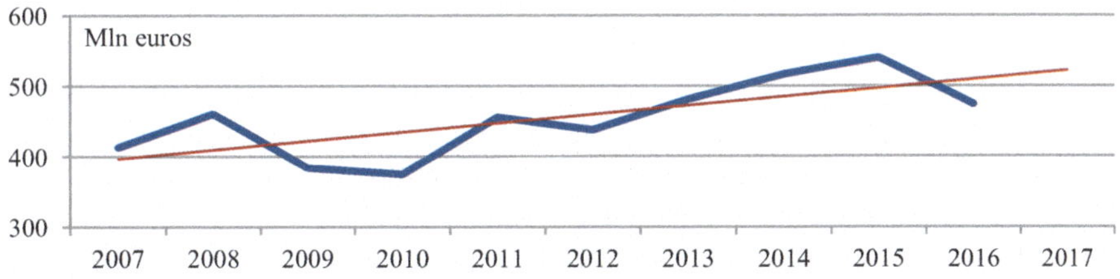

Fig. 9. Total expenditures during non-resident overnight travellers' trips.

These expenses of travellers make the Latvian tourism industry's incomes. The regression equation of the indicator under consideration

$$y = 12.54x + 385.22$$

with the determination coefficient R^2 = 0.5044 showing an unstable but positive trend. However, the fall of this indicator in 2016 is alarming. The dynamics of average daily expenses of a non-resident traveller during the post-crisis period from 2010 to 2016,

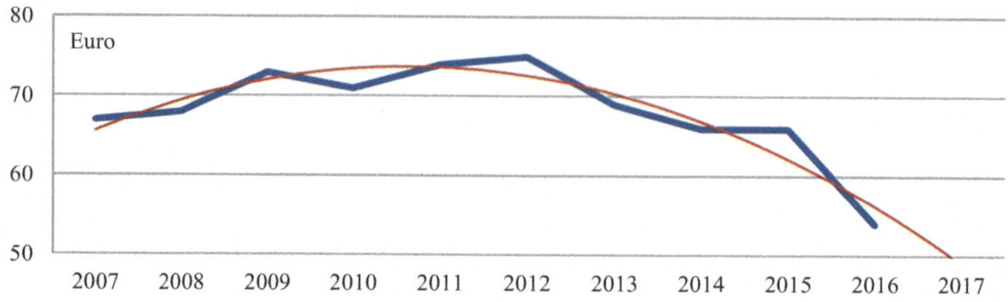

Fig. 10. Average daily expenditures per non-resident overnight traveller.

represented in Fig. 10 by the diagram (TUG12 [3], 3.2 [4], and the polynomial regression equation

$$y = -0.6098x^2 + 5.672x + 60.583$$

with $R^2 = 0.8793$, show a general trend of daily expense reduction by a non-resident traveller in Latvia.

2.8 General Conclusions to Sect. 2

The indicator analysis attests that despite the growth of absolute indicators of accommodation and an increasing flow of non-resident tourists in Latvia:

- the most important relative indicators practically do not grow—the average occupation of beds (significantly below 50%) and the average number of nights spent by non-residents in hotels of Latvia (2.03);
- the average daily expenditures per non-resident overnight traveller keep falling;
- "Rural tourism" is currently sluggish.

This indicates that the devised and implemented plans, as well as the suggested measures are not sufficient to considerably boost export of tourist services. It is necessary to think of other methods based on the latest information technologies. For this purpose, it is suggested using the New Distribution Capability (NDC) [6, 7] model in the framework of the International Air Transport Association's (IATA) project implemented on international airlines.

3 The Approach to Service Distribution in Latvian Tourist Industry: Ancillary Services in Passenger Air Transport Considering the IATA's NDC Model Adjusted to the Local Factors

3.1 Application Options of the NDC Model for Aggregating of Adjacent Transport and Tourist Services as Ancillary Air Transport Services in Latvia

In recent years ancillary services (AS), which are services of allied modes of transport, tourism, and other services [7, 8], have been being widely introduced in the world air transport. Tourist services can be understood as a package of accommodation services in "Hotels & others", transfers, cultural entertainment, recreation, visiting of sporting and entertainment events, and other services reserved by air travellers in conjunction with the air travel. These services are an important source of airlines' additional revenue.

In general, the NDC standard provides the following possibilities:

- aggregation of information on transport routes, availability of seats, availability of AS resources, fares and other information related to the air transportation;
- control over the implementation of air transportation and AS, formed in "One Order", at all stages of their implementation.

Using the unified information technology of NDC, it is possible to reserve and sell air transportation services and AS to customers through three channels:

- the direct channel "airline—customer";
- the two-level channel "airline—travel agent—customer";
- the three-level channel "airline—Content aggregators (Global Distribution System and the new systems)—travel agent—customer".

When working on the direct channel, the Content Aggregator is not only organizationally and technologically connected with travel agents or customers. The packages of services "are clung" by the airline's CRS (Computer Reservation System), aggregated in "One Order" and issued on the transport documents of the airline e-ticket and eMCO (electronic Miscellaneous Charges Order). This option is mostly suitable to arrange regional AS packaging and export (specific for a certain region, including such small countries as Latvia) for the air passengers of international airlines travelling to the region in question. Also this option can be used by the airlines of Global Distribution System (GDS) for reservation of air transportation and other non-regional AS.

3.2 Suggestions on Developing the Latvian Aggregator of Regional Ancillary Services (RAS)

Figure 11 represents a flowchart outlining the adjustment procedure of the previously described IATA's NDC model as the basis for setting-up the Latvian Aggregator (LA), as a system of regional AS aggregation and distribution through the newly established channel into the global market.

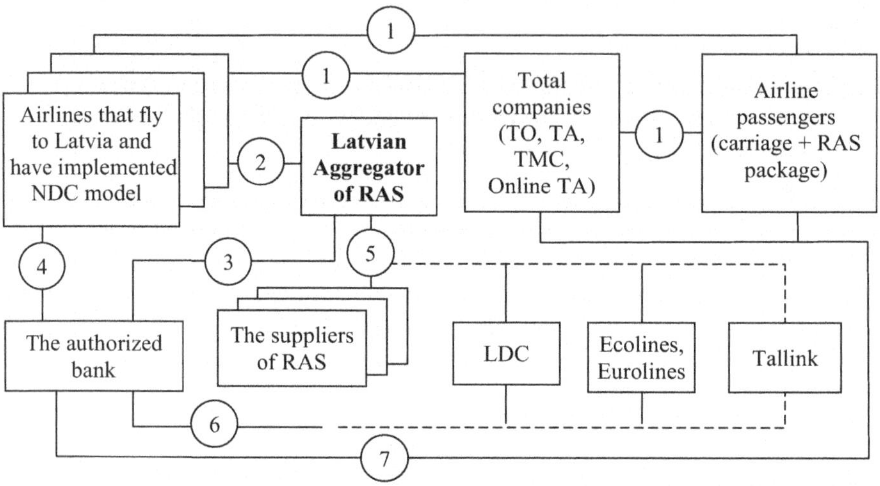

Fig. 11. The organizational framework of LA, as a new distribution channel for Latvian RAS distribution and export.

The digits in Fig. 11 indicate the following main technological procedures:

1. Reservation and processing of documents for the sold air transportation and complex RAS by "One Order";
2. The reserved RAS delivery; provision of customer support services throughout the entire term of the RAS delivery; execution of accounting operations for the sold RAS; work with claims;
3. Distribution of revenue between the suppliers for their sold RAS, payment for the LA services;
4. Receipt of the revenue and execution of accounting operations for the company's own services sold;
5. Aggregation and reservation of RAS for transmission to the airline CRS;
6. Execution of accounting operations for making the payment to the suppliers of the sold RAS;
7. Execution of accounting operations on the revenue for the aggregated and charged services in the "One Order".

The links represented in Fig. 11 also show the pending necessity of appropriate contractual relations between the interacting entities.

In the present framework the participants of the System are:

- traditional airlines (a/c) flying to Latvia, in particular, British Airways that has already implemented the NDC standard for "One Order" formation, and in the nearest prospect—Air Baltic, Lufthansa, Aeroflot—Russian Airlines, etc.;

- travel companies, reserving and selling air transportation and AS through the websites of airlines and GDS;
- suppliers of RAS;
- an authorized bank that provides payment instruments to the project participants and performs accounting/payment for services.

Potential members of the system for service export can be Latvijas Dzeļš Ceļš (LDZ), bus carriers Ecolines, Eurolines and the ferry carrier Tallink on the condition of the CRS interface provision with LA via the XML standard and acceptation of the respective airlines eMCO for their transport services.

The central framework participant performing the functions of RAS packaging, reserving, distributing and selling is LA, which, performing, in fact, the functions of an information online provider, must ensure, as a minimum:

- hosting of information resources of the RAS suppliers;
- on-line reservation and distribution of full or quota service content and fares of the RAS suppliers;
- supply of constantly updated information available for each product (characteristics, terms of delivery, standards, etc.);
- personalization of the RAS delivery with the identification and satisfaction of passengers' requests on an individual basis.

To perform the functions listed above, the Aggregator must:

- have the CRS providing a resource for storing the content of the RAS suppliers on equal competitive terms and interacting with the airlines' CRS when creating "One Order";
- process the "big data" strictly in accordance with the current legislation and other regulatory acts on personal data protection;
- keep mutual accounting and conduct claim work with the Aggregator members;
- implement mechanisms of sending NDC-transactions to information systems of distribution participants;
- provide data on RAS sales to their regional AS suppliers in an official form of files or messages for their back-office systems.

Design and implementation of the proposed System and RAS distribution model to the global market of air transport passenger services will provide:

- for air passengers: accessibility and guarantee of obtaining the regional tourist services; the opportunity to get a profitable personalized offer in the form of a AS package; actual support for the entire duration of the tourist trips;
- for the RAS suppliers: access to the global market of air transport passenger services; guaranteed payments for their marketed services; their increasing promotion in the tourist market;
- for the air carriers: expansion of the range of proposed AS; business development and improvement of financial indicators;

- for the country and its regions: developing export potential due to the growth of inbound tourism; domestic tourism development; reinforcement of the country's image as a regional tourist centre; the taxable base improvement.

4 Conclusion

1. The work presents an analysis of indicators and trends in the developing inbound and domestic tourism in Latvia, showing a modest and almost not growing engagement of hotel accommodation resources, a low average number of nights per stay spent by non-residents in Latvia, poorly growing incomes from inbound tourism, drop of the average daily expenses per non-resident and stagnation of rural tourism. Annually growing resources, both in terms of the number of accommodation establishments and the number of beds, remain underexploited even witnessing positive trends in inbound and domestic tourism. The redundancy of accommodation resources provokes a self-defeating competition and certain deterioration of business performance in this sphere.
2. The conducted analysis highlights the necessity to search for new approaches to boosting the performance of Latvian inbound and domestic tourism, not satisfying with the existing ones. One of such innovative approaches may become establishment of a new distribution channel for service export.
3. To qualitatively increase the flow of inbound tourists and occupancy indicators of accommodation establishments, it is proposed to modify NDC, a recently-developed distribution model. For this purpose, the researcher proposed the LA framework being regarded as a brand-new channel for Latvian RAS distribution and export, based on IATA's NDC adjusted to the local context. This kind of aggregator systems can also be exploited in other small countries, or as an integrated regional System for several countries, for example the Baltic States.

References

1. LR Ministry of Economics Latvijas tūrisma attīstības pamatnostādnes 2014–2020.gadam (2014). https://likumi.lv/doc.php?id=267332
2. Saeima of the Republic of Latvia: Sustainable Development Strategy of Latvia until 2030 (2014). https://www.cbs.nl
3. Central Statistical Bureau of Latvia: Hotels and other tourist accommodation establishments (2017). http://data.csb.gov.lv
4. Central Statistical Bureau of Latvia: Tourism in Latvia 2017. Collection of Statistics (2017). http://www.csb.gov.lv
5. Central Statistical Bureau of Latvia: Transport-Key Indicators (2017). http://www.csb.gov.lv
6. IATA: NDC-resolution-787 (2014). IATA e-Library
7. Hoyles, Y.: IATA-Led Simplifying the Business (StB). New Distribution Capability (NDC). Building Airline Retailing, pp. 9–44. IATA, Geneva (2014)

8. Mahareva, K.: The NDC model-based aggregation strategies on specific ancillary services offered by Latvian providers to air transport passenger. V International Scientific and Practical Proceedings. In: The Transformation Process of Law, the Regional Economy and Economic Police: The Relevant Economic and Political and Legal Issues, Riga, December 2016, pp. 188–195. Baltic International Academy, Riga (2017). ISBN 978-9984-47-143-3

Linkage Between Management of Long-Lived Non-financial Assets and Performance of Latvian Companies Listed on the Baltic Stock Exchange

Ieva Kozlovska(✉) and Irina Kuzmima-Merlino

Transport and Telecommunication Institute, 1 Lomonosova Street, 1019 Riga, Latvia
{Kozlovska.I,Kuzmina.I}@tsi.lv

Abstract. The research paper studies whether there is any influence or linkage between the management of long-lived non-financial assets and performance of Latvian companies listed on Baltic stock exchange within the system of corporate governance. The research paper is based on the analysis of theoretical literature and research papers within the area of information provided about companies' financial statements and the management of long-lived non-financial assets. Furthermore, the research focuses on investigation and analysis of mutual influence between the management of long-lived non-financial assets and performance of Latvian companies listed on Baltic stock exchange within the system of corporate governance. The aim of the research is to examine relationship between the management of long-lived non-financial assets and performance of Latvian companies listed on Baltic stock exchange. This research is limited to the performance of Latvian listed companies on Baltic stock exchange and research period is from year 2007 to year 2017. Results of this research might be of interest of academic researchers as well as educators and practitioners analyzing information provided in financial statements of listed companies on stock exchange as well as any other companies having significant proportion of long-lived non-financial assets from ones' balance sheet value. Conducted research enabled to define new recommendations relating the management of long-lived non-financial assets.

Keywords: Long-lived non-financial assets · Management · Financial performance

1 Introduction

Similar to other countries also the role of Latvian listed companies in economy of the Republic of Latvia is very important reflecting situation in the country. Specialists from European Central Bank maintain that "over longer horizons the growth rates of listed firms' aggregated earnings and of GDP tend to share similar trends" [1]. Total proportion of long-lived non-financial assets from total balance sheet values of Latvian listed companies is significant one. "In average this proportion in balance sheets of Latvian companies listed on Baltic stock exchange was 48%. In most of these

I. Kabashkin et al. (Eds.): RelStat 2018, LNNS 68, pp. 684–693, 2019.
https://doi.org/10.1007/978-3-030-12450-2_65

companies this percentage was higher and even up to 97%" [2]. "Tangible long-term assets occupy a large share in total company assets, especially in heavy industry despite the high rates of development of a service sphere and increase of a share of intangible assets under globalization" [3]. Therefore, processes of not only accounting and reporting, but also management of these assets of the companies listed on Baltic stock exchange should be accounting as very important issue.

The management of long-lived non-financial assets does have significant impact on company's profitability indicators. Use of return of capital employed, fixed assets turnover ratio plus inventory and debtors turnover ratios was concluded that the way how these assets are being managed does have significant influence on company's profitability [4]. Efficient management of long-lived assets is an integral component of the company's overall corporate strategy in order to create shareholders' value. Efficiency in asset management forces to increase owners' wealth [5].

The aim of the research is to examine relationship between the management of long-lived non-financial assets and performance of Latvian companies listed on Baltic stock exchange. Therefore, the research focuses studying the following issues:

1. Management of long-lived non-financial assets.
2. Influence/relationship between the management of long-lived non-financial assets and performance of Latvian companies listed on Baltic stock exchange.

The research paper has been structured as follows: in order to develop conceptual foundation of this research the literature review about the overview of the literature in the area of this research has been performed. Further, the methodology and findings of the research have been presented and discussed, followed by key conclusions at the end of the research paper.

2 Theoretical Framework of the Research

Long-lived non-financial assets management process is critical for further successes of the company increasing asset's overall performance and contribution to the company's overall performance. In most of cases companies invest significant financial resources to acquire and afterwards use these assets in company's basic operations. Company uses these assets with goal to create competitive advantage that further will increase company's value. Therefore, company's management and also shareholders influence various decision-making processes concerning these assets. In addition, this is not only before, but also after acquisition of tangible and intangible assets. "Lifecycle management of assets is essential for cost-effective maintenance and long-term economic viability. Properly maintained infrastructure provides significant economic advantages. Neglecting maintenance leads to lower productivity and imposes costs on users. Furthermore, delayed maintenance significantly increases total costs associated with repair or replacement" [6]. Further, most of companies' shareholders are more concerned with issues relating maximization of profit and their wealth not with issues relating strategies and management of these assets or "assets management". Therefore, also issue of ethics arises, especially relating intangible asset management [7]. Company's intangible asset management has been increasingly related also with its stakeholders' ethical concerns together with its corporate performance goals since intangible assets management is strongly related to company's superior performance. Ethical management of intangible assets requires a balance between traditional performance-based objectives (i.e. profit maximization, value creation) and broader business objectives that include maximizing stakeholder value and meeting societal obligations [7]. This is essential for companies in order to get reputation, license, etc. and to be able to operate [8, 9].

Importance of various issues relating company's long-lived non-financial assets management process has been supported by many researchers' idea that it plays significant rule in defining real value of particular company's assets reflected in its financial statements, therefore, also the real value of the particular company. Scientifically researches proves significant influence of effective long-lived non-financial assets management process on company's future successes, market value, share price, profit maximization, etc. Furthermore, effective management process of these assets is crucial aspect of improving company's operational performance ranging from financial management to maintenance plus assisting companies to come out of survival mode becoming successful and even thrive [10, 11].

Latvian listed companies listed in Baltic stock exchange has been adopted latest Corporate Governance Code from January 1, 2017. It has been developed by the team representing interests of investors and issuers, auditors and lawyers, public sectors, local financial supervisory, entrepreneurs, and others. Key goal of this documents is to ensure that company's management reach strategic goals, eliminating any discrepancies, increasing company's value and investors' (local and foreign) interest in ones. As a result, it is crucial the recognition and measurement processes of company's long-lived non-financial assets is correct, transparent, and in accordance with respective accounting and reporting regulations.

Authors came to the conclusion that nowadays functions of engineering, finances and outputs are joined staff that in combination with asset's life cycle and continues improving results in processes of asset management and assets governance.

Asset management is the process how tangible and intangible assets create and increase value to the company owning ones. Asset management is not just assets. It is the whole process how this particular asset can create value to the company from its acquisition to disposal. Furthermore, asset governance is the business strategy how particular company uses its long-lived non-financial assets in the most efficient way in order to increase growth indicators and the value of the company. In addition, there no any study performed about issues relating the management and the linkage of long-lived non-financial assets management and corporate performance of Latvian listed companies on Baltic stock exchange.

3 Research Methodology

The literature review conducted by authors in the section above allowed defining the research question (RQ): *does information relating long-lived non-financial assets owned by Latvian companies listed on Baltic stock exchange show/prove that these assets and management of these assets do have influence on companies' performance?*

This research is based on the secondary data—the author's collected data from financial statements/annual reports of these companies available at official home page of Nasdaq Riga [12]. In order to evaluate the relation between long-lived non-financial assets management and performance of Latvian companies listed on Baltic stock exchange the measurement of financial performance was used.

In order to answer research question defined above the analysis of financial statements of Latvian companies listed on Baltic stock exchange was performed for the time period of 11 years (from year 2007 to year 2017). In this case information publicly available at Nasdaq Baltic home page was used to study financial statements/annual reports of all 21 Latvian listed companies on Baltic stock exchange all three lists (Baltic Main List, Baltic Secondary List and Firth North Baltic Share List) as of August 10, 2018. The case companies were classified on 8 different industries:

- basic materials (code: 1000): 2 companies,
- industrials (code: 2000): 4 companies,
- consumer goods (code: 3000): 7 companies,
- health care (code: 4000): 3 companies,
- utilities (code: 7000): 1 company,
- financials (code: 8000): 2 companies,
- technology (code: 9000): 2 companies,
- telecommunications (code: 6000): 1 company.

Since the authors investigated whether there is any relation between long-lived non-financial asset management and performance of Latvian companies listed on Baltic stock exchange secondary sources of information available at the Nasdaq Riga website was used—annual reports of these companies. Analysis of financial performance covered two different periods within the time period of 11 years: year 2007 and year 2017. Within this period the authors selected in total 9 financial ratios/measures per time period of 11 years (year 2007 and year 2017 and in some cases also from year 2007 to year 2017) of Latvian companies listed on Baltic stock exchange.

The selection of the most appropriate ratios for analyzing efficiency and influence of company's long-lived non-financial asset management on overall corporate performance of Latvian companies listed on Baltic stock exchange was based on the research done by Tehrani [13] and also ones by view of the authors of this research paper.

Different methods of financial analysis had been applied—benchmarking or industry analyses and trend analyses by use of various different financial ratios as well as common-size financial statement analyses. The following ratios had been calculated and analyzed for this research: Liquidity, Return on Assets (ROA), Return on Equity (ROE), Return on average Assets (ROAA), Operating Cash Flow (OCF), Equity to Debt, Debt to Assets.

Further, in order to analyze mutual relation of long-lived non-financial assets and companies' overall corporate performance from the perspective of profitability was performed. It was done by use of various mathematical tools (averages, percentages, ratios) and simple linear regression analysis. For this analyzes intangible assets have been excluded. Since the proportion of the ones from the total balance sheet value of companies investigated in this research paper is not significant comparing to tangible ones. The method of selection of the most appropriate ratios for analyzing efficiency and influence of company's long-lived non-financial asset management on overall corporate performance of Latvian companies listed on Baltic stock exchange was based on the research done by Chakraborty [4, 14]. For this analyses only fixed or tangible assets owned by Latvian companies listed on Baltic stock exchange were used. The following ratios relating evaluation of efficiency of long-lived non-financial asset management within the system of corporate governance had been calculated and analyzed for this research:

- To evaluate efficiency of long-lived non-financial asset management: Fixed Asset Turnover Ratio, FATR = Net sales/Fixed assets;
- To evaluate profitability: Return on Capital Employed, ROCE (%) = Net operating profit/Capital Employed.

Furthermore, to ensure more precise and reliable results for the analysis of ratios mentioned above the rank test has been performed.

4 Research Results

Based on the information summarized and analyzed in the Table 1 below it can be concluded that during the time period of 11 years (year 2007 to year 2017) the average overall corporate performance of all Latvian companies listed on Baltic stock exchange (ones listed on the Baltic Main list and the Baltic Secondary List) has been improved considerably. Companies have been investing in long-lived non-financial assets. Therefore, average proportion of long-lived non-financial assets of total assets shows significant increase—companies of the Baltic Main List this average increase is 32% and the Baltic Secondary List—average increase of 31%. Therefore, it can be concluded that the management of long-lived non-financial assets of Latvian listed companies within *the system of corporate governance does have significant influence on the overall corporate performance of these companies.*

Table 1. Financial ratios analysis of corporate performance of Latvian listed companies by lists of Baltic stock exchange per year 2007 and year 2017

Ratios	Main list			Secondary list		
	2017	2007	±	2017	2007	±
LLA to total assets, %	36.83	27.82	+	59.67	45.54	+
Liquidity	2.70	2.14	+	3.24	4.57	+
ROE	0.14	0.06	+	0.26	0.02	+
ROA	0.09	0.05	+	0.09	0.01	+
ROAA	0.09	0.04	+	0.09	0.01	+
Operating CF, mil. EUR	9.7	2.6	+	2.7	5.1	–
Equity to debt	2.47	1.79	+	3.60	4.77	–
Debt to assets	0.66	0.51	+	0.51	0.60	–

Liquidity ratio shows that nevertheless in the research period the proportion of long-lived non-financial assets from the total assets has been increasing companies representing both of lists use their resources more efficiently. Also ROE and ROA proves the improvement of its overall performance and competitiveness of Latvian companies listed on Baltic stock exchange.

Comparing profitability of company's assets between year 2007 and year 2017 for the companies listed on the Baltic Main list shows that it is at the same level, but for the ones listed on the Baltic Secondary List it has even increased from 0.01 to 0.09. Therefore, it can be concluded that companies listed on the Baltic Secondary List use its assets more efficiently.

Analysing average operating cash flow per research period indicates very significant and positive trend for the Latvian companies listed on the Baltic Main

List—average increase of 268%. Therefore, it can be concluded that these companies have been generating cash flow from its basic business operations successful. Opposite it is for Latvian listed companies on the Baltic Secondary List—decrease of 47%.

Since average equity to total debt ratio for the companies of the both lists are above 1 it can be concluded the following—nevertheless there is slight fluctuations to both directions companies are stable enough to cover its short-term and long-term liabilities. This ratio is advisable to be at least 2. Therefore, it can be concluded since average equity to total debt ratio is above 2 for both of lists as in year 2007 as in year 2017 companies in average are stable enough to cover its debts from its own capital. Therefore, Latvian companies listed on Baltic stock exchange become less risky and in case of possible problems ones are able to set their financial obligations by selling assets ones own. Further, analyzing operating cash flow, ROE and ROA per the research period the authors of this paper came to the conclusion that average highest operating cash flow is for basic materials which also represents one of the highest ROE and ROA.

Further, analyzing operating cash flow, ROE and ROA the authors came to the conclusion that management of long-lived non-financial assets within the system of corporate governance do have significant influence on the companies' performance.

In order to find out is there any mutual relation between long-lived non-financial assets owned by Latvian companies listed on Baltic stock exchange and its overall corporate performance the authors performed analysis based on the performance of each company and not per industries represented by ones. According to the Table 2 below the following companies do have positive correlation between company's fixed (tangible) asset management process and company's overall profitability:

- high efficiency-high profitability: VEF, Grobiņa, Ditton pievadķēžu rūpnīca, Valmieras stikla šķiedra, Rīgas elektromašīnbūves rūpnīca, Latvijas jūras medicīnas centrs, Tosmāres kuģu būvētava, Latvijas gaze, HansaMatrix, VEF Radiotehnika RRR, Grindeks, Olainfarm;
- moderate efficiency-moderate profitability: Brīvais vilnis;
- low efficiency-low profitability: SAF Tehnika.

For the further research the author selected year 2007 and year 2017 and evaluate the mutual relation and correlation between fixed or tangible asset turnover ratio and return on capital employed. For the further research intangible assets have been

Table 2. Efficiency-profitability indicators of Latvian listed companies on Baltic stock exchange per time period from year 2007 to year 2017

FATR versus ROCE	Avg. ROCE: high ($\geq$25%)	Avg. ROCE: moderate (>15% but <25%)	Avg. ROCE: low ($\leq$15%)
Avg. FATR: High ($\geq$6)		Rīgas autoelektroaparātu rūpnīca; Rīgas kuģu būvētava; Pata Saldus; Kurzemes Atslēga 1	Grobiņa; Ditton; Valmieras stikla šķiedra; Rīgas elektromašīnbūves rūpnīca; Latvijas jūras medicīnas centrs; Tosmāres kuģu būvētava; Latvijas gāze; HansaMatrix; VEF Radiotehnika RRR; Grindeks; Olainfarm
Avg. FATR: Moderate (>2 but <6)		Brīvais vilnis	Siguldas CMAS; Rīgas juvelierizstrādājumu rūpnīca
Avg. FATR: Low ($\leq$2)		Latvijas Balzāms	SAF Tehnika

excluded since the proportion of these assets from the total balance sheet value of Latvian companies listed on Baltic stock exchange is not significant in comparison to fixed or tangible ones.

Further, in order to explore the mutual association between two variables—fixed asset turnover ratio (FATR) and companies' corporate performance or profitability (return on capital employed)—the author performed calculations and analyses of correlation coefficients shown in the Table 3 below.

Based on the calculations and analyses reflected in the Table 3 above the authors came to the conclusion that there is strong mutual correlation and influence between management of long-lived non-financial assets and companies' overall corporate performance.

The result of the analysis performed for the time period from year 2007 to year 2017 splitted per lists of Baltic stock exchange has been summarized in the Table 4 below. The correlation coefficient shows that there are is mutual correlation between FATR and ROCE. Furthermore, there is no any negative correlation coefficient indentified.

The result of the analysis performed for the time period from year 2013 to year 2017 splitted per lists of Baltic stock exchange has been summarized in the Table 5 below. The correlation coefficient shows that there are is stronger mutual correlation between FATR and ROCE of Latvian companies listed on Baltic stock exchange. Furthermore, similar to the analysis for the time period from year 2007 to year 2017 above there is no any negative correlation identified.

The results obtained show better performance of the correlation coefficient almost for all companies listed on Baltic stock exchange as it was for the research period of 11

Table 3. Analysis of mutual correlation between fixed asset turnover ratio (FATR) and return on capital employed (ROCE)

Companies	Correlation coefficient between avg. FATR and avg. ROCE (2007–2017)	Correlation coefficient between avg. FATR and avg. ROCE (2013–2017)
Rīgas autoelektroaparātu rūpnīca	0.41	0.75
VEF	0.86	0.97
Grobiņa	0.15	0.73
Ditton	0.46	0.55
Rīgas elektromašīnbūves rūpn.	0.45	0.62
Valmieras stikla šķiedra	0.22	0.92
Rīgas kuģu būvētava	0.06	0.97
Latviajs jūras medicīnas centrs	0.38	0.05
Tosmares kuģubūvētava	0.06	0.80
Brīvais Vilnis	0.08	0.93
Latvijas Gāze	0.39	0.57
Pata Saldus	0.28	0.50
Olainfarm	0.86	0.66
HansaMatrix	0.10	0.20
VEF Radiotehnika RRR	0.46	0.39
Grindeks	0.85	0.81
Kurzemes Atslēga 1	0.38	0.94
Siguldas CMAS	0.60	0.71
Rīgas juvelierizstr. rūpnīca	0.47	0.38
Latvijas Balzams	0.01	0.74
SAF Tehnika	0.61	0.18

Table 4. Analysis of mutual correlation between FATR and ROCE for Latvian listed companies splitted per lists of Baltic stock exchange

Correlation coefficient	From year 2007 to year 2017	
0–0.3	Baltic main list	HansaMatrix
	Baltic secondary list	Grobiņa; Valmieras stikla šķiedra; Rīgas kuģu būvētava; Tosmares kuģubūvētava; Brīvais vilnis; Saldus mežrūpniecība (AS Pata Saldus); Latvijas Balzāms
0.3–0.5	Baltic main list	–
	Baltic secondary list	Rīgas autoelektroaparātu rūpnīca; Ditton pievadķēžu rūpnīca; Rīgas elektromašīnbūves rūpnīca; Latvijas Jūras medicīnas centrs; Latvijas Gāze; VEF Radiotehnika RRR; Kurzemes Atslēga 1; Rīgas juvelierizstrādājumu rūpnīca;
0.5–0.7	Baltic main list	SAF Tehnika
	Baltic secondary list	Siguldas CMAS
0.7–1.0	Baltic main list	Olainfarm; Grindeks
	Baltic secondary list	VEF

Table 5. Analysis of mutual correlation between FATR and ROCE for Latvian listed companies splitted per lists of Baltic stock exchange

Correlation coefficient	From year 2013 to year 2017	
0–0.3	Baltic main list	HansaMatrix; SAF Tehnika
	Baltic secondary list	Latvijas Jūras medicīnas centrs
0.3–0.5	Baltic main list	–
	Baltic secondary list	VEF Radiotehnika RRR; Rīgas juvelierizstrādājumu rūpnīca;
0.5–.7	Baltic Main List	Olainfarm
	Baltic Secondary List	Ditton pievadķēžu rūpnīca; Rīgas elektromašīnbūves rūpnīca; Latvijas Gāze; AS Pata Saldus
0.7–1.0	Baltic Main List	Grindeks
	Baltic Secondary List	Rīgas autoelektroaparātu rūpnīca; Grobiņa; Valmieras stikla šķiedra; Rīgas kuģu būvētava; Tosmares kuģubūvētava; Brīvais Vilnis; Kurzemes Atslēga 1; Siguldas CMAS; Latvijas Balzāms

years (from year 2007 to year 2017). The only exceptions identified with opposite (decrease) performance of the correlation coefficient were the following companies:

- Baltic Main List: Olaifarm, Grindeks, SAF Tehnika;
- Baltic Secondary List: Latvijas Jūras medicīnas centrs and VEF Radiotehnika RRR.

Furthermore, it can be concluded that the higher FATR of Latvian listed companies on Baltic stock exchange the higher efficiency of management of long-lived non-financial assets within the system of corporate governance. Therefore, also companies earning capability increases and investors' wealth have been maximized accordingly as well.

5 Conclusions

Based on the research performed above the authors of this research conclude that there is mutual linkage between management of long-lived non-financial assets owned and performance of Latvian companies listed on Baltic stock exchange within the system of corporate governance. Therefore, not only management of these companies, but also shareholders (existing/potential) as well as other ones interested in the performance of the particular company and various organizations dealing with issues relating management of these assets have to take into account this finding when develop models and concepts to manage long-lived non-financial assets owned not only by Latvian companies listed on Baltic stock exchange, but also any other company owing these assets in its balance sheet.

References

1. European Central Bank: Economic and Monetary Developments (2017). https://www.ecb.europa.eu/pub/pdf/other/mb200709_focus05.en.pdf?93084b968732d1d3660b35b3669768a2 (2018)
2. Kozlovska, I.: The impact of long-lived non-financial assets depreciation/amortization method on financial statements. Copernic. J. Financ. Account. **4**(2), 91–108 (2015). http://www.apcz.pl/czasopisma/index.php/CJFA/article/view/CJFA.2015.018
3. Derun, I.: Tangible long-term asset accounting software selection in ukraine: case study. Manag. Theory Stud. Rural. Bus. Infrastruct. Dev. **35**(2), 185–198 (2013)
4. Chakraborty, K.: Linkage between efficiency of assets management and profitability during the post-liberalization era: a study on select companies of indian public sector. J. Inst. Public Enterp. **37**(3/4), 1–17 (2014)
5. Sur, D., Yadav, S.K.: Trends in asset manangement efficiency in Maharatna central public sector enterprises: a cross-sectional analysis. J. Inst. Public Enterp. **37**(3/4), 78–90 (2014)
6. Giglio, J.M., Friar, J.H., Crittenden, W.F.: Integrated lifecycle asset management in the public sector. Business Horiz. **61**(4), 511–519 (2018)
7. Gambetti, R.C., Melewar, T.C., Martin, K.D.: Guest editors' introduction: ethical management of intangible assets in contemporary. Organization **27**(3), 381–392 (2017)
8. Berrone, P., Surroca, J., Tribó, J.: Corporate ethical identity as a determinant of firm performance: a test of the mediating role of stakeholder satisfaction. J. Bus. Ethics **76**(1), 35–53 (2007)
9. Cornelissen, J.P., Christensen, L.T., Kunithia, K.: Bringing: corporate brands and identity: developing stronger theory and the call for shifting debate. Eur. J. Mark. **46**(7/8), 1093–1102 (2012)
10. Metka, R.: Asset management is key to improving performance. Can. Minig J. **136**(8), 32 (2015)
11. Roggi, O., Giannozzi, A.: Fair value disclosure, liquidity risk and stock return. J. Bank. Financ. **58**, 327–342 (2015)
12. Nasdag. Baltic Equity List: Baltic Regulated market (2018). http://www.nasdaqbaltic.com/market/?pg=mainlist&lang=en
13. Tehrani, N.G.: The effects of corporate governance system quality on company performance. Int. J. Manag., Account. Econ. **3**(7), 444–452 (2016)
14. Peavler, R.: Use Asset Management Ratios in Financial Ratio Analysis (2017). https://www.thebalance.com/use-asset-management-ratios-in-financial-ratio-analysis-393187

Business Continuity Assurance in Creative Industries

Jacek Woźniak([✉]) and Piotr Zaskórski

Military University of Technology, 00-908 Warsaw, Poland
{jacekj.wozniak,piotr.zaskorski}@wat.edu.pl

Abstract. The article presents selected aspects of operations of enterprises representing creative industries in Poland, particularly service companies were investigated. Moreover, there are combined two topics in the paper, i.e. the risk management as a basis for developing the objectives for the business continuity assurance system (for the core, i.e. innovative processes), as well as the functionality of the Business Intelligence systems. The aim of the article is to indicate, how the service enterprises of the creative industries in Poland assure the business continuity with the use of the BI systems' environment.

Keywords: Business continuity (BC) · Business intelligence (BI) · Risk · Creative services · Creative enterprise · Creative industries · Innovative processes

1 Introduction

The creative industries are starting to play an increasing role in the economies of many countries, also in Poland. Business entities that make up these industries (i.e. the creative enterprises, which use the creativity, invention and innovativeness of employees, as well as offer the creative goods and carry out creative processes/projects) have their peculiarity, that affects the way how they are managed. A special role in the operation of these enterprises is played by the risk management, which can be regarded as a chance to create the development opportunities or identification and threats reduction. Risk management is associated with the development of the basis for the business continuity assurance, which depends on the different factors.

The aim of the article is to indicate, how the service enterprises of the creative industries in Poland assure the business continuity with the use of the BI systems' environment. The research questions are: (1) What are the factors that are essential to the BC assurance in the service enterprises of the creative industries in Poland? (2) How those enterprises treat risk? (3) What is the approach of respondents to specify actions that are undertaken in the BC assurance? As well as (4) What is the meaning and role of BI systems in the process of BC assurance in those enterprises?

The article consists of four parts, which involve the definition and structure of the creative industries, role and importance of the BI systems in the BC assurance, methodology of the research and the research limitations—that show, which enterprises, how, and why have been surveyed in the context of the BC processes' assurance, as well as empirical results and conclusions—that emphasis, how the Polish

I. Kabashkin et al. (Eds.): RelStat 2018, LNNS 68, pp. 694–703, 2019.
https://doi.org/10.1007/978-3-030-12450-2_66

service enterprises of the creative industries treat risk and risk management as a basis for the BC assurance—according to the BI systems' functionality. The leading research method is induction (carried out on the basis of CATI interviews), supplemented by the critical literature analysis and synthesis.

The results of the empirical study indicate that, in spite of a holistic approach to risks identification, the undertaken actions are selective and random (it is not an integrated approach to the risk management). Risk is regarded primarily as a source of threats, and no benefits. However, the undertaken actions are the convenient environment for implementing and deploying the functionality of BI systems in the area of the business continuity assurance. There is a large potential in the surveyed enterprises, but there are also the key limitations, mainly they refer to the costs of implementing and maintaining the BI systems, the scale of action, and the age of the enterprises.

2 The Definition and Structure of the Creative Industries

Creative industries can be defined as "specific collections of business entities, created by creative companies that are market-oriented, and deal with the creation, production, distribution and/or spread creative goods and services through the media" [1]. The Department of Culture, Media and Sport (DCMS) from the United Kingdom defines creative industries as a set of activities that "are retrieved in individual creativity, skill and talent, and which have a reasonable potential to the creation of wealth and jobs for the next generation and exploitation of intellectual property" [2]—this is a so-called narrow approach to define the creative industries.

The document titled *Creative Industries in Berlin. Development and Potential* [3], adds other activities—next to the commercial activity—like: cultural services financed from the public funds, and the non-governmental sector organizations (non-profit) dedicated to activities in the area of culture. Another approach is offered by Kasprzak [4], who believes that the creative industries are a "sphere of social services, covering the area of the economic activity aimed to create and commercialize cultural products, which include various organizational forms of doing business". All these definitions give prominence to the activity of commercial nature, although with a varying force. It can be assumed that creative industries are mainly constituted by the enterprises. An essential complementation to commercial activities in creative industries are public and social organizations [5].

According to the classification presented in the publication titled *Creative Industries Mapping Document* [2], it can be assumed that creative industries include thirteen types of activities, grouped in the three key areas: (1) arts and crafts industries: performative arts, art and antiques markets, artistic crafts, (2) creative production industries: music, publishing, broadcasting, film and video industries, as well as computer games, video, and fashion ones, (3) creative services industries: advertising, architectural and urban design, software and computer services, design [4].

The study is followed by a focus on a narrow approach to defining the activities of creative industries—mainly for the reason that the creative industries are primarily created by the so-called "design" enterprises.

3 The Role and Importance of the Business Intelligence Systems in the Business Continuity Assurance

The business intelligence systems can be defined as a set of concepts and methodologies aimed at improving business decisions by using fact-based systems (according to the Gartner Group definition from 1989) [6]. In addition, the BI systems can be seen in a narrow approach (as an analytical tool), or a wide one (as a technology covering, besides the common application solutions, also the data warehousing, analytical processing technology, querying and reporting tools, as well as data mining) [7]. The BI systems can also be thought of as a management concept supported by information (IT) tools. These systems combine data from different sources, store historical data and efficiently provide data for analysis—in order to obtain decision-making information for users at all levels of management [8].

The BI systems are becoming not only the resource (an element of a technical infrastructure) but, above all, a specific "gateway key" to the global economy for creative enterprises. The use of the BI systems is especially important when a key role in the operation and development of the enterprise is played not only by the value and structure of financial and material resources, but also knowledge and the relationships with stakeholders—having an impact on the evolution of the enterprise's flexibility [9, 10]. One of the basic ways to use the potential (mostly the informational one) of an ambient of the enterprise (also from the creative industries) is to enable the different functionalities of the BI systems. However, it should be kept in mind that different types of enterprises can make use of the various features of this class's systems. The basic selection criteria for the functions of the BI systems can be such as: age, maturity and size of the enterprise (the number of employees or annual turnovers), the internal potential of employees and the financial capacity of the enterprises, the BI systems' efficiency in the core business of the company, the level of maturity of the industry, as well as the level of technological maturity of the local societies [11, 12].

The implementation of the BI systems in the enterprises improves the following processes: data acquisition and collection, data processing (e.g. aggregation, exploration, visualization), enterprise information integration, data and information sharing, as well as forecasting, and knowledge creation [13–18]. Therefore, it can be noted that the BI systems—in the processes of the business continuity assurance—are e.g. responsible for [19, 20]:

- supporting the processes of identification the information needs of the enterprise,
- providing an access to the information resources for relevant entities (employees),
- updating data and information (monitoring the key determinants of the BC),
- supporting the current communication and the knowledge diffusion,
- increasing the flexibility (agility) of the enterprise,
- the information and managerial/process integration of the enterprise,
- increasing the reliability of the decision-making processes,
- formalization of the process of the business continuity assurance,
- raising the "intelligence" of the enterprise by sourcing the knowledge repository.

Therefore, it can be assumed that the BI systems can be widely used in the business continuity assurance in modern enterprises. However, the scale and scope of the use of these systems in the business continuity management depend on various factors, and therefore the enterprises are not always able to make a full use of the functionality of BI systems. In this situation, a proper solution seems to be the implementation of the simpler systems/technologies, relating to the specific features and capabilities of the BI systems, e.g. using the cloud computing (CC) technology (without buying the technical infrastructure) [21–24] or the Internet of Things (IoT) (supporting the processes of data obtaining and sharing) [25, 26].

4 The Methodology of Research and the Research Limitations

The peculiarity of the Polish creative services' industries, the availability of the domestic and international research results, and the proximity of businesses, have made these industries to be chosen area for the empirical research. The specification of the research methodology is presented in Table 1.

Table 1. The methodology of research.

The components	Specification
1	2
Research scope	The scope and complexity of the risk management in the "design" (i.e. creative) enterprises from the industries of creative services in Poland as a basis for the BC assurance
Research tool	Computer Assisted Telephone-Interviewing (CATI)
Entity carrying out the study	Centre of Marketing Research INDICATOR LLC, Świętojerska 5/7 Str., 00-236 Warsaw
Period of study	2 months (August–September 2017)
Scope of study	Area of the whole country (16 voivodships in Poland)
Respondents	The owners or managers responsible for the risk management, environment or innovations (1 respondent per a business)
Criteria for selection of the research sample	The systematic random sampling (including the criterion of leading business profile) in the layers (corresponding to the enterprise's size)
Research sample	N = 104 creative enterprises Structure according to the criteria of: • average annual level of turnover: 0–10 PLN mln—41% of enterprises, 10–40 PLN mln—34%, 40–70 PLN mln—19%, 70 PLN mln and more—6% • size of the enterprise: micro—14% of enterprises, small—32%, medium—43%, big—11% • age of the enterprise: 1–10 years on the market—22% of enterprises, 11–20 years on the market—69%, 21 years on the market and more—9% • scale of the business: local—98% of enterprises, regional or domestic—89%, European—30%, international—4% (it was a multiple choice question)
Number of units (according to the numbers of the Polish Classification of Activity)	No. 62.01.Z (software processes): N = 27; No. 71.11.Z (architecture): N = 26; No. 73.11.Z (advertisement agencies): N = 25; No. 74.10.Z (special designing): N = 26

The study was carried out on a relatively small sample—therefore, the research can be regarded as a pilot study. It is difficult to carry out an inference on the research population. Conclusions can be applied only to the surveyed enterprises. The study focuses on the selected aspects of the BC assurance in terms of the risk management—the basic reason was the cost of the research in this case. The analysis of the results can be selective and limited. In the study, there were taken into account the opinions of respondents.

5 The Empirical Results

Based on the literature analysis, it can be concluded that the main risk areas are: the level of competences of managers, as well as competences of creative workers [27], systematic improvement of skills [28], recruitment processes and talent management [29], worktime and discontinuities in obtaining of income [28], transfer of the responsibility of employee to the employer or entrepreneur [30], as well as disposition of the creative employees [31]. The literature underlines the importance of factors such as: the global and national fundraising, creative activities' digitalization, as well as public awareness of designing processes, internationalization of creative work, and quality of the higher education system [32]. It should be also mentioned that an important role is played by the quality and richness of historical data, forecasting processes, as well as the use of information and communication technologies (ICTs) [33]. The key processes also include: the learning of enterprises and knowledge transfer, and evolution of network structures (clusters) [34–36]. On the basis of the empirical study it can be noted that the basic areas of risk are the finance management, markets' operation, human resources management, as well as regulations (Table 2).

In the business continuity assurance for the creative enterprises an important approach is to make an interpretation and treatment of risk. In the surveyed enterprises, there dominates above all the "classic" approach, e.g. regarding risk as a source of the potential losses or an inhibitory factor for the enterprise's development (Table 3). The risk management maturity is at a low (primary) level. There is a lack of the holistic and integrated approach to the risk management in the enterprises. Only in a small number of entities there can be found the symptoms of a relatively high complexity and integration of the risk management processes.

In view of the above, the organizational and analytical potential of the BI systems' environment in the surveyed enterprises might not be fully devoted in the designing and developing of the BC assurance processes. The classical approach to risk management advocates to implement only selected, individual functions and modules of these systems, e.g. related to the planning, budgeting, or control processes [37]. This is due to the fact that in the surveyed enterprises are e.g. non-integrated information that can be called as "islands" [38].

According to the above results, a reference can be made to the peculiarity of the actions undertaken by the enterprises in the framework of risk management—as the specific basis for the business continuity assurance in the creative enterprises. Based on the research, it can be concluded that (Table 4):

Table 2. The key areas of the risk management in the enterprises from the creative industries.

The main risk areas	Detailed risk areas
Finance management	Decrease in turnovers, high costs of employees' acquisition, too high material costs, too much financial effort, lack of adequate financial resources and financial liquidity, the need to increase the budget for innovation, and incur of additional unexpected costs
Functioning of the markets	The placing on the market the better services by competitors, withdrawal of the project partner or customer, incompetent and not paying contractors, forcing the adverse conditions by larger companies as customers, frequent market changes, "unstable" customers
Human resources management	The lack of qualified personnel, unstable job market, devaluation of the profession (loss of prestige), underestimation of the working time, an increase in the activity of freelancers who "spoil" the market, the instability of the workforce (impermanence of employment)
Regulations	Too many unexpected changes in legislation, changes in the law that enforce changes in the projects, increase in bureaucratic processes

Table 3. The general perception of risk (by owners and managers) in the enterprises.

Theoretical approach (the literature analysis)	Results of the empirical research
Risk is both the losses and potential benefits	The positive aspects of risk are not seen
Risk is a factor of development	The risk is associated with losses
The level of risk tolerance and risk appetite is at a relatively high level	The level of risk tolerance and risk appetite is at a relatively low level
The holistic approach dominates	The silo approach dominates

Table 4. The key actions undertaken in the business continuity assurance—the chosen aspects.

The predominant processes within the innovation management (core business activity):
There is taken place a high level of commitment of managers/owners and holding their own initiatives to create innovation in the enterprise In innovation processes, there is taken place a market and customer needs' analysis It can be observed a focus on the capital- and material-demanding innovation processes
The predominant processes in the area of efficiency management in core business activity:
The so-called total efficiency is taken into account, i.e. the efficiency of the entire enterprise The human resources & financial efficiency are taken into account
The predominant processes in the area of the risk management in the enterprise:
Identification of the different current types of risk, as well as the risk estimation Identification of risks both in the environment and inside the enterprise Identification of risks which the enterprise can manage and those outside its influence
The predominant processes of the risk management in the innovative processes:
Application of corrective/replacement actions, and use of different methods of risk analysis The use of the so-called event management (i.e. checking in the actually occurring risks) Designation of individuals/positions responsible for the risk management

- in the innovation management a great importance is played to the market analysis, as well as the involvement of managers or business owners, and paying attention to the financial and material aspects of the implementation of the innovation processes; therefore, in the BC assurance, there are inherently involved the different levels of management, but the decisive role has the highest level; in addition, in the BC assurance a particular significance have the processes related to the functioning of the enterprise on the market, as well as pursuing the relationships with customers (and other external stakeholders); one of the basic criteria in the BC assurance within the innovative processes are the costs and availability of resources;
- in the process of efficiency management in the innovative processes a great importance have the human and financial resources—these are the two areas of the evaluation of efficiency that are essential to the business continuity assurance for the core processes; in addition, in the surveyed enterprises, there is calculated a so-called total efficiency, which is critical in assuring the BC in the core processes;
- in the risk management across the whole enterprise—as an integrated system—dominate the processes of identification the current and key risk factors; in the case of the business continuity assurance it is important to identify the risk factors both within the enterprise and in its environment, as well as risk factors on which the enterprise has an impact, and those for which there is no enterprise's impact—it helps to e.g. develop the procedures for ensuring and maintaining the BC.

The above-mentioned actions constitute a convenient environment for the implementation and deployment of the risk management processes in the core (i.e. innovative) processes, as well as development of the organizational basis for the purposes of ensuring the BC. Here it can be indicated e.g. the implementation of corrective or replacement processes, event management, or creation of the posts responsible for the widely-understood risk management (Table 4).

On the basis of the above findings, it can be assumed that in order to ensure the BC in the surveyed creative enterprises the most important are: keeping the supplies of materials, ensuring the proper financing of the activities, as well as supporting the staff. It is also important to implement correctly the R&D processes—by promoting and supporting the creativity of employees. Therefore, the key areas in the BC assurance are: R&D, logistics, finance, and human resources management.

6 Conclusions

In the BC assurance in the surveyed enterprises, there is important a widely understood data analysis on the risk factors, as well as the processes of acquiring, collecting and sharing data and information. Having regard to the fact that ensuring the business continuity requires the continuous updating of the information resources and the development of the knowledge management system in the enterprise, it is reasonable to use of the specific ITCs. It can be used here, as an example, the BI systems, which in advanced are able to support e.g. the forecasting, planning, monitoring and control processes.

The above key areas of risk—identified both on a global and Polish scale—give rise to develop a holistic concept of the BC assurance in enterprises from the creative industries. It is worth to note the role of information resources, network structures and enterprises' cooperation, as well as knowledge management. Having regard to the progressivism and a high degree of internationalization of creative enterprises, it seems to be worth to enable BI systems, in the concept of ensuring the business continuity, to increase enterprise's efficiency, as well as shape the level of its security. The implementation of the BI system in the information and decision-making area in the enterprise often is an inefficient activity from a financial and organizational points of view—especially in the SMEs. Despite of the extensive processes of identifying the risk factors and understanding the role and importance of the information's analysis, the following trends can be observed, in the surveyed enterprises:

- the insufficient organizational and managerial foundations for in the BC assurance,
- the lack of orientation to take a full advantage of the potential of the BI systems, or no manifestations of the application of the BI systems.

On the other hand, the surveyed enterprises have hold great needs for the information resources' analysis, interaction with the different groups of stakeholders (e.g. in the form of the creative clusters, network or virtual structures, etc.), and monitoring the environment, as well as identifying and updating information on the risk factors. The implementation of such needs requires the use of ICTs' potential, especially in the functionality aspect of the integrated and analytical information systems.

Taking into account that the use of the full potential of this class of systems (BI) in the creative enterprises is difficult (and sometimes even impossible), it should be thought how—despite a number of the financial and operational obstacles—to enable the BI systems in the creative enterprises' development. In the case of creative enterprises it is reasonable to use simple and low-capital intensive ICTs as the cloud computing (CC). The CC technology gives the ability to use the potential of BI systems, increasing effectiveness of information and decision-making processes supporting business (design) processes, and reducing risk of losing valuable information resources. It is worth at this point to refer to the potential of the Internet of Things—especially in terms of managing the large data resources (Big-Data), and the communication between the different resources/factors in the process of designing and providing the creative services.

Despite the large financial and organizational constraints, as well as significant analytical and decision-making requirements or needs of the creative enterprises, even a partial utilization of the BI systems' functions can be helpful in the BC assurance in the innovative (i.e. core) process.

References

1. Mackiewicz, M., Michorowska, B., Śliwka, A.: Analysis of Needs and Development of Creative Industries. ECORYS Polska, Warsaw (2009)

2. Creative Industries Mapping Document: Background. https://www.gov.uk/government/uploads/system/uploads/attachment_data/file/183544/2001part1-foreword2001.pdf. Accessed 05 July 2016
3. Walther, I.: Creative Industries in Berlin. Development and Potential. Senate Department for Economics, Technology and Women's Issues (Projekt Zukunft), Berlin (2008)
4. Kasprzak, R.: Creative Industries in Poland: Circumstances and Prospects. Kamon Consulting, Warsaw (2013)
5. Markiewicz, J., Kotylak, S., Szczech-Pietkiewicz, E.: The Challenges of Creative Economy. Texter, Warsaw (2016)
6. Zaskórski, P.: Information Asymmetry in Processes Management. Military University of Technology, Warsaw (2012)
7. Januszewski, A.: The functionality of the IT management systems. T.2. Business Intelligence Systems. Wydawnictwo Naukowe PWN, Warsaw (2008)
8. Surma, J.: Business Intelligence: Business Decision Support Systems. Wydawnictwo Naukowe PWN, Warsaw (2009)
9. Bratnicka-Myśliwiec, K.: The impact of information technology on the dynamics of the creative entrepreneurial organization. e-mentor **5**(67), 40–45 (2016)
10. Wereda, W., Woźniak, J.: Risk criterion in the "agile" organization. Modern Manag. Syst. **10**, 61–87 (2015)
11. Nowicki, A.: Identification of information technologies. In: Nowicki, A., Turek, T. (eds.) Information technologies for economists. Tools. Applications, pp. 44–58. Publishing House of Wrocław University of Economics, Wrocław (2010)
12. Woźniak, J., Wereda, W.: Information and communication technologies (ICTs) in enterprises on the over the counter (OTC) markets in European Union: case study of Polish NewConnect market. In: Dima, A. (ed.) Doing Business in Europe, pp. 413–446. Springer, Cham (2018). https://doi.org/10.1007/978-3-319-72239-9_20
13. Chomiak-Orsa, I.: The importance of modern ICT in improving intra-organisational communication. Informatyka Ekonomiczna. Business Informatics **1**(39), 46–55 (2016)
14. Dohn, K., Gumiński, A., Matusek, M., Zoleński, W.: IT tools implemented in the information system supporting management in the range of knowledge management in mechanical engineering enterprises. Inf. Syst. Manag. **3**(1), 3–15 (2014)
15. Dziembek, D.: The objectives of the use of information technologies. In: Nowicki, A., Turek, T. (eds.) Information Technologies for Economists. Tools. Applications, pp. 38–44. Publishing House of Wrocław University of Economics, Wrocław (2010)
16. Lipińska, A.: Knowledge management of companies in the information society. Theory and practice. In: Sołek, C. (ed.) Management Dilemmas in the Information Technology Era, pp. 93–100. Military University of Technology, Warsaw (2012)
17. Pindelski, M., Mrówka, R.: Big data visualizations in the identification of management problems. Manag. Forum **5**(363), 18–28 (2014). Research Papers of Wrocław University of Economics
18. Woźniak, J. (ed.): Organization Design and Improvement in the Aspect of Its Security and Streamlining Decision-Making and Information Processes. Military University of Technology, Warsaw (2015)
19. Zaskórski, P.: Organization Management in Terms of Risk of Losing Information Continuity. Military University of Technology, Warsaw (2011)
20. Zawiła-Niedźwiecki, J.: Operational Risk as a Problematic Triad: Risk, Resource Security, Business Continuity. Edu-Libri, Cracow-Legionowo (2014)
21. Mateos, A.: Cloud computing: solutions for business. Helion, Gliwice (2011)
22. Nowicka, K.: Cloud Computing in Stakeholders Relationships Management. Przedsiębiorczość i Zarządzanie **17**(12), part 1, 267–277 (2016)

23. Parlinska, M., Petrovska, I.: Cloud computing and its benefits. Inf. Syst. Manag. **6**(4), 309–317 (2017)
24. Tvrdikova, M., Skyba, M.: Using Business Intelligence Tools through Cloud Computing Services. Studia Ekonomiczne/Uniwersytet Ekonomiczny w Katowicach **113**, 39–51 (2012)
25. Nitesh, D.: Abusing the Internet of Things. Blackouts, Freakouts, and Stakeouts. O'Reilly Media, Sebastopol (2015)
26. Sułkowski, Ł., Kaczorowska-Spychalska, M.: Internet of Things. The New Paradigm of the Market. Difin, Warsaw (2018)
27. Porfírio, J.A., Carrilho, T., Mónico, L.S.: Entrepreneurship in different contexts in cultural and creative industries. J. Bus. Res. **69**(11), 5117–5123 (2016). https://doi.org/10.1016/j.jbusres.2016.04.090
28. Hennekam, S., Bennett, D.: Self-management of work in the creative industries in the Netherlands. Int. J. Arts Manag. **19**(1), 31–41 (2016)
29. Ratalewska, M.: Entrepreneurship in creative sectors. Res. Pap. Wrocław Univ. Econ. **347**, 382–390 (2014). https://doi.org/10.15611/pn.2014.347.36
30. Lampel, J., Germain, O.: Creative industries as hubs of new organizational and business practices. J. Bus. Res. **69**(7), 2327–2333 (2016)
31. Patten, T.: "Creative?"… "Entrepreneur?"—understanding the creative industries entrepreneur. Artivate: J. Entrepreneurship Art **5**(2), 23–42 (2016)
32. Baculakova, K., Gress, M.: Cluster analysis of creative industries in the EU. Econ. Ann.-XXI **9–10**, 15–18 (2015)
33. Kembaren, P., Simatupang, T.M., Larso, D., Wiyancoko, D.: Design driven innovation practices in design-preneur led creative industry. J. Technol. Manag. Innov. **9**(3), 91–105 (2014). https://doi.org/10.4067/S0718-27242014000300007
34. Boix, R., Hervás-Oliver, J.L., De Miguel-Molina, B.: Micro-geographies of creative industries clusters in Europe: from hot spots to assemblages. Pap. Reg. Sci. **94**(4), 753–773 (2015). https://doi.org/10.1111/pirs.12094
35. Lee, M.: Fostering connectivity: a social network analysis of entrepreneurs in creative industries. Int. J. Cult. Policy **21**(2), 139–152 (2015). https://doi.org/10.1080/10286632.2014.891021
36. Olko, S.: Knowledge Management in Clusters and Networks in Creative Industries. CeDeWu, Warsaw (2017)
37. Sasin, R.: The evolution of the ERM concept. In: Kasiewicz, S. (ed.) Integrated Enterprise Risk Management in Poland: Directions and Tools, pp. 98–112. Wolters Kluwer, Warsaw (2011)
38. Zaskórski, P.: Information Strategies in the Management of Business Organizations. Military University of Technology, Warsaw (2005)

The Mechanism for Creating an Effective International Strategic Alliance in the Field of Air Transportation

Nataliya Kazakova[1(✉)], Elena Shuvalova[1], Alisa Chemarina[1], Artem Nikanorov[1], Irina Kurochkina[2], and Elizaveta Sokolova[3]

[1] Plekhanov Russian University of Economics, Moscow, Russia
{axd_audit, Eshuvalova}@mail.ru, alisa-chemarina@yandex.ru
[2] P.G. Demidov Yaroslavl State University, Yaroslavl, Russia
ipkurochkina@yandex.ru
[3] Financial University Under the Government of the Russian Federation, Moscow, Russia
sokolovaes15@mail.ru

Abstract. Today, international strategic alliances in the field of air transportation represent a progressive form of business integration that allows top management to implement a global strategy for the development of the industry. The result of the study is a developed by the authors' mechanism for creating an effective international strategic alliance, which includes a set of analytical procedures for assessing the key success factors, risks and synergistic financial effect of airline participation in the Alliance. The algorithm of formation includes the following stages: analysis of the current state of business and identification of the strengths and potential capabilities of the alliance from the perspective of strategic development; development of the alliance strategy; formation of the optimal structure of the alliance for the implementation of the strategy; evaluation of the effectiveness of airline participation in the Alliance.

Keywords: Air transportation · International strategic alliance · Efficiency

1 Introduction

Today, strategic alliances have become a powerful key factor in improving business competitiveness and a way of quick implementing of a global strategy. There is a single model of winning markets and increasing competitive positions in most fields of the economy. At the same time, air transport has unique features related to the aspects of globalization and competition, which requires an individual approach to the formation of strategic alliances in this industry and the assessment of their effectiveness. The analysis of the air transportation industry for 2014–2016 allows summarizing a complex of factors and trends that have a direct impact on the creation and development of strategic air hookahs. The sharing of passenger traffic on international routes is has a significant increasing trend every year, with 22% of the world's air traffic concentrated on only 300 routes. Each of them annually carries more than 1 million passengers, the

I. Kabashkin et al. (Eds.): RelStat 2018, LNNS 68, pp. 704–713, 2019.
https://doi.org/10.1007/978-3-030-12450-2_67

volume of world passenger traffic annually increases by 5% in average. The growing competition in the industry is largely stimulated by the development of low-cost carriers. Currently, they are more focused on the traditional markets of Europe and the United States, but Asia is considered to be the fastest growing and most competitive air transport market. The number of passengers in this region is increased by 9% yearly [1].

The formation of an effective international strategic Alliance is a complex and multi-stage process that requires instrumental and analytical support. In this regard, the purpose of this study is to develop the mechanism for creating an effective strategic alliance in the air transport industry taking into account benefits and risks for companies belonging to the alliance.

The tasks of the study are: (1) to propose the author's approach to developing a mechanism for an effective international strategic aviation alliance; (2) develop a methodology for evaluating the effectiveness of the participation of companies in international strategic aviation alliances; (3) to form a common conceptual model of the mechanism of an effective international strategic aviation alliance, taking into account all the advantages and risks for companies within the alliance. The subject of the research was the modern concepts of strategic alliance mechanism formation in aviation industry, methods and systems of performance indicators and performance management.

2 Factors for the Formation of Aviation Alliance and Strategies for Their Operation

Research and documents of international organizations consider various approaches to the formation of strategic alliances based on the goals of the participants, the sharing of their resources, their place in the system of inter-company relations, the system of corporate control and management (Table 1).

Summarizing the various approaches to the formation of international strategic alliances, the authors concluded that strategic aviation alliance is a form of long-term partnership of airlines that seek to jointly strengthen advantages over competitors by making optimal use of available resources, including increasing the value of brands, expanding market opportunities service, increase business profitability [4]. The model presented in Fig. 1 takes into account the macro and micro-factors that influence the creation of aviation alliances and the defining features of their strategy [5].

The strategy of the international strategic aviation alliance involves the establishment of certain requirements for the activities of partner companies, while maintaining the strategic flexibility of the alliance as a whole. The airlines participating in the alliance and their divisions are distinguished by a high level of integration in the areas of planning, ticket sales, loyalty programs, flight connections, code sharing, airport terminal sharing. In addition, they apply uniform principles of work, as far as possible, conduct common policies and use common procedures, cooperate in the technological sphere. This helps to reduce costs, allows you to share experiences, widely using benchmarking tools.

Table 1. Generalization of approaches to the formation of strategic alliances (developed by the authors based on [2, 3]).

Subject of approach	The most well-known authors	Description
The place and the role of the strategic Alliance in the system of inter-company relations	Garrett B., Dussauge P., M. Porter	The concept of intermediate inter-company relations, according to which firms are already acting in the common interest, sacrificing part of their economic autonomy, but their joint activities have not yet led to a complete merger of companies
Strategic objectives and resources of partners	A. Heck, Papenberg E., Backhaus K., Das T., Teng B.	Alliances are linked to the strategic objectives and interests of their members. Goals are considered strategic if they determine the movement of the company into the future and if they are associated with the creation and strengthening of its competitive advantage
Inter-company management and control structures	Gomez-Casseres B.	Alliances are the mechanisms of management and decision-making typical for inter-company consolidation
Operational and legal parts	UNCTAD, Bundes-kartellamt	Strategic alliances are considered regarding to the main strategies of transnational corporations, which determine the structural organization, the system of corporate control and management in the strategic Alliance
Management in the global competition	International air transport Association (IATA)	Strategic Alliance of companies is a form of long-term partnership between two or more companies that seek to jointly strengthen the competitive advantage by making optimal use of available resources, including increasing the value of brands and expanding the opportunities to seize market share, improving the quality of service to improve profitability

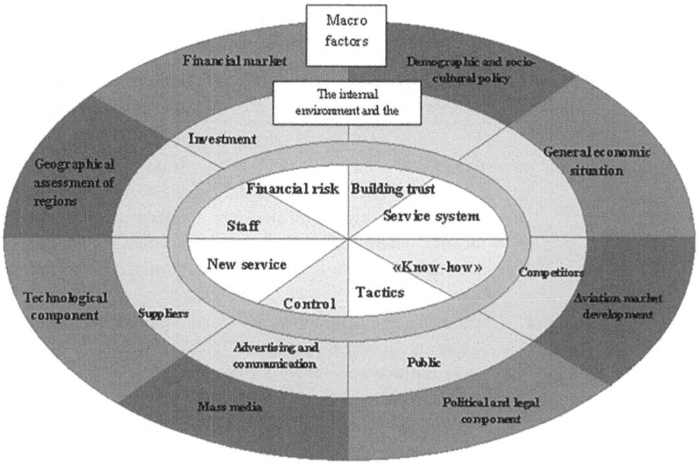

Fig. 1. Multi-factor model of the aviation Alliance strategy formation.

3 Analysis of Existing Approaches to Assessing the Effectiveness of the Company's Participation in Alliances

The most important condition for the formation of air alliance is to obtain a synergistic effect. The first developments in the field of assessing the economic viability of the participation of companies in alliances appeared in the 60s of the twentieth century. They are associated with the name of Igor Ansoff, an American mathematician-economist of Russian origin, who was the first to introduce the concept of "synergy" in relation to assessing the interrelationship of various types of activities within companies.

However, in the opinion of the author of the concept itself, the choice of the evaluation method directly depended on the type of synergistic effect. Therefore, in the future, other scientists developed the concept of Ansoff [6], proposing new improved assessment methods (Table 2).

The most well-known foreign methods based on econometric analysis methods are also associated with the names of such scientists as Ferrer, Orsag, McClure, Kinnunen, Xia and Xiuzhi [12–15]. The latter, in particular, draw attention to the need for risk analysis in assessing the synergistic effect, which is due to the likelihood of its zeroing due to the high level of risk.

Summarizing the results of the study, the authors make conclusion that joining and participating in the alliance provides companies with competitive advantages or key success factors. Benefits can be both qualitative (strengthening the image and brand development, reducing risks, increasing the attractiveness of services through the use of a general loyalty program for customers), and quantitative (increasing revenues due to commercial loading of flights, reducing the cost of services and transaction costs, including by tagging).

Table 2. Comparative analysis of methods for assessing the effectiveness of the organization's participation in alliances (developed by the authors based on [6–13]).

Method name	Authors	Brief description	Disadvantages limiting the spreading of the methodology
1	2	3	4
The concept of synergism	Ansoff [6]	Includes two methods of evaluation: determination of the level of cost savings for the same income; determination of profit growth at a certain level of investments	It does not provide a complete view of the degree of synergy effectiveness. It does not take into account the qualitative effect of participation in the alliance
Balance approach	Prahalad and Doz [7]	The balance model allows to determine in which functional areas positive and negative effects can be expected	Assumes only a qualitative assessment of the effect
The concept of profit multiplication	Naumov and Shubaeva [8]	The change in the profit of each participant of the system through the profit multiplication factor (Kp) is determined: $Kp = Ps/Po$, where Ps—profit as a result of synergy; Po—Profit before operational or organizational arrangements	Considers profit as the only evaluation criterion
EVA Model	Joel Stern and Bennett Stewart [9]	The EVA ("economic value added") indicator is the net operating profit after tax (NOPAT) minus the cost of capital. This indicator allows you to count the value of the company as a result of raising capital (by investments) in a certain year	There is no link with the non-monetary effect of participation in alliances. Can be applied only if the company has investments
The method of premium assessment, based on retrospective analysis	Davidenko [10]	To calculate the premium for a particular transaction, similar transactions that have already been committed by other companies earlier are analyzed. Based on several similar transactions, the most significant synergistic effects are identified, and a common portfolio of synergies for the transaction under analysis is created	The possibility of the lack of a sufficient number of similar transactions in the past to build a correct statistical portfolio
Methodology for determining the effect of company's participation in alliances	Polozov-Yablonsky [11]	It divides the effect of the participation of companies in alliances on non-economic and economic. The non-economic effect is determined on the basis of expert estimates by multiplying the factors that appeared after joining the alliance. Weights are not assigned to this factors	Difficulty of calculations, inaccessibility of information that is necessary for determining indicators

4 Research Results

4.1 Mechanism for the Creation of an Effective International Strategic Alliance in the Field of Air Transportation

Based on the study, the authors developed a model for the formation of an effective international strategic alliance in the field of air transportation, which includes four stages (Fig. 2).

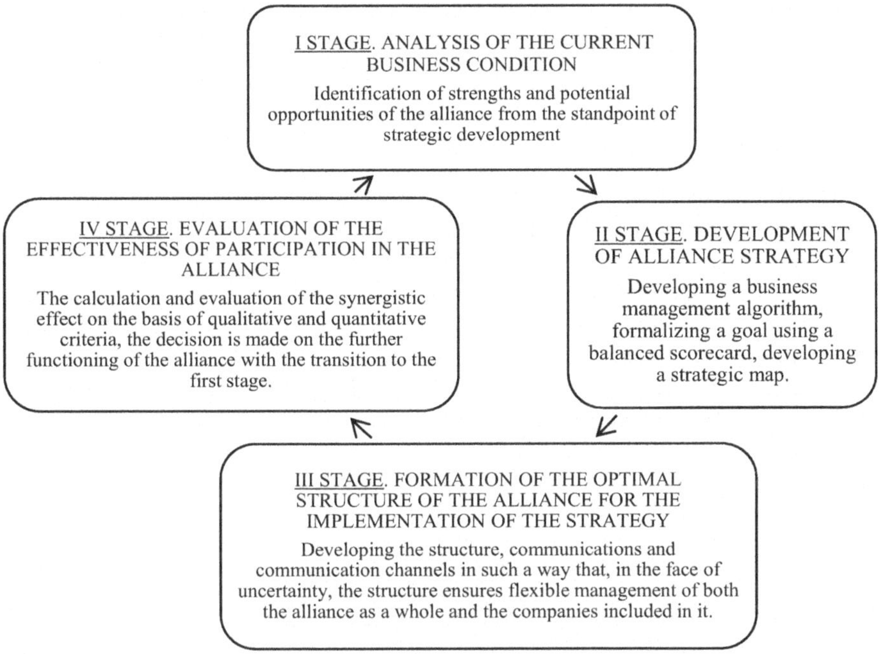

Fig. 2. Model for the formation of an effective international strategic Alliance in the field of air transportation.

The first stage "Analysis of the current state of the business and identification of the strengths and potential opportunities of the alliance from the standpoint of strategic development" provides an assessment of the company's market position, level of competition, search for opportunities and prospects for companies to participate in the Alliance.

The second stage "Development of an Alliance Strategy" involves the development of a business algorithm, including the definition of strategic goals and objectives of the alliance for groups of stakeholders; selection and justification of the directions of development of the alliance (revenue generation, cost reduction, the creation of a common structure). To formalize the goals and objectives of the alliance, taking into account the interests of its participants, we used a balanced scorecard system, including an assessment of financial resources and the effectiveness of their use; market and customer value assessment; indicators of monitoring the effectiveness of the organization and management of business processes; performance criteria for staff training and development. The stage ends with the development of a strategic map.

The third stage "Forming the optimal structure of the Alliance to implement the strategy" consists in developing the structure of the alliance, communications and communication channels within it so that, in conditions of uncertainty, the structure provides strategic flexibility for the alliance as a whole and for its companies. The mandatory elements of the structure are the internal audit and corporate control systems by various stakeholders.

The fourth stage "Evaluation of the effectiveness of participation in the Alliance" consists in calculating the economic benefits from cooperation, highlighting the qualitative and financial effects of the participation of companies in the alliance.

4.2 The Procedure of Assessing the Financial Effect of an Airline Company's Participation in Alliance

The method developed by the authors for evaluating the effectiveness of an airline's participation in an international strategic Alliance is implemented using the example of "Siberia Airlines".

Analysis of the qualitative effect is carried out on the basis of expert assessments. First, factors affecting the industry as a whole are identified, then factors affecting the alliance and key success factors. Next is the ranking and selection of the most significant factors for a particular participant, the assignment of weights and the choice of the rating scale, after which the total quality effect indicator is calculated. The ranking of the indicators is carried out according to the Fishburn weighting system [16, 17]:

$$K_1 = n/(1+2+\ldots+n), \tag{1}$$

where: K_1 is the weighting factor of the first significant indicator; n is the number of factors.

$$Ki = n{-}1/(1+2+\ldots+n), \tag{2}$$

where: Ki is the weighting factor of the any following indicator; n is the number of factors.

The proposed methodology of the analysis allows studying the financial benefits received and also quantifying the intangible effect of joining the alliance, which sometimes is more significant for companies operating on international markets.

Among the main driving forces, the most significant factors in the aviation industry are selected: growing globalization, the development of e-commerce, changes in government policy, changes in costs and profits, in particular, the level of prices for fuel and services. Among the key factors for the success of the international alliance "Oneworld", which includes "Siberia Airlines", should include: providing technological advantages, including know-how in customer service, high quality services, efficient use of aircraft, expanding the network of routes, reducing expenses, expansion of marketing services, including loyalty programs and guarantees for passengers; high professionalism of staff through investments in the development and training of employees; additional organizational capabilities, including optimization and improvement of business processes efficiency.

Evaluation of the financial effect of joining the Alliance is carried out on two components: on additional income and savings on expenses considered for two periods: a year before joining the alliance and a year later. Indicators are determined per flight. Based on the assessment of changes in selected indicators for the analyzed period, the savings on expenses and additionally received revenues are determined, which are summarized, turning into a quantitative financial effect from the company's participation in the alliance (Table 3).

Table 3. Calculation of the financial effect for the company "Siberia Airlines" from participation in the Alliance (in million rubles).

Line number	Indicators	Formula	Year before joining the Alliance	Year after joining the Alliance	Diff. ±
1.	Revenue per flight, S	S/number of flights (Q)	866.18	919	52.83
2.	Cost of sales per flight, COGS	COGS/Q	717.76	750.4	32.64
3.	Commercial costs per flight, SC	SC/Q	36.24	36.91	0.67
4.	Administrative expenses for one flight, AC	AC/Q	42.73	41.61	−1.12
5.	The total effect	Line 1-line 2-line 3-line 4	–	–	**20.64**

The proposed method makes it possible to assess not only the financial re-result, but also the intangible effect of joining the alliance, which is sometimes more significant for companies operating in international markets.

4.3 Summary: The Model of Formation of an Effective International Strategic Alliance in the Field of Air Transportation

The author's conceptual model of the formation mechanism of an effective international strategic aviation Alliance, presented in Fig. 3, demonstrates the technique of the management process from analyzing and evaluating the main trends and factors influencing the creation of an international strategic alliance, highlighting key success factors and assessing risks for companies in the alliance, to determining the synergy effect of joint activities. At the same time, analytical procedures at each stage of the author's mechanism are distributed in such a way as to ensure a complete, repeating cycle of performance management and development of the aviation Alliance.

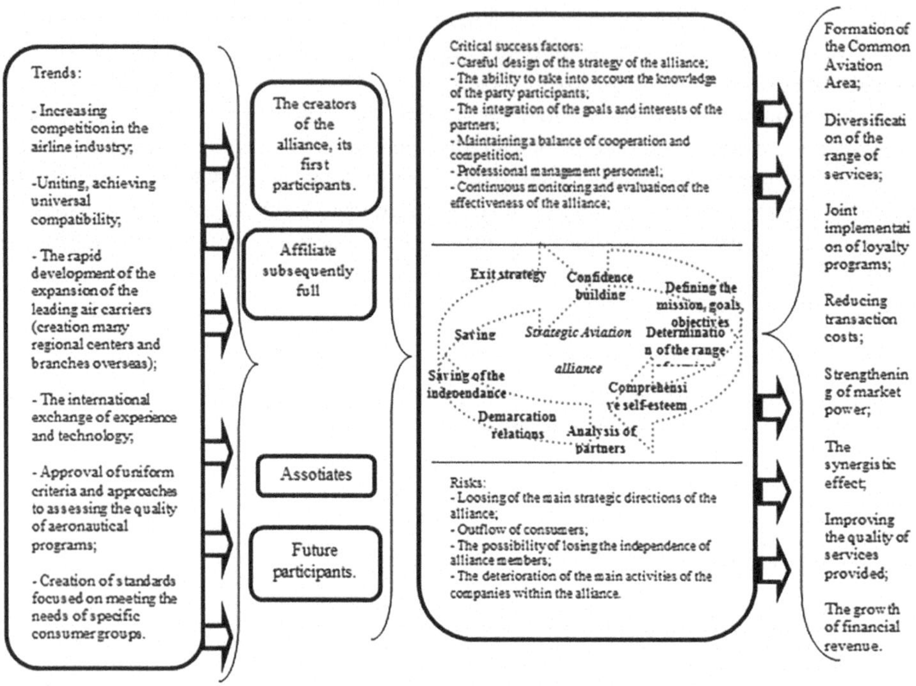

Fig. 3. Model of formation of an effective international strategic Alliance in the field of air transportation.

5 Conclusions

The scientific results of the study consist in the development of a mechanism for the formation of an effective international strategic aviation Alliance. The object of implementation of the research results is the company "Siberia Airlines", considered in terms of the impact on it of joining the "Oneworld" international strategic Alliance. Theoretical results of the study include the development of a conceptual model for the formation of an effective international strategic Alliance in the field of air travel, taking into account all the benefits and risks for companies within the alliance, as well as defining a set of analytical procedures as an applied toolkit for managing the performance and development of international strategic air alliance. The practical results of the study are to develop a procedure for evaluating the effectiveness of the participation of companies in international strategic aviation Alliances.

References

1. Information-analytical materials on the state and development of civil aviation in Russia, the CIS countries and around the world (2017). http://aviationweek.com/aviation-week-space-technology
2. Kazakova, N., Gendon, A., Sedova, N., Khlevnaya, E.: Estimation of factors and projected growth of the metallurgical branch of Russia under unstable market conditions. Metalurgija (Croatia) **57**(4), 360–362 (2018)
3. Thompson Jr., Arthur, A., Strickland III, A.: Strategic management: concepts and situations for analysis (2006)
4. Aaker, D.: Strategic Market Management, 7th edn. Wiley, Hoboken (2007)
5. Rangan, U.S., Yoshino, M.: Forging alliances: a guide to top management. Columbia J. World Bus. **31**, 6–13 (1996)
6. Ansoff, I.: New Corporate Strategy, p. 416. Wiley, Hoboken (1999)
7. Prahalad C.K.: Do You Really Have a Global Strategy? Harvard (1985)
8. Naumov, V., Shubaeva, V.: Strategic interaction of market entities in marketing systems, pp. 206–207 (2015)
9. Kazakova, N., Bolvachev, A., Gendon, A., Golubeva, G.: Value added analysis and trend forecasting in the manufacturing industry in Kaliningrad oblast. Stud. Russ. Econ. Dev. **28** (2), 160–168 (2017)
10. Davidenko, A.: Methods for the valuation of synergetic effects in mergers and acquisitions. The Youth Scientific Portal Lomonosov (2017). http://www.rsso.su
11. Polozov-Yablonsky, A.: Method of determining the effect of participation of the airline in alliances, p. 87 (2002)
12. Ferrer, R.: An empirical investigation of the effects of merger and acquisition on firms' profitability. Acad. Acc. Financ. Stud. J. **16**(3), 31–55 (2012)
13. Orsag, S., McClure, K.: Modified net present value as a useful tool for synergy valuation in business combinations. UTMS J. Econ. **4**(2), 71–77 (2013)
14. Kinnunen, J.: Valuing M&A Synergies as (Fuzzy) Real Options. Institute for Advanced Management Systems Research (2010). http://www.realoptions.org/papers2010/238.pdf
15. Xia, Z., Xiuzhi, Z.: Strategic analysis of synergistic effect on M&A of Volvo car corporation by Geely Automobile. I-Business **3**(1), 5–15 (2011)
16. Kazakova, N., Bolvachev, A., Gendon, A., Golubeva, G.: Monitoring economic security in the region based on indicators of sustainable development. Stud. Russ. Econ. Dev. **27**(6), 650–660 (2016)
17. Kampbell, K.: Strategic Synergies, p. 150 (2004)

The Principles of Creating a Balanced Investment Portfolio for Cryptocurrencies

Svetlana Saksonova[1] and Irina Kuzmina-Merlino[2](✉)

[1] University of Latvia, 19 Raina Blvd., Riga 1586, Latvia
saksonova.svetlana@gmail.com
[2] Transport and Telecommunication Institute, Lomonosova Street 1, Riga 1019, Latvia
Kuzmina.I@tsi.lv

Abstract. Despite their novelty, cryptocurrencies have by now acquired certain popularity due to convenience in making payments, high speed of transactions and the application of modern technology to ensure transaction security among other things. The aim of this research is to evaluate cryptocurrencies as an investment instrument. The tasks of the research are as follows: (i) to evaluate a hypothesis about potential profitability of the cryptocurrency investment portfolio, (ii) to analyse cryptocurrency investment profitability, (iii) to assess the attractiveness of creating an investment portfolio of cryptocurrencies, and (iv) to provide recommendations to a potential investor for creating an investment portfolio comprised of cryptocurrencies. The research methods used include statistical data analysis and econometric modelling. Applied results were obtained using correlation-covariance analysis yielding a balanced profitable investment portfolio according to the Markowitz model. The results of the research have allowed the authors to conclude that cryptocurrencies would be an attractive investment instrument; the following indicators can be used to assess the attractiveness of cryptocurrency: market capitalization, the objectives of currency creators (issuing tokens on different resources to currency holders, bounty programs, etc.), trend assessment, and the analysis of reproducibility and replicability of the platform on which the cryptocurrency is built.

Keywords: Cryptocurrencies · Balanced profitable investment portfolio

1 Introduction

Cryptocurrencies are a recent phenomenon with the paper outlining the proposal for the first cryptocurrency—Bitcoin published online in 2008 arguing that: “A purely peer-to-peer version of electronic cash would allow online payments to be sent directly from one party to another without going through a financial institution” [1]. A report by the European Central Bank [2] placed Bitcoin in context of a larger phenomenon—virtual currency schemes, that are classified in three types: (i) closed schemes (basically for

I. Kabashkin et al. (Eds.): RelStat 2018, LNNS 68, pp. 714–724, 2019.
https://doi.org/10.1007/978-3-030-12450-2_68

use in online gaming), (ii) schemes, which have a unidirectional flow with a conversion rate for purchasing virtual currency, which is mostly used for virtual goods and services, and, finally, (iii) currency schemes with bidirectional flows, where a virtual currency acts like any other convertible currency. This paper focuses on the virtual currencies of the last type.

The *aim* of this research is to evaluate cryptocurrencies as an investment instrument.

The *tasks* of the research are as follows: (i) to evaluate a hypothesis about potential profitability of the cryptocurrency investment portfolio, (ii) to analyse cryptocurrency investment profitability, (iii) to assess the attractiveness of creating an investment portfolio of cryptocurrencies, and (iv) to provide recommendations to a potential investor for creating an investment portfolio comprised of cryptocurrencies.

The *research methods* used include literature review, statistical data analysis and econometric modelling. The authors collected data on the historical exchange rates of cryptocurrencies from July 2017 to January 2018 and studied their correlations, covariance and the return on investment. This particular time period was chosen because there were significant changes in the exchange rate dynamics of different cryptocurrencies in the second half of the summer of 2017. In addition, data on certain cryptocurrencies was not available prior to this period. The cryptocurrencies chosen for the analysis were considered based on their statistical characteristics (price trend and correlation with other cryptocurrencies). The research was conducted using SPSS Statistics and Excel software. The authors analyzed the econometric models and instruments applied in the creation of an investment portfolio and determined the criteria to assess investment attractiveness of specific cryptocurrencies. Applied results were obtained using correlation-covariance analysis yielding a balanced profitable investment portfolio according to the Markowitz model. So, the paper outlines the application of modern portfolio theory to the cryptocurrency market and illustrates the creation of portfolios with the minimum risk and maximum yield, which can be recommended to a potential investor. Overall, therefore, cryptocurrencies can certainly be considered a legitimate and possibly attractive investment instrument. In this paper, the authors determine the optimal cryptocurrency investment portfolios calibrated both for maximum profit as well as minimum risk.

2 Literature Review

The European Central Bank argued that there are both advantages and disadvantages to virtual currency schemes. Among the advantages is the possibility of innovative payment solutions featuring lower costs, global reach, anonymity of the payer and speed of settlement. On the other hand, there are also several drawbacks and disadvantages for users, i.e. lack of transparency, clarity and continuity; high dependency on IT and on networks; anonymity of the actors involved; and high volatility [3]. In addition, users face risks associated with the anonymity of the payee, high volatility of the cryptocurrency exchange rates and the risk of fraud due to the lack of transparency. Currently, there is no technical possibility to eliminate these risks, as in other types of fintech [4] therefore, supervisory authorities in many countries have concerns about possible illegal operations, Ponzi schemes, and the possibility of tax evasion.

Recent literature on cryptocurrencies has typically focused on the most popular one —Bitcoin and arrived at a diverse set of conclusions. On the one hand, the Bitcoin market, for example, has been declared highly speculative [5]. A similar conclusion has been reached by Yermack [6] who argued that Bitcoin's daily exchange rates exhibit virtually zero correlation with widely used currencies and with gold, making bitcoin useless for risk management and exceedingly difficult for its owners to hedge. A few years later the sentiment had changed. Dyhrberg [7] argued that Bitcoin possessed some of the same hedging abilities as gold and could be included in the variety of tools available to market analysts to hedge market specific risk. Trimborn et al. [8] also argued that including cryptocurrencies can indeed improve the risk-return trade-off of the portfolio. Baur et al. [9] also found that Bitcoin was uncorrelated with traditional asset classes such as stocks, bonds and commodities and is mainly used as a speculative investment and not as an alternative currency and medium of exchange.

The formation of a profitable investment portfolio using the cryptocurrencies and its instruments is a relatively new task for financial market participants. Previous attempts to analyse this topic include Andrianto and Diputra [10], who analysed three cryptocurrencies together with stocks, commodities, and foreign currencies. They also found that cryptocurrencies can increase the effectiveness of an investment portfolio in part by lowering portfolio variance. On the other hand, Bouri et al. [11] found that Bitcoin is a poor hedge and is suitable for diversification purposes only.

This paper, in contrast to the literature discussed above, focuses on investment portfolios comprised exclusively of cryptocurrencies. Investment analysis of multiple cryptocurrencies has been conducted, for example, by Chen et al. [12], who constructed an investment index of 30 cryptocurrencies. Brauneis and Mestel [13] related the risk and return of different mean-variance portfolio strategies to single cryptocurrency investments and two benchmarks, the naively diversified portfolio and the index constructed by Chen et al. [12] finding that portfolios feature lower risk than single cryptocurrencies and that in terms of the Sharpe ratio and certainty equivalent returns, the 1/N (naïve portfolio) outperforms singles cryptocurrencies and more than 75% of mean-variance optimal portfolios.

Gandal and Halaburda [14] considered whether Bitcoin can be a beneficiary of network effects. The volatility of cryptocurrencies highlighted earlier has naturally raised questions about bubbles in cryptocurrency markets. Corbet et al. [15] concluded that cryptocurrencies exhibit periods of clear bubble behaviour. A similar conclusion has been reached by earlier by Cheah and Fry [16]. In this, however, cryptocurrencies are no different from other financial and non-financial assets.

Another challenge in considering the attractiveness of cryptocurrencies as an investment instrument is whether the market is efficient. Urquhart [17], for example, found that Bitcoin market is inefficient over the entire sample period, although it may be in the process of moving towards an efficient market. Bartos [18], on the other hand, found that Bitcoin is an efficient market at least in a sense that it immediately reflects publicly available information. Some degree of market inefficiency, however, is to be expected with such a new market and does not necessarily make cryptocurrencies a less attractive investment instrument compared to other financial instruments, which feature their own market inefficiencies.

3 Portfolio Theory and Selecting the Cryptocurrencies

In this section, the authors consider a model of modern portfolio theory, which is a mathematical formulation of risk diversification in investing, aimed at selecting a group of investment assets that collectively have a lower risk than any individual asset. Intuitively, this becomes possible, as the value of various assets often changes in opposite directions. As a rule, any investment portfolio makes a compromise between risk and profitability, with a high expected return on assets generally possibly with a high degree of risk. Portfolio theory makes the choice between these two considerations in an optimal manner. The foundations of modern portfolio theory have been formulated in a seminal contribution by Markowitz [19]. This is also a particularly attractive framework for our purposes, because as shown by e.g. Kroll et al. [20] or Levy and Markowitz [21], portfolio optimization using mean-variance criteria is almost equivalent to the expected utility maximization under non-normality of returns, which is likely to be the case with cryptocurrencies.

The process of formation and management of the investment portfolio involves the implementation of the following stages: setting goals and selecting an adequate type of portfolio, analysing investment objects, forming an investment portfolio, selecting and implementing a portfolio management strategy, and evaluating the effectiveness of decisions taken. When formulating the investment portfolio, one should adhere to the basic principles that are aimed at increasing profitability and minimizing risks:

1. Compliance with the goals of investment.
2. Balanced assets, risk diversification.

As already alluded to in Sect. 2, investors, in the first instance seek to diversify by different classes of assets (e.g. shares, bonds, cryptocurrencies, commodities, financial derivatives, etc.). Within the asset class one can also diversify, by e.g. different regions and countries, or different sectors of the economy within a country. In order to choose the best portfolio, the following steps are needed:

- Building a trend for the analysis of the prospects of cryptocurrencies,
- Correlation analysis,
- Optimization of the asset portfolio,
- Portfolio rebalancing.

These stages will help to make the investment portfolio the most optimized in the risk/return ratio. The universe of cryptocurrencies available for investment is large. In this paper, we will focus on the cryptocurrencies from the Top-30, for which data is available from the CryptoCompare website [22]. Specifically, we include Bitcoin (BTC), BitcoinCash (BCH), Ethereum (ETH), Ethereum Classic (ETC), Ripple (XRP), Litecoin (LTC), Tron (TRX), VeChain (VEN), Icon (ICX), Dash (DASH), IOTA (IOT), Neo (NEO), Monero (HMR), Stellar Lumens (XLM), Cardano (ADA), Lisk (LSK), OmiseGo (OMG), ZCash (ZEC), Cindicator (CND), Walton (WTC), BitcoinGold (BTG), Verge (XVG). Since the condition of a profitable investment portfolio is its liquidity, then for the analysis the cryptocurrencies with instant online conversion to the US Dollar were chosen (not all cryptocurrencies are converted to the US Dollar, the less popular ones can only be converted to Bitcoins). Such currencies from the Top-30 as EOS, BnbCoin, VibeCoin are not converted to the US Dollar, so these cryptocurrencies are not included in the portfolio.

4 Assessing Cryptocurrencies for the Inclusion in the Portfolio

The cryptocurrency market is characterized by high nonlinearity therefore the authors chose a polynomial trend line for a more accurate description of the rate change schedule. Polynomial approximation is used to describe quantities that alternately increase and decrease. It is useful, for example, for analyzing a large set of data on an unstable value. The degree of the polynomial is determined by the number of extrema (maxima and minima) of the curve. The polynomial of the second degree can describe only one maximum or minimum. The polynomial of the third degree has one or two extrema. The polynomial of the fourth degree can have no more than three extrema.

One also has to note the in-sample accuracy of the approximation. This can be measured by a simple R-squared coefficient. After the initial analysis of the cryptocurrencies from the list in Sect. 3, high-volatility and downtrend cryptocurrencies were excluded, specifically ZCash, Verge, Tron, BitcoinGold, Walton, Monero and Cardano. The remaining 15 cryptocurrencies should be analyzed using correlation analysis. Table 1 shows that in principle, the correlation between different cryptocurrencies is rather high. However, this is likely to be a spurious correlation caused by the presence of common trends (e.g. the tendency of all cryptocurrencies to appreciate in sync).

In order to get a clearer picture, one can remove the polynomial trend from the cryptocurrency exchange rate and repeat the correlation analysis. These results are summarised in Table 2. Based on the results in Table 2, we can exclude currencies which exhibit high correlation with others even after removing the trend. We set the threshold for “high” correlation at 0.8 to ensure that enough cryptocurrencies remain for portfolio construction. On this basis, we choose the following cryptocurrencies for the investment portfolio: Bitcoin (BTC), BitcoinCash (BCH), Cindicator (CND), EthereumClassic (ETC), Ethereum (ETH), Litecoin (LTC), Neo (NEO), OmiseGo (OMG), Ripple (XRP).

To create a balanced portfolio, this paper uses the Markowitz model, which is an approach based on the analysis of expected mean values and variations of random variables. In addition, the risk/yield indicators are calculated for each asset in the portfolio. As a criterion for a possible deviation of profitability, the condition of an economic recovery or recession is usually taken. The expected portfolio risk is the standard deviation of the possible yields.

To calculate yields and risk, it is necessary to set the time horizon of the portfolio. Since cryptocurrencies have high volatility, it is better to make the portfolio for a short period of time, for example, three weeks. One can thus calculate the mean yield and the standard deviation for the cryptocurrency in a series of three-week periods covering the time horizon of the paper. The results are summarized in Table 3.

Table 1. Cryptocurrency correlations.

	BTC	BCH	CND	DASH	ETC	ETH	IOT	LSK	LTC	NEO	OMG	VEN	XLM	XMR	XRP
BTC	1.00														
BCH	0.87	1.00													
CND	0.51	0.61	1.00												
DASH	0.92	0.97	0.65	1.00											
ETC	0.87	0.90	0.61	0.92	1.00										
ETH	0.78	0.86	0.80	0.87	0.89	1.00									
IOT	0.93	0.87	0.55	0.92	0.89	0.81	1.00								
LSK	0.73	0.87	0.85	0.87	0.82	0.95	0.77	1.00							
LTC	0.90	0.89	0.61	0.94	0.89	0.85	0.92	0.82	1.00						
NEO	0.64	0.75	0.77	0.74	0.76	0.94	0.68	0.87	0.71	1.00					
OMG	0.68	0.79	0.86	0.82	0.76	0.90	0.73	0.91	0.78	0.88	1.00				
VBN	0.60	0.70	0.84	0.70	0.74	0.94	0.64	0.89	0.67	0.93	0.83	1.00			
XLM	0.72	0.78	0.84	0.80	0.81	0.94	0.76	0.91	0.78	0.89	0.85	0.94	1.00		
XMR	0.91	0.95	0.72	0.98	0.93	0.92	0.93	0.91	0.94	0.81	0.87	0.78	0.86	1.00	
XRP	0.67	0.77	0.85	0.79	0.74	0.89	0.71	0.91	0.76	0.82	0.85	0.88	0.96	0.82	1.00

Table 2. Detrended cryptocurrency correlations.

	BTC	BCH	CND	DASH	ETC	ETH	IOT	LSK	LTC	NEO	OMG	VEN	XLM	XMR	XRP
BTC	1.00														
BCH	0.52	1.00													
CND	−0.04	0.26	1.00												
DASH	0.69	0.89	0.35	1.00											
ETC	0.68	0.76	0.32	0.83	1.00										
ETH	0.32	0.60	0.68	0.64	0.74	1.00									
IOT	0.86	0.67	0.20	0.81	0.76	0.55	1.00								
LSK	0.16	0.64	0.76	0.65	0.58	0.86	0.45	1.00							
LTC	0.74	0.72	0.31	0.87	0.75	0.64	0.81	0.57	100						
NEO	0.04	0.39	0.62	0.37	0.50	0.87	0.31	0.73	0.37	1.00					
OMG	0.12	0.50	0.77	0.56	0.50	0.79	0.41	0.81	0.52	0.75	1.00				
VBN	0.04	0.36	0.74	0.35	0.50	0.90	0.30	0.81	0.36	0.87	0.67	1.00			
XLM	0.29	0.49	0.73	0.55	0.61	0.88	0.50	0.81	0.52	0.77	0.70	0.88	1.00		
XMR	0.65	0.81	0.51	0.92	0.86	0.80	0.84	0.77	0.87	0.56	0.70	0.56	0.69	1.00	
XRP	0.23	0.53	0.76	0.57	0.49	0.80	0.43	0.84	0.53	0.67	0.71	0.79	0.94	0.66	1.00

Table 3. Yields of select cryptocurrencies (percent).

Date	BTC	BCH	CND	ETC	ETH	LTC	NEO	OMG	XRP
14.08.2017	56.62	0.00	0.00	−12.24	29.91	1.84	0.00	364.86	−15.71
04.09.2017	−1.40	106.62	0.00	35.83	18.82	73.75	−36.47	56.10	38.99
25.09.2017	−7.84	−30.94	−43.00	−45.72	−19.82	−39.90	−36.22	−19.93	−23.71
16.10.2017	46.44	−25.44	−17.77	14.89	19.10	38.06	41.83	−7.21	50.94
06.11.2017	20.84	99.24	−61.62	9.53	−11.99	−16.74	−5.17	−21.43	−24.70
27.11.2017	39.85	174.68	62.12	65.27	58.82	57.86	44.04	34.77	22.43
18.12.2017	94.92	6.32	125.08	57.13	52.56	270.73	76.18	84.14	191.12
08.01.2018	−21.06	38.47	352.38	4.23	55.73	−15.02	50.16	56.43	290.56

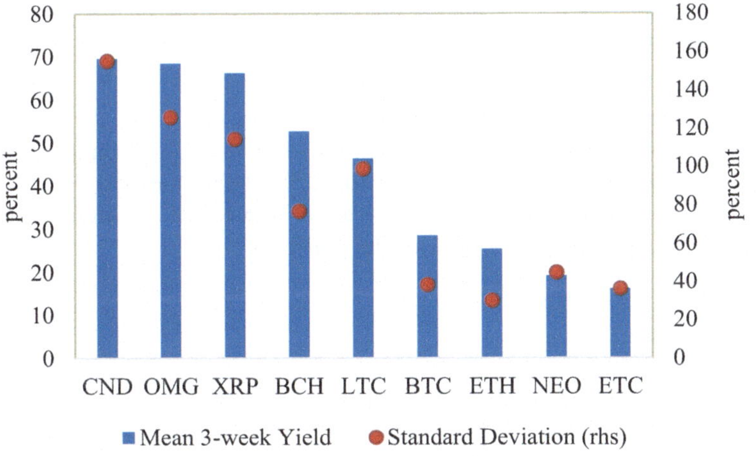

Fig. 1. Mean 3-week yield and standard deviation, (14.08.2017–08.01.2017, percent).

Figure 1 shows the calculated expected yield and standard deviation of the cryptocurrencies.

One can now calculate the overall standard deviation of portfolio returns (recall that in Markowitz model the optimal set of weights is the one that minimizes portfolio return variance, which is used as a proxy for risk). Total portfolio variance is given by:

$$\sigma_p = \sqrt{\sum\nolimits_{i=1}^{n} (w_i^2 \cdot \sigma_i^2) + 2\sum\nolimits_{i=1}^{n-1} \sum\nolimits_{j=i+1}^{n} (w_i \cdot w_j \cdot \rho_{ij} \cdot \sigma_i \cdot \sigma_j)}. \quad (1)$$

where σ_p is the standard deviation of portfolio returns, w_i is the weight of the i-th financial instrument, ρ_{ij} is the correlation coefficient of yields of the i-th and j-th financial instrument, and σ_i is the standard deviation of the yield of the i-th financial instrument. One can then use the formula to solve numerically for weights w_i such that the standard deviation of portfolio return is minimized or the weighted average return maximized. The corresponding solutions are shown in Fig. 2.

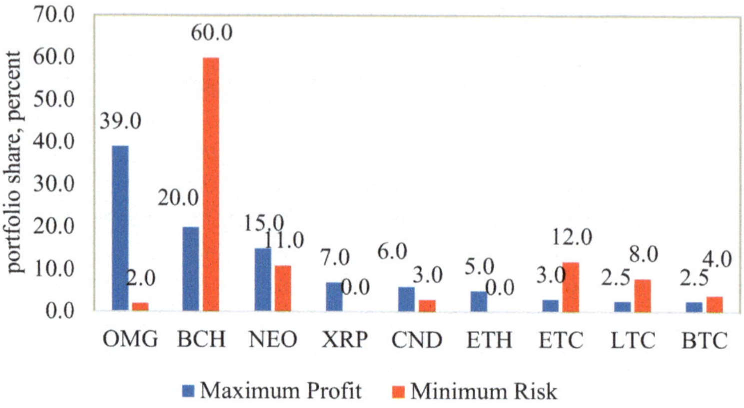

Fig. 2. Portfolio allocations for maximum profit and minimum Risk (percent total portfolio).

Thus, an investment portfolio targeting maximum profit (standard deviation of returns of 34%) is largely focused on two currencies: OmiseGo accounting for 39% of the portfolio, and BitcoinCash accounting for 20% of the portfolio. The investment portfolio with minimum risks (expected yield of above 30%) is largely focused on BitcoinCash, which accounts for 60% of the portfolio.

5 Conclusions

Cryptocurrencies emerged as a result of the dramatic innovation in finance. Over the past few years, the interest in cryptocurrencies has increased significantly, they are beginning to be considered an alternative to the usual money and a legitimate asset class. However, due to the high volatility of their returns, cryptocurrency investment portfolios will likely have to be monitored in real time and constantly adjusted.

In the short term, investing in cryptocurrencies yields can achieve substantial returns (in combination with substantial risk). These risks can be minimized if a prospective investor follows a few recommendations. First, there should be sufficient portfolio diversification. Second, cryptocurrencies comprising the portfolio should be liquid, that is, be freely convertible to fiat currency. Third, to the extent possible, uncorrelated cryptocurrencies should be included. Portfolio weights can then be determined by the application of the standard Markowitz model, taking into account the risk/yield ratio. This paper provided an illustrative demonstration of such an approach. It could easily be extended to more sophisticated applications by: (i) extending the time horizon considered and (ii) combining the cryptocurrencies with more traditional assets. Cryptocurrencies are a rapidly developing field, but it is likely that they will gradually become a traditional asset class to be used ever more widely.

The authors have also developed some recommendations for forming a balanced investment portfolio with cryptocurrencies. Any investment portfolio should be created considering investor objectives, and obey a logical relationship between risk and profitability. Therefore, cryptocurrencies should not be correlated; an investment portfolio should include seven to twelve cryptocurrencies—to safeguard against the risk of the loss of the entire portfolio; cryptocurrencies should be liquid, that is be freely exchangeable to a fiat currency on an appropriate exchange (e.g. dollars, euros). Investors should also consider regular rebalancing of the portfolio, which could help increase profitability.

References

1. Nakamoto, S.: Bitcoin: a peer-to-peer electronic cash system (2008). https://bitcoin.org/bitcoin.pdf. Accessed 15 July 2018
2. European Central Bank: Virtual Currency Schemes (2012). https://www.ecb.europa.eu/pub/pdf/other/virtualcurrencyschemes201210en.pdf. Accessed 13 July 2018
3. European Central Bank: Virtual currency schemes—a further analysis (2015). https://www.ecb.europa.eu/pub/pdf/other/virtualcurrencyschemesen.pdf. Accessed 11 July 2018
4. Saksonova, S., Kuzmina-Merlino, I.: Fintech as financial innovation—the possibilities and problems of implementation. J. Eur. Res. Stud. **XX**(3A), 961–973 (2017)
5. Baek, C., Elbeck, M.: Bitcoins as an investment or speculative vehicle? Appl. Econ. Lett. **22** (1), 30–34 (2015)
6. Yermack, D.: Is Bitcoin a real currency? An economic appraisal. NBER Working Paper 19747 (2013)
7. Dyhrberg, A.H.: Hedging capabilities of Bitcoin. Is it the virtual gold? Finance Res. Lett. **16**, 139–144 (2016)
8. Trimborn, S., Li, M., Härdle, W.K.: Investing with cryptocurrencies—a liquidity constrained investment approach (2018). SSRN: https://ssrn.com/abstract=2999782 or http://dx.doi.org/10.2139/ssrn.2999782
9. Baur, D.G., Hong, K., Lee, A.D.: Bitcoin: medium of exchange or speculative assets. J. Int. Finan. Mark. Inst. Money **54**, 177–189 (2018)
10. Andrianto, Y., Diputra, Y.: The effect of cryptocurrency on investment portfolio effectiveness. J. Finance Account. **5**(6), 229–238 (2017)
11. Bouri, E., Molnar, P., Azzi, G., Roubaud, D., Hagfors, L.I.: On the hedge and safe haven properties of Bitcoin: is it really more than a diversifier? Finance Res. Lett. **20**, 192–198 (2017)
12. Chen, S., Chen, C.Y., Härdle, W.K., Lee, T.M., Ong, B.: Econometric analysis of a cryptocurrency index for portfolio investment. Handb. Blockchain Digit. Finance Incl. **1**, 175–206 (2018)
13. Brauneis, A., Mestel, R.: Cryptocurrency-portfolios in a mean-variance framework. Finance Res. Lett. (2018, in Press). https://doi.org/10.1016/j.frl.2018.05.008
14. Gandal, N., Halaburda, H.: Competition in the Cryptocurrency Market. CESifo Working Paper 4980 (2014)
15. Corbet, S., Lucey, B., Yarovaya, L.: Datestamping the Bitcoin and Ethereum bubbles. Finance Res. Lett. (2017, in Press). https://doi.org/10.1016/j.frl.2017.12.006
16. Cheah, E.-T., Fry, J.: Speculative bubbles in Bitcoin markets? An empirical investigation into the fundamental value of Bitcoin. Econ. Lett. **130**, 32–36 (2015)
17. Urquhart, A.: The inefficiency of Bitcoin. Econ. Lett. **148**, 80–82 (2016)
18. Bartos, J.: Does Bitcoin follow the hypothesis of efficient market? Int. J. Econ. Sci. **IV**(2), 10–23 (2015)
19. Markowitz, H.M.: Portfolio selection. J. Finance **7**(1), 77–91 (1952)

20. Kroll, Y., Levy, H., Markowitz, H.M.: Mean variance versus direct utility maximization. J. Finance **39**(1), 47–61 (1984)
21. Levy, H., Markowitz, H.M.: Approximating expected utility by a function of mean and variance. Am. Econ. Rev. **69**(3), 308–317 (1979)
22. CryptoCompare website. https://www.cryptocompare.com. Accessed 14 July 2018

Management of Liquidity and Profitability in Commercial Banks

Natalia Konovalova(✉)

RISEBA University of Business, Arts and Technology, Meža str. 3, Riga LV-1048, Latvia
natalija.konovalova@riseba.lv

Abstract. Issues of liquidity and profitability management are currently highly topical for every commercial bank. Those banks, which operate with surplus liquidity, encounter the problem of profits that are less than optimal. Other banks, which conduct aggressive policy aiming to place all the available resources with maximum efficiency, are forced to seek solutions to the problem of locating additional liquid funds, in order to ensure timely fulfilment of liabilities. In this paper the author offers a methodology for evaluating the optimization of the balance between profitability and liquidity in commercial banks. This methodology is based on the analysis of bank's active and passive operations, which results in determining the amount of effective (profitable) resources, which is compared with the actual amount of profitable assets to discover the level of optimization of profitability and liquidity. The potential application and viability of the offered methodology is shown on the example of Latvian commercial banks. After analysing active and passive operations of Latvian commercial banks, the author develops recommendations to optimise profitability and liquidity.

Keywords: Profitability · Liquidity · Assets and liabilities management · Effective resources · Profitable assets

1 Introduction

Commercial banks attach great significance to the issues of profitability and liquidity management. To successfully operate on the market, each bank must be able to balance the two most important characteristics of its operations: on the one hand—to maintain the necessary level of liquidity, but on the other hand—to achieve the maximum possible level of profitability of its active operations. Because one of the characteristics of the bank's role of the financial intermediary is the use of funds attracted from clients, depositors and other creditors, the bank has to have a certain amount of highly liquid assets [1]. However, maximising highly liquid assets (cash, deposits on correspondent accounts at the central bank and other commercial banks, etc.) with respect to other assets comes at the expense of reducing profitability. That is, if the bank's liquidity is maximised, the bank's profits will be minimal [2]. At the same time maximising profitability requires using liquid funds to provide loans and make investments instead of keeping them on the balance sheet. Since that requires minimising cash balances and

I. Kabashkin et al. (Eds.): RelStat 2018, LNNS 68, pp. 725–736, 2019.
https://doi.org/10.1007/978-3-030-12450-2_70

balances on the correspondent accounts, the maximisation of profitability threatens the ability of the bank to fulfil its obligations to its clients without interruptions [3]. These trade-offs in banking operations give rise to the necessity of constant search for the optimal balance between profitability and liquidity.

Currently Latvian commercial banks have the ratios of liquidity, which many times exceed the necessary minimum requirements, that is they operate with surplus liquidity and, consequently, do not use the entire available reserves to obtain the maximum possible financial and commercial result. For example, on the middle of 2018, the liquidity coverage ratio (LCR) of Latvian banks has been 130–699% [4] instead of the required 100%. At the same time the indicator of the assets profitability was record-breaking low. During the last 5 years the return on assets (ROA) has been between 0.86 and 1.49% [5]. All of this is evidence about the irrational use of the available resources by the commercial banks, which leads to obtaining suboptimal profits.

Unfortunately, the issues of analysing profits forgone by the banks remain little explored.

Taking into account the above mentioned, the author attempts to develop the methodology to evaluate the optimisation of profitability and liquidity in commercial banks and approbates the possibility of its application in Latvian commercial banks.

The aim of this paper is to evaluate the effectiveness of managing banks' assets and liabilities using the analysis of profitability and liquidity of Latvian commercial banks and to develop suggestions for optimising the profitability and liquidity in the commercial banks.

Research tasks are:

- to analyse the profitability and liquidity in the Latvian banking industry;
- to point out the existing problems in managing profitability and liquidity of Latvian commercial banks;
- to propose a method of evaluating the optimisation of profitability and liquidity in commercial banks;
- to evaluate asset and liability management in Latvian commercial banks using the proposed methodology;
- to develop suggestions for optimising profitability and liquidity in Latvian commercial banks.

During the research the following analyzes are used: analytical, graphical, statistical and empirical research method. The theoretical and methodological basis for this research comes from specialised scientific literature, textbooks, the legislative provisions of the Republic of Latvia, regulations of supervisory authorities and the published annual reports of the banks.

2 Theoretical Aspects of Liquidity and Profitability

Studies by many authors show a close relationship between liquidity and profitability of commercial banks. Thus, American scientists M. Osborne, A. Fuertes, A. Milne on the basis of a study conducted in the US banking sector concluded that excessively high liquidity was a costly factor affecting the performance of banks, as it significantly reduced the profitability [6]. American scientists argue that high liquidity and lending are also interconnected since a reduction in lending on the one hand reduces the risk while on the other hand reduces the bank profitability and creates conditions for increasing liquidity [7]. The problems of excessively high liquidity are very dangerous for a bank since a reduction in profitability and a shortfall in profits under unstable conditions can lead the bank to insolvency and bankruptcy [8].

Earlier studies by M. Halling, E. Hayden, A. Demirgüç-Kunt and H. Huizinga [9, 10]. point to the inverse relationship between liquidity and interest margin: banks with high liquidity have a lower net interest margin. In the event of a liquidity crisis, the bank can lend money in the interbank market. However, this can also lead to a decrease in profitability since interbank loans during a crisis turn out to be expensive. The link between liquidity and profitability is indicated by the authors A.O. Agbada and C. Osuji [11]. They note that the presence of a large number of funds in comparison with attracted deposits does not allow banks to earn profits. However, if banks finance the reserved money in order to increase profits, they increase their risks and may face a serious problem, such as a liquidity crisis [11].

Liquidity management is important for both financial and non-financial enterprises. The level of liquidity allows to assess the financial condition of the bank [12]. M. Drehmann, K. Nikolaou characterize the liquidity as "the capability to immediately settle financial obligations". Liquidity is a method that is used by banks to convert assets into monetary form for making cash payments [13]. Liquidity is the liability of the bank to pay all financial obligations—short-term and long-term debts as well as other financial expenses, including taxes. On the other hand, if the amount of current assets is too high or when the bank operates with excess liquidity, then the return on investments in such a bank will be low [14, 15].

Liquid assets mainly include cash, easily marketable securities as well as funds on the accounts of the Central Bank and government short-term debt securities [16]. Liquidity is determined via the indicators and ratios that are especially important for banks. Banks work with a huge amount of customer moneys. And it is very important for banks to balance the incoming and outgoing cash flows. It is very good if liquid assets are represented by easily marketable securities because they are easy to convert into cash and to fulfill obligations to the customers [17].

However, it should be noted that liquidity risk management is the core of the risk management structure in the entire banking sector since it affects the profitability of banks and other financial institutions [18]. To avoid losses in a bank's operations, well-established control and management decisions are necessary in managing of liquidity and profitability [19]. J.A. Niresh believes that the liquidity is important for both internal and external analysts, and also extremely important for monitoring of daily business operations. He also emphasizes that the weak liquidity situation threatens the solvency and profitability of the company and is dangerous for doing business [20].

Issues of liquidity and profitability were considered in studies by H. Greuning and S. Bratanovic. Thus, in their opinion the bank liquidity represents the capability of a bank to effectively finance its transactions. Liquidity management can be viewed from at least two points of view: the management of insufficient or inadequate liquidity levels and management of excess liquidity. First, an inadequate level of liquidity can lead to the need for attraction of additional sources, resulting in additional costs that reduce the profitability of the bank. And this, in turn, can ultimately lead to bankruptcy. Second, excessive liquidity can lead to a decrease in the asset returns and, as a result, the bank will have poor financial performance. To avoid these two extremes—insufficient liquidity and excess liquidity—the bank must have the adequate liquidity potential. That is, to be capable of receiving the funds immediately and at a reasonable cost when they are needed. And to direct available resources to efficient profitable operations. In practice, achieving and maintaining the optimal liquidity and profitability is a true art of bank management [21].

A. Demirgüç-Kunt, E. Detragiache believe that the pursuit of high bank profitability can lead to a financial crisis [22]. In turn, U. Albertazzi, L. Gambacorta argue that the bank returns are certainly important. But with a view to assess the consistency and reliability of the financial and banking sector [23]. F. Heibati, M. Seid Noorani, S. Dadkhah define the profitability as the change between the expenditures and incomes during the financial year [24]. They consider that sufficient profitability will contribute to growth and expansion of activity. A.O. Agbada, C. Osuji in their studies pay a great attention to the planning of profit and profitability. Profitability planning is the main and complicated part of management, which allows the bank management to make decisions and develop strategies [11].

According to M.A.M. Kabajeh, S. Nuaimat, F.N. Dahmash, the profitability is represented by three alternative variables [25]. The most important is the return on assets (ROA). ROA shows the capability of a bank asset to generate profit. Another indicator is the return on equity (ROE), which shows to the shareholders the profit on their shares. Other important indicators of profitability also include the income from investments, or return on investment (ROI). This indicator measures the efficiency of a bank with the help of invested capital. According to authors J.A. Pearce, E. B. Freeman, R.B. Robinson, the group of important indicators of profitability includes the earnings per share (EPS) [26]. In turn, J. Darko, Z.A. Aribi, G.C. Uzonwanne state that net profit margin (NPM) and Tobin's ratio (Tobin Q) are also important factors for bank profitability [27]. Potential investors are interested in dividends and market evaluation of the company, so they pay a great attention to profitability indicators. In turn, managers are interested in measuring the operating rate of return. Low profit margins will witness an inefficient management and investors will not dare to invest in such a company [28].

E. Koskel & R. Stenbacka state that the profitability, liquidity and reliability are the main objectives of the monetary and lending policy. A commercial bank must earn a profit for its shareholders and at the same time must meet the needs of the clients in obtaining credit services and fulfilling its financial obligations to them. Therefore, the bank is always trying to achieve a dual ambivalent goal of liquidity and profitability by choosing a diverse and balanced portfolio of assets within the framework of instructions of the Central Bank and Supervisory Authorities [29].

3 The Methodology of Determining the Optimality of Profitability and Liquidity in Commercial Banks

In order to evaluate the level of optimality of profitability and liquidity, the author suggests using a methodology developed in this paper, which is based on analysing asset and liability management in the bank.

The method of optimising the profitability and liquidity includes the following stages.

Stage I. *Determining the total volume of bank resources based on the analysis of the bank's passive transactions.*

At this stage the passive transactions of the bank are analyzed and, taking into account the current and prospective market situation, the real volume of funding sources is determined. When analysing passive operations of the bank, it is necessary to pay particular attention to interbank loans, in order to avoid them turning into a constant source of creating bank resources [30].

Stage II. *Analysis of the allocation of bank resources in the bank's assets.*

The analysis of the resource allocation in the bank's assets presupposes the determination of each asset group share in total assets of the bank as well as the assessment of their profitability. At this stage, using the balance sheet data, the actual volume of earning assets is established. The main items of earning assets will include credit investments, leasing, factoring, investment in securities and interbank lending. When determining the earning assets of the bank, it is also necessary to take into account that overdue loans and other bad debts in the group of illiquid assets do not bring income to the bank and therefore will not be included in earning assets. Fixed assets, intangible assets and other assets withdrawn from the principal banking activities and playing a supporting role in carrying out the bank's transactions will not be regarded as earning assets either [30, 31].

Stage III. *Determining the volume of effective (earning) resources.*

When determining the volume of effective resources, it is necessary to remember that the liabilities section of the bank's balance sheet includes non-resource liabilities, such as deferred income and accrued expenses, provisions for the employees' holidays, deferred tax payments, accounts payable for business transactions as well as other non-resource liabilities. Therefore, when determining the volume of effective resources, it is necessary to take into account the difference between immobilized assets and non-resource obligations, i.e. the amount by which immobilized assets exceed non-resource obligations, which has to be subtracted from total resources of the bank.

On the basis of the above, the following formula is suggested for determining the volume of effective (earning) resources of the bank:

$$R(EF) = R - \Delta IM - (MRR \times RB) - (LCR - MRR) \times NLO$$

(the formula is developed by the author), where R (EF)—the amount of effective (profitable) bank's resources; R—total bank's resources; ΔIM—the excess of non-current assets over the liabilities that are not a part of bank's resources; MRR—minimum reserve requirements; RB—reserve base; LCR—liquidity coverage ratio = liquidity buffer/net liquidity outflows over a 30-calendar day stress period; NLO—net liquidity outflows over a 30-calendar day stress period.

Stage IV. *Assessment of the optimization of bank resources attraction and allocation.*

For optimizing the attraction and allocation of bank resources while ensuring the required level of liquidity and achieving the maximum possible level of profitability, it is necessary to strive for the equality of earning assets and effective (earning) resources.

If the volume of effective resources is substantially bigger than their effective (highly profitable) deployment in assets, the banks will get a *free volume of unprofitable resources*, which will indicate the inefficiency of resource management. In case the volume of earning assets is bigger than the available effective resources, the bank will be lacking resources, which will result in the shortage of liquid assets.

This concludes the main summary of the methodology of evaluating the profitability and liquidity of commercial banks developed by the author.

4 The Analysis of Profitability and Liquidity Management in Latvian Commercial Banks

The position in the banking sector of Latvia is characterized by excessively high level of liquidity and low level of profitability. At the middle of the year 2018 the average level of liquidity coverage ratio (LCR) in Latvian banking industry was 245.25% with the required norm of 100%. Changes in average liquidity ratios are given in Fig. 1.

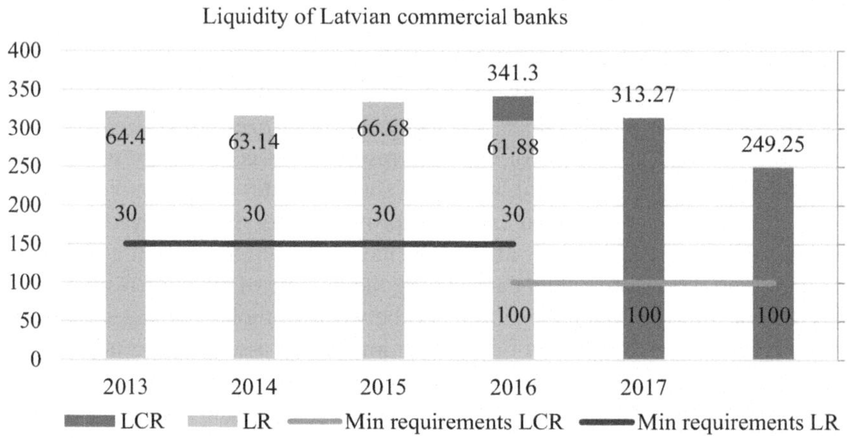

Fig. 1. The dynamics of average liquidity ratios in Latvian Banking industry, %. (developed by the author basing on Latvian Financial and Capital Market Commission data [4]).

The position of liquidity level in each Latvian commercial bank is demonstrated in Fig. 2. As we see from Fig. 2, liquidity coverage ratio (LCR) on 30.06.2018. is fluctuated from 130% in Luminor Bank to 699% in Rietumu banka. But minimum requirements of liquidity coverage ratio (LCR) according to Basel III is 100%. It demonstrates high excess liquidity in each commercial bank of Latvia.

Indicators of bank profitability over the time period considered (2013–2018) had a tendency towards deterioration. The changes of the profitability indicators are given in Fig. 3.

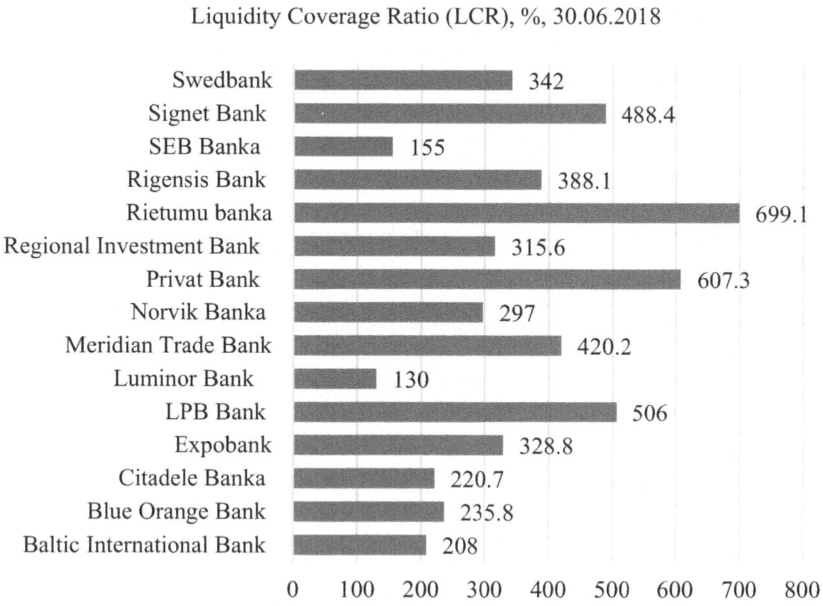

Fig. 2. Liquidity coverage ratio (LCR) in Latvian commercial banks (developed by the author basing on Latvian Financial and Capital Market Commission data [4] and Finance Latvia Association data [5]).

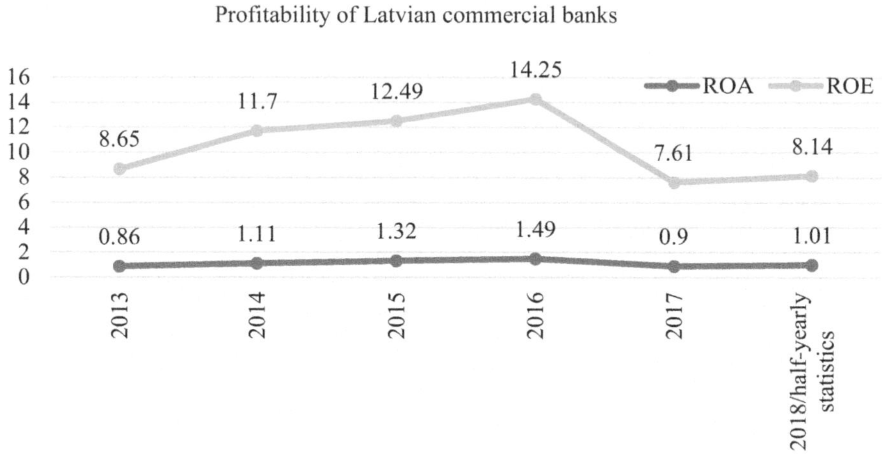

Fig. 3. The dynamics of average profitability indicators in Latvian Banking industry, % (developed by the author basing on Latvian Financial and Capital Market Commission data [4] and Finance Latvia Association data [5]).

Obviously, the profitability and liquidity indicators for each separate bank will differ from the similar indicators that are calculated over the entirety of the commercial bank system in Latvia. Therefore, it is necessary to conduct the analysis of profitability and liquidity considering each commercial bank separately.

Next, the author provides an analysis of profitability and liquidity in Latvian commercial banks, using the methodology described above.

Table 1 shows the calculations of effective (profitable) resources, the actual amount of profitable assets and the surplus of liquid funds for all Latvian commercial banks.

These calculations confirm that Latvian commercial banks operate with a surplus of liquidity. This is evidence for the fact that Latvian commercial banks do not rationally utilize resources at their disposal, as well as the fact that there are limited possibilities for Latvian banks to effectively place assets on the markets.

Table 1. Comparison of the amount of effective (profitable) resources with the actual amount of earning assets in Latvian commercial banks (%), 30.06.2018. (developed by the author basing on Latvian Financial and Capital Market Commission data [4], Finance Latvia Association data [5], and Latvian commercial banks data [32–46]).

No	Latvian commercial banks	Effective resources, R (EF)	Earning assets, A (EARN)	Surplus (+) of free resources	Return on Equity ROA	Return on Assets ROE
1.	Baltic International Bank	100.00	48.08	+51.92	–1.60	–18.00
2.	Blue Orange Bank	100.00	42.41	+57.59	2.00	21.20
3.	Citadele Banka	100.00	45.31	+54.69	1.10	10.40
4.	Expobank	100.00	26.19	+73.81	0.80	3.20
5.	LPB Bank	100.00	19.77	+80.23	2.80	25.80
6.	Luminor Bank	100.00	76.92	+23.08	0.90	8.40
7.	Meridian Trade Bank	100.00	23.81	+76.19	0.90	20.10
8.	Norvik Banka	100.00	33.67	+66.33	0.00	0.10
9.	Privat Bank	100.00	16.47	+83.53	–1.30	–7.60
10.	Regional Investent Bank	100.00	31.69	+68.31	1.30	12.40
11.	Rietumu banka	100.00	14.30	+85.70	1.80	8.70
12.	Rigensis Bank	100.00	25.77	+74.23	0.90	3.90
13.	SEB Banka	100.00	64.52	+35.48	1.40	12.20
14.	Signet Bank	100.00	20.48	+79.52	0.20	2.10
15.	Swedbank	100.00	29.24	+70.76	2.20	15.50

For comparison of the liquidity and profitability conditions for different banks, the relative indicators, which describe the share of surplus liquidity in the total volume of effective (profitable) resources for the bank, are the most appropriate. The smaller the share of liquidity surplus, the more effectively does the bank operate, and when the share of surplus liquidity increases, the effectiveness of bank's operations decreases. Consequently, the relative indicators of surplus liquidity determine the level of inefficient utilization of resources which determines the amount of bank profits forgone.

After comparing various banks, one can conclude that 8 commercial banks (Expobank, LPB Bank, Meridian Trade Bank, Privat Bank, Rietumu banka, Rigensis Bank, Signet Bank, Swedbank) work with the largest surplus of liquidity (ranging from 70 to 85%). 5 commercial banks (Baltic International Bank, Blue Orange Bank, Citadele Banka, Norvik Banka, Regional Investment Bank) are the banks that operate with the middle level of surplus liquidity (the range of surplus liquidity fluctuations for these five banks varies from 50 to 70%). The more moderate range of surplus liquidity is observed for two banks—SEB Bank and Luminor Bank (23–35%).

Results of analysis confirm, banks, which work with high level of surplus liquidity have low indicates of profitability (ROA, ROE). And banks, that operate with middle and low level of surplus liquidity, have indicates of profitability (ROA, ROE) higher (Table 1).

5 Conclusion

Thus, the results of the research allow formulating the following conclusions:

- The position in the banking sector of Latvia is characterized by excessively high level of liquidity and low level of profitability.
- Commercial banks in Latvia operate with surplus liquidity. Over the period from the year 2013 to the year 2016, the average indicator of current liquidity in Latvian commercial banks was fluctuating between 61.88% and 66.68%, whereas the minimum norm was 30%. The average liquidity coverage ratio (LCR), which was introduced into Latvian commercial banks in 2016, from the year 2016 to the year 2018 is fluctuating between 249.25% and 341.30%, whereas the minimum norm is 100%.
- The presence of surplus liquidity does not allow Latvian commercial banks to use resources at their disposal with maximum efficiency, which accordingly leads to getting a less than optimal profit.
- The minimum necessary reserve of highly liquid assets for commercial banks is determined by the mandatory reserve requirement, the indicator of current liquidity and liquidity coverage ratio.
- The analysis of the profitability and liquidity of Latvian commercial banks had shown that the largest surplus of liquidity can be observed in 8 commercial banks (Expobank, LPB Bank, Meridian Trade Bank, Privat Bank, Rietumu banka, Rigensis Bank, Signet Bank, Swedbank), which underutilize resources at its disposal by 70–85%.

- SEB Bank and Luminor Bank manage their profitability and liquidity in the manner that is closest to the optimal. With the average surplus liquidity indicator of 23–35%, the average return on equity—8.4–12.20%.

Based on the above conclusions and the obtained results, the author develops the following recommendations for the optimisation of the profitability and liquidity in the commercial banks.

1. The evaluation of the optimality of profitability and liquidity has to be conducted on the basis of the analysis of bank's asset and liability management.
2. The amount of effective (profitable) resources for the bank should be calculated using the method offered by the author.
3. One should continuously analyse bank's active operations with the objective of distinguishing the presence of profitable and unprofitable assets.
4. One has to control the relationship between the amount of profitable resources and the amount of profitable assets.
5. In case the amount of earning resources exceeds the amount of earning assets, to find opportunities for the efficient allocation of surplus earning resources.
6. In case the amount of earning assets exceeds the amount of earning resources, to regroup the assets increasing the share of highly liquid assets and to procure additional earning resources by increasing equity capital, attracting subordinated liabilities, attracting deposits, issuing debt instruments and using other sources.

References

1. Chistov, D.V.: Banking Information Systems and Technologies. Finance and Statistics, Moscow (2005)
2. Nikonova, I.A., Shamgunov, R.N.: Strategy and Cost of Commercial Bank. Alpina Business Books, Moscow (2005)
3. Scherbakova, G.N.: Analysis and assessment of bank activity, Top, Moscow, St. Petersburg (2016)
4. Latvian Financial and Capital Market Commission. http://www.fktk.lv/en/, http://www.fktk.lv/en/statistics/credit-institutions/quarterly-reports.html. Accessed 28 Sept 2018
5. Finance Latvia Association. https://www.financelatvia.eu/en/industry-data/. Accessed 18 Sept 2018
6. Osborne, M., Fuertes, A., Milne, A.: Capital and profitability in banking: evidence from US banks (2009)
7. Osborne, M., Fuertes, A.M., Milne, A.: In good times and in bad: bank capital ratios and lending rates. International Review of Financial Analysis (2016)
8. Arif, A., Nauman Anees, A.: Liquidity risk and performance of banking system. J. Finan. Regul. Compliance **20**, 182–195 (2012)
9. Halling, M., Hayden, E.: Bank failure prediction: a two-step survival time approach (2006)
10. Demirgüç-Kunt, A., Huizinga, H.: Determinants of commercial bank interest margins and profitability: some international evidence. World Bank Econ. Rev. **13**, 379–408 (1999)
11. Agbada, A.O., Osuji, C.: The efficacy of liquidity management and banking performance in Nigeria. Int. Rev. Manag. Bus. Res. **2**, 223–233 (2013)

12. Drehmann, M., Nikolaou, K.: Funding liquidity risk: definition and measurement. J. Bank. Finan. **37**, 2173–2182 (2013)
13. Diamond, D.W., Rajan, R.G.: Liquidity shortages and banking crises. J. Finan. **60**, 615–647 (2015)
14. Jagongo, A., Makori, D.: Working Capital Management and Firm Profitability: Empirical Evidence from Manufacturing and Construction Firms Listed on Nairobi Securities Exchange, Kenya (2013)
15. Van Horne, J.C., Wachowicz, J.M.: Fundamentals of Financial Management. Pearson Education, London (2008)
16. Duijm, P., Wierts, P.: The effects of liquidity regulation on bank assets and liabilities. Int. J. Central Bank. **12**, 385–411 (2016)
17. Carlin, B.I., Kogan, S., Lowery, R.: Trading complex assets. J. Finan. **68**, 1937–1960 (2013)
18. Majid, A., Rais, A.: Development of liquidity management instruments: challenges and opportunities. Paper Presented at the International Conference on Islamic Banking: Risk Management, Regulation and Supervision, Held in Jakarta Indonesia September (2003)
19. Merrouche, O., Schanz, J.: Banks' intraday liquidity management during operational outages: theory and evidence from the UK payment system. J. Bank. Finance **34**, 314–323 (2010)
20. Niresh, J.A.: Trade-off between liquidity and profitability. A Study of selected Manufacturing Firms in Sri Lanka. University of Jaffna (2012)
21. Greuning, H., Bratanovic, S.: Analysis and Risk Management Banking. Irecson Publishing House, Bucharest (2004)
22. Demirgüç-Kunt, A., Detragiache, E.: Monitoring banking sector fragility: a multivariate logit approach. World Bank Econ. Rev. **14**, 287–307 (2000)
23. Albertazzi, U., Gambacorta, L.: Bank profitability and the business cycle. J. Finan. Stab. **5**, 393–409 (2009)
24. Heibati, F., Seid Noorani, M., Dadkhah, S.: Evaluation of private banks compared to banks performance of countries around Arabic Gulf. J. Econ. **6**, 91–108 (2009)
25. Kabajeh, M.A.M., Al Nuaimat, S.M.A., Dahmash, F.N.: The relationship between the ROA, ROE and ROI ratios with Jordanian insurance public companies market share prices. Int. J. Hum. Soc. Sci. **2**, 115–120 (2012)
26. Pearce, J.A., Freeman, E.B., Robinson, R.B.: The tenuous link between formal strategic planning and financial performance. Acad. Manag. Rev. **12**, 658–675 (1987)
27. Darko, J., Aribi, Z.A., Uzonwanne, G.C.: Corporate governance: the impact of director and board structure, ownership structure and corporate control on the performance of listed companies on the Ghana stock exchange. Corp. Gov. **16**, 259–277 (2016)
28. Vieira, R.S.: The relationship between liquidity and profitability. An exploratory study of airline companies between 2005 and 2008. Master thesis 2010 Umeå University (2010)
29. Koskela, E., Stenbacka, R.: Is there a tradeoff between bank competition and financial fragility? J. Bank. Finance **24**(12), 1853–1873 (2000)
30. Konovalova, N.: Profitability and liquidity optimization in commercial banks: problems and solutions. Economy & Business International Scientific Publications 1., Science Invest—Bourgas, Bulgaria, pp. 235–247 (2007)
31. Konovalova, N.: Commercial Banks' Resource Management. University of Latvia. International Scientific Conference Finance and Accounting: Theory and Practice, Development and Trends, Proceedings (2009). ISBN 978-9984-45-106-0
32. Baltic International Bank. https://www.bib.eu/. Accessed 05 Oct 2018
33. Blue Orange Bank. https://www.blueorangebank.com/int/en. Accessed 05 Oct 2018
34. Citadele Banka. https://www.citadele.lv/en/. Accessed 05 Oct 2018
35. Expobank. http://www.expobank.eu/. Accessed 05 Oct 2018

36. LPB Bank. http://www.lpb.lv/en/. Accessed 05 Oct 2018
37. Luminor Bank. https://www.luminor.lv/. Accessed 05 Oct 2018
38. Meridian Trade Bank. https://www.mtbank.eu/en. Accessed 05 Oct 2018
39. Norvik Banka. https://www.norvik.eu/en/. Accessed 05 Oct 2018
40. Privat Bank. https://www.privatbank.lv/en/. Accessed 05 Oct 2018
41. Regional Investment Bank. http://www.ribbank.com/en/. Accessed 05 Oct 2018
42. Rietumu banka. http://www.rietumu.com/. Accessed 05 Oct 2018
43. Rigensis Bank. http://www.rigensis.lv/en/. Accessed 05 Oct 2018
44. SEB Banka. https://www.seb.lv/eng. Accessed 05 Oct 2018
45. Signet Bank. https://www.signetbank.com/. Accessed 05 Oct 2018
46. Swedbank. https://www.swedbank.lv/. Accessed 05 Oct 2018

Corporate Income Taxation: Challenges of E-Commerce Platforms

Māris Juruss(✉), Sigita Ragucka-Ragovska, and Monta Sandore

Riga Technical University, Riga, Latvia
Maris.Juruss@rtu.lv

Abstract. Sales turnovers, profits, hence taxes collected from businesses operating in digital economy are rising worldwide, including the EU. This arises additional challenges for different tax administrations on how to tax profits of such businesses as their operational models no longer fit in traditional taxation models. Consequently, the question is how to attribute fair amount of profits of such businesses to respective jurisdictions.

This is due the fact that businesses operating in digital economy have critical differences in their business models compared to the ones operating in traditional economy as further discussed in the research paper. Changing business models outline the need for new principles for taxation of such businesses to achieve fair taxation as recently pointed out by the European Commission.

Different businesses are operating in digital economy, such as, ride-to-hire companies, e-commerce retailers and social networks etc. This research paper focuses specifically on corporate income taxation of e-commerce platforms. The research paper in more detail outlines the methodology used to determine key profit drivers of e-commerce platforms. It should be specifically stressed that from upstream sources as per traditional business models, e-commerce platforms have changed its operations to downstream sources—i.e. customers. Subsequently, as finding suggested that the most appropriate method for determination of profits to be taxed by each jurisdiction is profit split method, application of the method was simulated. Similar ideas and approaches on how to tackle issues arising from corporate income taxation on businesses operating in digital economy also are supported by the Organization of Economic Cooperation and Development member states as well as researches.

Keywords: Digital economy · Corporate income tax · Profit split · Fair taxation

1 Introduction

Globalization process, artificial intelligence and technologies are becoming more popular, affecting economies around the world. The most successful businesses are operating in digital economy, allowing the Internet to become the main element to communicate to customers and use it for sale of goods and provision of services. In the EU 50% of the affiliates of digital businesses are foreign based, compared 80% for traditional businesses. Meaning businesses manage to have major sales turnovers in the EU, meanwhile the profits of such companies remain and are taxed not necessarily

I. Kabashkin et al. (Eds.): RelStat 2018, LNNS 68, pp. 737–746, 2019.
https://doi.org/10.1007/978-3-030-12450-2_71

within the EU. Such rapid development of digital businesses has created a major fiscal distortion: the effective tax rate for digital businesses is around 9.5%, compared to 23.2% for traditional businesses. This is due current international taxation concepts that aimed at traditional businesses leaving more room for harmful tax practices for digital businesses.

European e-commerce market is experiencing stable growth across all major countries, being the most dynamic and fastest growing industry in Europe. Online retail growth has continued in the 2017, with European e-commerce turnover increasing by 11% to 534 bEUR in the 2017 and forecasted to grow by 13% to 602 bEUR in the 2018. This represents an impressive growth rate from the 2013 figure of 307 bEUR. However, the levels of online shopping differ greatly across the region and represent a large potential audience with which to grow the consumer—base in the future. Countries such as the Netherlands, Sweden and the UK represent markets with mature e-commerce consumer-bases and are accustomed to ordering goods/services over the internet (82%, 81%, and 78% respectively) [1]. For instance, average annual revenue growth of the digital businesses was 14% compared to traditional multinational businesses growth that varies between 0.2% and 3% in 2017. During 2008–2016 the amount of businesses that are engaged in e-sales increased by 6 percentage points in the EU, increasing the turnover of e-sales respectively by 7 percentage points during the period. Thus, the interest in e-commerce is increasing from both consumers and companies increasing their investments in establishing an online presence in the market [2].

New technologies are in general bringing new challenges of taxing businesses on international level. Digitalization shapes new type of economy, meaning that not only businesses but also countries should adapt to the new system. Taking into account the developments of the concept of value creation as well as issues addressed by OECD have boosted the development of mind-set with regard of taxation of new business [3].

The aim of this research paper is to set conceptual approach for international taxation that allow fair taxation of digital businesses. In order to achieve the aim the following steps are to be taken: (i) Determination of expert findings and practical developments with respect to taxation of e-commerce platforms; (ii) Analysis of value creating factors of business models of e-commerce platforms and possible solutions for corporate income tax application; and (iii) Development of methodology for purpose of corporate income taxation on e-commerce platforms. For the purpose of the research comparative, data collection and systematization methods, as well as other scientific methods were used.

2 Literature Review

2.1 Concepts of E-Commerce Platforms

Technological innovations are the main driver for growth of the major economies. Due to new technologies and new products are developed helping to address such global challenges as environmental issues or health. Transformation of manufacturing processes as well as the way services are provided boosts productivity, creates jobs, as well as improves the global competitiveness of nation and the quality of life of citizens. The importance of innovation and its role in the growth has been much discussed in the economic literature. Also, it is confirmed by a number of empirical studies, e.g., Jacques Mairesse and Pierre Mohnen have highlighted the innovation role, that these are new production technologies or new products, in economic growth [4].

The term "digital economy" more often started to be used since 1990s, however there is still no unified definition of the term. The key aspects that differentiate businesses operating in digital economy from traditional ones are as follows:

1. Product offering—meanwhile traditional businesses offer standardized services and physical goods that require physical presence to provide or manufacture them, digital businesses offers customers digital services, which often do not require any presence in the country of the receiver of service;
2. Role of client—meanwhile in traditional businesses clients most probably would be offered standardized goods and services produced with no participation of customers, digital businesses would require users to be a part of their business operations in order to provide service or good, e.g., user participation on social networks;
3. Key assets used for generating the profit—meanwhile for the traditional businesses those would be manufacturing sites, equipment etc., digital businesses main asset would be information flows from customers.

According to definition proposed by the OECD "digital economy is an umbrella term used to describe markets that focus on digital technologies, which facilitate trade of goods and services through e-commerce" [5]. The definition itself already points out the types of companies operating in digital economy can be different—service oriented or goods oriented [6].

For the purpose of the paper, e-commerce platforms as part of digital economy will be analyzed. By definition "platforms are products, services or technologies that create value by enabling direct interactions between two or more independent customer groups" [7]. It is true that not all platforms operate the same, however, the core principles remain: (i) owners of platforms in control of intellectual property and governance; (ii) an intermediary serving as the platforms' interface with users; (iii) producers creating product offerings; and (iv) consumers using those offerings [8].

2.2 International Solutions for Taxation of E-Commerce Platforms

Corporate income taxation rules follow the general concepts as set in international tax law practice, which stipulate that the profits of international businesses are allocated between the country of residence and the country of source. Thus, the country of residence of the enterprise have the right to tax the worldwide income of the enterprise [9]. Meanwhile country of source of the profit of the enterprise poses only limited taxing rights on a territorial basis, i.e. profits arising from the income generated within the territory if significant economic presence within the territory takes place [10]. Significant economic presence is a principle developed in international taxation upon which non-resident enterprises are required to have a physical presence in the source country. It is defined either by means of a fixed place of business or of the presence of a person who is acting on behalf of the enterprise [5, 11].

Basing taxation on an enterprise conducting its activities at a fixed place is problematic for e-commerce platforms and companies alike. This has caused a loophole for tax planning, as countries of source of the income of e-commerce platforms are no longer able to trigger the significant economic presence for tax purposes. It is argued if the concept of significant economic presence should be re- characterized to minimize its association with the physical presence. Thus allowing the source country exercising its taxing rights in connection with activities effectively linked to and income derived from within its territory [3]. Several of new concepts are suggested, including defining significant digital presence, reconsidering current concept of significant economic presence, development of the concept of digital service permanent establishment [5, 12–14].

The idea of development of new concepts for e-commerce platforms is not new. As an example, an experience of the US should be noted. Already since 1967, the Supreme Court of the US has repeatedly outlined that significant presence of an enterprise should be linked to the physical presence of the enterprise rather than market exploitation or other economic considerations. However, the interpretations of the economic presence in the US had changed significantly. A great example of the change is the West Virginia Supreme Court decision on MBNA America Bank NA case in 2006, where it has adopted the most expansive interpretation of state authority to tax. The state Supreme Court decided that the physical presence was not necessary to create nexus for an out-of-state company that issued credit cards but conducted its business exclusively by mail and telephone from outside the state [15]. This was a breakthrough for the development of the economic presence nexus in the US, emphasizing the need to test both the quality and quantity of the company's economic presence. However, this concept only applies for the sales taxation purposes rather than corporate income taxation.

Nevertheless, on May 11, 2015, the Illinois Circuit Court of Sangamon County granted the Illinois Department of Revenue's cross motion for summary judgment in Capital One Financial Corporation v. Brian Hamer, Director of the Illinois Department of Revenue and held that Capital One had sufficient nexus with the state for corporate income tax purposes. The court adopted and applied the "significant economic presence" test [16, 17]. Meaning that with this recent ruling, the concept of significant economic presence was expanded and applicable for corporate taxation purposes. Similar developments for the expansion of the definition of significant economic presence are seen in other countries [5]. Thus, it can be concluded that significant economic presence for e-commerce platforms should exist when a large user base exists in the source jurisdiction, which is tested by applying different quantitative measures.

However, as studies have shown because of the rapid development of the e-commerce platforms the efforts for determination of significant economic presence for the platforms are recognized insufficient and to the larger extent income generated by businesses escapes taxation [13]. The failure to succeed on a common nexus for e-commerce platforms is mainly due to probable conflicts with supranational trade law and potential double taxation if the instrument is not introduced in a coordinated manner. Moreover, determination of the significant economic presence itself, does not solve the issue of fair corporate income taxation. Thus, when considered together, the broader tax challenges raised by the e-commerce platforms relate directly to the operation and interaction between two basic concepts that underlie the international tax rules: the rules for determining economic significant presence (nexus) and the allocation of profit [18].

When discussing how to tax and attribute respective profits to one or another jurisdiction, no universal solution can be developed. Taking into consideration interests of each and every jurisdiction it is least likely that a comprehensive solution exists. However, it is argued that the developments within the OECD within Base Erosion and Profit Shifting (hereinafter—BEPS) initiative actions as well as the latest guidance on profit attributions to permanent establishments can lead to a consensus between jurisdictions. Furthermore, on June 2018 the OECD had published the Revised Guidance on the Application of the Transactional Profit Split Method, which contains more in debt guidance of application of transactional profit spilt method (hereinafter—TPSM) [5]. The method is one of the five transfer pricing methods previously defined in the OECD Transfer Pricing Guidelines for Multinational Enterprises and Tax Administrations, most recently published in 2017 [11]. According to the guidance, TPSM focuses on analysis of value creation of the business in order to determine the main profit drivers of an enterprise. Therefore, the method is appropriate for determination of fair profit allocation of e-commerce platform for corporate income taxation purposes. This is supported by the developments in international practice, outlining that the profits of enterprises should be allocated to jurisdictions in which value creation occurs either on the supply or demand side [3, 14, 19].

There are two approaches of TPSM—contribution and residual analysis. Both approaches indicate that application of TSPM requires two steps: (1) determination of profit drivers of enterprises, and (2) weighting each profit drivers and consequently splitting profits according to the pre-defined weights [20]. Thus, to determine profit drivers for the enterprises, value chain should be analyzed. Porter had offered value chain framework that allows to analyze value creation of an enterprise. The framework suggests to identify activities of the company and further study the economic implications of those activities.

Under the value chain analysis, the primary activities and support activities are explored. Primary activities have a direct impact on value creation and involve the creation of physical products and include inbound logistics, operations, outbound logistics, marketing and sales, and service. Meanwhile support activities only impact on the performance of the primary activities [21, 22]. The support activities do not generate any revenue streams for the company, therefore authors do not consider any of these activities as profit drivers of e-commerce platforms.

2.3 Profit Drivers of E-Commerce Platforms

As noted to develop model for e-commerce platforms for corporate income taxation purposes, key profit drivers of e-commerce platforms shall be determined. It shall be noted that the patterns of defining value creation and profit drivers have undergone significant transformation. Western businesses are relying more in research and development, customer relations and services, boosting development of e-commerce platforms and shifting the value creation from physical presence to user participation. The value chain used to be linear and fit to models developed by notable economists, however now it has distorted into a circle, where customers are in the center of the value chain. From upstream sources as per traditional business models, e-commerce platforms have changed its operations to downstream sources—i.e. customers [23].

Although, that not all e-commerce platforms operate in the same manner, authors believe that general principles of operations remain the same, e.g. e-commerce platform owners are engaged in marketing and intellectual property development (e.g., IT system, webpage etc.); e-commerce platforms operate as a link between manufacturer and customer, offering a interface platform which to use for connection of the both sides.

Taking into account the two key functional aspects of e-commerce platform the profit is often created by combining algorithms for processing user data and knowledge how to use this data for sales (e.g. user can share their preference by liking page on social media, contributing value). This data from user later will be monitored, analyzed and used for advertising [24].

Additional emphasis on analysis from the demand side of value creation should be made. According to the definition the value is created where content is created and transactions are designed to generate revenue [21]. Therefore, activities that allow users to create content and the market side of interaction with users or customers need to be analyzed as they shape the revenues streams [19].

For e-commerce platforms customer orientation is critical, so one of the main tasks for such business is communication, reconfiguration and integration of its customers. That is the way how e-commerce platforms switch from product to customer approach. In addition, a core product of digital business is the generation and provision of digital content [25].

It is argued that IT infrastructure is tangible and intangible asset because products and services cannot be separated from IT infrastructure [26]. For instance, intangible assets are key value drivers for e-commerce platforms and they can take new forms other than copyrights or patents [27]. Moreover, it is critical to distinguish assets of ordinary character as they involve negligible risk between those assets with a larger contribution to value creation. According to transfer pricing ordinary asset involve minimal risk or is just internally used software or any other asset which does not generate income streams and is not used for business operations. For the e-commerce platforms data is form of an intangible asset and it is a key resource as well. Studies have revealed the importance of user base in e-commerce platforms. The main financial reason is that in e-commerce businesses, benefits occur from the use of the platforms according to network external factors [19, 17]. It is claimed that knowledge-based capital is becoming a predominant value driver in digitalization [28]. Thus, such kind of capital should have significant role in the profit-sharing substantial functional analysis.

3 Methodology

Taking into account the analysis of value chain and findings of various authors it can be fairly concluded that the significant presence of e-commerce businesses, is strived by two factors: intangible assets and users. However, under nowadays international taxation concepts e-commerce platform profits are usually only taxed in the country where intangible assets, e.g. advertising algorithms are made, rather the country of users. In thinking about a possible approach for solutions for taxation of digital businesses, the question is how much value these e-commerce businesses are deriving from user participation, and thus how much of their profit should be reallocated to jurisdictions in which users are located. The allocation keys used for determination of value of user participation is one extremely vital aspect on further development of guidance on profit split of e-commerce platforms.

In order to provide fair taxation for digital businesses TPSM should be applied. According to the findings presented previously, the following characteristics should serve as determinants for allocation of profits derived from e-commerce platforms: (i) intangible assets; (ii) user participation; and (iii) physical presence. Under contribution analysis of TPSM a weight is given to each profit driver, thus the following attribution of profits of e-commerce platforms is suggested:

1. Profit attributable to countries where intangible assets are owned, controlled and developed should amount to 50% of total profits derived by e-commerce platform. In cases where intangible assets are owned in various countries, profits should be split between those jurisdictions. For determination of the economic ownership of intangible assets guidance provided by the OECD should be used;
2. Profit attributable to countries where user participation is noted, should account for 30% of total profits derived from e-commerce platforms. In order to avoid administrative burden for taxpayers and tax administrations, only those countries where users or business partners in those countries exceed 5% of total income generated by the e-commerce platform are taken into account. In case various countries are entitled for those profits, they are split in a proportionate manner. The residual profits of the e-commerce platform are left in the jurisdiction of e-commerce platform;
3. Profit attributable to the country of e-commerce platform should account for 20% of total profit derived by e-commerce platform.

4 Results

As a result of these calculations, using variables as predefined before, corporate income tax taxable base can be calculated in various jurisdictions for e-commerce platforms.

Table below represents calculations of taxable for –ecommerce business that operates in 3 countries—X, Y, Z. The e-commerce platform is headquartered in Country X, meanwhile the main income generated by users divides as follows: (1) 40% of income is generated from country X; (2) 20% from country Y; and (3) 40% the rest of the world, where each country accounts for no more than 5% of total income streams.

Most of e-commerce business IT systems are developed and owned in country X. The total profit generated by the e-commerce platform is 100. Under current international taxation rules the e-commerce business is mainly taxed country Z, whose tax rate is 7%, resulting in tax of 7% for the group and effective tax rate of 7%. Tax rates in country X and country Y are 20% and 15% respectively.

Table below represents calculations of taxable base and tax payable according to the methodology as described above. Each country attributable share of profit is calculated taking into consideration three criterions, where total profits divided between countries should always equal to total amount of profits. After splitting the profits, tax payable is calculated for each country separately, thus resulting in fairer amount of taxable profit attributable to each country. Moreover, considering the results of calculation, effective rate of such e-commerce platform would increase to 11.61%. (Table 1).

Table 1. Paying balance between countries (by the authors).

Profit splitting criteria	Qualitative and quantitative indicators	Country X	Country Y	Country Z	Tax payable
Income generated by users (A)	40% of users	20	0	0	4
	20% of users	0	10	0	1.5
	Other users	0	0	20	1.4
	Total A	20	10	20	6.9
Ownership of core IT systems (B)	IT systems protection	3		3	0.81
	IT system development	8			1.6
	IT system maintenance		6		0.9
	Total B	11	6	3	3.31
Country of e-commerce business (C)	Total C			20	1.4
Balance (D)		31	16	43	11.61

5 Discussions

So far only very few jurisdictions have developed guidance on taxation of e-commerce platforms, hence leaving a space for development for other jurisdictions. Nevertheless, no consensus has been reached between countries on common criterions for taxing e-commerce platforms or application of TSPM. Moreover, various authors argue that traditional transfer pricing methods developed by OECD are not the most appropriate to economically more justifiable results in digital world. A model where different companies in the digital economy engage in bargaining and consider the network effects that the activities of each other would have on their financial situation in suggested. In conclusion it is emphasized more profits should be allocated to that countries where users and customers are located even if their assets and functions are limited from a traditional perspective [8, 19, 29].

In advance TPSM application would require businesses to be aware of all taxation models worldwide in order to apply each portion of profits under appropriate tax. Moreover, taxation deadlines may differ across countries. Thus, it reveals the main issue of the model—it is not efficient unless common rules for corporate income taxation are developed. Such approach is commonly discussed by the EU member states under the initiative of implication of Common Consolidated Corporate Tax Base [24, 30].

Moreover, not only corporate income taxation should be considered when talking about e-commerce platforms. Various countries have also started to implement various sales taxes on e-commerce platforms, which constitute so called value added tax as already applicable for traditional businesses. However, the main issue with respect sales tax would be how to define the service provided by e-commerce platforms. Furthermore, how to avoid double taxation arising in cases where sales tax is applied to the same income. This would require critical change of mindset and agreements between various jurisdictions [3].

6 Conclusions

In today's developing economy, economic theory as well as practice should encompass that value creation of e-commerce business occurs in a modern, less physical sense. Thus, by allowing not only to ensure the transparency of such businesses but also tax profits of such businesses in proper manner. For the purpose of common taxation of businesses methodology as described above can be used, which relies on a combination of three key factors: (i) IT system development and usage by e-commerce businesses, (ii) user participation, and to less extent, (iii) physical presence of business.

However, application of such model would require a common set of rules on multilateral level, rather unilateral. Therefore, common approaches for implementation of methodology should be analysed in more detailed, leading to changing taxation principles on a multilateral level.

References

1. European B2C ecommerce still growing fast, with national markets moving at different speeds. https://www.ecommerce-europe.eu/press-item/european-b2c-ecommerce-still-growing-fast-national-markets-moving-different-speeds/. Accessed 31 Jan 2019
2. European Commission: E-sales and turnover from e-sales, 2008 to 2016, EU28. http://ec.europa.eu/eurostat/statistics-explained/index.php/File:V1_Esales_and_turnover_from_esales,_2008_to_2016,_EU28_(%25_enterprises,_%25_total_turnover).png. Accessed 30 Aug 2018
3. Hongler, P., Pistone, P.: Blueprints for a new PE Nexus to tax business income in the era of the digital economy, pp. 10, 14, 24–25. IBFD White Papers (2015)
4. Constantinescu, L.M., Panagoret, I.: Digital economy a bridge to Romanian competitiveness from regional to global. J. Sci. Arts **17**, 491–502 (2017)
5. www.oecd.org

6. Kulcsár, T.: EU Update, Commission. 54 Eur. Taxn. 8 sec. 1.1. J. IBFD (2014)
7. Hagiu, A.: Strategic decisions for multisided platforms. MIT Sloan Manag. Rev. **55**, 92–93 (2014)
8. Van Alstyne, M.E., Parker, G.G., Choudary, S.P.: Pipelines, platforms, and the new rules of strategy. Harvard Bus. Rev. **94**, 54–62 (2016)
9. Schreiber, U.: International Company Taxation. Springer Nature America, Inc (2013)
10. Cavelti, L.U., Jaag, C., Rohner, T.F.: Why corporate taxation should mean source taxation: a response to the OECD'S actions against base erosion and profit shifting. World Tax J. IBFD **99**(3), 353 (2017)
11. www.arcoabogados.es
12. Requena, J, González, S.: Adapting the concept of permanent establishment to the context of digital commerce: from fixity to significant digital economic presence. Intertax, Int. Tax Rev. [Serial Online] **45**(11), 732–741 (2017)
13. Baéz Moreno, A., Brauner, Y.: Withholding taxes in the service of BEPS action 1: address the tax challenges of the digital economy, p. 4 (2015)
14. Olbert, M.: International taxation in the digital economy: challenge accepted? https://www.sbs.ox.ac.uk/sites/default/files/Business_Taxation/Events/conferences/2016/Doctoral_mtg_2016/olbert.pdf. Accessed 30 Aug 2018
15. Gall, M., Kulwicki, L.: West Virginia Supreme Court holds that "significant economic presence" creates Nexus for MBNA. J. State Taxn. [Serial Online] **25**(3), 5–51 (2007)
16. BPQ Partners LLC: Illinois circuit court holds that significant economic presence is proper test for corporate income tax nexus. https://gbq.com/illinois-circuit-court-holds-that-significant-economic-presence-is-proper-test-for-corporate-income-tax-nexus-2/. Accessed 31 Jan 2019
17. Wilkie, J.S.: Intangibles and location benefits (customer base). 68 Bull. Intl. Txn. 6/7 J. IBFD, 352–360 (2014)
18. www.deblislaw.com
19. www.ibfd.org
20. Ilhan, C.B.: The use of value chain analysis in a profit split. Issue: Int. Trans. Pricing J. **25**(4) (2018)
21. Amit, R., Zot, C.: Value creation in E-business. https://www.uazuay.edu.ec/bibliotecas/e-business/Value_Creation_in_E-Business.pdf. Accessed 30 Aug 2018
22. www.kppod.org
23. Biacabe, J-L., Vadcar, C.: Key points of the study creating value in a digital word changing economic action. https://www.institut-friedland.org/sites/default/files/atoms/files/etude-synthese-creationdevaleur-numerique-va-1701_0.pdf. Accessed 30 Aug 2018
24. European Commission: Fact sheet. http://europa.eu/rapid/press-release_MEMO-18-2141_en.htm. Accessed 30 Aug 2018
25. Laudon, K.C., Laudon, J.P.: Management Information Systems—Managing the Digital Firm, p. 415 (2014)
26. Pagani, M.: Digital business strategy and value creation: framing the dynamic cycle of control points. MIS Q. **37**(2), 619 (2013)
27. White, S.: Intangibles drive value in the digital age. J. Account. **221**, 21 (2016)
28. Tavares, R. J. S., Owens, J.: Human capital in value creation and post-BEPS tax policy: an outlook. 69 Bull. Intl. Txn. 10 J. IBFD (2015)
29. Pellefigue, J.: Transfer pricing economics for the digital economy. Int. Trans. Pricing J. IBFD **22**(2), 99 (2015)
30. European Commission: Questions and answers on a fair and efficient tax system in the EU for the digital single market. http://europa.eu/rapid/press-release_MEMO-18-2141_en.htm. Accessed 30 Aug 2018

Education and Training in Engineering

On Higher Education Realities: Supporting Education and Research

Yulia Stukalina[1] and Milan Pol[2(✉)]

[1] Transport and Telecommunication Institute, 1 Lomonosova Street, 1019 Riga, Latvia
Stukalina.J@tsi.lv
[2] Masaryk University, 1 Arna Nováka Street, 602 00 Brno, Czech Republic
pol@phil.muni.cz

Abstract. Providing quality higher education is a central goal related to the creation of the European Higher Education Area. Cost Action CA15221—*Advancing effective institutional models towards cohesive teaching, learning, research and writing development*—addresses the challenge of creating synergy among the increasingly more specialised and centralised supports for the above activities, offering the most advantageous models and practices for supporting teaching, learning, writing and research (TLWR).

The aim of the paper is to investigate basic elements of integrated support for TLWR in a higher education institution. To achieve the above aim the following objectives were set by the authors: (1) studying the forms and methods of centralized support for research, writing, teaching and learning, which can be used in a modern university; (2) examining what educational services, designed to facilitate effective teaching, learning, and research, are offered to students, academic and attending staff; (3) defining basic processes and knowledge transfer schemes involved in the centralized model of support for research, writing, teaching and learning employed in a contemporary higher education institution; (4) analysing how modern technologies and communication platforms allow for greater collaboration between all study process participants in the agenda of creating the common ground in terms of shared purposes, processes, knowledge, values and skills among centralized institutional supports for the four key activities in order to capitalise on their synergies.

The following research methods were used by the authors: (1) analysis of technologies and communication platforms used for supporting teaching, learning, research and writing (TLWR); (2) analysis of the main documents regulating centralized support for TLWR strategic plans, documents related to quality assurance in higher education, relevant publications, and study courses descriptions; (3) interviews with representatives of the management, research and academic staff of the university in focus.

Keywords: Educational management · Teaching · Learning · Research · Writing

I. Kabashkin et al. (Eds.): RelStat 2018, LNNS 68, pp. 749–758, 2019.
https://doi.org/10.1007/978-3-030-12450-2_72

1 Introduction

As higher education institutions (HEIs) of the EU play a key role in generating new knowledge, "shaping critical thinkers and problem-solvers", the best teaching and learning environments should motivate students to cultivate confidence in their talents, promote strong community engagement and understanding that learning is a "lifelong phenomenon" [1]. European HEIs have to contribute to the progress towards more "coherent and holistic lifelong guidance provision" [2]. This is critical for providing economic growth in the EU in the agenda of transition to the knowledge-based economy [3–6].

Cost Action CA15221—*Advancing effective institutional models towards cohesive teaching, learning, research and writing development*—addresses the challenge of creating synergy among the increasingly more specialized and centralized supports for the above activities, offering the most advantageous models and practices for supporting research, writing, teaching and learning. As there is a lack of professional conversations around the shared territory of support for the above areas, corresponding research across higher education institutions would illuminate intersections and contribute to institutional transformation based on complementary, coherent and integrated provision. The aim of the paper is to investigate basic elements of integrated support for TLWR in a higher education institution. To achieve the above aim the following objectives were set by the authors:

1. Studying the forms and methods of centralized support for research, writing, teaching and learning, which can be used in a modern university.
2. Examining what educational services, designed to facilitate effective teaching, learning, and research, are offered to students, academic and attending staff.
3. Defining basic processes and knowledge transfer schemes involved in the centralized model of support for research, writing, teaching and learning employed in a contemporary higher education institution.
4. Analysing how modern technologies and communication platforms allow for greater collaboration between all study process participants in the agenda of creating the common ground in terms of shared purposes, processes, knowledge, values and skills among centralized institutional supports for the four key activities in order to capitalize on their synergies.

The following research methods were used by the authors:

1. analysis of technologies and communication platforms used for supporting teaching, learning, research and writing (TLWR);
2. analysis of the main documents regulating centralized support for TLWR strategic plans, documents related to Quality Assurance in higher education, relevant publications, and study courses descriptions;
3. interviews with representatives of the management, research and academic staff of the university in focus. The university in focus is of a larger size (more than 30 thousand students), relatively well-established for almost a century, located in the city of almost 400 thousand people. It is relatively complex in its focus providing education in more than 200 programmes.

2 Supporting Education and Research in a Higher Education Institution

The citizens of Europe are very much interested in the quality of higher education provision, since the graduate who has received high quality teaching is expected to be more flexible, innovative, self-confident and employable [1]. Quality assurance in education is vital for supporting the creation of the European Higher Education Area [7–9]. Thus, the need for policies and appropriate systems for increasing the quality of is widely recognized in Europe; enhancing the quality of education is a central concern in the policy debate on education in the EU at different levels [10].

Innovative approaches to learning, teaching, writing and research made possible by new technologies, can complement, consolidate, support and further advance these efforts. New emerging technologies (including on-line solutions), modes of teaching and learning, individualized pedagogical approaches are now being implemented to increase the quality of European higher education and to make it more accessible in the agenda of modernising higher education [11]. The main emphasis is placed on student-cantered learning and teaching-learning processes [12]. Key higher education activities also include research, and contribution to the society [13, 14]. The research and education symbolize the two important aspects of the mission of higher education institutions, research being a "kind of the original investigation" necessary for generating new knowledge, and education being connected with the teaching and learning process aimed at developing competences and skills [15].

It should be mentioned that the "processes of effective teaching and learning" are directly linked to the "processes of effective research", research-based study being associated with innovative teaching and experiential active learning in the agenda of life-long learning [16]. The synergy between teaching and research can be considered as a joint enhancement in quality and effectiveness: teaching should be associated with the research added value; in turn, research should be associated with the teaching added value [17].

Linking research and education is a complicated process; the evolution of higher education to a great extent depends on accomplishing synergy between research and teaching-learning activities relationship, and such relationship presupposes attaining "biunique effects" [15]. In this context, there is a need of developing effective institutional models towards cohesive teaching, learning, research and writing development to address the challenge of creating synergy among the increasingly more specialised and centralised supports for teaching, learning, research and writing (TLWR) that would contribute to institutional transformation based on matching, coherent and integrated provision. Quality enhancement is supposed to be a result of using shared, high-quality learning materials and applying more creative and individualised pedagogical approaches [11].

Using new modes of learning and teaching in the agenda of creating synergy among the increasingly more specialized and centralized supports for TLRW is closely associated with the current challenges faced by higher education managers. The challenges include:

- Developing an institutional vision and leadership for more fully engaging academic staff in the potential offered by the new modes of learning and teaching [11].
- Ensuring that both academic staff and students possess digital skills needed for using modern technologies [ibid.].
- Ensuring that the quality assurance system is in place and it is applied to all types of credit-awarding provision [ibid.].
- Developing applicable tools for a methodical analysis of the research-education synergy effects [15].
- Promoting innovative teaching-learning methods to contribute to the enhancement of the research skills of students and academic staff [ibid.], etc.

3 Discussion: Investigating Basic Elements of Integrated Support for TLWR in a Higher Education Institution

3.1 Main Issues Discussed in Relation to Integrated Support for TLWR in a HEI

For investigating basic elements of integrated support for TLWR in the university in focus, the following issues were discussed in relation to the objectives set by the authors of the paper.

1. *Issues discussed in relation to Objective 1:*

   - How is integrated support for TLWR performed (at what level—University/faculties/departments, etc.)?
   - What kind of documentation regulates the process—strategic plans, guidelines, quality manuals, standards, etc.?
   - What kind of staff is involved in the process?
   - What kind of training and development programmes is offered for the academic staff in the context of TLWR support?
   - How are training needs identified?
   - What is the role of study programme and its director in the process?
   - How are synergies between specific areas of support created?

2. *Issues discussed in relation to Objective 2:*

   - Is there a special unit supporting TLWR?
   - What areas are covered (international education and grants; guidance and information on the University's study programmes and courses; University's strategy and policies; engagement in research activities; staff issues; University-wide communication; career guidance and counselling services, etc.)?
   - How synergy is created through the established communication models in relation to TLWR?
   - How are students with any type of physical, sensorial, psychological disabilities supported?

3. *Issues discussed in relation to Objective 3:*

   - What communication models and knowledge transfer schemes are used at different levels—University/faculty/departments/individuals?
   - What kind of structured (planned) and spontaneous knowledge transfer schemes (staff and faculty meetings, newspapers, newsletters, social spaces, mentoring, social media, etc.) are employed for supporting TLWR?
   - How is feedback from students/staff obtained?
   - How is collaboration between faculties/departments/teams/individuals ensured to create synergies in relation to TLWR?
   - How are decisions made at the University? Is this process evidence-based?
   - How is necessary information collected and stored?
   - Are TLWR strategies and policies in line with institutional policies and strategies?
   - Is there a balance between central provision and departmental autonomy?
   - What is the role of the Quality Management System in supporting research, writing, teaching and learning?

4. *Issues discussed in relation to Objective 4:*

   - What is the role of the University's Learning Management System (LMS) in supporting TLWR activities?
   - What is the role of the University intranet in managing communication within the University and inspiring social negotiation as an important part of these activities?
   - Does technology play an essential role in supporting TLWR at different levels?

The results of the discussion performed in the framework of the research are presented below.

3.2 Documents Regulating Integrated Support for TLWR in a HEI

Today, the need for long-term planning and creation of strategies for higher education is widely acknowledged in the EU countries [18]. The strategic approach to managing a HEI allows developing an educational organization holistically [19], which enables creating synergies among various organizational units.

The research conducted by Stukalina [20] shows that strategic planning in a HEI involves developing strategic goals and area-specific strategic objectives aimed at achieving excellence in different functional domains. These domains include research, education, staff and infrastructure [21]. Strategic objectives are reflected in strategic plans (developed by the HEI and its units), and documents related to Quality Assurance. Contemporary universities focus their efforts on sustainability, innovation, internationalization, lifelong learning, dissemination of research, inter-institutional and international cooperation, community engagement, establishing internal quality assurance systems, etc. [20]. Thus, support for TLWR is integrated in their strategic plans; this support is carried out at different levels and across different units of a HEI. Some strategic priorities and related measures and activities aimed at supporting TLWR are presented in Table 1.

Table 1. Strategic priorities as identified in the strategic plan of a HEI: integrated support for TLWR (based on the Strategic Plan of the University in focus).

Strategic priority	Institutional measures and activities (University's level)	Decentralized measures and activities (Faculty's/Department's level)
1. Quality in education	1. Introducing an internal quality assurance (QA) system Develop advanced communication tools to ensure interactive teaching	1. Introducing technological innovation in teaching activities and promoting the use of IT tools Introducing new forms/methods of education for diverse target groups
2. Research excellence and relevance/research efficiency and a stimulating research environment	2. Developing internal grant schemes supporting teams focusing on research priorities Developing project support and associated services Optimizing the collection of electronic information sources accessible online Securing information services support for using research grants	2. Introducing support tools for improving research potential utilization Developing project support and associated services Implementing tools designed to stimulate the research performance of individuals and teams Promoting knowledge sharing and mutual awareness
3. Organizational culture based on shared values	3. Developing an internal communication concept using modern communication tools	3. Implementing cultural, social, sports and leisure-time programs for students and employees
4. Inspiration and social responsibility	4. Popularizing scientific knowledge and organizing educational activities for the general public (a museum, "University for children", Support centre for students with special needs, etc.)	4. Creating an electronic newsletter for the academic community, innovating the university website, employing digital marketing tools for informing the public about university activities
5. Personnel management and employee development	5. Developing and implementing an internal employee education program Developing instruments for assessing the work performance and quality of both academic and non-academic employees	5. Establishing a unit dedicated to supporting the development of educational/other competences Providing adequate counselling and additional services for teachers when introducing technological teaching innovations
6. Information systems and IT support	6. Creating a shared work/communication platform Continuing the digitization of administrative processes and streamlining internal document flow	6. Developing a long-term document/data storage concept Developing a suitable sensitive data processing environment and identifying critical IT infrastructure components

3.3 Integrated Support for TLWR: Creating Synergy Across a HEI

As seen from the above table, support for TLWR should encompass students and academics engaged in the relevant activities throughout the HEI.

The following elements of centralized support for TLWR in the university in focus are integrated in the university's management scheme.

- *Evidence-based support for TLWR*

The activities associated with TLWR are among strategic priorities of the university. Support for TLWR is implemented in an evidence-based manner in the agenda of Quality Assurance, the completion of an internal quality assurance system being a key challenge for the administration and academic staff. The support is based on the empirical data collected from students and staff on a regular basis (including staff self-evaluation reports). The obtained information is stored in the organizational intranet, and is available to academic staff and administration).

- *Keeping balance between central provision and departmental autonomy*

There should be a balance between central provision and accountability, and departmental autonomy. This is very important for creating the so-called "systemic synergy". Sufficient cross-departmental linkages are established to support university-wide communication, collaboration and knowledge exchange in the agenda of supporting TLWR. These linkages are established at different levels of the HEI (university/faculty/department) through formal structures that span departmental boundaries. In addition to these formal structures, informal means of communication and collaboration facilitate cross-departmental linkages.

- *Knowledge transfer models used for creating synergetic effects*

Structured and spontaneous knowledge transfer models supporting integrated TLWR activities include University/faculty/department/Senate meetings, social spaces, social media, etc. Tight cross-departmental linkages allow creating synergetic effects throughout the HEI. Synergetic overlaps between the different functional units and areas of support are supposed to be beneficial for all stakeholders of the University. In the University, there are centralized offices, where specialized staff with expertise in TLWR is working. Such structures provide support for academics in the TLWR related fields, as well as guidance and counselling to students. These units are as follows:

- Centre of Academic Staff Development,
- Quality Assurance Centre,
- Research and Development Centre,
- Centre for Information Technologies,
- Central Library,
- Experimental Humanities Laboratory,
- Centre for the Advancement of Research in the Liberal Arts,
- Office for Doctoral Studies and Advanced Master's State Examinations,
- Career Centre,
- Support Centre for Students with Special Needs,
- Office for Studies,
- Centre for International Cooperation, etc.

- *Staff's professional and individual development*

Academics assigned with both teaching and research, should be provided with the possibility of individual development and growth regarding their competencies. To ensure sufficient proficiency of the academic staff regular staff training is organized at the university level in cooperation with every faculty; it includes qualification improvement courses. So qualifications of faculty members are supposed to be developed continuously throughout their academic careers, publishing in internationally recognized scientific journals (indexed in Scopus, WoS, etc.) and presenting at international scientific conferences being supported by the faculties' budgets.

- *Incorporating modern technologies in the educational process*

Support for TLWR is also implemented through incorporating modern technologies in the educational process. The role of the Learning Management System and intranet (developed by the university's IT specialists) in supporting communication, interdisciplinary learning, transdisciplinary cooperation and related TLWR activities can not be underestimated. Modern technologies and solutions enable accessible, blended and collaborative learning and research environments for both students and academic staff. They include a wide range of features and functionalities for supporting synchronous and asynchronous communication; this combination seems quite efficient to optimally support student and staff engagement.

4 Conclusions

This paper has discussed some issues related to ensuring integrated support for TLWR in a HEI in the context of providing quality higher education. The research conducted in the frame of Cost Action CA15221—*Advancing effective institutional models towards cohesive teaching, learning, research and writing development*—has allowed the authors to identify and describe essential elements of integrated support for TLWR in a HEI.

These elements include, among other things, evidence-based support for TLWR, keeping balance between central provision and departmental autonomy, using knowledge transfer models to generate synergetic effects across the university, academic staff's professional and individual development, incorporating current technologies in the educational process.

The support for TLWR addresses both educational and research activities in the frame of quality assurance in higher education. It is aimed at accomplishing synergistic effects across the university, and is reflected in the regulatory documents that define the basis of the relationship between different functional units. The coordinated support is data-driven, and tracks the needs of students and academic staff. The university is supposed to be committed to supporting students' academic and personal development throughout their studies, as well as staff's professional development. Integrated support for TLWR is associated with the creation of a constructive educational environment and enhancement of academic work through attaining synergy between teaching/learning and research, cultivating personalized and customized teaching and learning practices, improving students' academic achievement and staff's performance.

This study has some limitations mainly related to the research base. A future study with a broader research base is recommended to verify and generalize the findings.

Acknowledgements. This research is part of a COST project (COST Action) that is surely carried out according to ethical policies.

References

1. Report to the European Commission on improving the quality of teaching and learning in Europe's higher education institutions, High Level Group on the Modernisation of Higher Education (2013). http://ec.europa.eu/dgs/education_culture/repository/education/library/reports/modernisation_en.pdf. Accessed 18 May 2018
2. Lifelong Guidance across Europe: Reviewing Policy Progress and Future Prospects: Working Paper No. 11. Cedefop: Luxembourg, Publications Office of the European Union (2011). http://www.google.lv/url?sa=t&rct=j&q=&esrc=s&source=web&cd=3&ved=0CC8QFjAC&url=http%3A%2F%2Fwww.cedefop.europa.eu%2Ffiles%2F6111_en.pdf&ei=3edZVfa4IouWsgHXiI-GoDA&usg=AFQjCNFKUbK4euUYXvM7cR7A9CK33DEjdQ&bvm=bv.93564037,d.bGg. Accessed 12 May 2018
3. COM (2003) 58 final: The role of the universities in the Europe of knowledge. Communication from the Commission of 5 February 2003 (2003). http://europa.eu/legislation_summaries/education_training_youth/lifelong_learning/c11067_en.htm. Accessed 10 May 2018
4. The European Higher Education Area in 2012: Bologna Process Implementation Report (2012). http://www.ehea.info/Uploads/(1)/Bologna%20Process%20Implementation%20Report.pdflast. Accessed 10 May 2018
5. Report to the European Commission on Improving the Quality of Teaching and Learning in Europe's Higher Education Institutions (2013). http://ec.europa.eu/education/library/reports/modernisation_en.pdf. Accessed 11 May 2018
6. Modernisation of Higher Education in Europe: Access, Retention and Employability: Eurydice Report. Education, Audiovisual and Culture Executive Agency (2014). http://eacea.ec.europa.eu/education/eurydice/documents/thematic_reports/165EN.pdf. Accessed 07 May 2018
7. Bologna Declaration. Joint Declaration of the European Ministers of Education (1999). http://www.ond.vlaanderen.be/hogeronderwijs/Bologna/documents/MDC/BOLOGNA_DECLARATION1.pdf. Accessed 15 Apr 2018
8. London Communiqué. Towards the European Higher Education Area: Responding to Challenges in a Globalised World (2007). http://www.ond.vlaanderen.be/hogeronderwijs/bologna/documents/MDC/London_Communique18May2007.pdf. Accessed 17 May 2018/03/17
9. Standards and Guidelines for Quality Assurance in the European Higher Education Area. EURASHE, Brussels (2015). http://www.enqa.eu/wp-content/uploads/2015/11/ESG_2015.pdf. Accessed 11 May 2018
10. European Commission/EACEA/Eurydice, Assuring Quality in Education: Policies and Approaches to School Evaluation in Europe. Eurydice Report. Publications Office of the European Union, Luxembourg (2015). http://eacea.ec.europa.eu/education/eurydice/documents/thematic_reports/178en.pdf. Accessed 21 May 2018

11. Report to the European Commission on New Modes of Learning and Teaching in Higher Education, High Level Group on the Modernisation of Higher Education (2014). http://ec.europa.eu/dgs/education_culture/repository/education/library/reports/modernisation-universities_en.pdf. Accessed 15 May 2018
12. Tremblay, K., Lalancette, D., Roseveare, D.: The rationale for an AHELO: higher education in the 21st century context (Chap. 1). In: Assessment of Higher Education Learning Outcomes. Feasibility Study Report. Design and Implementation, vol. 1. OECD (2012). http://www.oecd.org/education/skills-beyond-school/AHELOFSReportVolume1.pdf. Accessed 11 May 2018
13. Watson, D.: Managing Strategy. Open University Press, Buckingham (2000)
14. Veugelers, R., Del Rey, E.: The contribution of universities to innovation, (regional) growth and employment. EENEE Analytical Report No. 18, January 2014 (prepared for the European Commission). European Expert Network on Economics of Education (EENEE) (2014). http://www.google.lv/url?sa=t&rct=j&q=&esrc=s&source=web&cd=3&ved=0ahUKEwiJ2Mrc-7vOAhUMM5oKHdUrAXkQFggpMAI&url=http%3A%2F%2Fwww.eenee.de%2Fdms%2FEENEE%2FAnalytical_Reports%2FEENEE_AR18.pdf&usg=AFQjCNEKQqRHJLqj8vXOsxkj4b2o3_kkOQ. Accessed 19 May 2018
15. Helerea, E., Popescu, M., Coman, C.: Enhancing Academic Research & Education Synergy—Transilvania University (Chap. 9). In: Talaba, D., ten Thij, H. (eds.) Teaching and Research Synergy in the Context of University-Industry Cooperation, pp. 76–83. ZkP—Chevalier de Seyn Publishers, Eindhoven (2007)
16. Fat, S.: Designing the Curriculum—transmitting research based academic knowledge (Chapt. 8), In: Talaba, D., ten Thij, H. (eds.) Teaching and Research Synergy in the Context of University-Industry Cooperation, pp. 66–74. ZkP—Chevalier de Seyn Publishers, Eindhoven (2007)
17. Talaba, D.: Teaching-Research Synergy and University-Industry cooperation in the knowledge based society (Chap. 1). In: Talaba, D., ten Thij, H. (eds.) Teaching and Research Synergy in the Context of University-Industry Cooperation, pp. 2–13. ZkP—Chevalier de Seyn Publishers, Eindhoven (2007)
18. Higher Education Governance in Europe: Policies, structures funding and academic staff. Eurydice (2008). http://eacea.ec.europa.eu/education/eurydice/documents/thematic_reports/091EN.pdf. Accessed 25 May 2018
19. Kettunen, J.: Competitive strategies in higher education. J. Inst. Res. **11**(2), 38–47 (2002)
20. Stukalina, Y.: Some issues associated with strategy formulation and strategic planning in a contemporary university. In: The European Proceedings of Social & Behavioural Sciences, vol. XXXIV, pp. 1–11 (2017). http://dx.doi.org/10.15405/epsbs.2017.12.02.1
21. Stukalina, Y.: Strategic management of higher education institutions. Manag. Org.: Syst. Res. **70**, 80–90 (2014)

Developing a Marketing Strategy for a Higher Education Institution in the Agenda of Customer–Driven Education

Olga Zervina(✉) and Yulia Stukalina

Transport and Telecommunication Institute, 1 Lomonosova street, LV–1019 Riga, Latvia
{zervina.o,Stukalina.J}@tsi.lv

Abstract. Modern higher educational institutions (HEIs) offer exciting, creative programs for thousands of students worldwide. The higher education industry is designated by development in creativity, training and employability for qualifications and degrees. Without using competitive marketing strategies, student enlistment numbers often fight to meet and surpass expectations. That is the reason advanced education organizations are employing creative marketing experts to maintain their foundations and studies stand-out and attractive. Universities are moving past such traditional methods as magazines and leaflets; they are rather pushing assets towards speaking to the always developing computerized advertising scene. There, they can have more extensive reach and more noteworthy potential in quickly expanding their program's comprehension. There's some specific challenges higher education marketers face today, for instance, limited marketing resources, media overloading, image based on previous information, etc. Moreover, every year university applicants start a new lifecycle. Consequently, higher education market experts have to face a new target audience. That is why it is important to define the latest marketing strategies and trends, while re-evaluating the foundations to ensure the university sends the right message to the right audience. In order to create a competitive market position for a higher education institution in such circumstances it is necessary that it adopts marketing concept and philosophy and creates its strategy and activities in terms of market performance. The aim of the paper is to examine some marketing strategies to be applied in a higher education institution in the agenda of customer-driven education. The following objectives were set by the authors: (1) to identify modern trends and challenges in the higher education industry in the context of developing an efficient marketing strategy; (2) to discuss some marketing strategies to be used in a contemporary university in the agenda of customer-driven education. The following research methods were used by the authors: (1) review of theoretical literature and EU documents on the research topic; (2) analysis of the open-source marketing reports of some universities.

Keywords: Higher education · Competitive position · Marketing tools · Quality

I. Kabashkin et al. (Eds.): RelStat 2018, LNNS 68, pp. 759–767, 2019.
https://doi.org/10.1007/978-3-030-12450-2_73

1 Introduction

Contemporary higher educational institutions (HEIs) offer exciting, creative programs for thousands of students worldwide. The higher education industry is designated by development in creativity, training and employability for qualifications and degrees. As stated in *UNESCO Tertiary Education Data* [1], the number of tertiary students doubles every 15 years. That is why, according to Driscoll and Wicks, many universities and colleges practice a "market-in" approach, with the satisfaction of student wants and needs being paramount; students are increasingly being referred to as customers, and education is the product that is being offered [2]. This is what the term "customer-driven education" presupposes. And without using competitive marketing strategies, student enlistment numbers often fight to meet and surpass expectations. Under the efficient marketing strategies one can understand the strategies that lead to the maximum profit potential. That is the reason advanced education organizations are employing creative marketing experts to maintain their foundations and studies stand-out and attractive. Remarkable advanced education promoting methodologies help bring issues to light while enrolling new students and attracting more resources. Universities are moving past such traditional methods as magazines and leaflets; they are rather pushing assets towards speaking to the always developing computerized advertising scene. There, they can have more extensive reach and more noteworthy potential in quickly expanding their program's comprehension. There's some specific challenges higher education marketers face today, for instance, limited marketing resources, media overloading, image based on previous information, etc.

Moreover, every year university applicants start a new lifecycle. Consequently, higher education market experts have to face a new target audience. That is why it is important to define the latest marketing strategies and trends, while re-evaluating the foundations to ensure the university sends the right message to the right audience. In order to create a competitive market position for a higher education institution in such circumstances it is necessary that it adopts marketing concept and philosophy and creates its strategy and activities in terms of market performance. The most important aspect of an effective strategy is making sure it aligns with the overall marketing and recruiting goals; a good strategy should also include a combination of digital activities [3].

The aim of the paper is to examine some marketing strategies to be applied in a higher education institution in the agenda of customer-driven education. The following objectives were set by the authors:

1. To identify modern trends and challenges in the higher education industry in the context of developing an efficient marketing strategy.
2. To discuss some marketing strategies to be used in a contemporary university in the agenda of customer-driven education.

The following research methods were used by the authors:

1. Review of theoretical literature and EU documents on the research topic.
2. Comparison of the open-source marketing reports of some universities. Five most popular and top-ranked American and European universities were randomly selected.

2 Modern Trends and Challenges in the Higher Education Industry in the Context of Developing an Efficient Marketing Strategy: Focus on Quality

The citizens of Europe are willing to invest into their education more than ever in the history. Moreover, European educational institutions tend to become lifelong service providers. Lifelong learning has become an essential element of the European Higher Education Area [4–7]. According to Eurostat report [8], participation in lifelong learning education has increased for 3.7% in 2002–2016 (Fig. 1). The EU's ET 2020 framework [9] also includes a benchmark that aims to raise the share of adults participating in learning to at least 15%.

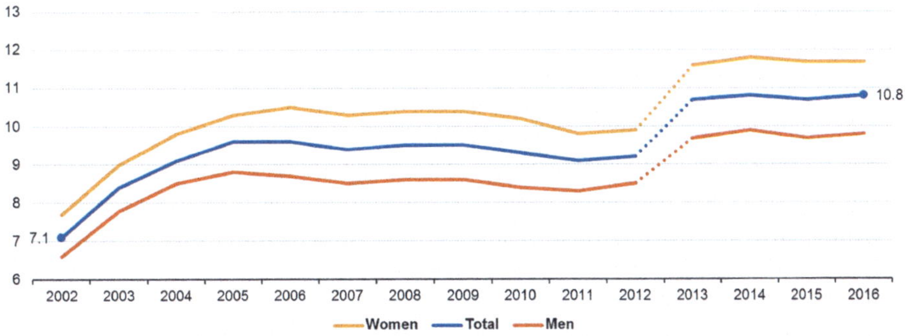

Note: Breaks in time series in 2003, 2006 and 2013. Lifelong learning refers to persons aged 25 to 64 who stated that they received education or training in the four weeks preceding the survey (numerator). The denominator consists of the total population of the same age group, excluding those who did not answer to the question 'participation in education and training'.

Fig. 1. Participation in lifelong learning, EU–28, 2002–2016 [9]

The role of the universities in the "Europe of knowledge" is closely related to increasing public responsibility for higher education systems across the EU, stimulating equity and access, as well as delivering excellence at different levels: local, regional, national and international [10]. In a fast changing higher education sector, the number of higher education providers (profit and non-profit) is growing [11]. Thus, the main concern is on the quality of higher education provision [12, 13], quality assurance in higher education being in the focus of educational managers' efforts [6, 14–16]. The value and reputation of higher education are based on quality and applied standards [11].

In view of the aforesaid, to show the quality of the education should be the top priority of the universities' marketing strategies.

Before making a decision what school to choose students are subject to a precise marketing politics on the educational field. Today, the advertising activities manager demand, supply and cost of products and services. It likewise takes care of the clients' wishes. Promoting is additionally a wellspring of data for both the organizations and buyers. It gives essential related data with respect to request, supply, and rivalry of items other than victories and disappointments of item in the market. In light of that the organizations' exercises are arranged in better way. Showcasing work force serves the customers in proficient way and deal with them by giving administrations through dispersion, estimating, credit, client benefits resultantly mean consumer loyalty. The organizations procure more benefits as a result of better advertising procedures. The advertising methodologies additionally guarantee smooth coordination of business in the chosen field. But it is still based firstly on the market analysis, as the marketing strategy "lays out target markets and the value proposition" to be offered on the basis of "an analysis of the best market opportunities" [17]. Marketing strategies used in a contemporary university in the agenda of customer-driven education are aimed at improving the competitive position of a higher education institution in the modern turbulent environment.

3 Marketing Strategy Development in Higher Education

Educational marketing can be analysed from different perspectives including a social perspective and a "services marketing perspective", which will "lead to specific ways of marketing strategies implementation" [18].

However, it should be noted that, in the higher education sector, marketing strategies (as all other marketing-related activities) are subject of the combination of marketing activities also known as marketing mix. It consists of four Ps (can be also referred as four Cs): Product (customer value), Price (Cost), Place (Convenience) and Promotion (Communication) [17].

In the rapidly changing marketing environment, advertisers face two main challenges: sky-rocket speed of e-marketing growth and globalization. That means changes and challenges to the traditional marketing mix for higher education institutions. Such product as educational programs is evaluated by increasing numbers of international students. They want to see the education that can be adoptable internationally and not for the local market only.

Price is the most conservative part of the educational marketing mix. Though, it also undergoes some changes: students can evaluate the opportunity costs such as giving up job or moving to a university area. To avoid that they can choose e-learning instead of full or part time studies.

The place element in higher education is related to creating service availability and accessibility [19]. Place or premiums part of marketing mix in the field of higher education consists of convenient and promising location, university premises, affordable accommodation, international exchange opportunities, and technological facilities available for classes. Each point from this list can be evaluated by prospective students online and be compared with other universities.

Promotion in higher education is used to stay in touch with all relevant stakeholders —students, employers, academics, etc. [18]. Promotion in the higher education sector has been influenced by the online era the most. Traditional communication with prospective students such as press advertising, personal publicity was changed almost in every aspect. Instead of direct mail and press publishing we use electronic media. Personal communication is now using streaming video, online chats and social media.

Higher education marketing strategies include the identification of consumer needs concerning the type and structure of study programmes, as the identity of a contemporary university is created based on both the nature and quality of its programmes [18]. Developing a marketing strategy modern higher education institution may go through the following stages [20]: (1) collecting necessary data by means of a few simple marketing tools such as questionnaires and interviews, and creating a special database; (2) carrying out the market segmentation and assessing the market capacity; (3) assessing the potential of the university and its competitors; (4) analyzing the external factors (macro-environment factors) influencing the university's operation.

4 Comparison of Marketing Plans of Some Modern Universities

The research performed by the authors includes the comparison of marketing plans developed by five modern universities operating in different parts of the globe; the research is based on their open-source marketing reports. Table 1 describes strategic priorities reflected in these marketing plans.

In Table 2, various marketing procedures and tools used by the five universities are described.

Table 1. Strategic objectives as reflected in marketing plans of five modern universities [21–25]

Kennesaw State University	University of Chester	South Dakota State University	University of Manchester	Stockholm University
–Improving reputation of the University –Identifying and targeting key internal and external customers –Positioning the University as an innovative institution –Creating awareness to the business community and establishing a funding opportunity for all University's programs	–Focusing on reputation management –Intensifying market research –Focusing on stakeholders relationship building –Market research underpinning –Benchmarking, monitoring and review	–Promoting academic excellence through quality programs and innovative teaching/learning environment –Encouraging innovations and promoting creativity to contribute to the public good –Extending the reach and depth of the university through strategic programs and collaborations –Securing sufficient human and fiscal resources	–Enhancing academic reputation –Generating "pride and engagement" –Focusing on outputs of the educational process	–Strengthening research and academic excellence –Improving international/national collaboration –Focusing on continuing professional development –Ensuring administrative and operational support at all levels

Table 2. Marketing procedures and tools as reflected in marketing plans of five modern universities [21–25]

Kennesaw State University	University of Chester	South Dakota State University	University of Manchester	Stockholm University
–Creating a magazine/a series of videos/a newsletter/events website calendar –Launching a student social media campaign –Creating a campaign to increase awareness of innovative programs by publishing informative articles –Submitting media alerts to select outlets to promote program recognition –Sub–scribing to Google analytics to maximize search engine recognition –Developing a speakers bureau –Implementing pilot program "corporate access to student talent" –Implementing the "corporate tuition assistance program" to increase enrolment	–Enhancing comprehensive qualitative/quantitative market research and creating a strong evidence base for developing marketing plans –Collecting alumni feedback –Providing a web-site/marketing materials for encouraging target audiences to engage with the University's brand identity and values –Applicant tracking/Collecting applicant feed-back –Collecting competitor data –Tracking digital activity and response rates –Providing individual course market information –Creating a photo library –Performing statistical analyses	–Employing news promotions, publications, web and other electronic communications –Creating and supporting a positive campus experience to encourage student enrolment and retention –Developing effective research tools for engaging and motivating target audience –Maintaining and enhancing the university's brand identity –Utilizing research-based decision-making processes to inform media and message decisions for promoting –Exploring and adopting new technologies and project management systems to maximize available resources and allocating funding streams	–Communicating the three core goals (more distinctive offering): world-class research/outstanding learning and student experience/social responsibility –Enhancing the power of stakeholders (better engagement) –Content marketing and digital marketing (more effective content production and channel selection) –Campus as University's "image"(more effective use of our physical assets and spaces) –Professional communications and marketing (more effective use of resources)	–Stimulating the appointment of student representatives to various bodies –Signing additional central collaboration agreements –Supporting newly arrived international students –Monitoring the quality and cost of the University's housing service –Creating pilot programmes in English at the bachelor's level –Developing criteria and a review process for promotions –Developing and monitoring the recruitment of academic leaders at various levels –Improving the technological infrastructure –Evaluating the University's financial model

As seen from the Tables 1 and 2, similar strategic objectives are formulated in marketing strategies of different universities in the agenda of accomplishing competitive advantage in the international education market, though they are set with regard to the national or regional context. Contemporary universities all care about reputation, brand identity, quick responsiveness to market changes, use of online resources, developing their students' professional awareness.

To reach strategic objectives universities choose the following tools: message of connection, visual identity for brand recognition, specially designed programs (e.g. "CAST—Corporate Access to Student Talent"), a model that guarantees the core operations' involvement in the administration of University-wide IT systems, tracking digital activity and response rates.

It should be mentioned that all five universities show common awareness about brand recognition, online resources development, study outcome track and evidence-based decision models. They are planning to invest in qualitative and quantitate market research, campus development, training programmes, improving IT-based infrastructure, digital marketing, social media campaigns, etc.

Every university provides its information in the English language. It is an obvious message that these universities welcome international students. Some universities have also claimed supporting newly arrived international students as a strategic goal.

5 Conclusions

This paper has examined current marketing strategies to be implemented in a higher education institution in the agenda of customer-driven education. Modern higher education systems have to meet some competitive requirements. Contemporary universities must "stay closer" to their customers responding to their needs, proposing high-quality educational services, improving service delivery, and building long-term relationships For this purpose, adopting high quality standards is vital.

The authors have identified noteworthy trends in the higher education industry in the context of developing an efficient marketing strategy, sky-rocket speed of e-marketing growth and globalization being among the most influential ones. The impact of these challenges cannot be underestimated as it affects the choice of a marketing strategy directly. That influence is reflected in the comparison of marketing plans of five modern universities.

There are various marketing techniques and tricks incorporated into a marketing strategy of a modern higher education institutions for promoting their educational services. The marketing tools used by the universities under review are all based on above mentioned trends; first and foremost, they include focus on digital resources, brand name recognition and international target audience.

References

1. UNESCO Tertiary Education Data (2014). http://uis.unesco.org/en/topic/higher-education
2. Driscoll, C., Wicks, D.: The customer–driven approach in business education: a possible danger? (2010) https://www.tandfonline.com/doi/pdf/10.1080/08832329809601663?needAccess=true

3. Kendall, D.: Higher Ed Marketing Leaders Share Their Top Digital Strategy Tips (2014). https://www.workzone.com/blog/digital–marketing–higher–education/
4. The Bologna Process 2020—The European Higher Education Area in the New Decade. Communiqué of the Conference of European Ministers Responsible for Higher Education, Leuven and Louvain–la–Neuve, 28–29 April 2009 (2009). http://media.ehea.info/file/2009_Leuven_Louvain–la–Neuve/06/1/Leuven_Louvain–la–Neuve_Communique_April_2009_595061.pdf. Accessed 01 June 2018
5. Bucharest Communiqué: Making the Most of Our Potential: Consolidating the European Higher Education Area (2012). http://media.ehea.info/file/2012_Bucharest/67/3/Bucharest_Communique_2012_610673.pdf. Accessed 10 June 2018
6. London Communiqué. Towards the European Higher Education Area: Responding to Challenges in a Globalised World (2007). http://www.ond.vlaanderen.be/hogeronderwijs/bologna/documents/MDC/London_Communique18May2007.pdf. Accessed 10 June 2018
7. Lifelong Guidance Across Europe: Reviewing Policy Progress and Future Prospects: Working Paper No. 11. Publications Office of the European Union, Luxembourg Cedefop (2011). http://www.google.lv/url?sa=t&rct=j&q=&esrc=s&source=web&cd=3&ved=0CC8QFjAC&url=http%3A%2F%2Fwww.cedefop.europa.eu%2Ffiles%2F6111_en.pdf&ei=3edZVfa4IouWsgHXiIGoDA&usg=AFQjCNFKUbK4euUYXvM7cR7A9CK33DEjdQ&bvm=bv.93564037,d.bGg. Accessed 12 May 2018
8. EU Strategic Framework – Education and Training 2020 (2009). http://ec.europa.eu/education/policy/strategic–framework_en. Accessed 12 May 2018
9. Eurostat. Participation in lifelong learning, EU–28. 2002–2016 (2017). http://ec.europa.eu/eurostat/statistics–explained/index.php?title=File:Participation_in_lifelong_learning,_EU–28,_2002–2016.png&oldid = 344834. Accessed 17 May 2018
10. COM (2003) 58 final. The role of the universities in the Europe of knowledge, Communication from the Commission of 5 February 2003 (2003). http://europa.eu/legislation_summaries/education_training_youth/lifelong_learning/c11067_en.htm. Accessed 10 May 2018
11. How Do We Ensure Quality in an Expanding Higher Education System? University Alliance, London (2014). https://www.unialliance.ac.uk/wp–content/uploads/2014/05/UA_Quality_Final.pdf. Accessed 11 June 2018
12. The European Higher Education Area in 2012: Bologna Process Implementation Report (2012) http://www.ehea.info/Uploads/(1)/Bologna%20Process%20Implementation%20Report.pdflast. Accessed 05 June 2018
13. Report to the European Commission on Improving the Quality of Teaching and Learning in Europe's Higher Education Institutions, High Level Group on the Modernisation of Higher Education (2013). http://ec.europa.eu/dgs/education_culture/repository/education/library/reports/modernisation_en.pdf. Accessed 11 June 2018
14. Bologna Declaration. Joint Declaration of the European Ministers of Education (1999). http://www.ond.vlaanderen.be/hogeronderwijs/Bologna/documents/MDC/BOLOGNA_DECLARATION1.pdf. Accessed 05 June 2018
15. Standards and Guidelines for Quality Assurance in the European Higher Education Area. EURASHE, Brussels (2015). http://www.enqa.eu/wp–content/uploads/2015/11/ESG_2015.pdf. Accessed 11 June 2018
16. Standards and Guidelines for Quality Assurance in the European Higher Education Area. EURASHE, Brussels (2015). http://www.enqa.eu/wp–content/uploads/2015/11/ESG_2015.pdf. Accessed 02 June 2018
17. Kotler, P., Keller, K.: Marketing Management, 14th edn. Pearson, London (2011)
18. Filip, A.: Marketing theory applicability in higher education. Procedia—Soc. Behav. Sci. **46**, 912–916 (2012)

19. Kotler, Ph, Fox, K.: Strategic Marketing for Educational Institutions, 2nd edn. Prentice-Hall, Englewood Cliffs (1995)
20. Cherkunova, N.G.: The formation of marketing strategy of the higher educational institutions to increase their efficiency. Int. J. Econ. Financ. **6**(S2), 37–42 (2016)
21. University of Chester Marketing Strategy 2013–2015. http://www.chester.ac.uk/sites/files/chester/Document9_0.pdf
22. South Dakota State University. Impact 2018. https://www.sdstate.edu/sites/default/files/AD314%20–%20ImpactBook.pdf
23. Manchester University. Communications and Marketing Plan. http://documents.manchester.ac.uk/display.aspx?DocID=25256
24. Kennesaw State University. Strategic Marketing Plan. http://uc.kennesaw.edu/resources/resources_docs/uc_marketing_plan_2015_2017.pdf
25. Strategies for Stockholm University 2015–2018. https://www.su.se/polopoly_fs/1.220898.1496237800!/menu/standard/file/Strategies.pdf

E-Learning and E-Teaching Effectiveness: Academic Staff Perception

Jekaterina Bierne[1], Anna Svirina[2], and Jelena Titko[1(✉)]

[1] University College of Economics and Culture, Riga LV1019, Latvia
katrina.bierne@gmail.com, jelena.titko@eka.edu.lv
[2] Kazan National Research Technical University named after A.N. Tupolev, Kazan 420111, Russia
anna_svirina@list.ru

Abstract. Implementation of e-learning practices requires from teachers not only digital skills, but also motivation to work in e-environment. The overwhelming goal of the planned large-scale research is to determine e-teaching value perceived by the academic staff members representing higher education institutions in Latvia and other countries. The goal of the current paper is to reflect the results of the authors' conducted pilot survey, using the authors' developed questionnaire. The authors focused on the factors affecting e-teaching effectiveness, perceived by the academic staff members. 38 academic staff members representing the University College of Economics and Culture participated in the survey. The respondents were offered to evaluate nine predetermined factors according to their perceived importance. Data has been processed in SPSS 24.0 environment with application of Multilayer Perceptron (MLP) analysis. The analysis revealed that the most important characteristics that has an influence on respondents' perception of factors effecting e-teaching effectiveness was educational background. On the opinion of the most respondents, the factor contributing to effective e-teaching is a regular training in the field of Information and Communication Technologies (ICT).

Keywords: E-Teaching · Survey · Higher education · Latvia

1 Introduction

The application of information and communication technology in education has been gaining attention worldwide. E-learning is increasingly essential in higher education institutions and other organizations that provide education and training services. E-learning is an instrument of open education and the digitalization of the education is one of the priorities stated by the European Commission. E-learning has been continuously present in current educational discourse, thanks to technological advances and public and organizational policies. Educators and an administration staff of higher education institutions (HEI) are interested in possibilities offered by new technologies to enrich learning environment and to enhance teaching and learning, and academic staff is playing the main role in the integration of virtual learning environments like Moodle in teaching-learning process.

I. Kabashkin et al. (Eds.): RelStat 2018, LNNS 68, pp. 768–778, 2019.
https://doi.org/10.1007/978-3-030-12450-2_74

The goal of the current paper is to reflect the results of the authors' conducted pilot survey. The given pilot study follows the first one conducted by the authors in 2016 [1]. It was performed using the authors' developed research instrument (questionnaire) at the University of Economics and Culture (EKA). The survey goal was to investigate the perception of e-teaching experience and to determine a Moodle platform value perceived by the academic staff. The authors focused on Moodle (Modular Object Oriented Dynamic Learning Environment) in their research, because it is the most popular IT-based tool used in study process in Latvia [2]. The underlying goal of the first pilot study was to test the research instrument in terms of its appropriateness for further usage in the large-scale research. The survey results indicated several problems: (1) on the opinion of the respondents, many of questions required adjustments and clarification, they were not properly formulated; (2) the list of statements about e-teaching perception did not include some important e-teaching aspects; (3) some important factors affecting e-teaching efficiency were not mentioned; (4) on the other hand, the list of statements for evaluation was too large and consisted of too many less important statements for the investigated issue.

The initial questionnaire was amended, discussing the questions within the group of experts. According to the Gorbans and Bierne [3] perception of e-teaching experience could be divided into two dimensions—pedagogical and managerial. Thus, the criteria for the academic staff members to be invited as experts were: (1) a long-term e-teaching experience, and/or (2) doctoral degree in Pedagogy.

The research instrument used in the current pilot survey involved three sections: (A) respondent profile questions (B) perception of e-teaching (B_1 "General statements about of e-teaching" and B_2 "Factors affecting e-teaching effectiveness") and (C) e-teaching perceived experience. 38 representatives of EKA academic staff participated in the survey. The respondents' answers on only the B_2 questions were analysed in the current paper. Multilayer Perceptron (MLP) analysis has been performed by SPSS 24.0 version.

The hypothesis stated for the research is: *Socio-demographical characteristics of the academic staff members, such as age, gender and education background, have an influence on their perception of the factors affecting e-teaching effectiveness.*

The analysis yielded four neural network models: (1) "Age-gender"—"Layout variables", (2) "Field-gender"—"Layout variables", (3) "Level-gender"—"Layout variables", and (4) "Level-age"—"Layout variables".

2 E-Learning and E-Teaching Effectiveness

In turn, implementation of e-learning practices requires from teachers not only digital skills, but also motivation to work in e-environment. Teachers' integration into e-learning is influenced by organizational factors, attitude towards technology and some other factors [4]. Despite the interest in e-learning in higher education, little is known about how lecturers working in conventional universities perceive e-learning, what are they conceptions and beliefs on it and how these ideas associates with such parameters as age, gender and other similar factors [5]. In mentioned research interviews with the academic staff were analyzed from the grounded theory perspective, giving rise to some emerging themes. Evidence showed that it is a complex area with many influences, some of which can be attributed to personal, social and cultural factors.

Advancements in web technologies are leading to new and innovative ways of learning. These technologies enable activities that allow users to be active learners participating in the e-learning process. E-learning success determinants and determinants of user perceived satisfaction are examined in some empirical studies with wide variety of examined variables and research standpoints. The acceptance and use of e-learning can be influenced by different factors. Typically studies are examining the factors that are having an impact on students' perceptions about the use and acceptance of Moodle, using the Technology Acceptance Model (TAM) as an underlying theory [6–8]. Technology Acceptance Model developed by Davis in 1989 aims to explain the effects of system factors that have influence on user acceptance of information system through user perception [9]. Some researches are focused on the impact of human factors on the decision to accept e-learning, using Theory of Acceptance and Use of Technology providing valuable insights on how to promote use and continuance of e-learning. Recognizing human components make it possible to improve e-learning experience of users, still the focus remains mostly on students' perception and acceptance of it [10–12], including contribution of personal and social factors [13]. Determinants of adoption of Moodle by students get explored in the range of researches [14]. The findings revealed that adoption of learning management system and e-learning is affected by perceived usefulness, perceived ease of use and computer self-efficacy [6, 15]. In general, perceived usefulness was found as the strongest predictor of attitudes toward using Moodle [16]. Collaboration quality, information quality, users perceived satisfaction explain e-learning use, while the drivers of user perceived satisfaction are information quality and instructors attitude [17]. Perceived effectiveness of e-learning studied from different viewpoints, still the main focus remains learner related dimensions, such as digital literacy, self-reported internet skills and previous knowledge [18, 19].

Currently e-learning is considered an essential part of the everyday learning process for both teachers and students in higher education. Some researches are conducted in order to compare students and teachers perceptions about perceived advantages, challenges and needs of their e-learning experiences, concentrating mostly on student-teachers interaction aspects. Measuring teachers and learners' perceptions of the quality of their online learning experience showed that while teachers perceive collaborative learning variables as crucial, learners are more concerned with their own learning benefits [20, 21]. Such aspects as relevant training, peers support and careful design of study course appeared significant [22]. Information quality, task-technology fit, usefulness and some other factors appears as significantly influencing satisfaction and perceived effectiveness of e-learning [23].

Teachers' perception about use of e-learning in educational activities in terms of effect of the use of technology on their perceived motivation, perceived usefulness and perceived use examined and discovered in some researches [24], as well as teachers' attitude to e-learning in association with cultural perceptions, digital literacy, background training [25]. Still there is relatively few data about factors, which are influencing teachers' perception of e-learning dimensions, such as gender, age, teaching experience or education level. Schiller [26] found that male teachers have got more positive self-reported attitude toward usage of technology in educational process and generally concluded that personal characteristics such as educational level, age, gender educational experience can influence the adoption of technology.

Whilst teachers' perceptions of e-learning itself are relatively well covered, hypothesis about association of these perceptions with teachers' age, gender and experience are not especially tested. As different from previous studies, in current studies some hypotheses will be tested about teachers' personal factors such as age, gender, professional experience and education level connection with their perception of e-teaching effectiveness factors.

3 Research Methodology

For the research purposes, the research instrument (questionnaire) was developed and tested. The final version of the questionnaire included three sections: A section—respondent profile questions with pre-determined response alternatives (A1—affiliation, A2—gender, A3—age, A4—field of education, and A5—level of education); B section (B1—20 statements regarding the perception of e-teaching has the structure, and B2—9 statements regarding the factors affecting e-teaching effectiveness); C section—14 statements regarding the staff perceived experience in e-teaching.

The only B2 section statements were offered to the respondents for evaluation, using the 5-point scale, where "1" meant "not important factor" and "5" meant—"critically important factor".

For the purposes of this study, we have used Multilayer Perceptron (MLP) analysis performed by SPSS 24.0 version. The procedure was aimed to reveal the difference in perception of e-teaching effectiveness factors by respondents with different socio-demographic characteristics, such as age, gender, and educational background (field of education and level of education).

In the sample, we had 37 valid cases (1 case excluded) where 29 were used for training on neural network, and 8—for testing the model. As it is a relatively low number of cases, the results are preliminary, though still even this sample size allowed us to create some understanding on the issue. Interpreting of the achieved results is based on mathematical approach for neural networks assessment.

The neural network graphs indicate a few characteristics. First, the number of intermediary figures (H1:1, H1:2) reveal if the interrelation of independent variables when they influence dependent one, is different. In case only one interrelation type can be present one will see only H1:1, but if the situations can be different, there will be more than one intermediary figure. Second, the most important ties are pictured as thick lines, while thin lines indicate low level of influence. Labels used in Figures representing the yielded models are described in the Table 1.

Table 1. Explanation of the figure labels used in the paper.

Label	Explanation	Label	Explanation
Age = 1	18–25 years	Field = 1	Economics and management
Age = 2	26–45 years	Field = 2	Social sciences and law
Age = 3	45–62 years	Field = 3	Humanities and art
Age = 4	Over 62 years	Field = 4	Pedagogy
Gender = 1	Male	Field = 5	Natural sciences, engineering and ICT
Gender = 2	Female	Field = 6	Engineering, production and construction

4 Results

For the purposes of this study we have used neural networks to estimate the relationship between four independent variables (age, gender, field of original studies and education level). The independent entry factors were analyzed in pairs, while the set of output factors was always the same and included 9 variables.

The first pair of entry factors used for analysis were age and gender—these characteristics are natural and cannot be changed by the respondent. A neural network

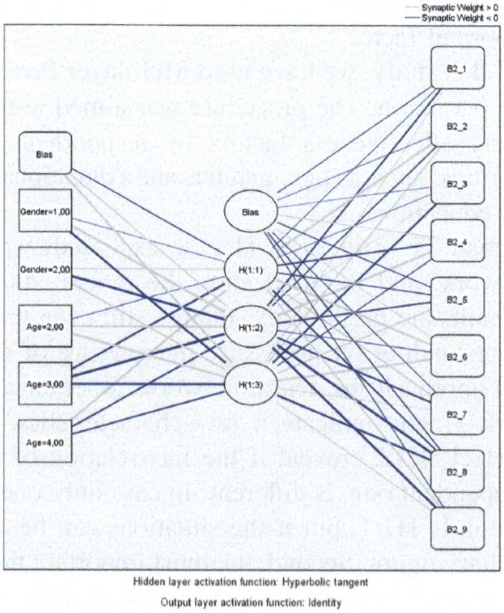

Fig. 1. Neural network model "Age-gender"—"Layout variables".

model with entry factors "age" and "gender" is shown in Fig. 1.

Second, we have analyzed the influence of field-gender set of characteristics on the same set of output variables. In this case 37 valid cases were used for analysis, including 25 used for training of the model and 12—for testing (see Fig. 2).

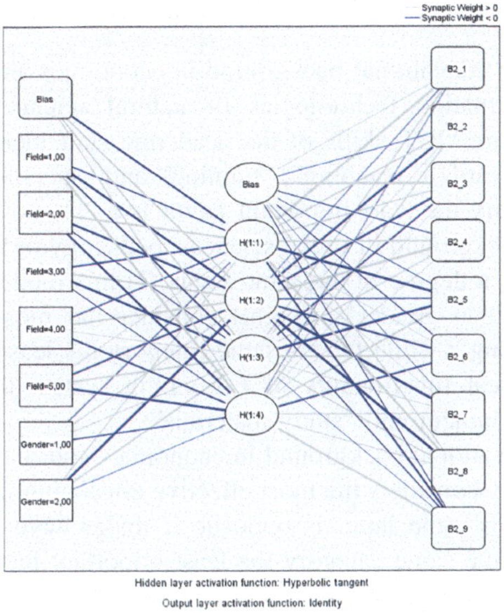

Fig. 2. Neural network model "Field-gender"—"Layout variables".

The layout variables that witness the most significant influence by age-gender set of independent variables allow us to make the following conclusions:

- males from 45 to 62 years considers the most effective e-teaching factors as "financial motivation from HEI administration for engagement into e-teaching process" ("financial motivation") and the level of the technical functionality of E-teaching system;
- males from 45 to 62 years considers the least effective e-teaching factors as ICT skills of the academic staff members (the same conclusion applies to female teaching staff from 26 to 45) and a "qualitative and regular training at the working place in order to help staff members to acquire skills for working in e-environment" (in this case it is the opposite for female teaching staff from 26 to 45);
- female teaching staff above 45 considers the most effective e-teaching factors as motivation to use ICT-based solutions, ICT skill level of academic staff; active engagement of students in e-learning process; specifics of study course and existence of study materials which can be used in e-environment ("study materials");
- female teaching staff above 45 considers the least effective e-teaching factor as strategic orientation of higher educational institutions towards e-teaching (opposing males of the same age in this issue);
- male teaching staff above 45 considers the most effective e-teaching factors as strategic orientation of HEI towards E-teaching and specifics of a study course;
- male teaching staff above 45 considers a "qualitative and regular training at the working place in order to help staff members to acquire skills for working in e-environment" ("training") as the least effective e-teaching factor.

The main conclusions that are derived from the Fig. 2 are the following:

- for females with an educational background in economics and management, social sciences and information technologies or natural sciences the most effective e-teaching factors are “ICT skills of the academic staff members”, “strategic orientation of HEI towards E-teaching”, “training” and “specifics of a study course”; for the same category the least important factor is a “financial motivation”;
- females with the background in social sciences, pedagogy or natural sciences, alone with males holding a degree in the same fields (in this model the field’s weight is two times higher than gender’s weight) consider the most effective e-teaching factors to be “training”; while for the same category the least important factors are level of the technical functionality of E-teaching system that allows effectively working in e-environment and “study materials”;
- males with the educational background in economics and management, humanities and art, or pedagogy considers the most effective e-teaching factors to be “training” and “study materials” (the latter is opposite if males have some technical background); while for the same category the least important factors are motivation of the academic staff members and the use ICT-based solutions in study process and ICT skills of the academic staff members;
- for respondents with a background in humanities and arts (despite their gender) the most effective e-teaching factor is the “level of the technical functionality of E-teaching system that allows effectively working in e-environment”; while the least effective factor is “training”.

Third, we have analyzed the gender-level model. 26 cases were used for training, and 11—for testing to create a valid model (see Fig. 3). In this model Level = 1 stands for Bachelor degree, level = 2 stands for a Master degree and level = 3 stands for a Doctor degree.

- The main conclusions from Fig. 3 are:
- the level of studies appears to be much more significant than gender; also, in this model bias is created by the alternative factors not used as an input set of variables. The bias appears most significant for specifics of a study course factor;
- the teaching staff with a Master degree, despite of the gender, considers the most effective e-teaching factors to be “financial motivation” and “training”, while the least effective factor for them strategic orientation of HEI towards E-teaching;
- for respondents with a Doctor degree the most effective e-teaching factors are “training” and “specifics of a study course”, while the least important is and “study materials”. This category is more used to developing their own study materials, but at the same time need more training in e-environment to accept this way of studies as a regular part of the course.

It is important to see how the level of studies becomes important for the studied field as the teaching staff earns their master’s or doctoral degree (for bachelors there appeared to be no valid neural networks model); also, financial motivation is significant for Master’s, but less significant for the teachers with higher education level.

The last bit of analysis was performed for age-level set of variables. In this case training took 30 cases while testing took 7 cases (see Fig. 4).

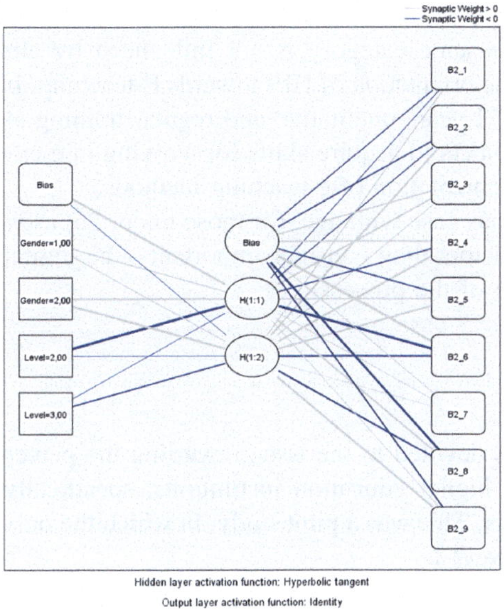

Fig. 3. Neural network model "Level-gender"—"Layout variables".

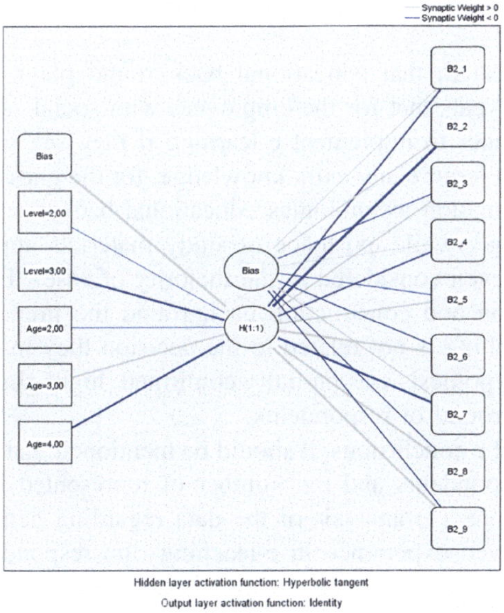

Fig. 4. Neural network model "Level-age"—"Layout variables".

From Fig. 4 it is seen that for the respondents with a Doctor degree above 62 the most effective factors to motivate them for e-teaching are ICT skills of the academic staff members and active engagement of students into e-learning process—so actually in this case one can only enrol those ones who already see value in e-teaching and are familiar with ICT; the same category is not influenced by almost any other factor, especially—by strategic orientation of HEI towards E-teaching. In turn, for the teaching staff between 45 and 62 years qualitative and regular training at the working place in order to help staff members to acquire skills for working in e-environment seems to be a must to ensure implementation of e-teaching methods.

Again, as in previous case with age, for those under 26, as well as bachelors, there are no effectiveness factors they consider important—the digital generation considers e-teaching regular part of the process.

5 Conclusions

The current paper was devoted to the issue regarding the perception of e-teaching by the academic staff of higher education institutions, specifically the factors affecting e-teaching effectiveness. This was a pilot study, in which the only staff members of one Latvian HEI were surveyed.

In the result, no specific patterns were found for both genders in case respondents' age was below 25, and same applies to males from 26 to 45. To explain this, we can conclude that those categories are very familiar with e-tools, and do not need to be specifically motivated to implement e-learning, they just see it a part of normal study process. For the others different approaches should be used based on their age-gender characteristics.

The findings indicated, that educational background plays more important role, than gender; it also reveals that for the employees with social science and humanities education are more eager to implement e-learning if they see strategic value and the technical side does not require any extra knowledge; for the ones with natural sciences, engineering and information technologies educational background the most inspiring motivation is "training", while existence of study materials appropriate for using in e-environment is not even considered in the majority of cases. Finally, for those with engineering, production and construction background the implementation of misimplementation of e-teaching is not related to the decision they make on using the tool.

Thus, the tested hypothesis was partially confirmed, highlighting the importance of the educational background of respondents.

As for validity of the conclusions, it should be mentioned that the study was limited by the number of respondents and the number of represented institutions. The next steps of the research are: (1) analysis of the data regarding general statements about e-teaching and perceived experience in e-teaching (the respondents' answers on the questions from these sections were not reflected in the current paper); (2) dissemination of the questionnaire among representatives of other Latvian higher education institutions; (3) dissemination of the questionnaire among the academic staff members of the partner institutions in foreign countries.

References

1. Bierne, J., Titko, J.: Perception of e-teaching experience in latvian higher education institutions. In: IBIMA Conference Proceedings, pp. 1723–1730. IBIMA Publishing (2016)
2. Birzina, R.: E-Learning for lifelong learning in Latvia. ASEM white paper e-learning for lifelong learning. http://priede.bf.lu.lv/grozs/Didaktika/ASEM/2_White_paper.pdf. Accessed 20 June 2018
3. Gorbans, I., Bierne J.: The E-course Management within Moodle in Higher Education: Technological and Pedagogical Issues. In: Proceedings of the International Scientific Conference "Society. Integration. Education", pp. 45–60. Rezeknes augstskola, Rezekne (2015)
4. Chen, C.H.: Why do teachers not practice what they believe regarding technology integration? J. Educ. Res. **102**(1), 65–75 (2008)
5. McConnell, D.: E-learning in Chinese higher education: the view from inside. High. Educ. **75**, 1–15 (2017)
6. Sánchez, R.A., Hueros, A.D.: Motivational factors that influence the acceptance of Moodle using TAM. Comput. Hum. Behav. **26**(6), 1632–1640 (2010)
7. Escobar-Rodriguez, T., Monge-Lozano, P.: The acceptance of Moodle technology by business administration students. Comput. Educ. **58**(4), 1085–1093 (2012)
8. El-Masri, M., Tarhini, A.: Factors affecting the adoption of e-learning systems in Qatar and USA: extending the Unified Theory of Acceptance and Use of Technology 2 (UTAUT2). Educ. Technol. Res. Dev. **65**(3), 743–763 (2017)
9. Marangunić, N., Granić, A.: Technology acceptance model: a literature review from 1986 to 2013. Univ. Access Inf. Soc. **14**(1), 81–95 (2015)
10. Alsabawy, A.Y., Cater-Steel, A., Soar, J.: Determinants of perceived usefulness of e-learning systems. Comput. Hum. Behav. **64**, 843–858 (2016)
11. Asoodar, M., Vaezi, S., Izanloo, B.: Framework to improve e-learner satisfaction and further strengthen e-learning implementation. Comput. Hum. Behav. **63**, 704–716 (2016)
12. Nganji, J.T.: Towards learner-constructed e-learning environments for effective personal learning experiences. Behav. Inf. Technol. **37**, 1–11 (2018)
13. Olasina, G.: Human and social factors affecting the decision of students to accept e-learning. Interact. Learn. Environ. 1–14 (2018)
14. Kauffman, H.: A review of predictive factors of student success in and satisfaction with online learning. Res. Learn. Technol. **23**(1), 26507 (2015)
15. Kumar, D., Samalia, H.V.: Learning management system adoption in universities-perception about moodle as a pedagogical tool in a private university in India. Int. J. Manag. Behav. Sci. **6**, 169 (2015)
16. Kilic, E.: Determining the factors of affecting the Moodle use by using TAM. The story of a university after a destructive earthquake. Hacettepe Üniversitesi Eğitim Fakültesi Dergisi, 29 (29–1) (2014)
17. Cidral, W.A., Oliveira, T., Di Felice, M., Aparicio, M.: E-learning success determinants: Brazilian empirical study. Comput. Educ. **122**, 273–290 (2018)
18. Algahtani, A.: Evaluating the effectiveness of the e-learning experience in some universities in Saudi Arabia from male students' perceptions. Doctoral dissertation, Durham University (2011)
19. Burin, D.I., Irrazabal, N., Ricle, I.I., Saux, G., Barreyro, J.P.: Self-reported internet skills, previous knowledge and working memory in text comprehension in E-learning. Int. J. Educ. Technol. High. Educ. **15**(1), 18 (2018)

20. Emelyanova, N., Voronina, E.: Introducing a learning management system at a Russian university: students' and teachers' perceptions. Int. Rev. Res. Open Distrib. Learn. 15(1) (2014)
21. Gómez-Rey, P., Barbera, E., Fernández-Navarro, F.: Measuring teachers and learners' perceptions of the quality of their online learning experience. Distance Educ. **37**(2), 146–163 (2016)
22. Pihlajamaa, J., Karukka, M., Ålander, H.: Comparison of higher education student and teacher perceptions of e-learning. In: European Conference on e-Learning, p. 782. Academic Conferences International Limited, October 2016
23. Al-Samarraie, H., Teng, B.K., Alzahrani, A.I., Alalwan, N.: E-learning continuance satisfaction in higher education: a unified perspective from instructors and students. Stud. High. Educ. **43**, 1–17 (2017)
24. Barberà, E., Gómez-Rey, P., Fernández-Navarro, F.: A cross-national study of teacher's perceptions of online learning success. Open Learn. J. Open Distance and e-Learn. **31**(1), 25–41 (2016)
25. Albirini, A.: Teachers' attitudes toward information and communication technologies: the case of Syrian EFL teachers. Comput. Educ. **47**(4), 373–398 (2006)
26. Schiller, J.: Working with ICT: perceptions of Australian principals. J. Educ. Adm. **41**(2), 171–185 (2003)

A Logistics Management Game for Actors of a Geographically Distributed Supply Chain

Tobias Reggelin(✉) and David Weigert

Otto von Guericke University Magdeburg, Universitätsplatz, 2, 39106 Magdeburg, Germany
{tobias.reggelin,david.weigert}@ovgu.de

Abstract. Management games are a common method for training employees in production and logistics, both in expert knowledge and in soft skills. Typically, all participants are located in one room while playing the game, which allows for a face-to-face interaction among the participants and between the participants and the trainer. This facilitates the typical learning loop of a management game: play, analyze and improve. For companies with supply chains covering faraway geographical locations and different time zones, the approach of having a management game seminar with all relevant participants (actors from different locations of the supply chain) in one room leads to high costs due to traveling and lost time for traveling. Multiplayer computer games played through the Internet could enable companies to conduct management games with actors of geographically distributed supply chains without spending money and time on travelling. The paper conceptually compares board games, multiplayer computer games played distributed over the internet and multiplayer VR games. Furthermore, the paper describes a prototype implementation of a computer game on supply chain management suitable for actors of a geographically distributed supply chain.

Keywords: Supply chain management · Logistics · Game-based learning · Serious games · Business games · Management games · Board games · Computer games · Multiplayer games

1 Introduction

The first management games used in business are closely associated with military games [1], where officers had "an opportunity to practise decision making creatively in a myriad of hypothetical yet true-to-life competitive situations. Moreover, they are forced to make decisions in areas outside their own speciality; a naval communication officer, for example, may play the role of a task force commander" [2]. One example is the management game "Top Management Decision Simulation" developed by the American Management Association (AMA) in the 1950's [2]. The probably best-known business simulation game on supply chain management is the Beer Distribution Game developed at MIT in the 1960's, played in many versions, both as a board game and a computer-based game, and described in many publications, for example in [3–8].

I. Kabashkin et al. (Eds.): RelStat 2018, LNNS 68, pp. 779–788, 2019.
https://doi.org/10.1007/978-3-030-12450-2_75

Players of the Beer Distribution Game experience the Bullwhip Effect in a supply chain and lean about its causes.

Today, management Games are a common method for educating and training students and employees in production and logistics, both in expert knowledge and in soft skills. Vlachopoulos and Makri [9] review the use of games and simulation in higher education. The usefulness of business games was shown in several studies [10–15], which are all analyzed and summarized by Riedel et al. in [16].

When training people with a management game in form of a board game, all participants are located in one room while playing the game, which allows for a face-to-face interaction among the participants and between the participants and the trainer. This facilitates the typical learning loop of a management game: play, analyze and improve (Fig. 1).

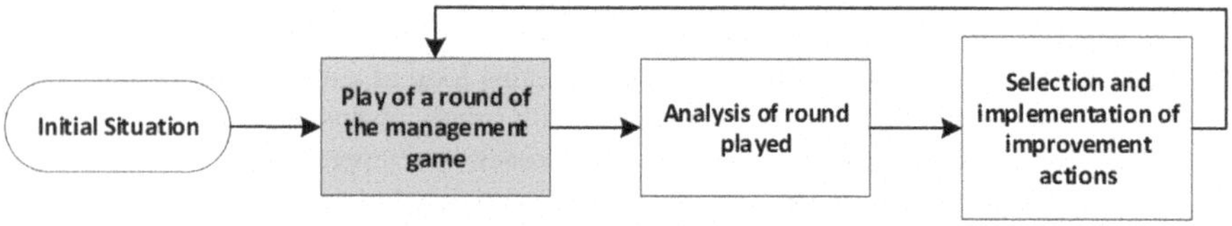

Fig. 1. Learning loop of a management game

For companies with supply chains covering different geographical locations and different time zones, the approach of having a management game seminar with all participants in one room leads to high costs due to traveling and lost time for traveling.

Multiplayer computer or VR games played through the Internet could enable companies to conduct management games with actors of geographically distributed supply chains without spending money and time on travelling. Before transferring a board game into a multiplayer computer game with players in different locations and time zones, a game developer needs to consider the following main aspects:

- The role of the trainer (facilitator):
  - Will there be a trainer or not?
  - How will the trainer communicate with the participants?
- The interaction and communication among the players:
  - Will there be any audio, video, chat interaction among the players
- The process of analyzing and improving the logistics processes within the learning loop:
  - Will the learning loop be similar to the learning loop in a classical board game?
  - How will the players interact?
  - What will be the role of the trainer?
  - How flexible can the players implement improvement measures?
- The degrees of freedom within the game:
  - How flexible can the trainer and the players change the structure and processes of the game?

This all comes down to two main questions:

- How can the game developer transfer the flexibility of a board game in terms of changing the structure and processes of the supply chain into a computer-based game?
- How and to which extend will the game allow for an interaction among the players and between the players and the trainer in order to facilitate the learning loop depicted in Fig. 1?

Why do the authors not use an already existing internet-based implementation of the Beer Distribution Game or of another game? The authors could not find an implementation of the Beer Distribution Game in the Internet that satisfies the requirements of the authors, especially in terms of flexibility, interaction of the participants and involvement of a trainer. Most of the games are "are 'automated' versions of the traditional game, primarily intended to make the game run faster" [8]. Klemke et al. [17] state that, "little work is reported that focus on multi-user learning situations".

2 Supply Chain Management Game

Figure 2 shows the game on supply chain management, which the authors intend to transfer into a computer game. Ten players take different roles within a supply chain. The customer orders a product in four different variants. Each day every player has to decide how many products to order from the preceding actor in the supply chain. The two main goals of every player are to ensure a high delivery reliability and to minimize the inventory.

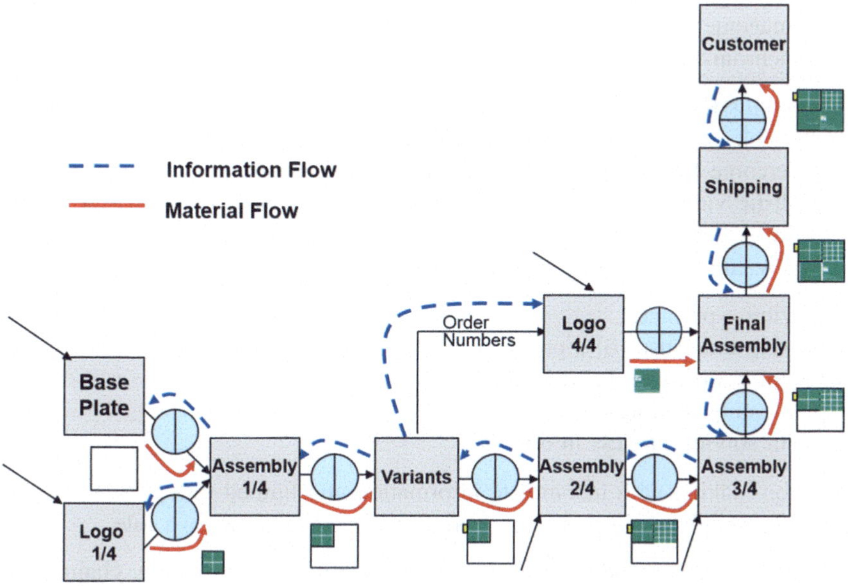

Fig. 2. Supply chain management game: structure, material and information flows

The game resembles the Beer Distribution game, but does not have the weaknesses of the traditional Beer described by Kaminsky and Simchi-Levi [8], "the Beer Game does not necessarily provide students with insight on how to better manage the supply chain." The players experience the bullwhip effect and have to create and implement strategies to avoid the bullwhip effect and deliver all ordered products on time while minimizing the holding costs.

The participants play the game in rounds of ten days of playing time. After every round of ten days, the trainer moderates the process of analyzing the game and improving the processes. Usually the participants have improved the supply chain after four rounds of ten days each in a way that the customer gets all the products on time and the bullwhip effect does not occur anymore.

The main Learning goals and learning content of this board game on supply chain management are:

- Logistics Knowledge:
  - How and why does the bullwhip effect develop in a supply chain?
  - How to avoid the bullwhip effect?
  - Controlling a supply chain with methods like ConWIP [18];
  - Increasing logistics performance and customer satisfaction:

1. Reducing inventory and consequently holding costs;
2. Reducing lead times and consequently delivery times;
3. Increasing delivery reliability.

- by applying the following methods:

  1. Improving information flows;
  2. Using logistics control loops like ConWIP;
  3. Management of variants (postponement);
  4. Synchronization of parallel processes.

- Other Skills/Knowledge:
  - To look beyond one's own nose/Think out of the box;
  - Overcome function orientation and think process oriented;
  - Holistic view on the overall process (from a worm's eye view to a bird's eye view);
  - Analyzing and problem solving skills:

1. Analyzing a problem;
2. Analyzing the causes of a problem;
3. Developing measures to solve the problem;
4. Implementing the measures;
5. Problem solving process in a team.

- Decision making with incomplete information and limited time;
- View the supply chain processes from other roles than the own role.

The authors have conducted the described board game on supply chain management worldwide as trainers with both university students as part of their study program and employees of companies as part of company trainings. The game was also adapted to specific supply chains of real companies in order to use the game for company-wide trainings of employees as described in [19]. The authors always played the game as a board game with all the participants and the moderator in the same room. A more detailed description of the game can be found in [20].

During the last time, the authors noted that companies show interest in using Internet-based games to train their employees located in different locations and time zones and do not seem to find suitable games. The reason for this could be that many computer games lack a proper interaction between the participants and an adequate involvement of a trainer in order to support the learning loop presented in Fig. 1.

An additional learning content arises because the actors of worldwide supply chains work in different locations and time zones, which often leads to communication problems and information delays. A classical board game with all participants in one room does not realistically represent this problem. However, playing a computer game from different locations and time zones allows for a quite realistic representation of this problem.

3 Conceptual Comparison of Board Games, Computer Games and VR Games

Table 1 shows the conceptual comparison of the game on supply chain management, described in the section before, implemented as a board game, a multiplayer computer game and a multiplayer VR game. The authors included a VR game into the conceptual comparison since it seems to be an interesting option for a future implementation of a management game. Another comparison of a board game and a mobile game can be found in [17].

The number of participants deduces from the ten roles in the supply chain represented in the described game (Fig. 2). Playing the board game gives the trainer flexibility. One player could play two of the roles with less workload at the same time or two players can play one role together, since the main aspect of the game is not to perform the mechanic work that is associated with each role, but to understand, analyze and improve the processes represented in the game.

The same applies for the computer game, where two players play one role of the supply chain together. This flexibility gets lost when playing the game with a VR headset.

The supply chain that serves as a generic base for the conceptual comparison has actors in Germany and the USA with a time difference of six hours. For this reason, the playable time shrinks to three hours a day, when the players play the game at the same time. An iterative playing of the game becomes necessary when the time difference gets bigger. In this example, the duration of a management game seminar can take up to one week with three hours of playing each day, compared to half a day up to two days with a board game.

All three kind of implementations allow for a job rotation (participants play different roles of the supply chain in the course of a game) which is important to fulfill the

Table 1. Comparison of the game on supply chain managements (a board game), a multiplayer computer game playable on a standard PC and a multiplayer VR game played with VR headset

	Board game	Computer game	VR game
1	2	3	4
Number of participants	8–20	10–20	10
Suitability for geo-graphical distributed participants	No	Yes	Yes
Geographic locations and possible time slots for playing	One location	Example: 5 participants in the USA (GMT-5): 8 – 11 am 5 participants in Germany (GMT + 1): 2 – 5 pm 3 h of simultaneous playing time a day. Participants of one location can be together in a conference room or separately in their offices	
Duration of a seminar	0.5–2 days	1–5 days Each day consists of 3 h playing the game	
Participants play different roles of the supply chain	Yes	Yes	Yes
Communication of participants beyond the actual computer game	Face-to-face (all participants in one conference room)	Video, audio, chat communication, virtual Whiteboards etc. for the process of analyzing and improving the game's processes	Virtual communication. Audio: through VR headset. Visual: avatar/pictures of participants with simple gestures (nodding, movement of the hand)
Trainer	In one room with participants	Trainer can be either together with the participants at one location or at a different location; interact with the participants by video, audio, chat communication	Trainer uses also a VR work station
Interaction of participants during the actual game	Hand over products, parts and information (paper) like in a board game Face-to-face (all participants in conference room)	Hand over products, parts and information in a computer game (mouse and keyboard)	Hand over products, parts and information in the VR world Possible Input devices: Multifunction trackpad, grip buttons, dual-stage trigger, system button, menu button

(*continued*)

Table 1. (*continued*)

	Board game	Computer game	VR game
1	2	3	4
One-time costs	Game development or license cost vs. no cost with standard game and external moderator	Game development costs or license cost vs. no cost with standard game and external moderator	Game development costs, costs for VR headsets and suitable PC for each participant
Costs for playing with geographical distributed participants	Travel and accommodation cost. Cost for moderator vs. own game with own moderator	Cost for moderator vs. own game with own moderator	Cost for moderator vs. own game with own moderator

learning goal of seeing the processes through the eyes of other roles than the own role in the supply chain.

There is probably nothing better than face-to-face communication as in the board game in order to jointly analyze the overall processes, find improvement measures and implement them. However, this does not represent the actual daily communication in a geographically and timely distributed supply chain very well. Thus, the disadvantage of only using audio, video and chat communication in a distributed computer game even resembles the daily experience of the participants better and consequently could lead to a more realistic game scenario compared to a board game.

Beside the aspect of the communication among the participants, the involvement of a trainer who facilitates the learning process in the game could play a critical role for the success of a computer game with players in different locations. The authors suggest that the trainer should at least be included in the game by video, audio and chat communication.

In terms of cost, a computer game can save lot of money and resources because cost and time consuming travel as for the joint board game is not necessary anymore.

4 Prototype of a Multiplayer Computer Game on Supply Chain Management

The authors let students develop and test a first prototype of a computer game on supply chain management based on the board game described above within a project seminar of the study program Industrial Engineering with focus on Logistics at Otto von Guericke University Magdeburg [21]. The students implemented the game with MS Excel and VBA. Figure 3 depicts a screenshot of the user interface.

A shared master Excel file serves as the central database for the game to which the distributed files, which the players use, send their data at the end of each playing day. At the beginning of each playing day, the participants' files receives the data on the orders from the succeeding actor in the supply chain and the material deliveries from the preceding actor in the supply chain from the master file.

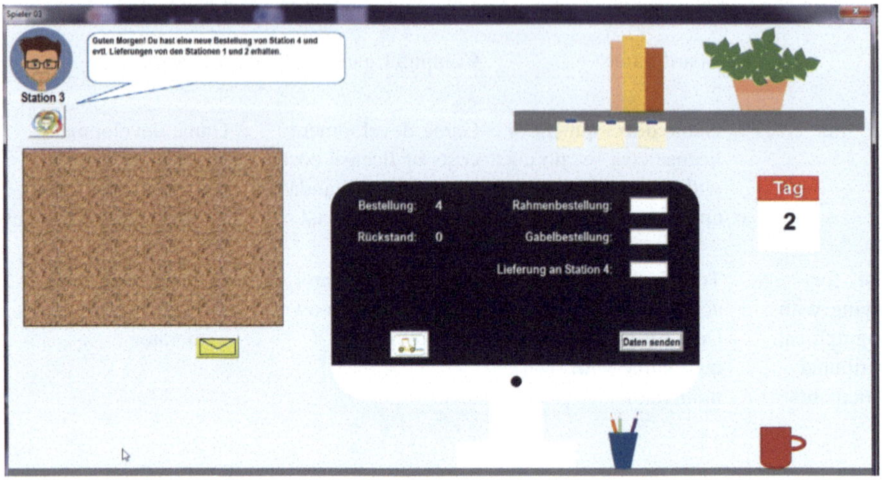

Fig. 3. Screenshot of the prototype of a multiplayer computer game on supply chain management

This setup allows geographically distributed players playing the game as long as their files have access to the master file. The trainer also uses an Excel file that allows for interventions to the game, for example changing the demand of the customer.

There are two main differences compared to the board game:

1. The computer game automatically logs all the data to the orders of each player on each day, the inventory of each player on each day and the backlog of each player on each day. This makes the analyzing much faster than in the board game.
2. In the computer game, the players cannot implement flexible improvement measures as in the board game. They have to choose from already implemented improvement measures. The prototype computer game provides the options:

– to postpone the building of variants;
– to use a control loop (ConWIP) to control the ordering in the supply chain and;
– to make all the available information of the game (orders, inventories, backlogs) accessible to every player.

5 Conclusion and Future Work

The authors plan to implement the described game on supply chain management as an Internet-based game based on the prototype game described above together with a company that has a worldwide operating supply chain. The main goals for the implantation are

- to address all relevant learning goals described in this paper,
- to combine the game with communication tools for audio, video and chat communication and with tools for a virtual analyzing and creative process, like virtual whiteboards and brainstorming tools,
- to keep the flexibility of the current board game in terms of flexible implementation of improvement measures, mainly in terms of changed information flows and
- to keep the role of a trainer to guide the participants throughout the learning loop of playing, analyzing and improving which is essential to achieve the stated learning goals.

If the game will meet these requirements, it can combine the advantages of a board game (mainly flexibility and direct interpersonal interaction) with the advantages of an Internet-based digital game (playing with geographically distributed participants, automated recording of players' decisions, mainly in terms of order amounts, automatically logged inventories and backlogs).

Another option will be the implementation as a VR-based game in order to increase the immersion into the game. However, the authors are not sure yet whether or not the learning and the motivation will increase by the use of a VR-based game. It even seems that the proposed implementation as an Internet-based game played on a PC resembles the actual day-to-day situation of the employees better than a VR-scenario. Furthermore, the technical requirements and costs to conduct a VR-based game are higher.

References

1. Eilon, S.: Management games. J. Oper. Res. Soc. **14**, 137–149 (1963)
2. Ricciardi, F.M., Marting, E.: American Management Association: Top Management Decision Simulation: the AMA Approach. American Management Association (1957)
3. Sterman, J.D.: Modeling managerial behavior: misperceptions of feedback in a dynamic decision making experiment. Manage. Sci. **35**, 321–339 (1989)
4. Sterman, J.D.: Teaching takes off—flight simulators for management education. OR/MS Today **35**, 40–44 (1992)
5. Sterman, J.D.: Instructions for running the beer distribution game. In: System Dynamics Group Working Paper D-3679, MIT, Sloan School of Management, Cambridge (1984)
6. Machuca, J.A.D., del Pozo Barajas, R.: A computerized network version of the beer game via the Internet. Syst. Dyn. Rev. **13**, 323–340 (1997)
7. Jacobs, F.R.: Playing the beer distribution game over the Internet. Prod. Oper. Manage. **9**, 31–39 (2000)
8. Kaminsky, P., Simchi-Levi, D.: A new computerized beer game: a tool for teaching the value of integrated supply chain management. Global Supply Chain Technol. Manage. **1**, 216–225 (1998)
9. Vlachopoulos, D., Makri, A.: The effect of games and simulations on higher education: a systematic literature review. Int. J. Educ. Technol. High. Educ. **14**, 173 (2017)
10. Cook, D.M.: The effect of frequency of feedback on attitudes and performance. J. Acc. Res. **5**, 213 (1967)
11. Trieschmann, J.S.: Teaching insurance with an insurance management computer game. J. Risk Insur. **43**, 43 (1976)

12. Wolfe, J.: Effective performance behaviors in a simulated policy and decision making environment. Manage. Sci. **21**, 872–882 (1975)
13. Wolfe, J., Luethge, D.J.: The impact of involvement on performance in business simulations: an examination of Goosen's "Know Little" decision-making thesis. J. Educ. Bus. **79**, 69–74 (2003)
14. Pourabdollahian, B., Taisch, M., Kerga, E.: Serious games in manufacturing education: evaluation of learners' engagement. Procedia Comput. Sci. **15**, 256–265 (2012)
15. Ben-Zvi, T.: Using business games in teaching DSS (2006)
16. Riedel, J.C.K.H., Feng, Y., Azadegan, A., Romero, M., Usart, M., Baalsrud Hauge, J.: Measuring the commercial outcomes of serious games in companies—a review. In: Ma, M. (ed.) Serious Games Development and Applications, SGDA 2014., vol. 8778, pp. 176–191. Springer, Cham (2014). Proceedings of 5th International Conference, Berlin, Germany, 9–10 Oct 2014
17. Klemke, R., Kurapati, S., Lukosch, H., Specht, M.: Transferring an educational board game to a multi-user mobile learning game to increase shared situational awareness. In: Zaphiris, P., Ioannou, A. (eds.) Learning and Collaboration Technologies, 9192, pp. 583–594. Springer, Cham (2015)
18. Hübl, A.: Conwip. In: Hübl, A. (ed.) Stochastic Modelling in Production Planning. Methods for Improvement and Investigations on Production System Behaviour, pp. 87–107. Springer, Berlin (2018). Fachmedien Wiesbaden, Wiesbaden. s. 1
19. Wojanowski, R., Schenk, M., Blötz, U.: Das maßgeschneiderte Planspiel (The taylored business game). Planspiele in der beruflichen Bildung: Auswahl, Konzepte, Lernarrangements, Erfahrungen–aktueller Planspielkatalog (2008)
20. Schenk, M., Wojanowski, R.: SILKE—SCM hautnah erleben und gestalten. In: Beckmann, H. (ed.) Supply Chain Management, pp. 123–143. Springer, Berlin, Heidelberg (2004). Strategien und Entwicklungstendenzen in Spitzenunternehmen
21. Aberle, S., Brand, H., Dittrich, J., Puhst, A.: Konzeptionierung, Gestaltung und prototypische Umsetzung eines rechnergestützten Simulations-Planspiels aus dem Bereich Produktion und Logistik. Magdeburg (2016)

Application of Process Approach to Knowledge Management in Education Institutions: Competence Centres

Natalia Salienko(✉) and Victoria Klyueva

Bauman Moscow State Technical University, 2-ya Baumanskaya str. 5/1, Moscow, Russia
verno555@mail.ru, vklyueva@emtc.ru

Abstract. In recent years, Russia has carried out reforms in the vocational education system. In order to improve the reforms efficiency leading engineering universities of the country are organising Competence centres. In technical universities this tendency is actually to provide training for teachers, professors, experts in hi-tech industries. The article considers the main functions of competence centers as structures aimed at active search, transfer and accumulation of new knowledge and unique experience, based on the students and employers needs. It is suggested to apply process approach for university centres design, while in the framework of this approach it is possible to optimize the acquisition, storage, and knowledge management process.

Keywords: Competence centres · Process management · Educational institutions

1 Introduction

Today the world is on the verge of the sixth technological paradigm forming, which contours are already visible in advanced countries with a high level of innovative high-tech products [1]. The core of the new technological paradigm is NBIC (nano-, bio, info-, cognitive) technologies: convergence of nano- and biotechnologies, cognitive technologies and information and communication technologies (for example, quantum computers, etc.). A number of studies show that the commercialization of products based on NBIC technologies will start in the near future [2].

The transition to a new paradigm is possible only with large-scale investment, not only in the means of production and modernization of the economy on their basis, but also in the education system. With the development of technologies of the 6th technological paradigm, there will be a demand for highly qualified specialists in the field of the NBIC (nano-, bio, info-, cognitive technologies). Consequently, special attention should be given to the training system (especially to the innovation sphere). According to the Atlas of emerging jobs 186 new professions will appear [3], it is vital to restructure the training system for the requirements of society and time. In the opinion of the Atlas developers, system thinking, project management, lean manufacturing, inter-industry communication belong to the over-professional skills and abilities of the future.

I. Kabashkin et al. (Eds.): RelStat 2018, LNNS 68, pp. 789–797, 2019.
https://doi.org/10.1007/978-3-030-12450-2_76

At the moment, the modern education system is unable to prepare highly qualified personnel that will meet the requirements of employers, industry and society as a whole. Consequently, it is necessary to reconsider the organization of educational system in universities. One of the possible options for improving the quality of education is the process approach to the management of educational activities.

The aim of the article is to exam competence centers as the basis of the process approach in education in order to improve the quality of educational processes management.

Research tasks were the following:

- to describe the nature of competence centre in educational organization;
- to examine students' and future employers' attitude to the competence centres creation;
- to develop the model of process approach for competence centres.

Theoretical part of the article introduces the concept of the Competence Centres for the process approach in education and examines the status of Competence centers in Russia at the moment. In the practical part the requirements for the Competence Centers were proposed survey basis. Finally, the model showing how the process approach can be applied at the case university within the framework of the Competence Centers.

2 Theoretical Framework

2.1 The Concept of Competence Centres Application Within Process Management Approach in Education

The concept of process knowledge management begins to be applied in educational institutions [4, 5]. The advantage of the process approach is the continuity of control that it provides at the interface of individual processes, as well as their combination and interaction. According to this concept, the activity of the organization is considered as a set of processes.

At the moment, a lot of problems that arise in the activities of any organization are related to the linear-functional management structure. These are problems of interaction, when a structural unit or an official body is not responsible for the overall end result, does not understand how the results of his work are used by other employees of the organization. This model also complicates quality control.

According to the standard ISO 9001:2000, process is a set of interrelated and interacting activities that transform inputs into outputs.

The process definition is quite broad, which facilitates the independent choice of options for constructing a process model leading to the intended goal. Taking into account the above definitions, the general scheme of the process can be represented in the following form (Fig. 1).

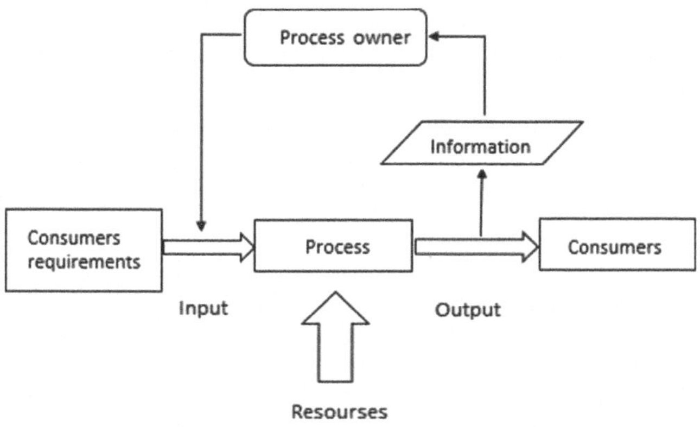

Fig. 1. Process general scheme (developed by the authors)

The process owner is an official or an organization that has at its disposal the necessary resources, information about the process, controls the progress of the process and is responsible for the progress, results and effectiveness of the process. Resources transformed into the process outputs during the process are considered as the process inputs. The output of the process is the result (product, service) of the process. The consumer is the entity that receives and uses the result of the process, the degree of satisfaction of which is the criteria for assessing the process results quality.

One of the key ideas representing the essence of the process approach to managing the educational organization activities is to model, transform, optimize, improve the processes implemented in it for the best result at the output of the educational system, i.e. improving the quality of education and customer satisfaction.

The process model provides clear representation of processes sequence and interaction, the organization activities, both in horizontal and vertical directions. The process approach implies the implementation of many small educational processes within the framework of one large process. For these purposes, there should be a single platform for interaction between different departments and divisions of the university, monitoring the scientific activities of students, stimulation of interdisciplinary projects development (it is necessary to create conditions for the knowledge and competencies exchange on discussion platforms between representatives of different departments, universities, enterprises); assistance in the project technological implementation (simplified access of students to partner enterprises production sites and University laboratories should be organized, and it is also proposed to consider the creation of pilot plants for student scientific communities).

The basis for the process approach implementation in universities could be the Competence centers—structures or structural subdivisions of knowledge-based enterprises aimed at finding new knowledge, their active transfer, as well as the training of qualified specialists with a wide range of professional competencies and innovation skills. The main types of competence centers are the following: corporate, university, regional, sectoral and international [6].

An adequate approach to the creation of competence centers has been developed abroad. Large companies independently form and take an active part in the works of these structures. Corporations perceive them as missing parts of the innovation chain—the point of new practices development based on advanced technologies that ensure the competitiveness of the company [7].

In this article university Competence centres will be considered as a training tool for the new technological paradigm, since in universities there is an established infrastructure for knowledge generation and transfer. Such competence centers can be considered as an option for the education quality improvement.

2.2 Competence Centres in Educational Organizations

In Russia, such centres often perform as marketing structures with blurred functions at universities or technological parks. Activities of such centers usually have no practical value, are not involved organically in the innovative business process, and therefore are of interest to large enterprises. In order to improve the current situation, the Competence centres of National Technology Initiative are created, which task is to train leaders in new technologies development, to promote new products and services through the implementation of educational programmes. The Competence centres of National Technology Initiative create and implement educational programs of the engineering profile. The programmes are meant to create teams (researchers (teachers) and students) in different scientific fields and different structural divisions of the association members. Cooperation between teams of different research units provides a synergistic effect and systemic implementation of its research programs.

The creation of Competence centres on the basis of universities seems to be the most promising. University Competence centres can be created on the basis of a department or scientific and educational center; faculty or their association for the development of complex projects; partnership of the University with academic science (the Russian Academy of Sciences)—for joint implementation of breakthrough fundamental research projects and the introduction of innovations in the national economy.

There are 2 directions of competence centers specialization: the fields of professional knowledge (natural sciences, technology, humanities) and the fields of professional activity [8]. The Competence centres of the first direction implement short-term intensive training programs for hi-tech industry personnel. A group of trained specialists is formed in the Competence centre, teachers are appointed, a schedule of lessons is drawn up and training is conducted, oriented to the certain competencies acquisition. The second direction Competence centres are oriented to practical innovative solutions that are important for this industry: selection of best practices, identification and systematization of effective practices, transfer of new knowledge, development of technology standards, maintenance of various projects, and initiatives related to knowledge management, etc. Such Competence centres are designed to interact with the following educational institutions:

1. Schools (acquaintance with the basics of various types of professional activity);
2. Specialized centers of technical, cultural creativity of youth;
3. Secondary vocational education;
4. Research institutions.

3 Research Methods

According to the authors, it seems logical to re-engineer university Competence centres, relying on the needs of students and future employers (who are potential university process consumers). Bauman Moscow State Technical University (BMSTU) has a number of research and educational centers [9] and teaching methods [10–13], which give students the opportunity to acquire important knowledge and skills and to be attractive to future employers. Based on the results of a theoretical study conducted by the authors in the previous part of the article, the authors formulated the following research questions:

- RQ1: What are difficulties in project implementation?
- RQ2: In your opinion, does university have an opportunity to support your project?

In order to answer these questions, survey was conducted among students of Bauman Moscow State Technical University. During the research, 500 students of engineering faculties of the university and the faculty of business and management were surveyed. Information was collected by questioning.

It is important to take into account the need of modern industrial and scientific organizations in highly qualified personnel. In order to understand which subjects should be focused on while organizing competence centers in universities, a survey of 45 companies from the industry was conducted. The aim of the study was to find out what the graduates lack in terms of employers.

In order to conduct a survey the following questions were formulated:

- RQ3: What qualities do graduates lack for a successful start?
- RQ4: In your opinion, what is the level of graduates qualification?

4 Research Results

4.1 BMSTU Student Survey

As shown by survey (500 students-respondents) conducted in Bauman Moscow State Technical University (BMSTU), many students are interested in participating in research and in technological projects development, but they have certain financial difficulties (34% of respondents), as well as lack of competence and knowledge (35% of respondents) to implement the idea. For 20% of respondents the main barrier to develop a project is absence of technical facilities; 13% of respondents require a project team with like-minded people. Nevertheless, the answers to the question "In your opinion, does university have an opportunity to support your project?" have demonstrated that the university has all the required facilities for the implementation of scientific and technical projects (70% of respondents); only 15% of respondents think that university can't provide all the required facilities. Thus, during the Competence Centers forming, it is necessary to take into account the importance of facilitating students' access to the technological base of the university or partner organizations.

The function of the Competence Centers is also to find potential investors for student projects, assistance in the development of business plans to students, etc.

4.2 Employer Survey

According to the survey of Russian high-tech companies and start-ups (45 companies surveyed) future employers 27% of employers 27% of employers are not satisfied (Fig. 2) with the practical skills and knowledge of graduates.

The survey (Fig. 3) has also demonstrated that graduates are lack of practical skills (45% of respondents), basic knowledge (23% of respondents), responsibility (21% of respondents), desire to work (11% of respondents).

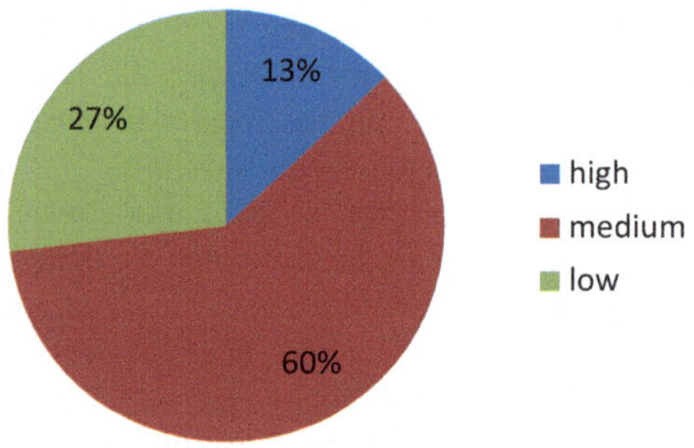

Fig. 2. Respondents' answers on the question "In your opinion, what is the level of graduates qualification?"

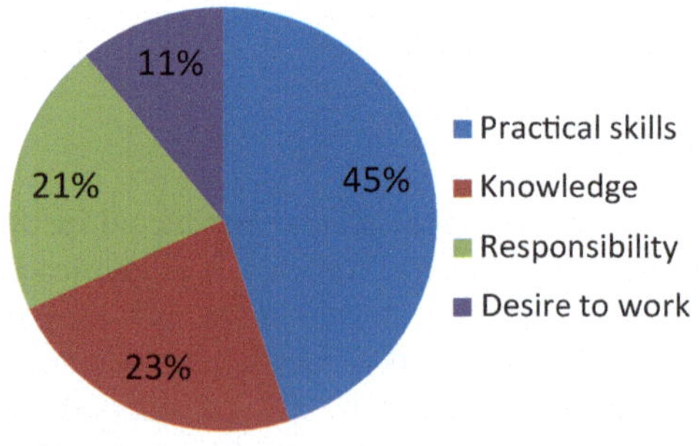

Fig. 3. Respondents' answers on the question "What qualities do graduates lack for a successful start?"

The results of the survey (Figs. 2 and 3) show that the current system of higher education does not meet the requirements of employers. Most graduates demonstrate an average level of training, which is not enough to solve the unconventional problems of the modern high-tech industry.

Consequently, the tasks of the Competence Center are to prepare students to the level required by future employers. Lean manufacturing methods training, acceleration programmes, entrepreneurship fundamentals studying (business planning, marketing, public relations), technological entrepreneurship courses should be organised in the Competence Centres.

4.3 Designing Model of the Process Approach in Competence Centres

Thus, on the basis of the competence center, knowledge and competencies should be exchanged between representatives of different departments, universities, and enterprises on discussion platforms. The Center should support the project technological implementation: simplified access of students to production sites and laboratories of partner enterprises and the University should be organized. On the site of the center, students will study methods of lean manufacturing and the Fundamentals of technological entrepreneurship.

In order to improve the effectiveness of educational activities, it is proposed to consider the activities of the University Competence Center as a knowledge management process. The advantage of the process approach is the continuity of control and individual process combination and interaction. An offered model of the knowledge management process in the Center is presented in Fig. 4.

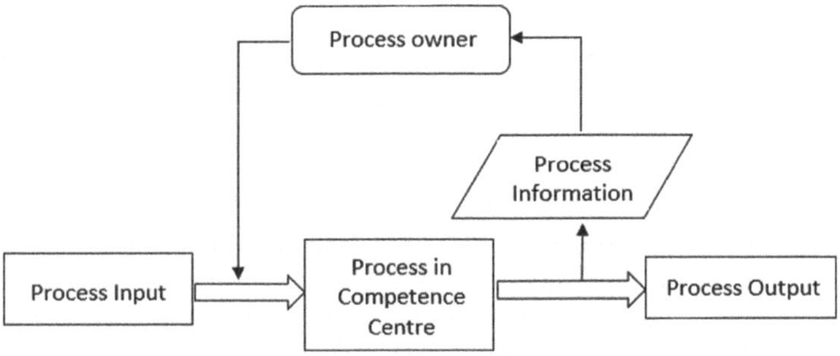

Fig. 4. Model of the knowledge management process in the Competence centre (developed by the authors)

The University will be the process owner, where the Competence Center will be organized. The process consumers are students, postgraduate students, future employers, society. At the input of the process, there is the necessity for students to acquire knowledge and skills that enable them to implement and promote high-tech projects, as well as to increase students competitiveness. In this respect, students require improved accessibility to the technological equipment of university scientific and research centres.

The best option is to create students experimental plants equipped with modern machines and measuring equipment and tools. In order to help students, mentors should be attached, which will supervise the students' research activities. To promote products on the market, students should study the fundamentals of entrepreneurship. On the Competence centre basis, students who receive education in different areas can enter into a dialogue, discuss problems in the project implementation, and create interdisciplinary teams. "Process output" are students who received new knowledge and implemented the project, future employers who received new highly qualified staff. The resources of the process are the personnel, technical, scientific base of the university, investors.

Consumers play a significant role within this approach to the education quality management. Customer satisfaction monitoring requires an assessment of information on consumers perception of their requirements fulfilment. This information, including process parameters, product parameters, and customer satisfaction, will be collected at the output of the process and analyzed by the Process Owner, who will take measures to improve the quality of output indicators.

5 Conclusion

The need to match the activities of the Competence centre with the main social and technological processes that affect the changing environment (automation, increasing complexity of management systems, increasing competition, increasing demands for environmental friendliness) necessitates a knowledge management approach in the educational institutions.

It was suggested to introduce the process management in the Bauman Moscow State Technical University on the basis of the Competence Centres. In order to reeingineer the Competence centres according to the requirements of process consumers (surveyed students and companies) the Competence Centres are to facilitate students' access to the technological resources of the university and to bring together the representatives of various scientific and industry branches to develop interdisciplinar projects. The Competence Centers are also to provide a set of educational courses conducted by the best specialists of academia and business community.

The importance of competence centers creation reflects the general professionalization, specialization and interdisciplinarity tendency in solving the tasks of the innovation economy and the new technological paradigm formation. The management of competence centers should cultivate special values and an internal environment that motivates and encourages the ideas and creativity realization.

This research result is the beginning of further studies series. Subsequently, it will be necessary to identify which subjects are most important for teaching in the centers of competence. The obtained data will form the basis for educational programmes development. The survey of companies will also be continued in order to identify the format of cooperation between enterprises and university competence centres. As a result of this research, we would like to draw the attention of the public to the creation of competence centers.

Acknowledgements. This study is surely carried out according to ethical policies.

References

1. The Bloomberg Innovation Index (2015). http://www.bloomberg.com/graphics/2015-innovativecountries/. Accessed 3 May 2018
2. Kutov, N.M., Galiakhmetov, R.A.: Personnel training in development of the sixth technological order. Bull. Kalashnikov ISTU **3**(71), 26–28 (2016)
3. Atlas of emerging jobs. http://atlas100.ru/. Accessed 3 May 2018
4. Llamosa-Villalba, R., Mendez-Aceros, S.E.: Process management model for higher education: improvement of educational programs in software quality. In: 2010 IEEE Proceedings of Education Engineering (EDUCON), 14–16 April, pp. 1955–1963. IEEE, Madrid (2010)
5. Ewy, R.W., Gmitro, H.A.: Process Management in Education: How to Design, Measure, Deploy and Improve Organizational Processes, 120 p. Quality Press, Seattle (2009)
6. Gitelman, L.D., Kozhevnikov, M.V.: Competence centers—a progressive form of organization of innovation activities. Innovations **10**(180), 92–98 (2013)
7. Berdnikov, R., Kholkin, D.: Innovative competence centers in the electric power industry. Energymarket No 4 (2012)
8. Aksenova, M.A.: Competence center as an educational resource for the training of highly qualified specialists. Vocat. Educ. Market Labor **4**, 18–24 (2017)
9. Mayorova, V., Grishko, D., Leonov, V.: New educational tool for teaching high-school students at university scientific-educational centres. In: Proceedings of the International Astronautical Congress, IAC (2016)
10. Smirnova, E.V., Dobrjkov, A.A., Karpenko, A.P., Syuzev, V.V.: Mentally structured educational technology and engineers preparation quality management. In: Communications in Computer and Information Science, vol. 754, pp. 119–132 (2017)
11. Karpov, A.O.: Generative learning in research education for the knowledge society. Math. Educ. **11**(6), 1621–1633 (2016)
12. Egorova, O., Omelchenko, V., Kuzchbah, A.: An efficient educational methodology for teaching “theory of machines and mechanisms” discipline. In: 2015 Proceedings of the IFToMM World Congress, IFToMM 2015 (2015)
13. Mayorova, V., Zhdanovich, O.: Innovative collaborative educational programs for space systems engineers. In: The Proceedings of the International Astronautical Congress, IAC (2016)

Conflict Management in the Educational Process at the University

Oksana Pozdnyakova(✉) and Anatoly Pozdnyakov

Transport and Telecommunication Institute, Lomonosova 1, Riga LV 1019, Latvia
Pozdnakova.O@tsi.lv

Abstract. This article is devoted to the management of the conflicts between academic staff and students. The progress of the whole pedagogical process will largely depend on the preventation, regulation and resolution of the conflict faculty-students. In order to improve conflict management system in high school, the authors of the article conducted an empirical study of this type of conflict interaction. The aim of this study was to investigate the causes that gave rise to the conflicts and the positions of the conflicting parties, as well as to identify the problems underlying the emergence of disagreements. The study involved the lecturers and students of the Transport and Telecommunication Institute. The main methods of collecting information were questionnaire survey with closed and semi-closed questions (for students) and interviewing (for faculty). Based on the received data, the measures for managing the conflict between faculty and students are proposed. The findings of this article should be useful for academic and administrative staff of the university.

Keywords: Conflict · Conflict management · Academic staff · Students

1 Introduction

Conflicts are an indispensable attribute of any organization, since the productive life of the collective is impossible without the clash of ideas, life positions, the goals taking place between its individual members or the groups of employees with the manager.

To the full extent, this provision is applicable to higher education institutions. Conflicts at the university can develop in different directions: among students, among the academic staff, between faculty and students, between employees and the administration, etc.

The literature devoted to the conflicts in organizations emphasizes that it is unambiguously wrong to evaluate this phenomenon from the negative point of view only [1]. It also refers to the university. Due to the conflicts, the problems and contradictions taking place in the educational process are revealed. So, the conflicts can be considered as a means of diagnosing the work of the university. At the same time Ciuladiene and Kairiene [2] assert that the constructive results of the conflict will depend to a large extent on the way how the conflict has been resolved and what strategies for its resolution have been used. On the other hand, as the researches show

I. Kabashkin et al. (Eds.): RelStat 2018, LNNS 68, pp. 798–807, 2019.
https://doi.org/10.1007/978-3-030-12450-2_77

[3, 4], in any case, conflicts hinder taking the effective solution for the basic education processes—learning and teaching.

Therefore, in order to maintain the stability within the educational environment of the university, to reduce the frequency of conflicts between its subjects, it is necessary to predict and prevent the emergence of conflicts, to diagnose them at an early stage of development, to regulate their flow, to facilitate their resolution, i.e. conflicts need to be managed.

To make the conflict management constructive and having the desired result, it is necessary to conduct a thorough investigation of the causes that gave rise to the conflicts, to study the positions of the conflicting parties, to reveal the real and imaginary problems underlying the emergence of disagreements. On this purpose, authors of this article conducted an empirical study of conflicts between faculty and students of Transport and Telecommunication Institute, and the measures for managing this type of conflict interaction are proposed. Although, as it has been noted above, the conflicts at the university can arise in different directions, nevertheless, the main field of the struggle, according to the authors of the article, unfolds between the main subjects of the pedagogical process—educators and trainees, and this kind of conflicts is the most dangerous. The point is that interaction along the line "faculty member-student" is one of the main ones within the social and pedagogical process. The whole range of knowledge, information, attitudes, value orientations is accumulated in the higher education system and is directly transferred, and it happens in this process exactly. And the conflicts between these parties can disorganize the normal course of the educational process for a long time. Moreover, the progress of the whole pedagogical process will largely depend on the course and resolution of the conflict between the academic staff and students [5]. That is why this kind of pedagogical conflict has been chosen for the study.

2 Methodology

The students of Transport and Telecommunication Institute participated in the research; they were the students of all years and forms of training, as well as a number of leading faculty of the university. All participants were informed on the purpose of the study and agreed to participate.

The main methods of collecting information were questionnaire and interviewing. The main form of students' survey was a questionnaire, which allowed attracting a large number of respondents (113 people) to the study. The questionnaire was conducted anonymously and contained closed and semi-closed questions aimed at identifying the presence of conflicts between academic staff and students, studying the causes of conflicts, their frequency and the possibilities of their resolving. For obtaining more complete information, the fifth part of all participating students, who agreed to express their detailed opinion on the issues proposed in the questionnaire, were interviewed. The interviews as the most flexible method of gathering information were preferred for collecting the faculty's opinion on the nature and causes of conflicts with students. 20 members of the university academic staff, male and female and of various qualifications took part in the survey. Interviewed respondents gave answers in free form, which were fixed.

3 Analysis of Conflicts between Academic Staff and Students (TTI Example) and Management of these Conflicts

The scientific literature pays much attention to the conflicts arising in the organizations, and their management. There are also detailed descriptions and investigation of conflicts developing at schools. The problem of the conflict in the higher school institutions, its prevention and resolution is also covered in the scientific literature. The researchers mainly focus on the study of conflict resolution strategies (their description and analysis) [3, 6–10]. The researches of conflicts between academic staff and students [11, 12], conflicts among faculty [13], the causes of conflicts in student groups [14, 15], studies referring the issues of training academic staff to get out of conflict with colleagues or students [3, 5, 16] have also been presented. However, in general, the researches, devoted to the conflicts at the Universities and its management, are very limited.

Less attention of researchers to conflicts in high educational school, in particular, to conflicts between academic staff and students, probably due to the fact that the pedagogical process at the university is less conflict-related compared to the similar process taking place at school. The organization of the educational process at the university is different; moreover, the status of university faculty and university students differ from the status of school teacher and school student. The university academic staff has a high professional status at the university environment. This is the most qualified part of the staff of the university. Students, in turn, are also directly involved in the learning process; they are able to evaluate the offered knowledge and competences consciously. On the other hand, the university faculty, working with already established personalities, may not adapt to his listeners as schoolteachers; vice versa, the students must adapt to them. Moreover, the students, especially the senior students, have other psycho-physiological characteristics, in comparison with schoolchildren. They are adults, and they are less impulsive, demonstrate less of maximalism and categoricity, which are so natural, for example, for adolescents. In turn, due to the greater regulation and formalization of professional activity, as well as the existence of a test-examination system, and as a consequence, a greater predictability of the situation, the university academic staff is less exposed to long-term stress compared to the school teachers. Accordingly, they are not in such a strong mental tension, contributing to the rapid deterioration of the intellectual and emotional spheres of the individual.

Nevertheless, the universities cannot function without conflicts. The subjects of the educational process are bearers of various social statuses, differ in their roles and functions. Moreover, the pedagogical process is characterized by a socio-professional hierarchy; for some participants the pedagogical process is a professional activity, for others—it is educational activity. One category of participants is attributed to the role of the carrier of socially significant experience and socially valuable qualities and abilities, while the role of the potential recipient of this experience is fixed after another category.

It is also noted by [3] that difficulties caused by ethno-sociocultural, age differences, differences in perceptions and values may arise in the process of communication between a lecturer and a student. In addition, the parties are connected not only with formal relations. By direct interacting, they also build certain interpersonal relationships. As a result, disagreements and conflicts between faculty and students may be due to the difference in their individual psychological characteristics.

The task of the empirical study was to determine how often (in the opinion of students and the academic staff), there are conflicts between TTI students and faculty. It turned out that only 6% of the surveyed students noted the presence of frequent conflicts with academic staff. 28% of interviewed students may have conflicts with faculty, but they are occasional, not on regular basis. The majority of students (66%) claimed that they did not directly participate in conflicts with faculty.

There can arise the conflict not only with the individual student, but also with a group as a whole. The answers received to the question: "Have you ever been involved in a conflict between your study group and the lecturer?", were distributed as follows: often—6%; episodically—25%; never—69%. Thus, the students witnessed mostly episodic manifestations of the conflict between the group and the teacher.

To correct these indicators, the question was presented to the students with a slightly different formulation: "Have you ever witnessed a conflict between another student and the lecturer during the time of study at the institute?" The answers were as follows: often—10%; occasionally—48%; never—42%.

In their turn, all the interviewed lecturers indicated that they had occasional conflicts with students in their pedagogical practice.

Thus, it can be assumed that the conflicts on the faculty-student line take place from time to time, and therefore they require attention. Since they are an inevitable phenomenon, it is senseless to struggle with them; they must be managed.

Conflict management should be preceded by the stage of its diagnosis, i.e. the identification of the main components of the conflict, and its causes [17]. Therefore, in the course of the empirical study conducted by the authors of the article, various factors were suggested to the attention of the students, presumably ones which could provoke conflict with the faculty. These factors were suggested to the students for ranking (number 1 is the most conflict-generating factor, i.e., the most often causing conflict, number 10 is the most non-conflict-generating factor).

The authors of the article assumed that the students suppose that the main reasons for conflicts with academic staff is a violation of their disciplinary requirements, such as compulsory attendance of classes, the performance of tasks on time, etc., due to disagreement with them, or the complexity of teaching methods used by faculty. It was also assumed that neither ethno-social factors, nor worldview differences, nor political views would have any importance for the emergence of conflicts between lecturers and students. The assumption was based on the fact that the ethnic and social environment of the university is relatively homogeneous though there are various political forces and relatively disunited society in Latvia; the serious political and ideological battles between faculty and students have no ground.

The interviewing of students confirmed the hypothesis expressed by the authors of the article that ideological and age contradictions infrequently cause conflicts between faculty and students of the TTI (Table 1).

As it can be seen from Table 1, interviewed students believe that the most common cause of lecturer-student conflict is inadequate or unfair assessment of their knowledge.

Table 1. The causes of conflicts "Faculty-students" (students' opinion)

The causes of conflicts between faculty and students	Level of conflict (from 1 to 10)
A biased evaluation of the student's knowledge by the lecturer	1
Excessive demands of the faculty towards students	2–3
Illogical (incomprehensible) explanation of educational material, dryness, complexity of the language of lectures	2–3
Authoritarian style of lecturer's behaviour	4
Subjectivity (differential treatment of students)	5
Fuzzy system of requirements and criteria for assessing knowledge	6
Failure of the student to comply with the standards of education (compulsory attendance of lectures and practical classes, performance and delivery of tasks on time, etc.)	7
Non-observance of the rules of behaviour in the classroom by the student (being late, chattering, dealing with things not connected with lecture)	8
Disrespectful attitude of the lecturer to students, lack of tact	9
Differences in ideological, political issues	10

This factor as a source of conflicts is typical enough, and it is supposed to be the most common reason of conflicts by a number of researchers of pedagogical conflicts [2, 17].

The next reasons of discontent and then conflicts by their importance are as follows:

- Pickiness, overstated requirements of lecturers towards students.
- Methodological flaws in the work of lecturers, though, in general the students have no complaints on the professionalism of faculty.

Thus, the assumptions of the authors of the article, that students' lack of discipline is one of the main causes of conflicts, according to this table have not been confirmed. Nevertheless, when the students were asked during the interview whether they have been participants or witnesses of the conflict between the faculty and students, caused by these factors (Table 2), the students named the violation of the norms of instruction and discipline in the classes as the most common cause of conflicts with lecturers.

Table 2. Actions of the student causing the conflict with the lecturer (students' opinion)

Actions of the student causing the conflict with the lecturer	Level of conflict (from 1 to 6)
Low discipline in lectures	1
Disagreement with the grades	2
Student's violation of educational standards	3
Cheating during the exam (test)	4
Demonstrating a point of view different from the lecturer's one	5
Manifestation of rudeness, tactlessness	6

The authors of the article asked the interviewed faculty to comment on the students' position regarding the causes of conflicts and to propose their own version. Confirming that students from time to time express dissatisfaction with the assessments, the academic staff noted that the overestimated self-esteem of students and unreasonable claims for a higher mark often contribute to the emergence of conflict.

Agreeing that the inexperience of a young lecturer or weak pedagogical skill can lead to methodological flaws and even cause criticism from students' party, and even lead to conflicts, faculty pointed out that violating the rules of behaviour in the classroom often prevents student from full comprehension of educational material. Considering, that outside conversations during the classes and other violations of discipline on the part of the student disturb both groupmates and the lecturer, the overwhelming majority of the interviewed faculty (90%) confirmed that this factor can also lead to conflict interaction. Moreover, it was noted that in any student group there may be one or more conflicting personalities, which destabilize relations not only between the groupmates, but also between students and faculty. It should be noted that there may be a conflict person among the faculty as well.

Subjectivity in the requirements towards students is also mentioned by respondents as a reason for the emergence of a conflict situation and even conflict (the fifth place). Faculty, while not denying that the problem of differentiated attitudes toward students exists, nevertheless, name the unpreparedness of students for university studies: poor mastery of the school curriculum, especially in the field of sciences, (primarily among first-year students), as well as laziness, irresponsibility, lack of independence and slovenliness of some students as one of the main causes of conflicts.

If to build a hierarchy of the main causes leading to conflict, basing on the opinion of faculty, the situation will be as follows:

- laziness, irresponsibility, lack of independence and discipline of students (60%);
- overestimated students' self-esteem and unreasonable claims for a higher mark (30%);
- incorrect behaviour of students (10%).

Thus, the assessment and ranking of the causes of conflicts between faculty and students vary, which certainly does not contribute to the normalization of relations between different parties.

It should also be noted that subjective factors, equally important for students and faculty, can contribute to the conflict just "on an equal footing". Among them can be mentioned: chronic fatigue caused by overload (the lecturer performs an increased teaching load, student studies after a working day); financial or family problems; the student or a lecturer has a strong nervous tension, stress.

In the course of the research the following task was set: to find out what the participants of the conflict think about its main initiators. It turned out that when listing the causes of conflicts between students and faculty, the other party was automatically recognized as the culprit of the conflict. On a direct question "Who is more likely to create a conflict situation?" the lecturers confirmed their position and answered unequivocally that the students are the main conflict generating social group at the university. The answers of the students were more diverse: 37% agreed with the opinion of the academic staff, 8% felt that the faculty often initiate the conflicts, and

55% indicated that both are equally to blame for the emergence of conflicts. However, in the process of additional interviews when they offered to describe the most memorable conflict with the faculty, students, as a rule, talked about the conflicts that happened, in their opinion, precisely because of the lecturer's fault. Such a mutual shifting of the blame for creating a conflict situation and developing it into an open confrontation must be alarming, since non-recognition of one's fault, a harboring sense of resentment, hinders finding the most optimal solution of the conflict—a compromise between the parties.

In the course of the empirical research both students and faculty were asked about the assessment of the solvability of conflicts. Students were asked, in whose favour, according to their observations, the conflict between the lecturer and the student (or the student group) is most often resolved; the answers were as follows: 33% claimed that in favour of faculty, 4%—in favour of students, 26%—solutions divided approximately equally in favour of both groups, and only 30% felt that conflicts were resolved fairly, to mutual benefit. The recognition of a fair resolution of conflicts among faculty was much higher. Although conflicts are far from always resolved to mutual satisfaction, more than half of them (65%), according to the interviewed lecturers, can be considered as fully resolved. Nevertheless, even if we take these percentages closer to reality than those named by the students, still the remaining 35% indicate an insufficient regulation of the conflict, which is quite dangerous, since the losing side, even if it has reconciled itself to the defeat in the conflict, remains negatively inclined in relation to opposite party. It, in turn, gives grounds for a new round of escalation and might provoke the development of a new conflict situation and conflict.

Also, all participants of the study were asked if any conflict between faculty and students can be resolved. While answering it, opinions were divided: a small part (9%) of both parties believes that all conflicts between academic staff and students are perpetual, because they are caused, among other things, by objective factors. However, the majority of the respondents, including the authors of this article, tend to think that the rational behaviour of participants makes possible to resolve the conflict in the full sense.

The authors propose the following measures to manage conflicts (their prevention, resolution and regulation). First of all, the problem of the conflicts cannot be hushed up. The situation is complicated by the fact that not all lecturers, unlike students, willingly admit the existence of conflicts with their wards. Most faculty still have a cautious attitude to the very word "conflict", that's why they negatively assess any conflict as a phenomenon. The authors of the survey, while surveying the academic staff, in half cases faced the opinion that allowing the emergence of a conflict situation, or creating it and bringing this situation to an open conflict testifies to the pedagogical incompetence of the lecturer.

Students, in turn, are mostly ready to discuss the conflicts with lecturers, and expect the same openness from their opponents. In [2] interesting facts are demonstrated: the students were more interested not even in the resolution of the conflict, but in feedback from the faculty. As [3] show, strategies such as communicating respect, clarifying course goals, involving students in solving the problem, and encouraging a sense of classroom community received the highest effectiveness scores.

The study conducted by the authors of the article demonstrated that the typical reasons for the conflict interaction, called by the students, were a biased evaluation of the students' knowledge and unclear criteria by which the task was evaluated. Therefore, as a prevention of the conflict, it is recommended to create a clear system of requirements and criteria for assessing knowledge for each task, as stipulated in the curriculum, and for each training course as a whole.

A great help to both faculty and students can become the special normative documents regulating the actions of academic staff and students in cases of conflict, prepared by the administration of the university. Only in the case when both parties are well aware of all the norms, rules for conducting and ending the conflict, and accepting them, there is a great possibility that even a preserved negative emotional background does not serve as a basis for the formation of a new conflict.

The separate question is "Who is responsible for resolving the conflict?" According to [18] and the opinion of the interviewed academic staff, the faculty plays the main role in preventing and resolving the pedagogical conflict. As it has already been noted, the lecturer is usually more experienced than the student because of life and professional experience; moreover, the faculty involved in conflicts with students are not in a hurry to "wash the dirty linen in public" and try to find a compromise solution. Nevertheless, in order to exit the conflict adequately, the lecturer needs certain knowledge on pedagogical conflicts and their characteristics, as well as skills of using them. Then the lecturer is able to predict, to prevent conflict, or make the course of the conflict more constructive.

To increase the conflictological competence of academic staff, it is desirable to devote a complex of pedagogical seminars and trainings for the analysis of diverse conflict situations with the search for solutions to specific problems. According to the authors of the article, the participation of students in these trainings is also desirable. This allows identifying the conflict situations associated with the actions of the faculty, which they may not even perceive as the beginning of the conflict; trainings also allow better understanding the feelings of students.

As it has already been mentioned, there is always a possibility that there is one or several conflict persons in the student group, capable of creating chaos in relations with colleagues and the academic staff. Consequently, the administration faces the task of identifying conflict persons and conflict student groups and providing psychological assistance to faculty working in such groups to prevent conflict situations and prevent conflicts.

Nevertheless, there are such situations when the lecturer, due to certain circumstances (for example, the cause of the conflict is not clear or when the conflict has gone too far and has taken a destructive emotional-personal character), cannot resolve it considering the mutual interests. In this case, the finding of an exit from the conflict between the lecturer and students falls on the shoulders of the Dean, as the immediate head of the lecturer. However, the Dean, being an interested person, is not always able to assess mutual claims of opponents objectively, and as the students point out, they often resolve the dispute in favour of a colleague. Therefore, it is desirable to resolve all particularly difficult disputes and conflicts by an independent and disinterested person, guided by the principle of fairness, the norms of legislation and the Charter of the university. This role can be given to the ombudsman of the Institute.

4 Conclusion

As a result of research carried out at the Transport and Telecommunications Institute, the following patterns have been revealed:

- Most of the students of the TTI consider the university faculty as highly qualified specialists and respect them.
- Nevertheless, 58% of the surveyed students noted the presence of at least episodic conflicts with the academic staff (by participating in these conflicts directly or by watching fellow students in conflict with the faculty). Also, all interviewed lecturers indicated that there were conflicts with students in their pedagogical practice.
- According to the students' opinion, conflicts were caused mainly by either bias and unjustifiably high demands on the part of lecturers, or by methodological flaws in teaching.
- According to the academic staff's opinion, the most frequent causes of conflicts are the overestimated self-esteem of some students and their unpreparedness for studying at the university.

The authors suggest the following measures for managing the conflicts:

- not to hush up the problem of conflicts;
- to elaborate the clear system of requirements and criteria for assessing knowledge for each task provided for in the curriculum and for the training course as a whole;
- to prepare and familiarize faculty and students with the special normative documents regulating the actions of academic staff and students in cases of conflicts;
- to dedicate a complex of pedagogical seminars and trainings (with the participation of both faculty and students) to analyze various conflict situations with the search for solutions to specific problems;
- to establish the position of an independent arbitrator—the ombudsman of the institute for the resolution of conflicts between faculty and students, taking destructive character.

The authors in the article analyzed only one type of the conflicts—the conflict between academic staff and students. However, it should be taken into account that the conflicts at the university can arise in different directions. Since they take place at the same organization, they are often interrelated and have influence on each other.

Acknowledgements. This study is surely carried out according to ethical policies.

References

1. Kirschner, S.: Conflict and dispute management system design: overview. ADR Bull. **1**(5), Article 1 (1998). http://epublications.bond.edu.au/adr/vol1/iss5/1
2. Ciuladiene, G., Kairiene, B.: The resolution of conflict between teacher and student: students' narratives. J. Teacher Educ. Sustain. **19**(2), 107–120 (2017). https://doi.org/10.1515/jtes-2017-0017

3. Meyers, S.A., Bender, J., Hill, E.K., Thomas, S.Y.: How do faculty experience and respond to classroom conflict? Int. J. Teach. Learn. High. Educ. **18**(3), 180–187 (2006)
4. Rimm-Kaufman, S., Sandilos, L.: Improving students' relationships with teachers to provide essential supports for learning (2011). www.apa.org/education/k12/relationships.aspx#
5. Isaacson, A.: Resolving a teacher-student conflict: an intrinsic case study. J. Educ. Learn. **5** (1), 68–77 (2016). https://doi.org/10.5539/jel.v5n1p68
6. Barsky, A.E.: Conflict Resolution for the Helping Professions, 2nd edn. Brooks/Cole, Belmont (2007)
7. Dincyurek, S., Civelek, A.H.: The determination of the conflict resolution strategies of university students that they use when they have conflicts with people. Behav. Anal. Today **9** (3–4), 215–233 (2008). https://doi.org/10.1037/H0100660
8. Harrison, T.R.: My professor is so unfair: student attitudes and experiences of conflict with faculty. Confl. Resolut. Q. **24**, 349–368 (2007). https://doi.org/10.1002/crq.178
9. Pasque, P.A., Chesler, M.A., Charbeneau, J., Carlson, C.: Pedagogical approaches to student racial conflict in the classroom. J. Divers. High. Educ. **6**(1), 1–16 (2013). https://doi.org/10.1037/a0031695
10. Skelton, A.: Value conflicts in higher education teaching. Teach. Learn. High. Educ. **17**(3), 257–268 (2012). https://doi.org/10.1080/13562517.2011.611875
11. Cothran, D.J., Ennis, C.D.: Students and teachers' perceptions of conflict and power. Teach. Teach. Educ. **13**, 541–553 (1997)
12. Tantleff-Dunn, S., Dunn, M.E., Gokee, J.L.: Understanding faculty-student conflict: student perceptions of precipitating events and faculty responses. Teach. Psychol. **29**(3), 197–202 (2002). https://doi.org/10.1207/S15328023TOP2903_03
13. Hearn, J.C., Anderson, M.S.: Conflict in academic departments: an analysis of disputes over faculty promotion and tenure. Res. High. Educ. **43**(5), 503–529 (2002). https://doi.org/10.1023/A:1020197630478
14. Argon, T.: The development and implementation of a scale to assess the causes of conflict in the classroom for university students. Educ. Sci. Theor. Pract. **9**(3), 1033–1041 (2009)
15. Borg, M., Kembro, J., Holmén Notander, J., Petersson, C., Ohlsson, L.: Conflict management in student groups—a teacher's perspective in higher education. Högre utbildning **1**(2), 111–124 (2011)
16. Meyers, S.: Strategies to prevent and reduce conflict in college classrooms. Coll. Teach. **51** (3), 94–98 (2003). https://doi.org/10.1080/87567550309596419
17. Ozgan, H.: The usage of domination strategies in conflicts between the teachers and students: a case study. Educ. Res. Rev. **11**(4), 146–152 (2016)
18. Ciuladiene, G., Kairiene, B.: The resolution of conflict between teacher and student: teachers' narratives. In: Proceedings of the International Scientific Conference Society. Integration. Education, 25–26 May 2018, Rezekne, Latvia, pp. 235–245 (2018)

The Quintuple HELIX Innovation Model: Cooperation for E-Services Development and Education of Society. A Case Study in Latvia

Ina Gudele(✉)

Institute of Regional Development, Latvian University of Applied Sciences, Svetes iela 18, LV 3001 Jelgava, Latvia
ina.gudele@riseba.lv

Abstract. The Helix innovation model has been in development for over 20 years and has gone through several development stages, from the Triple Helix innovation model at the beginning of the nineties, the Quadruple Helix innovation model, and then to the Quintuple Helix model as a driver for innovation in several fields of development. The author of this article analyses the usage of the Helix models for the development of innovative services in one project in Latvia, where the successful cooperation between two universities from Latvia and Lithuania, a small innovative enterprise, local and national media, a municipal organization, and a professional non-governmental organization, brought about excellent results. The article, analyses the main methods of cooperation for the creation of innovation products and the education of society in order to use the created electronic services developed for the inhabitants of a regional town in Latvia. The author describes the main roles of the partners in the project and the communication between the partners, the methods used for communication, and highlights the points of societal education and the involvement of all parties in this process. As the results of this project were being finalized, recommendations were made for cooperation between municipal and commercial enterprises. Together, with the participation of universities, non-governmental organizations and the mass media, the development of services would be an important process for the rural regions of Latvia, to become more active in the knowledge base and in the development of the smart economy. Each of the partner organizations had a specific role in the project, and a positive influence on the successful result of the project and of the benefits to everyone, and society in general.

Keywords: Innovation · Quintuple Helix · Electronic services · Education of society · E-tools for education

1 Introduction

In this paper, the authors analyze the development of the different Helix innovation models and the influence of political, economic and societal processes. The paper consists of four chapters.

I. Kabashkin et al. (Eds.): RelStat 2018, LNNS 68, pp. 808–817, 2019.
https://doi.org/10.1007/978-3-030-12450-2_78

The primary aim was to find the best method of cooperation between the interested parties in the development of electronic services for society. The process of innovation usually involves several sections, and the process of cooperation between different organizations with different structures and stakeholders can be difficult. Several countries look for the best solutions in order to find the most appropriate ways for government institutions and municipalities to find the best partners for the development of public e-services for citizens, and to create innovative cooperation with research organizations and small enterprises. All of those processes are strongly linked with the OECD defined Knowledge-based Economy processes and have a global impact on economic growth and development [5]. The main part of this paper—a case study of the project in Latvia, consists of several parts and mainly focuses on the education of society in order to use innovative services.

Part-Participants and roles- describing participants of project working using the Quintuple Helix innovation model and each role in project, connection between organizations in education process using electronic environment. Part-Results of project demonstrate all results of the project for several beneficiaries, usage of the results and improvement for each of them. In part—Main conclusions and recommendations collected main findings and recommendations from all participants for usage in next projects in Latvia and other countries. As the project was successful this positive experience can be disseminate to any other municipality to improve process of education of society.

2 Theoretical Discussion

With the development of technology and the effects of globalization, there was a need to look for more effective methods of cooperation in the development of new technologies and products. The main principle of the concept for cooperation between universities, governments and industry was initiated in the 1990s by Etkowitz and Leydesdorff [1, 2]. The Triple Helix thesis shows the potential for innovation and economic development from a Knowledge Economy perspective, which played a much more important role in the hybridization of elements from university, industry and government in order to generate new institutional and social formats for production, the transfer and application of knowledge with value added to the economy. Those principles relate very closely with the Knowledge Economy principles defined in 1996 by OECD [5], and which are still very actual in the global economy. The Knowledge Economy as a basement theory has been used in the development of all the most important policy documents of different states and municipalities. The same principles are included in the main economic development policy document, "National Development Plan: Republic of Latvia for Medium Planning Period from 2014–2020" (NDP20202) highlighting knowledgebase economy development as a main strategy for development in this period [4].

"In science and research, cooperation among partners is crucial, and there is a need for the creation of larger and consequently, more competent and robust associations, thus inducing joint and private-sector investment in research and development" [4].

In the mid-nineties, procurement for new research and innovation was mostly undertaken by government departments or big industrial companies, and for this reason the development of the Triple Helix innovation model used the government as the procurer, a university as the researcher and industry as the implementer. The concept of the Triple Helix model of innovation [2] was recently introduced as an analytical framework that synthesizes the key features of Triple Helix interactions into an 'innovation system' format, defined according to the systems theory as a set of components, relationships and functions between partners. Among the components of the Triple Helix model, a distinction is made between R&D and non-R&D innovators.

With the process of globalization and changes in the economy structure, the role of participants in the innovation processes was changed as well. Municipal institutions to reflect regional and local interest and needs are now replacing large governmental institutions more and more. The bigger industries have started to be replaced by medium and small sized businesses that are investing in new development and in cooperation with research institutions. As small and medium enterprises do not have the capacity to organize big projects they started to cooperate and join up with professional non-government organizations. Democratic processes in society were more active than before and it positively influenced the development of non-government institutions. This in turn became active processes in the development of the economy and innovation.

"Triple Helix innovation model transfer to Quadruple Helix and by additionally adding the helix (and perspective) of the natural environments of society. The Quadruple Helix already encourages the perspective of the knowledge society and knowledge democracy for knowledge productions and innovation" [6] (Fig. 1).

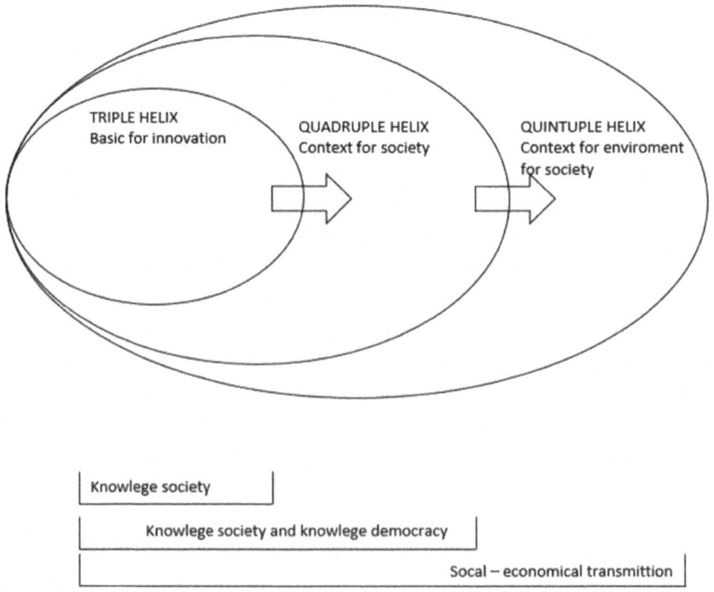

Fig. 1. Development of Helix innovations models and its influence on the transformation of the economy and society. *Source* authors, based on Carayannis, Bart and Campbell [6].

This model more involve in innovations process wide organizations and providing sociological transaction.

The development of the information society, knowledge-based society and knowledge-based economy continue transforming and developing innovation processes as well influencing participants and involve new partners in the process. For the sciences it is important not just to disseminate the latest discoveries and to exchange information for scientific discussions, but to involve new participants for practical tests and expertise. Governments, municipalities and entrepreneurs need to communicate with them auditory and it is space for one more important stakeholder in next model-Quintuple Helix innovation model- media. The Media has an important place in any innovation process as a messenger between the wider society and processes potential auditory and implementers. It is a key to the success of any practical implementation.

This is what makes the Quintuple helix model for innovation an interesting and valuable model for the analysis of innovation ecosystems, and to explain the processes of the knowledge exchange, that takes place in such collaborative innovation ecosystems. Any stakeholder in this ecosystem complements each other. The same principles are possible to use not just for the dissemination of research and development, but to share knowledge with focus groups in order to use the results of the development and so educate society about the benefits of those projects.

3 Research Results and Discussion. Case Study. The Development of Remote Control and Electronic Services in the City of Ventspils, Latvia

3.1 Participants and Roles

Within the EU Latvian-Lithuanian Cross border Development Program, the project, "Remote Control of Ventspils Water Supplying System and Smart Metering Services" was completed in May 2016. The project was organized using the Quintuple Helix innovation model with several participants.

The Municipality of Ventspils was the main beneficiary of the project, which was to create a remote control and smart metering system for a water supplying system in Ventspils city and to educate society in how to use the new electronic services, and to improve levels of cooperation between the residents and the municipal utility companies.

The project had several tasks, which were shared between the project participants:

Ventspils University College and Kaunas Technological University had to develop a new type of electronic equipment for reading data and transmitting it from water network nodes, software to organize all the data from the networks and to create e-services from based on that data.

Entrepreneurs such as the small enterprise, Smart Meter Ltd and Ventspils UDEKA (municipal enterprise of water supply) had the role of testing equipment, implementing services and maintaining all the systems after the end of the project.

The non-governmental professional institution, Latvian Internet Association (LIA), was very interested in the implementation of this kind of project because it positively influenced the development of e-services in the country, and contributed to the education of society and enlarging the potential for other entrepreneurs. The main goal of this NGO is to educate society in the use of internet technologies and to improve the business environment for SMEs in Latvia, acting as a mediator between government and entrepreneurs. The role of LIA was to prepare education materials for citizens about the available e-services and to help the municipality to organize the introductory seminars about the services and new communication possibilities. LIA was one of the dissemination channels and organizers in working with the other stakeholder- the media.

As the Quintuple Helix innovation model has not been very widely used in Latvia, experience in this type of cooperation was limited, especially in attracting traditional media. Local media has an important role in communication with the local community and any project success depends on successful communication with the potential audience. The local newspaper, "VentasBalss" was invited to take part in the project together with the usual social media channels which were used by all the project partners.

As the project partners were located in different parts of Latvia and Lithuania, and the distance between them was a few hundred kilometers, it was impossible to organize regular meetings. During the entire three-year project period, the project partners communicated using a specially designed platform for project management and communication. As a major part of the project was the organization of communication with the local community, several electronic tools for communication with the project target group were chosen and used (Table 1).

Table 1. Partner roles in the project and way of communication

Partner	Role in the project	Cooperation model and method of communication
1	2	3
Municipality of Ventspils city (*government*)	The main contractor for the project	Cooperation with all involved partners, communication with citizens via mass media
Ventspils University College Kaunas Technological University (*Academic*)	Research and development of remote control system and electronic for e-services	Close cooperation between academic and SME partners in special electronic communication platform, installing and demonstrating services to partners, provision of information to other partners
Smart Meter Ltd. (*SME*)	Production of electronic equipment and system of e-services	
Ventspils UDEKA (*SME*)	Installation of equipment on premises and network, organization cooperation with citizens	
Latvian Internet Association (*NGO*)	Creation of education materials for citizens about usage of e-services	Receiving information form academic and SME partners and preparing interactive materials for studies about e-service usage, close cooperation with mass media
Newspaper "VentasBalss" (*Mass media*)	Distribution of information about the project and usage of e-services	Cooperation with all partners, receiving information and distribution information to citizens about the project, results and benefits

Source authors own project materials (2016)

The system was created by both universities in two years, and was tested immediately after the development of each system element. The project organization was managed by specialists from LIA and Ventspils University. The project team worked in three different cities—Riga, Ventspils (Latvia), and Kaunas (Lithuania). The total project time with the creation of electronic services and organization of the training seminars for citizens of Ventspils was three years.

3.2 Project Results

All project partners successfully finished the project and the planned goals were achieved. As a result, further experience of working in the social-ecological field for the creation of innovative products and services was gained (Fig. 2).

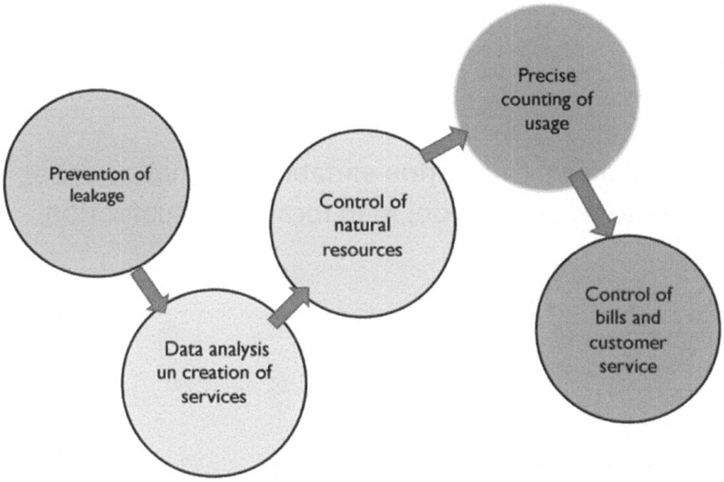

Fig. 2. Results of project. *Source* authors own used project materials [3]

The main goals and achievements of this project were to create for Ventspils City several improvements for their water supply system, mostly technological, but also greater experience in cooperation.

The system was created by both universities in two years, and was tested immediately after the development of each system element. The project organization was managed by specialists from LIA and Ventspils University. The project team worked in three different cities—Riga, Ventspils (Latvia), and Kaunas (Lithuania). The total project time with the creation of electronic services and organization of the training seminars for citizens of Ventspils was three years.

More detailed results from the project, see Figs. 3 and 4, gave the water supply company of Ventspils Municipality, "UDEKA" a lot of financial and human resource savings. Comparing the figures for preventing leakages of drinking water between 2012 and 2015 on trunk pipelines, the situation improved by 35%. If statistics from previous years could be more precise it will be possible to calculate even more precisely the

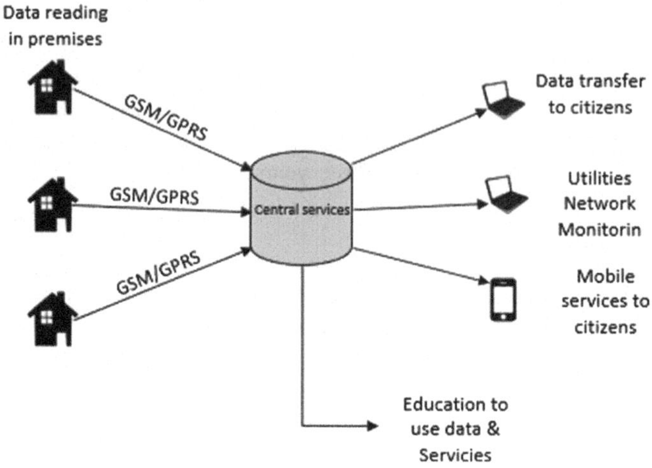

Fig. 3. General results of the project. *Source* authors own, used project materials

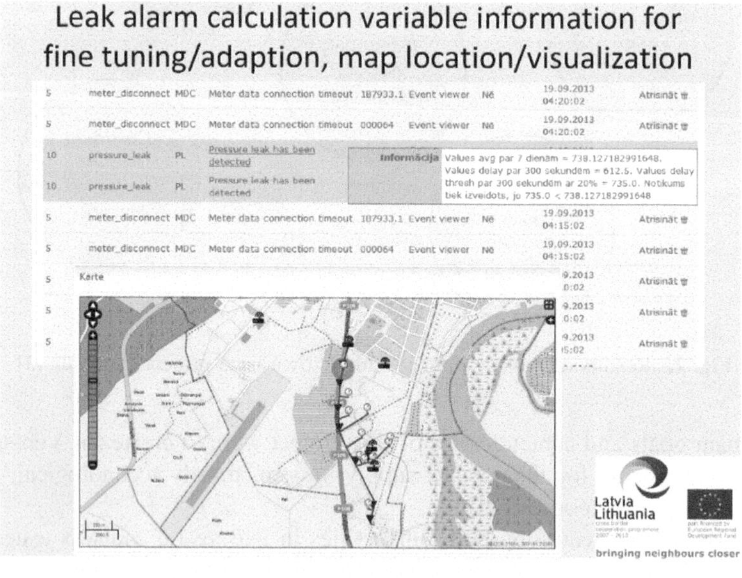

Fig. 4. Electronic services for utilities. *Source* authors own, using project materials

results of improvement. After the installation of the equipment, leakages are possible to detect within 1–3 h, whereas before, it could take several days to detect.

Currently, UDEKA does not have sufficient human resources but in two years it plans to redirect existing specialists to more specialist work because monitoring of pipelines is possible to organize electronically. This means that there will be a better use human resources as well.

The results of the project can be divided into three groups:

1. Direct results of the project—remote control system for the water supply system for utilities and end users.
2. Indirect results for the project partners—cooperation model to create new products and services between project participants from different groups—municipal, academic, non-government, small or medium enterprise and mass media.
3. Communication with society—municipality made project process transparent for society and organised education opportunities for the local community about the new available e-services using the existing municipal portal and local mass media.

The system gives the possibility of monitoring not just the pipelines but all the water meters used in the city in general and in each building. The system has the possibility to avoid problems with incorrect meter readings, can detect water leakages, can calculate alarms in use with the system, and can visualize any problems on the city map and with other different services used for utilities. One of the biggest problems to overcome was with the difference between the use of the central meter system and the use of individual meters in apartments. The biggest benefit for citizens was the reduction of bills for water and a fairer price for consumption.

The biggest achievements of this project were the electronic services for citizens giving them the possibility to follow water usage and to participate in the nature recourse saving program. It is also important to follow the financial account and to receive some additional information about other services and issues including water leakages. The Water Supply Company or entrepreneur in the short-term cannot achieve such big results and implement so many services as was done by this project.

Citizens have possibilities to follow online the usage of water in their apartments using the system created during this project, and can analyze data and find possibilities of conserving water as a natural resource, see Fig. 5.

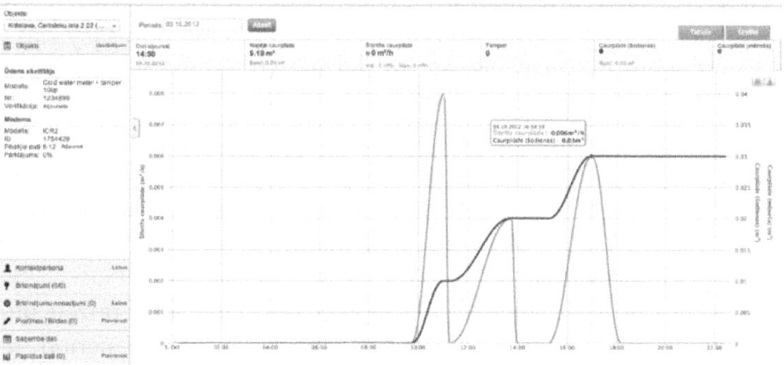

Fig. 5. Diagram for personal water usage in apartment (online). *Source* authors own, using project materials

The next stage of the development of services will provide citizens with mobile services and with information about water leakages in apartments, or where there is too much water consumption in a household, and to show the possibility of how to save natural resources and improve cooperation between residents and municipal enterprises. At this stage, it was very important to organize the distribution of information about the newly created services and to start using those services. The Latvian Internet Association prepared the content for training purposes and provided material to newspapers and portals to inform citizens about the possibilities of how to use e-services for the remote control use of water consumption and other resources.

The electronic system available online provides citizens with easy access, and for this reason has become popular especially with the younger generation. They need more and more new services. One of these services is to inform customers about the amount of water usage or potential leaks in an apartment by mobile phone.

4 Main Conclusions and Recommendations

The case study of the project "Remote Control of Ventspils Water Supplying System and Smart Metering Services" shows that by using the Quintuple Helix innovation model, it is possible to successfully create not just a lot of new products and services, but to provide some additional social benefits for society. Of course, technical achievements are very important, as are innovative technologies to improve the flow of utilities, and safe financial, technical and human recourses, but no less important are the non-material measured benefits for society.

Universities have the possibility to work on new generation electronic remote control equipment and to improve technologies, give scientists and students opportunities to develop new knowledge and experience.

NGOs can improve communication between society and industry, expand the field of electronic services and acquire new knowledge in communication practice with new partners such as academic institutions.

Entrepreneurs can make use of equipment and services but now they can gain much more knowledge about the management of the workflow of development and information, communication with society and potential audiences.

The Media has the possibility to monitor processes of development from the inside, to participate in its own right, and to be part of something new. The best and the easiest way to introduce e-services is to get information about water consumption in premises is through the local media, newspapers, and municipal portals because people have easy access to this information.

The Municipality was the biggest beneficiary receiving modernized utilities with improved structures and services and satisfied citizens. In a short period, the citizens started to use e-services because the information was made easily available as was training about how to use the newly created services.

An important result of the project was information for the local community. Whereas some innovations are created just for capital cities or a region this project showed how the self-awareness of local communities can be increased, and as a consequence enable communities to be stronger and self-confident particularly for economic development.

Experience of this project shows that the Quintuple Helix innovation model works not just in a big organization but also in small regional communities and with cross border projects as well. It is important to organize the correct information exchange and cooperation between stakeholders.

The Quintuple Helix Innovation model is still valid for innovation development and can be a useful method in crossbred projects. Using modern electronic tools for communication, it is possible to organize effective communication between partners and more importantly to provide information and knowledge to the target group of project—the citizens of a municipality. The case study of the project shows the usage of the Helix innovative model as an accelerator for the development of several e-services and the creation of an environment for the development of smart economy in rural regions with the involvement of SMEs and local society. This positive example shows that it is possible to replicate it in other Latvian regions and countries.

Acknowledgements. This paper was supported by Latvian State Research project EKOSOC 5.3.2.

References

1. Etzkowitz, H., Leydesdorff, M.: Triplex Helix Research Group Reports Stratford University (1995). http://triplehelix.stanford.edu/3helix_concept
2. Etzkowitz, H., Ranga, M.: Spaces: a triple helix governance strategy for regional innovation, pp. 2–3. Routledge, London (2013)
3. Gudele, I.: ICT tools and methods implementation for natural resources and energy saving in Baltic countries in Smart and Arrowhead projects. RTU (2015)
4. National Development Plan 2014–2012 Republic of Latvia, pp. 21–23 (2014). http://www.varam.gov.lv/lat/pol/ppd/ilgtsp_att/?doc = 13858
5. OECD: The knowledge-based economy. Head of publication services, CEDEX, pp. 21–23 (1996)
6. Carayannis, E., Campbell, B.: J. Innov. Entrep. A Syst. View Across Time Space **1**, 2 (2012)

Author Index

I. Kabashkin et al. (Eds.): RelStat 2018, LNNS 68, pp. 819–821, 2019.
https://doi.org/10.1007/978-3-030-12450-2

The manufacturer's authorised representative in the EU is Springer Nature Customer Service Centre GmbH, Europaplatz 3, 69115 Heidelberg, Germany. If you have any concerns regarding our products, please contact ProductSafety@springernature.com

Printed and bound by CPI Group (UK) Ltd, Croydon, CR0 4YY

05/10/2026

02228147-0006